D0946795

SUBMARINE GEOLOGY

SUBMARINE GEOLOGY
Third Edition

FRANCIS P. SHEPARD, 1897–
Scripps Institution of Oceanography

HARPER & ROW, PUBLISHERS
New York, Evanston, San Francisco, London

To Kenneth O. Emery and Robert S. Dietz, with appreciation for the fact that these former students have taught me more than I ever taught them.

SUBMARINE GEOLOGY, Third Edition

Copyright © 1948, 1963, 1973 by Francis P. Shepard.

Standard Book Number: 06-046091-1

Library of Congress Catalog Card Number: 72-80127

Contents

PREFACE TO THE THIRD EDITION ix

1: INTRODUCTION AND HISTORY 1

2: METHODS AND INSTRUMENTATION USED BY MARINE GEOLOGISTS 9
INTRODUCTION 9 • POSITIONING AT SEA 10: *Bearings and Horizontal Sextant Angles 10, Navigation from Chart Soundings 11, Electronic Position Finders 12* • SOUNDING METHODS 12: *Echo Soundings 13, Correction of Soundings 13, Deep-Tow Methods 15, Side Scanning 16* • SAMPLING METHODS 16: *Grab Samplers 18, Coring Devices 19, Dredging 25, Techniques in Coring and Dredging 26, Methods of Handling Samples 28, Deep-Sea Drilling (JOIDES) 28* • SEA-FLOOR OBSERVATIONS BY GEOLOGISTS 31: *Scuba Diving 31, Deep-Diving Vehicles 33* • BOTTOM PHOTOGRAPHY 34 • BOTTOM CURRENT MEASUREMENTS 37 • SOIL MECHANICS METHODS 39 • GEOPHYSICAL METHODS 40: *Seismic Reflection Profiles 40, Seismic Refraction 40, Gravity Measurements 41, Magnetic Anomaly Measurements 42, Heat-Flow Measurements 42*

3: WAVES AND CURRENTS 43
INTRODUCTION 43 • WAVES GENERATED BY THE WIND 44: *Shoaling Transformations and Breakers 45, Wave Refraction, Diffraction, and Reflection 49, Nearshore Currents Related to Waves 50* • CATASTROPHIC WAVES 53: *Tsunamis 53, Storm Surges 56, Landslide Surges 57* • TIDES 59 • OCEAN CURRENTS 60: *Wind-Drift Currents 61, Permanent Currents and Prevailing Winds 61, Bottom Currents Related to Permanent Currents 61, Cascading Currents on the Continental Slopes 64, Tidal Currents 64, Turbidity Currents 66, Internal Waves and Other Types of Bottom Currents 66*

4: OCEAN SEDIMENTS AND SEDIMENTATION 68
INTRODUCTION 68 • SEDIMENT SIZE 69: *Grade Scales 69, Measures of Size Distribution 70* • CONSTITUENTS OF SEDIMENT PARTICLES 75: *Coarse-Fraction Analysis 76, Effective Density of Constituents 78* • SHAPE OF SEDIMENT PARTICLES 78 • PACKING, POROSITY, AND DILATATION 81 • PERMEABILITY 83 • FLOW OF LIQUIDS AND THE TRANSPORTATION OF GRANULAR MATERIAL 84: *Fluid Flow over a Granular Bed 84*

5: PLATE TECTONICS (SEA-FLOOR SPREADING AND CONTINENTAL DRIFT) 88
INTRODUCTION 88 • PLATE TECTONICS 89 • EVIDENCE FOR PLATE TECTONICS AND SEA-FLOOR SPREADING 94: *Bilateral Symmetry of Magnetic Bands Bordering Oceanic Ridges 94, Earthquake Epicenters and First Motion 94, Basement Age Under Ocean Floors 94, Fit of Opposite Oceanic Margins 96, Aseismic Ridges and Hot Spots 98, Paleomagnetism 98, Heat Flow From Ocean Floors 99, Permo-Carboniferous Glaciation 99, Evaporite Distribution in History 101*

6: SEACOAST CLASSIFICATION AND ORIGIN 102
INTRODUCTION 102 • NOMENCLATURE FOR COASTAL FEATURES 103 • MAJOR COASTAL CLASSIFICATIONS 103 • EMERGENCE, SUBMERGENCE, AND NEUTRAL CLASSIFICATIONS 106 • CLASSIFICATION BASED ON LAND-SEA MORPHOLOGICAL RELATIONS 109 • GENETIC COASTAL CLASSIFICATION 109 • CLASSIFICATION OF COASTS 111

7: BEACHES AND SHORE PROCESSES 123
INTRODUCTION 123 • BEACH AND COASTAL TERMINOLOGY 124 • BEACH CLASSIFICATION 126: *Arctic Beaches 129* • MECHANICS OF BEACH DEVELOPMENT 129 • COMPOSITION OF BEACH SAND 132 • SOURCES OF BEACH SAND 133 • BEACH CYCLES 134 •

PERMANENT LOSS OF BEACH SAND 137 • BARRIER BEACHES AND BEACH RIDGES 141: *Chenier Ridges 145* • HURRICANE AND TSUNAMI EFFECTS ON BEACHES 147 • CUSPATE SHORELINES 148: *Cuspate Forelands 148, Cuspate Spits 150, Giant Cusps 152* • EFFECTS OF ENGINEERING STRUCTURES ON BEACHES 152 • BEACH STRUCTURES AND MARKINGS 156: *Beach Stratification 156, Beach Cusps 157, Beach Ripples 157, Miscellaneous Features 161*

8: DELTAS, LAGOONS, AND ESTUARIES 162

INTRODUCTION 162 • DELTAS 164: *Mississippi Delta 164, Niger Delta 167, Ganges–Brahmaputra Delta 168, Orinoco Delta 169, Yukon Delta 169, Po Delta 169, Colorado Delta of Texas 172, Alaskan North Slope Deltas 172, General Characteristics of Deltas 172* • COASTAL LAGOONS 174: *Laguna Madre 175, Lagoons of Central Texas Coast 176, Barataria Bay, Louisiana 176, Lagoons of Northwest and West Florida 180, Florida Bay 180, Gulf of Batabano, Southwest Cuba 180, Lagoons of Central East Coast, United States 182, Lagoons of Western Baja California 182, Lagoons of the California Coast 182, Wadden Sea of The Netherlands 185, Lagoons of Northern Gulf of Guinea 185, General Characteristics of Coastal Lagoons 187* • ESTUARIES 188: *Chesapeake Bay 188, Estuaries of Southern New England 188, Columbia Estuary 190, Yaquina Bay, Oregon 190, San Francisco Bay 190, Tampa Bay, Florida 192, Gulf of Paria 192, Estuaries of Western France 192, The Wash, Eastern England 193, The Rias of West Galicia, Spain 193, Inland Sea of Japan 193, General Characteristics of Estuaries 194*

9: CONTINENTAL TERRACES 196

INTRODUCTION 196 • DEFINITION OF TERMS 197 • EFFECTS OF GLACIATION AND SEA-LEVEL CHANGES 198 • TECTONIC TYPES OF CONTINENTAL TERRACES 199 • DESCRIPTION OF CONTINENTAL TERRACES 200: *Eastern North America 200, Labrador 200, Newfoundland 201, Gulf of St. Lawrence 202, Nova Scotia 203, Gulf of Maine and Georges Bank 204, Nantucket Island to Cape Hatteras 209, Cape Hatteras to Straits of Florida 212, Blake Plateau —Straits of Florida, and the Bahama Banks 214, Gulf of Mexico 218, West Florida 218, Mississippi–Alabama 221, Greater Mississippi Delta 222, Texas–Western Louisiana 224, Eastern Mexico 226, Northern South America 227, Eastern South America 229, Western South America and North America to Gulf of California 231, Western Baja California and California 233, Northernmost California, Oregon, and Washington 240, British Columbia and Southern Alaska 241, Bering Sea and East Asia 242, Australia 252, Southern Asia 255, East Africa 258, West Africa 259, The Mediterranean 263, Western Europe 265, The Arctic 273, The Antarctic 275*

10: ORIGIN AND HISTORY OF CONTINENTAL TERRACES 276

INTRODUCTION 276 • CONTINENTAL SHELF CHARACTERISTICS 277: *Shelf Topography 277, Shelf Sediments 278* • CONTINENTAL SLOPE CHARACTERISTICS 279 • BASIC TYPES OF CONTINENTAL TERRACES 279 • PROCESSES INFLUENCING SHELF TOPOGRAPHY 284: *Shelf Glaciation 284, Effects of Changing Sea Level 284, Coral Growth on Shelves 286, Sand Ridges on Shelves 286, Nondepositional and Relict Sediment Zones on the Continental Shelves 288* • EXPLANATION OF WIDE SHELVES 292 • MAXIMUM SEA-LEVEL LOWERING DUE TO GLACIATION 292 • CAUSE OF CONTINENTAL SLOPE IRREGULARITIES 294: *Mass Movements on the Slopes 294, Faulting and Folding on the Slopes 296, Slope Erosion 297* • UNDERLYING CAUSES OF CONTINENTAL SLOPES 298 • EVOLUTION OF CONTINENTAL RISES AND THEIR FUTURE 300 • ECONOMIC RESOURCES OF THE CONTINENTAL TERRACES 300

11: SUBMARINE CANYONS AND OTHER MARINE VALLEYS 304

INTRODUCTION 304 • TYPES OF MARINE VALLEYS 305 • DESCRIPTION OF SUBMARINE CANYONS AND THEIR FAN VALLEYS 306: *La Jolla and Scripps Canyons 306, San Lucas Canyon, Baja California 312, Monterey Canyon 315, Astoria Canyon 315, Bering Sea Canyons 319, Tokyo Canyon 321, Congo Canyon 321, Canyons of West Corsica and the French Riviera 321, Canyons off the Northeastern United States and Nova Scotia 323, Great Bahama Canyon 325* • ORIGIN OF SUBMARINE CANYONS 327: *Discredited Hypotheses 328, Subaerial Erosion as a Contributing Cause 329, Turbidity–Current Erosion of Canyons 330, Combination of Submarine Processes as Canyon Origin 331, Upbuilding of Canyon Walls 333* • DELTA-FRONT TROUGHS 334: *Ganges Trough 334, Indus Trough 334, Mississippi Trough 334, Origin of Delta-Front Troughs 336* • SLOPE GULLIES AND THEIR ORIGIN 337 • INTERMEDIATE TYPES OF VALLEYS 338 • FAULT VALLEYS OF THE SEA FLOOR 339 • FOLD VALLEYS 340 • UPLIFTED SUBMARINE CANYONS 341

12: CORAL REEFS 342

INTRODUCTION 342 • BACKGROUND INFORMATION 343: *Organic Reefs Defined 343, Types of Reefs 343, Reef Ecology 346* • ATOLLS 348: *General Character 348, Unusual Atolls 352, Drowned and Emerged Atolls 354, Slopes to Sea Floor Outside Atolls 354, Geophysical Prospecting 355, Borings into Atolls 355* • ISLAND BARRIER REEFS 358 • TERRACES AND BASINS OF ATOLL AND ISLAND BARRIER LAGOONS 360 • ORIGIN OF ATOLLS AND ISLAND BARRIER REEFS 361: *Effects of Glacially Controlled Sea Levels 363* • QUEENSLAND GREAT BARRIER REEF AND OTHER SHELF REEFS 366

13: DEEP-OCEAN FLOOR TOPOGRAPHY 369

INTRODUCTION 369 • TOPOGRAPHIC FEATURES DEFINED 370 • GENERAL SHAPE OF OCEAN BASINS 371 • THE ATLANTIC 371: *Mid-Atlantic Ridge and the Rift Valley 372, Atlantic Fracture Zones 372, Deep-Sea Channels of the Atlantic 376, Trenches of the Atlantic 376, Effects of Concentrated Bottom Currents 377* • THE PACIFIC 379: *East Pacific Rise 380, Pacific Fracture Zones 380, Channels and Abyssal Plains of the Northeast Pacific 382, Volcanic Ridges and Guyots of the Central Pacific 383, Arcuate Ridges and Trenches of the Western Pacific 388, Trenches of the Pacific 389* • INDIAN OCEAN 393: *Indian Mid-Ocean Ridges 393, Aseismic Ridges and Plateaus 395, Bengal Fan 395* • MEDITERRANEANS 395: *Arctic Ocean 397, Gulf of Mexico and the Caribbean 397, The Mediterranean and Black Seas 402*

14: DEEP-SEA DEPOSITS AND STRATIGRAPHY 404

INTRODUCTION 404 • SURFACE SEDIMENTS OF THE DEEP SEAS 405: *Sources of Deep-Sea Sediments 405, Classification of Deep-Sea Deposits 406, Classification of Deep-Sea Sediments 407, Abyssal Clays 407, Quartz and Feldspar 408, Zeolites (Phillipsite) 408, Manganese Nodules and Crusts 409, Cosmogenous Components of Sediments 410, Biogenous Deep-Sea Sediments 410, Volcanic Sediments 411, Deposits of Turbidity Currents and Other Ocean-Floor Currents (Contourites) 411, Glacial Marine Sediments 412, Sands on the Sea-Floor Highs 412, Erosional and Nondepositional Areas 412, Distribution of Deep-Sea Sediments 413* • QUATERNARY SEQUENCE FROM LONG CORES 415 • TRANSOCEANIC SEISMIC PROFILES 418 • DEEP-SEA DRILLING (JOIDES) 419: *The Atlantic 420, Gulf of Mexico 424, Caribbean Sea 425, Mediterranean Sea 425, Pacific Ocean 426, Summary of JOIDES Results 429*

REFERENCES 431

CONVERSION TABLES 482

GAZETTEER 483

INDEX 491

Preface to the Third Edition

In the second edition of *Submarine Geology,* I noted "there has been a surprising growth of information in what was until recently a very small field of investigation." Now it could be said without exaggeration that the marine studies in geology constitute a large field in which both marine operations and reports are increasing on an exponential curve that looks like that of the population explosion. Trying to keep abreast of the important new information in the field reminds me of the old movies of Charlie Chaplin rushing up a rapidly descending escalator.

I had not planned on writing a third edition so soon after the second appeared, but the rapid development of plate tectonics and its many applications to all phases of marine geology and the flood of new information coming from the deep-sea drilling program (JOIDES) left the older version sadly inadequate. New chapters were desperately needed and old ones had to be rewritten almost completely.

In virtually starting over again, I decided to eliminate several sections, including two chapters which did not fit in well with the rest of the book. The extensive discussion of wave mechanics and mechanics of sedimentation has been considerably simplified to come into line with the rest of the text. Both these subjects require books by themselves, as does the omitted chemistry of marine sediments. The relation between modern sediments and ancient sediments has also become so well documented that it, too, should be treated in a separate book. These account for omissions. On the other hand, a chapter on plate tectonics seemed required near the beginning of the book, and many people had suggested that there should be more treatment of deltas, estuaries, and lagoons; therefore a chapter of these is added. Because of the tediousness of making two trips around the continents for separate chapters on continental shelves and continental slopes, these have been combined into one chapter on continental terraces. The vast amount of new information has caused this chapter to grow beyond where it was intended, but it is difficult to omit many of the valuable additions that have come since writing the second edition.

An attempt is made here to refer to many new articles and books without discussing them extensively. It is hoped that this will provide the reader with a source of additional material on any subject of interest to him in the field.

Much help in preparing this edition has been given me by many friends. Critics of the various chapters include: Andre Rossfelder and Neil F. Marshall (2), Robert S. Arthur (3), D. L. Inman (3, 4), Sir Edward Bullard and R. S. Dietz (5), H. W. Menard (5, 13), F. B Phleger (7, 8, 14), R. E. Stevenson (8), Elazar Uchupi (9), J. R. Curray (9, 10), R. F. Dill, P. R. Carlson, and C. H. Nelson (11), W. A. Newman (12), J. W. Hawkins and S. M. Smith (13), and G. G. Arrhenius, Dorothy Echols, Frances L. Parker, and E. L. Winterer (14). A section of Chapter 2 was contributed by E. L. Hamilton. Information on the Indian Ocean and on the deep dredging methods was supplied by R. L. Fisher. Help with the discussion of deep-ocean drilling was provided by M. N. A. Peterson and E. L. Winterer. Numerous helpful suggestions on the style of the manuscript have been made by Margaret R. Miller. My wife Elizabeth has also been helpful with the manuscript.

Francis P. Shepard

Chapter 1
INTRODUCTION
AND HISTORY

Geologists have always known the importance of investigating the world's vast submerged territory, but until the 1940s it remained virtually a terra incognito. Finally, the world's worst war, along with all its horror, had among a few redeeming features the persuasion of naval scientists and their civilian advisors that naval operations needed a background knowledge of the ocean and its floor. The geological study of this ocean realm was first called *submarine geology* but is now more commonly referred to as *marine geology,* or less commonly as *geological oceanography.* The three names can be used synonymously. Included in this field are the study of the coasts and beaches as far as they are related to marine processes; of the continental terraces, consisting of broad shallow-water platforms surrounding the continents and the relatively steep slopes that extend to the abyssal depths; and of the deep-ocean floor. The last, by far the largest area of the oceans, is perhaps the most interesting portion, although the continental terraces certainly contain the greatest economic resources.

As information has increased, it has become evident that processes influencing the sea floor have far less relation to depth of water than was believed to be the case before bottom photography and drilling in the deep oceans exposed much new evidence. An important phase of marine geology is the study of sediments from the sea floor. The relation of these sediments to topography, both on the continental terraces and in the deep ocean, is of special interest. The developments of electronic methods in sound transmission through the ocean bottom and in determining magnetism, gravity, and heat flow in the ocean crust have collectively been of enormous importance to the rapidly developing field.

Marine geology is only a recent science. Among the first indications that the future study of the sea floor had great possibilities were the speculations and observations of Charles Darwin on coral reefs (1842). Somewhat later, a considerable backlog of information came from the history-making voyage of H.M.S. *Challenger* (1872–1876). Although no geologists were included in the personnel, the large collection of sea-floor samples was studied by geologists, including A. F. Renard (Murray and Renard 1891). Some geological information also came from the explorations of the U.S. Fisheries steamer *Albatross* from 1888 to 1920. The resulting samples were studied by Murray and Lee (1909), Louderback (1914), and Trask (1932). Another large collection was made by the Netherlands ship *Siboga* in 1899 and 1900, and these samples from the East Indies were studied by the geologists Böggild (1916) and Molengraaff (1916, 1922, 1930).

Early in the twentieth century, the collections made by the German steamers *Edi Stephan* and *Planet* in European waters were studied by Andrée (1920) and led him to write his pioneer book on marine geology. Also, the German South Polar Expedition of 1901–1903 produced cores studied by Philippi (1912). The Germans continued to play an important role in marine geology with Böggild's (1916) report of the *Siboga* Expedition. Then, shortly after World War I, the German South Atlantic *Meteor* Expedition showed for the first time that the ocean floors have irregularities as great as the lands (Stocks 1933). Early studies of the European continental shelf were made by Dangeard (1928). The *Snellius* Expedition to the East Indies, 1929–1930, led to important reports by Kuenen (1942).

Around the turn of the century, geologists first became interested in the strange valleys of the sea floor. Spencer (1898) made his extravagant claim that these valleys with their great depths were proof that the continents had been uplifted several kilometers, causing the widespread Pleistocene glaciation. Other geologists, including Lawson (1893), had previously reached the interpretation that the valleys were caused by diastrophism, and Smith (1902) attributed them to submarine currents. For many years these different hypotheses were hotly debated.

The development of echo soundings (an offshoot of hunting for submarines with sonic methods in World War I), prompted marine surveyors, especially the U.S. Coast and Geodetic Survey, to construct very detailed charts of the continental shelves and slopes. These made possible bathymetric contour charts along the East Coast (Veatch and Smith 1939) and along the California coast (Shepard and Emery 1941). These gave a much better picture of the submarine canyons off both coasts and showed the significance of the fault scarps and fault troughs off the California coast.

Studies of the bottom character notation of the coastal charts of the world led to the discovery that sediments do not grade seaward across the continental shelves from coarse to fine (Shepard 1932). The extensive sample collection of Trask (1932) provided the first clear indications of the types of marine environments in which petroleum is accumulating. The long cores taken in the Atlantic in the 1930s showed that it was possible to recognize the times of Pleistocene glaciation in the sediments of the sea floor (Stetson 1939; Phleger 1939; Bramlette and Bradley 1940).

Marine geology programs in the United States were begun in the 1930s at Woods Hole Oceanographic Institution, Scripps Institution of Oceanography, and Lamont Geological Observatory (now Lamont-Doherty Geological Observatory). These programs have expanded enormously and since World War II have been supplemented by active geological studies of the sea floor at Texas A & M and the universities of Miami, Washington, Oregon State, Rhode Island, and Southern California.

The U.S. Geological Survey for many years contributed to marine geology studies, but it started its first large project in cooperation with Woods Hole in 1962 and in 1966 began a concerted project for the study of the Pacific at the Menlo Park, California, office. The U.S. Navy initiated its first marine geology studies during World War II and continued with a large group at the Hydrographic Office in Washington and in the Navy Electronics Laboratory in San Diego (now Naval Undersea Research and Development Center).

Another American source of information in marine geology has come from funds of the National Science Foundation, which recently overtook the military in scientific expenditures, particularly since it initiated the Deep Drilling Project in 1965. The American Petroleum Institute has contributed heavily, especially in the study of sediments of the northwest Gulf of Mexico (Shepard et al. 1960) and in the Gulf of California (van Andel and Shor 1964).

In other countries somewhat smaller programs have nevertheless been of great importance to the science. The French have contributed active programs from the University of Paris, under the initiative of Jacques Boucart, Louis Glangeaud, and André Guilcher, studying the sea floor off the Mediterranean and western France. The influence of Jacques-Yves Cousteau, a former French Navy officer, has been very great since he initiated the use of scuba diving and had much to do with deep-diving vehicle development, both providing important visual information to geologists. The British have contributed through their National Oceanographic Institute with the incentive of Morris Hill, E. C. Bullard, and Arthur Stride, and its large sea-going vessels. The Canadians have active programs, both in the East, centered around Dalhousie University in Nova Scotia, and at the University of British Columbia in Vancouver. The Germans had been an important factor until World War II, as has already been indicated, but were set back for many years thereafter and only recently have resumed investigations under the leadership of E. Seibold of Kiel University. Sweden has contributed largely through the efforts of Hans Pettersson and his renowned world-encircling Swedish Deep Sea Expedition in 1947–1948. The geological reports were mostly by Arrhenius (1950), Phleger and others (1953), and Olausson (1960). Japan has an active program connected with the University of Tokyo and the Japanese Institute of Fisheries.

Among the ships that have helped make history in marine geology are the old *Atlantis* of Woods Hole (Fig. 1–1), the *E. W. Scripps* of Scripps Institution (Fig. 1–2), and the *Vema* of Lamont-Doherty (Fig. 1–3). More modern successors include *Atlantis II* of Woods Hole (Fig. 1–4), the *Melville* of Scripps Institution (Fig. 1–5), and *Discovery III* of the National Institute of Oceanog-

Fig. 1–1 The 175-ft motor ketch *Atlantis,* which conducted most of the investigations of Woods Hole Oceanographic Institution prior to World War II. Photo by Woods Hole Oceanographic Institution.

Fig. 1–2 The 104-ft Scripps Institution motor schooner *E. W. Scripps* on which most of the Institution work in marine geology was accomplished prior to World War II.

Fig. 1–3   The 202-ft motor schooner *Vema* belonging to Lamont-Doherty Geological Observatory and used extensively in operations throughout the North and South Atlantic and in the Indian Ocean. Courtesy of B. C. Heezen.

Fig. 1–4   *Atlantis II,* one of the modern fleet of Woods Hole Oceanographic Institution. Courtesy of K. O. Emery.

Fig. 1–5 The *Melville*, a modern ship of Scripps Institution of Oceanography. Photo courtesy of University of California at San Diego.

Fig. 1–6 *Discovery III* of England's National Institute of Oceanography. Photo by J. R. Curray.

Fig. 1–7   The *Lomonosov*, one of the large oceanographic ships of the Soviet Union.

raphy (Fig. 1–6). The large fleet of oceanographic ships operated by the Soviet Union (Fig. 1–7) outclasses that of all other countries.

The first large-scale development in marine geology came directly after World War II, when the U.S. Navy began extensive subsidizing of the studies at various oceanographic institutions, both through the Office of Naval Research and the Bureau of Ships. This made it possible to map widely the bathymetry of the oceans, producing the well-known maps by Heezen and Tharp of the Atlantic, Pacific, and Indian oceans, which appeared in the *National Geographic* (Oct. 1967, June 1968, Oct. 1969), and those of Menard (1964), which have revealed the character of the Pacific. The most recent additions are of the Indian Ocean by Fisher and others. (1971). Mapping of the oceans has also been pursued on a large scale by the Soviet Union scientists, who have issued valuable bathymetric maps, particularly those around Antarctica. The Soviet pioneering work was done by Klenova (1948), and, more recently, leaders in the field have included A. P. Lisitzin, A. V. Zhivago, and V. P. Zenkovich.

In the 1950s, an important development in marine geology was the initial use of seismic reflection profiles, showing the character of the structure and formations underlying the ocean floor (McClure et al. 1958). Many of the continental slopes of the world have been investigated by this method and innumerable profiles have crossed all the oceans, providing an enormous fund of information during the past twenty years.

The 1960s saw the development of a revolutionay change in thought in marine geology and, in fact, in the entire science of geology. The idea of *sea-floor spreading,* subsequently termed *plate tectonics,* had its seeds in the Wegener (1912) hypothesis of continental drift, but its modern and very different form was the result of independent but almost simultaneous articles by Dietz (1961) and Hess (1962). Since the idea was proposed, it has been documented in many ways by a group of geophysicists. The magnetometer tests are said to provide almost irrefutable confirmation (Vine and Hess 1970), and the earthquake motion studies have been equally favorable (Sykes et al. 1970). Heat flow probes into the ocean floor are of direct support (Bullard 1970). What seemed to be the most needed test of the theory has been made by the deep drilling into the ocean floor (JOIDES) from an essentially stationary platform on the ship *Glomar Challenger*. Begun in 1965, the project has collected samples from more than 300 holes (March 1972), and drilling is still continuing. The results to date appear to confirm the general idea of the spreading plates moving out from mid-ocean ridges where lavas are rising.

The interval between the second and third editions of this book has seen only two brief summary books on marine geology, Ottmann (1965) and Keen (1968). However, the appearance of volumes 3 and 4 of *The Sea*, edited, respectively, by Hill (1963) and Maxwell (1970), covered so many subjects in marine geology and related geophysics that it has made the intervening decade highly productive.[1]

In writing the second edition of *Submarine Geology*, I seriously considered converting units to the metric system but reluctantly decided that fathoms were more useful because most charts were using them and most oceanographic ships were taking soundings in fathoms. Since then, the swing toward the metric system has been accentuated and many bathymetric charts are now given in meters. Accordingly, the metric system is adopted in this edition except where old contour charts in fathoms have been repeated. Also, current velocities are given largely in centimeters per second and in knots. For rough conversion from meters to feet one should multiply by 3.28, and from meters to fathoms, divide by 1.83. To convert centimeters per second to knots, divide by 50.

Because of the large number of geographic names, especially in Chapter 9, a gazetteer is provided as an appendix. This includes latitude, longitude, and figure number of map where one is available (see Appendix).

---

[1] Since completing this new edition, two important books have emerged. The publication of *The Face of the Deep* (Heezen and Hollister 1971) and the submission for publication of *The Western North Atlantic* (Emery and Uchupi, in press) appear to be monumental developments which, unfortunately, cannot be included in this text.

Chapter 2
METHODS AND INSTRUMENTATION USED BY MARINE GEOLOGISTS

INTRODUCTION

The 1960s was a period of rapid development in techniques employed by marine geologists. The most spectacular results have come from the new methods of drilling into the deep ocean floor from a stationary ship. The improvement of seismic methods for determining the structure under the ocean bottom was also remarkable during this decade. For lowering instruments, tapered wires of high-strength steel now allow heavy weights to be supported at greatest depths. The new designs of deep-diving vehicles enable geologists to obtain visual examination of many bottom features previously seen only in pictures taken by remotely controlled cameras. Sonic instruments towed near the bottom produce far more accurate profiles of the topography than those of surface ships. The new method of side scanning from deep-towed instruments or surface ships covers a wide path on either side, adding a third dimension. Considerable advancement toward acquiring greater position point accuracy during operations has occurred with such innovations as satellite navigation and submerged buoy patterns with sonic devices (transponders) allowing triangulation. Various other electronic positioners have also been used with great success. Shipboard computers and digitizing data reduction systems are now widely used.

In sampling, we have seen the development of the unattached free instruments that are recovered with the help of flashing lights, radio, and radar. They are now important time-saving methods and allow taking multiple cores or grab samples during a traverse. The distortion of sediment structures inherent in coring operations is greatly reduced by the use of the large box corers that bring up samples with only slight disturbance of their structure, often showing, also, the character of the bottom from which they were collected. Using the free-instrument method, it is now possible to obtain good current-meter records from the bottom at all depths.

POSITIONING AT SEA

The improved methods of navigation have eliminated much of the difficulty encountered in locating positions in the early days of marine geology. Electronic devices, usually operated by the ship's officers, take care of most problems. In general, plotting relies on intersection of two lines of position, two circles, or hyperbolas. However, it is well for the scientists to keep careful check on the positions provided by the bridge, because many ship officers do not appreciate the importance of scientific accuracy. Also, in operation on smaller coastal boats, it is often necessary for the scientists to use the older methods of bearings and horizontal sextant angles. In fact, these often provide a good check on the electronic methods and, in some instances, may prove more accurate.

Bearings and Horizontal Sextant Angles Working near shore in an area with accurately located land objects or with buoys and beacons allows easy and accurate locations, either by taking bearings or horizontal sextant angles. Most oceanographic ships are equipped with a pelorus mounted on the repeater compass on each wing of the bridge. If taken accurately, two bearings will establish a position, but it is always advisable to take three or more bearings to avoid errors. The swinging of the compass in a rough sea may result in a false reading, because the compass is apt to be slow in adjusting to a heading deflection produced by the sea. If the vessel is under way and only one bearing can be taken at a time, another source of error can come from the time necessary to complete all bearings. With gyro compasses, the bearings do not need to be corrected unless there is a known gyro error. With magnetic compasses, the readings must be corrected for the regional magnetic variation and the deviation of the compass due to remnant magnetism in the ship.

Considerable accuracy is possible using a sextant or a hydrographic circle for measuring the horizontal angles between three well-located objects. Unless the angles are small or the objects used for the angles are on or close to the circumference of a circle in which the vessel is also located (Fig. 2–1), sextant angles will provide great accuracy. These are plotted by means of a three-arm protractor. Using two sextants that are "shot" simultaneously adds additional value to the method, because there is no lag between the time of the observation and the mark placed on the sounding record. In taking sextant angles, be careful to hold the sextant approximately horizontal. If one position

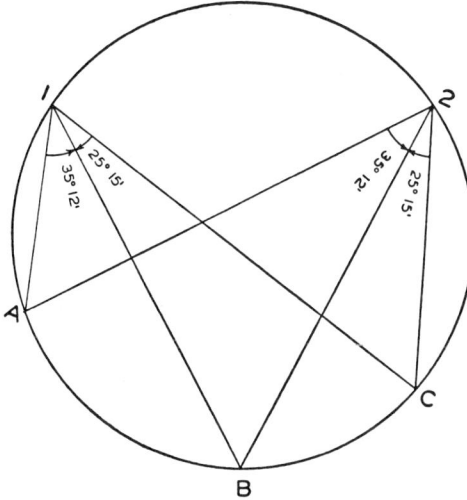

Fig. 2–1 A "swinger" where the three points, A, B, C, used in taking horizontal sextant angles, are on the arc of a circle that also includes the ship. In this case, the same sextant angles will be obtained at any point on circle, as at 1 or 2.

point is at a higher elevation than the other, choose an object below the higher point so that the two points will be on the same level; otherwise the angle will be too large.

Navigation from Chart Soundings Where good sounding charts exist (and this applies to all the echo-sounding surveys off the coasts of the United States despite any minor inaccuracies), it is frequently possible to obtain positions by running sounding lines between work stations. If these soundings are plotted on strips of paper to the same scale as the chart and the strips are superimposed onto the chart, a fit can be made by moving the paper strip around in the general vicinity, keeping it oriented along the course direction. An example is illustrated in Fig. 2–2. The fix often will not be accurate, because the speed and course may not be properly estimated owing to water currents. In this case, the soundings can be stretched or compressed somewhat, and a fit may then be discovered. The method is totally impractical, of course, where horizontal bottom is encountered. Navigators familiar with

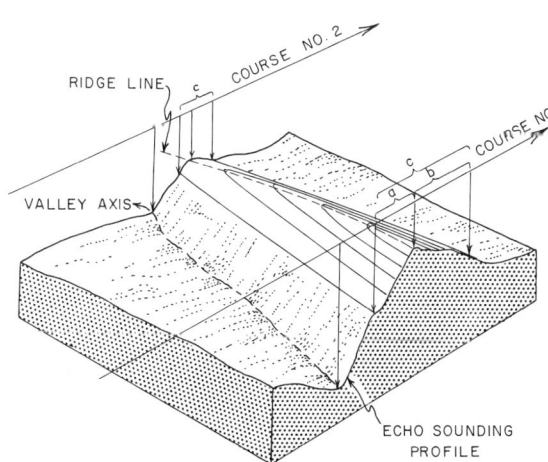

Fig. 2–2 A position can often be obtained by the depth of a crossing of a valley and ridge. From Adams (1939).

other methods may object at first to this innovation and prefer to use such methods as radio bearings, but comparisons of results off Southern California, where the topography has considerable relief, have shown that the sounding method is superior in most cases except in close proximity to radio stations. If the echo sounder is not working, it may be helpful to take wire soundings at intervals as frequent as practical in order to maintain a position.

Electronic Position Finders In modern ship operations, virtually all positions are acquired by using electronically operated instruments. In work near a well-charted coast, it is possible to use radar with which fixes are obtainable for about 25 km from low coasts and about 55 km from land margins with steep bluffs or large buildings. Radar (an acronym for radio direction and range) operates at 3000–10,000 Mc/sec with a wavelength of 3–10 cm. This is a line-of-sight method, which, under favorable refraction conditions, will go beyond the horizon and will invariably carry much farther than visual fixes. It is important to check the radar instruments frequently; they tend to go out of adjustment very easily. Because of errors, it is best to use at least three points on the radar and to plot both distance and bearings, although the latter are generally less accurate. Three good distances will provide an accurate position by drawing three arcs of circles set on a large pair of dividers at the scale of the chart. Where the arcs intersect is the location. If the arcs do not exactly intersect, use the center point of the triangle formed by the three-line crossings, but give consideration to the bearings and to the expected position of the ship, obtained by using dead reckoning (course and speed). Often one arc may be eliminated because it is clearly out of line with the expected position.

Radar can also be used for detailed surveys of submarine relief at considerable distances from land where absolute position is uncertain. With a good radar reflector or radar transponder mounted on a buoy, located, for example, on top of a seamount, lines can be run in relation to the buoy. Care should be taken to account for the tide, wind, or current shift of the buoy position if it is at all slackly moored. With this method, good relationship between lines can be obtained even if the exact latitude and longitude cannot be determined.

Various other electronic position finders are available with "slave stations" on land, some of them very accurate, that is, less than a ship's length in a distance of 150 km from the slave stations. These include Moran, Raydist, Decca, Map, and others, all very expensive and hence not available to most oceanographic institution vessels, but they are used by petroleum companies and the United States government services. For oceanographic institution ships, by far the most practical method for navigation at points beyond good radar control is the use of satellites (Talwani 1970B). This method produces an accuracy of about 200 m, and using a computer system it is possible to obtain a good fix within a few minutes of each satellite pass. The Omega network has now been completed for positioning in the Northern Hemisphere and will soon be available in the Southern Hemisphere.

SOUNDING METHODS

Until 1920, all depth soundings were made by lowering a weight on the end of a line (wire was used after 1870 to eliminate stretching of the line) and determining the amount of line necessary to reach bottom. To get a true depth, this method requires maintaining the wire nearly vertical. To reach great depths, this has always been a tremendous undertaking. The first deep-ocean sounding was taken by Sir James Ross in 1840. He let out almost "12,000 feet" (3660 m) to reach the bottom of the Atlantic, and because he used hemp rope, which was subject to stretch, his depth measurement was far from accurate. Wire line sounding is very difficult, even using high-speed winches and highly maneuverable vessels that can maintain approximately vertical wire angles. The time consumed for deep soundings runs into hours Consequently, only some 15,000 deep-ocean soundings had been taken until 1923.

Echo Soundings Almost all soundings are now obtained by sound impulses, known as echo sounding. The devices are ordinarily set at a speed of 1460 m (800 fm)/sec. For precise echo sounding, it is absolutely necessary to have a frequency-regulated power source. Otherwise, slight variations in frequencies and fluctuations in voltages will cause the speed of the synchronous drive motor to vary and result in relatively large errors in the recording of the travel time between outgoing and incoming sound impulses. These errors appear as sudden drops or rises on the echogram. Most survey echo sounders now have an internal frequency standard that provides stable drive voltages to the drive motors, producing accuracies in the range of .5 to 1.0 percent. However, at sea, the scientist must be constantly aware that these devices may fail, and he should be able to distinguish a power frequency shift from a topographic change on the sea floor. Many, perhaps most, of the lines of soundings taken across the ocean until about 1950 were obtained without a frequency regulator, so that the soundings are confused because of errors and are therefore quite useless in constructing good charts. Only since the constant frequency devices have been acquired has it become possible to get reasonably good profiles of the deep ocean.

Another difficulty with the earlier versions of echo sounders has been that they record on too small a scale and fail to show minor irregularities. The older instruments generally had a large scale for shallow depths and a small scale for great depths. The precision depth recorders, Raytheon (PFR), Gifft (PSR), and Ocean Sonics (OSR), show details of the bottom with about 1 m of relief at any depth. The older precision depth recorders (PDR) are still useful, particularly for great depths. All these recorders change automatically when they reach the limit of a scale (Fig. 2–3) . Wet and dry recording papers are used, and many recorders now have programming circuits so that an operator may select optimum signal returns for highest quality visual display. The method does not give absolute depths with this accuracy, because the speed of sound is not well known throughout the water column (see Matthews' tables, 1944). However, with return to the same position and use of the same frequency, the same depth should be obtained except for such minor differences as may come from variations of temperature or salinity and, of course, height of tides. As far as we know, significant changes of temperature and salinity occur only in near-surface water. Sound travels faster the higher the salinity, the higher the temperature, and the higher the pressure, but the differences are not very great.

Correction of Soundings One of the problems confronting both the surveyor and the scientist regarding echo soundings is whether or not to apply corrections to give true depths rather than the depths recorded using a velocity of 1460 m/sec. The actual depths of the deepest parts of the ocean should be known to leaders of expeditions investigating these depths so that they can apply corrections for sound velocities. On the other hand, to make soundings useful for navigation for vessels operating with echo sounders set

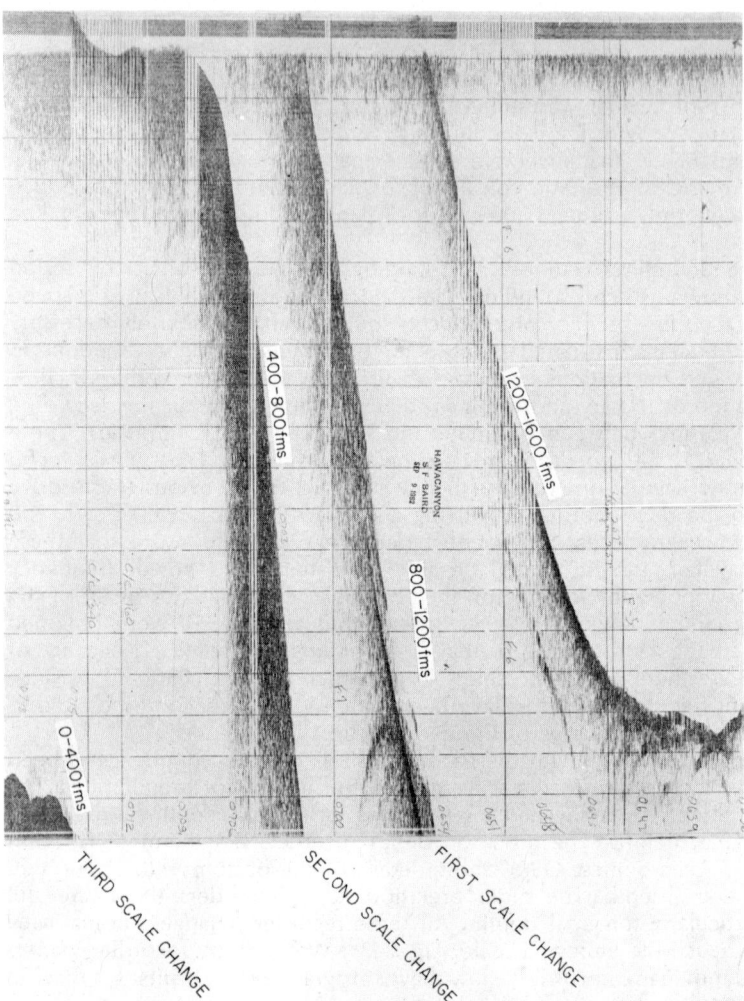

Fig. 2–3 An example of the scale changes in a fathogram during the crossing of a slope. It is important to label each scale on the fathogram.

for 1460 m/sec, much can be said in favor of publishing uncorrected soundings. For example, in crossing a valley, trough, or ridge with a relatively even axial slope, a position can be obtained by comparing the depth at which the axis is crossed with the chart (Fig. 2–2). However, if the published depths are corrected so that they do not agree with the fathometer of the vessel, the navigator may be thrown off position. As an example, the corrected depths in crossing San Diego Trough (Fig. 2–4) are offset by more than a kilometer from the uncorrected depths; hence, a vessel using a good survey for this approach would be thrown off course by this distance unless the navigator laboriously applied corrections for sound. Mariners should bear in mind that many charts, such as those of the U.S. Coast and Geodetic Survey, apply these corrections.

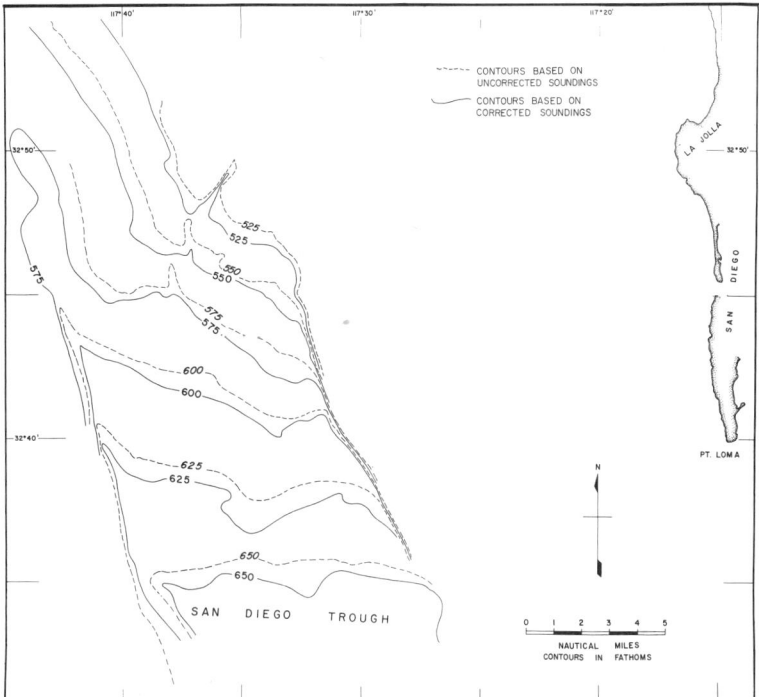

Fig. 2–4 Contours may be considerably shifted by correcting depths for speed and sound. In this case, the navigator crossing San Diego Trough and using a ship's fathogram and corrected chart soundings for his location would be a kilometer or more out of position.

Actually, however, there are few places where the corrections will interfere seriously with echo sounder navigation.

Another inaccuracy in echo soundings comes from the fact that the sound beam sent out by the transducer or sound head has a spread in most cases of about 30° from the vertical on both sides. As a result, where there is a steep submarine slope, the echo usually comes from the nearest point on the slope, rather than from the bottom directly beneath the vessel. This can make a large difference. However, in crossing a valley, it is often possible to get the true valley floor depth by observing a curved surface that often rises as an arc beneath the echoes coming from the walls (Fig. 2–5). This may come in with a strong echo when the center of the valley is crossed. Narrow-beam echo sounders between 5° and 7° are now being used on a few oceanographic ships (Krause and Kanaev 1970). They are of special value in delineating small sea-floor features. Work is also under way in using tridimensional bathymetry with a sweeping array of transducers.

Deep-Tow Methods It is possible to avoid the inaccuracies in echo-sounding profiles by towing an instrument near the bottom and using a narrow-beam echo sounder combined with an upward directed sounder (Spiess and Mudie 1970). This eliminates most of the side echo confusion. Towing speeds are about 1 m/sec (2 kn). Locations are obtained by acoustic transponders fixed on the sea bottom. The information is provided by telemetering

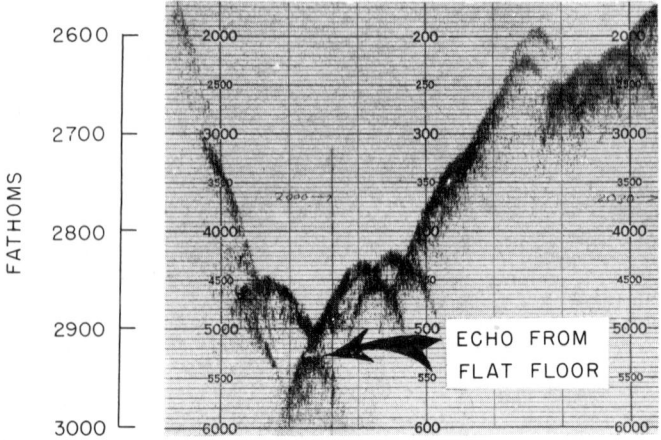

Fig. 2–5   The inverted V that often shows the true bottom of a submarine valley. Note that it is 20 fathoms (36 m) deeper than indicated in the fathogram.

through the towing cable. Using this method, minor irregularities on the deep-sea floor, which are entirely missed by echo soundings from surface ships, can be identified.

Side Scanning   Another method for obtaining a picture of the bottom on either side of a vessel is to use the asdic transducer, similar to that developed for hunting submarines and directed down from the surface by a few degrees (Fig. 2–6A). This method, called *side scanning,* was devised by Tucker and Stubbs (1963) and has been used extensively by Stride (1961A, B) and for deep tow (Mudie et al. 1970). It shows such features as narrow ridges, folds, faults (Fig. 2–6C), and outlines of sandy and rocky bottom (Fig. 2–6B), and even distinguishes fine and coarse sand.

SAMPLING METHODS

In the early days of oceanographic exploration, most bottom sampling was by dredging, and the primary purpose was to collect marine life. Small coring devices have been available for almost a century, and samples were obtained even earlier by placing tallow in the hollow cup at the base of a sounding lead. Except in recent years, all samples were procured by lowering attached instruments to the bottom and bringing them back with, or often without, a sample. In deep water this consumes hours of time and, because most oceanographic ships cost thousands of dollars a day to operate, the time devoted to sampling is an important consideration. In order to conserve time, free instruments have been developed that are cast loose from the ship, and after reaching the bottom, take a sample, photograph, or record data over a period of time and then return to the surface after dropping a weight that leaves them buoyant. The method was envisioned by M. Ewing and Vine (1938), and several successful types of free instruments have been developed subsequently (Isaacs and Schick 1960; D. G. Moore 1961B; Sachs and Raymond 1965; Walthier, Rossfelder, and Schatz 1971). The method is now used

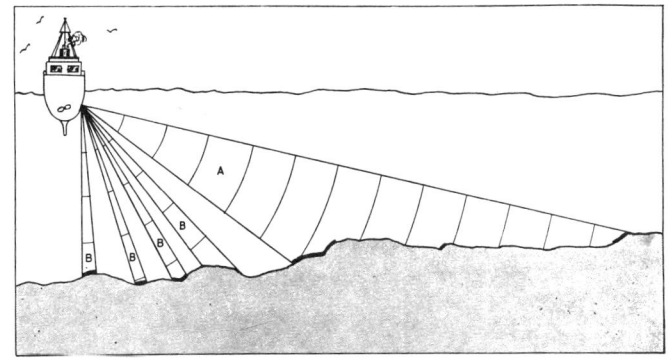

A

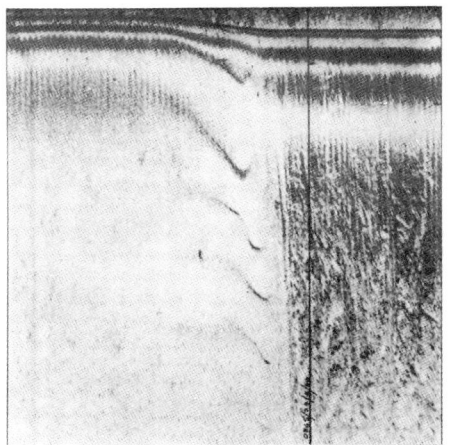

B

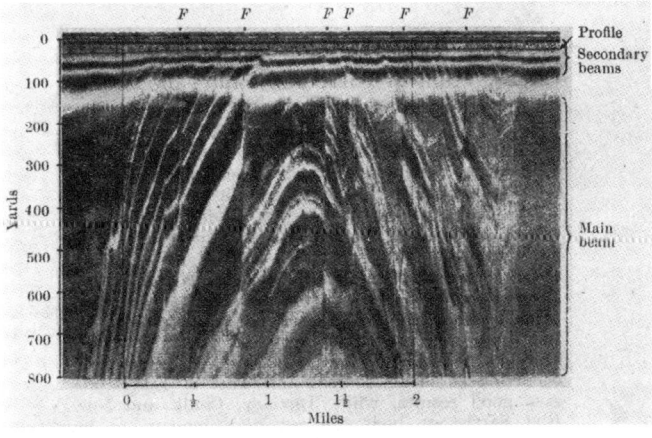

C

Fig. 2–6   A, method of side scanning. From Stride (1961B). B, side scanning showing contact between rock floor (right) and sand bank (left) with sand waves at edge of bank. From Stride (1961A). C, side scanning showing rock outcrops, gentle fold, and small faults. From Stride (1960).

extensively and patterns of samples are taken in areas, arranging time so that they are dropped and recovered in sequence. However, two difficulties exist: The samples are generally not as good as in attached devices; and some samplers are not recovered, either because the release fails to work or because the instruments are not seen when they surface. Attached lights, radio, and radar targets help in the recovery (Martin and Kenny 1971). From a large number of samplings, C. E. Schatz (personal communication) found losses on grab samplers of less than 10 percent, and in view of the time saving this means about 10 times more material collected for the same cost of ship time.

For the ordinary types of samplers, a general survey of the field has been made by Hopkins (1964) and Bouma (1969). A total of 55 samplers are included with references dating from 1911 to 1961. Some of these will be discussed here along with several more modern devices.

Grab Samplers The first bottom samples were procured by "arming" the cup in the bottom of the sounding lead with tallow. This generally brings a trace of sediment back to the surface after each sounding. Near the beginning of the century, a more effective method was devised, consisting of steel jaws held open during descent of the weighted sampler and then, upon

Fig. 2–7 Large Campbell grab. From Emery and Schlee (1968).

striking the bottom, the jaws are closed by a spring or other type of release. These samplers vary from small telegraph snappers that hold only a mass that would fill an egg cup to huge jaws that will hold approximately a cubic meter of material. The orange-peel buckets (Shepard 1963, Fig. 10A) have a canvas top to inhibit the stirring of the sediment during the return to the surface, but if the sample is taken at a depth greater than 400 m, water is likely to circulate sufficiently to leave a much disturbed sample. The large samples in the Petersen grab (Shepard 1963, Fig. 10B) may be so little disturbed that on returning to the surface it is possible to obtain good stratification by opening the top and inserting a tube to get a short core. The large Campbell grab (Emery and Schlee 1968) (Fig. 2–7) proved very successful in the extensive operations on the continental shelf off the East Coast of the United States, carried on by Woods Hole and the U.S. Geological Survey. It even contains a camera, allowing bottom photography prior to sampling.

Small snappers often bring up a very unrepresentative sample of the bottom, because a piece of gravel or a shell may get caught in the jaws and hold them open sufficiently for the finer material to wash out while surfacing. Also, the jaws of the smaller grab sampler are likely to become bent on encountering rock bottom, preventing the sampler from making a good seal when closed.

Coring Devices Cores of the sea bottom are in many ways superior to grab samples. Cores make it possible to distinguish between the most recent sedimentation and that of an earlier period, which is often quite different. Although the large grab samplers may also show this stratification, it is usually somewhat disturbed, making it difficult to tell what is recent and what is older material. Coring techniques have greatly improved in recent years and many new methods have been devised. Gravity corers are still used rather commonly despite the considerable advantage of piston coring. The Phleger corer (Fig. 2–8) is a small practical means of obtaining short core samples and can be used manually from small boats. Another light coring device was developed by Mackereth (1958). This makes use of "hydrostatic pressure operating on a cylindrical anchor chamber embedded in the sediment." It can be used only in relatively shallow water. The core is obtained in a plastic liner which allows easy storage. Most small gravity corers take a core of 2.5-cm diameter, but 5- or 6.5-cm core barrels can be used just as easily and will ordinarily take somewhat longer cores. Good corers should include: (1) large water escape valves; (2) a firmly closed core retainer; (3) a positively closed top check-valve; (4) ratios devised by Hvorslev (see Rosfelder and Marshall 1967).

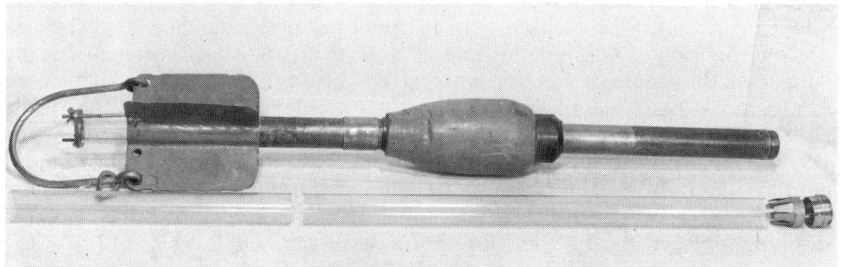

Fig. 2–8 Phleger bottom sampler used to obtain short cores. The core liner and core catcher (right) help prevent escape of sediment.

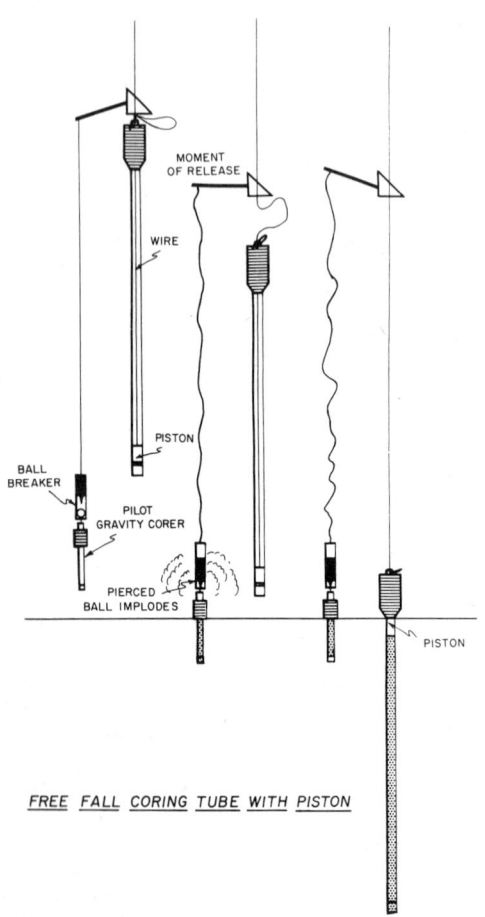

Fig. 2–9 Operation of the free-fall method of coring and the action of the piston in producing a long core by reducing friction on the inside of the core tube. Ball breaker is also included. Ball breakers have now been discontinued as impractical.

Because of the danger of fouling wire, it is unwise to lower a core tube at high speed. For this reason, the free-fall method of coring (Hvorslev and Stetson 1946) has proved helpful in increasing the velocity and momentum of the core tube prior to hitting bottom (Fig. 2–9). The device is lowered to the bottom rather slowly, where a suspended weight hits first, releasing a lever arm that allows the core barrel to fall freely for a few meters and sink rather deeper into the bottom than if it were directly attached to the evenly descending cable. Another advantage is that the free-fall corer avoids the danger of the operator not recognizing the bottom when an ordinary corer is lowered in deep water. The free fall gives a much clearer indication on the winch tensiometer than does the lowering of an ordinary corer to the bottom. The use of a *pinger* gives another indication when all types of instruments get near the bottom. The pinger is attached to the wire above the instrument. The echo is recorded on the fathogram aboard ship and shows the height of the pinger above the bottom during the approach.

The piston corer, invented by Kullenberg (1947), also falls freely after being released by the same method as the Hvorslev and Stetson corer. The piston in the inside of the core tube remains attached to the wire so that the core tube falls past the piston and hence avoids serious friction in taking the core (Fig. 2–9). Numerous modifications of the Kullenberg device have been made.

A serious difficulty in piston coring develops whenever the core barrel is not completely filled when the tube stops penetration. In such cases, the lifting of the apparatus by pulling on the piston tends to raise the core in the barrel and suck in bottom material or water from below. Thus the core is deformed by being pulled up through the barrel, and material of unknown length is added to the bottom. A common result is that the lower portion of the core has vertical flow lines indicative of the flow of material into the barrel during retraction. Core material may also be sucked into the tube when the device hits hard bottom and tips over, pulling up the piston and drawing in surface sediment (Bouma and Boerma 1968). Consequently, many cores have a deceptive appearance. The nature of the top of the core can often be determined by a study of the material obtained in a small gravity corer used as a pilot weight. The two cores may be comparable provided the piston had been pushed to the top and the tube was filled before retraction. In any case, the pilot core ordinarily will be considerably shorter than the piston core (Fig. 2–10); the difference in length is caused by frictional compaction of the sediment in the pilot corer, which has no piston (Emery and Dietz 1941), whereas the piston corer causes only minor changes of length.

Fig. 2–10 An example of the shortening of layers in a gravity core (right) compared to a piston core (left). These cores were taken, respectively, with a pilot core and the piston corer, as illustrated in Fig. 2–9. It is unusual that a piston corer shows as much of the top layer as indicated in this core by the Lamont-Doherty Geological Observatory. Photo from Lamont-Doherty Geological Observatory.

To overcome the difficulty of sucking up the core into a partially filled tube, it is necessary to have a device that will lift the core tube without pulling up the piston. Various methods have been devised to deactivate the piston after the core has ceased to enter the bottom of the tube. Those requiring external deactivators encounter some difficulties that are avoided in a device developed by Woodruff (1970), which makes use of a ball valve inside the piston.

One difficulty with piston coring has been the extraction of the core from the tube. If the hydraulic jack is used, the core is generally very disturbed. Core liners that fit inside the tube avoid this difficulty but can increase internal friction, making cores shorter. One method has been to use strong plastic tubes for the core barrel, which are subsequently removed and stored after each operation so that the core can be extracted in the laboratory by cutting the tube (Inderbitzen 1963).

The upper portion of a piston core often does not represent the surface material, particularly where soft mobile mud at the surface overlies sand. Comparing numerous Phleger gravity cores with piston cores, Ross and Riedel (1967) showed that the former gave a truer representation of the surface layer. This depends on the sediment and operation techniques. Some piston cores show surface material adequately (Fig. 2–10).

Kullenberg (1955) maintained that it is possible to obtain extremely long cores in soft bottom by the use of a piston coring device that trips only after it has penetrated for some distance. The friction that stops a core barrel from penetrating comes very largely from the walls, rather than from the penetration of the lower end of the tube. Therefore, by building a very heavy corer without great length, it is possible for a corer to penetrate until the sediment becomes sufficiently compact to stop the descent. A rope unwinds as the core tube sinks until a given depth is attained. At that point, the piston is stopped and the core tube keeps falling, so that the piston action allows a core to be taken at a depth below the bottom roughly equivalent to the length of the rope. In this way, only a deep section will be obtained in the core tube, but this can be supplemented by taking an ordinary piston core above or by using a shorter length of rope.

Reports from the Soviet Union (Bezrukov and Petelin 1960) indicate that cores up to 34 m in length have been obtained on the *Vitiaz* "without any serious distortion." Their device is somewhat like that of an earlier unsuccessful model by Varney and Redwine (1937). It makes use of an evacuated chamber at the top that is opened from below when the corer hits bottom, allowing the core to move up into the core tube to replace the low pressure in the chamber above. The great problem in such a device must come from sucking in material from the sides due to the vacuum effect. Presumably this has been overcome by the Soviet scientists, but no further use of this corer has been reported. Rosfelder (1966A) has devised other uses of hydrostatic pressure for marine operations, such as actuating rotary drills, firing core barrels, or for anchoring into the deep-sea floor. The Norwegians have developed another type of instrument for driving a piston corer deeper into the bottom than attained by a free fall (Andresen et al. 1965). After penetration stops, a rocket propellant is ignited, driving the tube deeper into the bottom.

Obtaining cores from a hard surface with packed sand generally has proved impossible by either gravity or piston coring. This may be overcome by the use of a vibrator that operates on the corer after it has hit bottom. Such a device has been used by the Russian scientists in coring (Kudinov

1957). The vibration acting like a jackhammer forces the corer into the sand. It may be necessary to have the device on a tripod that will hold it vertically during the operation on the bottom. Van den Bussche and Houbolt (1964) described a streamlined coring device that was successful in obtaining sand cores 1.5 m in length. The sand is held in by the inversion of the corer after it is pulled out of the bottom.

Another problem in coring has been to prevent the loss of the core by leakage from the bottom on the way to the surface. The building of a good core catcher has not proved easy. The common type consisting of plastic or metal leaves (Fig. 2–8) that fold in to prevent the core from escaping is fairly successful in mud cores but does not make a perfect seal. Sandy sediments are very likely to flow out through the leaves during the ascent of the core tube to the surface. Using a piece of cloth attached to the catcher on the inside is somewhat helpful but not entirely successful. A core retainer consisting of a nylon sleeve that is twisted by a torsion spring set into operation after the sampler has stopped penetrating (Rosfelder 1966B) serves the purpose very well. Also, the sphincter core retainer (Burke 1968) operates on the principle of pursing a nylon sock with a drawstring.

Rock has been cored largely where the core tube passed through the entire sediment thickness and still has had sufficient velocity to take a very small core of the underlying rock. Usually, the core nose is damaged and often the barrel is bent. Off the California coast, however, rock has been cored intentionally by Southern California oil companies without damage, because the instruments were designed to support the impact. Here, very short core barrels of hard steel are used and a few inches of rock obtained. Another type of jet sampler was described by Barr (1951). It is useful for taking fragments of bedrock beneath a mud sediment cover. For greater penetration, a drilling rig is necessary, and great progress has been made in this field (see p. 29). For short drill holes, bottom-standing corers are used, such as the French "flexoforage" (Delacour and Moulin 1965), the Canadian Bedford Institute drill, and the Dutch Geodoff drill.

Orienting of cores is helpful in determining current directions when the sediments were deposited and also in magnetic studies. Numerous devices have been developed for more than a century, as described by Rosfelder and Marshall (1966). Core orientating is possible by locking a compass before withdrawing the corer from the bottom, or by smashing an inexpensive compass in its set position before retracting the corer. Petroleum companies have various devices for orienting samples in oil-well drilling operations that are adaptable to marine geological investigations.

Perhaps the most useful device for obtaining undisturbed cores has been the box corer (Kastengreifer) (Fig. 2–11). This was devised by Reineck (1958) after a device by Ekman (1911). The boxes have a variable length and a diameter of 24 x 30 cm or larger, which would ordinarily result in the cores sliding out, but upon retraction a spade is rotated over the snout in such a way as to make a tight seal. Because of the large diameter, very little distortion occurs in the structure of the sediment, and often rock fragments are brought up in the bottom of the core. This corer has proved satisfactory for penetrating the hard sand floors of submarine canyons where piston corers often fail to procure a sample. The good preservation of bottom trails and ripple marks often found on the top of the core is also informative. Long rectangular box corers of 10 cm diameter have also been used (Kögler 1963; Rosfelder and Marshall 1967, Fig. 4). These are successful in soft bottom but apparently not very good for sand.

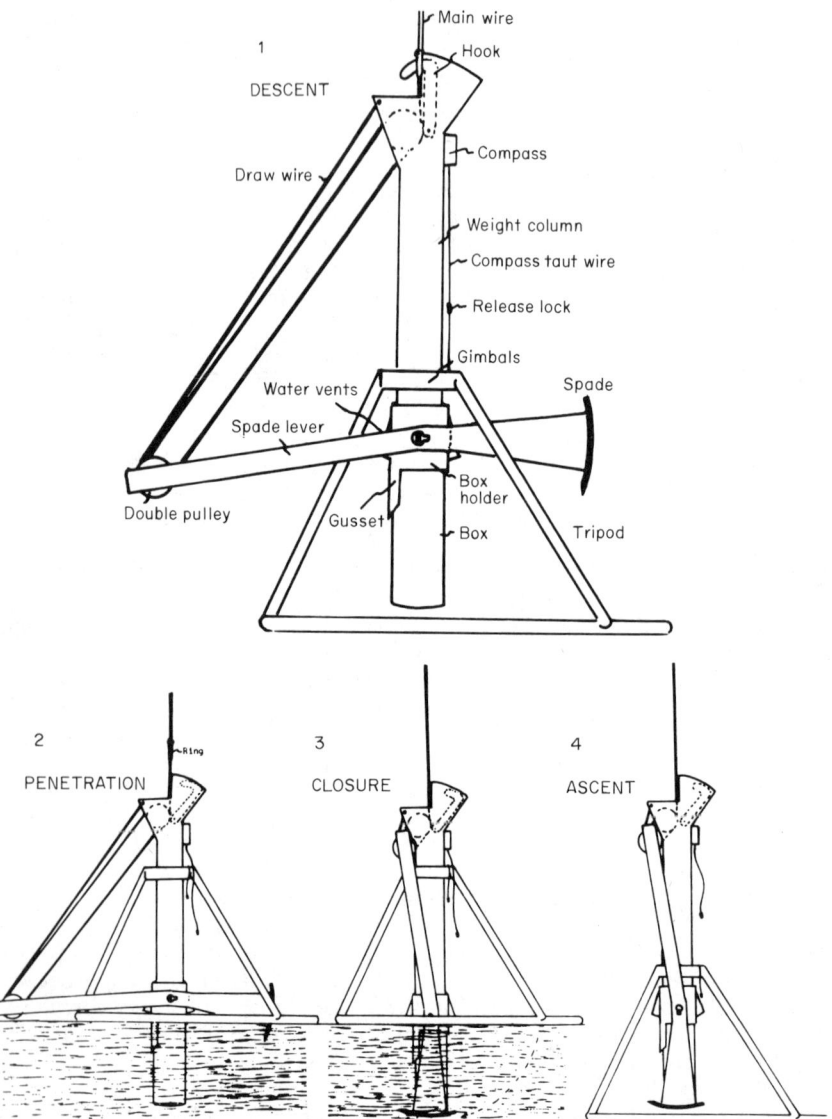

Fig. 2–11 The method of taking a box core and holding the core in place by covering it with the spade before raising above the bottom. From Rosfelder and Marshall (1967).

It is often helpful to have closely spaced cores to determine the continuity of sediment layers and to show details of sediment structures. One system is to use dual corers attached to the same descending line (Blackman and Anderson 1969). Even triple corers have been attempted. These are given free fall by the same type of release (a suspended weight below the corers) as used for piston coring (Fig. 2–9).

A useful investigation of piston coring involves mounting cameras and other instruments within a corehead (McCoy and Von Herzen 1971). The photographs indicate that the corer rotates during descents and develops about 6° deviation from the vertical during free fall. The mud mark on the outside of the recovered core barrel is a good index of penetration. No consistent relationship exists between the penetration and core recovery, mostly because of (1) poor core nose design with insufficient water escape at the top valve during penetration; and (2) sediment is sucked out (poor core retainer) during extraction.

Dredging Most rock from the sea floor has been obtained by dredging. The two types most commonly used by oceanographic ships are the frame dredge and the pipe dredge (Fig. 2–12). The frame dredge with a chain bag has a larger capacity and is considered more practical for work on the deep oceanic ridges. Because of its large width, it is possible for one of these dredges to get hooked onto a large firm outcrop, requiring tremendous strain

A   B

Fig. 2–12 A, the pipe dredge used successfully in coring crystalline rock from the walls of submarine canyons. In the foreground the barrel of a piston corer can be seen. B, a large frame dredge with a chain bag used for catching rock samples broken from the sea floor.

to break loose. However, with a strong cable, such as the commonly used ⁹⁄₁₆-inch (14-mm) wire, this is rarely a problem. With smaller wire, it may prove advisable to use the pipe dredge, which is rarely lost because it cannot catch as large a rock and tends to pivot under strain. On the other hand, these dredges are likely to tip over and lose their samples, although using a chain bag at the base helps to alleviate this problem.

Dredging in depths of less than about 1000 m is somewhat different from dredging at great oceanic depths. For the former, the usual procedure is to get into a position downslope from the desired locality, stop and lower the dredge about 50 m, and then attach a weight. Continue lowering to about the depth desired on the slope, and then move the ship slowly toward the slope, letting the wire trail behind the ship as you approach until you have a length of about 1.5 times the depth. Then set the winch brake enough to hold a moderate strain, and maintain an upslope speed sufficient to produce a 45° wire inclination. When the tensiometer on the winch jumps, indicating contact with the bottom, reduce speed but maintain sufficient headway to keep the dredge from falling away from the slope. After there have been strong pulls for several minutes, stop the engine and pull in the wire.

If dredging a canyon, chose the wall on the side where the current will carry the ship back into deep water when the wire is being reeled in. Otherwise, the dredged samples may come from a very broad area, because the dredge will continue to hit bottom as the wire is coming in slowly and the drift is carrying the dredge into shallower water, so that there may be no net gain in clearing the bottom for a long period.

Before dredging on a deep slope,[1] the current direction should be established and the ship stopped upcurrent from the slope. The weight should be attached at about 275 m, and it is important to attach a pinger another 90 m above the weight. When the tensiometer indicates a loss of strain, which shows that the dredge has hit bottom, decrease the descent rate to 5 to 10 m/sec. When the pinger indicates it is 20 to 40 m above the bottom, stop lowering and drift. Drifting should be continued until the tensiometer shows sudden strong indications that the dredge is biting into the bottom. Ordinarily, one should drift for an hour or more. Then start the winch but pull in the wire very slowly. The ship may be anchored and pulled back against the current as the wire comes in. When the pinger is about 250 m off the bottom, the strain should have decreased as the dredge has presumably broken free. However, continue to bring in the wire slowly until it is quite clear that the tensiometer shows a steady strain. Then increase winch speed gradually for another 100 m, because the dredge may hit a projecting ledge higher up the slope that is not indicated by the pinger. It is, of course, important to slow the winch when getting near the 365-m level to take off the pinger, and later at the 275-m level for the weight. Instruments are often lost or damaged by suddenly surfacing and ramming into the winch sheave.

Various types of large dredging systems have been suggested for mining placers and manganese nodules from the sea floor (Mero 1965, pp. 243–280). It seems quite possible that the near future may see some of these devices in operation for obtaining manganese nodules, particularly through two approaches, (1) using a pumpline from the sea floor, or (2) a continuous bucket conveyor belt.

Techniques in Coring and Dredging   A few simple techniques have proved very helpful in shipboard operations involving coring and dredging.

---

[1] Résumé of a manuscript by Robert L. Fisher.

Relative to coring, it is often important to put the corer in an exact position. For example, the center of a canyon may be very narrow so that to obtain a core on the canyon floor it is necessary to have the ship directly over the center when the corer hits the bottom. Unless electronic position control is available, this is difficult because of the problem of determining the location of the canyon bottom and because of drifting during the lowering of the device. The best way to find the canyon bottom is to proceed across the canyon, start up the far wall and then reduce speed and make a turn first right and then left (Williamson turn) so as to come back into the wake of the earlier crossing (Fig. 2–13A). In this way, the depth of the canyon bottom as shown on the fathogram from the first crossing will be known, and the speed should be so slow when the axis is reached that reversing the engines will bring the vessel to a halt at the deepest spot or, if there has been an overrun, a small amount of backing will bring the vessel into the desired locality. Backing, however, puts bubbles under the sound head and often stops the echo returns for a brief period.

If there is sufficient drift to throw the vessel out of position during the lowering, the winch should be stopped when the corer is fairly near the bottom, and then the vessel should be maneuvered back into position. This is relatively easy if the direction of drift can be determined by a fix on land objects or by noting the wire angle. If the wire is leading over the stern, as is true of the large winch wire in most oceanographic vessels, the ship is slowly pushed ahead (never backed unless wire is well free of stern because of the danger of tangling the cable in the propeller) and turned if necessary in order to bring the ship to the desired spot. After getting into position again, the engines are cut, and the wire is allowed to settle to approximately a vertical position before lowering is continued. If the wire leads over the side of the vessel, the maneuver is much more difficult. However, even with such a lead it may be possible to bring the ship very slowly back into position, turning only toward the direction in which the wire leads, that is, to port if the wire is on the port side or to starboard if on the starboard side. This will prevent the wire from getting under the vessel or, even worse, into the

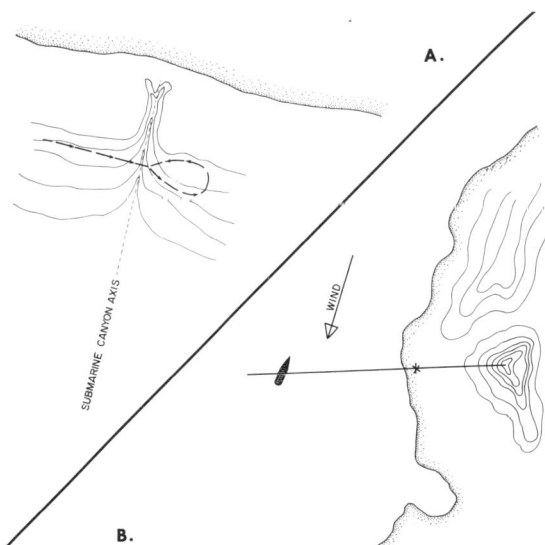

Fig. 2–13 Positioning methods for sampling. *A*, a submarine canyon is crossed and the ship is turned in such a way as to get back into the ship's wake; then the ship proceeds very slowly until the same axis depth obtained previously is reached. *B*, a position may be maintained by heading the vessel into the wind and steaming just sufficiently to keep shore points on range.

propeller. Where good ranges are visible on shore, it is often possible to keep the ship in position during the lowering by heading slowly into the wind so as to counteract wind drift and keep the same range (Fig. 2–13B).

Dredging is much easier over the stern than over the side, because dredging from the side of the vessel may get the wire under the hull, so that the wire has to be pulled against it and wears off paint or even gets caught on some projection on the hull. This will occur if the wind is blowing from the side opposite that on which the wire is let out or if the current is coming from the direction opposite the wire.

The letting out of wire from the side of the vessel, however, may have one considerable advantage. In coring, it is possible to build a cradle that can be used for housing the coring device and making it easier to operate than by bringing a heavy corer in over the stern. The coring operations on Woods Hole's *Atlantis I* and Lamont-Doherty's *Vema* were conducted from the sides of the vessels, and the coring operations were thereby speeded, but maneuvering was certainly hindered during operations. A very large A-frame on the stern allows the raising of the core barrels sufficiently to get the corer on deck without difficulty but only after many of the weights have been removed unless a movable cradle is maintained at the stern.

Methods of Handling Samples We are fortunate in having a book by Bouma (1969), which gives explicit details on many modern methods for preserving and handling samples. The taking of peels is useful in modern cores and in studying ancient formations. Impregnation of core samples is a great help in preserving the structures for study. Radiography brings out many structures that would not be visible without use of the method. Other methods, including drying of unconsolidated sediments, staining, sand blasting, infrared photography, imbedding of samples in a clear medium, and cementing of broken samples, are all valuable.

Peels are used largely for preserving undisturbed samples from unconsolidated material. Peels can be obtained from cores by cutting them in two longitudinally by wire or by an electroosmotic core cutter (Chmelik 1967), smoothing the cut with a spatula, and, after drying slowly for muds or lacquering for sand, placing a bandage over the core and covering this with thick lacquer. After the lacquer is dried the peel is detached.

Impregnation is a very useful technique in studying box cores, but can also be applied to other cores. Injection of a cementing agent has to be conducted so the arrangement of the grains will not be disturbed. Unsaturated polyesters are used for the purpose, with particularly good results when the resin Vestopol-H is employed (Altemüller 1962). Various methods of procedure are given by Bouma (1969, pp. 84–139). Polished surfaces and thin sections can be obtained from the impregnated samples for petrographic studies.

X-ray radiography can be applied to core sections to bring out special sedimentary structures. This was first developed by Hamblin (1962). Microfaults and -folds are often indicated in the x-ray films. Also, various fossils, both flora and fauna, are indicated that are not visible without the x-ray method. An example of an x-ray film compared with ordinary photography is seen in Fig. 2–14.

The preparation of microfossils from cores has been discussed extensively (see particularly Kummel and Raup, eds., 1965). It is important to follow the suggested procedures for the best results from samples submitted to micropaleontologists.

Deep-Sea Drilling (JOIDES) The petroleum industry, as a result of their great success in drilling on the continental shelves, has gradually been developing techniques for drilling from an unanchored ship. Through grants

Fig. 2–14 X-ray photographs (A) bring out features not shown in ordinary photographs. (B) slice of box core from Santa Maria Submarine Canyon, Baja California, 670-m depth. Arrow shows downcanyon direction. From Bouma and Shepard (1964).

from the National Science Foundation, five of the United States oceanographic institutions have developed a long-range program for drilling into the deep ocean basins (National Science Foundation 1970; Graham et al. 1970). This program, with headquarters at Scripps Institution in La Jolla, California, is called JOIDES (Joint Oceanographic Institutions for Deep Earth Sampling). The project, using the Global Marine Company's *Glomar Challenger* (Fig. 2–15), has had excellent cooperation between geologists and petroleum engineers with results that have proved amazingly successful (see pp. 419–430).

The most important factor in drilling from a ship is to maintain the same position. This is accomplished by dropping a sonic beacon at the desired location. When the beacon reaches the bottom it sends out acoustic signals that are picked up by hydrophones located in the ship's hull. The difference in time of arrival of the pulses at the different hydrophones makes it possible to triangulate and determine the direction and distance necessary to maintain the ship in exactly the right position. With the help of computers, the direction and amount of movement necessary are determined, and four "thrusters," two on each side, move the ship easily in any direction. Gyroscopically controlled roll stabilizing keeps the ship remarkably steady except in very high seas.

The 43-m derrick for the drilling is located over a 6.1 × 6.7-m hole that extends through the bottom of the ship. The drill pipe, suspended from the derrick, passes through a structure that distributes the stress to prevent excessive bending due to roll and pitch of the ship. The drill string is lowered with a bit at the end. Several types of bits are used, depending on the type of bottom expected. A tungsten roller-type bit is used for deep penetration but has less core recovery. The core barrel and drill collars are of the same type as in ordinary drilling for petroleum. The drilling string is rotated from the drilling derrick.

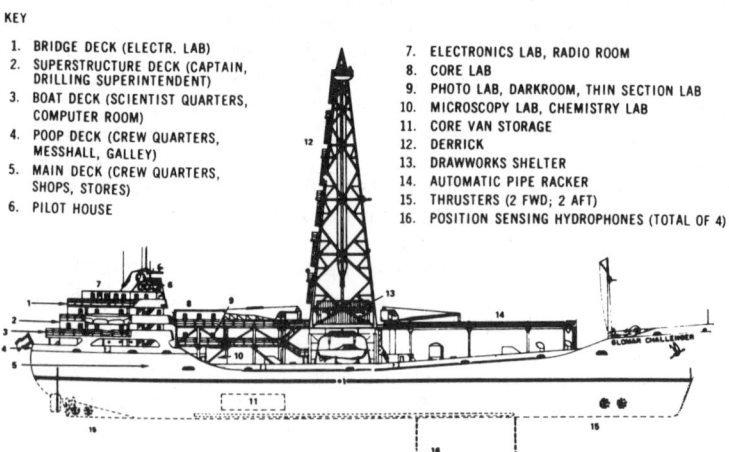

KEY

1. BRIDGE DECK (ELECTR. LAB)
2. SUPERSTRUCTURE DECK (CAPTAIN, DRILLING SUPERINTENDENT)
3. BOAT DECK (SCIENTIST QUARTERS, COMPUTER ROOM)
4. POOP DECK (CREW QUARTERS, MESSHALL, GALLEY)
5. MAIN DECK (CREW QUARTERS, SHOPS, STORES)
6. PILOT HOUSE

7. ELECTRONICS LAB, RADIO ROOM
8. CORE LAB
9. PHOTO LAB, DARKROOM, THIN SECTION LAB
10. MICROSCOPY LAB, CHEMISTRY LAB
11. CORE VAN STORAGE
12. DERRICK
13. DRAWWORKS SHELTER
14. AUTOMATIC PIPE RACKER
15. THRUSTERS (2 FWD; 2 AFT)
16. POSITION SENSING HYDROPHONES (TOTAL OF 4)

Fig. 2–15   The *Glomar Challenger* used for deep-ocean drilling operations (JOIDES).

During the first two years of JOIDES, drilling of many holes had to be terminated because bits were worn out, particularly from boring through chert. In 1970, a system was perfected for reentering the hole after a dulled bit had been brought to the surface for replacement. A reentry funnel is left on the sea floor and sonar scanners lowered on the new bit assembly locate the funnel, and the ship maneuvers the bit over the funnel.

The 12.5-cm pipes used for drilling are racked forward of the derrick and an automatic pipe-racking device handles the pipes and stacks them, thus avoiding danger to personnel. If the operation were manual, the risk would be great in heavy-sea conditions.

By keeping the ship headed into the prevailing seas, drilling operations have been able to continue with winds up to almost 50 kn and with seas up to 4.5 m in height. During drilling, roll and pitch averages less than 3°. As a result, there is almost no time lost because of sea conditions, and the ship has actually been used for drilling more than 50 percent of the time. Another 38 percent of the time has been necessary for cruising to stations and ports. Less than 1 percent has been breakdown time. This is clearly a very efficient operation.

To July 1971, 60 percent of the coring interval has been represented by cores. However, the core lengths do not necessarily agree with the distance drilled when the cores are being taken. Sometimes soft sediments, such as deep-sea oozes, squeeze into the core barrel faster than drilling proceeds. This results in considerable distortion. Hard-rock cores are the least disturbed. With new improved bits, drilling into hard rock is fairly fast, although it was very slow with the old bits during the first two years of operation.

SEA-FLOOR OBSERVATIONS BY GEOLOGISTS

The early marine geology work was largely done by using remote instrumentation, and direct observation played a very minor role. Now, with the advance of scuba techniques and the great improvement of deep-diving vehicles, opportunities are expanding for geologists to observe the ocean floor and choose the samples they wish to collect, rather than to make blind grabs that were often ineffective. Although visibility along the bottom is generally quite limited in distance, it is remarkable how many features have been discovered by diving.

Scuba Diving The old "hard hat" method of diving with a hose supplying air from a surface ship has now been virtually abandoned except in heavy construction or salvage operations and is seldom used by geologists (Klenova 1948, p 31; Shepard 1949). Although experiments with rudimentary self-contained free-diving equipment date back to the 1880s, the first practical scuba (with air tanks strapped to the diver's back and a demand regulator) was developed in 1943 by Jacques-Yves Cousteau and Emile Gagnan. A more complete history including earlier experiments is given in Larson (1959). The method was first applied to geological problems in 1950 (Dill and Shumway 1954) and has been used extensively since that time, especially on the narrow shelf off Southern California. It has proved very practical in mapping the geology, using methods similar to those of land geologists. Inclinometers, compasses, hammers, slates, and air lift bags make up the standard equipment (Fig. 2–16). Because scuba divers are weightless, they can examine outcrops of rock on cliffs without the difficulty encountered on land. By using towed diving sleds (Sigl et al. 1969), diving geologists can cover a large territory in a short time. Cousteau and others use motor-driven sleds for the same purpose.

Fig. 2–16 Scuba geologists at work with geological hammer and lift bag to bring rock samples to the surface by filling bag with air. Photo by N. F. Marshall.

According to R. F. Dill, geological mapping off Southern California has aided in the development of at least three new oil fields. Observations by scuba divers have shown the importance of mass movements on the sea bottom; creep is indicated by motion of stakes in sea canyons, landslide scars have been observed in many places, and large cracks are seen at the edge of terraces (Dill 1964). Also, sand flows have been observed on the canyon walls (Fig. 11–7A).

Depth of scuba diving has generally been limited to 50 m, but experiments with nitrogen and other gasses for divers' tanks has shown that it is quite possible to operate at much greater depths. By sending the divers down in dry chambers in which they can live and become gradually adjusted to greater pressure and to breathing various mixtures of gasses, the divers can emerge through traps in the floor at depths as great as 300 m. The U.S. Navy program that made such good progress in Sealab II conducted at 64 m has been indefinitely postponed after one of the participants in Sealab III was killed. Continuing physiological experiments in France, England, and the United States are developing techniques to take man well below 300 m and possibly to 1500 m.

Deep-Diving Vehicles The development of small submarines in which observers can view the bottom made much progress in the 1960s (Terry 1966). The earliest deep-diving vehicle was the bathyscaph *FNRS-I,* built for Auguste Piccard in 1948 and later taken over by the French Navy. This was made like a balloon with a steel ball for the cabin attached to a large gasoline-filled gondola. A second improved bathyscaph, *Trieste,* built by Piccard in Italy, was taken over this time by the U.S. Navy. This second deep-diving vehicle actually dove in 1960 to the bottom of the deepest known depression in the world, the Challenger Deep, 10,915 m (Piccard and Dietz 1961). Although both the Italian-built bathyscaph *Trieste* and the more recent third bathyscaph, the French-built *Archimede,* have made many dives and have obtained interesting information, this type of submarine has not proved very practical because of excessive cost, difficulty of operations, and lack of maneuverability. Somewhat better results have come from Cousteau's *Soucoupe* (diving saucer) and similar submarines that are small enough and maneuverable enough to operate in the rugged areas of the sea floor. The *Soucoupe* is limited to 300-m depth, but *Deep Star,* built by Westinghouse in collaboration with Cousteau, has made many dives to 1200 m with impressive results. Woods Hole's *Alvin* (Fig. 2–17), built by the U.S. Navy, can dive to 1800 m and has been of great value to geologists. After operating for several years, the *Alvin* was sunk during a launching in rough sea conditions, but after a year was brought back to the surface and is again in operation. The *Aluminaut*

Fig. 2–17 The *Alvin,* the Woods Hole Oceanographic Institution deep-diving vehicle. From Terry (1966).

of Reynolds Submarine Services is the largest of the deep-diving vehicles, can descend to 4500 m, and can stay submerged for many days. In 1970, the *Ben Franklin*, of similar design, drifted for 60 days along Blake Plateau investigating the Gulf Stream. Various companies have built small two-man submarines that dive to about 100 m, notably the Perry Company *Cubmobile* and the General Oceanographics' *Nekton*.

BOTTOM PHOTOGRAPHY

The earliest report of bottom photography was made by Boutan in Algeria in 1893, who, in a diving suit, took photographs with a protected camera in shallow water. Shortly before World War II, an early model of the remote camera devices for photographing the bottom at greater depths was developed (M. Ewing et al. 1946). This operated in water less than 200 m deep by setting off an unprotected flash bulb and synchronously taking a photograph from a Robot camera in a brass case mounted in front of a glass porthole. Now there are many underwater cameras that have pressure housings strong enough to operate at all depths (Fig. 2–18). They have electronic flash units

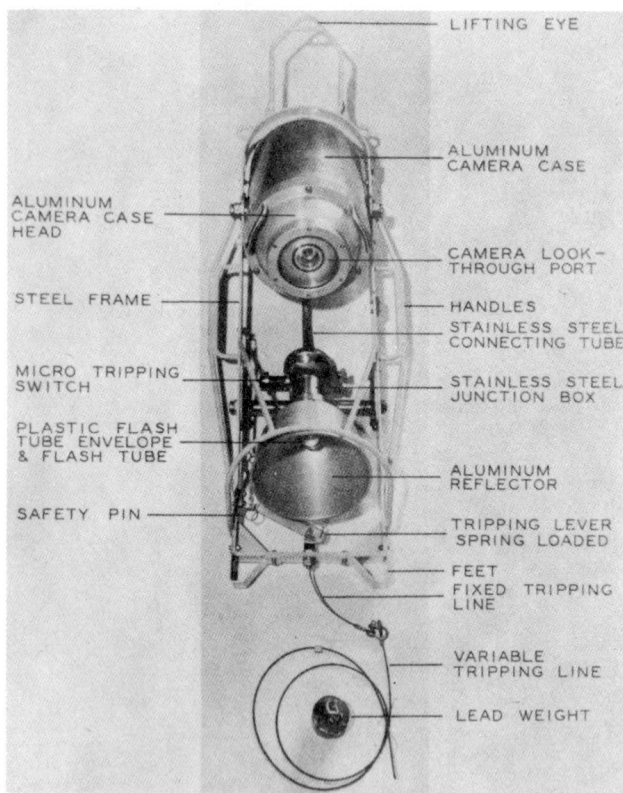

Fig. 2–18 The U.S. Navy Electronics Laboratory Type III deep-sea camera designed and constructed by Carl J. Shipek.

and automatic film advancement for taking a large number of photographs during a single lowering. They maintain height above the bottom with a pinger. The increased chances of losing the camera by taking multiple photographs as the camera moves over the bottom are offset by the great amount of time saved. Taking photographs going down a slope rather than going up a slope considerably diminishes the chances of loss.

The use of wide-angle lenses on the cameras permits a large area to be photographed. Stereo pairs are useful in showing bottom relief (Fig. 2–19).

Numerous photographs were taken along with samples in the giant Campbell grab sampler in the recent work off the East Coast of the United States (Emery et al. 1965). The method made it possible to compare the character of the bottom as shown by both the photographs and the samples. Another useful method for obtaining photographs over a large area is to pull a sled (*troika*) with mounted camera, designed by Cousteau and Gagnan (Laban et al. 1963), over the bottom (Fig. 2–20). Repeated flash pictures show the character of the terrain over which the sled has traveled. Free-instrument camera systems are now in use to record bottom topography on a gross scale.

Thousands of bottom photographs are now available and only a small number have been published. Two books (Hersey 1963; Heezen and Hollister 1971) represent an attempt to coordinate and summarize some of the significance of this vast source of scientific material.

Underwater photography by scuba divers has become very successful (Frey and Tzimoulis 1968). Color photography in clear shallow water with the new 35-mm cameras is no problem, even for amateurs. These cameras are waterproofed and do not require the difficult mounting of a camera inside a plastic case. However, colors change rapidly with depth beyond about 3 m because of the fading of the red end of the spectrum. Therefore, to get true color reproduction requires a light source, which adds to the bulk of equipment to be handled. In clear water, natural light allows good results in black and white films down to about 40 m. Exposure meters, also somewhat combersome, are very helpful. Photographs taken near the surface may be rather blurred because of wave action unless high-speed film is used or the swimmer steadies his camera on the bottom. A simple method is to hang onto a projecting rock or coral.

BOTTOM CURRENT MEASUREMENTS

In the second edition of this book, it was noted that much progress had been made in measuring surface and intermediate depth currents, but that bottom currents had been badly neglected. Unfortunately this is still true. We know more about currents from deep-ocean photographs (Heezen and Hollister 1971) that often show ripple marks, current streaks, and grooves around rocks and nodules than from actual measurements. Also, sequence shots from corehead cameras give the approximate character of currents by showing the rate and direction of spreading of sediment clouds set up by the coring operation (M. Ewing et al. 1967).

Free-instrument Savonius rotor current meters are now being used; these can be dropped to the bottom to make a record for a number of days and then return automatically to the surface (Isaacs and Schick 1960). Using explosive releases to drop the anchor weight, they will surface at exact times, or, by using magnesium releases, the time of return can be estimated within a number of hours. Attached radio transmitters operate when the instrument

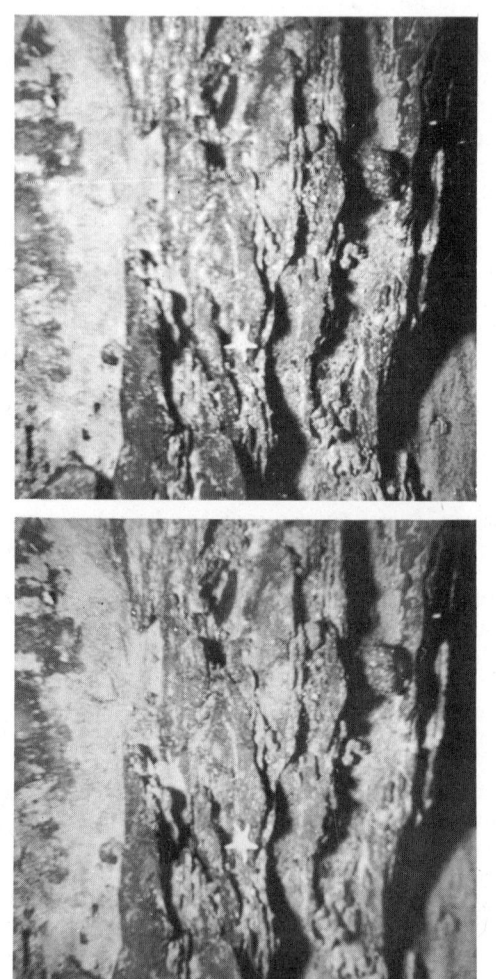

Fig. 2–19 Stereo pair of photographs obtained on board a Soviet vessel, *Vitiaz*, by N. L. Zenkevitch. Photograph obtained 200 miles north of New Zealand at a depth of 2040 m. Note dipping rock formation.

reaches the surface. The records are inscribed on pressure-sensitive paper driven by a precision motor and indicate current direction and speed (Fig. 2–21). However, speeds beyond 75 cm/sec (1.5 kn) cannot be recorded because the velocity marks on the recording paper of the meter are so closely spaced as to be blurred.

The Isaacs–Schick current meter has been used for many records of the currents along the bottom of submarine canyons off California and Baja California (Shepard and Marshall 1969 and 1973). The same device is equally successful in measuring the currents of the continental shelf and has also been used for currents of the deep-ocean floor (Isaacs et al. 1966).

A current meter placed in the shallow head of a canyon and attached by cable to the Scripps Institution pier has shown electrical records of currents up to 180 cm/sec (3.6 kn) (Inman 1970). At this speed, the wire parted and the meter was lost. Similarly, an Isaacs–Schick current meter was put into Scripps Canyon just before a storm and then disappeared, suggesting that there had been a very strong current (see p. 66). To monitor the high-speed currents that apparently exist, N. F. Marshall has devised a low-cost current pendulum (Fig. 2–22), which is released from a bottom weight when the current gives it a certain inclination. If washed ashore and recovered, it shows that the current that released the instrument reached a velocity of at least the amount for which it had been set. While on the bottom, the current pendulum is surrounded by a protective screen that prevents kelp or other debris from being pushed against it. Using the same principle, pendulum meters have been used to measure the amount and direction of bottom currents by a sequence of films taken by a camera attached to the device. These show the nature of the pendulum deflection (Saski et al. 1965).

Other simpler and less expensive devices that are useful in shallow

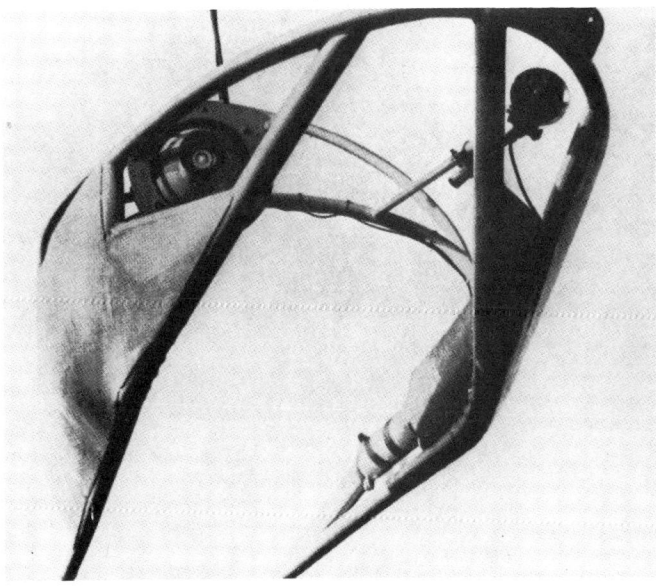

Fig. 2–20 Troika (sled) with mounted camera. From Laban, Pérès, and Picard (1963).

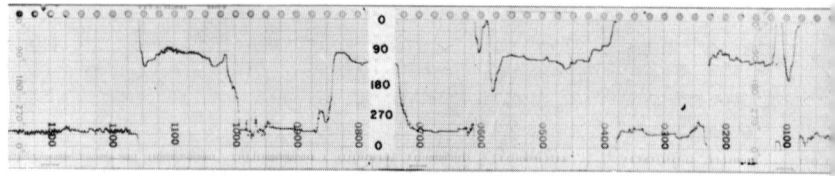

Fig. 2–21 A current meter record taken in the axis of La Jolla Submarine Canyon. Note the direction of current shown by height of line above base, and velocity by closeness of tick marks. This record shows alternating upcanyon (90°) and down-canyon (270°) lines.

water include vertical current crosses suspended from horizontal surface floats that show the movement of the current near the bottom because of the greater resistence of the crosses. If the current carries the cross into shallow water it strikes bottom, and if it moves seaward, it gets too far above the bottom; but considerable information can be learned from such a device.

CURRENT PENDULUM

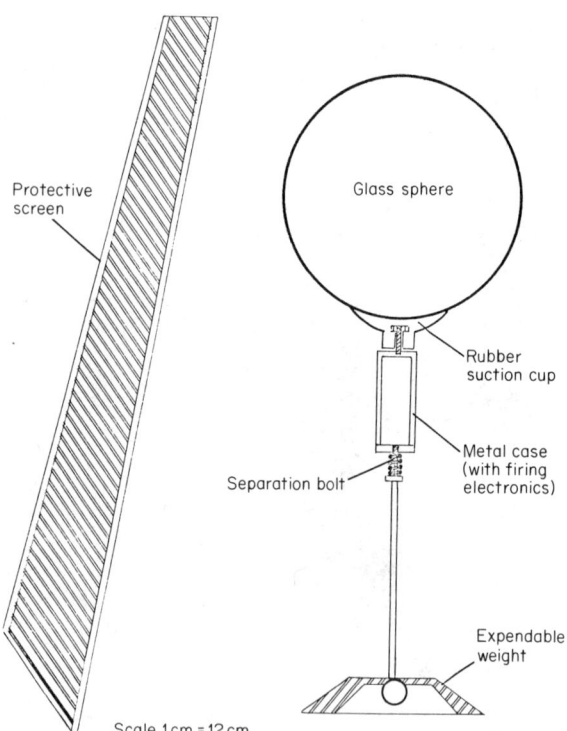

Fig. 2–22 Pendulum current meter device is released when current reaches a definite speed and then surfaces. Guarded from kelp by screen that surrounds it. Designed by Neil F. Marshall.

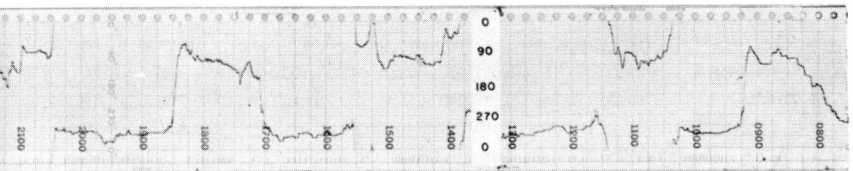

Carruthers (1958) devised another simple system for use by net fishermen. It consists of a plastic bottle into which a hot solution of powdered gelatin is poured, filling it to the halfway mark, and a floating compass is placed on top. The bottle is covered and a twine attached to the top and to a weighted stick that will anchor the device to the bottom. Another line extends to the drifting ship and is payed out for several minutes, allowing the gelatin to solidify before hauling in the line. The amount of tilt will vary according to the current, and a table is given by Carruthers that enables the observer to make a good estimate from the angle of inclination at the top of the jelly.

Strain gauges, doppler shift, thermal generation, and electromagnetic fields are now being used to measure currents. The advantages of these techniques are a relative insensitivity to fouling, a lack of moving parts, and a near-instantaneous response to current velocity; whereas with the Savonius rotor, a considerable operational threshold exists. Interesting new data will eventually be available with these new devices in studies of orbital motion in waves and current flow around stationary objects.

SOIL MECHANICS METHODS[2]

When any object, equipment, or structure is placed on the sea floor it is necessary for the scientist or engineer to consider the sediment properties of importance in soil mechanics and foundation engineering. These properties determine the extent to which any object placed on the sea floor will sink into the sediment as the result of failure of the sediment in shear, and the slow sinking of the object because of the adjustments of the sediment structure to overlying pressure (*consolidation* in soil-mechanics terminology). In addition, the thickness and properties of the sediment determine slope stability; and hence whether an object placed on the sea floor will cause a slope failure, or will cause failure of a slope above the object, burying the installation. The sediment properties required by the engineer for a foundation study include all of those commonly studied by geologists and geophysicists, with special emphasis on shear strength, consolidation, and thicknesses of various layers. These properties are best determined in situ, for example, from a deep-diving submersible, by scuba divers, or by special sampling, or measuring devices dropped or lowered from a surface vessel.

Soil-mechanics literature includes significant contributions to the fol-

---

[2] This section was provided by E. L. Hamilton of Naval Undersea Research and Development Center.

lowing subjects of importance to marine geologists and geophysicists: (1) sediment structure, (2) consolidation, or the response of a sediment body to effective overburden pressure, (3) sediment strength under imposed loads, (4) penetrability of sediment to instruments, (5) dynamic rigidity, and the velocity and attenuation of compressional and shear waves, (6) the collection of least disturbed samples, and (7) the stability of sediment slopes. Recent articles containing longer discussions of these fields are by Hamilton (1959, 1970A, 1971) and Noorany and Gizienski (1969). The methods of correcting laboratory measurements to in situ values were recently discussed by Hamilton (1970B).

GEOPHYSICAL METHODS

Geophysical methods in marine operations have greatly improved in the past decade and their use has been an important factor in revolutionizing geological thought, as discussed in several of the following chapters.

Seismic Reflection Profiles The nature of seismic profiling was well described in *The Sea,* Vol. 3, by Hersey (1963), but because of the great advances in instruments and of the quantity of new information, J. Ewing and M. Ewing (1970) summarized the subject again in Vol. 4 of *The Sea,* showing many of the new deep-ocean profiles made by Lamont-Doherty Geological Observatory. Seismic profiling is similar to echo sounding, but instead of high-frequency it uses low-frequency sound impulses. These are less attenuated in traveling through the sediment and rock of the sea floor. Interfaces between sediment layers make good profiles below the bottom, showing increase in printing density (Fig. 2–23). This reveals the structure of the underlying sediments and rock layers and allows one to estimate the depth to the basement and other important horizons.

The sound impulse is emitted from a towed electric transducer, such as a sparker (arcer) that produces an electric spark at intervals, a boomer that produces an acoustic impulse by electrically forcing two metal plates apart, or an air gun that emits a bubble impulse. The echoes are received in a hydrophone towed behind the transducer. In order to conduct profiling while the ship is moving at speeds up to 12 kn, it has been necessary to tow an array of hydrophones in flexible tubing with the hydrophones connected electrically in a series. Inactive sections of tubing are included in front of and behind the hydrophone array to improve streamlining. The graphic recorders are diverse in character, but all show "a cross sectional display of reflection time versus distance along a traverse" (J. Ewing and M. Ewing 1970).

Seismic Refraction The seismic refraction method used for determining the nature of crustal layers is also discussed in Vols. 3 and 4 of *The Sea* (J. Ewing 1963; Shor 1963; Hill 1963; Ludwig et al. 1970). The principles of refraction shooting are described in various textbooks on exploration geophysics (see Dobrin 1960). Explosives are set off at regular intervals from a moving ship and are recorded by a hydrophone or geophone on a buoy or a stationary ship as the distances separating the shot and detector increases or decreases. Under favorable circumstances, the travel-time plot from the stationary hydrophone will show the depth of discontinuities below the bottom where there is an increase in velocity of the elastic waves. This, in turn, may show the velocity in the layer underlying each discontinuity and hence, by comparing known velocities in various types of rock, may make it possible to predict with some reliability the nature of the rock in the various

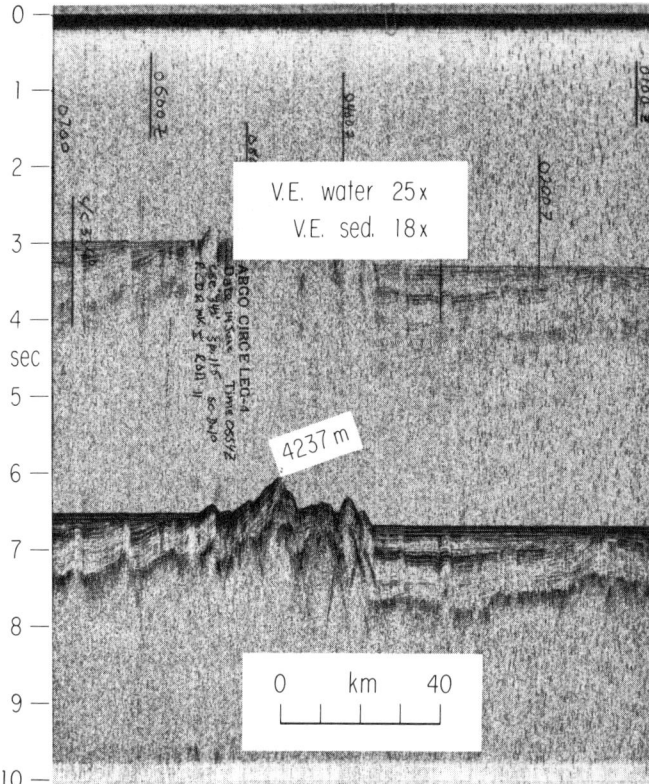

V.E. water 25x
V.E. sed. 18x

4237 m

0   km   40

Fig. 2–23 Seismic reflection profile, showing recently folded hills rising locally above the Bengal Fan in the Indian Ocean. Courtesy of J. R. Curray and D. G. Moore.

underlying layers. Also some idea of the dip of the surface of the discontinuities may be obtained.[3]

Seismic refraction is not commonly used now in crustal studies because the seismic reflection method gives greater continuity and is much less expensive, as it takes far less time, both in the field and in the laboratory. However, the reflection technique is still not capable of reaching the deepest layers, and the study of the Mohorovičić discontinuity requires refraction shooting or earthquake studies.

Gravity Measurements At first, gravity at sea was measured only from submarines because they could be kept stable when submerged, and the multiple-pendulum system devised by Vening Meinesz (1929) could be used only with this stability. Now, surface ships have taken over all gravity measurements. The study of acceleration of gravity below the ocean floor has also been discussed in Vols. 3 and 4 of *The Sea* (Worzel and Harrison 1963; Talwani 1970A). During the interval between these two articles, much progress had been made in surface-ship gravimeters (LaCoste et al. 1967). The new

[3] For a more complete discussion of the method, see Shepard, *Submarine Geology,* 2nd ed., pp. 28–31 by G. G. Shor, Jr.

instruments are being used on stable platforms. Talwani (1970A) also describes a vibrating string and force-balance type that represents further improvement. With the general acceptance of some type of convection current in the mantle (see Chap. 5), interest has increased in the study of both long-wave-length anomalies, the former obtained by satellite observations. The importance of gravity measurements has increased with the new methods.

Magnetic Anomaly Measurements The inexpensive method of towing a total intensity magnetometer from a ship or airplane[4] has become of enormous importance to marine geologists and geophysicists during the past decade, because the linear magnetic anomaly belts have provided the most important method for verifying and measuring sea-floor spreading (see p. 94). In the same two volumes of *The Sea*, good summary articles on magnetism and magnetic anomalies are given (Bullard and Mason 1963; Heirtzler 1970). The alternating negative and positive anomalies are found in general to be parallel to the mid-ocean ridges where lavas are rising. By matching the anomalies, displacements of large order are found that are, in turn, related to plate tectonics. Most oceanographic expeditions now make magnetometer measurements during all of their voyages.

Heat-Flow Measurements Because of the almost constant temperatures along the floors of the deep oceans, an ideal opportunity exists for measuring thermal flux from the earth's interior. In *The Sea*, we again find summary material (Bullard 1963; Langseth and Von Herzen 1970). Gradient-measuring probes are forced into the sea floor. The probe contains two or more temperature-measuring elements and their readings are recorded in a pressure-tight container. Many of the measurements are taken along with cores, the probe being attached to the corer. Fins are used to keep the temperature from being disturbed by the coring action. The temperature-sensitive element is either a thermocouple or a thermistor. The method is well described by Bullard.

According to Langseth and Von Herzen, approximately 3000 heat-flow measurements had been made up to July 1968. The measurements have proved very important in understanding the volcanic activity in mid-ocean ridges that is involved in sea-floor spreading. Consistent results have been obtained in relation to the ridges.

[4] This method is described in Shepard, *Submarine Geology*, 2nd ed., pp. 34–36 by Victor Vacquier.

Chapter 3
WAVES AND CURRENTS

INTRODUCTION

Although ocean waves and currents are primarily related to physical oceanography, they have played a very important part in developing coastal forms, in modifying continental shelves, and in the transport and deposition of sediments, and are therefore closely related to the geological phases of oceanography. The problem of beach maintenance and the effect of engineering structures on beaches and harbors can only be interpreted with a background of wave mechanics and the hydraulics of water currents. However, in order to avoid long technical discussions and conform with the rest of this book, the present edition omits much of the quantitative treatment in the second edition, and the reader interested in obtaining such material may find it from other sources, such as the second edition (pp. 50–78), Kinsman (1965), and Phillips (1966).

The study of waves and currents dates back at least to Leonardo da

Vinci, who became interested in the nature of wave motion in 1480. In 1513, Ponce de Leon first discovered the powerful Gulf Stream. Benjamin Franklin was the first to produce a map of this great current. During the past century, both waves and currents have been given considerable attention by mathematicians, physicists, and engineers. During World War II, it became important to understand propagation of waves from storm centers for the Allied landing operations, and this need resulted in a special burst of investigations. Since then, military engineers and oceanographers have continued the research, along with many publications from the Coastal Engineering Research Center of the U.S. Army Engineers (formerly the Beach Erosion Board). Ocean currents have also been an important aspect of the work at all oceanographic institutions.

WAVES GENERATED BY THE WIND

When the wind blows over a water surface it transfers energy to the water by tangential stress exerted in an irregular fashion because of the turbulent and gusty nature of the flow of the wind. Initially, the wind sets up ripples on the water surface and these grow into waves with a sheltering effect on their lee sides. This in turn leads to propagation of the waves in the direction in which the wind is blowing.

There are various types, called Airy waves, Stokes waves, and solitary waves, which have forms that are, respectively, sinusoidal (circular or elliptical), trochoidal (like a particle motion on the hub of a wheel), and solitary (with isolated crests and broad troughs). The dimensions of waves are given as heights (H), meaning the difference in elevation between the crest and the trough (Fig. 3–1A), and length (L) or distance between two adjacent crests. The significant height is defined as the average height of the highest one-third of the waves during a stated time interval, commonly 20 minutes. The time between passage of two wave crests past the same point is called period (T) expressed in seconds. The significant period is the average of the periods of the highest one-third of the waves. The wave velocity (C) or the rate of propagation of the wave form is equal to the wave length divided by the period, that is, $C = L/T$.

Considering the Airy wave, the simplest type, which is sinusoidal in profile, particles in deep water complete a circle during one wave period. In shallow water, waves are more typically trochoidal, and their profile resembles a car wheel that is spinning on a slippery surface and advancing only a small portion of the distance over which the wheel rotates (Fig. 3–1B). The result of the forward motion is that, in general, a floating object will move toward the beach or in the open sea in the direction of the wind that caused the wave.

In a storm center, the waves are very irregular and appear to be moving in various directions, although roughly parallel to the direction of the wind. The waves in or near a storm center are called *sea*. These waves move outward from the center and, when they are not under the influence of the wind, become much more smooth in shape and resemble sine waves in profile. Also, the longer period waves run faster than short period. The smooth waves beyond the storm area are called *swell*. Sea waves during storms have been measured to heights of as much as 30 m, and swell has been estimated as high as 15 m.

Wave motion decreases rapidly with depth below the surface, accounting for the smoothness of motion in a deeply submerged submarine. The

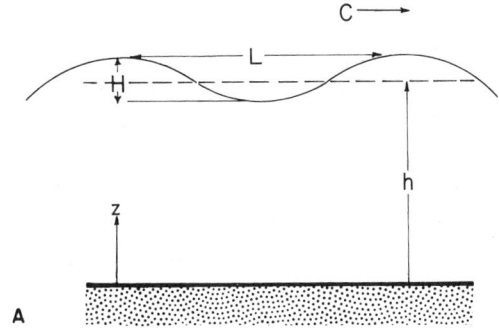

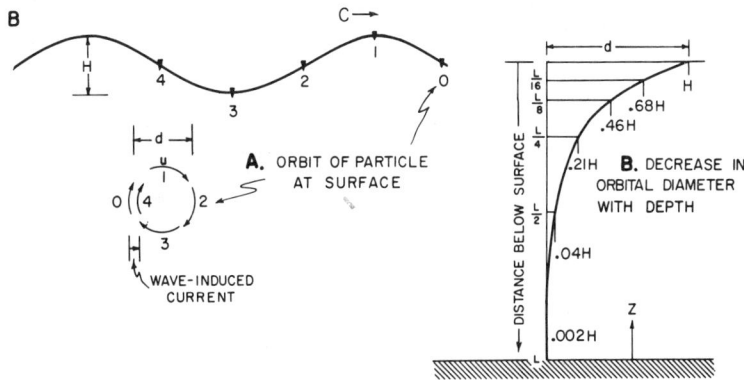

Fig. 3–1   A, definition of terms. Schematic diagram of a sinusoidal wave of height *H*, and length *L*, traveling with phase velocity *C* in water of depth *h*. *B,* orbital velocity and diameter of an Airy wave traveling in deep water. (A) trajectory described by particle at surface as successive parts of wave travel past point *O*. Note that at the surface the diameter of the orbit is equal to the wave height, *d* = *H*. (B) decrease in orbital diameter with depth for a wave traveling where depth of water is equal to the wavelength *L*.

motion becomes negligible at a depth equal to about one-half the wave length but is still significant at a depth of a few hundred feet in the case of the long waves that outrun great storms The most amazing thing about waves from a storm center is their ability to continue for thousands of miles over deep water. Their loss of energy is negligible after leaving the storm area Loss of height in the first thousand miles is commonly not more than half. Storm waves generated in the Antarctic can travel to the West Coast of the United States and break with sufficient force to cause serious beach erosion. In some areas, notably Southern California and Morocco, large breakers along the shore often show little relation to the local weather because they have come from distant storms. Using weather reports from ships, it is possible to make fairly accurate predictions of breaker height and time of arrival of maximum surf from storms thousands of miles from the coasts on which the waves will ultimately break.

Shoaling Transformations and Breakers   The typical profiles of ocean swell in deep water are long and low in slope, approaching sinusoidal curves.

Fig. 3–2 *A*, plunging breaker with hollow front. Distant wave source. Photo courtesy of Ron Church. *B*, spilling breaker during local storm. Photo by author. *C*, surging breaker with steep beach, Kaanapali, Maui. Photo by D. L. Inman.

However, as the waves enter shallow water, where the depth is equal to about a quarter of the deep-water wave length, the shallow water causes a decrease in wave velocity and length, and, as energy is conserved, wave height increases and the crests steepen to sharp peaks separated by relatively flat troughs. The heights of breaking waves may reach several times that of waves in the open sea. The increase in height is particularly pronounced for the long-period swell from a distant storm, whereas the increase is only slight for the shorter wavelengths associated with local storm waves.

When the water depth is about equal to the wave length, the wave collapses and particles of water rush forward together and move up onto the shore. In swell waves on a gently sloping bottom, the typical relation of wave height (*H*) to water depth (*h*) at the outer edge of the breakers is $H/h = 0.78$; whereas in the short-period steep waves, the relation is about $H/h = 0.6$. Therefore we see storm waves breaking farther out to sea than long-swell waves of the same height. For example, in La Jolla, California, large swells rarely break at the 5-m depth at the end of the Scripps Institution pier, but storm waves often break well outside the pier.

While coming through the surf in small boats, one should remember this relationship of breaker height to water depth, but even more important is to watch the changes in the train of breakers. There is a common saying that every seventh wave is the large one. Actually, there may be groups of large and small breakers at somewhat irregular periods, but the best time to go through the breakers is after a series of large waves has been followed by a relatively small wave. The grouping of high and low waves is called *surf beat*.

The types of breakers include: (1) *plunging breakers* (Fig. 3–2A), which have a hollow (concave upward) front and generally come from long swell approaching a gently sloping beach; (2) *spilling breakers* (Fig. 3–2B), which

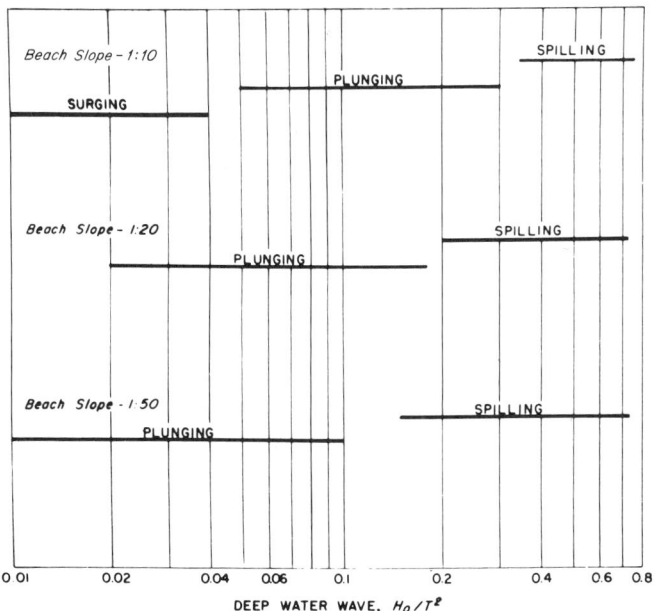

D

DEEP WATER WAVE, H_0/T^2

Fig. 3–2D Diagram showing relation between breaker type, beach steepness, and steepness of open sea waves. From Iversen (1953).

50   0   1 0

MILES   KM

Fig. 3–3 Oblique approach of wave crests to the beach at Atlantic City, New Jersey. Photo by U.S. Coast and Geodetic Survey.

have a steep front but no hollow and accompany most short-period wind waves; and (3) *surging breakers* (Fig. 3–2C), which do not actually break but surge up steep beaches. The most dangerous breakers are, as would be expected, those with the hollow fronts because the water drops vertically with great force.

Tank experiments by Iversen (1953) indicated a relationship of breaker type between a measure of deep-water wave steepness (H/T^2, where H is height of open-sea waves in feet and T is the wave period) on the one hand and beach slope on the other (Fig. 3–2D). Thus the boundary between spilling and plunging breakers occurs at a higher ratio between open-sea wave steepness and wave length for a steeply sloping beach than for a gently sloping beach. To produce spilling breakers on a steep beach, the open-sea wave steepness must be relatively high; and conversely, to produce plunging breakers on a gently sloping beach, the open-sea wave steepness must be relatively low. The same diagram shows also a relation between plunging breakers and surging breakers. It will be noted that the gentler sloping beaches do not have surging waves.

Wave Refraction, Diffraction, and Reflection   Another effect of the comparatively even ocean waves coming diagonally into shallow water is a bending of the wave crests as the result of the slowing of the advance. The wave trains over a smoothly sloping bottom curve in order to become more parallel to the coast (Fig. 3–3). The wave energy per unit length along the crest is decreased because of the spreading of the wave rays as the crest is turned. If the ocean floor is irregular, the wave crests tend to conform somewhat to the submarine contours (Fig. 3–4). This process is called *wave refraction*. The presence of a valley on the sea floor spreads the wave rays even more markedly by *divergence*, and the breakers at the valley head are considerably reduced. Over an adjacent ridge the rays are packed together in what is called *convergence*, and the breakers are greatly increased in height. It is not unusual for this combination of divergence and convergence to produce breakers ten times as high inside a ridge as inside the adjacent valley (Fig. 3–5). As a result, landing through the surf is far easier where there are sea-floor valleys extending in close to the coast. Fishing ports like Nazare on

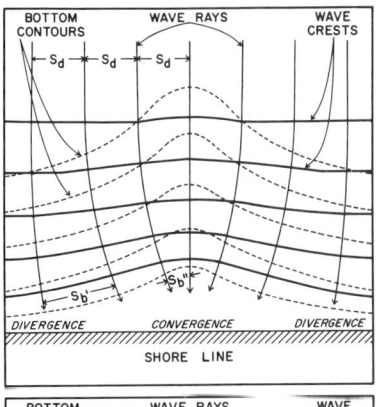

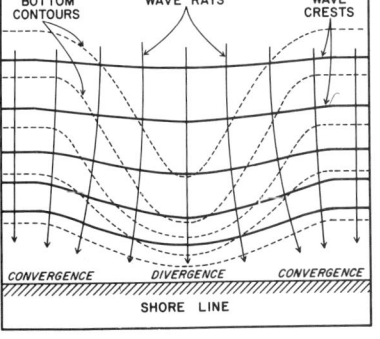

Fig. 3–4 Wave refraction diagram showing: *A*, the convergence of wave rays over a ridge producing higher breakers inside the ridge, and *B*, the divergence of wave rays over a valley producing smaller breakers inside the valley head.

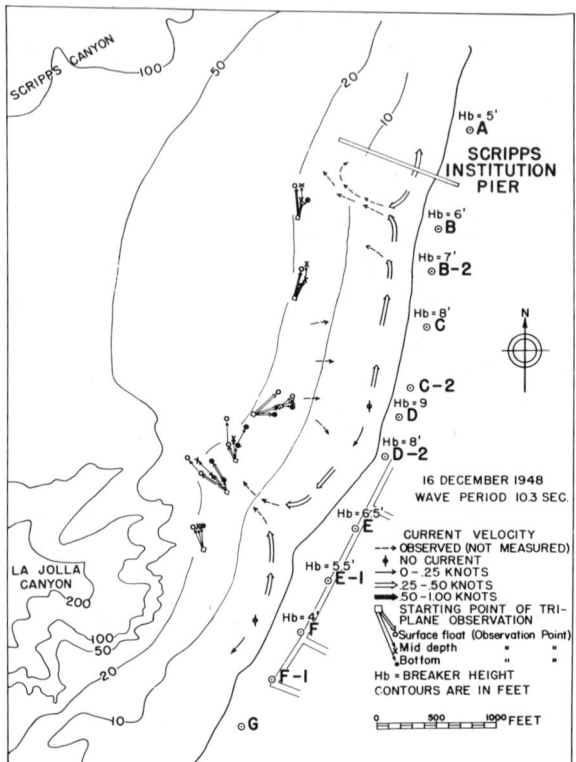

Fig. 3–5 Cell-like features of the circulation pattern during relatively long-period waves along the beach south of Scripps Institution. The rotary currents are related to the wave convergence between canyons and divergence at canyon heads. Velocities were measured by tracing movement of suspended and surface floats. Letters refer to ranges used in the work. From Shepard and Inman (1950).

the west coast of Portugal and Cap Breton on the southwest coast of France owe their existence to this valley effect, allowing the fishermen to launch their boats inside the adjacent valley heads.

Where a wave crest passes the breakwater at a harbor entrance, the train is interrupted, but some of the energy from the waves is transmitted at right angles to the direction of wave advance, allowing small waves to extend into the shadow of the barrier (Fig. 3–6). This is called *diffraction*.

A wave approaching a vertically walled coast undergoes little steepening during the approach. In such cases, most of the energy is reflected and the reflected wave travels seaward like a billiard ball from a cushion. Actually, there are few vertical sea-floor slopes at the shore, so the reflected waves are usually greatly reduced in size. Where a reflected wave hits an incoming breaker a large spout of water may result. Figure 3–7 shows a wave reflected from the precipitous north coast of Kauai encountering an approaching wave and sending a column of water more than 30 m into the air. Another type of wave travels along the beach and is called an *edge wave*.

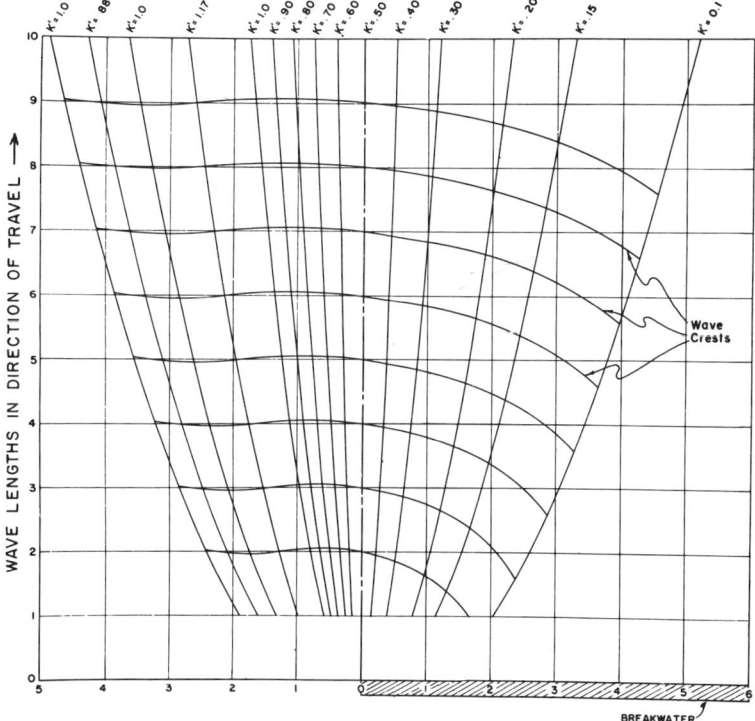

Fig. 3–6 Generalized diffraction diagram for waves passing a breakwater with uniform depth shoreward of the breakwater (after Dunham 1951, Fig. 6; and *Beach Erosion Board Tech. Rept. 4,* Fig. 21). Each unit on the grid is equal to one wavelength of the undiffracted wave; curved crest lines show the position of successive wave crests after diffraction. The values of K' give the proportion of the height of the undiffracted wave that is to be found in the lee of the breakwater.

Nearshore Currents Related to Waves   The approach of waves diagonal to the shore sets up currents inside the breakers, which play an important part in distributing the sandy coastal sediments. Because of the net forward motion of a water particle in trochoidal waves, water is carried into the shore. The beach, however, represents an almost impenetrable obstacle that must reduce to zero the pileup of water against the shore. The superelevation of the water surface produces a normal outward force that must create a balancing outward movement. If the direction of wave travel is diagonal to the shore, the return flow has a net longshore component, which contributes to the longshore current. This is usually confined to the surf zone. The velocity of this current inside the surf zone would continue to increase with distance along the shore were it not for the existence of outward-flowing rip currents at discrete places (Figs. 3–5, 3–8). After passing through the surf, often with a velocity too great to allow ordinary swimmers to progress against it, the rip tends to spread out into what is known as the *head* and, beyond this, the water again is returned to the breaker zone by ordinary wave motion.

A very special type of circulation is set up in relation to the refraction that accompanies alternating valley and ridge topography adjacent to the shore (Fig. 3–5). The convergence on the intervalley ridges causes a higher

Fig. 3–7 Reflected wave encountering an oncoming breaker off the Napali coast of Kauai. Photo by author from helicopter at 30-m height.

water level along the adjacent shore, and this in turn causes water to flow away to either side as a *longshore current.* These diverging currents encounter a weaker current in the opposite direction that is set up by divergence inside the valley head. As a result, rip currents, often at very pronounced speeds, move seaward near the valley. In some cases, the rip currents move directly into the area over the valley and, losing velocity, cause a considerable deposition of sediment.

It has often been reported that breaking waves are accompanied by *undertow* that carries water below the oncoming waves and results in drowning of bathers. Actually, this erroneous interpretation probably results from the fact that the strong backwash, such as that on steep beaches, knocks unwary bathers off their feet. The backwash is reversed into uprush as soon as the next wave arrives, and, during uprush, the bottom current moves upslope, contrary to what was supposed in the undertow myth. In some circumstances, however, the actual measurement of flow may locally show a net movement seaward along the bottom; and, in rip currents, the current moves seaward along the bottom as well as at the surface. The bottom-flowing current in a rip does not usually extend out much beyond the surf zone, but at the surface the flow may continue seaward in extreme cases for as much as a kilometer. If a drowning swimmer has swallowed sufficient water to make his body heavier than water, he may be carried out below the surface or even near the bottom in a rip current.

Fig. 3–8 A typical rip current carrying muddy water seaward. Note foam lines on the front of each seaward surge in the head of the spreading rip.

CATASTROPHIC WAVES

In addition to ordinary wind waves with their accompanying sea and swell, there are other types of waves that may produce serious disasters to coastal communities. Popularly they are all called tidal waves; actually they have little relation to the tides. These waves include *tsunamis* (seismic sea waves), *storm surges* (storm tides), and *landslide surges*. They all represent a piling up of water onto the shore for a period and, depending on the cause, they may come singly or in a group with intermediate periods of lower sea level.

Tsunamis The word *tsunami*, taken from a Japanese word that means high water in a harbor, is used by the Japanese for the rising surge that carries water onto the shore, both in bays and on the open coast, as the result of a movement of the sea bottom or a volcanic disturbance. Historically, these waves have raised the water level as much as 41 m (Johnson 1919, p. 41). Of particular interest have been the tsunamis of Japan that have caused many drownings and much loss of property. Since 1946, tsunamis have been particularly destructive in the Hawaiian Islands and in southern Alaska. Although the Japanese and Alaskan disasters have been related to movements of the nearby sea floor (Van Dorn 1965), the destructive tsunamis in Hawaii have come almost entirely from great distances, even as far as the west coast of South America. The only disastrous tsunami waves that have hit the conterminous (the 48 states) portion of the United States have also come from great distances. In fact, we had experienced no serious damage until waves from the March 1964 Alaska earthquake struck Northern California and Oregon. The Atlantic has had few severe tsunamis, although that of Lisbon in 1755 swept up the Tagus with waves up to 12 m and destroyed the lower part of the city. This was due to a fault movement outside the coast and affected areas as far as Gibraltar and the nearby Azores Islands. These islands have had many other tsunamis throughout history. In the West Indies, Port Royal, Jamaica, was partly destroyed in 1692 by a tsunami coming from a local fault movement. The destructive tsunamis during the 1964 earthquake in Alaska were mostly of rather local origin, the result of large slumps in the bays. Offshore faulting and tilting of a large block to the southeast sent waves down to the California coast and caused serious damage to Crescent City (Van Dorn 1964). In the Alaska bays, water rose as much as 52 m locally, probably from slumps on the sea floor rather than from fault displacements (Plafker et al. 1969).

Most tsunamis are probably the result of sudden movements of the sea floor along the deep-ocean trenches. These movements, which are accompanied by world-shaking earthquakes, are thought to be caused by either a sudden deepening or a sudden shoaling over a wide area, resulting in the cascading of the displaced water mass into the depressed area or out from the elevated area. The elevated or depressed area is generally elongate, and the waves move out with greatest energy normal to the axis of this elongation (Van Dorn 1965). This explains why the destructive waves from the 1946 and 1957 earthquakes coming from an east–west section of the Aleutian Trench damaged Hawaii and not the West Coast of the United States; whereas the waves from the 1964 earthquake coming from a northeast–southwest elevated area on the shelf off Alaska (Van Dorn 1964) caused destruction along the West Coast of the United States and did not damage Hawaii. Examination of a globe will show that displacements in the deep trench off Chile are capable of directing waves toward the Hawaiian Islands and account for the recurrent disaster at Hilo in 1960 (Eaton et al. 1961), but a much greater disaster occurred along the coast of Chile (Sievers 1961).

While traveling across the deep ocean, tsunamis, because of their great length, advance at much higher speeds than ordinary wind waves. This velocity has been measured as averaging from 725 to 800 km per hour in crossing the Pacific. The equation $C = \sqrt{gh}$ (where C is the velocity and h the depth of water) makes it possible to estimate the speed. The wave height varies inversely as the fourth root of the water depth. Accordingly, the wave is very low in crossing the deep ocean, so it is not observed by mariners; and it rises only as it is slowed by encountering a shallow shelf and approaching the shore or moving up a narrow funnel-shaped bay. The waves may come in with a steep front like a tidal bore or may rise like an ordinary tide. The period between waves is always long but is frequently between 10 and 30 minutes.

A study by three scientists of the 1946 tsunami in Hawaii (Shepard et al. 1950), during which the writer observed the waves, has led to some tentative conclusions. Figure 3–9 shows the heights to which the highest waves rose on the various islands. It will be seen that the greatest heights were, as would be expected, on the north coasts because the waves came from that direction. Other conclusions from the study include the following:

1. The waves in the Pacific that follow earthquakes are probably the result of movement along a line rather than in a narrow zone, such as might have been due to submarine landslides (Shepard et al. 1950, p. 394).

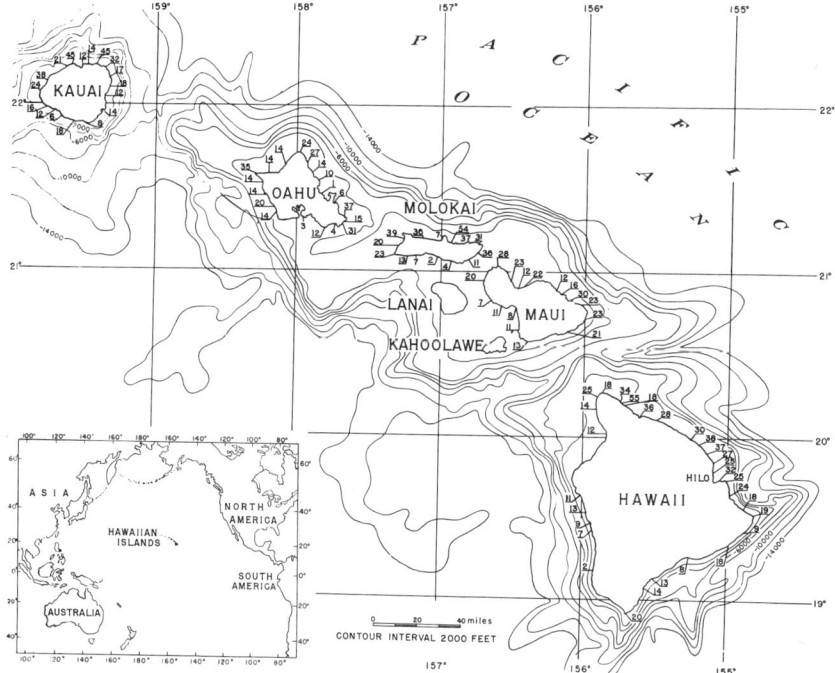

Fig. 3–9   Maximum heights reached by waves from the 1946 tsunami in the Hawaiian Islands. Heights expressed in feet. From Shepard et al. (1950).

2. The above indicates that faulting is the cause. According to Van Dorn (1965), the "absence of a bore and the leading trough suggest a source aligned with the axis of the Aleutian Arc, with downthrust on the south side."

3. Tsunamis can travel for thousands of miles and retain energy enough to rise for a score or more feet on an unprotected coast facing the direction of wave approach.

4. A withdrawal of the water from the coast is likely to be the first substantial manifestation of approaching waves, although a slight crest often precedes

5. The first wave is not likely to be the largest.

6. In general, the waves are relatively high inside submarine ridges and relatively low inside submarine valleys, provided these features extend into deep water.

7. The waves may be small where they hit a projecting point that is bordered by deep water.

8. The waves are greatly decreased by the presence of coral reefs bordering the coast.

9. The waves have a funneling effect in some small bays but are generally small at the heads of long estuaries.

10. The waves can bend around a circular island without great loss of energy but are considerably reduced in height when encircling an elongate angular island.

Coastal changes often accompany the inundations of tsunamis. Scarps are cut at levels well above those of the ordinary waves. Also, boulders and other debris are often carried up onto the coastal lowlands and left in a confused mass (Fig. 3–10).

Storm Surges During great storms, such as hurricanes and typhoons, broad coastal inundations often occur. The sea level in and near the low-pressure area is considerably raised, due largely to winds blowing toward shore, and the sea piles up onto the shore with one or more surges that usually develop at or near high tide. If there is a violent onshore wind at the same time, great destruction may result. When Hurricane Camille hit the coast of Mississippi in August 1969, the water rose 9 m, accompanied by pounding waves. Gusts of wind were estimated to be as high as 370 km per hour. Virtually all man-made structures in the inundated area were destroyed and hundreds of lives were lost. A smaller wave height was an even worse killer in 1900 when the city of Galveston was wiped out with almost its entire population of 7000. Hurricane tracking has now enabled advance warnings to be given, so these waves are somewhat less destructive to life, but sometimes the warnings are too late or not heeded soon enough to be very effective. Also, sea walls, as at Galveston (constructed after the 1900 disaster), have saved property and lives. Most hurricanes start in ocean areas, and there is now hope that seeding their centers with chemicals may help break them up before they hit the coast.

Geologically, hurricanes are responsible for even greater coastal changes than tsunamis. Since 1940, it has been the practice of several government agencies to take overlapping strips of aerial photographs along the coasts of the United States. This has made it possible to discern graphically

Fig. 3–10 Destruction on coastal plain of northern Kauai by tsunami of April 1, 1946. Coral boulders and uprooted trees are scattered over the denuded terrace. From Shepard et al. (1950).

the nature of these hurricane changes, especially along the beaches and low sand islands of the East and Gulf Coasts (results are described in Shepard and Wanless 1971). Hurricanes often cut inlets into long sand islands (Fig. 3–11). Other inlets that have been closed by sand deposition across their mouths are reopened. Beaches and coastal dunes are greatly eroded by the high sea levels and violent waves during the passing of the eye of a hurricane. Escarpments left in the dunes may later be covered by deposition of sand. Beaches also are usually rebuilt rapidly after the waves have ceased. Another common effect of a hurricane is to sweep sediment through inlets and deposit it as tidal deltas on the inside (Fig. 3–12).

Landslide Surges It is difficult to distinguish between tsunamis and local waves coming from submarine landslides. This was particularly the case for the 1964 Alaska earthquake. However, waves generated by rockfalls initiated above sea level are in a distinctly different category from tsunamis. A large rockfall from a cliff into an embayment or lake may cause an uprush of water to tremendous heights. As a result of an earthquake in 1958, 40 million cubic yards of rock fell from an altitude of 900 m into Lituya Bay, southern Alaska, and caused an impulsive wave that crossed the kilometer-wide fiord and surged up to 530 m onto the mountain on the opposite side, cutting off all the trees up to that level (Fig. 3–13). This was investigated by Miller (1960). The water swept down Lituya Bay to the entrance at 180 km

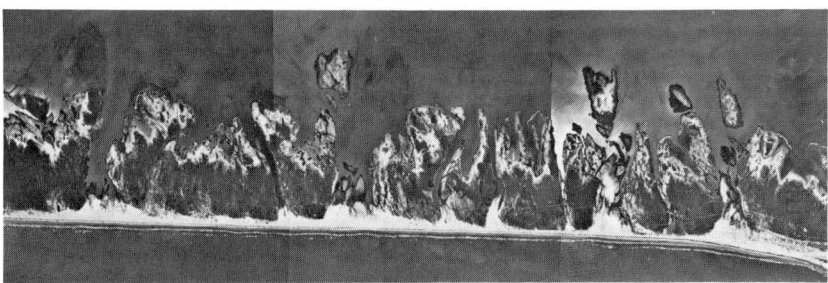

A

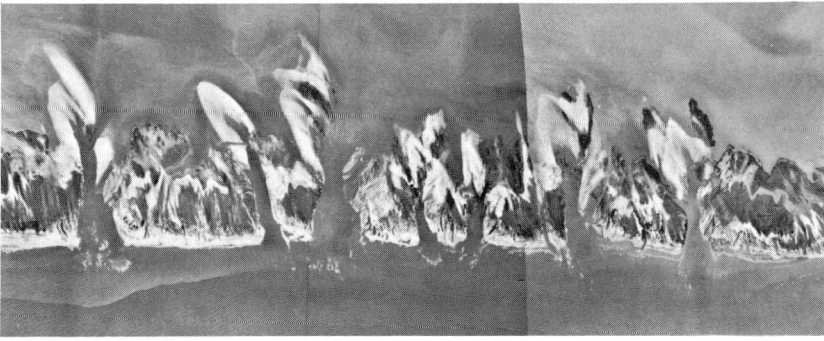

B

Fig. 3–11 A, the continuous barrier of Matagorda Peninsula before Hurricane Carla, 1961. *B,* same area six days after the hurricane with newly opened sluiceways and washover deltas formed inside the barrier. Photos by U.S. Coast and Geodetic Survey.

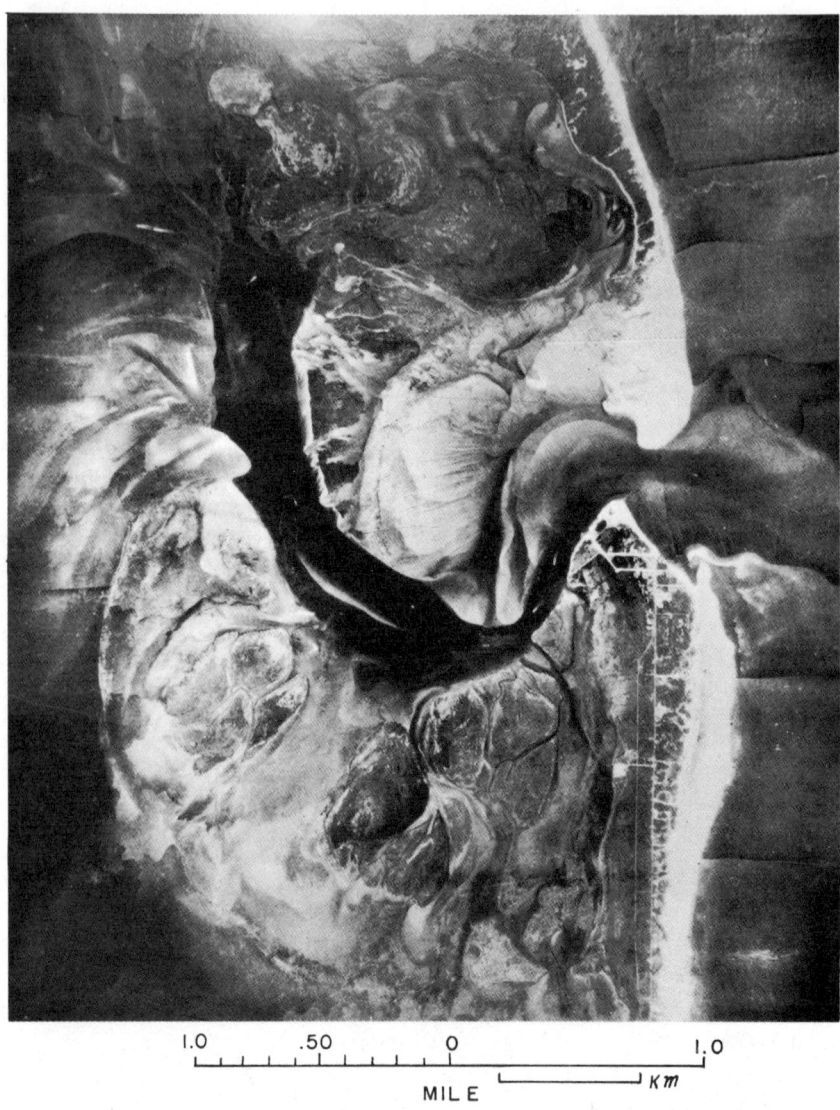

Fig. 3–12 An aerial mosaic of Barnegat Inlet, New Jersey, shows tidal delta in early 1930s. Photo by Airview Inc., Long Beach, New Jersey.

per hour and carried a boat across the entrance spit so high that the occupants could look down on the treetops while passing over them. The boat sank but the people managed to escape in their skiff. None of the landslide surges has produced a wave that could be traced seaward for any distance. Although the landslide surge at Lituya Bay killed only a few boatmen who were in the bay, other rockfalls have been more destructive. According to Miller (1960), the worst rockfall wave on record occurred in 1792 in Shimbara Bay of Kyushu Island, Japan, where a fall at the time of an earthquake pro-

Fig. 3–13   Trimline caused by giant wave of July 9, 1958, which cut off all vegetation on the cliff opposite the huge rockslide up to 530 m. The front of Lituya Glacier (right) was cut back 400 m by the slide. From D. J. Miller (1960).

duced three large waves that drowned 15,000 people along the edge of the bay. In 1936 at Loen Lake, Norway, surges up to a height of 70 m were due to rockfalls without an earthquake as the initiating cause. Icefalls into fiords or lakes from hanging glaciers or from the cliff front of a tidal glacier are also known to produce large surges and have at times destroyed small boats.

TIDES

The ocean tide, a long-period type of wave, results from the gravitational attraction of the moon and, to a smaller extent, the sun. If the earth were covered with a uniform thickness of water, the tides would be high when the moon passed overhead. On the opposite side of the earth, the high tides would be the result of the moon's pulling the earth away from the water. The tides are far more complicated, however, as the result of the interference generated by the continents, islands, and shallow continental shelves. The direct relation of the moon and sun to the tides is shown by the highest and lowest tides occurring when the sun and moon are in conjunction, that is, on the same side of the earth (new moon) or on directly opposite sides (full moon). These are called *spring tides,* although there is no seasonal significance. The times of least tidal range are when the moon and sun have a right-angle relation to the earth (first and third quarter). These are called *neap tides.*

The times of high tide vary as a result of nodal points of small tide that develop in the middle of large oceans and the tidal oscillation around

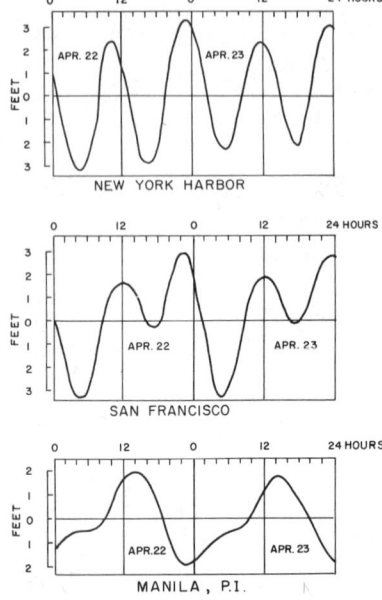

Fig. 3–14 Three principal types of tidal curves; semidiurnal (New York), mixed (San Francisco), and diurnal (Manila).

these nodes, causing the tidal wave to move up and down the continental margins. Long bays and inlets usually have high tides later than on the open ocean. Funnel-shaped bays, like the Bay of Fundy and the English Channel, have the greatest tidal ranges (up to 18 m). The smallest tides occur in inland seas, such as the Mediterranean and the Gulf of Mexico, and on the islands that are near nodal points, like the Hawaiian group.

The tides are waves commonly with periods averaging 12 hours and 26 minutes between successive high or low tides (due to earth rotation and the motion of the moon around the earth). Tides are either semidiurnal, mixed, or diurnal (Fig. 3–14), depending on the natural period of the ocean basin and the declination of the moon. Atlantic tides are mostly semidiurnal, whereas Pacific tides are more mixed and even diurnal along some of the island coasts. Currents associated with tides will be discussed later.

OCEAN CURRENTS

It was formerly believed that most of the abyssal ocean had very low-velocity currents, not greater than 1 cm/sec. We now know that almost all of the water is in a constant state of movement, and complete stagnation exists only in a few deep basins with shallow sills. Even with these exceptions it is doubtful if stagnant conditions occur over long periods. In most places, water flow is very slow, but in some places, the velocity is as high as in the large rivers on the continents. Aside from the currents related to the breaking waves in the surf zone, ocean currents result from (1) wind stress on the surface of the ocean, (2) tidal forces, and (3) differences of water density. The last may be due to evaporation, differential heating, freezing of sea water, melting of sea ice, introduction of river water into the ocean, and development of dense sediment-laden water as the result of stirring by waves of bottom sediments or slumping of slope deposits.

Wind-Drift Currents   The local wind produces frictional drag at the surface and sets up currents. Near the surface in the open sea, the Coriolis force causes these flows to move at approximately 45° to the right of the wind direction in the Northern Hemisphere and 45° to the left in the Southern Hemisphere. These angles increase to 90° below the surface. Along the coast, however, the currents are restricted in direction of movement so that the Coriolis deflection is slight and has little influence. In long narrow bays, the wind currents roughly conform to the direction of the bay margins.

The depth of wind-induced currents (aside from the permanent currents discussed below) depends on the velocity and constancy of the winds, and the currents become negligible at depths of about 50 m. Wherever strong steady winds occur, the warm surface water of equatorial and temperate latitudes is mixed downward, and thick isothermal layers may be produced.

Permanent Currents and the Prevailing Winds   The large permanent ocean currents, like the Gulf Stream, are all related to the density distribution of the oceans but are maintained primarily by the prevailing winds.

These great streams of water have a very gentle surface slope downward to the left in the Northern Hemisphere and to the right in the Southern Hemisphere. The water on the higher side is always warmer and hence is less dense. The major subtropical gyral currents turn to the right in the Northern Hemisphere and to the left in the Southern Hemisphere (Fig. 3–15). The westward movement at low latitudes is the result of the trade winds, and the eastward movement the result of the west winds at higher latitudes. The most concentrated currents occur along the east side of the continents of Asia and North America because of the pileup of water due to the trade winds. The level is raised sufficiently in the Pacific Ocean to cause a countercurrent, not only to the north but also to the south of the equator. Within a narrow band along the equator (2°N to 2°S) an even stronger but subsurface eastward flow, the Cromwell or Pacific Equatorial Undercurrent (Knauss 1961), was discovered by Cromwell, Montgomery, and Stroup. These undercurrents are also well developed in the Atlantic and seasonally in the Indian Ocean. At high latitudes on the east side of the Atlantic and Pacific, a branch of the east-flowing current, which is driven by the westerly winds, extends along the west coasts of Europe and North America with considerable warming of the adjacent lands. Cold currents come down along the northeast coasts of Asia and North America as far as New England (Labrador Current) and northern Japan (Oyashio Current). Similarly, cold currents in the Southern Hemisphere extend well north along the west coast of South America (Humboldt Current) and Africa (Benguela Current).

Some of these great streams actually move hundreds of times as much water as do the Mississippi or Amazon rivers. For example, the Gulf Stream off eastern and southern Florida has a flow of 26 million cu m/sec. It is about 170 km wide and flows at the surface with maximum speed of about 1.5 to 3.0 m/sec (3 to 6 kn). The flow has sharp boundaries that shift from time to time and throw off large eddies after the Stream leaves the coast at Cape Hatteras.

Fast-flowing currents, such as the Gulf Stream, develop countercurrents along the sides that have an important influence on the coastal configuration of Florida and the Carolinas, where there is a series of capes and concave bights. Similar features along the east coast of Japan may have some connection with the countercurrents from the Kuroshio.

Bottom Currents Related to Permanent Currents   Until recent years the general belief was that the deep circulation related to the permanent currents moved over the ocean floor only at very slow speeds and thus was

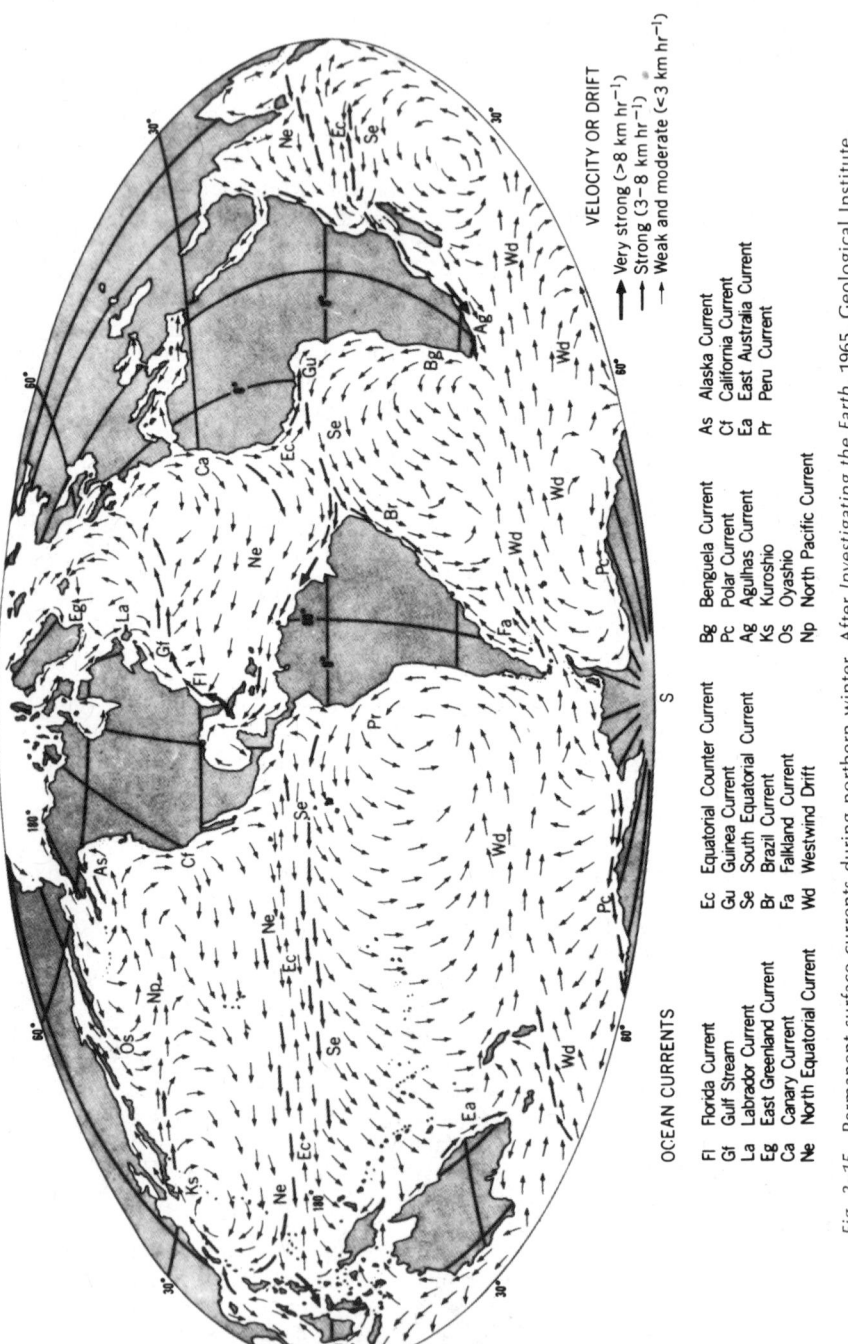

OCEAN CURRENTS

Fl	Florida Current	Ec	Equatorial Counter Current	Bg	Benguela Current	As	Alaska Current
Gf	Gulf Stream	Gu	Guinea Current	Pc	Polar Current	Cf	California Current
La	Labrador Current	Se	South Equatorial Current	Ag	Agulhas Current	Ea	East Australia Current
Eg	East Greenland Current	Br	Brazil Current	Ks	Kuroshio	Pr	Peru Current
Ca	Canary Current	Fa	Falkland Current	Os	Oyashio		
Ne	North Equatorial Current	Wd	Westwind Drift	Np	North Pacific Current		

VELOCITY OR DRIFT

→ Very strong (>8 km hr⁻¹)
→ Strong (3–8 km hr⁻¹)
→ Weak and moderate (<3 km hr⁻¹)

Fig. 3–15 Permanent surface currents during northern winter. After *Investigating the Earth*, 1965. Geological Institute.

not important in either erosion or deposition. The first indications that this supposition was not correct came from the discovery of hard bottom on portions of Blake Plateau at 700 to 1100 m off the southeast coast of the United States. Similarly, an abundance of notations of rock bottom along the Japanese coast under the Kuroshio to depths of about 1300 m have suggested that there might be substantial currents in this area. With the advent of bottom photography and the recent observations from deep-diving vehicles, there has been a gradual buildup of information showing that currents strong enough to produce ripple marks, approximately 15 cm/sec (0.3 kn), are common on the tops of seamounts and oceanic ridges (Heezen and Hollister 1964, 1971) and on the floors of submarine canyons at all depths (Shepard and Dill 1966). They are found also in passageways on the ocean floor where return circulation from deep descending currents is concentrated, as in the Straits of Florida and Drake Passage (Heezen and Hollister 1964, 1971). Less frequently, ripples have been found with other evidence of currents along the contours of the continental rise off the East Coast of North America (Heezen et al. 1966A), and ripples are seen in a few photographs of the floors of elongate basins. Aside from the current ripples in submarine canyons, it seems probable that most of the other ripples in deep water areas are related to the permanent currents. Much water from the surface sinks in the polar regions and returns along the bottom. In general, these bottom currents are probably very slow, but over wide areas they are concentrated sufficiently to produce ripples (Fig. 3–16) or at least to transport fine sediment. In some places, they may even be strong enough to erode the bottom (Rona et al. 1967).

This evidence of relatively strong currents does not agree entirely with physical oceanographers' flow estimates that generally indicate somewhat slower speeds. However, these estimates are averages from flow information well above the bottom and do not take into account local concentration or temporary acceleration. Wüst (1958) thought that western Atlantic currents probably have speeds along the bottom sufficient to transport sediment. Also, M. Ewing and Thorndike (1964), using a nephelometer, found murky sediment-laden water, a nepheloid layer, near the bottom in the North Atlantic; the same discovery was made in the Arctic (Hunkins et al. 1969).

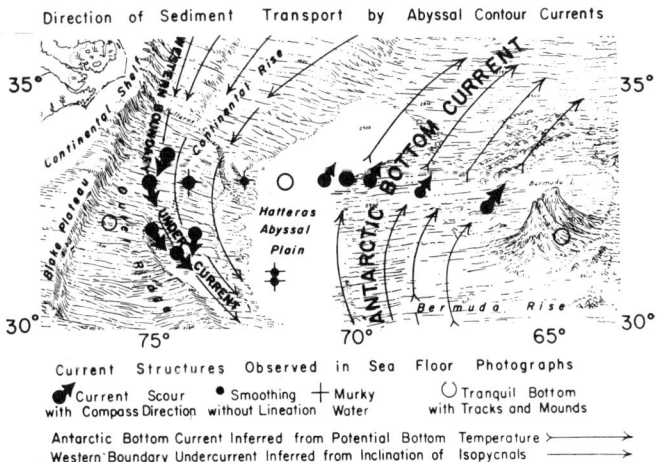

Fig. 3–16 Contour currents along the floor of the Atlantic. From Heezen et al. (1966C).

Cascading Currents on the Continental Slopes A special type of current that is related to the deep circulation of permanent currents is referred to as *cascading* (Cooper and Vaux 1949). This action has been described relative to the Celtic Sea south of Ireland and the slopes off the Norwegian coast. Cold water of winter forming along the shore will flow down across the shelf and may move with considerable velocity down the slopes beyond. Bougis and Ruivo (1954) found that even in the Mediterranean Sea, at the west end of the Golfe du Lion, local cooling waters near the coast reduced the temperature sufficiently (11° C) to cause cascading into a submarine canyon that is located off Banyuls. No velocities have been given for such currents, but the coarseness of sediments on the outer shelf in cold areas and in the canyon off Banyuls may be attributed to such flows. It was even suggested by Iselin (1963) that the disappearance of the U.S. submarine *Thresher* may have been due to a special type of cascading related to a storm circulation. This has never been confirmed.

Tidal Currents The currents produced by the tides have a somewhat greater effect on the shallow sea floor and coastal configuration than do the permanent currents. Because of the difference of times of tides in bays from those on the open coast, the water flows with considerable velocity through the bay entrances. At narrow entrances, velocities of several knots are common, and these currents even occur in the Mediterranean and the Gulf of Mexico despite the low tidal ranges. Thus, the water in the Strait of Messina is subject to whirlpools due to the tide (Charybdis of Greek mythology), and the entrance to Barataria Bay in Louisiana has currents that have scooped out a hole 50 m deep. In seas with greater tidal range, much higher velocities are found. For example, the Seymour Narrows in British Columbia had currents up to 7.5 m/sec (15 kn) prior to the blasting out of a huge rock from the entrance. Tsugaru Strait between Honshu and Hokkaido islands, Japan, has tidal currents up to 3.5 m/sec (7 kn) (Fig. 3–17). These currents are apparently responsible for the deep holes at the entrance to many bays. At Bungo Strait, the entrance to the inland Sea of Japan, there is a hole with depths of 390 m, and at the entrance to San Francisco Bay, the Golden Gate has a 110-m depression.

Unlike both wind and permanent currents, where the velocity decreases rapidly with depths, the velocity of tidal currents is essentially the same from top to bottom in a water column, although near the floor the bottom friction reduces the flow considerably. Thus we measured near-bottom currents in the Golden Gate of 1.5 m/sec (3 kn) when the surface currents were flowing at 3 m/sec (6 kn).

The general belief has been that tidal currents have not been important in contributing to sedimentation in the ancient seas. It is true that tidal currents are not very pronounced in the broad inland seas where so much of the ancient sedimentation appears to have occurred. However, at the entrance to these seas, narrow straits may have had tidal currents that introduced a considerable amount of sediment. Also, tidal currents in shelf seas, where ancient sediments were deposited, are often detectable and are important to sedimentation. Tides moving up into fingerlike mouths of a partially drowned delta can certainly play an important role in redistributing the deltaic sediments. Furthermore, the tides may have been much more important in the past if the moon was at one time much closer to the earth than at present (Munk 1968). It is quite possible that in the late Precambrian there were great tidal forces that may account for the scarcity of life in some of the ancient deposits and for the abundance of coarse sediments. At present this is only speculation.

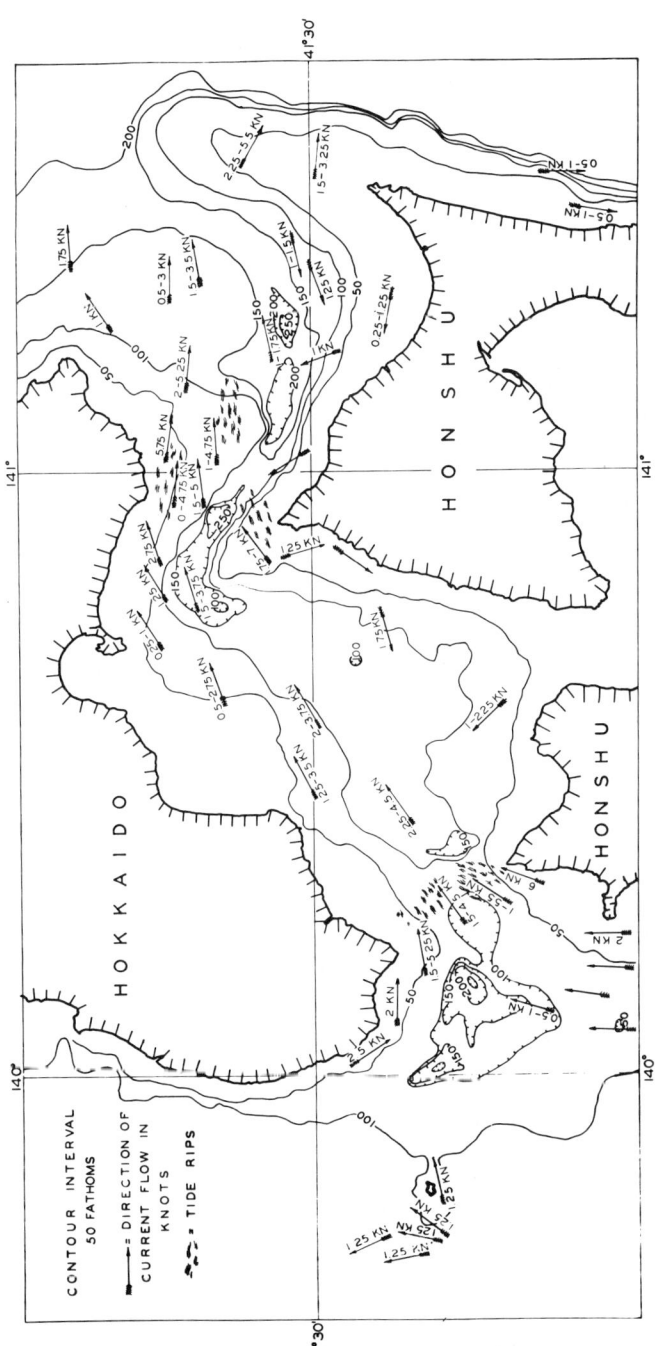

Fig. 3-17 The surface currents of Tsugaru Strait, Japan, and the deep holes in the two narrows.

Turbidity Currents   When sediment is suspended in water, the combination is heavier than the adjacent clean water. The resulting sediment-laden water will descend a slope displacing the clean. Experiments in the laboratory show that currents produced in this way are turbulent as well as turbid, so that the name turbidity currents has significance for two reasons. In nature, turbidity currents were first observed in Lake Geneva where the Rhone River entering with sediment-laden water is sufficiently heavy to descend under the relatively clear water of the lake. Turbidity currents caused by the Colorado River have been traced along the length of Lake Mead to Hoover Dam. Their velocities are of the order of 25 cm/sec (0.5 kn), generally slower in the lower end of the lake (Gould 1951). However, turbidity currents are not likely to be formed in this way in the ocean because entering river water is usually lighter than sea water, and river water mostly spreads out over the sea surface beyond the river mouth.

Daly (1936) was the first geologist to recognize the importance of turbidity currents in the ocean as a means of transporting sediment into deep water and as an agency of erosion of the slopes of the sea floor. Ph. H. Kuenen is largely responsible for documenting Daly's suggestion and demonstrating the role of turbidity currents in producing many of the puzzling coarse deep-water sediments found among the ancient sedimentary rocks (see especially Kuenen 1951, 1959, 1964, 1965, and Kuenen and Migliorini 1950).

Causes for turbidity currents in the ocean include waves that stir up sediments from the bottom in shallow water, which develop a relatively heavy mass of water. Also, slides along the sloping sea floor may carry considerable sediment into suspension. Unfortunately, we have seen only slow-flowing turbidity currents during diving operations, so that we have to rely largely on indirect evidence, such as the successive cable breaks on the continental slope off eastern Canada that followed the Grand Banks earthquake of 1929 (see p. 331). Because many characteristics of sediments, particularly in deep water, appear to demand substantial turbidity currents, their existence in the ocean is generally accepted. Once the currents are initiated they have the ability to continue flowing indefinitely down slopes by stirring up more sediment along their path. Their importance relative to submarine canyons is so great that they will be given a major discussion in the chapter on canyons and sea-floor valleys (pp. 304–341).

Internal Waves and Other Types of Bottom Currents   Measurements of currents in submarine canyons have shown that there are alternating up and down flows that are related neither to turbidity currents nor to the permanent currents of the ocean. Their periods are largely nontidal. They may be the result of *internal waves,* which have long been known to exist. They were first reported occurring in the oceans by Ekman (1904), who observed the "dead water" in the Arctic, where a thin layer of meltwater overlies a more saline mass. Measurements showed that an oscillation of the boundary occurred without any accompanying surface waves. During World War II, Ufford (1947) used the bathythermograph to study these internal waves and found periods as frequent as 5 minutes. The more recent work by LaFond (1961) off Mission Beach, Southern California, demonstrated that the internal waves have their greatest wave height at intermediate depths and the greatest horizontal velocities at the bottom. The observed horizontal velocities are capable of moving sediment or perhaps even of producing erosion. It is possible that internal waves may be significant on the tops of seamounts, and they may account for the alternating up- and downcanyon flows of water in submarine canyons.

According to Emery (1956A), internal waves cause bottom currents that are responsible for stirring bottom sediments in the deep basins off the Southern California coast, and also aerate the basins, replacing the oxygen that is used by organisms, and thereby prevent stagnation.

The alternating currents in submarine canyons may have some cause, not yet apparent, other than internal waves. The study of bottom currents in the ocean is still in its infancy.

Chapter 4
OCEAN SEDIMENTS
AND SEDIMENTATION[1]

INTRODUCTION

Prior to discussing beaches and the sea floor it is necessary to have a rudimentary understanding of the physical and chemical properties of sediments as well as of the mechanics of sedimentation. However, a quantitative discussion of sediment mechanics requires a lengthy mathematical treatise. Although such information is important, this chapter will conform with the rest of the book and leave the mathematical and physical treatment to others. The best references now available to these phases are Bagnold (1954, 1963), Inman and Bagnold (1963), and Inman (1963) in the second edition of *Submarine Geology*. The present chapter will deal largely with the description, characteristics, and methods of analysis of sediments.

[1] Portions adapted from Shepard, *Submarine Geology*, 2nd ed., Chapter V by D. L. Inman.

Certain physical properties of sediments appear to be fundamental to the marine geologist in his study of the classification of sedimentary deposits, as well as to the student of sedimentary mechanics in his attempt to understand the natural dynamics by which the sediment was transported and ultimately deposited. Density, size, and size distribution are controlling parameters in almost all the physical properties of sediments. Settling velocity, while fundamental to sedimentation and suspension, is also an important measure in determining the size of fine sediments; in this discussion, the "equivalent size" is that of a quartz sphere, having the same settling velocity as that of the less spherical natural grains. Packing and permeability become important considerations in sediment being picked up by waves and currents, and the slopes of beach faces are, to a certain extent, controlled by the permeability of the beach sand. Permeability, in turn, is determined by the packing, size, and size distribution of the sediment grains.

SEDIMENT SIZE

Sediments are frequently referred to as sand, silt, or clay. These terms refer to the *size* of the sediment particles, which is one of the fundamental physical properties of any sediment. The size distribution or the range in size of the particles that make up the sediment aggregate is also of importance. Size and size distribution are frequent variables in determining most sediment properties; i.e., they are related to basic concepts of settling velocity, transportation, permeability, sorting, and so on. The purpose of this section is to discuss size, the concept of size scales, and measures for describing the distribution of sediment size.

Grade Scales   Early in the study of sediments it was found that a linear or arithmetic scale was not the most convenient for treating sediments. Most properties of sediments, such as settling velocity and permeability to fluid flow, were found to vary as some power of sediment size rather than directly with size.

If the size distribution of a sediment aggregate is considered on an arithmetical scale, it is found that the distribution is very skewed,[2] so that the bulk of the sediment occurs in one of the fine sizes of an arithmetic distribution. Thus, a representation of this type is difficult to use because sediments differing widely in physical characteristics appear somewhat similar. From the nature of the size distribution curve and from the settling relations, discussed later, it is apparent that a logarithmic or geometric scale is better suited for describing sediment distributions.

A geometric series is a progression of numbers of such a nature that there is a fixed ratio between successive elements of the series. The first geometric scale for sediment size to receive extensive use was developed by Udden (1898), who used powers of 2 mm in his scale. In 1922, Wentworth extended Udden's scale and gave names to the various elements of the series. This scale is now in common use in the United States. Wentworth's scale as well as others are illustrated in Table 1.

Krumbein (1936) used the exponents, that is, the powers of two in the Wentworth series, as the basis for a logarithmic scale of sediment size. Because most sediments are finer than 1 mm and thus would have a negative

---

[2] *Skewed* means bunching of grains on one side of the median and tailing out on the other.

Table 1 Grain Size Scales for Sediments

GRADE SCALES					
WENTWORTH (1922) after Udden (1898)	Phi $\phi = -\log_2$ (m.m.)	(m.m.)	MICRONS μ	U.S. BUREAU OF SOILS	
BOULDER					
	−8	256			
COBBLE	−7	128 100			
	−6	64			
	−5	32		LARGE	GRAVEL
PEBBLE	−4	16 10			
	−3	8			
	−2	4		MEDIUM	
GRANULE	−1	2			
SAND VERY COARSE	0	1	1000.	FINE	
COARSE	+1	½	500.	COARSE	SAND
MEDIUM	+2	¼	250.	MEDIUM	
FINE	+3	⅛ 10	125.0	FINE	
VERY FINE	+4	1/16 20	62.5	VERY FINE	
SILT COARSE	+5	1/32	31.3		
MEDIUM	+6	1/64	15.6	SILT	
FINE	+7	1/128	7.8		
VERY FINE	+8	1/256 200	3.9		
CLAY COARSE	+9		1.95		
MEDIUM	+10	1/1024	0.98		
FINE	+11		0.49	CLAY	
VERY FINE	+12	1/4096	0.24		
COLLOID					

exponent which is cumbersome to use in computations, he used the negative logarithm to the base of 2. Krumbein defined a phi unit as the negative logarithm to the base of 2 of the particle diameter in millimeters; $\phi = -\log_2$ $(D$ in millimeters). This notation has the disadvantage that large sizes have negative values and small sizes are positive. On the other hand, it is convenient in that each whole phi unit designates the boundary between a single Wentworth size; thus, fine sand, which includes sizes from 0.250 to 0.125 mm, is contained between +2.0 and +3.0ϕ. Also, the phi scale considerably simplifies plotting of frequency distribution and cumulative frequency curves of sediments, and helps the computation of standard deviations and skewness of sediment distribution. A conversion chart from millimeters to phi units is given in Fig. 4–1 (for a more detailed conversion table, see Paige 1955).

Measures of Size Distribution Numerous descriptive measures obtained from the grain-size analysis of sediments are used in the literature to indicate the salient features of the size-frequency distribution of a sediment. These measures range in complexity from those obtained by computing the various moments[3] of the sample distribution about some central measure of

---

[3] In mathematical statistics the moment is used to denote the product of the frequency of occurrence within a given (size) class and the difference (or the difference raised to some power) between the class size and that of some origin, such as the means of the distribution. The sum of the products for all classes divided by the number of classes is the moment of the distribution. The first four moments, when converted to the proper form, are referred to as the mean, standard deviation, skewness, and kurtosis.

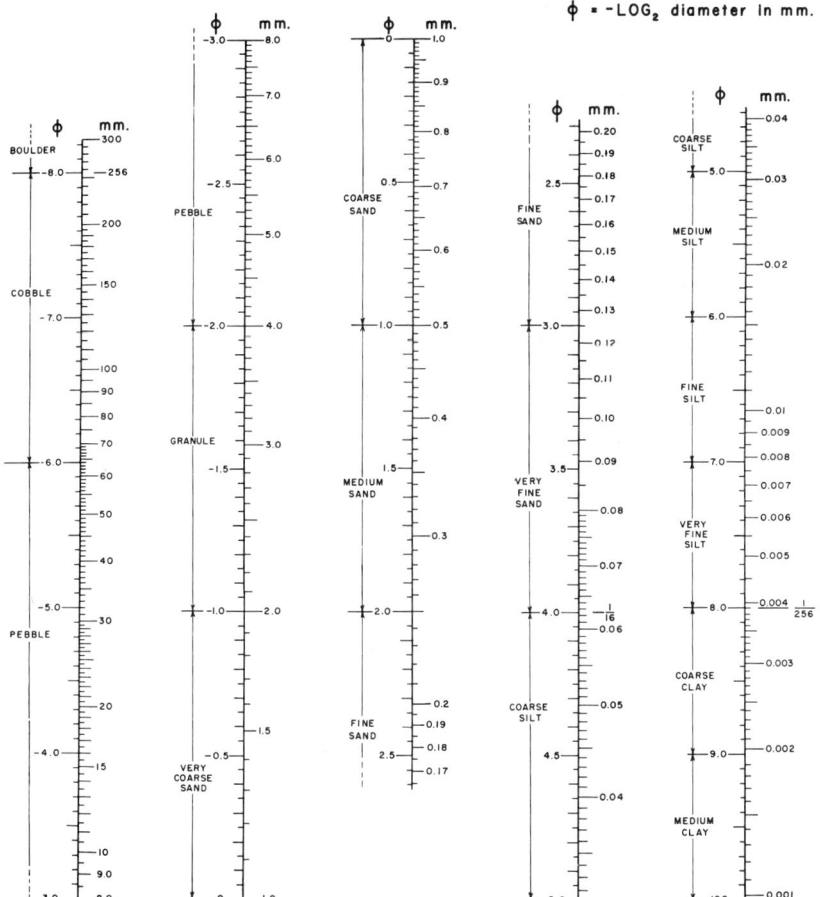

Fig. 4–1 Conversion chart for diameters in phi units and millimeters. From Inman (1952).

sediment diameter to those obtained by selecting a few points from the central 50 percent of the cumulative frequency curve of the sediment.[4]

In the preceding section it was noted that the size-frequency distribution curves of sediments become more symmetrical when the logarithm of the diameter instead of the diameter is plotted as the independent variable. In many cases, these curves approach normal distribution or one of the family of curves derived from a normal distribution.[5] This is advantageous because it allows the properties of the distribution to be described in terms of devia-

[4] Trask (1932) defined the sorting coefficient as $So = \sqrt{Q_1/Q_3}$, and the skewness coefficient as $Sk = Q_1Q_3/Md$, where Q_1 and Q_3 are the diameters in millimeters corresponding to the 25th and 75th percentiles, respectively, of a cumulative percent (coarser) curve, and Md is the median diameter.

[5] A normal distribution is one whose frequency distribution curve is "bell-shaped" and symmetrical. A more complete description may be obtained from any text on statistics, such as Dixon and Massey (1951, p. 47).

tions or anomalies from the normal distribution, a procedure well established in mathematical statistics.

Measures, such as the mean, standard deviation, and skewness, obtained from moments provide a convenient and concise notation, which yields the general features of the size-frequency distributions of sediments. However, the dependence of moment measures on the entire distribution is a limitation to their practical application to sediments, because mechanical analyses of sediments frequently result in open-ended curves and thus do not give the coarse and fine limits of the distribution. Another disadvantage of moment measures is the complex and time-consuming procedure required to compute them, although, where available, modern computer techniques make the handling of large quantities of data quite rapid.

In a normal frequency distribution curve, approximately 68 percent of the population occurs between plus and minus one standard deviation either side of the mean (or median, since these two measures are equal for a normal distribution), and 95 percent occurs between two standard deviations either side of the mean. Thus, the 16th and 84th percentiles of a normal sediment cumulative frequency curve represent diameters one standard deviation either side of the mean.

The above considerations led Inman (1952) to develop expressions for

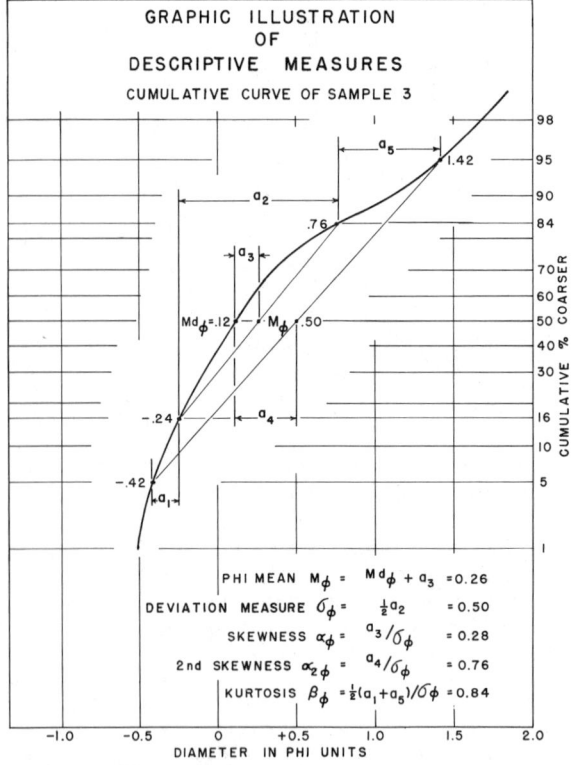

Fig. 4–2   Graphic illustration of the descriptive measures for the size distribution of sediments as obtained from the cumulative frequency curve. From Inman (1952).

the particle size distributions of sediments in terms of phi measures that serve as approximate graphic analogies to the moment measure of the mean diameter, standard deviation, kurtosis, and two measures of skewness, the second being sensitive to skew properties of the "tails" of the sediment distribution. The latter are computed from five percentile diameters obtained from the cumulative size-frequency curve of the sediment and are defined in Fig. 4–2 and Table 2.

The phi median diameter Md_ϕ is a measure of central tendency. It is sometimes used in preference to the mean diameter $M\phi$, because it may be obtained directly from the cumulative curve without interpolation and because it is less influenced by extreme values of skewness than the mean. The phi deviation measure, σ_ϕ, is a measure of sorting or spread and is approximately equal to the standard deviation of statistics. Since one phi unit is equivalent to one Wentworth division, the phi deviation measure gives the standard deviation of the sediment distribution in terms of Wentworth units.

In a symmetrical distribution, the mean and the median coincide, but if the distribution is skewed, the mean departs from the median, and the extent of this departure is a measure of skewness. The phi skewness, α_ϕ, gives the departure of the mean from the median in terms of the phi deviation measure and is therefore a dimensionless measure of skewness, independent of the spread or deviation of the distribution.

Table 2   Descriptive Measures

Measure		Nomenclature	Definition
Central tendency	1[a]	Phi median diameter	$Md\phi = \phi_{50}$ $= M\phi - (\sigma\phi\alpha\phi)$
		Phi mean diameter	$M\phi = \frac{1}{2}(\phi_{16} + \phi_{84})$ $= Md\phi + (\sigma\phi\alpha\phi)$
Dispersion (sorting)	2	Phi deviation measure	$\sigma\phi = \frac{1}{2}(\phi_{84} - \phi_{16})$
Skewness	3	Phi skewness measure	$\alpha\phi = \dfrac{M\phi - Md\phi}{\sigma\phi}$
	4	2nd phi skewness; Measure	$\alpha_2\phi = \dfrac{\frac{1}{2}(\phi_5 + \phi_{95}) - Md\phi}{\sigma\phi}$
Kurtosis (peakedness)	5	Phi kurtosis measure	$\beta\phi = \dfrac{\frac{1}{2}(\phi_{95} - \phi_5) - \sigma\phi}{\sigma\phi}$
Diameter in phi units corresponding to a given percentage		5th percentile diameter	$\phi_5 = Md\phi - \sigma\phi + \sigma\phi(\alpha_2\phi - \beta\phi)$
		16th percentile diameter	$\phi_{16} = Md\phi - \sigma\phi + (\sigma\phi\alpha\phi)$
		50th percentile diameter	$\phi_{50} = Md\phi$
		84th percentile diameter	$\phi_{84} = Md\phi + \sigma\phi + (\sigma\phi\alpha\phi)$
		95th percentile diameter	$\phi_{95} = Md\phi + \sigma\phi + \sigma\phi(\alpha_2\phi + \beta\phi)$

[a] In reporting data, it is sufficient to list only one measure of central tendency, because the second may easily be computed from the other parameters.
Source: From Inman (1952).

The phi skewness measure is zero for a symmetrical size distribution. If the distribution is skewed toward smaller phi values (coarser diameters), the phi mean is numerically less than the median and the skewness is negative. Conversely, α_ϕ is positive for a distribution skewed toward higher phi values. Thus, in Fig. 4–2 the skewness is positive as there is a tailing out of the fine constituents.

The secondary skewness, $\alpha_{2\phi}$, has the same form as the primary skewness, but it is based on the 5th, 50th, and 95th percentile diameters. Although the primary skewness is sensitive to skew properties occurring in the bulk of the particular size distribution, the secondary skewness is most sensitive to the distribution within the tails of the sediment. Also, $\alpha_{2\phi}$ serves as a check on the continuity of skewness indicated by α_ϕ. Because $\varnothing_5$ and $\varnothing_{95}$ are 1.65 standard deviations from the mean, $\alpha_{2\phi}$ divided by 1.65 would equal α_ϕ if the distribution were normal. The inclusion of the second skewness, together with the other measures, allows five significant points ($\varnothing_5$, $\varnothing_{16}$, $\varnothing_{50}$, $\varnothing_{84}$, and $\varnothing_{95}$) of the cumulative curve of a sediment distribution to be obtained from the phi measures without resort to the original mechanical analysis of the distribution.

The phi kurtosis measure, β_ϕ, is a parameter sensitive to the relative lengths of the tails of a distribution compared with the spread of the central portion, and is thus a measure of peakedness. This measure may be regarded as the ratio of the average spread in the tails of a distribution, that is, the average value of $\varnothing_{16} - \varnothing_5$ and $\varnothing_{95} - \varnothing_{84}$, to the phi deviation measure, σ_ϕ. For a normal distribution, β_ϕ has a value of 0.65. If the tails have a greater spread than in the case of a normal curve, β_ϕ is greater than 0.65. Conversely, lower values of β indicate that the tails have less spread than for a normal curve with the same deviation measure. The limitations and the departure of the graphic phi measures from their moment equivalents is discussed by Inman (1952).

In reporting sediment size distributions, it is sometimes convenient to convert the mean and median to their equivalent micron or millimeter value, because these metric units are directly applicable in sediment transport relations. However, the phi deviation, skewness, and so on should remain in the phi notation, because they give sorting and skewness directly in Wentworth units and have no meaningful linear equivalents. Also, such relations as the Krumbein and Monk (1942) expression for permeability use σ_ϕ directly (see section on permeability). The phi deviation measure can be computed directly from a cumulative frequency curve plotted in millimeter or micron notation by the relation

$$\sigma_\phi = 1.66 \log_{10} \left[\frac{D_{16}}{D_{84}} \right]$$

where D_{16} and D_{84} are the 16th and 84th percentile diameters in mm or microns and $D_{16} > D_{84}$.

The procedures and measures of size distribution discussed above provide a convenient notation, which in turn yields a reasonably accurate description of the mean, standard deviation, and skewness of the distribution of sediment sizes, that is, the broad outline of the spectrum of sediment size. This provides no assurance, however, that such measures are necessarily the most relevant in terms of sediment transport. For example, it may be that a minor mode obscured within the coarse portion of the distribution is the more relevant measure insofar as the threshold of grain movement at the bed is concerned.

CONSTITUENTS OF SEDIMENT PARTICLES

Perhaps of equal interest to the geologist is the nature of the constituents of a sediment. In sands, the grains may be predominantly *terrigenous,* derived from the weathering and erosion of the rocks found mostly on the lands; *biogenous,* the product of marine organisms; or *authigenic,* deposited on the sea floor as the result of chemical reactions that are not directly attributable to organisms.

Among the terrigenous sands, the mineral quartz is the most common, but large percentages of feldspars and ferromagnesian minerals, including hornblende and pyroxene, are also common. In addition, mica, tourmaline, zircon, and garnet may be very abundant. Many terrigenous sands have a large percentage of rock fragments. Depending on the percentages of these constituents in a sand, a sediment may be classified as an *orthoquartzite* if predominantly quartz; an *arkose* if largely feldspar and sand-sized igneous rock fragments; or a *graywacke* if predominantly mica and metamorphic rocks (Fig. 4–3).

The biogenous sands consist mostly of the skeletal material of calcareous-secreting organisms, principally molluscan shells or shell fragments, including foraminifera tests, ostracods, bryozoans, and echinoids. In tropical areas, large quantities of coral and algal fragments are found in the sands. Siliceous organisms are usually smaller than sand size, but the sands may contain some of the larger diatoms and radiolarians. The biogenous sands are generally referred to as *calcarenites* if the constituents are lime carbonates; *diatomites* or *radiolarites* if predominantly siliceous diatoms or radiolarians. Sands often contain wood fragments, particularly off river mouths where

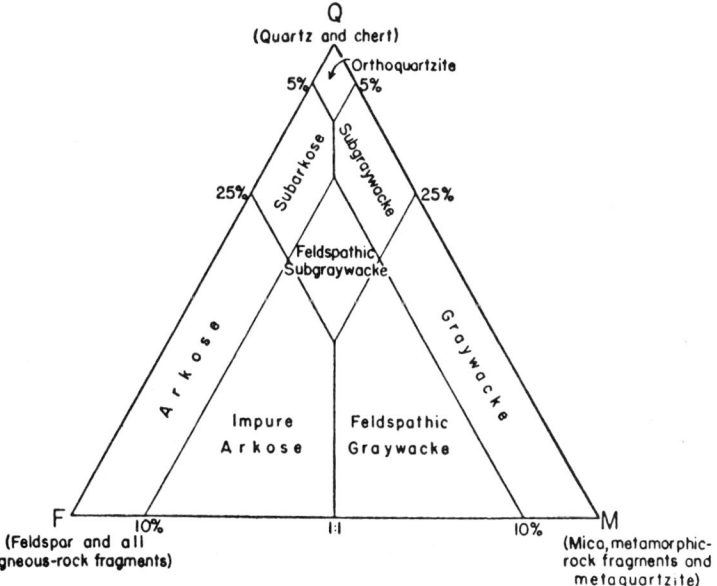

Fig. 4–3 Eight types of terrigenous sediments as defined by mineral composition of detrital silt-sand-gravel fraction. From Folk (1954).

wood is introduced in large quantities from the land. In some lakes and marshes, the sediment may be predominantly woody material, which constitutes a peat or, if consolidated, a coal.

The authigenic constituents of sands include principally glauconite, pyrite, and various forms of manganese and phosphatic nodules. Most of the nodules are coarser than sand size. There is some uncertainty whether oölites, the oval calcareous grains with concentric shells, are entirely authigenic or partly biogenous; they may be a combination of the two (Newell et al. 1960).

Coarse-Fraction Analysis It is often helpful in determining the depositional environment of old sediments to estimate the percentages of constituents of the coarse fraction, that is, of the grains coarser than 62 microns. Present-day sediments usually show their environment relations with very distinctive assemblages of minerals and faunal types.

To estimate the constituents of the coarse fraction, the most practical method appears to be to sieve the sand fraction into the standard sizes: > 1 mm, 1–0.5, 0.5–0.25, 0.25–0.125, and 0.125–0.062 mm. Each fraction is then weighed, and a cut is taken from each sieve and spread over a gridded tray. Then 100 or more grains are classified; bias is minimized by choosing the grains nearest grid intersections. The cut can be made with a sample splitter, but it has been proved by careful tests in a Scripps Institution laboratory that a method just as effective and one consuming much less time is to spread the sample over a large tray and take small scoop samples from various parts of the tray. The counting of the grains is accomplished very satisfactorily by using a Denominator blood counter of the type with eight keys that can be operated by the touch method without taking one's eyes from the field of the binocular microscope. When 100 grains have been counted, the device stops automatically and the figures give the percents of eight or fewer types. If larger numbers are required, another counter can be added, but each grain recorded on the additional counter should be included on the right-hand key of the Denominator in order to get the count to stop at 100 (or a multiple of 100 if desired).

When the counts have been made, the percents of constituents in each size should be multiplied by the weight percent for the respective grade sizes; thus, the percent of each constituent in the whole sample can be determined by using any standard computer. The average for a particular environment can be obtained by adding percentages of all analyzed samples in that environment.

Results can be plotted in various ways. Figure 4–4 illustrates two of the methods that were used for certain U.S. Gulf Coast environments. In the pie diagrams, the percent of fines is included to give another means of distinguishing the several environments, but the pie diagrams can also be made by using only the coarse fraction. The results of each sieve size are given in the middle bracket, and the upper bracket shows the weight percents in each size.

The choice of the constituents to be counted should, of course, depend on the capability· of the investigator and on the relative importance of the various constituents in the area under investigation. The groups that have proved most useful in the work at Scripps Institution have included terrigenous minerals (often separating these into a quartz-feldspar group, ferromagnesians, and mica); rock fragments; authigenic minerals (often separating these into glauconite, phosphatic grains, and pyrite); foraminifera (separating, if possible, into planktonic and benthonic); echinoids; ostracods; diatoms; bryozoans; and aggregates (describing these in terms of their physi-

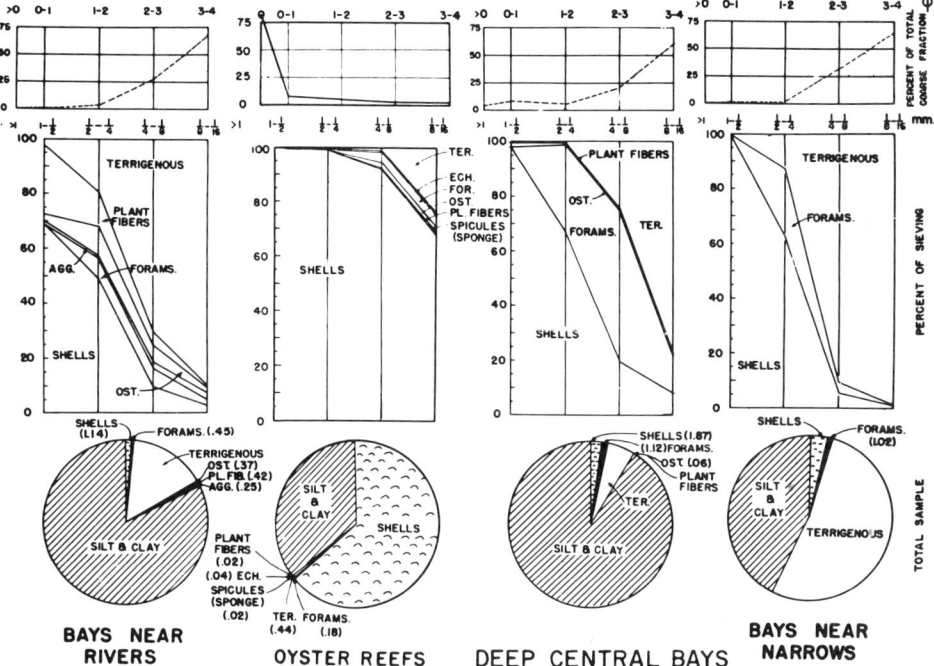

Fig. 4–4 Methods of indicating coarse fraction according to different environments in a Texas bay. This shows average percentage of material in different sizes of coarse fraction, percentages of constituents in each sieve size, and percentage of materials in total sample (shown in pie diagram, along with percentage of silt and clay).

cal properties). Although the above list exceeds eight and hence the capability of the counter, in most samples a group of eight seems sufficient for the purpose. The number of aggregates will depend considerably on the type of disaggregation that is practiced in preparing the sample. The use of sodium oxalate 67 gm with sodium carbonate 10.6 gm per 100 liters, or sodium hexametaphosphate is helpful in breaking up most of the aggregates, although some are very persistent and may have genetic significance. For some purposes it is best not to try to disaggregate the sediment, because certain features that may be preserved in sedimentary rocks, such as fecal pellets, are usually disaggregated by a good peptizer. Preparing two samples for study, one with disaggregation and the other with only washing in distilled water, will help solve this difficulty. Percentages of less than 3 are deceptive unless many samples are averaged. In some cases, it is better to observe the whole field and estimate by eye the percents of the rarer constituents.

The coarse-fraction method is definitely not highly accurate; it suffers somewhat because it gives discontinuous distributions that are arbitrarily cut off at 62 microns and because values are given as weights but are based in large part on numbers of individuals in the counts. Counts of mineral or organic species that have modes near the cutoff of 62μ could be very misleading when compared with similar samples. The grains of constituents such as mica or foraminifera obviously weigh less than terrigenous grains, and much less than magnetite, but this can scarcely be avoided and the results appear to justify the means. In fact, when experts in various fields have been asked to diagnose conditions in which the sample sediments were deposited, without their knowledge of the environments, the results coming from coarse-fraction analysis rank at least as well as from the others, such as microorganisms and macroorganisms.

The recognition of some of the constituents may be facilitated by referring to Fig. 4–5. Here, typical samples of various minerals and life forms are included to show how they appear under the binocular microscope.

Effective Density of Constituents It is obvious that the nature of the current or wave motion necessary to account for the erosion, transportation, or deposition of a sediment cannot be deduced entirely from the grain size. For example, a sediment consisting largely of shells could be deposited in the absence of current if the animals lived where the shells accumulated. The fluid stresses required to transport solid mineral grains are greater than those necessary to transport hollow foraminiferal tests of the same size and shape. An even greater stress is needed to transport heavy minerals, such as magnetite. Also, the settling velocity varies with the excess density $(\rho_s - \rho)$ of the grain where ρ_s is the density of the solid grain and ρ is the density of the fluid. Thus the excess density of a quartz sphere $(\rho_s = 2.65)$ settling in water $(\rho = 1.00)$ is 1.65 g/cm^3, while that of a hollow foraminifer with, say, one-half of the effective density $(\rho_s = 1.32)$ would have an excess density of only 0.32 or approximately one-fifth. In the range of viscous settling where diameters are less than about 0.2 mm, such a foraminifer would fall with a velocity of only one-fifth that of the quartz sphere.

SHAPE OF SEDIMENT PARTICLES

The shape of the mineral grains is also an important characteristic of granular sediments. Spherical grains are easily dislodged or set in motion on the sea floor, whereas angular grains may be more difficult to dislodge, and the flat tabular grains may offer considerable resistance where they are resting on a smooth surface or partly sunk into a finer-grained matrix.

Once dislodged, the spherical grains are not likely to be transported as far as are the platy grains, because, for example, mica settles slowly once it has been carried into suspension. Mica is carried away from the shore by rip currents (see p. 51) and in offshore moving currents of muddy water.

The shape of grains of sand size is particularly important in sedimentation because of the wide range in shape from thin flat grains, such as mica, to spherical foraminiferal tests. Somewhat less significant are the differences between spherical and cubic or other angular grains with three roughly equivalent axes.

Numerous articles discuss methods of determining roundness and sphericity. In general usage, *roundness* refers to the ratio of radius of curvature at the corners of a solid to the radius of curvature of the maximum inscribed sphere (Waddell 1932). *Sphericity*, however, refers to the degree to which the shape of a fragment approaches that of a sphere. Thus a perfect cube has high sphericity, whereas the corners are sharp, giving it poor rounding.

Because of the difficulty of developing good three-dimensional methods of comparison of shapes of grains, most methods of evaluating grain shapes have been based on two-dimensional analysis (for example, Waddell 1932; Krumbein 1941; Riley 1941). The scale proposed by Powers (1953) is primarily

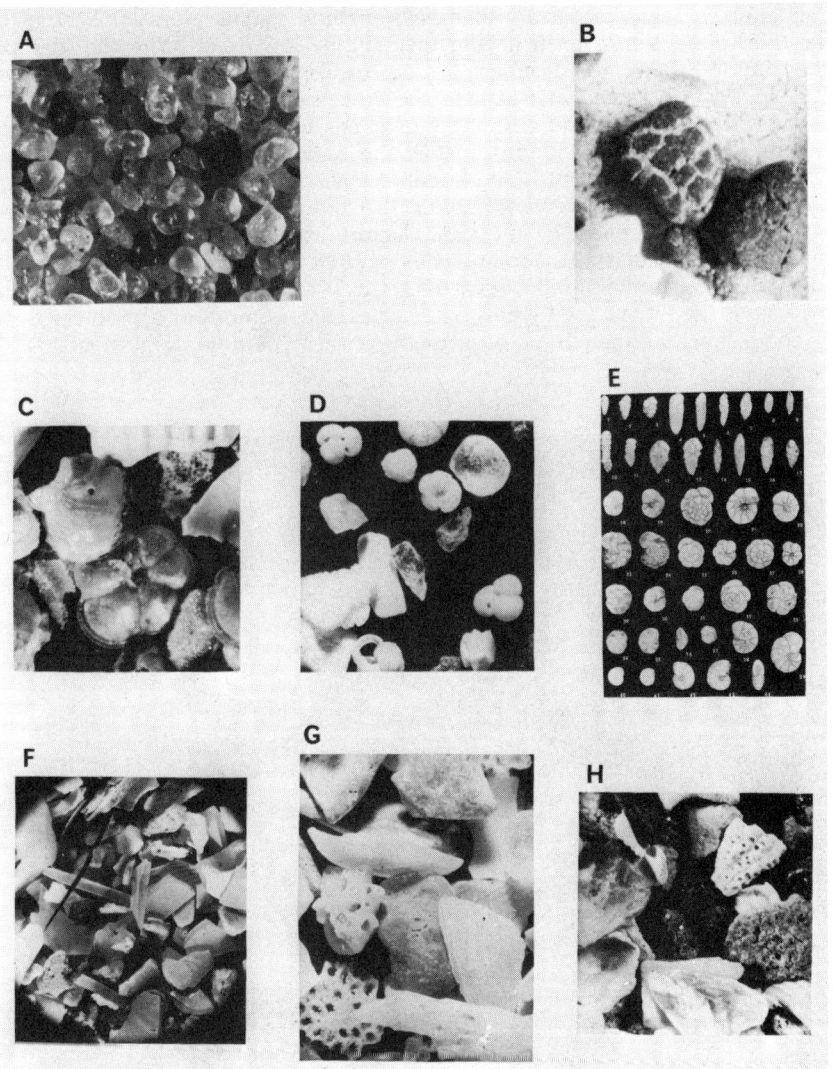

Fig. 4–5 Examples of constituents that can be counted in coarse-fraction analysis. Microphotographs by Ruth Young Manar (except E). *A,* quartz (white, clear), ferromagnesian (dark). *B,* glauconite. *C,* planktonic foraminifera (oval), echinoderms (porous), shell fragments (upper right). *D,* planktonic foraminifera (intersecting globes), echinoid (left), shell (lower left corner), angular quartz (center), well-rounded quartz (upper right). *E,* benthonic foraminifera, shallow-bay types. From Parker et al. (1953). *F,* angular shell fragments; hair from laboratory brush. *G,* shell fragments, bryozoan (holes in stalk) (lower left). *H,* bryozoan (upper right), shells (lower).

two-dimensional, but the photographs of his models give some indication of the third dimension. Because this is obviously of great importance in sedimentation, it seems necessary to have a method that is more applicable to the third dimension even if it is mathematically less exact. With the idea of checking the roundness of grains, a series of classes was developed (Shepard and Young 1961) somewhat akin to the Powers models.

The analysis of the roundness of grains using the models in Fig. 4–6 can be accomplished in approximately the same way as that described for coarse-fraction constituents. One hundred grains of the same mineral, quartz being preferred, are classified under the binocular microscope, and the numbers are multiplied by the same factors used by Powers, that is, 0.14 for the very angular; 0.21 for angular; 0.30 for subangular; 0.41 for subrounded; 0.59 for rounded; and 0.84 for well rounded. For some studies it may be advisable to count grains of all types. Bias is more of a problem than in coarse-fraction analysis because of the ease of putting marginal grains in

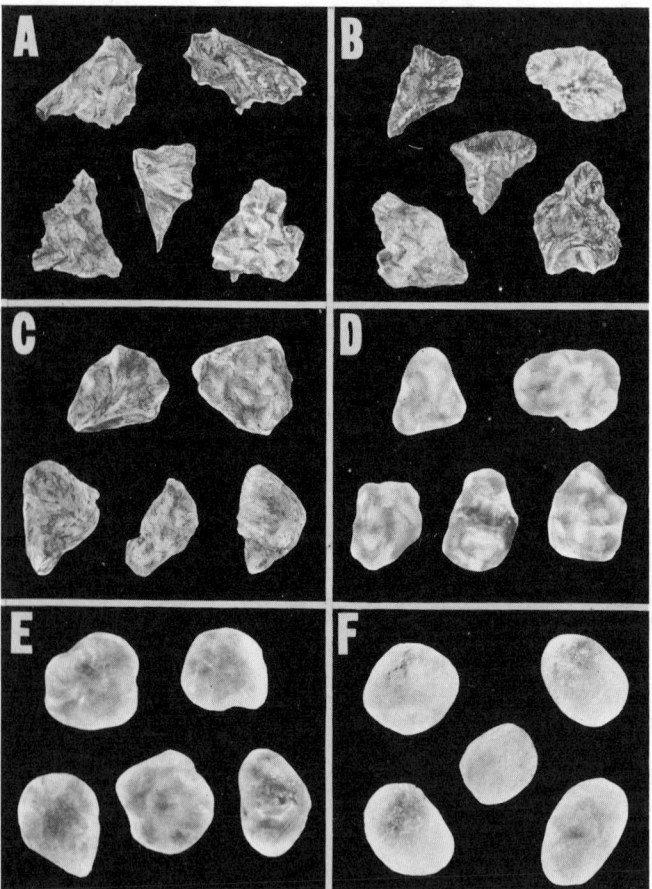

Fig. 4–6 Examples of the six classes used for roundness determinations. Modified from Powers (1953). A = very angular, B = angular, C = subangular, D = subrounded, E = rounded, F = well rounded. Photographs by Ruth Young Manar, retouching by J. R. Moriarty.

a higher or lower bracket, following a subconscious desire to find a sample rounder or more angular in order to fit some preconceived hypothesis. As a result, it is very important to study all samples as unknowns using a code system. It is also advisable to have a second operator to offset a common tendency to call grains more rounded on certain days than on others. When two operators show distinct differences in results, the sample should be rerun to see if a closer agreement can be reached.

The use of shape analysis lies in its application to differentiating sedimentary environments. The application to beach and dune studies has been repeatedly demonstrated (MacCarthy 1935; Beal and Shepard 1956; Shepard and Young 1961), but little attempt has been made to apply the method to other environments where there well may be important correlations. Shape may also tell something about the distance of transportation of grains, although roundness does not necessarily increase with distance of travel and may decrease, as shown by R. D. Russell (1939) and Kuenen (1958).

PACKING, POROSITY, AND DILATATION

There are numerous possible spatial arrangements of neighboring grains in a granular mass at rest, and the aggregate property is referred to as *packing*. The type of packing becomes important in such considerations as fluid stress on a granular bed, spontaneous liquefaction of granular masses, porosity, and permeability. Graton and Fraser (1935) showed that there are six cases of simple systematic packing in spheres, in addition to several haphazard cases. Systematic packing of uniformly sized spheres ranges from the loosest possible or cubic packing to the densest or rhombohedral packing (Fig. 4–7). In cubic packing, the centers of the spheres form the eight corners of a cube, an arrangement producing a maximum of pore space between spheres. Rhombohedral packing is the most stable and the most compact possible arrangement of spheres. It is an arrangement where the centers of the spheres are situated at the eight corners of a regular rhombohedron. The volume concentration, N, of the solids in these two cases—that is, the ratio of solid to whole space—is 0.524 and 0.740 for cubic and rhombohedral packing, respectively. The porosity, the ratio of pore space to whole space, in the above two cases is $1 - N$ (volume concentration) or 0.476 and 0.260, respectively.

In nature, packing is complicated by the occurrence of nonspherical shapes and by nonuniformity of size, factors that produce a considerable

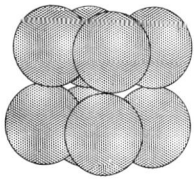

A. CUBIC PACKING

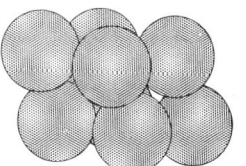

B. RHOMBOHEDRAL PACKING

Fig. 4–7 Two extreme cases of systematic packing of uniformly sized spheres, ranging from (A) the "loosest" possible or cubic packing to (B) the "densest" or rhombohedral packing. After Graton and Fraser (1935).

degree of disorder in packing. Departure from a spherical shape may produce an increase in porosity, whereas nonuniformity of size decreases porosity. In general, the greater effects of gravity on the grains at the moment of final settlement in relation to the effects of fluid flow over the bed, the looser the packing and the greater the porosity. Thus, the denser rhombic packing tends to be found where the bottom bed is actively subjected to current or wave stresses, as on the shallow portions of sandy continental shelves and on the beach face. Loose packing is more common where sediments settle from suspension into still water.

Chamberlain (1960, Table 7), working with fine sand from the beach face, measured an in situ porosity of about 0.42 as compared with a porosity of 0.40 for the same sand when compacted by vibration to its densest packing. However, sand settled into the less active waters of an adjacent submarine canyon head had in situ porosities of about 0.47, whereas micaceous sand from the canyon had porosities as great as 0.73. Hamilton and others (1956) found similar ranges in porosity for sediments from the shelf and bay in the vicinity of San Diego.

When a dense grain aggregate is at rest, the packing arrangement cannot be changed without moving and rearranging the grains. Because grains at rest are in contact with their neighbors on all sides, rearranging them requires that there be at least a temporary expansion or dilatation in the volume of the aggregate. Reynolds (1885) first investigated the "dilatancy" of a granular mass and pointed out that it is a property not possessed by known fluids or solids. He showed that a change in bulk occurred when the shape of the granular aggregate was changed. In other words, whenever a granular mass is sheared, a change in volume is produced, and hence a change in porosity.

The instantaneous "drying" of the surface of a wet beach when stepped upon is a good example of the dilatation of a densly packed sand due to shear. The dilatation produces a sudden increase in pore space and a local deficiency in pore water. Capillary action causes water to flow toward the area of dilatancy, and an excess of water is observed when the weight causing the shear is removed.

If a loosely packed granular mass is subjected to a shear or shock, a slight dilatation causes each grain to be separated from its neighbor, and the entire mass may collapse toward a denser packing arrangement. If the loosely packed granular mass is water saturated, the sudden collapse produces a dense granular suspension, and the entire mass may flow. This sudden collapse and flow of rigid granular structure is called *spontaneous liquefaction* and is probably associated with many avalanchelike phenomena in nature, including some turbidity currents. Spontaneous liquefaction may also be caused by an increase in pore water pressure rather than from an external mechanical stress or shock (Terzaghi and Peck 1948, p. 100). For example, an underground spring issuing into a sand body may produce a local excess in pore water pressure greater than that due to hydraulic head of water immediately above the sand. An excess pore pressure acts as a dispersive grain pressure, permitting grains to flow in response to other sources, such as gravity.

PERMEABILITY

The resistance of the bottom bed to the discharge of liquid through it in part determines the dissipation of energy of waves moving over the bed, and enters into considerations of sediment transport by waves. Permeability is a major factor in determining the slopes of the foreshores of beaches.

The discharge rate of a liquid through a porous granular bed depends on grain and liquid properties as well as on the impelling force of pressure head producing the flow. The complete expression for the discharge is known as Darcy's law, and is valid for all directions of flow and for all velocities small enough that forces of inertia are negligible compared to those of viscosity. Following the presentation of Hubbert (1958), the volume discharge, q, of fluid crossing a unit area in a unit time is given by

$$q = (GD^2)(\rho/\mu)(-gdh/dz)$$

where G is a dimensionless factor of proportionality depending on the geometry of the pore space; D is a measure of size of the pore space, such as the mean grain diameter; ρ and μ are the density and dynamic viscosity, respectively, of the liquid; g is the acceleration of gravity; and dh/dz is the rate of change of the pressure head in the direction of flow. If the flow were through a cylinder, dh/dz would be the decrease in pressure as indicated by fluid height in the manometers divided by the flow distance between manometers.

Of the three factors on the right of equation (1) GD^2 depends only on properties of the porous media, and ρ/μ are properties of the liquid only. The combined geometrical factor GD^2 is often expressed by the symbol K and is called the permeability of the porous media. The permeability has the units of length squared and is sometimes given in darcies, where one darcy equals approximately 10^{-8} cm^2.

Krumbein and Monk (1942), experimenting with water flow through sands of differing sizes and standard deviations, obtained the following empirical expression for the permeability K expressed in darcies:

$$K = 760 \, D^2 e^{1.31\sigma_\phi}$$

where D is the geometric mean diameter expressed in millimeters, e is the base of natural logarithms, and σ_ϕ is the deviation measure in phi units. From this relation it is seen that the permeability increases with the square of the grain size diameter and decreases exponentially with increasing standard deviation. Thus, for a given mean size, well-sorted sands are more permeable than poorly sorted sands.

The permeability of bed material may be expressed, independently of the scale of the impelling force, as the perpendicular water discharge through a unit area of a bed of thickness equal to the perpendicularly applied head of water. This gives a value of unity to dh/dz and permits the discharge of any given liquid to be expressed in centimeters per second as a function of the permeability. This discharge, using the relation of Krumbein and Monk (1942) for the permeability, is graphed in terms of sediment size and phi deviation measure in Fig. 4–8. The relatively low discharge through sands of less than 1 mm mean diameter as compared with coarser material is readily apparent. A typical beach sand has a phi deviation measure of about ¼ ø.

FLOW OF LIQUIDS AND THE TRANSPORTATION OF GRANULAR MATERIAL

The transportation of granular solid material has been widely studied by hydraulic engineers, and it has been generally assumed that the laws of fluid flow could be applied directly to sediment transport. However, as

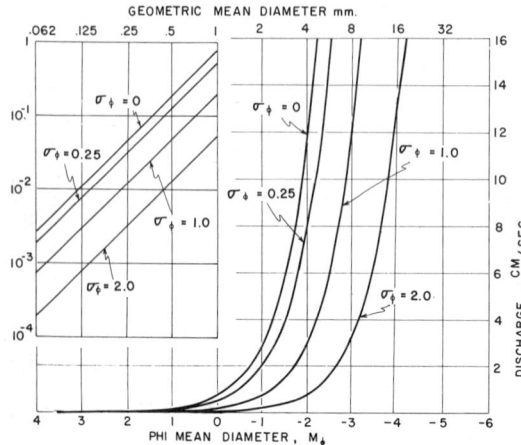

Fig. 4–8 Perpendicular discharge *q* in centimeters per second through unit area of the bottom bed of thickness equal to the applied head of water. Computed from the relation of Krumbein and Monk (1942, eq. 17) for sea water at 15° C, salinity $35^0/_{00}$, and for sediments with phi deviation measures of $\sigma\phi$ = (single size), $\frac{1}{4}\phi$, 1ϕ, and 2ϕ. The very low discharge through sands less than 1 mm is shown on the logarithmic scale in the insert.

Reynolds (1885) and Bagnold (1954) have shown, a mixture of granular and fluid material has properties differing quite widely from that of the fluid alone. Thus, the laws of fluid flow can apply only when the concentration of granular matter is very low, as in a very dilute suspension of solids, or to the granular bed boundary directly, before many grains are set in motion.

In by far the majority of cases—when appreciable solid granular material is in motion—the laws of the fluid alone are inadequate.

Fluid Flow over a Granular Bed   The nature of the motion in a moving fluid can be either *laminar* (that is, composed of thin fluid elements each moving uniformly relative to its neighbors) or *turbulent* (that is, each fluid element moving at random in relation to its neighbors). If the initial motion of the flow is gradual and steady, and the flow is shallow, its motion will probably be laminar throughout. Turbulence results whenever there is tangential shear at a fluid boundary and spreads to other parts of the fluid if the Reynolds number[6] of the flow is sufficiently large.

In a shallow flume, turbulence is commonly initiated when an eddy or vortex forms in the lee of some bottom irregularity, such as a large granule or pebble. If the eddies are sheared and carried upward into the fluid, they transport their motion to other layers of the fluid. If the bed is sufficiently rough to produce turbulence, the turbulence is transferred to successively higher layers until at some downstream point the entire flow becomes turbulent. Motion in large rivers and in ocean currents is inherently turbulent because of the very large Reynolds numbers of the flow and the presence of numerous tangential stresses exerted at the surface boundary by winds and at the bottom boundary by flow over bed irregularities.

At this point it is helpful to give an elementary description of the nature of the grain movement and flow conditions observed in a simple flume as the velocity of the water is gradually increased. Consider a flat-bottom bed of sand, as in a river or flume, with water flowing above at a velocity so low

---

[6] The Reynolds number is a dimensionless criterion expressing flow intensity. It is given as $R = \bar{u}h/v$ where the $\bar{u}$ is the mean velocity of flow, h is the depth, and v is the kinematic viscosity equal to the molecular viscosity μ divided by the density of the fluid. For water at 20° C, v has a value of about 0.01 cm^2/sec. For the flow where the width is large compared with depth, turbulence can become general throughout the flow when R is 600 or greater.

that none of the bed grains is moved. If the flow rate is gradually increased, the following sequence of sediment movement will be observed.

The first grain movement will consist of an intermittent rolling and sliding of individual grains. This motion, intermittent in time and space, indicates the random nature of the turbulence in the fluid flow and the irregularities in grain packing on the bed. This first incipient sediment motion is referred to as the initiation or threshold of grain motion (Fig. 4–9).

As the flow rate is increased, the number of particles moved increases, and some particles are lifted off of the bed, executing short trajectories or saltations before falling back. Saltation is an indication that the transportation of material on the bed as bedload has reached a more advanced stage. If the flow is turbulent, some of the grains may be lifted higher above the bed by the random motion of the water, and the grains are then said to be in suspension. In well-developed suspension, the sediment grains deviate from the even low-angle "cannon ball"-shaped trajectory, the mark of true saltation, and partake in the more random motion of the turbulent water. Thus, turbulence is a necessary condition for suspension.

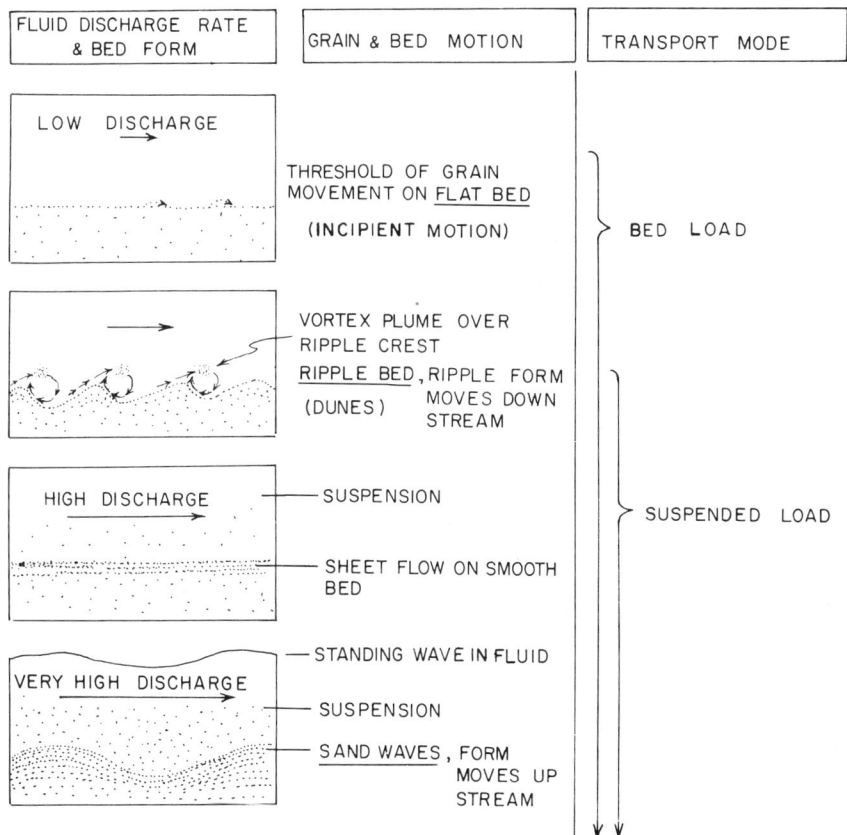

Fig. 4–9 Schematic representation of the various modes of transport and types of grain motion as observed through the glass window port of a flume as the liquid velocity is gradually increased.

As the velocity and flow rate are increased further, the amount of material transported in suspension increases, and, if the sediment is fine, suspension becomes an important mode of transport. If one looks down on a flume where suspended transport is well developed, it will be observed that the suspended material moves in clouds or billows in a random manner, rather than in a uniform and constant concentration. Here again, the more turbulent nature of the fluid flow is reflected in the motion of the suspended particles.

If grain motion continues, the flow will produce the familiar bed irregularities known as sand ripples. At low velocities, the grains roll up the gentle upstream face of the ripple and slide down or are deposited on the steeper lee face. At somewhat higher velocities, the sand may jump or saltate from one ripple to the next. In the regimen of bedload transportation the velocity of the sand is always slow compared with the velocity of the fluid, and the form of the ripple moves in the direction of the current. In deep rivers, a much larger bottom feature, termed a *dune,* is eventually formed under flow intensities similar to those forming ripples. Aqueous sand dunes are analogous to the dunes of wind-blown sand and differ greatly in scale from ripples. Ripples are the small-scale features that can be observed moving up the windward face of sand dunes.

Ripples attain a maximum length and height at some intermediate velocity range. If the flow is increased beyond this range, a condition will be reached where the ripples decrease in height and vanish, and the bed will become smooth again. Because the bottom is nearly flat after the ripples vanish, this flow condition is sometimes referred to as *sheet flow.* The entire bottom is then in motion, and it can be observed that the motion extends several grain layers into the bed. The intense bed shear produces a marked increase in sediment transport. In this sediment regime, the laws of fluid flow alone are inadequate to describe the sediment transport, and it is necessary to introduce the grain-to-grain shear forces.

At still higher flow rates, undulations again form on the bottom if the flow is shallow. Unlike the ripples, the new undulations, which are called *sand waves,* are more sinusoidal in profile than ripples, and their form moves upstream instead of down. This is because the sand that is eroded from the lee face of the sand wave is deposited on the upstream face of each succeeding wave crest. Sand waves were termed *anti-dunes* by Gilbert (1914) and are usually accompanied by waves at the water surface above the sand wave. However, there is considerable doubt if sand waves occur in deep rivers; the large-scale features observed in such cases are more appropriately called dunes.

Certain fundamental similarities and differences are worth noting between the phenomena occurring at the granular bed boundary under unidirectional flow and that occurring under oscillatory motion due to wave action. In both flows, there is a threshold below which grain motion does not occur; only under wave action is the threshold further complicated by the presence of acceleration and deceleration. In both types of motion, after local turbulence develops behind ripple crests, there is a phase when the turbulent vortex produces a scour in the ripple trough moving grains in counter direction to the flow over the crest (Inman and Bowen 1963). The convergence of flow at the ripple crest causes a plume of sediment to rise in the vortex above the ripple. In the oscillatory case, the sediment plume forms suddenly at the end of the forward or backward motion. This is because deceleration of flow enhances the formation of turbulence, whereas acceleration inhibits its formation. Also, oscillatory flow tends to inhibit the upward spread of turbulence from the bottom, because the oscillation does not allow sufficient time for the turbulence to spread. In both unidirectional and oscillatory flow, providing the orbital diameter of the oscillation is large compared to the ripple wavelength, intense shear causes ripples to disappear and the bed becomes smooth. The similarities in boundary layer phenomena between the two types of flow suggest that the same basic principles apply to both.

Chapter 5
PLATE TECTONICS
(SEA-FLOOR SPREADING
AND CONTINENTAL DRIFT)

INTRODUCTION

The 1960s saw the establishment and wide acceptance of a theory that is an outgrowth of the old Wegener (1912, 1924) hypothesis of continental drift. In the first edition of *Submarine Geology* (1948), the Wegener hypothesis was discussed with emphasis on the arguments against the Atlantic rift concept. In 1961, when the second edition was written, the first presentation of sea-floor spreading had appeared (Dietz 1961[1]) and the idea was discussed briefly, but the arguments against the drifting of continents were still given precedence. Since that time, the new concept of plate tectonics including sea-floor spreading has been supported by so much substantial evidence that the old objections to migration of vast tracts of the earth's crust have somewhat disappeared. Many of the difficulties related to the Wegener hypothesis

[1] A somewhat similar approach was made by Holmes (1945, p. 505).

do not apply to the new concepts. A few scientists still voice serious objections, but these do not appear to refute the vast body of substantiating evidence.[2]

In the earlier editions, continental drift was considered in chapters related to the deep-ocean basins, but plate tectonics clearly has an impact on most aspects of marine geology and should therefore be discussed near the beginning of the book. Here, only the fundamental ideas and a brief exposition of the great mass of documentary evidence that has appeared in recent years will be included along with some of the apparent inconsistencies. Further evidence will be presented in the chapters on the continental terrace and on the deep ocean.

PLATE TECTONICS

Prior to introducing plate tectonics, we should consider briefly the difference between ocean and continental crust. As a result of refraction shooting (p. 40) and earthquake studies, it has been learned that there is a major discontinuity under both the oceans and continents below which velocities of sound are slightly greater than 8 km/sec, with distinctly lower velocities above. This is called the *Mohorovičić discontinuity* or *Moho*. It is considered to be the juncture between the earth's crust and the underlying mantle. Under the continents, the Moho lies at an average depth of 35 km, but, under the ocean, the crust has only an average thickness of 6.5 km (Fig. 5–1). The continental crust is believed to be largely granitic and the oceanic crust mostly basaltic. The mantle is thought to be ultrabasic rock, such as peridotite. The *lithosphere* is the rigid zone in the crust and in the upper mantle. The *asthenosphere* is a layer below the lithosphere where plastic movement takes place. The contact between the two is estimated to be found at about 100 km.

The plate tectonics theory as it stands at this writing is a compilation of the work of numerous scientists. Among the outstanding contributors have been Dietz (1961, 1968), Hess (1962), Wilson (1965), Vine (1966), Sykes (1967), Le Pichon (1968), Morgan (1968, 1971), Bullard (1969), and Menard (1969). Many articles by joint authorship could be included if space permitted. Symposia containing much of the important literature include: Blackett et al. (1965), Runcorn (1962), Kay (1969), and Johnson and Smith (1970).

The theory of plate tectonics and sea-floor spreading is in agreement with Wegener's hypothesis in considering that the continents were once united as Pangaea (Laurasia and Gondwana) and later separated to form the six continents of the present (Fig. 5–2) However, the mechanics of drift are quite different. As suggested by Dietz (1961) and Hess (1962), the basic concept of sea-floor spreading is the formation of tension cracks in the earth's crust, which are filled with ascending currents of lava from the mantle. Where the newly formed dikes develop in a continent, they produce a heavy mass that because of isostasy will sink below its surroundings, forming a rift valley. Examples of this early stage of rifting are found in the Gulf of California, the Red Sea, and the rift valleys of Africa. If the tension that forms the crack

---

[2] Since the above was written, papers by A. A. Meyerhoff and H. A. Meyerhoff (1972 A, B, C) have appeared, giving even more critical coverage of the opposition to plate tectonics than was given in the earlier Meyerhoff (1970) and Meyerhoff and Teichert (1971) papers. It is too late to review critically this mass of data and no rejoinders have yet been made, even to the earlier Meyerhoff papers.

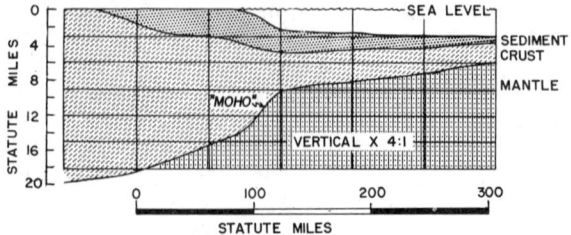

A

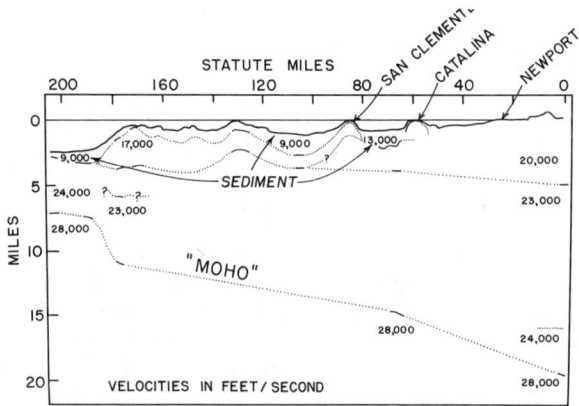

B

Fig. 5–1 A, generalized diagram showing the relation of the over-lying sediment (layers 1 and 2), the crust (layer 3), and the Moho to the continental shelf, continental slope, and deep ocean basin off the East Coast. From Worzel and Shurbet (1955). *B,* cross section of the sea floor off Newport, Southern California, showing the thickness of the sediment under the basins and the travel velocities in feet per second in the underlying rocks. Note the progressive increase in the depth to the Moho in approaching the continent. From Shor and Raitt (1958).

continues, new openings of the crust will occur and an embryonic ocean will be formed and gradually increase in size. This is thought to have happened to Pangaea (Fig. 5–2), with the initial rifts forming during early Mesozoic. A major crack developed between what are now the Americas and Africa and Europe, forming the Atlantic Ocean. Other cracks opened at approximately the same time between East Africa, India, Australia, and Antarctica, producing the Indian Ocean. The subcontinent of India was moved northward by the expanding Indian Ocean until it encountered Asia. Australia had a somewhat similar ride, coming into close contact with Southeast Asia. The Atlantic opened gradually to its present width and, in the Cenozoic, Greenland sepa-rated from North America, forming Baffin Bay. An eastward continuation of the cracks producing the Indian Ocean extended across the South Pacific and thence up along the west coasts of the Americas. Movement of crust away from this oceanic ridge appears to be well documented since Early Cretaceous.

As the cracks widened, the newly emplaced crustal material moved away on both sides. Near the spreading axis, the crust was hot and therefore

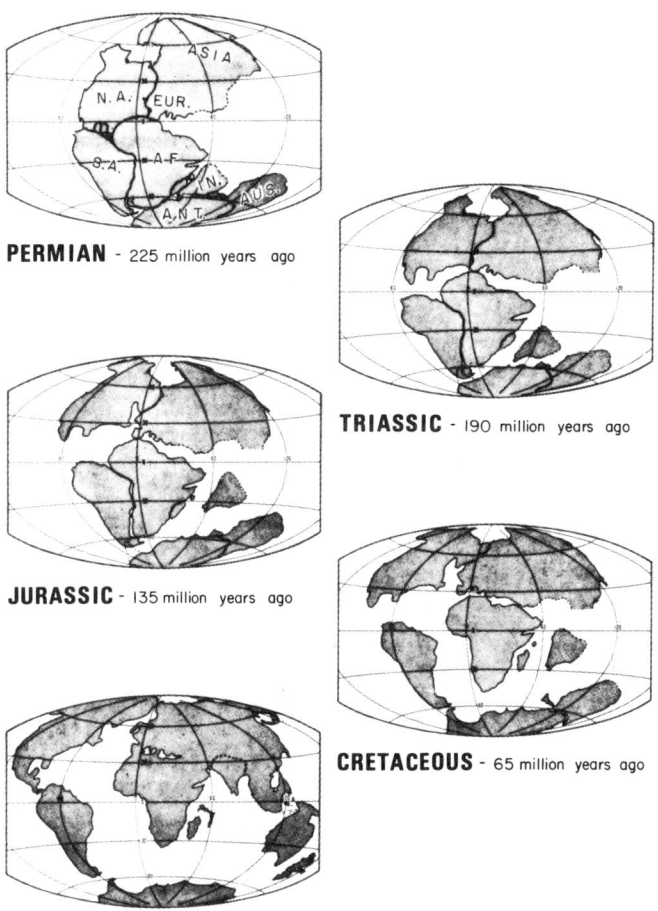

PERMIAN - 225 million years ago

TRIASSIC - 190 million years ago

JURASSIC - 135 million years ago

CRETACEOUS - 65 million years ago

CENOZOIC - Present

Fig. 5–2 One suggested sequence of breakup of the old supercontinent, Pangaea. From Dietz and Holden (1970).

stood above the rest of the expanding ocean, which had cooled as it moved away from the center. In this way the mid-ocean ridges were formed, such as the Mid-Atlantic Ridge, the East Pacific Rise, the Antarctic Rise, and the Carlsberg Ridge of the Indian Ocean. The spreading away from the oceanic ridges has taken place at rates from about 1 to 10 cm/yr, averaging about 2 cm in the Atlantic and about 5 cm in the Pacific. As a result of this spreading, which has been continuing at least since the Jurassic, young crustal material has spread entirely over ocean basins, pushing older crust under the continental margins. This explains the absence of pre-Jurassic sediments, both in oceanic dredgings from escarpments and from the cores of the JOIDES drillings (see pp. 419–430).

The original cracks were probably discontinuous and the present ridges show large offsets. As spreading took place, ridges and troughs, called *fracture zones*, developed between the offset ends, extending approximately

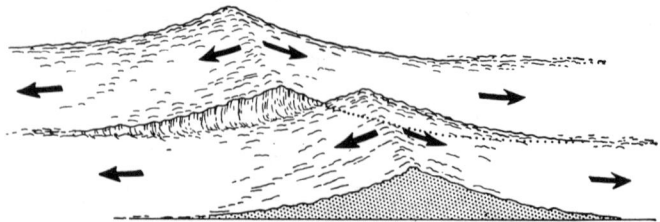

Fig. 5–3   Transform faults resulting from spreading of offset ridge crests. From Menard and Atwater (1968).

at right angles to the ridges. As a result of the spreading away from the various sectors of the ridges, displacements take place along strike-slip faults with the movement in the opposite direction from what one would suppose from the relation of the disjointed ridges (Fig. 5–3). These are termed *transform faults*. The fracture zones can be traced for thousands of kilometers across the oceans and in some places extend into the adjacent lands. The San Andreas Fault is now interpreted by many as a transform fault.

   According to plate tectonics, a series of relatively rigid plates are being moved over the interior away from the axial cracks. Although there is no complete agreement on the number, six major plates are recognized as active at present (Fig. 5–4). About 20 small plates are said to exist. The con-

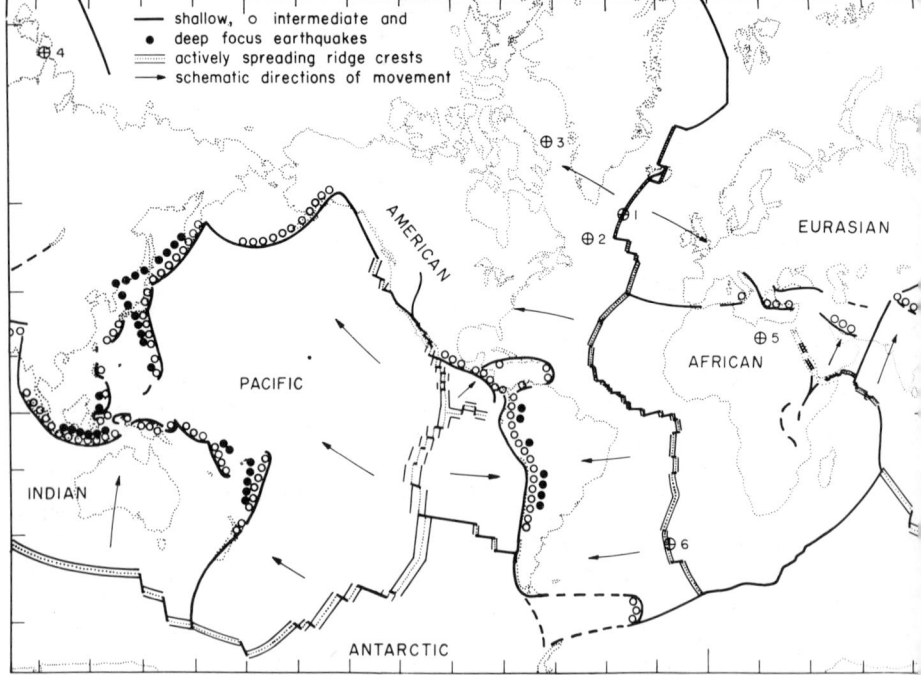

Fig. 5–4   The six major plates, the spreading ridge crests, the depth of earthquake hypocenters, and the direction of plate movements. From Vine and Hess (1970).

tinents are not thought to move as a unit, like a ship in a sea, as suggested by the Wegener hypothesis, but are tied to portions of adjacent oceans; for example, the two portions of the American plate include the western half of the Atlantic joined with the adjacent American continents and are moving away from the Mid-Atlantic Ridge as a unit.

The leading edges of the plates are in collision with other plates and are therefore localities where the crust is being lost, either by *subduction* associated with the deep oceanic trenches, where the crust from one plate moves down under the other plate (Fig. 5–5), or by crumpling of the crust that forms mountain ranges. Subduction is said to take place along the west coast of South America and along the arcuate islands of eastern Asia and southwestern Alaska. Crumpling of the crust is well exhibited in the Himalayas where the Indian plate is encountering the Eurasian plate. Since there is no subduction of the Atlantic margins, the basin is growing wider at about 4 cm/yr, 2 cm for each side.

The moving plates are thought to have thicknesses of the order of 100 km. It is believed that they extend down to where the underlying mantle is heated to such an extent that it is a quasi-liquid and yields readily to the movements of the plates above it. Alternatively, the movement of a liquid layer in the mantle resulting from a convection current is carrying with it the overlying crust. In turn, the crust develops cracks at points where the rising convection currents flow apart, and subduction is taking place where the convection currents are descending.

Here we have discussed plate tectonics in relation to movements that have taken place in the Mesozoic and Cenozoic. However, there is no reason to believe that the activity did not exist at earlier periods. The formation of

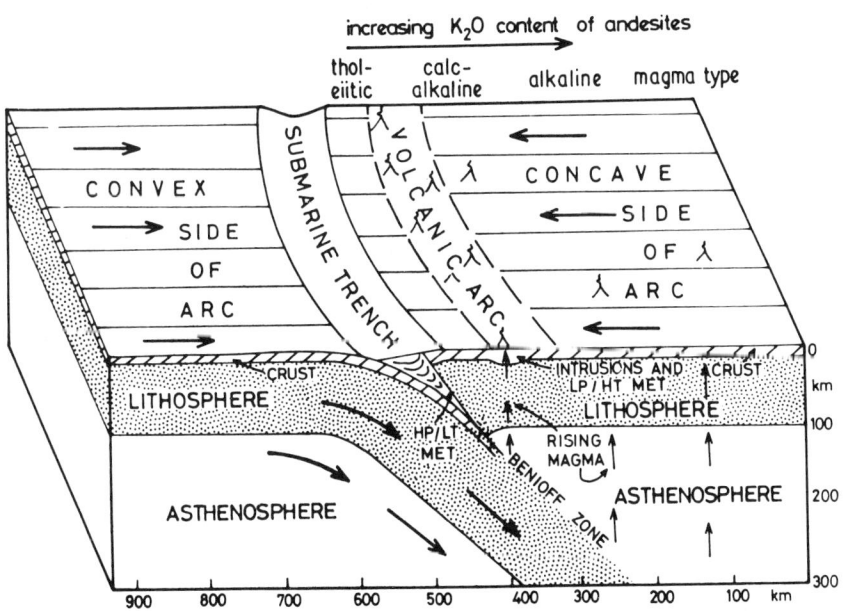

Fig. 5–5 The subduction of the lithosphere forming the Benioff zone landward of the trenches and the relation of subduction to volcanism. From Mitchell and Reading (1971).

Paleozoic mountain ranges can be explained by movements at the edge of an old Atlantic Ocean, and the formations of the Urals between Europe and Asia can be explained by two plates moving together and closing an ancient sea.

EVIDENCE FOR PLATE TECTONICS AND SEA-FLOOR SPREADING

The preceding has outlined the status of what is becoming a widely held theory. We can now discuss the nature of the supporting evidence. It is well to remember that scientists are apt to be carried away by any hypothesis that appears to be very successful in providing answers for many puzzling features. Sometimes this has led us in the wrong direction, and it is advisable to be critical, where possible, of the arguments that have caused any such revolutionary change in the philosophy of geology, as has plate tectonics.

Bilateral Symmetry of Magnetic Bands Bordering Oceanic Ridges Vine and Matthews (1963) have suggested that if sea-floor spreading has taken place, the periodically reversing polarity of the earth's magnetic fields should be indicated by alternating bands of negative and positive polarity parallel to the mid-oceanic ridge axes that formed a mirror's image of each other on the two sides. This has now been well authenticated for the Pacific (Pitman and Heirtzler 1966), the Atlantic (Heirtzler and Hayes 1967; Pitman and Talwani 1972), and the Indian Ocean (Le Pichon and Heirtzler 1968). The matching goes at least as far back as about 78 million years (anomaly 32), and the last part of the reversal time scale is partly checked by ages of lava on land and magnetism in cores from deep-ocean drilling. It seems impossible to explain these extraordinary matches of magnetic belts (Fig. 5–6) other than by sea-floor spreading. With only this evidence there would be good reason to believe in the new theory. On the other hand, Meyerhoff and Meyerhoff (1972B) give examples of rather poor matching.

Earthquake Epicenters and First Motion An examination of the epicenters of earthquakes (Sykes et al. 1970) shows that they occur mainly in the trenches where, according to plate tectonics, the crust is being subducted under the adjacent land masses, along the ridge crests where the crust is being formed, and along the portion of the fracture zones that form the offsets between the crests. The slip vectors determined by study of seismograms (Fig. 5–7) indicate that the motion at trenches is away from the oceanic plates and hence in agreement with plate tectonics. The vectors also indicate that spreading of the ridges and the direction of movement along the fracture zones is that of transform faults (Fig. 5–3), counter to what would exist if the offsets in ridges were due to progressive separation of the two sectors of the ridge.

Earthquakes also support the subduction of the ocean crust at trenches because shallow-focus earthquakes occur in the trenches and deep-focus earthquakes take place under the adjacent lands, as, for example, beneath the Andes. Underthrusting at the continental margins and on the ocean side of the island arcs would account for this greater hypocenter depth landward from the trenches.

Basement Age Under Ocean Floors If the oceans have existed more or less in their present form since early geologic time, as was believed by those who advocated the permanency of ocean basins, drilling into the ocean floor should have lead to the recovery of Paleozoic and even Precambrian sediments above the basement rocks. However, if the basement has been

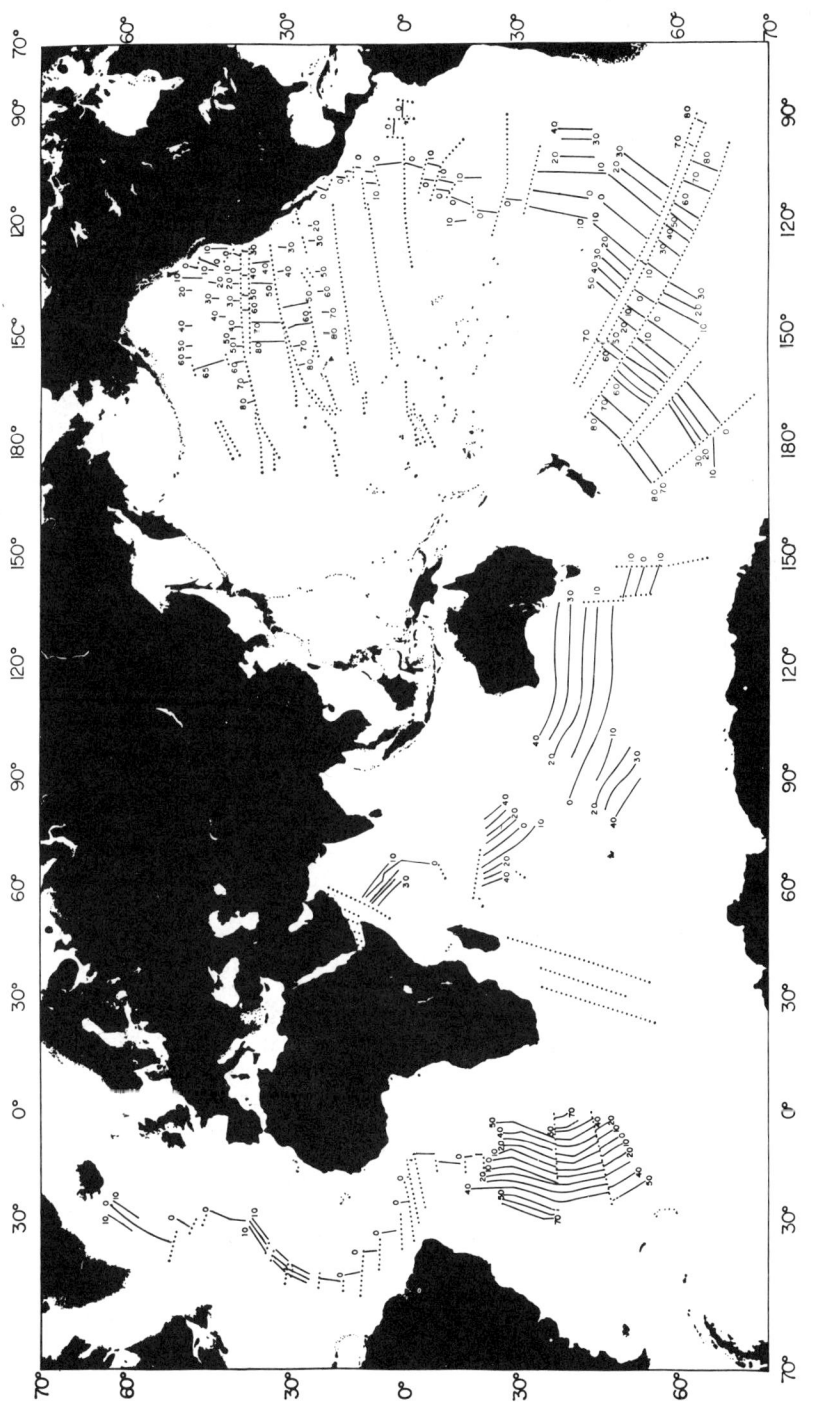

Fig. 5-6 Matching of magnetic belts on the two sides of the spreading ocean ridges. Numbers are age in millions of years. Dotted lines are fracture zones. From Heirtzler et al (1968).

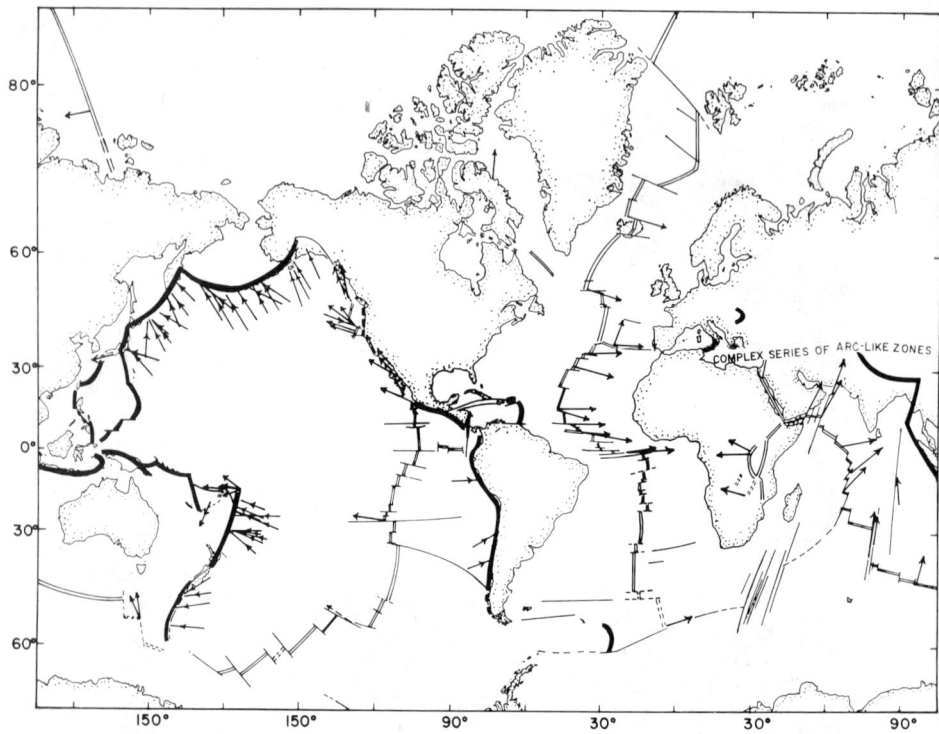

Fig. 5–7 Vectors of slip during major earthquakes and their relation to plate tectonics. From Isacks et al. (1968).

the result of sea-floor spreading from the mid-ocean ridges, and if the rates have been one or more cm/yr, the drillings should show almost no ancient sedimentary formations and should indicate that the sediments directly over-lying the basement are youngest over the ridges and grow progressively older away from the ridges. The latter has been well demonstrated by the JOIDES operations (see pp. 419–430). At the ridge crests, the sediment cover is either absent or very young. In general, the oldest sediments are found farthest from the ridge crests. The agreement between the ages of the basement covers and the age of the magnetic anomalies again provides what seems to be good evidence of the spreading sea floor.

The basement cover gives further indication of spreading by showing a sequence of deep-sea deposits that are suggestive of deepening water. As the sediment cover grew, the area moved farther away from the mid-ocean ridge where the basement was formed. Thus, the sequence upward from calcareous oozes to siliceous oozes and then to brown clay is found in many of the cores from drillings into deep basins. We know that organic oozes are limited in depth by solution and siliceous oozes occur somewhat deeper than the calcareous, but brown (red) clay is deposited at the greatest depths.

Fit of Opposite Oceanic Margins The earliest versions of continental drift made use of the fact that the two sides of the Atlantic show a very good fit. Actually, if one takes the continental slopes rather than the coast as the margin, the fit is considerably improved. This was demonstrated by Bullard and others (1965) in a map partly reproduced in Fig. 5–8. There are slight gaps and overlaps. Some of the overlaps are due to outbuilding of the continents since the continents were separated; for example, the growth of the Niger Delta. The Bahama region may also have been a deep-sea floor (Dietz et al. 1971; Uchupi et al. 1971) and may have grown up to the surface since the continents were separated. Gaps may also be due to independent spreading, as at the entrance to the Caribbean. Other examples of good margin fits are between Antarctica and southern Australia (Sproll and Dietz 1969),

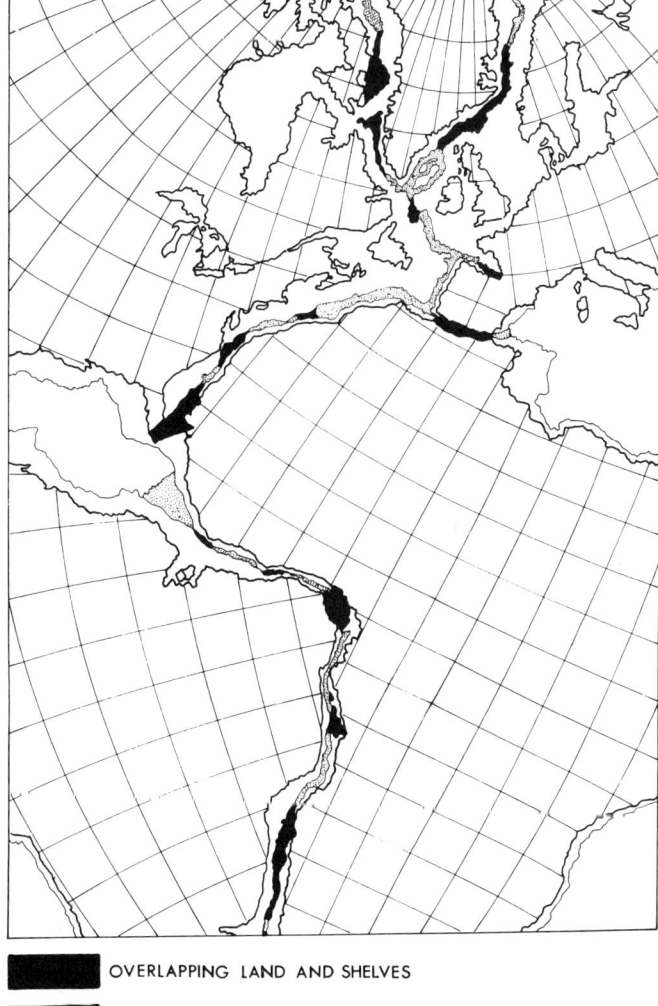

■■■ OVERLAPPING LAND AND SHELVES

▢ GAPS

Fig. 5–8 The remarkable fit of the two sides of the Atlantic suggests the rigidity of plate movements. Some of the small gaps and overlaps are explained in text. From Bullard et al. (1965).

and between Antarctica and South Africa (Dietz and Sproll 1970). Meyerhoff and Meyerhoff (1972), however, show that some good fits may be coincidental.

If the continents have drifted apart, this should be revealed by comparison of the old mountain ranges on the two sides, particularly those ranges that terminate at the coasts. The fit of the Appalachian system with the Varsican and Caledonian ranges of Europe is impressive, and some of the large strike-slip faults of the Canadian provinces appear to fit with those of Scotland (Wilson 1962). South America and Africa have late Paleozoic mountains that would match if the continents were rejoined, including the Cape Range of South Africa and the Sierra de Pillahuinco of Argentina. The Precambrian rocks in West Africa and Brazil are also well matched chronologically (Hurley, de Almeida, et al. 1967).

Aseismic Ridges and Hot Spots In addition to the great ocean ridges where sea-floor spreading is taking place, there are various long volcanic ridges, particularly common in the Pacific, which are essentially aseismic, or at least their earthquakes are only related to volcanic activity, and such activity is taking place almost entirely at one end of the ridge. In the Hawaiian Ridge, it is known that the time of active volcanism increases in age, at least as far as Midway, where it was Miocene (24.6 $\pm$ 2.5 million years) (Funkhouser et al. 1968; Jackson et al., 1972). Wilson calls attention to the significance of these aseismic ridges. It is thought that the heat for the volcanism comes from a long-persistent source below the asthenosphere, and that, as the moving asthenosphere carries the overlying lithosphere over this stable hot spot, new volcanoes develop and the old are moved in the direction that the convection currents transport the crust. Morgan (1971, 1972) has noted that three of the Pacific aseismic ridges—Hawaiian–Emperor, Tuamotu–Line Island, and Austral–Gilbert–Marshall—have had parallel movement and even show the same change in trend (Fig. 5–9) in their westerly portions. This suggests rigid plate movement over three hot spots and a change in direction of the motion. If future studies verify this increasing age in the ridges away from the hot spots it will be difficult to think of a better interpretation. The Walvis and Rio Grande ridges in the South Atlantic trend, respectively, northeast and northwest away from a hot spot near Tristan da Cunha Island. This, according to Dietz and Holden (1970), indicates that there has been northward movement with respect to a hot spot, but the spreading sea floor has produced net movements diagonally on either side of the Mid-Atlantic Ridge.

Morgan (1971, 1972) refers to 20 hot spots and notes the contrast between the lavas at the hot spots and those of the spreading ridges. He suggests that the former represent primordial material coming directly from the deep mantle, whereas the ridge basalts represent the filling of the cracks from the moving asthenosphere. He proposes that the hot spots may provide the driving force for the plate motion.

Paleomagnetism Magnetic minerals crystallized during consolidation of recent molten rocks show a relationship to the magnetic poles of the earth. However, it has been known for many years that the "fossil" magnetism of ancient rocks is often different from the direction that a compass needle would point today in the locality in which the rocks are now found. Because the magnetization assumed by such minerals as magnetite and maghemite is related to the magnetic pole and hence probably to latitude and longitude of the time at which they were crystallized, it is possible to make a crude estimate of the position of the magnetic poles when the rocks crystallized. Furthermore, it is probable that the magnetic poles have always been near the geographic poles. Therefore, a study of the magnetism of ancient rocks

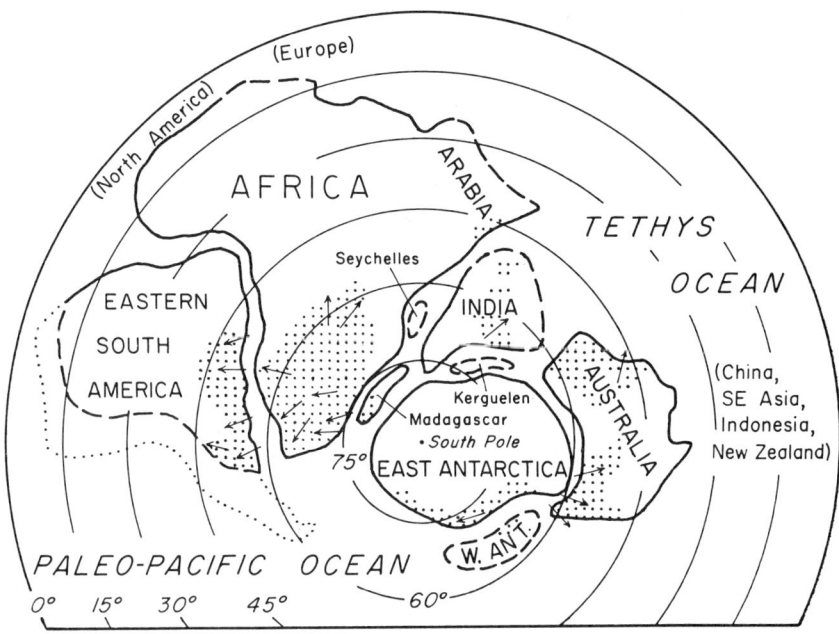

Fig. 5–9 Glaciation of Gondwana at the beginning of the Permian shows directions of ice movement. Solid lines are margins of present continents that have changed little in shape since the Permian. Dotted line is outer margin of present South America. Note many localities where ice was coming from what are now separated continents. From Hamilton and Krinsley (1967).

can give a moderately good indication of the relative positions of continents, such as the Americas, Europe, and Africa.

The accuracy of this method has been debated (Meyerhoff 1970). Most authorities agree, however, that paleomagnetism indicates that the Americas have separated from Africa and Europe, and that there have been large-scale movements in a general northerly direction that have carried South America, Africa, India, and Australia north away from the South Pole where these continents were fitted together with Antarctica during the late Paleozoic. The paleomagnetism of the rocks in the base of the deep drillings also supports a general northward drift, particularly in the Pacific (see p. 429).

Heat Flow from Ocean Floors The rate of heat flow is far easier to measure from the ocean bottom than from the continents because of the stable temperature of the water directly above the bottom, in contrast to fluctuating temperature of the atmosphere in contact with the lands. The probes driven into the sea floor with thermistors have indicated that heat flow is consistently greater from the crests of the extensive system of oceanic ridges than from the flanks, and the average values of the 3.12 μcal per cm^2/sec from the crest of the East Pacific Rise and 2.92 from the Mid-Atlantic Ridge contrast with the world mean from land and sea floor of 1.5. This suggests that hot mantle rock is moving up into the ridge crests, which is in accordance with the other evidence for spreading ocean floors.

Permo-Carboniferous Glaciation Glaciation is known to have been widespread during the Permo-Carboniferous in South America, South and Central Africa, India, Australia, and Antarctica. This knowledge was used by

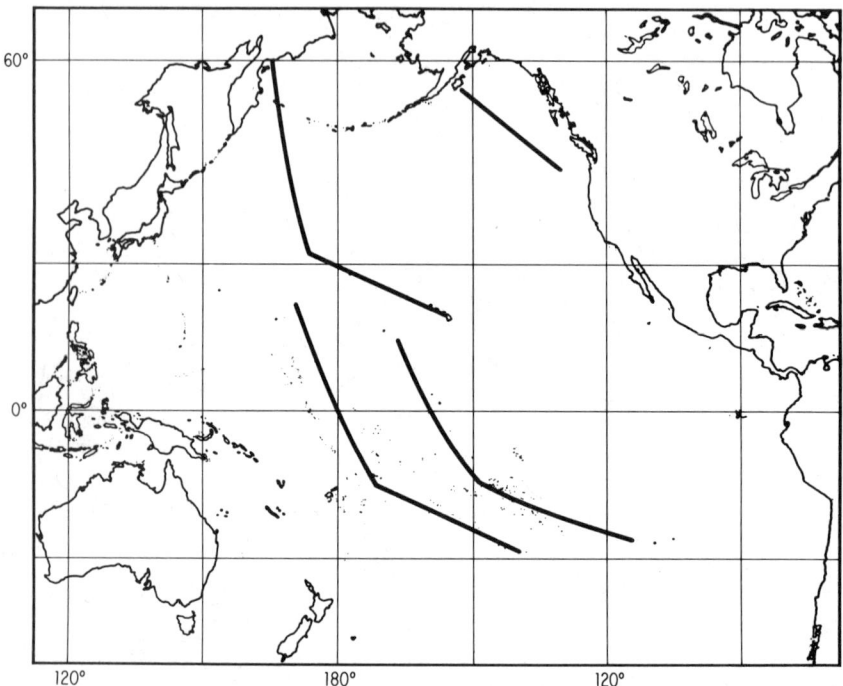

Fig. 5–10 Hot-spot migration in the Pacific. Youngest volcanoes are to the southeast. Note similarity of directions, all indicative of the northwest movement of the Pacific plate and of an earlier more northerly migration. From W. J. Morgan (1972).

Wegener and many others to support continental drift. The grouping of these localities together in the high latitudes of the Southern Hemisphere, as is indicated by the preceding discussion, removes much of the difficulty in explaining glaciation in many places which are now located in low latitudes. Furthermore, the direction of ice movement suggests that much of the ice was coming onto the present land margins (Fig. 5–10), which presumably would mean that it was derived to a considerable extent from the grouped continents of Pangaea (Crowell and Frakes 1970A; Hamilton and Krinsley 1967). Thus, ice appears to have moved onto Argentina and South Africa from the east and onto South Australia from the south. This is also shown in Frakes and Crowell (1970B, Figs. 1, 2), and in Meyerhoff and Teichert (1971, Figs. 7, 13).

However, Meyerhoff and Teichert consider that glaciation could not have occurred in much of eastern South America, South Africa, India, and Australia if they had been joined together, because they believe that little moisture could have reached much of the area. Actually, there is little precipitation in much of central Antarctica, but it has a thick ice cap. Also, the Pleistocene continental ice centers were approximately as far removed from an ocean source as those of glaciated territory in the Pangaea reconstruction (see Flint 1971, Fig. 25–1). Without the Andes barrier to moisture on the west and with a good source from the southwesterly moving water currents in the ancient Tethys Sea, which probably advanced closely to the glaciated areas, there appears to be no serious objection to glaciation of Pangaea. It would be far more difficult to understand how there could have been such extensive glaciation if the continents had maintained essentially their present arrangement. If, as claimed by Meyerhoff (1970), the glaciation in low latitudes was all of the high-mountain type, it seems amazing that so much of the glacial pavements and other evidence would have remained. Obviously, erosion in mountain ranges would have left little trace for geologists to collect after an interval of some 250 million years.

Evaporite Distribution in History   Less certainty concerning the shifts in latitude comes from studies of evaporites. At present, salt, gypsum, and other saline deposits that constitute the evaporites are being deposited largely near 30°N and 30°S latitudes in the high-pressure areas of low rainfall. If the plates have moved north away from the South Pole, there should have been a corresponding shift of the dry belts. If the areas formerly constituting the 30°N dry belts have moved north, these areas would have been carried into the higher rainfall zone of the westerlies, whereas the equatorial rainy belts would have moved into the latitudes of aridity close to 30°. The maps of Lotze (1963) appear to show such a shift for the lands in the Northern Hemisphere. His evaporite belts shifted progressively to the south, as would be expected from the north motion of the plates. Lotze believes that this movement indicates a polar shift, but it could equally well be related to the sliding plates or to both. Furthermore, the Jurassic opening of the Atlantic would have greatly modified the zones of low rainfall in the center of the continents; this appears to have been the case.

A very different conclusion is reached by Meyerhoff (1970). His paleogeographic maps show a general similarity to those of Lotze, but Meyerhoff interprets the changing boundaries of the evaporite belts as due to variation in width of the arid zones during the various periods. Also, he points to the fact that almost all localities where ancient evaporites were deposited are outside present-day areas of high rainfall. However, saline lakes, precursors to evaporites, are found even today over wide areas including Antarctica (McLeod 1964). The problem seems to be rather unsettled and awaits more study, but this seems to be a minor point, and we do have substantial evidence for the new theory. Also, the effects of this crustal spreading must have been profound in relation to the development of the relief features of the oceans and of the continents. On the other hand, many aspects of the theory are likely to be changed as investigations continue. It is only the general principle that appears to be rather well established.

Chapter 6
SEACOAST CLASSIFICATION
AND ORGIN

INTRODUCTION

Just as it was important to discuss the newly established theory of plate tectonics and sea-floor spreading prior to consideration of the various provinces of the sea floor, it is equally important to discuss the development of various types of coasts and their classification prior to considering beaches. Also, the shallow seas surrounding the continents are greatly influenced by the nature of the adjacent coastal areas, both relative to coastal relief and to coastal climate and drainage.

In the second edition of this book, I may have put too much emphasis on what appears to be a practical method of classifying coasts from a study of large-scale ground and aerial navigational charts and photographs. This method has been criticized, partly because it does not include coasts of emergence (Valentin 1952; Cotton 1954; King 1959, pp. 236–240) and partly because it does not give the major subdivisions, such as active mountain coasts and those of coastal plains (McGill 1958; Inman and Nordstrom 1971). I realized the importance of coastal emergence but hesitated to include such a class because of complications resulting from eustatic sea level changes and river and glacial erosion during glacial stages of low sea level. This made interpretations of emergence difficult to diagnose, especially from charts and photographs. Similarly, a major scale classification was omitted because I was not satisfied with the various methods that had been proposed. However, we now have an interesting suggestion for a major classification by Inman and Nordstrom (1971) based largely on plate tectonics. Also, Bloom (1965) has given a good proposal for including the interplay between sea level changes, uplift of land, and erosion and deposition. Nevertheless, my earlier classification still appears to be useful for most purposes and so will be included with some additions and more discussion of diastrophic changes.

NOMENCLATURE FOR COASTAL FEATURES

Definitions relative to beach terms are given in Chapter 7, pp. 125–126, but a few terms will be discussed here. Much of our nomenclature on coastal and shoreline features dates back to Johnson's *Shore Processes and Shoreline Development* (1919). Some revisions have been suggested more recently (Price 1951, 1955, and 1956; Shepard 1952; Beach Erosion Board 1954), and we are indebted to Wiegel (1953) for a complete glossary of terms and definitions relating to shores and shore processes. The *shoreline* is generally defined as "the line where land and water meet," whereas the *shore* is the zone from mean low tide (or lower low tide) line to the inner edge of wave-transported sand. The *coast* is the broad zone directly landward from the shore. The coast includes the sea cliffs and elevated terraces as well as the lowlands inside the shore. Because the shoreline is dependent on the state of the tide, the direction and force of the wind, the height of the breakers, and the effect of seiches, as well as the slope of the shore, it is only a line for any particular moment and may migrate over a belt miles in width, as at Mont-Saint-Michel of the Normandy coast of France. Also, high tide or mean tide line is not constant but an average position for a certain period. The interplay between shore and coast makes it impractical to classify them separately, therefore the term *coast* will include shore in this classification.

Other terms that should be defined in the present chapter are *bars* and *barriers*. The word "bar" has been used rather indiscriminately. Among mariners, a bar is a slightly submerged ridge, usually of sand or other type of sediment. This usage seems desirable. On the other hand, the term "offshore bar" for an elongate sand island or peninsula as used by Johnson (1919) is confusing. Some of these sand masses have lengths of 100 km or more, widths of more than 20 km, and heights above sea level of 50 m or more, hardly suitable for the term "bar." The word "barrier" seems far more applicable for the feature, and the combined terms *barrier island, barrier spit,* or *bay barrier* (Fig. 6–1) can be used where appropriate. *Barrier beach* is suggested where the sand mass consists only of a beach, lacking both the dunes and marsh flats which form a part of the larger barriers.

MAJOR COASTAL CLASSIFICATIONS

In the first edition of *Submarine Geology* (p. 78), a brief classification of the following major coastal features was included:

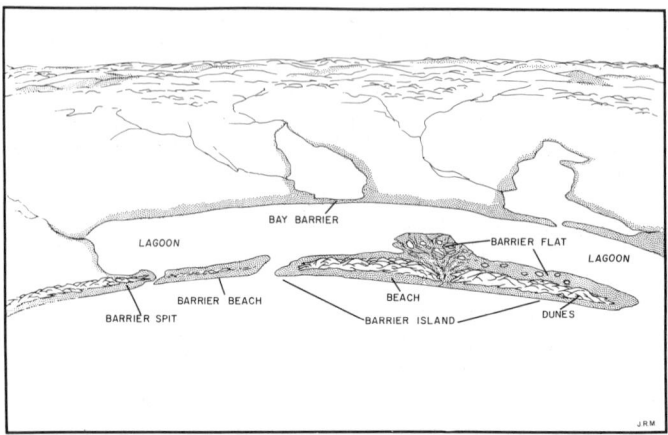

Fig. 6–1   Various types of barriers and the component parts of a barrier island.

Coasts with young mountains
Coasts with old mountains
Coasts with broad plains
Glaciated coasts

Later, the classification of McGill (1958) included similar divisions with the addition of various types of coastal plateaus, along with many of the subdivisions in other classifications. McGill also made a map of the world with an attempt to depict these subdivisions.

The most recent classification of major features (Inman and Nordstrom 1971) makes use of plate tectonics, subdividing the coasts into (1) *collision coasts*,[1] including those on the collision or subduction side of continents and island arcs; (2) *trailing-edge coasts,* including those on the sides of continents and islands that are moving away from the rising and spreading oceanic ridges; and (3) *marginal seacoasts* that lie on the protected side of island arcs. Inman and Nordstrom further subdivided the trailing-edge coasts into (a) *neo-trailing-edge coasts,* where new rifts have recently opened and only a small amount of separation has taken place, as in the Red Sea and the Gulf of California; (b) *Afro-trailing-edge coasts,* where both east and west coasts are trailing, as Africa and Greenland; and (c) *Amero-trailing-edge coasts,* where the other side of a continent has a collision coast, as on the East Coast of the Americas. The world distribution of these types, according to Inman and Nordstrom, is shown in Fig. 6–2.

As indicated, the young mountain coasts of the world are located primarily or perhaps entirely along the collision coasts. They are characterized by narrow continental shelves and are high in seismicity and volcanism. The neo-trailing-edge types mainly have precipitous coasts and few or no continental shelves. They are subject to some volcanism and seismicity. The Afro-trailing-edge coasts have a predominance of plateaus alternating with

[1] Subduction coasts might be preferable.

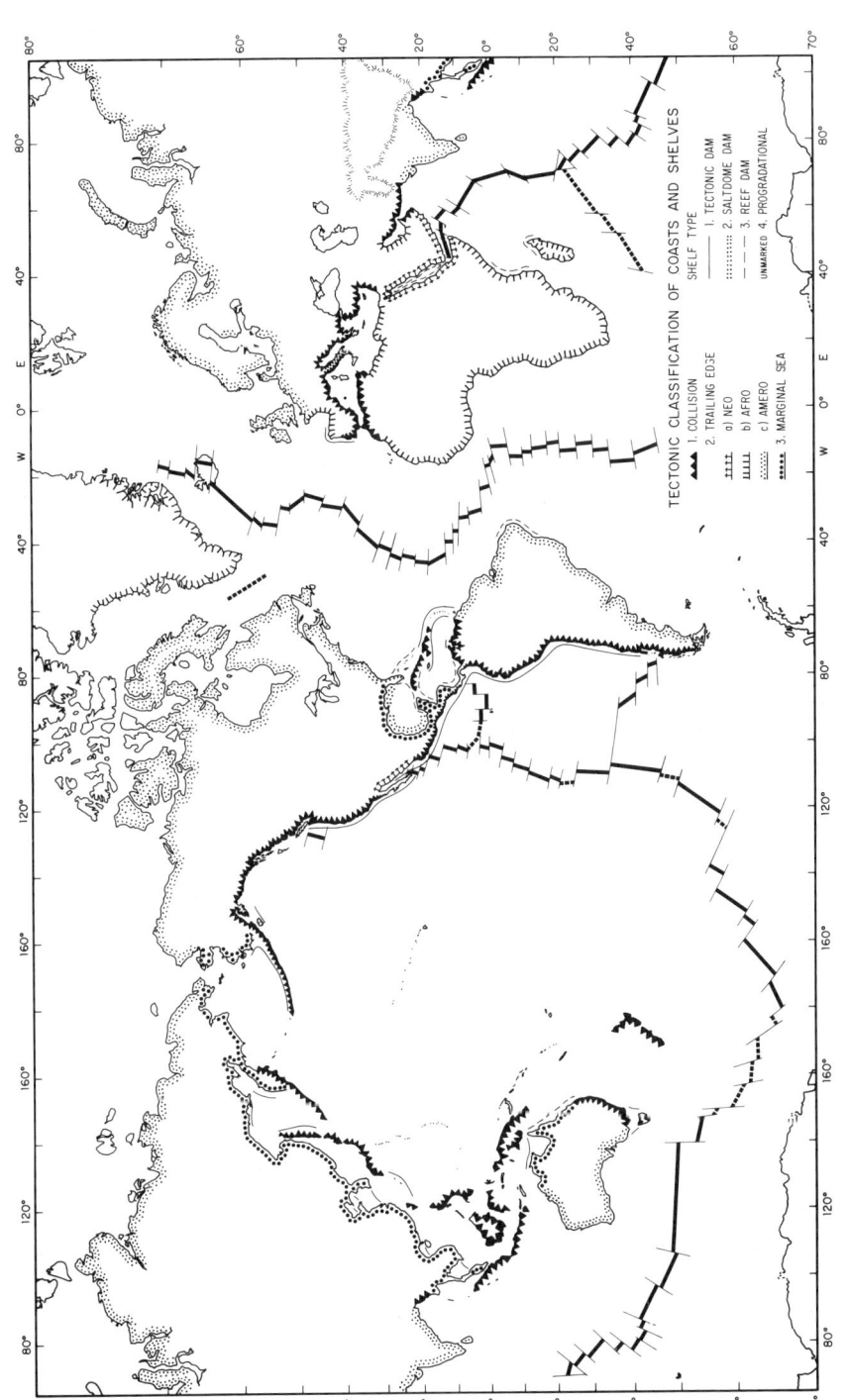

TECTONIC CLASSIFICATION OF COASTS AND SHELVES

COLLISION
1. COLLISION
2. TRAILING EDGE
  a) NEO
  b) AFRO
  c) AMERO
3. MARGINAL SEA

SHELF TYPE
1. TECTONIC DAM
2. SALTDOME DAM
3. REEF DAM
4. PROGRADATIONAL
UNMARKED

Fig. 6–2 A tectonic classification of coasts in conformity with plate tectonics. *Collision* coasts might be better called *subduction* coasts. From Inman and Nordstrom (1971).

plains and have relatively narrow continental shelves. The Amero-trailing-edge type is notable for broad coastal plains and equally broad continental shelves. It should be noted that exceptions are found in the narrow shelf off central Brazil south of Cabo São Roque and off southeastern Florida. Both the Afro and Amero types have low levels of volcanic and seismic activity.

EMERGENCE, SUBMERGENCE, AND NEUTRAL CLASSIFICATIONS

The long-accepted Johnson (1919) classification had as its principal basis the recent history of coasts relative to local sea level, including shore-lines of *submergence, emergence,* or *neutral,* where no change had occurred. In 1937, I called attention to what has now generally been recognized as an unfortunate feature of the Johnson classification, namely, that the postglacial rise of sea level continued into such recent times that all coasts are potentially "coasts of submergence," and that most coasts may even be considered as coasts of emergence, since the sea level was higher than at present some 80,000 to 150,000 years ago during an interglacial episode (Fig. 6–3). Thus the typical drowned valleys of Johnson's coasts of submergence are likely to be found wherever uplift in the past few thousand years has not exceeded the postglacial rise in sea level. Such features as raised wave-cut terraces and raised coral reefs may also be found on any coast that has not undergone downwarping to an extent greater than the height of the interglacial and preglacial high stands of sea level. A sudden melting of the Antarctic ice cap would raise the sea level about 66 m (Theil 1962). As a result, most relatively stable coasts could be classified as either coasts of emergence or submergence, depending on whether their raised terraces or their drowned valleys happen to be the most evident or least modified.

Another objection to the Johnson classification came from the use of what he considered as the important indications for coasts of emergence—straightness and barrier islands (called *offshore bars* by Johnson). We now know that straight coasts are principally those that have been straightened either by wave erosion or wave deposition after the virtual termination of rise of sea level during deglaciation. Also, the extensive drilling in barrier islands has shown that they have been growing upward as they submerged along various coasts of the world (Shepard 1960; Hoyt 1967). Furthermore, the chief examples used by Johnson for neutral coasts are deltas, and here again there is now clear evidence that instead of being neutral most large deltas are actually subsiding, as has been shown in so many areas (Fisk et al. 1954; Russell 1958; Scruton 1960).

Valentin (1952) developed a classification much more realistic than that of Johnson, but again with the major divisions advancing (vorgerückte) and receding (zurückgewichene) coasts. He divides the advancing coasts into emerging coasts, where the sea bottom has been exposed by uplift, and upbuilt coasts, where organisms, such as mangroves and corals, have built the coast forward or where the coast has been prograded by deltas or marine deposition. The receding coasts are subdivided into coasts that have been submerged (including drowned river valleys, fiords cut out by glaciers, and downfaulted coasts) and eroded coasts where waves have cut back the shore and formed cliffs. He has constructed an interesting map showing his types for the entire world. Some of the objections to Johnson's classification could also be applied to that of Valentin. He shows few areas where he has called the coasts emerging; his examples include the south side of the upper Gulf

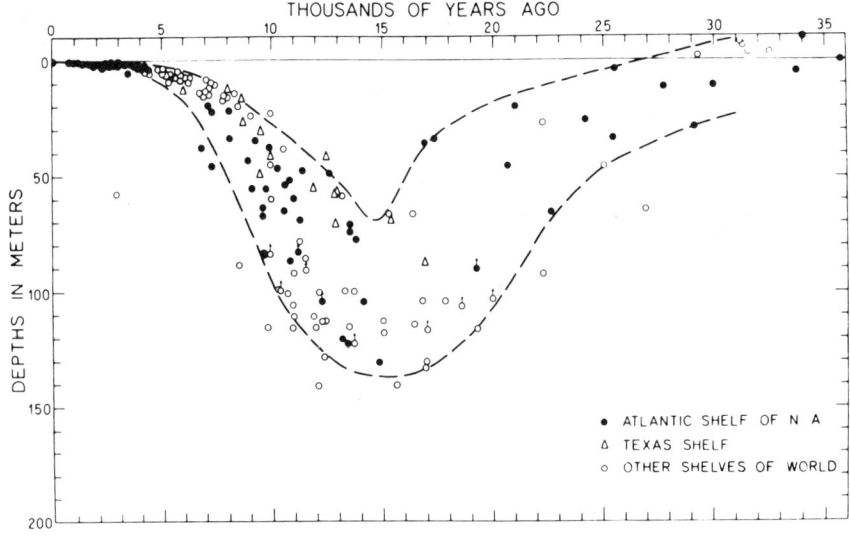

A

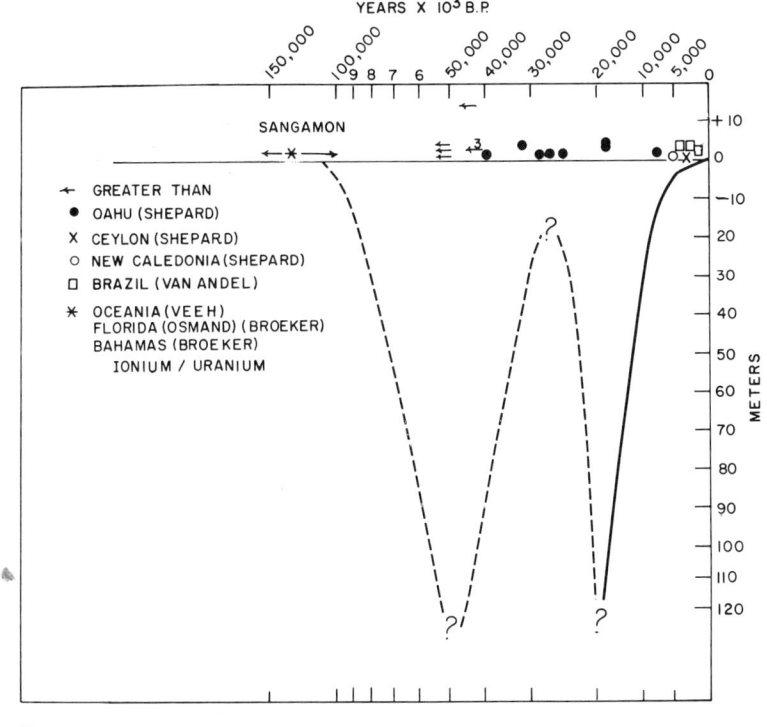

B

Fig. 6–3   A, carbon-14 dates and depths at which coastal fossils were found on the continental shelves. From Emery et al. (1971). B, approximate sea-level curve for the past 150,000 years, based on various methods. From Shepard and Curray (1967).

of St. Lawrence, the southwest side of Hudson Bay, and the east side of the Gulf of Bothnia. Although these have emerged by postglacial rebound, so have virtually all glaciated areas. One wonders why differentiate these particular coasts from their neighbors, which are reasonably classified as glacial erosion coasts. Many of his "faulted coasts," which he includes as a subdivision of "submerged coasts," are from the point of view of recent information much better included under his "emerged coasts" as are his "cliffed coasts" as, for example, the west coasts of the Americas from the state of Washington to Chile (with a few gaps) and the north coast of Morocco. Nevertheless, the continental portion of his world map contains much useful information.

Perhaps the most successful attempt to refer to coasts as "advancing" and "retreating" has been made by Bloom (1965). He developed a graph (Fig. 6–4) in which the history of a coast in past time is given in relation to sea level changes. Thus, for New England this takes into account the rise in sea level due to deglaciation and the upwarping of the land due to recovery from

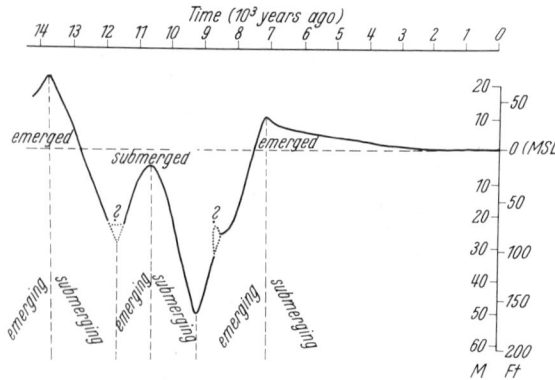

Fig. 6–4   Coastal changes in southwest Maine related to postglacial sea-level rise and postglacial rebound. From Bloom (1965).

ice load. Part of the time during melting episodes, the sea-land border of southwest Maine stood inland from the present coast because upwarping had exceeded sea level rise; whereas, at other times, the coast stood seaward of the present location because rising sea level had exceeded upwarping. This example, which lacks the further complication that may come from erosion and deposition, illustrates how difficult it is to determine whether a coast is one of emergence or submergence, and how completely impossible it is with the present state of our knowledge to make any map that would so classify the coasts of the world. However, where information is available, it is certainly of great interest to determine what the history of emergence and submergence of a coast has been in relation to time.

Because of all these implications, I feel that it is not practical to make a coastal classification that uses emergence and submergence (or advancing and receding) as a major factor. On the other hand, where information is available, as along the Alaska coast after the 1964 earthquake (Fig. 6–5) or the Japanese coast of Sagami Bay raised at the time of the 1923 earthquake, certain local areas can be referred to properly as coasts of emergence or submergence because the position of the shoreline was known before the earthquake.

Fig. 6–5 A wave-cut terrace on Montague Island, Alaska, uplifted during the 1964 earth-quake exposing 0.5 km of sea floor. From Plafker, U.S. Geological Survey (1969).

CLASSIFICATION BASED ON LAND-SEA MORPHOLOGICAL RELATIONS

An interesting type of coastal classification was suggested by Ottmann (1962). In this, the different relations between the profile of the coast and of the adjacent sea floor are used as a basis (Fig. 6–6). His **A** type has a steep land slope that continues seaward of the shoreline with no continental shelf. Subdivision **A-1** consists of a series of fault scarps extending beneath the sea, and **A-2** is a volcanic slope. His second type, **B**, consists of a sea cliff with an abrasion platform on the seaward margin. Type **C** has a steep land slope that continues down to a slightly submerged continental shelf. He includes in this **C-1**, a glacial fiord. Type **D** consists of coasts where the continental relief continues in somewhat modified form beyond the shoreline. Type **E** consists of coastal plains that continue beyond the shoreline as gentle slopes. This last group includes deltas, barrier and lagoon margins, dune coasts, and low coasts inside coral reefs. The classification is almost entirely morphological and only incidentally genetic. It should prove to be useful where little evidence is available concerning the history of coastal development.

GENETIC COASTAL CLASSIFICATION

An effective system is to classify coasts according to the geological processes that have been operating on them in recent times. For such a classification to be practical, especially for students, it should be so divided that it is possible to determine the subdivisions from a study of coastal charts and contour maps with supplementary use of ground and aerial photographs.

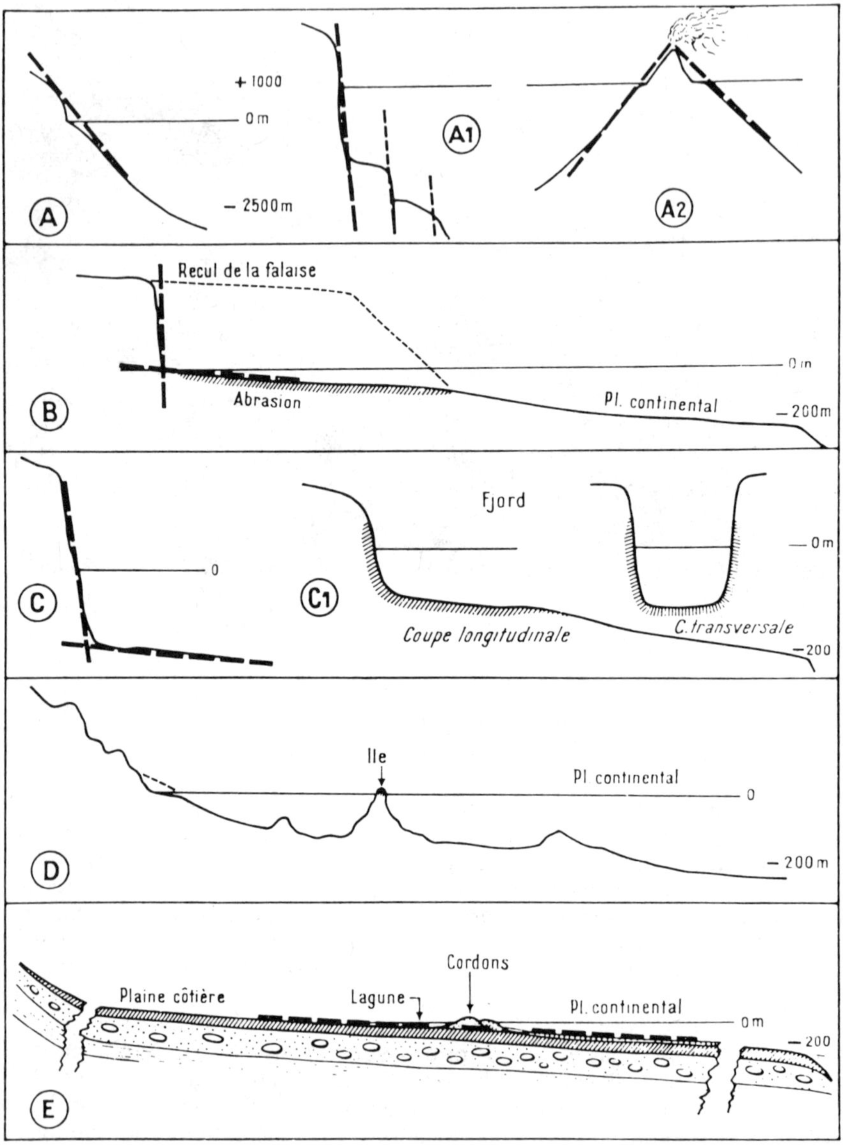

Fig. 6–6 Coastal classification of Ottmann (1962).

Many years of teaching have convinced me that the classification that follows, while still in need of refinement, is generally satisfactory. It is, of course, possible to misinterpret some areas, but even from the use of coastal charts alone a considerable degree of success may be attained. Because so much difficulty results from applying such terms as coast of emergence and submergence, as has been seen, these terms are only given as subsidiary to some of the types.

The classification to follow has many points of similarity with others that are also genetic. In my earlier attempts, many of Johnson's types were included. As the classification has evolved, use has been made from many of the other schemes of dividing coastal types. The numerous excellent papers by Sir Charles Cotton have been particularly helpful, and the worldwide classification with many illustrations developed by Putnam and others (1960), along with the world map by McGill (1958), have proved stimulating. The discussion of classifications by Bird (1969) contains many good suggestions. The large-scale coastal types of Price (1955) and Tanner (1960) coming from their studies of the Gulf and East Coasts of the United States, with emphasis on wave attack and degree of equilibrium, were also given important consideration.

The classification here has two major subdivisions: (1) *primary coasts*, essentially unmodified by marine processes, their character due to the sea level coming to rest against a land form that was the result of terrestrial agencies (erosion or deposition), volcanism, or earth movements; and (2) *secondary coasts*, where present-day marine processes (erosion or deposition) or growth of marine-living organisms has been responsible for the character of the coast. Primary and secondary thus correspond to the youthful and mature coasts of Johnson. Further subdivisions are made according to which specific agent, terrestrial or marine, has had the greatest influence on coastal development. It should be realized that numerous gradational types exist, which are difficult to classify in one single category, so a dual or triple classification may be desirable. However, most coasts appear to show only one dominant influence as the cause of their major characteristics. Difficulty also arises in separating primary and secondary coasts because marine processes may begin to have an effect almost immediately after the sea has come to rest at any particular place. On the other hand, most coasts appear to be either predominantly the result of terrestrial or marine processes.

CLASSIFICATION OF COASTS

I. *Primary coasts* Configuration due to nonmarine processes.
   A. *Land erosion coasts* Shaped by subaerial erosion and partly drowned by postglacial rise of sea level (with or without crustal sinking) or inundated by melting of an ice mass from a coastal valley.
      1. *Ria coasts (drowned river valleys)* Usually recognized by the relatively shallow water of the estuaries which indent the land. Commonly have V-shaped cross section and a deepening of the axis seaward except where a barrier has built across the estuary mouth (1 in Fig. 6–7).
         (a) *Dendritic* Pattern resembling an oak leaf due to river erosion in horizontal beds or homogeneous material.
         (b) *Trellis* Due to river erosion in inclined beds of unequal hardness (Fig. 6–8).
      2. *Drowned glacial erosion coasts* Recognized by being deeply indented with many islands. Charts show deep water (commonly more than 100 m) with a U-shaped cross section of the bays and with much greater depth in the inner bays than near the entrance. Hanging valleys and sides usually parallel and relatively straight, in contrast to the sinuous rias. Almost all glaciated coasts have bays with these characteristics.

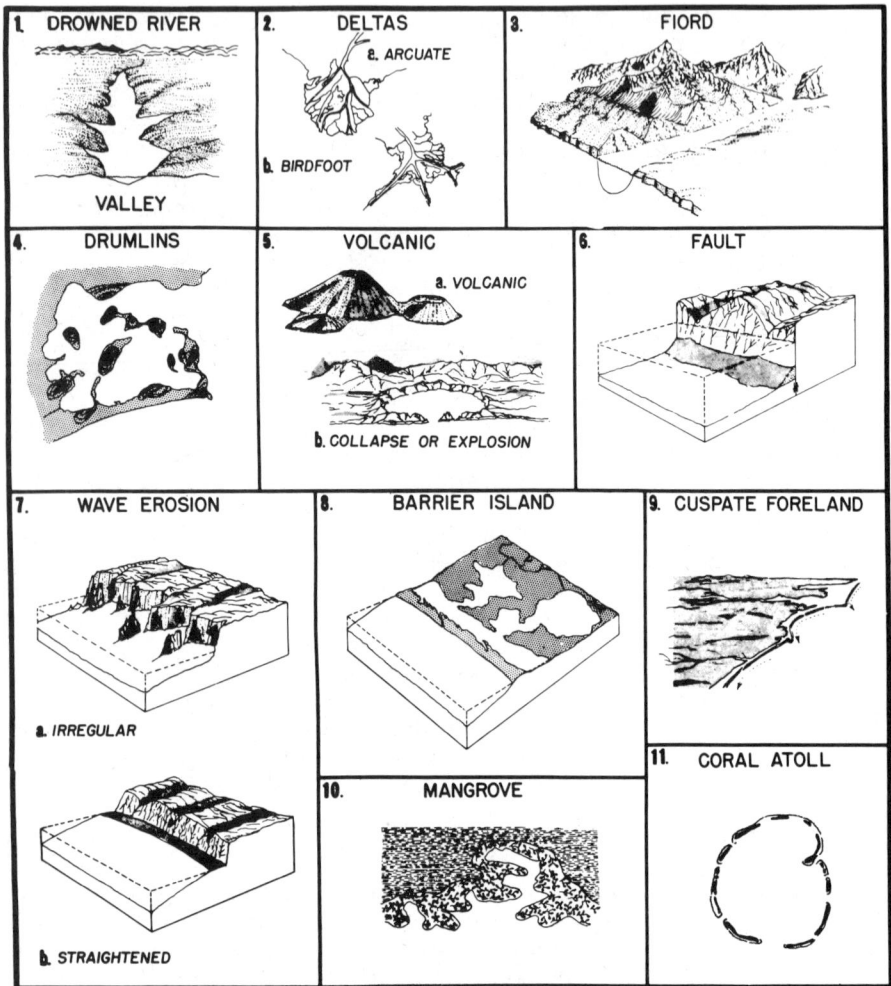

Fig. 6–7   Various types of coasts included in the suggested classification.

(a) *Fiord coasts*   Comparatively narrow inlets cutting through mountainous coasts (3 in Fig. 6–7).

(b) *Glacial troughs*   Broad indentations, like Cabot Strait and the Gulf of St. Lawrence or the Strait of Juan de Fuca.

3. *Drowned karst topography*   Embayments with oval-shaped depressions indicative of drowned sinkholes. This uncommon type occurs locally, as along the west side of Florida north of Tarpon Springs, the east side of the Adriatic, and along the Asturias coast of North Spain.

B. *Subaerial deposition coasts*

1. *River deposition coasts*   Largely due to deposition by rivers extending the shoreline since the slowing of the postglacial sea level rise.

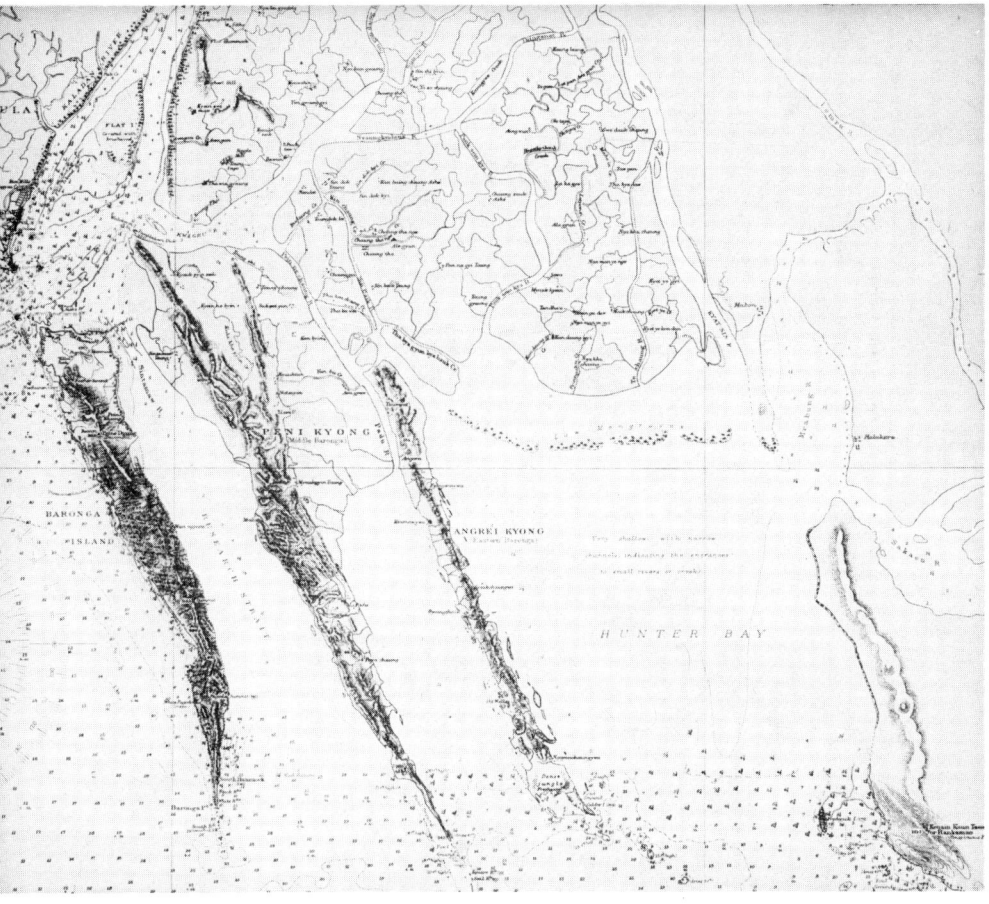

Fig. 6–8 Trellis type of erosion coast caused by wave erosion of inclined layers of varying resistance. From U.S. Oceanographic Office Chart 3705 of Burma coast at 20°N latitude.

(a) *Deltaic coasts*
   (i) *Digitate (birdfoot)*, the lower Mississippi Delta (2b in Fig. 6–7).
   (ii) *Lobate*, western Mississippi Delta, Rhone Delta.
   (iii) *Arcuate*, Nile Delta (2a in Fig. 6–7).
   (iv) *Cuspate*, Tiber Delta (Fig. 6–9).
   (v) *Partly drowned deltas* with remnant natural levees forming islands (Fig. 6–10).
(b) *Compound delta coasts* Where a series of deltas have built forward a large segment of the coast, for example, the North Slope of Alaska extending east of Point Barrow to the Mackenzie (Fig. 6–11).
(c) *Compound alluvial fan coasts straightened by wave erosion* (Fig. 6–12).

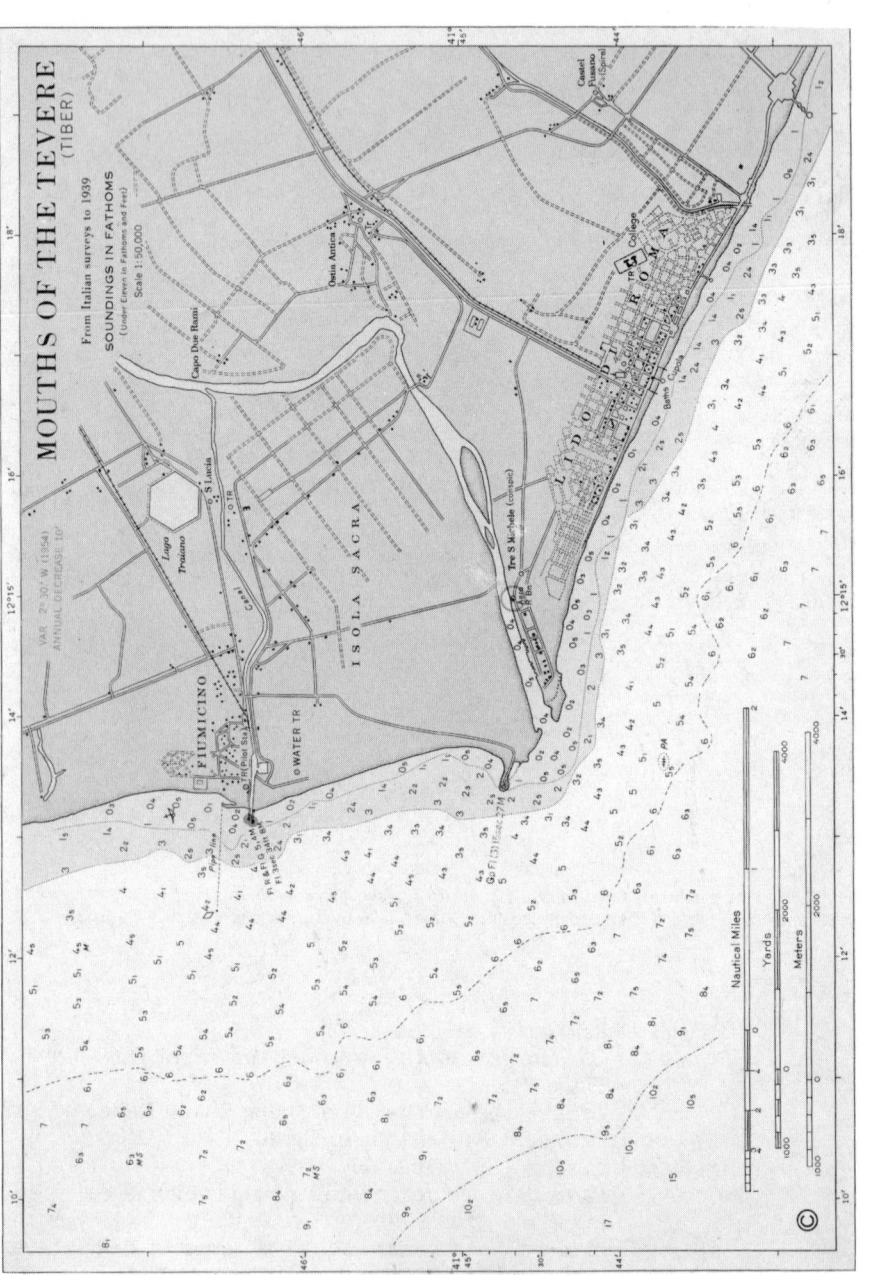

Fig. 6–9 The cuspate delta of the Tiber River in Italy. From U.S. Oceanographic Office Chart 3493.

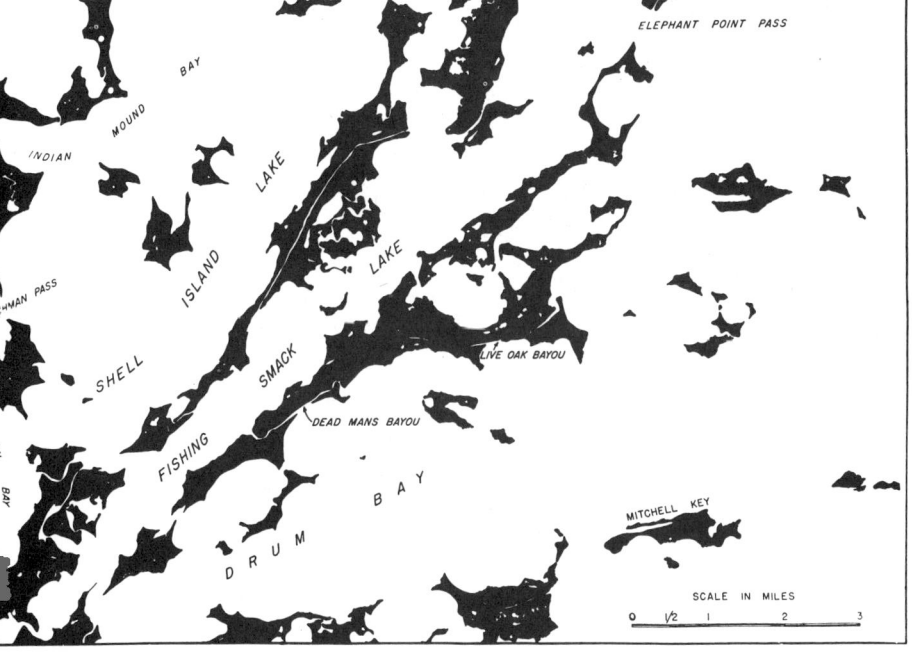

Fig. 6–10 Natural levees of partly submerged old abandoned channels of the St. Bernard subdelta of the Mississippi River. From Russell (1936).

2. *Glacial deposition coasts*

(a) *Partially submerged moraines* Usually difficult to recognize without a field study to indicate the glacial origin of the sediments constituting the coastal area. Usually modified by marine erosion and deposition as, for example, Long Island.

(b) *Partially submerged drumlins* Recognized on topographic maps by the elliptical contours on land and islands with oval shorelines, for example, Boston Harbor (4 in Fig. 6–7) and West Ireland (Guilcher 1965).

(c) *Partially submerged drift features*

3. *Wind deposition coasts* It is usually difficult to ascertain if a coast has actually been built forward by wind deposition, but many coasts consist of dunes with only a narrow bordering sand beach.

(a) *Dune prograded coasts* Where the steep lee slope of the dune has transgressed over the beach (Fig. 6–13).

(b) *Dune coasts* Where dunes are bordered by a beach (Fig. 6–14).

(c) *Fossil dune coasts* Where consolidated dunes (eolianites) form coastal cliffs (Fig. 6–15).

4. *Landslide coasts* Recognized by the bulging earth masses at the coast and the landslide topography on land (Fig. 6–16).

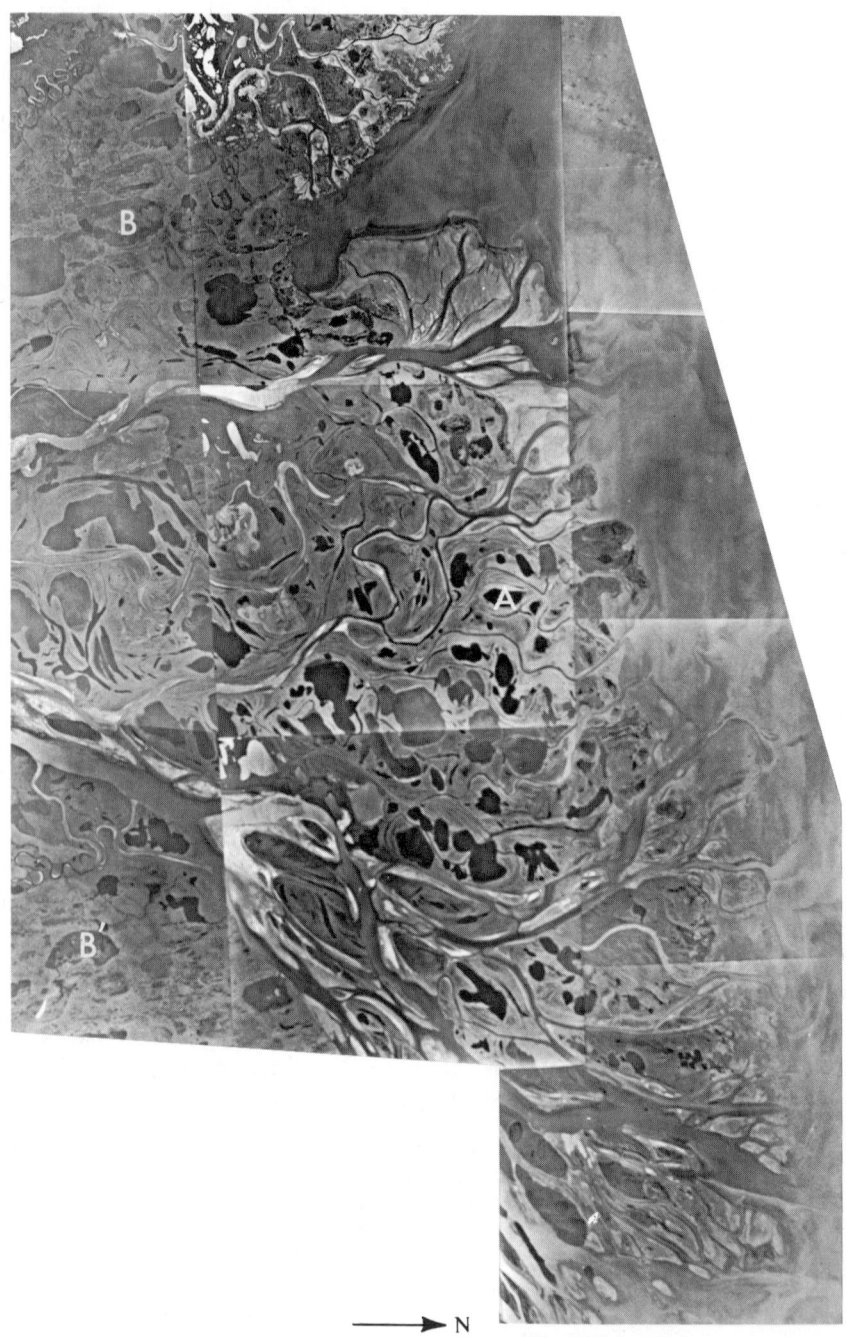

→ N

Fig. 6–11 The delta of the Colville, the largest river on Alaska's Arctic Slope. This delta, like that of the Yukon, consists of channels interlaced with sand bars and numerous delta lakes. Because permafrost action is more intense here than in the Yukon Delta, there are abundant ice polygons (*A*) on the delta flats. Along the landward margins of the delta, oriented lakes (*B, B'*) are visible. Photo by U.S. Geological Survey (1955).

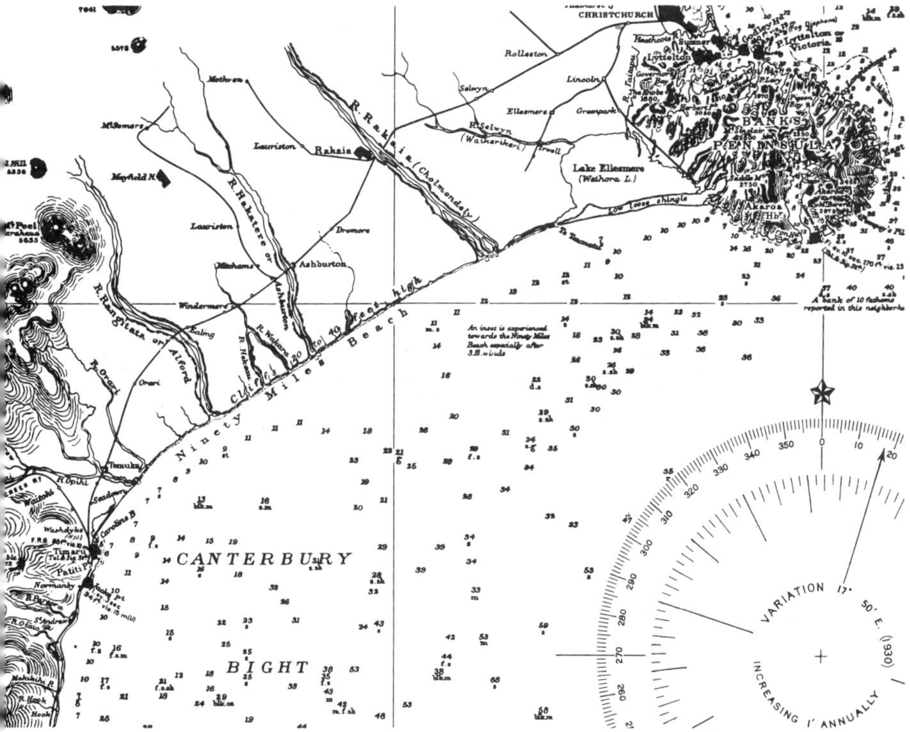

Fig. 6–12 An alluvial fan coast on South Island, New Zealand, straightened by wave erosion.

Fig. 6–13 Oblique aerial photo of San Miguel Island, Southern California, illustrating a wind-prograded coast where the dunes are advancing locally into the sea despite wave erosion. Naval Research and Development Center photo.

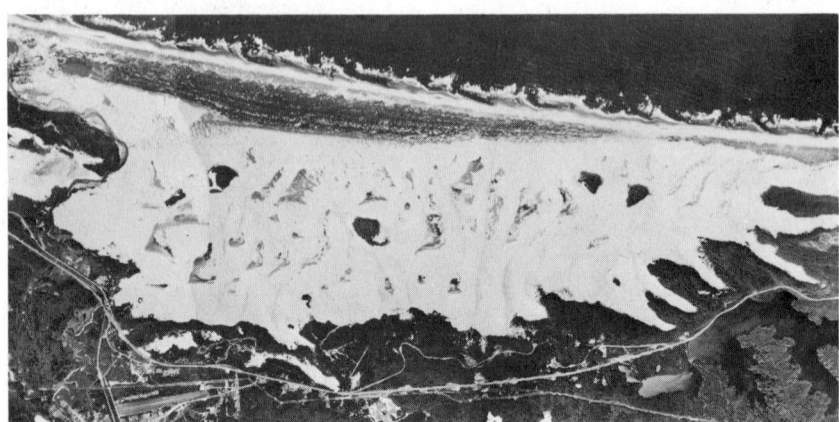

Fig. 6–14   A dune coast with bordering sand beach, north of Coos Bay, Oregon. Some of the dunes have blocked valleys, forming lakes. From U.S. Department of Agriculture.

C. Volcanic coasts

1. *Lava-flow coasts*   Recognized on charts either by land contours showing cones, by convexities of shoreline, or by conical slopes continuing from land out under the water. Slopes of 10° to 30° common above and below sea level. Found on many oceanic islands.

Fig. 6–15   Fossil dune coast. This is an eolianite ridge with a small nip caused by the waves of a lagoon. From Kaye (1959).

Fig. 6–16 A large rockslide that has built out the coast at Humbug Mt., Oregon. Photo by University of California Engineering Laboratory.

      2. *Tephra coasts* Where the volcanic products are fragmental. Roughly convex but much more quickly modified by wave erosion than are lava-flow coasts.

      3. *Volcanic collapse or explosion coasts* Recognized in aerial photos and on charts by the concavities in the sides of volcanoes (5b in Fig. 6–7).

  D. *Shaped by diastrophic movements*

      1. *Fault coasts* Recognized on charts by the continuation of relatively straight steep land slopes beneath the sea. Angular breaks at top and bottom of slope.

        (a) *Fault scarp coasts* For example, northeast side of San Clemente Island, California (6 in Fig. 6–7).

        (b) *Fault trough or rift coasts* For example, Gulf of California and Red Sea, both being interpreted as rifts.

        (c) *Overthrust* No examples recognized but probably exist.

      2. *Fold coasts* Difficult to recognize on maps or charts but probably exist.

      3. *Sedimentary extrusions*

        (a) *Salt domes* Infrequently emerge as oval-shaped islands. Example: in the Persian Gulf.

        (b) *Mud lumps* Small islands due to upthrust of mud in the vicinity of the passes of the Mississippi Delta (Fig. 6–17).

Fig. 6–17 Mud lump islands off Pass a Loutre, Mississippi Delta. These islands rise rapidly and soon disappear because of wave erosion. Photo courtesy of J. P. Morgan.

       E. *Ice coasts*   Various types of glaciers form extensive coasts, especially in Antarctica.

  II. *Secondary coasts*   Shaped primarily by marine agencies or by marine organisms. May or may not have been primary coasts before being shaped by the sea.

       A. *Wave erosion coasts*

           1. *Wave-straightened cliffs*   Bordered by a gently inclined sea floor, in contrast to the steep inclines off fault coasts.

              (a) *Cut in homogeneous materials* (7b in Fig. 6–7).

              (b) *Hogback strike coasts*   Where hard layers of folded rocks have a strike roughly parallel to the coast so that erosion forms a straight shoreline.

              (c) *Fault-line coasts*   Where an old eroded fault brings a hard layer to the surface, allowing wave erosion to remove the soft material from one side, leaving a straight coast.

              (d) *Elevated wave-cut bench coasts*   Where the cliff and wave-cut bench have been somewhat elevated by recent diastrophism above the level of present-day wave erosion (Fig. 6–5).

              (e) *Depressed wave-cut bench coasts*   Where the wave-cut

bench has been somewhat depressed by recent diastrophism so that it is largely below wave action and the wave-
cut cliff plunges below sea level (Fig. 6–6C).
2. *Made irregular by wave erosion* Unlike ria coasts in that the
embayments do not extend deeply into the land.
(a) *Dip coasts* Where alternating hard and soft layers intersect the coast at an angle (Fig. 6–18); cannot always be
distinguished from trellis coasts.
(b) *Heterogeneous formation coasts* Where wave erosion
has cut back the weaker zones, leaving great irregularities.
B. *Marine deposition coasts* Coasts prograded by waves and currents.
1. *Barrier coasts* (Fig. 6–1).
(a) *Barrier beaches* Single ridges.
(b) *Barrier islands* Multiple ridges, dunes, and overwash flats.
(c) *Barrier spits* Connected to mainland.
(d) *Bay barriers* Sand spits that have completely blocked
bays.
(e) *Overwash fans* Lagoonward extension of barriers due to
storm surges.
2. *Cuspate forelands* Large projecting points with cusp shape.

Fig. 6–18 West coast of Italy, near Sestri Levante. A dip coast with irregularities due to difference of resistance of tilted layers that strike normal to the shore. From Putnam et al. (1960).

Examples include Cape Hatteras (Fig. 7–18) and Cape Canaveral (9 in Fig. 6–7).

3. *Beach plains*   Sand plains differing from barriers by having no lagoon inside.

4. *Mud flats or salt marshes*   Formed along deltaic or other low coasts where gradient offshore is too small to allow breaking waves.

C. *Coasts built by organisms*

1. *Coral reef coasts*   Include reefs built by coral or algae. Common in tropics. Ordinarily, reefs fringing the shore and rampart beaches are found inside piled up by the waves.

(a) *Fringing reef coasts*   Reefs that have built out the coast.

(b) *Barrier reef coasts*   Reefs separated from the coast by a lagoon (Fig. 12–11).

(c) *Atolls*   Coral islands surrounding a lagoon (11 in Figs. 6–7, 12–2).

(d) *Elevated reef coasts*   Where the reefs form steps or plateaus directly above the coast (Fig. 6–19).

2. *Serpulid reef coasts*   Small stretches of coast may be built out by the cementing of serpulid worm tubes onto the rocks or beaches along the shore. Also found mostly in tropics.

3. *Oyster reef coasts*   Where oyster reefs have built along the shore and the shells have been thrown up by the waves as a rampart.

4. *Mangrove coasts*   Where mangrove plants have rooted in the shallow water of bays, and sediments around their roots have built up to sea level, thus extending the coast (10 in Fig. 6–7). Also a tropical and subtropical development.

5. *Marsh grass coasts*   In protected areas where salt marsh grass can grow out into the shallow sea and, like the mangroves, collect sediment that extends the land. Most of these coasts could also be classified as mud flats or salt marshes.

Fig. 6–19 Aguijan Island, Mariana Islands. The low plateau is an elevated coral reef. From Putnam et al. (1960).

Chapter 7
BEACHES AND
SHORE PROCESSES

INTRODUCTION

The term *beach* is applied to a shore with a covering of sand or gravel. This excludes muddy shores, such as are found around some deltas, and the rocky shores where no appreciable amount of sediment exists. In contrast to rocky shores, beaches are very unstable. During stormy weather and times of high tide, beaches tend to retreat or even disappear, whereas during times of small waves and small tidal fluctuation, beaches tend to prograde. The works of man along the shore have also had a great influence on the stability of beaches. Because beaches are of great economic importance, both for recreational use and as a buffer zone to protect property from wave attack, engineers and geologists have studied extensively the causes of beach erosion.

In the second edition of this book, attention was called to the important work of the Beach Erosion Board of the U.S. Army Corps of Engineers and of the Council on Wave Research of the University of California at Berkeley. This work has continued with many important results. The Coastal Studies Institute of Louisiana State University has become a major contributor to the study of beaches, particularly in the southern United States. Also, the Marine Institute of the University of Georgia has issued important papers, and the Coastal Research Group of the University of Massachusetts is concentrating on the New England beaches with interesting results. The coastal studies group under D. L. Inman and the Hydraulic Laboratory at Scripps Institution of Oceanography has made much progress in the study of wave mechanics and sediment movement along the shore. The reports by Moberly and collaborators (1963, 1964, 1965) give detailed information about Hawaiian beaches.

Supplementing the older important works, such as D. W. Johnson (1919) and C. A. M. King (1959),[1] several new books with major emphasis on beaches have become available since the second edition of this book. The most important is the monumental work by Zenkovich (also spelled Zenkovitch) (1967), an English translation from the 1962 Russian edition. Although there is no reference to Soviet work subsequent to 1961, one gathers the opinion that at least up to that time the Soviets were ahead of the rest of the world in their coastal investigations. Bascom's book (1964) is largely popular and has interesting examples and ideas coming from his extensive experience in studying beaches, particularly on the West Coast of the United States. Bird (1969) has contributed a book with special attention to Australian beaches. Beaches of the United States are described in a recent book (Shepard and Wanless 1971). Many new articles on beaches have appeared, notably from England, Holland, France, Germany, and Australia.

BEACH AND COASTAL TERMINOLOGY

As a result of coordinated efforts by engineers and geologists, the nomenclature of beach features is now quite well established (Fig. 7–1). This

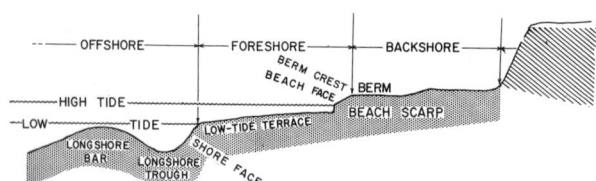

Fig. 7–1   The principal subdivisions of beaches and of the adjacent shallow-water area.

has been important in connection with legal practice arising from disputed ownership of beaches and lawsuits over the erosion of beach property. Among the terms, the following (in alphabetical order) appear to be the most pertinent:[2]

[1] Since writing this, a second edition (King 1972) has appeared with many new features.
[2] For more complete definitions see *Beach Erosion Board* (1954), Wiegel (1953), and *Glossary of Geology* (1972 edition).

Accretion   The building up of a beach, either by natural processes or by artificial works of man.

Backshore   The zone of the beach lying between the foreshore and the coastline (Fig. 7–1).

Backwash ripples   Low-amplitude ripple marks formed on fine sand beaches by the backwash of waves. Commonly about 45 cm apart (Fig. 7–28A).

Bar   An elongate slightly submerged sand body. May bare at low tide (Fig. 7–1).

Barrier   A sand beach (barrier beach), island (barrier island), or spit (barrier spit) that extends roughly parallel to the general coastal trend but is separated from the mainland by a relatively narrow body of water or a marsh (Fig. 6–1).

Beach   The zone of unconsolidated material extending landward from the mean low water line to the locality where there is a change in material or physiographic form as, for example, the zone of permanent vegetation, a zone of dunes, or a sea cliff. The upper limit of a beach usually marks the effective limit of ordinary storm waves.

Beach face   The sloping section of the beach below the berm normally exposed to the wave uprush (Fig. 7–1).

Beach ridge (storm beach, chenier)   A low lengthy ridge of beach material piled up by storm waves landward of the berm. Usually consists of coarse sand, gravel, or shells. Occurs singly or as a series of more or less parallel ridges (Fig. 7–14). Should not be confused with dune ridges that form particularly where the sand is fine and resemble beach ridges (Fig. 7–16).

Beach scarp   An almost vertical slope along a beach. It is caused by unusually large waves or developed where a beach is retrograding under moderately large wave attack.

Berm (beach berm)   The nearly horizontal part of a beach inside the sloping foreshore (Fig. 7–1).

Berm crest (berm edge)   The seaward limit of a berm (Fig. 7–1).

Bight   A slight indentation of the shoreline with a crescentic shape.

Chenier   A beach ridge usually built upon swamp deposits (Fig. 7–17).

Cusp   One of a series of short ridges on the foreshore extending transverse to the beach and occurring at more or less regular intervals with spacing increasing with wave height (Fig. 7–9A).

Cuspate foreland   A large sandy cusp-shaped projection of the coast (Figs. 7–18, 7–19).

Cuspate sandkey   A cusp-shaped sand island.

Cuspate spit   A sandy cusp-shaped projection of the shoreline, found on both sides of some lagoons (Fig. 7–20).

Feeder beach   An artificially widened beach serving to nourish downdrift beaches by littoral currents.

Foreshore   The sloping part of the beach lying between the berm and the low water mark (Fig. 7–1).

Groin (groyne, British)   A short wall built perpendicular to the shore for the purpose of trapping littoral drift (Fig. 7–26).

Jetty   A structure built out into open water with a greater length than a groin, and designed to prevent the shoaling of a channel by confining stream or tidal flow (Fig. 7–22).

Longshore bar (ball or ridge)   A sand ridge or ridges, extending along the shore outside the trough, that may be exposed at low tide or may occur in the offshore below the water level (Fig. 7–1).

Longshore trough (runnel or low)   An elongate depression or series of depressions along the lower beach or in the offshore zone inside the breakers (Fig. 7–1).

Mole   In coastal terminology, refers to a massive solid-filled structure (generally revetted) of earth, masonry, or large stones.

Offshore   The breaker zone directly seaward of the low tide line (Fig. 7–1).

Overwash   The portion of the uprush that carries water over the crest of the berm.

Revetment   A facing of stone, concrete, etc., built to prevent shore erosion.

Rill marks   Small drainage channels forming in the lower portion of a beach at low tide.

Rip channel   Channel cut by seaward flow of rip current, usually crosses longshore bar.

Sand domes   Small domed-up surfaces of the beach due to entrapment of air. Erosion of the domes usually produces ring structures.

Shingle   Defined variously as a beach with flattish pebbles or consisting of smooth well-rounded pebbles.

Shore face   The narrow sloping zone seaward of the low tide shoreline over which the beach sands and gravel oscillate most actively (Fig. 7–1).

Swash mark   The thin wavy line of fine sand, mica, or fucus left by the uprush along a beach (Fig. 7–30A).

Tombolo   A sand zone above ordinary high tide level that connects an island or rock to the mainland or another island.

Uprush (swash) The rapid flow of water up onto the beach face following the breaking of a wave.

BEACH CLASSIFICATION

Beaches may be classified in various ways. A differentiation can be made between the variety of features found in beach profiles (Fig. 7–2). The profile may show a continuous slope, may have one or more berms, or may have a low tide terrace with or without bars and troughs that are exposed at low tide. The same beach, however, may develop all these characteristics at different times, so this does not serve very well as a fundamental classification of beach types. More striking differences exist between the gravel or shingle beaches, coarse-sand beaches, and fine-sand beaches. The typical gravel beach has a beach ridge on the inside where the waves have piled up the gravel sometimes as much as 6 m above normal high tide. In some areas,

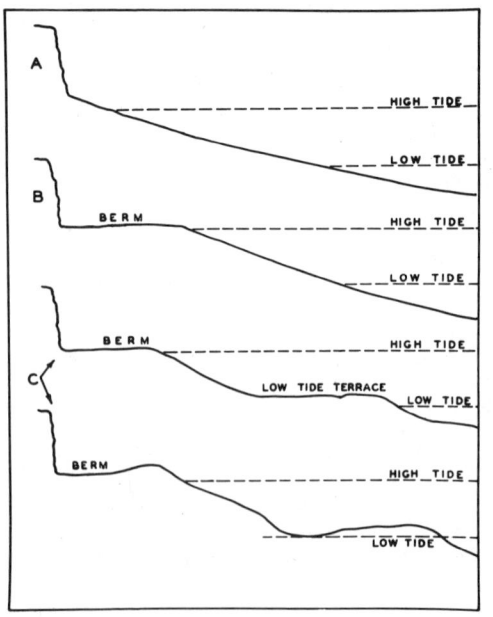

Fig. 7–2   The principal types of beach profiles.

the ridge consists of shells, such as the cheniers or the Gulf Coast (Russell and Howe 1935; Gould and McFarlan 1959). Ordinarily there is no appreciable berm in the gravel beaches and the foreshore slopes continuously seaward. However, often the slope is interrupted by a step near the low tide line. In many places this step is sand covered.

Coarse sand beaches may have berms but these berms slope landward, often at considerable angles. The foreshore is steep, although somewhat less so than in gravel beaches (Table 3).

Table 3 Average Beach Face Slopes Compared with Sediment Diameters

Type of Beach Sediment	Size	Average Slope of Beach Face
Very fine sand	1/16– 1/8 mm	1°
Fine sand	1/8– 1/4 mm	3°
Medium sand	1/4– 1/2 mm	5°
Coarse sand	1/2– 1 mm	7°
Very coarse sand	1– 2 mm	9°
Granules	2– 4 mm	11°
Pebbles	4– 64 mm	17°
Cobbles	64–256 mm	24°

Source: Data compiled by D. L. Inman and F. P. Shepard.

The slope of the beach face results from a dynamic equilibrium between the run-up (or swash) of water up the beach face and the return flow (or backwash) of water down the face. Although primarily related to the beach permeability, the slope shows a strong correlation with grain size because the latter determines the permeability. It will be noted that the loss of run-up because of discharge into the beach is ten times greater for beach grains 4 mm in diameter than for 1 mm, and that the discharge through sand less than 1 mm in diameter is almost negligible for times as short as a wave period (see Fig. 4–8). Thus, there is almost as much water in the backwash of a fine-sand beach as in the run-up, and the resultant beach face cannot stand at a steep angle except where a scarp has been cut recently. All coarse sand beaches are soft, making them equally poor for walking and for traffic, and the offshore commonly has bars and troughs.

Fine-sand beaches differ from the others chiefly in having very gentle foreshore slopes. The sand is generally packed hard on the foreshore so that one can walk along the beach without sinking in and, unless the tide has recently retreated, the foreshore is likely to be hard enough to support an automobile or to land a small plane.

Beaches also vary somewhat according to the amount of tide and their exposure to wave attack. In tideless or nearly tideless seas, beaches are often bordered by a series of longshore bars and troughs, whereas seas with large tidal ranges are likely to have broad terraces, each with at most one large bar exposed at the low tide level. During neap tides, bars and troughs are generally more evident on the lower beach. In some areas, especially in tideless seas, the bars adjacent to the beach are crescentic (Fig. 7–3) with the points between the crescents faced toward the land. Where the same type of sand is exposed to larger waves along one portion of a beach than an-

N ←

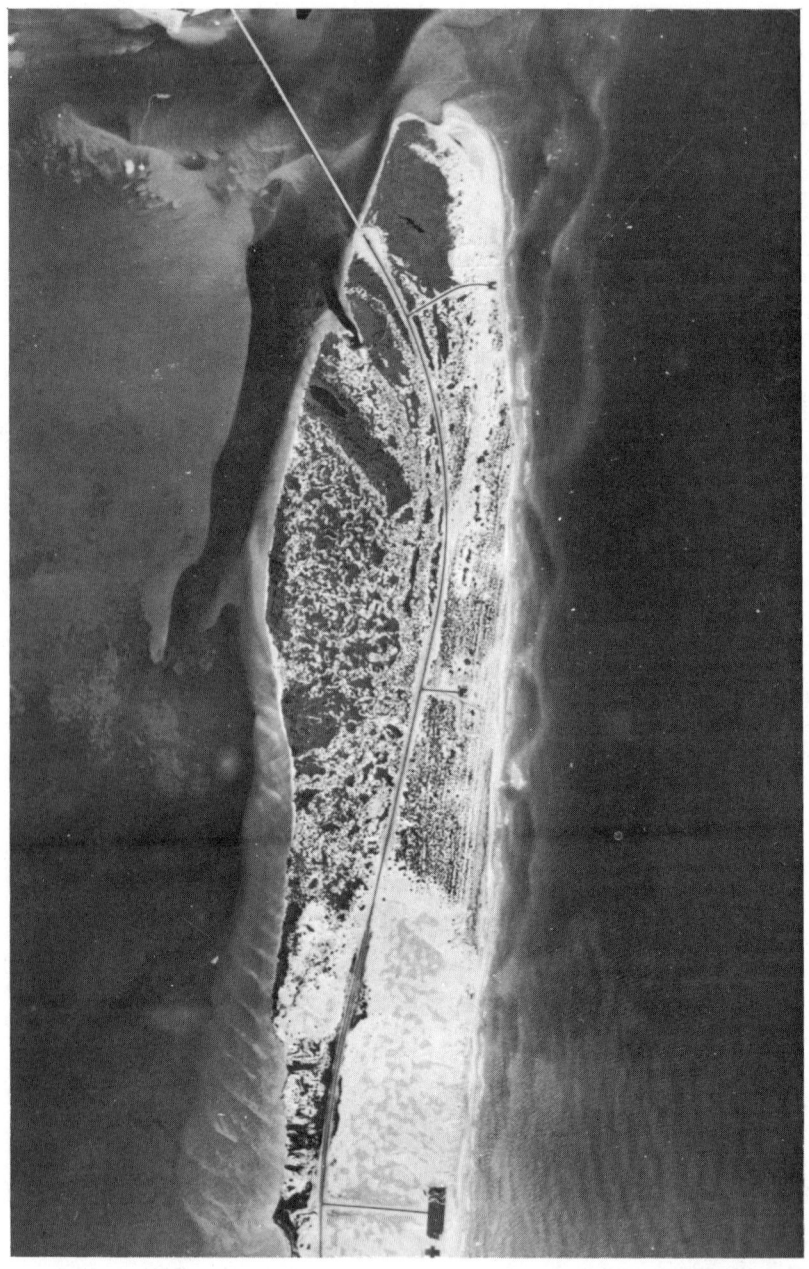

Fig. 7–3 The submerged crescentic sand bodies, cuspate bars, found principally in seas with small tides. For straight longshore bars, see Fig. 7–16.

other, the inclination of the foreshore is lower inside the zone of large waves. This difference is more striking in coarse-sand beaches than in fine. Another fundamental difference in beaches is related to their ground plans. Thus the long continuous and relatively straight beaches of the barrier type are in marked contrast to the short crescentic beaches in the coves between rocky headlands (Fig. 7–4). Along some cliffed coasts, as in much of Southern California, long, relatively straight beaches extend between widely separated rocky points. Of these three types, the cove beaches are likely to contain the coarsest sand because it has been locally derived from the sea cliffs, whereas the long beaches are more commonly fine-grained because here the sand has come largely from streams (the coarser products of the streams do not usually reach the coast).

Arctic Beaches Because of the intense frost action, the almost continuous sea ice along the coast, and the glacial tongues that enter the sea, the beaches of the high latitudes in the Arctic and the Antarctic have some marked differences from those of warmer climates (Nichols 1961; G. W. Moore 1966, 1968; Greene 1970A). The pack ice along the shore allows wave action only for short periods of time, and even when the sea next to the coast is open, the offlying ice pack prevents heavy wave action, and therefore high storm beach ridges are rare. However, low ridges mostly gravel, are very common because of the greater importance of physical than of chemical weathering in high latitudes. Some of the beaches have a covering of silt that comes from the solifluction flows during the summer thaws. Ice floes driven along the coast by the wind while the ice pack is breaking up may be thrust over the beach, making chaotic piles of ice, and may leave behind boulders and gravel that was frozen into the bottom of the ice (Fig. 7–5). In winter, a flat rampart, consisting of alternating layers of sediment and ice (called a *kaimoo*) (Fig. 7–5), often develops along the arctic beach shores. This may form a rampart that protects the beach from the disruption caused by moving ice floes.

In the Antarctic, many gravel beaches rest on ice. When the ice melts, it leaves very irregular surfaces with many collapse depressions in the gravel. The beaches are likely to terminate abruptly against snow drifts. The beach gravel in the Antarctic is in general more angular than elsewhere because the sea has not had much chance to act on it.

Elevated beaches are common features in northern and eastern Canada where the land depressed during glaciation has been recovering its former elevation since the ice melted. These beaches, consisting mostly of low concentric gravel ridges, often follow the contours of the elevated areas (Fig. 7–6).

MECHANICS OF BEACH DEVELOPMENT

Experiments in hydraulic laboratories and instrument measurements in the surf zone have established the general mechanics of beach development and beach erosion. Studies of orbital velocities in the surf zone (Inman and Nasu 1956) have confirmed data from tank experiments (Beach Erosion Board 1941) that velocities of onshore motion are greater under advancing wave crests than velocities of offshore motion under troughs. Studies have also shown that during periods of low waves, differential velocity is sufficient so that sand tends to move onshore except in zones of rip currents. Onshore migration is particularly large during long-period waves when there is more opportunity for sand to be deposited. Conversely, with high waves of short

Fig. 7–4 A cove beach showing typical summer (upper) and winter (lower) conditions at Boomer Beach in La Jolla, California. Some of the sand in winter has shifted along the beach to the right of the photo.

Fig. 7–5 A kaimoo with ice-cemented gravel cover (foreground) and a jumbled pile of ice pushed up by wind-driven pack ice during a period of about one hour, observed and photographed by George W. Moore.

Fig. 7–6 Elevated gravel beach ridges formed in northern Canada by recovery of the land from the great weight of the ice sheets. From Washburn (1947).

period, which keep sand in suspension, the beach retreats largely because sand washed off the foreshore by the backwash does not settle until it is carried into a rip current and has moved seaward to relatively deep water.

Beaches develop their well-sorted condition due partially to the common presence of oscillation ripples in the surf zone (Inman 1957). The turbulence and resultant lifting force is greatest at the crest of the ripples; therefore only the coarsest sediment is deposited there. This material tends to move landward during the passage of a wave crest. The fine material of the troughs is carried into suspension only during large waves and hence is carried seaward in the rip currents.

COMPOSITION OF BEACH SAND

Beaches along continental shores consist predominantly of terrigenous minerals, the residue of disintegration of coarse-grained igneous rocks. Because quartz is the most stable of the common igneous minerals, it is the most abundant in sands, and in some areas where there has been extensive chemical weathering, beaches are almost entirely quartz. Notable examples are the beaches of northwestern Florida. Most continental beaches also have an admixture of feldspar and ferromagnesian minerals. These groups are particularly abundant in semiarid regions, like Southern California, where chemical weathering has been slight. Mica is plentiful in some beaches and layers of mica are quite common in beach sands. The heavy minerals magnetite and ilmenite are also interbedded in many beach sands or concentrated in layers, especially at the base of the sand. Heavy minerals of economic importance, such as gold, tungsten, scheelite, and wolframite, are locally of sufficient quantity in beach sands to be mined on a small scale. At Nome, Alaska, large amounts of gold have been mined from the present beach and from older beaches above and below the present sea level.

Biogenous constituents, mostly calcareous, are found in most beaches and often predominate in tropical and subtropical areas, including oceanic islands. Shell beaches are rare in higher latitudes but a few occur where terrigenous material is scarce, as around some islands (Raymond and Stetson 1932). In the tropics where there are offlying reefs from which the sand is derived, the beaches have an abundance of coral fragments, but usually the foraminifera greatly outnumber the coral grains and the algae *Halimeda* is very common, especially in the Bahamas (Lowenstam and Epstein 1957). In the Hawaiian Islands, Moberly and others (1956) found that quantitatively the order in calcareous beach sands is: foraminifera, mollusks, algae, echinoids, and then coral. They suggest that it is a misnomer to refer to "coral sands" for the beaches in the coral-reef areas. In tropical areas without reefs, foraminifera are usually less abundant and mollusk fragments are often the largest constituent in the sands, as for example, the coquina beaches of eastern Florida. In northern Florida, the shells are mostly old, but near Miami, mostly modern (Rusnak et al. 1966). The east Florida coast shows an alternation of terrigenous beaches and shell beaches (Martens 1931), the former have hard sand with low foreshore slope and the latter are steep and soft so are poor for walking. Daytona Beach, of fine quartz sand, is so hard packed that it has been used for auto races.

Oceanic islands have many beaches with volcanic sand. In the well-studied Hawaiian beaches, grains of basaltic lava are common (Moberly et al. 1965). Volcanic glass is plentiful near areas of recent eruption, producing glistening black-sand beaches. Some green-sand beaches are the result of a concentration of olivine crystals weathered from the lavas. The beaches of Kauai have more volcanic sands on the dry side of the island and more calcareous beaches on the wet side (Inman et al. 1963), because the active weathering on the wet side decomposes most of the sand. Hence, the streams do not transport as much volcanic sand to the shore as on the dry side.

SOURCES OF BEACH SAND

The direct source of almost all beach sand is the shallow sea floor, although previously the bulk of the shelf sand was carried into the ocean by runoff from the land. A somewhat smaller source comes from wave erosion of sea cliffs. In tropical areas, much of the sand is derived from the remains of marine organisms; therefore the beach sands here may not even indirectly have a terrestrial source. Beaches may also receive their supply from sands that were deposited on the continental shelf during the Pleistocene stages of low sea level when the shelves were largely dry land. Hence, here the terrestrial source may be remote in time.

Where estuaries formed by the postglacial rise of sea level have been filled, as in parts of Southern California, flood conditions carry large quantities of sand into the ocean, and longshore currents move this along the coast until reaching an area where conditions are favorable for the formation of a beach. If estuaries are shallow and of small length, flood conditions may produce currents that carry sand to their mouths, providing an adequate source for beaches along the open coast. Longer estuaries trap most of the sand supply and, unless there is a plentiful source of Pleistocene sand on the shallow adjacent shelf, there will be few beaches on the open coast in the immediate vicinity.

The source of some beach sands is remote. The beach sands of Israel are derived mainly from the Nile River and are perhaps the only gift from Egypt to the Israelis. The beach west of the Santa Barbara breakwater in Southern California, according to Trask (1952), has contributions from the Santa Maria and Santa Ynez rivers, which enter the ocean well north of the turn in the coast, more than 100 km from Santa Barbara.

Cliff erosion is an important source of beach sands where cliffs of unconsolidated material are subject to wave attack. Such cliffs often retreat at rates of a meter or more a year (D. Johnson 1919, p. 295, 1925, p. 318; Shepard and Grant 1947; C. King 1959, pp. 294–361; Zenkovich 1967, p. 165; Shepard and Wanless 1971, pp. 43, 543). Glacial outwash deposits are a source of much beach sand. A notable example is the cliffed portions of Cape Cod, Massachusetts, where retreat has been 54 m in 70 years (Zeigler et al. 1964). The sand is carried north to form the beaches of Provincetown and south to form extensive beaches that are constantly changing (Shepard and Wanless 1971, pp. 41–50).

Cliffs of more consolidated rock are generally a poor source of sand, because retreat is slow. D. Johnson (1919, pp. 184–195) cited examples of glacial striations along hard-rock coasts that extend down to the sea, showing that there has been no erosion since the glaciers retreated, even in some places where the rock is exposed to the full attack of the sea. The general absence of beaches in hard-rock areas where glaciers have scoured off the unconsolidated cover is explained in this way. On the other hand, small cove beaches often receive their supply of sand from relatively hard rock cliffs.

Evidence that old shelf deposits constitute the source of beach sand is found in areas where streams are not introducing any appreciable amount of material and where there are no alluvial or soft-rock cliffs to supply the sand; yet the beaches are forming actively, and the sand is migrating along the coast. A good example is furnished by the beaches of the barrier islands along the coasts of Alabama and Mississippi, west of Mobile Bay (J. C. Ludwick, unpublished manuscript). It is almost certain that the great bulk of the sand brought in by the Mobile River, at the head of Mobile Bay, is deposited at the bayhead delta, because the lower bay has predominantly muddy sediments. Cliffs are virtually nonexistent on the open coasts of Alabama and Mississippi, which are bordered by sand islands. These islands are migrating westward under the influence of the easterly winds (Otvos 1970), but new supplies keep re-forming at the east ends of the islands. This sand may come from the extensive sand shelf deposits to the south and east. Trask (1952) presented evidence that some of the sand accumulating on the west side of the Santa Barbara breakwater during the recent dry years of little runoff must have come from a shallow offshore area. There had not been sufficient sediment from the intermittent streams to account for the sand budget at the breakwater, and the small quantity of hornblende in the recent sand indicated that little sediment had been derived from streams north of Point Conception. In his study of the minerals in the beach sands of New Jersey, McMaster (1954) concluded that the southern beaches must have had their source sand from the continental shelf. From a study of the beaches between Capes Hatteras and Lookout, Pierce (1968) concluded that approximately one-third of the sand was derived from offshore. From work on the mineral assemblages of the northwest Gulf of Mexico, van Andel and Poole (1960) found that the Texas beaches had received much of the sand from erosion of the old sediments on the adjacent continental shelves, evidence that beaches can receive their sediments largely from offshore sources.

BEACH CYCLES

Almost all beaches are subject to frequent large fluctuations in size and shape. Some small beaches disappear completely during periods of large waves but are usually rebuilt during the periods with small waves that follow (Fig. 7–7). Other beaches show a reduction in berm width during stormy seasons, along with a reduction of their foreshore slope and often an increase in the height of the remaining berm. Here, also, one usually finds a return of the prestorm conditions during intervals of small waves. The periods of large waves are most common in winter, although this is not necessarily true in all areas, because winter winds along some coasts, such as northwest Florida, are mostly offshore. Also, hurricanes and typhoons with their enormous cutting power usually occur in late summer and early fall, in both the Atlantic and Pacific Oceans. The nature of the beach cycles has been determined by numerous profiles. Among the more recent additions to the literature we can include: Zeigler and Tuttle 1961; Sitarz 1963; Dolan 1965; Gorsline 1966; and Sonu and Russell 1966.

Examples of the winter and summer cycles are shown in Fig. 7–7. It will be seen that the berm is cut back or disappears entirely during the stormy season. As a result, the beach foreshore becomes more gently sloping, although on some occasions a beach scarp may form on the foreshore because of excessive cutting at one level. The offshore ordinarily develops rather deep

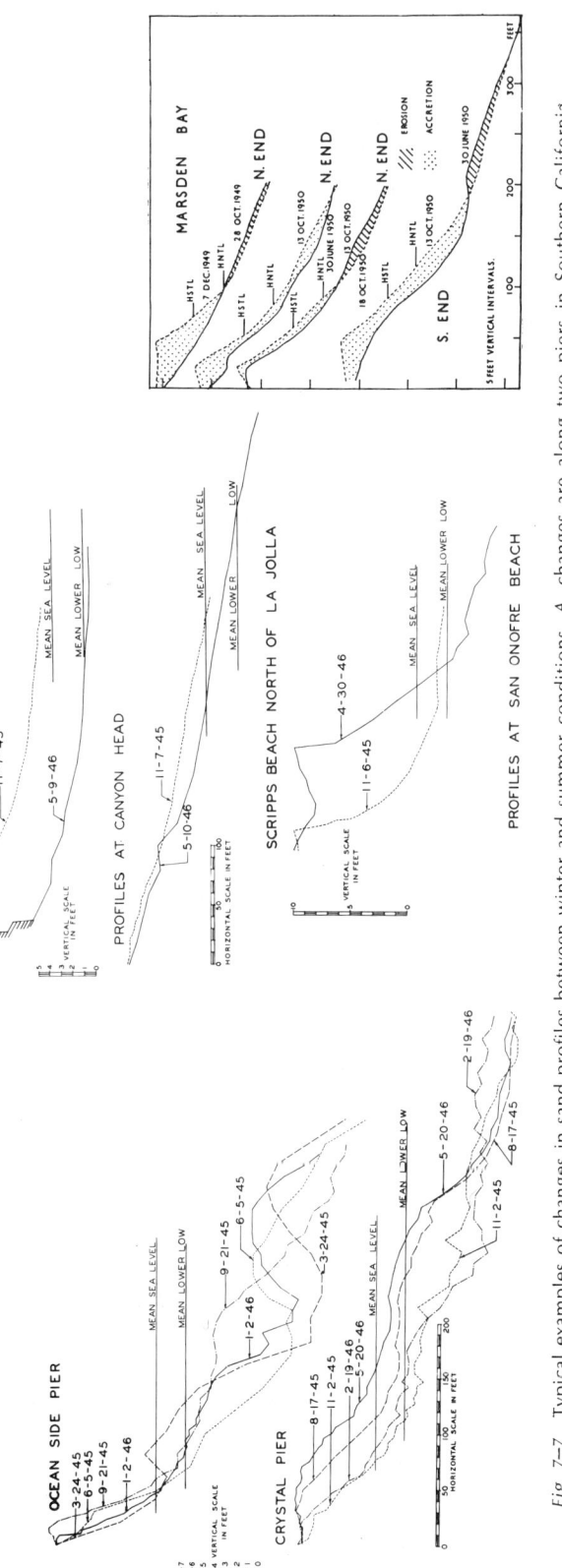

Fig. 7–7 Typical examples of changes in sand profiles between winter and summer conditions. A, changes are along two piers in Southern California. B, the difference in the changes at La Jolla, California, between points of high wave convergence and of wave divergence (at submarine canyon head). Shows also a change at San Onofre Beach, Southern California, where a considerable lateral shift has built up the beach at the south end as a result of northwest approaching waves. C, profiles of Marsden Bay, County Durham, on the English coast, showing the development of berms. From King (1959, Fig. 8–1).

channels due to strong longshore currents and rip currents. Bars are built outside because of the combined effect of offshore creep of sand inside the breaker line and onshore creep outside. The net effect on the offshore is deposition that takes care of most of the loss of sand on the inside.

The return to small-wave conditions tends to reverse the sand shift so that the longshore bars migrate shoreward and fill the inner trough, and the sand begins to pile up on the beach to form a new berm that is almost always lower than any remnant of an old one. The new berm will ordinarily build up gradually to the level of the old berm crest and carry the sand up to the higher level left by the old berm.

In addition to the winter and summer cycles, there are cycles of shorter period. These may come from an isolated unseasonal storm followed by small waves, or they may develop because of the change from neap to spring tides. The storm cycle is similar in its effects to seasonal cycles. The tidal cycle is the result of the change of elevation of sites of erosion and deposition on the upper beach. Ordinarily, deposition occurs at the highest level reached by the waves, because at this point most of the water sinks into the sand and leaves behind some of the products of the uprush. Thus, given waves of the same magnitude, deposition will occur at a certain level during neap tides, but during spring tides the waves will reach a higher level, and in the zone where deposition had occurred during neap tides, the backwash will tend to cut away the deposits of the neap-tide period. The small sardinelike fish called *grunion* make use of this cycle by laying their eggs in shallow holes they dig in the sand at the upper levels reached by high tides at the end of a spring-tide period. This allows the waves to deposit sand over the eggs and keep them covered during the next fortnight, after which the eggs hatch during a period of erosion so that with the sand removed, the young can return to the

Fig. 7–8   Pt. Dume, California, with beach built on upcurrent side of a projecting land mass. Photo by D. L. Inman.

sea (Walker 1952). The nature of the instinct that enables the grunion to be familiar with a process only recently discovered by man is not yet clear.

Another important effect of seasonal changes is the shifting of sand along the beach as a result of a change in wave approach. Beach shorelines tend to form as closely as possible at right angles to the direction of wave approach (C. King 1959, p. 169). This is the most stable beach form, because longshore drift is at a minimum under this condition. Consequently, a wide beach develops on the side of the embayment facing the wave approach (Fig. 7–8). If the wave approach changes, the wide beach will tend to shift to the other end of the embayment. A good example of such a shift is found at La Jolla Point in Southern California where Boomer Beach (Fig. 7–4), on a north–south coast, builds up in summer as a result of the southerly approach of the waves. The sand on this beach disappears entirely during the fall or winter because the large waves usually approach from the northwest. At this time much of the sand of Boomer Beach goes to the south end of the small embayment and forms a beach at that point.

Seasonal cycles are sometimes very erratic. As an example, I had observed for many years that the sand beach just south of Scripps Institution became denuded during the winter and for a considerable period the underlying gravel was exposed (Fig. 7–9A). Since 1947, however, the gravel condition has only once developed as it did in former years, and, except for small patches of gravel that have been uncovered locally, the sand cover has persisted (Fig. 7–9B). Also, the extensive exposure of rock directly north of Scripps Institution pier that formerly occurred every winter has been greatly reduced in area in recent years.

The fact that, except in 1951, the beach near Scripps Institution has maintained much of its sand every winter since 1947 is not easy to explain, particularly because the area rainfall has been very light since the March flood of 1938. As a result, far less sediment should have been brought into the ocean to supply the beach from the north. A minor source of sand coming from the retreat of the alluvial cliffs on either side of Scripps Institution has also decreased, partly because of a new sea wall and partly because the rodents that were causing a large amount of the erosion have been controlled. Possibly decreased wave activity may explain these years of beach maintenance. The erosion of fine-sand beaches at low tide by the occasional heavy rains may have been partially responsible for the former winter denudation. Also, the more saturated condition of the beach, which, as indicated by Grant (1948), tends to increase erosion because of less absorption of the backwash, may explain the erosion during the former higher rainfall periods.

PERMANENT LOSS OF BEACH SAND

Because beaches receive new supplies from runoff and from cliff erosion and because the waves are effective both in setting up currents that move sand along the shore and in bringing sand up onto the land to form the beaches, it should follow either that beaches would grow continuously wider or that some means exists of disposing of the excess sand. Sand is disposed of in several ways and is lost to the beaches either temporarily or permanently. One such means is evident from aerial photographs taken of the downcurrent extremity of some elongate beaches. Dunes can be seen in many land valleys inland from the downcurrent end of these beaches (Fig. 7–10), often extending in from the shore for many miles. The sand may even-

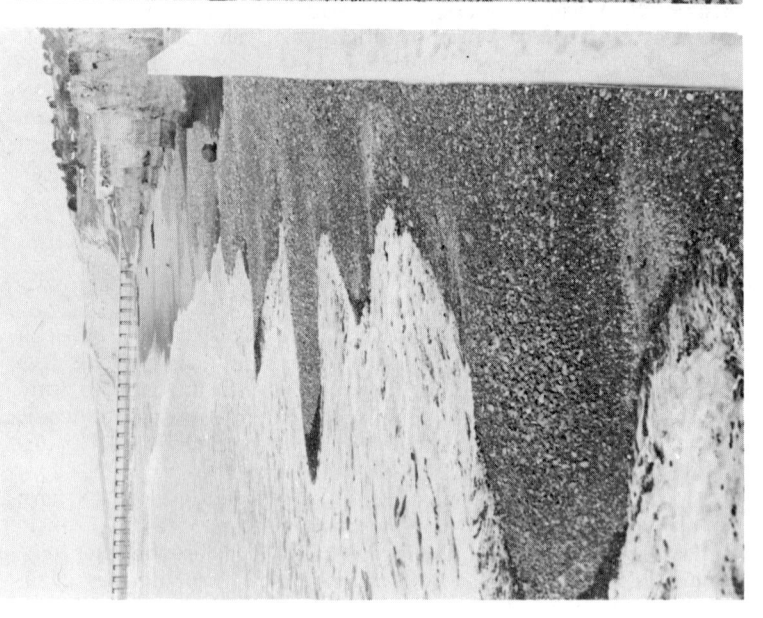

A

B

Fig. 7–9 A, gravel cusps that formerly appeared every winter at La Jolla, California. B, the same beach after sand has covered the gravel. Since 1947 the exposure of the gravel has occurred only once. The alluvial cliffs formerly retreated 25 cm a year but have now become essentially stable. Photos by U. S. Grant IV, 1936.

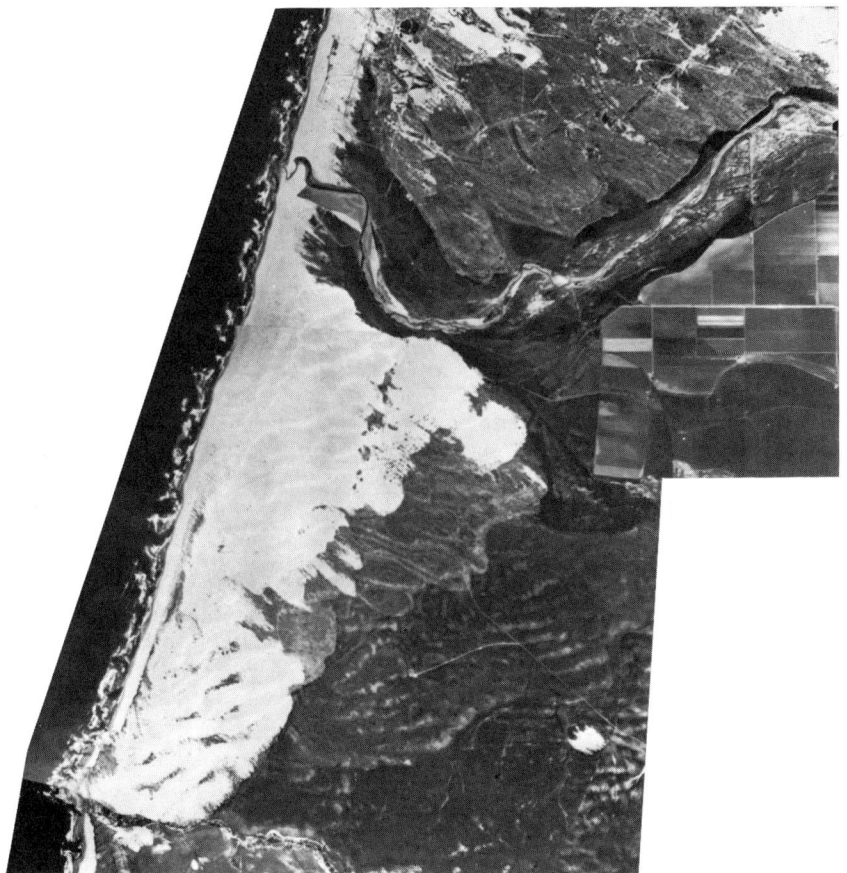

Fig. 7–10 Dunes north of Pt. Sal formed inside the beach at the mouth of the Santa Maria River, central California. Photo by U.S. Department of Agriculture, 1967.

tually cross divides and return to the shore downwind from the dunes by stream transportation. Alternatively, it may be carried inland into desert basins and hence become lost to the beaches.

Some of the sand that is carried downcurrent along the shore is deflected seaward at the lower end of the beach where a point of land causes strong rip currents. If carried seaward far enough across the shelf by such a current, it may not return to the shore. However, studies by Trask (1952, 1955) show that some sand can even bypass major rocky points and continue along the shore, provided that the water off the points is not very deep near the shore. Thus, much of the sand coming down the California coast passes Point Conception, turns at a large angle, and continues on down to the Santa Barbara breakwater, where it was temporarily trapped until a pumping operation set up by a U.S. Army Corps of Engineers project now allows the sand to continue moving down the coast. The nature of the wave refraction at a projecting point is such that the sand is deflected and swept around the point rather than continuing seaward unless caught in a rip current. Trask found that very active bypassing occurs in the zone to a depth of 9 m off Southern California, and that some transfer takes place to a depth of about 18 m. This agrees with the changes observed in measurements made on stakes off La Jolla (Inman and Rusnak 1956).

A very important loss of sand occurs where submarine canyon heads extend in toward a beach. Studies of profiles of the canyon heads along the California coast (Shepard 1951; Chamberlain 1960; Inman and Frautschy 1966) have shown the rapid fill that takes place in these localities. Computations have indicated that the rate of this fill is approximately equal to the amount of sediment supply being carried down the coast from the runoff in the entire area that lies between canyon heads. The sand would fill the canyons rapidly and hence eliminate the canyon heads if it were not for currents that carry the sediment out along the canyon axes, presumably in the form of sand flows and turbidity currents (see p. 308).

According to Inman and Frautschy (1966), there are a number of essentially closed beach compartments along the California coast. At the south end of each compartment, a submarine canyon heads close to the coast and catches most of the sand supply that is drifted toward it from the north (Fig. 7–11). Directly south or a short distance to the south of each canyon head is a rocky shoreline, usually in the form of a projecting point. Farther south, the beaches re-form and become progressively wider because of an increasing supply from entering streams. These beaches persist with few, if any, interruptions until a canyon head is reached. In Southern California, one compartment extends essentially from Point Conception to Hueneme and

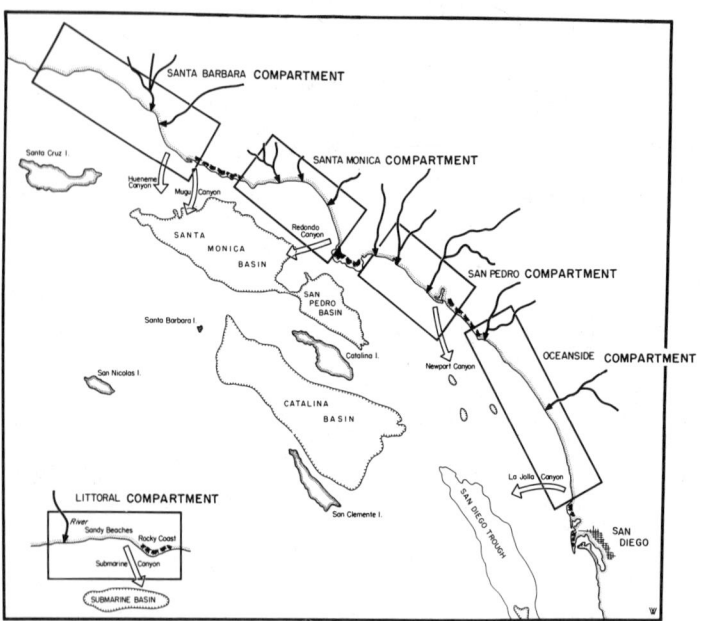

Fig. 7–11 The compartmentalization of beaches along the Southern California coast with loss of sand into the heads of submarine canyons at the southern end of each compartment. From Inman and Frautschy (1966).

Mugu canyons. This compartment receives most of its supply from the Santa Ynez Mountains along the Santa Barbara coast, although, as shown previously, some sand comes from the north around Point Conception (Trask 1952). During times of small runoff, as in recent years, the continental shelf has supplied much of the sand for the beaches (Trask 1955). In the second compartment, from Mugu Canyon to Redondo Canyon, the sand comes largely from the runoff of the Santa Monica Mountains and bypasses Point Dume because the submarine canyon at this point does not extend in very close to the coast.

The third compartment extends from the Palos Verdes Hills to Newport Canyon. Here, a beach exists as a spit for a short distance beyond Newport Canyon, forming Newport Harbor. This spit, according to Inman (personal communication), remains almost static in size because after periods of small runoff of the Santa Ana River, the chief source of supply, most of the sand bypasses the canyon head, whereas after periods of heavy runoff the canyon head receives an excess supply, and hence the beach beyond does not grow. It seems possible that bypassing of the Scripps Canyon head, at the southern end of the next compartment, may be regulated in somewhat the same way, or that even more sand goes past the canyon head during periods of low runoff, and so this also may account for the increased sand on the beach south of Scripps Institution during the drought of recent years.

BARRIER BEACHES AND BEACH RIDGES

The most common type of beach is located along the seaward side of barriers (formerly called *offshore bars*). These are found along most of the lowland coasts of the world. Approximately 47 percent of the United States coasts are bordered by some type of barrier. They are almost continuous along the Gulf Coast from Florida to New York on the East Coast. Other areas with very long barriers include arctic Alaska, Holland, west-central Africa, southern Brazil, and southeastern Australia.

Among important recent reports on barriers are those by Allen (1965A), Bernard and LeBlanc (1965), Bird (1965), van Straaten (1965A), Hoyt (1967), Zenkovich (1967, pp. 236–240), Hails and Hoyt (1968), Weidie (1968), Curray et al. (1969), and Otvos (1970). With this new information, which includes many drilling logs, the knowledge of barriers becomes amazingly complete. We can now test the various hypotheses of barrier origin with ample facts.

The early idea of de Beaumont (1845) was that erosion of the bottom in the breaker zone piled up sand into a ridge at some distance from shore and thus created a bar and later a barrier (Fig. 7–12[?]). Gilbert (1885, p. 87)

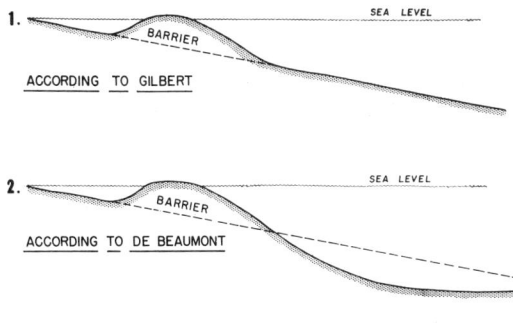

1.

SEA LEVEL

BARRIER

ACCORDING TO GILBERT

2.

SEA LEVEL

BARRIER

ACCORDING TO DE BEAUMONT

Fig. 7–12 The hypotheses of Gilbert and de Beaumont on the origin of barrier islands. Gilbert believed the barrier, introduced by longshore currents, was simply added to the slope; de Beaumont believed that the waves excavated the sea floor on the outside and built the barrier on the inside.

A

B

Fig. 7–13 The development of a new barrier island along the west coast of Florida between 1939 and 1951. Photo *A* by U.S. Coast and Geodetic Survey; photo *B* by D. L. Inman.

was convinced that the barriers were due to longshore drift carrying sediments and depositing them as a spit along the shore (Fig. 7–12[1]). Later, according to Gilbert, gaps were cut through the split, producing barrier islands. D. Johnson (1919, pp. 348–392) favored de Beaumont's hypothesis because his profiles suggested erosion, as in Fig. 7–12[2]. Johnson thought that conditions were made favorable for barrier development largely by the emergence of the inner portion of the very gently sloping continental shelf. As we have seen (p. 106), the emergence idea is opposed by information obtained from virtually all of the drillings into barriers, which suggest that there has been submergence and that the barriers have grown upward with the late Holocene rise of sea level. Hoyt (1967) has offered substantial evidence from drillings into barriers and lagoons, which indicates that most barriers have formed neither according to the de Beaumont nor the Gilbert hypotheses. If either idea were correct, beach and neritic sediments (primarily sand) should have appeared along the former shore in the zone now occupied by the lagoon before the barrier formed. The deposition of mud would have become possible only where there was a protected area on the inside of the new barrier. The drillings give little, if any, indication of such a prebarrier condition under lagoonal deposits. Therefore, Hoyt believed that the barriers were originally beaches or dune ridges that were separated from the shore by regional submergence. This appears to satisfy most of the evidence. However, at least small barriers, mostly arcuate, are occasionally formed by wave action on gently sloping sea floors. This is illustrated in Fig. 7–13, where aerial photos taken in 1939 and 1951 show that a cuspate barrier has formed in the 14-year interim. Also, as Hoyt admits, barrier islands are sometimes formed by spit growth across large bays and subsequent truncation of the spits during storms.

After having been formed by these various methods, barriers are subject to seaward and lagoonward growth. Many barriers have grown seaward, as shown by the numerous ridges on their seaward side (Fig. 7–14). Old maps confirm some of the seaward growth. In areas where beaches are of coarse sand, gravel, or shells, it is possible for a ridge to form along the shore as a storm beach (Fig. 7–15). As an example of how a barrier may grow seaward with a series of storm beach ridges and intervening low ground, we can refer to the Tabasco area of Mexico southeast of Yucatan Peninsula. Here, according to Psuty (1965, 1966), broad summer berms are built out into the Gulf of Mexico, and then winter storms with higher sea level rework portions of these berms into storm beaches landward of the summer beach berms. These ridges often develop on either side of a river channel that supplies the sediment.

If the beach sand is fine, as is true of most beaches, the sand cannot be piled up appreciably above the berm level because storm waves sweep it inland across the barrier. Most of the ridges along barriers of fine sand are probably dune ridges that form just inside many fine-sand beaches. This origin was established by studies along the Texas coast (Shepard and Moore 1955, p. 1562). The sands in the ridges are more typical of dunes than of beaches, as shown by their roundness and their higher content of black minerals (Shepard and Young 1961).

The truncation of barrier beaches, which often occurs during hurricanes, allows sediment to be swept through into the inner lagoon. This has led to the development of fanlike projections into the lagoon, known as *tidal deltas* or *overwash fans* (Fig. 7–16). These protrusions are at first low marshy areas but may be built up to form dry land. Wind deposition has covered some of the marshy areas on the lagoon side of barrier islands (Fig. 7–16).

Fig. 7–14 The Nayarit coast of western Mexico illustrates the seaward progradation of a group of parallel beach ridges and their subsequent truncation (foreground). Photo courtesy of J. R. Curray.

Fig. 7–15 Cobble storm beach on the seaward side of Marblehead Neck, Massachusetts. Photo by author, 1950.

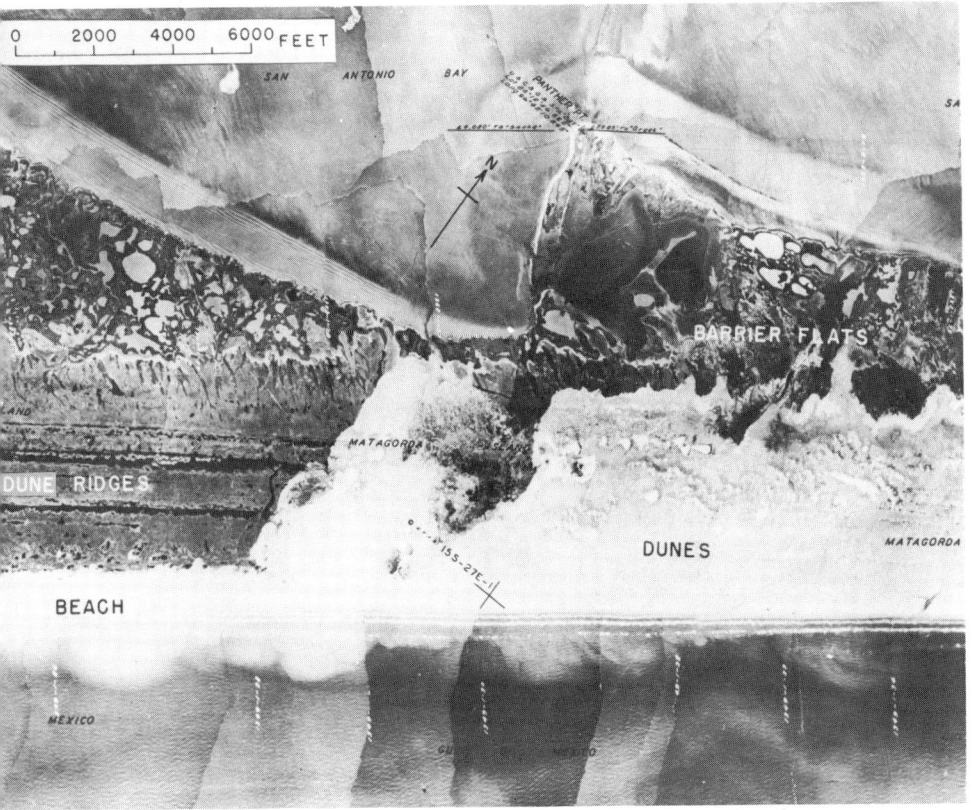

Fig. 7–16 The major physiographic divisions of a large barrier island. The foreground shows the Gulf of Mexico and the background a bay shore along the Texas coast. Two types of dunes are illustrated: stabilized longitudinal dune ridges and transverse migrating dunes. Longshore bars are indicated by the breaker lines, and, on the inside, bars can be seen through the temporarily clear water. The barrier flats are overwash fans. Photo courtesy of Edgar Tobin Surveys.

This growth of the barriers in both directions has resulted in land masses that are often several kilometers wide. Except where there has been an uplift, all of the highest portions of the barriers are dunes.

Barriers are very common around deltas. They form mainly where old deltaic lobes are temporarily abandoned by the active distributaries, and the lobe has been sinking slowly because of compaction and other causes (Russell 1936; Lankford and Shepard 1960). A classic example is the Mississippi Birdfoot Delta where there are extensive barrier coasts both to the west and east. The great arc of the Chandeleurs to the east is the remnant of the St. Bernard subdelta; and Grand Isle, Dernieres, and Timbalier Islands are remnants of the Lafourche subdelta (Kolb and Van Lopek 1958). Extensive barriers are also located along the north coast of Alaska where inactive deltas have had the same subsidence.

Chenier Ridges A special type of beach ridge, called *cheniers* (from the French *chêne*, meaning oak), can be seen along the coastal marshlands of

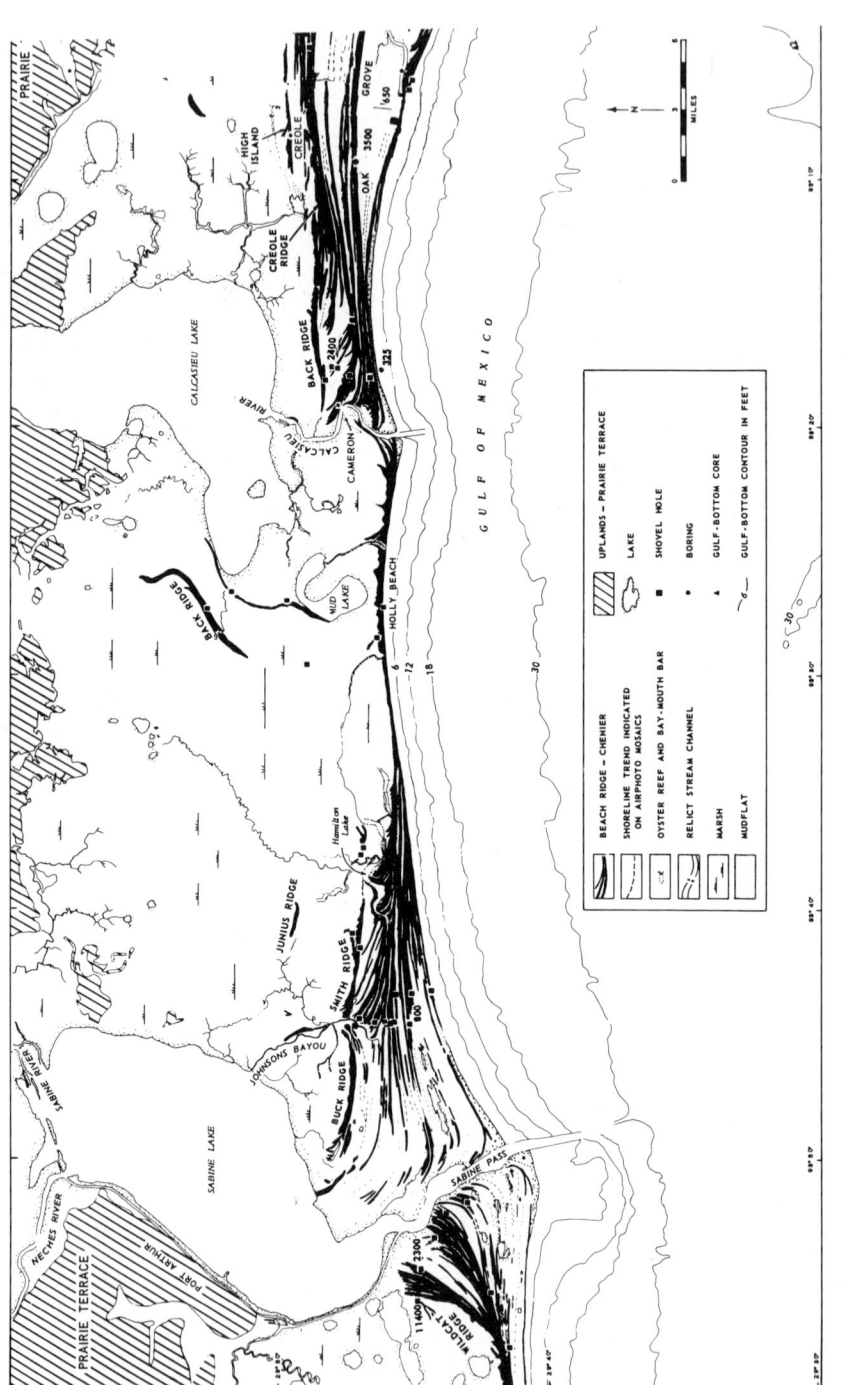

Fig. 7–17 Map showing distribution of cheniers along a portion of the western Louisiana coast. From Gould and McFarlan (1959).

southwestern Louisiana (Fig. 7–17). They consist of sand and shells. Many papers have described them; perhaps the most comprehensive are those of Gould and McFarlan (1959) and Byrne and others (1959). The Louisiana cheniers contrast with ordinary beach and dune ridges in resting on muddy marsh deposits and then being overlapped by more of the marsh deposits. Gould and McFarlan show that they range in age from 2800 to 300 years B.P. (before the present), and hence were formed after postglacial seas had reached close to the present level. When sediment supply from the Mississippi was abundant, the shore built forward rapidly as a marsh-capped mud flat; but when the supply was slight, wave attack cut back these marshes and concentrated the coarse sediment with its abundance of shells into the beach ridges. Because the periods of progradation were greater than those of erosion, the coast advanced during the 2800 years for a total of approximately 16 km, leaving the chenier ridges at intermediate points.

HURRICANE AND TSUNAMI EFFECTS ON BEACHES

Large changes in beaches are produced by the catastrophic waves of hurricanes and tsunamis. The coastal rise in sea level that accompanies the waves from storms and from underwater faulting generally cuts back the beaches. Some low sandy coasts have retreated as much as 1 km during a hurricane. Scarps are frequently cut into the dunes inside beaches. Although the scarps may persist for long periods, the beaches usually are rebuilt within a few months after the largest waves have ceased. The immediate result of catastrophic waves is to concentrate debris on the denuded beaches. Boulders, trees, and blocks of rock are generally scattered along the shore (Fig. 3–10).

Along the East and Gulf coasts of the United States, hurricane surges have cut through coastal barriers at numerous places (Fig. 3–11). These cuts are often sealed off from the ocean within a few months after the storm by new barrier beaches, which usually persist until the next large waves reopen the inlet.

Storm-surge effects have been studied from the unusual storm of January–February 1953 that funneled water into the North Sea, raised the sea level a maximum of 3 m, and flooded the coastal lowlands of Holland and England (C. King 1959, pp. 283–312; Robinson et al. 1953). Along the coast of Suffolk, cliffs 2 m high were cut back as much as 27 m, and cliffs 12 m high regressed 12 m. At the same time, many breakthroughs occurred in beach ridges with flooding of the marshes inside. The effects of hurricanes along the East and Gulf coasts of the United States have been widely studied (for example, J. Morgan et al. 1958, Shepard and Wanless 1971, Chaps. 3–9). The raising of sea level as much as 5 m has caused numerous breakthroughs in barriers and has eroded the dunes behind beaches along the coasts where these storms have occurred. Lower beaches have been built up by some of these hurricanes because of removal of great quantities of sand from high levels followed by deposition at the low levels. In June 1957, Hurricane Audrey striking southwestern Louisiana spread sand from many of the beaches widely over the marshlands and allowed later wave attack to cut back the coast for great distances (J. Morgan et al. 1958). The sheltering effect of well-developed shell-sand ridges in the western part of the storm area prevented important changes, but temporary inlet channels were developed across the ridges that were soon sealed by littoral drift. Large mud arcs were produced locally from the eroded mud-flat material.

CUSPATE SHORELINES

Cusp-shaped points are common along beaches. These sand points grade in size from small beach cusps (discussed on p. 157) to the huge cuspate forelands, like Cape Hatteras. Although somewhat similar processes are involved in cusp formation of all sizes, the problems are rather different, therefore in this section we will consider only the larger features—cuspate forelands, cuspate spits, and giant cusps.

Cuspate Forelands The huge projections, many of them scores of kilometers across, were first described by Gulliver (1896), who gave them their appropriate name. The most remarkable example of a series of cuspate forelands is located along the coast of the Carolinas between Hatteras and Cape Romain (Fig. 7–18) (Shepard and Wanless 1971, Chap. 5). The intervals between these capes are from 110 to 160 km, and they protrude 25 to 40 km

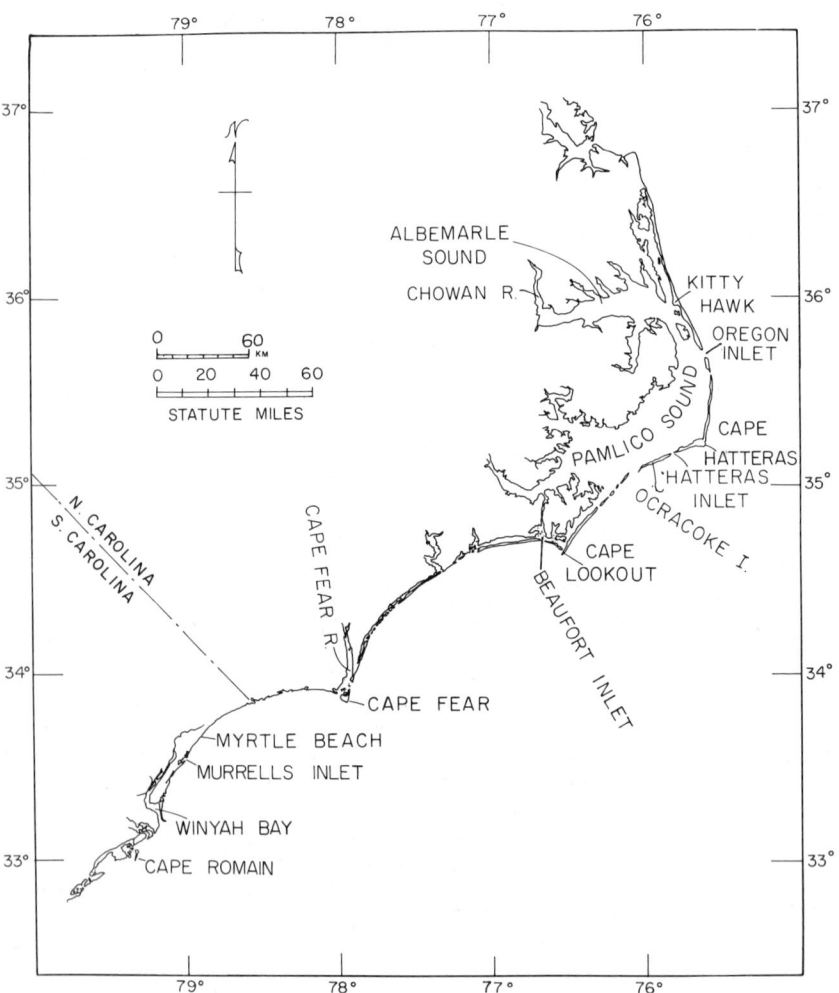

Fig. 7–18 Cuspate forelands and barriers along the Carolina coastal plain.

seaward of a straight line on the inside. Farther south, Cape Kennedy (former-ly Cape Canaveral) is as large as the Carolina examples but is isolated; around the west side of the Florida coast, a series of smaller cuspate forelands cul-minate to the northwest at Cape San Blas, which has major proportions. The only other comparable cuspate forelands exist in Alaska, north of Bering Strait. Here, Cape Espenberg, Cape Krusenstern, and Point Hope are somewhat isolated, and then Icy Cape, Point Franklin, and Point Barrow constitute a group similar to that in the Carolinas from Cape Hatteras to Romain. Within the northern Alaskan group, each point is about 130 km apart. Point Hope (Fig. 7–19) is somewhat unique because it protrudes 25 km from a narrow base. Cape Krusenstern is of special interest because archaeologists, using carbon-14 dating, have traced the history of its beach ridges back for about 5000 years (Giddings 1966) and thus give an idea of how the cuspate foreland grew.

Where soundings are available, one or more elongate shoals have been found extending seaward off the cuspate forelands. The reason for more than one shoal appears to be that shoals, developed at former positions of the point, have been left on the sea floor after the point migrated to a new posi-tion.

Although most cuspate forelands have been built as the progradation of a series of beach and dune ridges, occasionally there are large lagoons

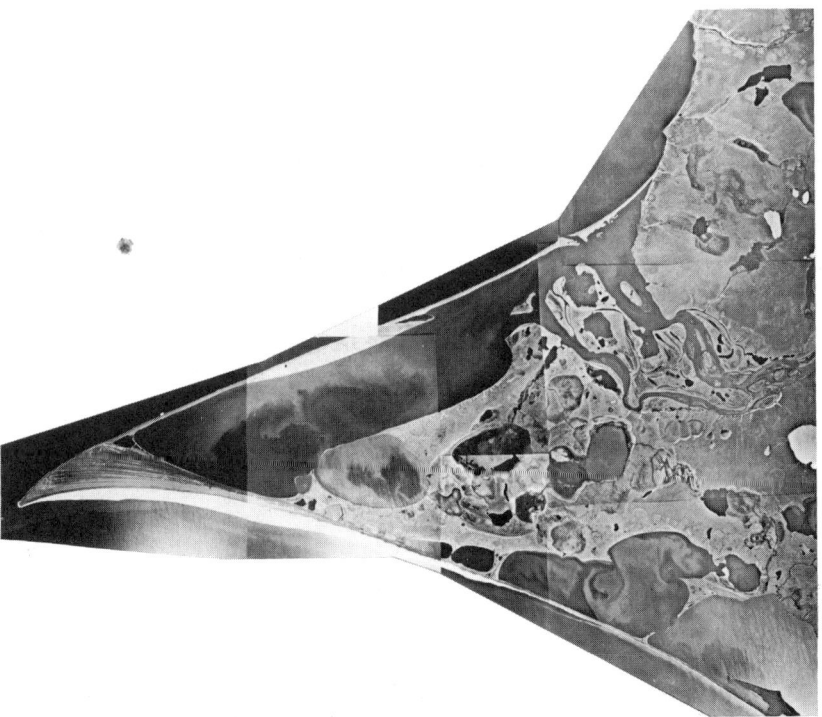

Fig. 7–19 Point Hope, a cuspate foreland protruding into the Chukchi Sea. The Kukpuk River built a delta at this point. Sand barriers grew outside the delta, forming the spectacular point. The spit on the south side shows many parallel beach ridges. From U.S. Coast and Geo-detic Survey.

inside the points, as for example, Cape Hatteras, which shelters the broad Pamlico Sound (Fig. 7–18). Other forelands have marshes or sand plains between the ridges but no open water.

Most cuspate forelands probably are deposits in the slack-water zone between two coastal eddies. The eddies commonly develop inside a major current that flows along the coast, such as the Gulf Stream north of the Straits of Florida. Whether this explanation also applies to the capes north of Bering Strait is not clear, although there is said to be a north-flowing current along the coast with velocities up to 50 cm/sec (1 kn) (McCulloch 1967). Possibly at some time during the past this current may have been more powerful and developed strong eddies. McCulloch observed that the somewhat raised Sangamon interglacial shoreline lacks cuspate forelands. Also, active erosion of the cliffs occurred prior to the growth of the barriers south of Point Barrow. Thus, these cuspate forelands can be considered as the result of a temporary condition.

Some small cuspate forelands are clearly due to progradation of beaches on the inside of small islands or other obstructions. Zenkovich (1967, Fig. 192) shows a cuspate beach on the landward side of a wreck. If such points extend far enough seaward they form a tombolo, or land-tied island. An example is Cape Verde, along the west coast of Africa. Often the tombolo has two arms connecting the island to the shore with open water on the inside, as seen in Mont Argentario, north of Rome.

Cuspate Spits   The cusps that form on both sides of lagoons inside barriers (Fig. 7–20) have been discussed by many authors (particularly, Shaler

Fig. 7–20   Cuspate spits in the east arm of Pensacola Bay, Florida. Note the underwater bars extending out beyond the spits. Cuspate bars are shown in the offshore area (foreground). Photo by U.S. Coast and Geodetic Survey in 1945 (C-3258).

1889; D. Johnson 1925, p. 445; Fisher 1955; Price and Wilson 1956; Zenkovich 1967, pp. 514–524). These spits are somewhat more common on the barrier island side but often are as well developed next to the mainland. Like cuspate forelands, they may have a solid form or may have only barrier beaches with a pond on the inside. Where the water is clear in the lagoon, a submarine ridge can often be seen extending out from the cuspate point.

Johnson explained the cuspate spits on Nantucket Island as the result of the elongation of the barrier spit on the outside with the development of recurved spits at a succession of points. He thought that these hooks were later modified by the eddies of tidal currents. Fisher suggested that the breaching of barriers by storms carries sediment over into the lagoons, forming a deposit on the inside. This deflects the currents moving along the lagoon and results in further deposition. Price and Wilson, however, considered that standing wave oscillations or seiches in a lagoon develop the paired spits at nodal points. Zenkovich attributed the formation of spits to shore drifting processes in the lagoon but thought that they may represent a remnant of slightly lower sea conditions. Each of these more or less overlapping ideas may explain the cuspate spits as different processes operating in different localities. Gierloff-Emden (1961) has observed that all of these lagoonal cuspate spits occur in areas with small tidal range. Aerial photographs of the Nantucket cuspate spit show that their ridges are more indicative of current

Fig. 7–21 Typical giant cusps along the outside of Cape Cod, Massachusetts, near Highland Light.

action in the lagoon than of the other suggested causes. Most of them show growth lines that could not be due to former hooked spits nor to overwash fans, but are suggestive of eddy currents within the lagoon. This agrees with Shaler's very early interpretation of the Nantucket cuspate spits, which he explained as the result of the tide running along the lagoon.

Giant Cusps Along many open beaches, projecting points are spaced at intervals of a hundred or more meters. Elsewhere these cusps occur as solitary points (Fig. 7–21). An explanation for one of these cusps has been determined. Currents diverge from the zone of wave convergence north of La Jolla Submarine Canyon and are turned seaward where they encounter weaker currents that are due to the diagonal approach of the waves away from the head of the canyon (Fig. 3–5). The back eddies from these seaward-moving masses cause deposition that produce a point. Other points are probably related to zones where similarly conflicting currents turn seaward.

EFFECTS OF ENGINEERING STRUCTURES ON BEACHES

The building of jetties or other types of projecting walls to develop harbors along straight coasts has had disastrous effects on many beaches. The usual result is illustrated in Fig. 7–22, where a pair of jetties has been constructed seaward from Lake Worth Inlet at Palm Beach, Florida. The prevailing current is from the north so that the sand has been trapped next to the north jetty and greatly widened the beach. Directly south of the south jetty, the beach has a somewhat normal width because of wave refraction with resulting north drift; but farther south, the beach has been lost and property undermined because of absence of the normal supply of sand from the north. Numerous similar examples have developed in Southern California, notably at Santa Barbara and Redondo. At Santa Barbara (Fig. 7–23), the sand built up along the entire west side of the jetty and then built around the end to form an island in the entrance to the harbor. Pumping this sand has been necessary in order to nourish the beaches to the east, which were being robbed of their supply. The Army Engineers first deposited the sand as a bar separated from the shore by water with a depth of 5 m. This proved ineffective because the sand did not migrate toward the shore, so it had to be dumped in shallow water.

At Santa Monica, a breakwater was built parallel to the shore (Fig. 7–24), hoping that the sand would be carried through the harbor and hence supply the downcurrent beaches. This system did not work because the wave shadow caused by the jetty allowed the sand to build up on the inside, very much like a cuspate foreland landward of an island. This threatened to fill the harbor and caused loss of beach downcoast. Accordingly, pumping had to be resorted to here, too.

The old pier at Redondo Beach, on the south side of Santa Monica Bay, suffered disastrous effects following construction of the breakwater. The pier is located directly inside a submarine canyon head. As at similar locations, the canyon head produces a convergence of waves along the shore to the north and south. When the southward-curving breakwater was constructed north of the canyon head, access of normal beach sediment coming from the supplies to the north was halted. However, the breakwater did not have much effect on the wave convergence just north of the canyon. Consequently, during storm periods the beach was cut back, and no sand was available during quiet periods to allow the beach to be replenished. As

Fig. 7–22 The effects of building jetties at Lake Worth Inlet, east Florida. The current approaches from the north and has built up the sand on that side, whereas erosion is occurring on the south side of the jetty (lower right). Photo by U.S. Coast and Geodetic Survey in 1945 (C-1511).

erosion continued, the city lost an entire block of valuable property in this area. Only after the breakwater had been extended for some distance to the south in front of the eroded section was erosion stopped and the shore stabilized.

Durban, South Africa, is largely built on a beach that apparently grew as a spit in the shelter of a bluff on its south side (Fig. 7–25). The situation is almost identical to San Diego (Shepard and Wanless 1971, Fig. 10–2), with the bluff comparable to Point Loma and the beach to Coronado. The prevailing current is north at Durban, and sand was formerly carried around the point and along the shallow bar at the entrance to the bay then deposited on the shore, due to an eddy that forms in the current lee of the point. However, according to C. King (1953), the deepening of the channel and the building of jetties on both sides prevented the natural supply of sand from reaching the beach; so the beach has retreated extensively and has been preserved only by pumping sand into the area. At San Diego, the situation is

Fig. 7–23 The effect of the Santa Barbara, California, breakwater. Note the large accumulation of sand on the near side of the breakwater due to the wave approach from that side; the sandbar has built into the harbor after the sand has bypassed the end of the breakwater. Photo by J. H. Filloux in 1959.

somewhat different because the chief source of sand is from the south. Although the dominant current is to the south along the Southern California coast, a countercurrent develops along the Coronado Strand, bringing sand north from the mouth of the Tia Juana River to supply the Coronado beaches, a source now partly stopped by river dams.

Not all jetties cause beach problems. If they are put in at points where rocky headlands project on the downcurrent side, no local beaches will be

Fig. 7–24 The parallel breakwater built at Santa Monica, California. The wave shadow has caused deposition of sand on the inside, which threatened to fill the harbor until it was removed by pumping to the beaches on the downcurrent side. Note the groins at various points along the coast. Photo by U.S. Coast and Geodetic Survey in 1934 (E-5753-22).

Fig. 7–25 Durban, South Africa, beach built as a barrier caused by an eddy on the down-current side of a projecting rocky point. Current flows to left (north) along this coast. Situation duplicates Point Loma and Coronado Strand in San Diego. From U.S. Oceanographic Office Chart 5267.

Fig. 7–26   The undermining of beach cottages at Newport Beach, California, as a result of storm surge on October 10, 1934. Note the large rip current carrying sediment seaward from the area being undermined. The short walls extending into the sea are groins. Photo by U.S. Air Force.

ruined, and the effect on beaches considerably farther south will not be very great. This is the situation, for example, at Newport Harbor south of Long Beach, California.

Groins (groynes) (Fig. 7–26) cause much less trouble than jetties. Constructed along many beaches where erosion threatened, they stabilize the sand rather than causing the building out of the broad beaches, characteristic of the upcurrent side of jetties. Beaches downcurrent from groins may have slight losses, but the effects can often be largely offset by erecting groins at these places also. A special form of groin, suggested by Rivière and Laurent (1954), leaves a gap between the inner and outer portions, allowing a flow of current between the two; this may prevent excessive building out of the sand.

BEACH STRUCTURES AND MARKINGS

The structures and special markings characteristic of beaches were described in detail by D. Johnson (1919, pp. 457–550) and more recently have been the subject of much observation and experiment by McKee (1950, 1962), McKee and Sterrett (1961), van Straaten (1953), Potter and Pettijohn (1963), and Rosalsky (1964). Knowledge of these features can prove useful in studying ancient sediments and attempting to determine environments of deposition.

Beach Stratification   In areas where a wide assortment of minerals are available for beaches, stratification is usually visible in trenches cut through the beach (Fig. 7–27). Alternating dark and light colored layers underlying the foreshore are usually due to changes in wave intensity. The dark layers often

Fig. 7–27   Typical beach stratification with dark layers due to heavy minerals. At the right, the disturbance in the stratification is probably produced by a hoofprint of a horse. Foot rule is included for scale.

have an abundance of iron minerals as well as ferromagnesians and are formed as the result of large waves that concentrate the heavier minerals by attrition. Layers consisting largely of mica are indicative of very small wave conditions, which may also concentrate small shells. Small folds in the stratification (Fig. 7–27) may be the result of the escape of trapped air (Stewart 1956) or a hoof print (van der Lingen 1969). Where old berms have been eroded and new foreshore sediments deposited, cross-bedding may develop.

Beach Cusps   Small cusps with points facing the sea and rounded embayments in between (Fig. 7–9A) occur frequently on coarse-sand and gravel beaches, whereas they are more sporadic on fine sand beaches. The cusps are usually rather evenly spaced, separated by distances from about 0.5 m to 100 m. The spacing of cusps is clearly related to the height of the waves when they were produced. This is shown by tracing cusps from the zones of high waves at a convergence along the Scripps Institution beach to zones of low waves at the divergence formed by a submarine canyon (Fig. 3–5). The cusps show a progressive decrease in spacing toward the canyon. It is often possible to find widely spaced cusp remnants on beaches left from storm waves at a high level where normal waves do not reach, whereas the more closely spaced cusps at lower levels are the result of smaller waves. Cusps along the shores of bays and small lakes or the lagoon side of coral reefs, where there is little fetch, are very closely spaced.

Cusps in fine-sand beaches show a strong tendency to develop during neap tide periods, where there is less tidal range, and to disappear during spring tides. On the long beach in the north part of La Jolla, the cusps are most common during the fall when the beach berm is first being cut back.

D. Johnson (1919, pp. 457–486) explained cusps as the result of normal approach of waves to the coast and thought their destruction was caused by diagonally approaching waves. Although this contention appears to be fairly well supported, observations at La Jolla show that intersecting patterns of waves approaching the beach diagonally from both directions may be accompanied by cusp development. Strong longshore drift destroys the cusp shape. The cusps are evidently due to piling up of water on the berm or on the upper beach with backwash concentrated along rather evenly spaced channels. The interchannel ridges are sharpened into points, and the channel floors are rounded by the waves. Directly seaward of the indentations there are usually small rises on the beach where the material from the indentations is deposited. Diagonally approaching waves are much more effective in destroying cusps on fine-sand beaches than on coarse.

Beach Ripples   At low tide, most beaches exhibit ripple marks of several types left by the receding water. These include the following. On the

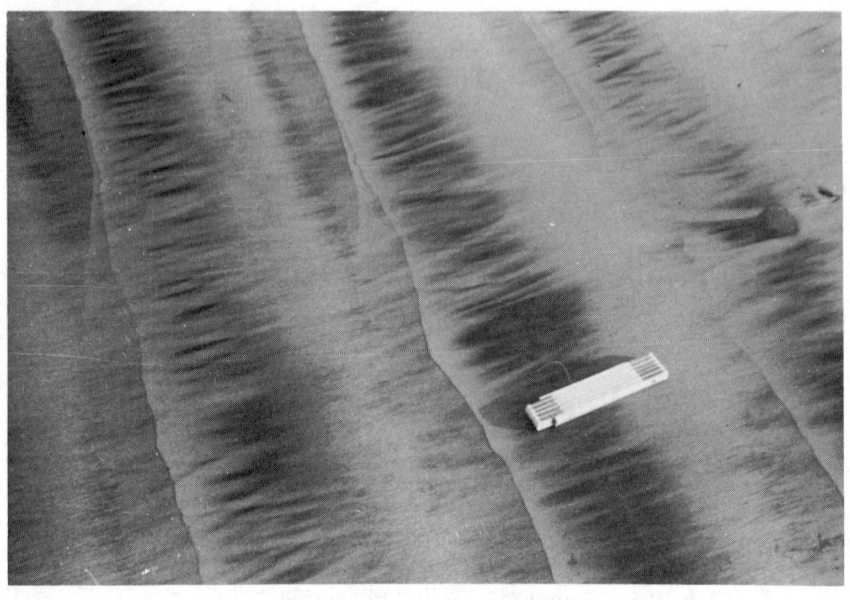

A

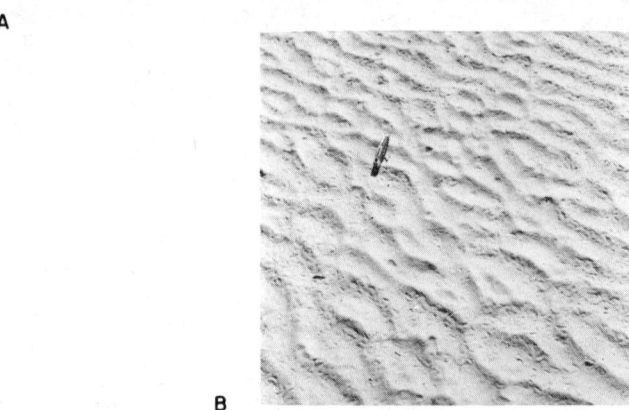

B

Fig. 7–28   A, backwash ripples with a typical 45-cm wavelength. These develop only on fine sand beaches. *B,* typical current ripples produced in a channel along the lower foreshore by longshore currents. Wavelength approximately 7 cm.

foreshore, backwash ripples (Fig. 7–28A) develop on fine-sand beaches as a result of the backwash of the waves setting up a turbulent motion. These ripples usually are about 50 cm apart and extend parallel to the contours of the beach. They are low in amplitude and frequently have concentrations of mica in the troughs, whereas heavy black sands are more common on the crests. At lower levels, where rip current channels are exposed at low tide, current ripples are seen that trend roughly normal to the beach (Fig. 7–28B). These ripples have wavelengths of a few centimeters in fine-sand beaches and up to 30 cm in coarse. In some places, giant ripples (also called *dunes*) that are several feet across are found in the larger channels in areas of strong currents (Fig. 7–28C). Intersecting patterns of ripples are often developed as the result (1) of longshore currents moving parallel to the shore, producing current

C

Fig. 7–28C Giant ripples on one side of a tombolo along the Oregon coast. The wavelength is approximately 2 m. Photo by Ph. H. Kuenen.

ripples, and (2) of normally approaching waves producing oscillation ripples. As indicated by Trefethen and Dow (1960), superposed current ripples may intersect an earlier generation of ripples where one set is formed by a strong flood tide and a superposed set is due to weaker cross drainage currents during the ebb (Fig. 7–29).

Fig. 7–29 An example of cross ripples formed where the strong flood tide has caused the major ripples (roughly normal to the photograph base), and lateral drainage into an estuary during the ebb tide has caused the smaller ripples (roughly parallel to the base). From Trefethen and Dow (1960).

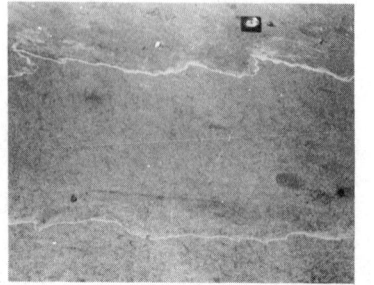

A

B

C

Fig. 7–30 A, swash marks, debris left at the highest point reached by a wave along a beach. *B,* large backwash marks produced by the tsunami of 1946. For scale note the object in the upper right which is a displaced bridge. Smaller backwash marks are shown in Fig. 7–28. Aerial photo by U.S. Navy. *C,* trail left by an echinoid crawling along a beach. Photo by D. L. Inman.

Miscellaneous Features Various other features characterize beach surfaces. Where water has been entrapped at high tide and then returns to sea at low tide, it cuts miniature channels, called *rill marks*. Where the top of a wave comes to a halt, the water sinks into the sand and leaves behind lines of beach debris or mica, called *swash marks* (Fig. 7–30A). These are destroyed by each wave that reaches a higher level. Another type of marking develops where the backwash encounters obstacles, such as pieces of gravel, shells, or occasionally the projecting antennae of little burrowing sand crabs (*Emerita*). These obstacles divide the flow and often leave streaks of dark sand in a diagonal pattern, called *backwash marks* (Fig. 7-30B). Where the sand is very porous, pockets of air often form. The air may coalesce and raise the overlying sand surface into a minute dome, called a *sand dome* (Emery 1945). Erosion of the top of these domes often develops ring structures because of the alternating black and white sand layers. Where the air has escaped, small holes with minute craterlike rims can be seen.

Organisms traversing or burrowing under beaches make distinctive markings. The craters made by digging crabs are particularly common in the tropics. Sand hoppers make raylike patterns (Emery 1944), and trails are caused by invertebrates moving along the beach (Fig. 7–30C). Shore birds also leave tracks and may leave small holes as evidence of their digging for worms.

Special features sometimes develop along foreshore slopes and beach scarps (Rosalsky 1964). *Sand drips* are rounded masses that are the remnant of a sand flow on the slope when the backwash water became absorbed into the underlying sand. These are usually not preserved. Rosalsky also describes *patterned sand,* alternating firm and loose sand areas of rectangular shape on beach scarps due to "windows of loose dry sand in the wetted and firm surfaces" left by thin sheets of water rolling off a slope and wetting only portions of it.

Numerous other features of interest can be discovered on beaches. Many of them have not yet been described or explained.

Chapter 8
DELTAS, LAGOONS, AND ESTUARIES

INTRODUCTION

Just as beaches are related both to the land and to the sea, deltas, lagoons, and estuaries are also marginal between the two. Deltas, mainly low marshy areas that are slowly subsiding, become flooded by the sea when rivers cease to provide sediments to keep them built above sea level. There have been unfortunate failures by many authors to differentiate between lagoons and estuaries. For geologists, the easiest and most natural distinction can be made by classifying *lagoons* as the shallow bodies of water extending along the inside of barriers (Fig. 8–1A, B) and *estuaries* as the bays that deeply indent the coastline because of drowning of river valleys (Fig. 8–1C). Wide river mouths may be included in the latter. Thus, Chesapeake Bay is an estuary and the long body of water inside Padre Island in southern Texas is a lagoon. Some embayments are a combination of lagoon and estuary, as for example, San Antonio Bay in Texas (Fig. 8–13A).

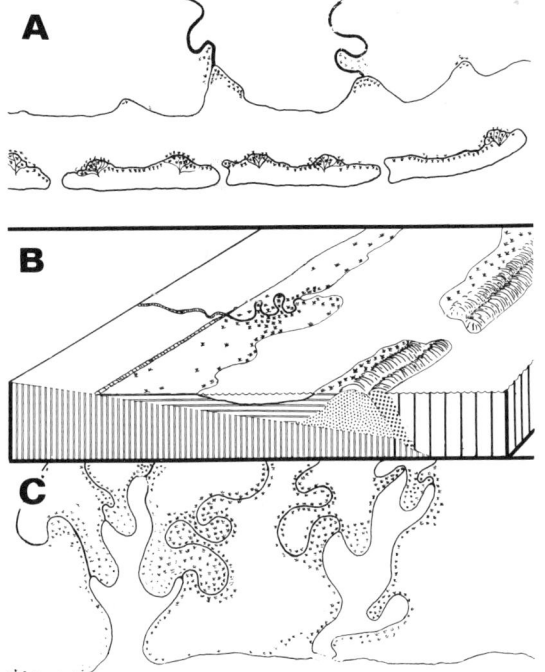

Fig. 8–1 A, typical lagoon. B, cross section of lagoon and barrier island. C, typical estuaries.

This distinction between estuary and lagoon is not entirely in agreement with the definitions given elsewhere (Hedgpeth 1967; McHugh 1967; Fairbridge 1968), because these authors use the original Latin *aestus,* meaning boiling and hence tidal effect, as an important part of their definition along with a mixing of salt and fresh water due to the tides entering a bay. However, a very similar tidal effect takes place in both an estuary and a lagoon (as defined here). For example, Bruun (1969) describes the tidal effects for both types of bays without differentiation. We certainly should not limit an estuary nor a lagoon to a body of water with a salinity of water 0 to 35°/00, as suggested by McHugh. The use of the name estuary for glacially excavated fiords is debatable because the fiords are, in a sense, drowned valleys; but they have characteristics and conditions very distinct from unglaciated drowned river valleys. The very deep water of the fiords and usual lack of strong entrance currents and of barrier spits at their mouths make tides far less significant and often result in stagnant euxinic basins within the fiords. Therefore, only shallow bays with narrow entrances will be considered in this chapter. Fiords are related to glacial troughs of the open shelf and will be discussed in Chapter 9.

The three main features of this chapter have been covered in recent symposia. Modern and ancient deltas were the subjects of a 1965 symposium in New Orleans (J. Morgan, ed., 1970). Coastal lagoons were featured in a 1967 conference in Mexico City (Castañares and Phleger, eds., 1969). Estuaries were the subject of a conference held in 1964 at Jekyll Island, Georgia (Lauff, ed., 1967). Also, the engineering aspects of estuaries were discussed in another symposium (Ippen, ed., 1966). Both estuaries and lagoons were covered

earlier in *Treatise on Marine Ecology and Paleoecology* in an article by
Emery and Stevenson (1957).

DELTAS

Deltas were given their name by the Greek Herodotus because of their
roughly triangular shape. For generations, deltas have been recognized as the
locus of very active deposition, particularly in bays protected from the attack
of the waves of the open sea. Many bays are being filled rapidly by the
incursion of deltas, necessitating frequent remapping. Along unprotected
coasts with low-energy waves, deltas are building seaward wherever a large
supply of sediment is being introduced, as at the mouth of the Mississippi.
Most large deltas that are not building into enclosed estuaries and lagoons
are located at heads of broad embayments, such as the Niger in the Gulf of
Guinea, the Ganges in the Bay of Bengal, the Indus near the head of the
Arabian Sea, the Rhone in the Gulf of Lyons, and the Yukon in Norton Sound.
Other large deltas are building along the open arctic coasts where the pro-
tection of the ice pack allows forward building.

Mississippi Delta   The Mississippi River has built one of the largest
deltas of the world and is certainly the best known (Gould 1970; J. Morgan

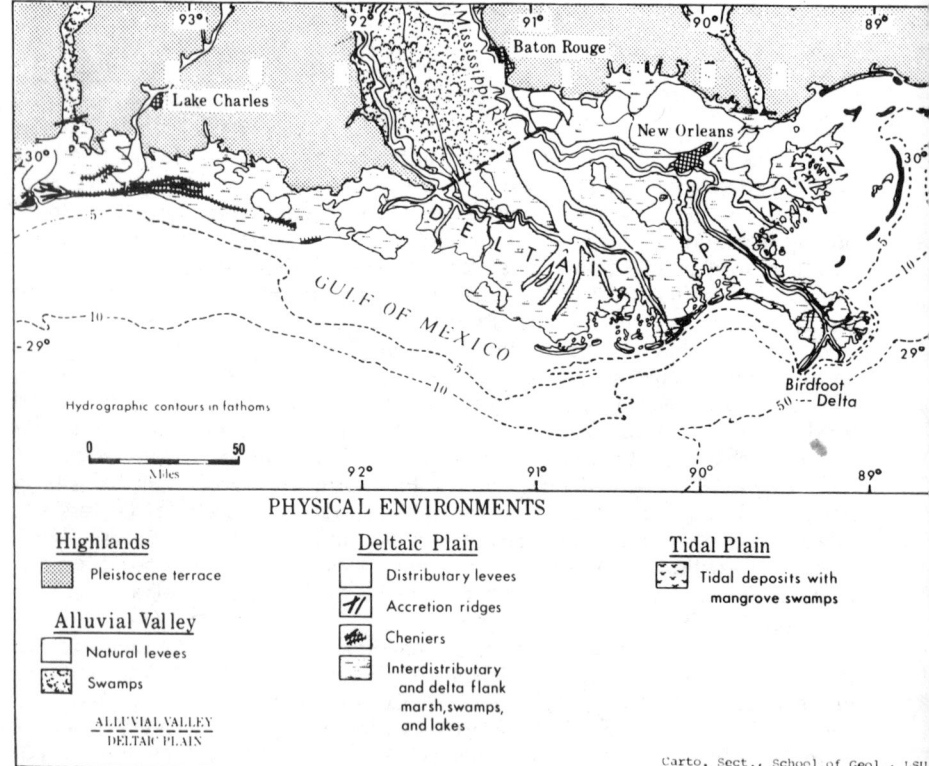

Fig. 8–2   The major environments of the Mississippi Delta. From J. Morgan (1970).

1970; Fisk 1952; Fisk et al. 1954; Fisk and McFarlan 1955; Kolb and Van Lopik 1958; Scruton 1960; Shepard 1960; Coleman and Gagliano 1964; Russell 1967; Shepard and Wanless 1971, Chap. 8). In the early stages, the delta was initiated at the head of the present alluvial valley near Cairo, Illinois. At the lower end of this plain is the postglacial deltaic complex of numerous distributaries consisting of old abandoned mouths and of the Birdfoot Delta, which is still actively building into the Gulf of Mexico (Fig. 8–2). The Atchafalaya, the main distributary, is depositing most of its sediment into large deltaic depressions. The nature of the land in the alluvial valley and the deltaic plain is indicated in Fig. 8–2. It consists mostly of swamp and marsh with almost the only dry portions found on the natural levees formed by overflow from the distributary channels. The river meanders across this swampy channeled area.

The Mississippi Delta has developed as a result of overlapping and laterally displaced sediment lobes that are slowly subsiding (Fig. 8–3). Where the distributaries have been abandoned, as have the Cocodrie, Sale Cypremont, Teche, and St. Bernard, much of the delta is below sea level and only portions of the levee are left, except where the waves have built up the sand from the old subdeltas into barriers, like the Chandeleur Islands and Grand Isle at Barataria Bay.

The development of what has been called *bar fingers* (Fisk et al. 1954; Fisk 1961) is illustrated in Fig. 8–4. Crescentic sand bars form at the mouth of the distributaries and are subsequently breached during flood stages and

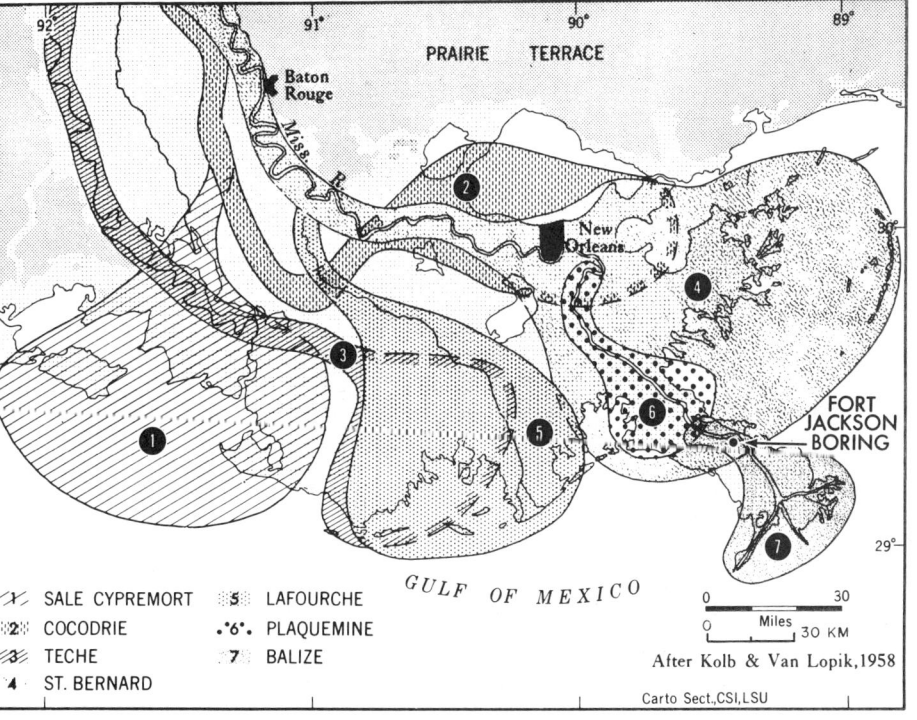

Fig. 8–3 Succession of lobes of the Mississippi Delta. The Balize lobe is also called Birdfoot Delta. From Kolb and Van Lopik (1958A).

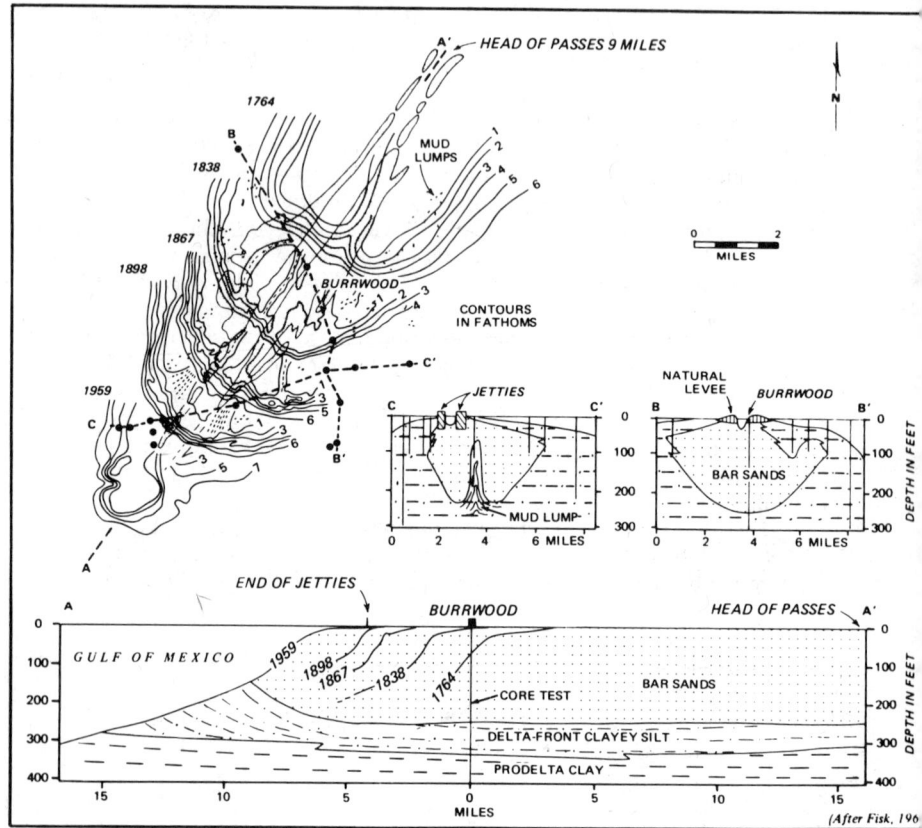

Fig. 8–4 Historic development of the bars at the mouth of Southwest Pass. The depth of the "bar fingers" in the profile is questionable. See text. From Gould (1970).

new bars form farther seaward. In this way, the bar crest at Southwest Pass has advanced 15 km in 195 years. Sampling during the American Petroleum Institute Project 51 showed that sediments with high sand content along the front of the bar only extend down to about 8 m (Shepard and Lankford 1959). This suggests that the 60-m depths for the bar sand shown in Fisk's profiles are open to question, and a series of borings into levees conducted by Project 51 failed to find any such thick sand fingers, nor do the borings shown by Fisk confirm his contention. It seems much more likely that local relatively thick sand masses are formed by point bars, such as exist in the present deep channels, especially on the inside of bends. Other thick masses of sand may represent the barrier islands of earlier delta stages (Lankford and Shepard 1960). However, relatively thin bar fingers may exist along the advancing channels, and local thickening of the sands may be due to compaction of the underlying muds (Gould 1970). The advance of a pass and its later abandonment should leave important deposits of relatively clean sand of river point bars and bar mouth origin with a shape roughly like that of the bar fingers.

The Mississippi Delta is unique in having built local lobes completely across the comparatively wide continental shelf, so that the mouths of the

Birdfoot Delta are now pouring sediment onto the continental slope. This has only been true for a short period. About 1500 years ago, the main mouth of the river was at Head of Passes, where three large distributaries diverge (Fisk 1961). Despite this rapid growth, the entire mass of the Mississippi Delta has retreated more than it has advanced since accurate charts have become available (J. Morgan 1963).

Niger Delta Another of the large deltas of the world has been built by the Niger River of West Africa. Although built into the open sea, the Niger Delta represents a type that contrasts in many ways with the Mississippi (Fig. 8–5). Instead of having the various subdelta lobes, the Niger Delta has a single symmetrical arc with concentric sediment facies (Allen 1965B, 1970). The river has a group of distributaries that appear to have built forward into the Gulf of Guinea at about the same rate and to have advanced the margin of the continental shelf so that it conforms with that of the Delta margin. Thus, this can be considered a classical delta, like the Nile. Otherwise, the Niger Delta has many of the same facies as are found in the Mississippi and includes channels, levees, swamps, bay-mouth barriers, and barriers at the margin (Fig. 8–5). Studies by Weber (1971) show that the Delta has growth faults caused by gravity of sedimentation.

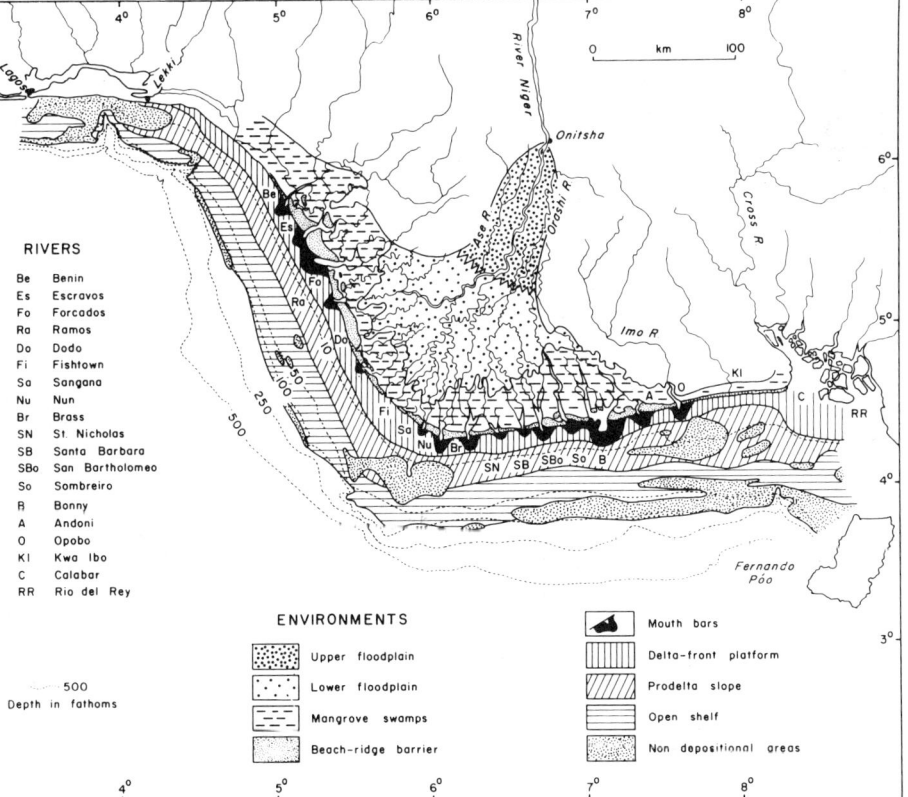

Fig. 8–5 The arcuate Niger Delta with concentric zones built around the margin. See also Fig. 9–36. From Allen (1970).

Ganges–Brahmaputra Delta   Another large delta is located at the combined mouths of the Ganges and Brahmaputra Rivers in Bangladesh and India (Fig. 8–6). The rivers in the broad flood plain differ from the Mississippi in having a braided pattern, and, unlike both the Mississippi and Niger, in having a broad tidal plain, the result of the high tides at the head of the Bay of Bengal. This plain is subject to extensive flooding by storm surges, causing great loss of life to the fishing population living on the higher sand bars. The tides scour out deep channels and produce much overlapping of sediment on the inside. Huge quantities of sediment are brought in by the rivers from the Himalayas and other high mountain ranges to the north. Many of the braided channels in the flood plain have been abandoned, leaving large masses of sand. The huge city of Calcutta is located in one. According to J. Morgan (1970), old channels should be ideal places for oil reservoirs if such deposits have been preserved in ancient rocks. The Ganges plain is subject to tectonic activity and small recent faults are found, some of them producing basins (J. Morgan and McIntire 1959).

The Ganges–Brahmaputra Delta is apparently not growing appreciably, and probably most of the sediment is being carried into the Bay of Bengal (see p. 395).

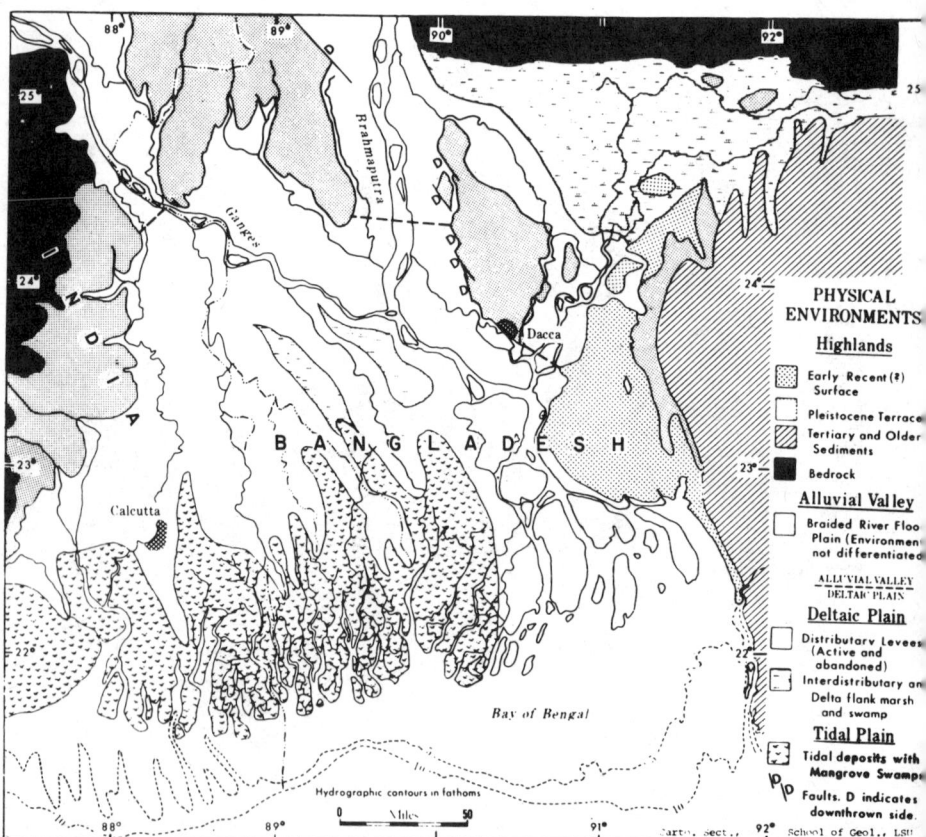

Fig. 8–6   The environments of the Ganges–Brahmaputra Delta. From J. Morgan (1970).

Orinoco Delta   The largest delta in South America is at the mouth of the Orinoco (Fig. 9–20). According to van Andel (1967), the Orinoco Delta has been built mostly of fine-grained sediment from a broad lowland area and has had large contributions from the north-flowing North Equatorial Current, which brings an abundance of fine sediment from the Amazon. The northern, actively building portion of the Delta advances by accretion of mud flats and a few developments of sand cheniers. The mud-flat growth is possible because of the absence of large waves. Farther south, the older abandoned mouths remain as gaping estuaries, which receive their sediments from the east. The inner delta has an alluvial plain of sand and clay, and the outer portion, covered with marsh and swamp clays, has a scattering of cheniers around all except the south margin.

Van Andel (1967) and van Andel and Sachs (1964) have traced the postglacial history of the Orinoco Delta and its relation to the Gulf of Paria (Fig. 9–20). The modern delta appears to have been initiated about 8000 B.P., developing the Amacuro plain to the southeast and advancing at about 2 km/century. The main river mouth now enters the Gulf of Paria and, according to van Andel, should be filling that body of water in about 5000 to 10,000 years.

Yukon Delta   The combined Yukon and Kuskokwim deltas extend for about 800 km along the Bering Sea coast of Alaska and receive the major drainage from that mountainous state. The Yukon Delta has been studied in part by the U.S. Geological Survey (Hoare 1968; Hoare and Condon, 1966), but it is not as well known as the previously discussed deltas. Like the Mississippi, the Yukon greater delta appears to have developed a series of different lobes that changed position frequently. A sequence suggested by Shepard and Wanless (1971, p. 461) is (1) main mouth at the present Kuskokwim Estuary, (2) mouth shifted to Baird Inlet, with Kuskokwim mouth beginning to sink and form an estuary, (3) new mouths forming on both sides of the Askinuk Mountains, (4) new lobe developing on the south side of Norton Sound, forming the present drainage system (Fig. 8–7). In building this large deltaic plain, several mountainous islands became connected to the mainland. Aerial views show many interesting features of the Delta (Shepard and Wanless 1971, Figs. 14.6 to 14.8). The myriads of lakes related largely to permafrost and thawing are a characteristic of the area, differing from those of other more temperate deltas. The oldest charts of the area, dating as far back as 1898, show little change in coastline, in contrast to the Mississippi Delta growth.

Po Delta   The Po Delta is interesting because it is being built into the shallow protected head of a gulf and has a large load of relatively coarse sediment coming from the Alps. The work of Nelson (1970) provides recent information concerning the growth of the Delta. Historical records, which are very complete for the area, indicate that the margin has advanced as much as 25 km since 1000 B.C. (Fig. 8–8). The modern works of man have caused much faster building today than in the past. This growth has occurred despite a sinking of the area that reaches a maximum of 13 cm/yr at the apex of the Delta, decreasing on either side but amounting to 2 mm/yr at Venice, which is located inside a barrier 50 km north of the Delta. A salt wedge intrudes under the fresh water at the mouth of the river, like that at the Mississippi passes, and extends up the river at high tide but is flushed out at low tide. This plays an important part in sediment deposition. The sediments are predominantly sand and silt.

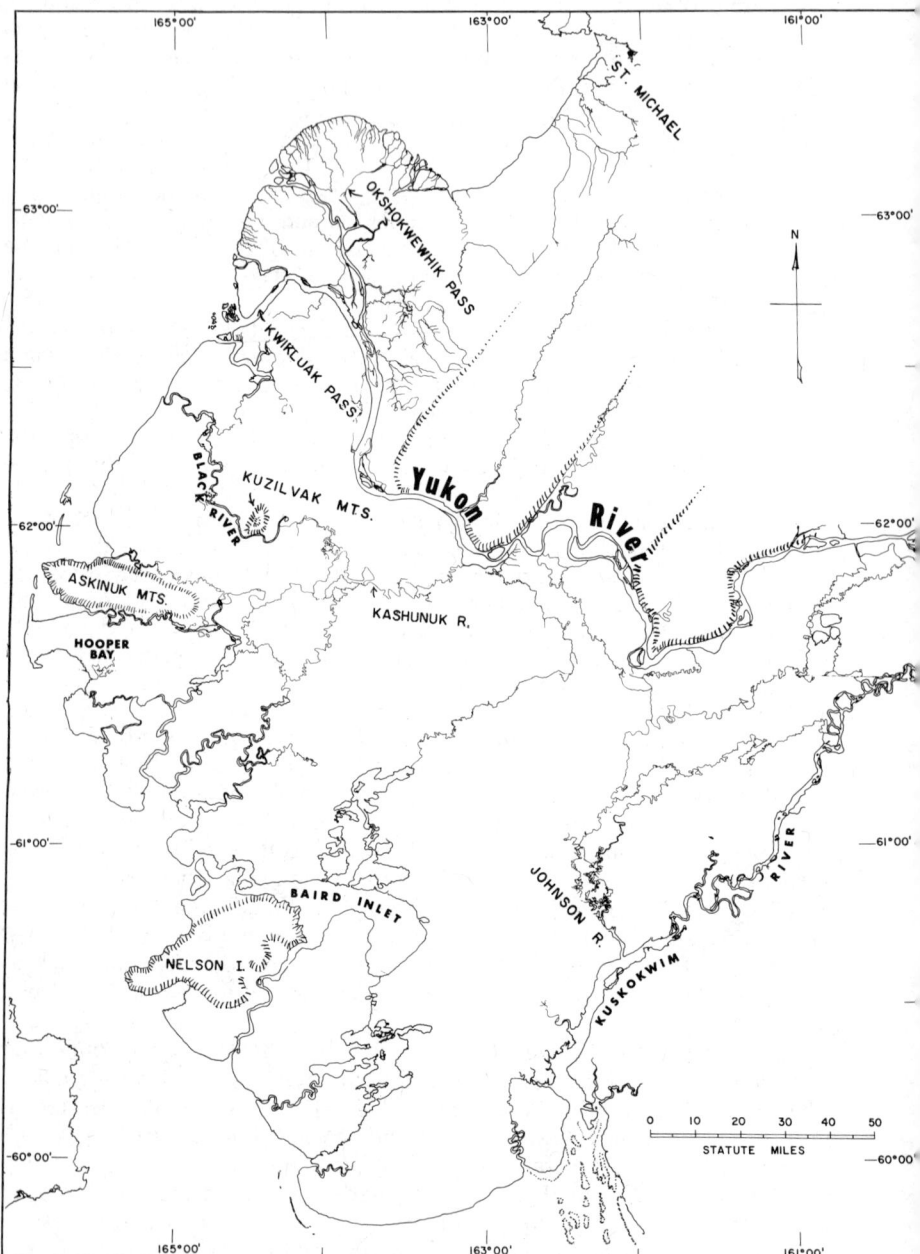

Fig. 8–7 Outline map of the Yukon and Kuskokwim deltaic plain with suggested sequence of drainage changes of the Yukon River (A to L), and present river drainage into Norton Sound to the north. The early outlets appear to have been along the Kuskokwim, which now enters an estuary. From Shepard and Wanless (1971).

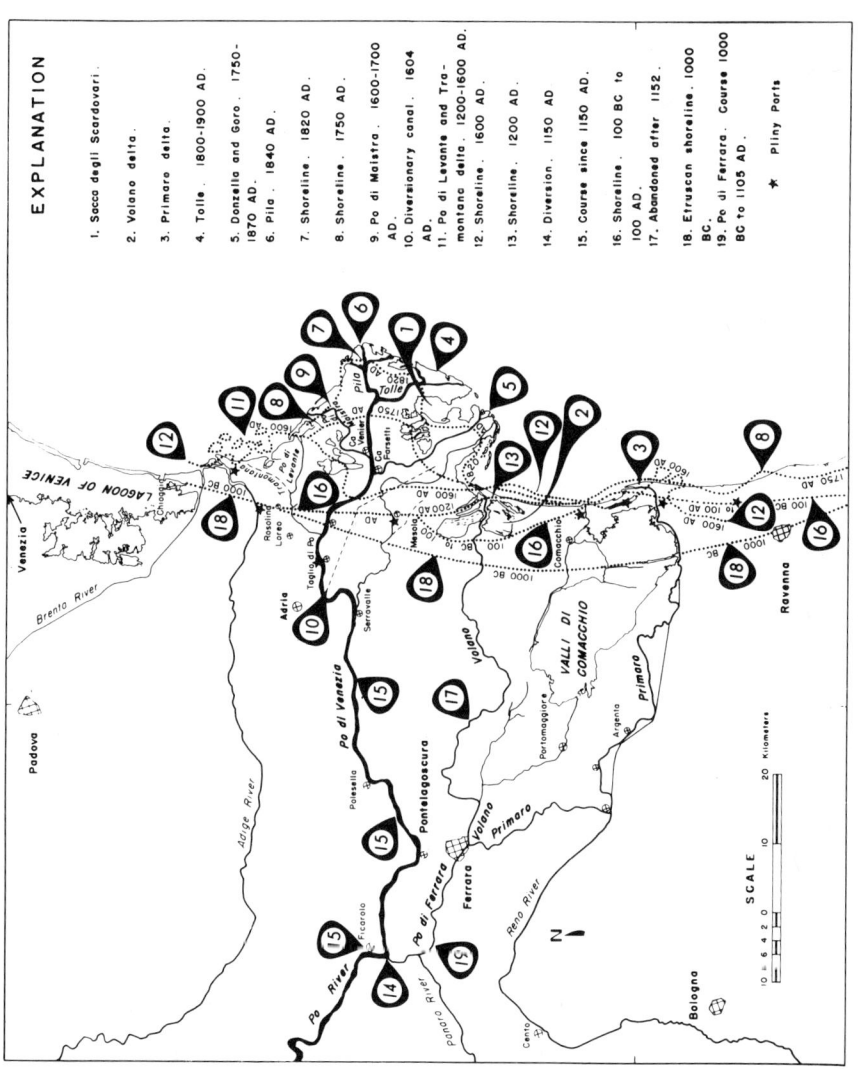

Fig. 8–8 Showing the history of the Po Delta since 1000 B.C. Old shorelines are dated approximately. From Nelson (1970).

Colorado Delta of Texas   Spectacular delta growth often occurs where a river empties into a well-protected lagoon (Fig. 8–9). Thus, the Colorado, the fourth largest river of Texas, built its delta almost entirely across the 7-km-wide Matagorda Bay between 1929 and 1956 (Bouma and Bryant 1969; Kanes 1970; Shepard and Wanless 1971, p. 236). Prior to several hundred years ago, the mouth of the Colorado was farther north and joined with the Brazos River in filling a large estuary near the present city of Freeport. It then developed a new mouth and entered a small estuary in Matagorda Bay; normal growth was impeded by a logjam that was broken up by a flood in 1929, starting the rapid growth. In 1936, the U.S. Army Corps of Engineers cut a canal across Matagorda Peninsula, and the Colorado now enters into the open Gulf of Mexico, where delta building has not occurred because of sufficient beach drifting to bypass the entering sediment.

The sediment in the rapidly growing delta was found to consist mostly of sand at the base of the fill with gradual increase of silt and of wood fibers near the surface (Shepard and Moore 1960). The sandy base was due to the rapid flow when the logjam broke up, and the increase in silt and wood fibers shows the more usual sediment of rivers crossing a broad plain.

Alaskan North Slope Deltas   One of the longest deltaic coasts in the world consists of a series of compound deltas along the Arctic Coastal Plain of northeastern Alaska and continuing into Yukon Territory with a total length of about 900 km. An example is seen in Fig. 6–11, showing the delta of the Colville, the largest of the North Slope rivers (for other examples, see Shepard and Wanless 1971, Figs. 2.5, 14.28). The Brooks Range is the source of most of these rivers. The deltas have masses of distributaries and are highly braided. Many of the distributaries have been abandoned, and barriers have formed in front of their mouths as the result of the occasional summer storms that produce erosion of the deltaic lobes, concentrating the sand along the coast. The active streams flow only during a few months of the year (Arnborg et al. 1967), and about half of the flow of the Colville River was measured by Arnborg during a few weeks. The pack ice is piled against the coast except during a few summer months, but part of the time the water of the large rivers flows under the ice pack into the Arctic Ocean. Large permafrost polygons are common on the surface of the deltas.

General Characteristics of Deltas   The preceding description of some of the large deltas of the world has consisted largely of their sediments, their surficial characteristics, and their rates of growth. Almost all of these large deltas have grown rapidly, although some of them are not now advancing appreciably. Most of the deltas are sinking, either because of compaction of their sediments with their high water content or because they are formed in areas of subsidence. As a result, delta formations alternate between subaerial and submarine facies.

As most deltas build forward, they develop foreset beds that slope outward from their source. The larger deltas have mostly low inclinations in their foreset beds, commonly from 0.5 to 3°, but small deltas built into steep-sided lakes may have foreset beds that are as steep as 30° or even 40°. Such steep beds are usually much deformed by slumping. Creep or sliding also occurs on many of the gentler slopes when deposition has been rapid (Terzaghi 1956).

On top of the foreset beds, braided and meandering streams deposit essentially horizontal formations. These are likely to be sand in the channels and silt and clay in flood plain overflows.

In almost all deltaic sediments, one finds a great abundance of wood

Fig. 8–9 Stages in growth of the Colorado Delta (Texas) across Matagorda Lagoon. Rapid growth caused by breaking of logjam. From Shepard and Wanless (1971). Outline from 1930 shown by black line.

fibers, mica, and ferruginous aggregates. The levees may have considerable
sand, but most of them consist of finer sediments. In the channel sand bars,
cross-bedding is common. The sand bodies should provide excellent reservoir
sands for petroleum accumulation, particularly as they are usually covered
with fine sediments when the channels shift and the sea encroaches onto the
deltas.

COASTAL LAGOONS

The importance of shallow lagoons inside sand barriers can be appreciated when it is realized that about 47 percent of the coasts of the United
States are bordered by such barriers with lagoons on the inside, although the
lagoons are in various stages of filling and usually have broad marshes.
Lagoons are particularly common along lowland coasts subject to low or
relatively low-wave energy. The Gulf of Mexico is a prime example (Fig. 8–10)
with several lagoons extending for more than 200 km. Lagoons are also very
common along the central part of the West African coast.

In the very complete coverage of coastal lagoons (Castañares and
Phleger 1969), the bays discussed fit mostly into the definition given here

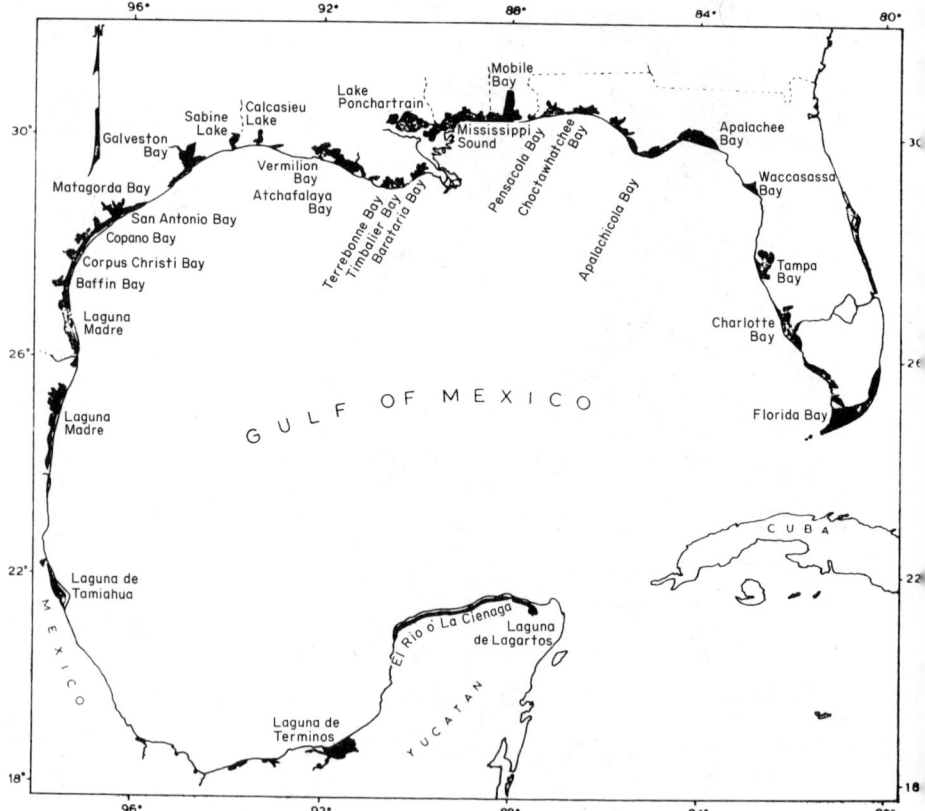

Fig. 8–10   The extensive lagoons of the Gulf of Mexico. From Gunter (1969).

(Fig. 8–1A), although a few bays that indent the coast (called *inner lagoons*) might come under the definition of estuaries, even though they are protected by the same type of barrier as the adjacent lagoons. Also included in the lagoon volume is the type of lagoon inside encircling atoll reefs, which is discussed in Chapter 12 covering coral reefs.

Laguna Madre One of the most studied lagoons in the world is Laguna Madre, which extends for 200 km from Corpus Christi Bay, Texas, to the Rio Grande deltaic plain (Fig. 8–11) (Fisk 1959; Rusnak 1960). The central section is usually dry but is covered with water during northers. Ordinarily, the lagoon has a short northern arm and a long southern arm. The lagoon sediments show the influence of a semiarid climate and of overwashes from the narrow barrier island. The numerous borings give a clear idea of the

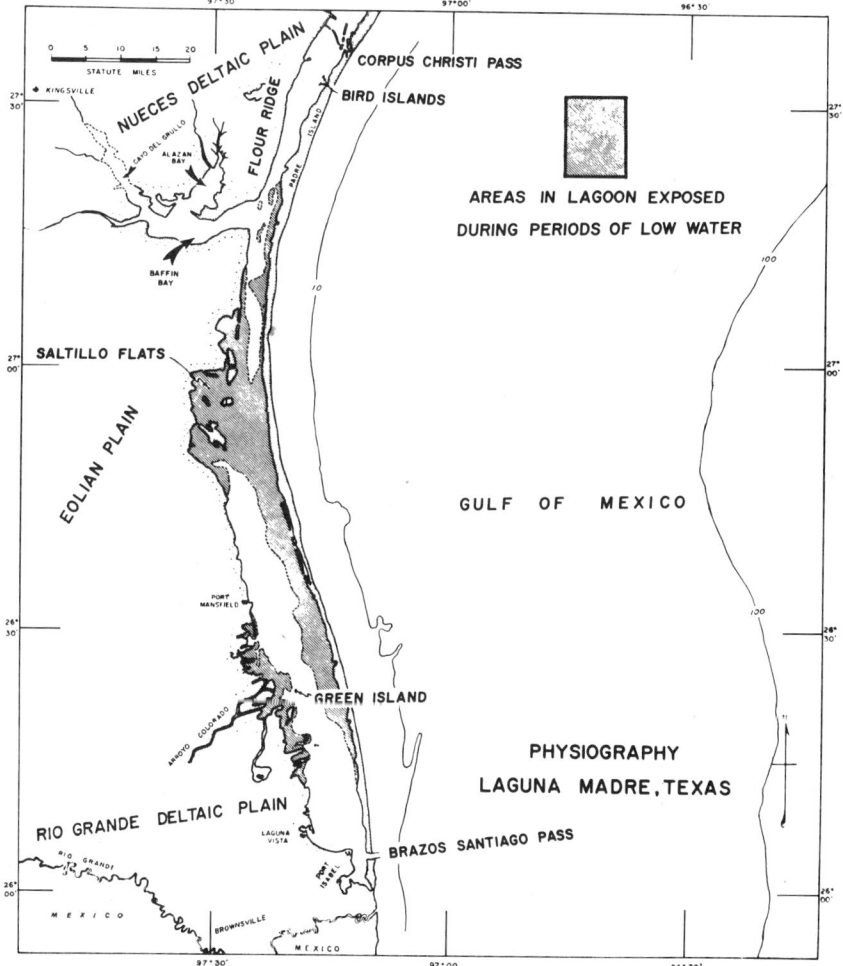

Fig. 8–11 Physiographic divisions of Laguna Madre, Texas, with normally exposed flats in the middle. From Rusnak (1960).

sedimentation history in late glacial and postglacial time (Fig. 8–12). The bay deposits are alternations of clays and sand with abundance of shallow-water organisms, except in the upper layers where the connections with the Gulf had become restricted and salinity had increased because of evaporation. Gypsiferous deposits and oölites began to form at this time, and washover sand from the barrier island became more important, being ultimately replaced by dune sand next to the barrier island.

The recent lagoon deposits show a variation in structure with distance from the inlets at the two ends. Near the inlets, the sediments are mottled due to benthic fauna mixing, whereas at greater distances in the high-salinity portions, they are laminated because of lack of benthic fauna (Rusnak 1960). In the very shoal or exposed middle section, the surface is predominantly an algal mat. This covers much of the barrier-flat sands and is found interbedded with the underlying sands. Large drying cracks develop in the algal mat surfaces (Fisk 1959). The mats, usually brown, turn red when covered with water on rare occasions.

Lagoons of Central Texas Coast From Corpus Christi north to beyond Galveston, a series of lagoons extend inside the almost continuous barrier islands. Despite their proximity, these lagoons contrast with Laguna Madre because they are in a more humid climate with progressive increase in rainfall to the north (Shepard and Rusnak 1957). Also, unlike the Laguna Madre, they are all fed directly or indirectly by entering rivers. As a result, the salinity is mostly below that of the open Gulf. The bays around Rockport (Fig. 8–13A) have been studied in detail (Shepard and Moore 1955, 1960) as a part of the American Petroleum Institute Project 51. The sediments differ considerably from those of Laguna Madre. The deeper Rockport bays have silty clay sediments with silty sand near the inlets where the tides have brought in the sand from the barrier islands and shallow-water Gulf. The sediments are rarely laminated, except near river mouths, but show a mottling because of the stirring of the actively burrowing benthic organisms. Extensive oyster reefs are found in the bays, making navigation difficult, even in small boats, because the reefs grow almost to the surface. The bay sediments differ from those of the open Gulf by lack of glauconite and echinoid fragments, which are both common in the nearby shelf sediments. Study of the coarse fractions of the bay sediments showed that various distinct environments exist (Fig. 8–13B). Thus, the plant fibers near the river mouths, the shell content near the oyster reefs, and the high content of foraminifera in the central bays away from oyster reefs are characteristic. The bay foraminifera are largely benthonic and there are far fewer species than on the open shelf.

Rates of sedimentation in the bays are determined partly by comparison of old sounding surveys with new (Shepard 1953). An average shoaling of 32 cm per century is indicated for central and northern Texas. A much faster rate occurs near the deltas. The actual rate of bay filling is different from that shown by depth changes as the muddy sediments undergo compaction and there is some evidence of bay subsidence, although this may be only local and caused by the removal of oil and water from the underlying layers.

Barataria Bay, Louisiana An example of a lagoon inside the barrier islands around a submerged delta lobe is afforded by Barataria Bay, on the west side of the Birdfoot Delta (Krumbein and Aberdeen 1937; Krumbein and Caldwell 1939). Here the coarsest sediments (3.3 phi) were found in the deep entrance channels where the currents are strongest, and the finest sediments (6 phi) in the zones fringing the low marshy islands within the Bay. The organic carbon content showed an inverse relation to the diameters of the

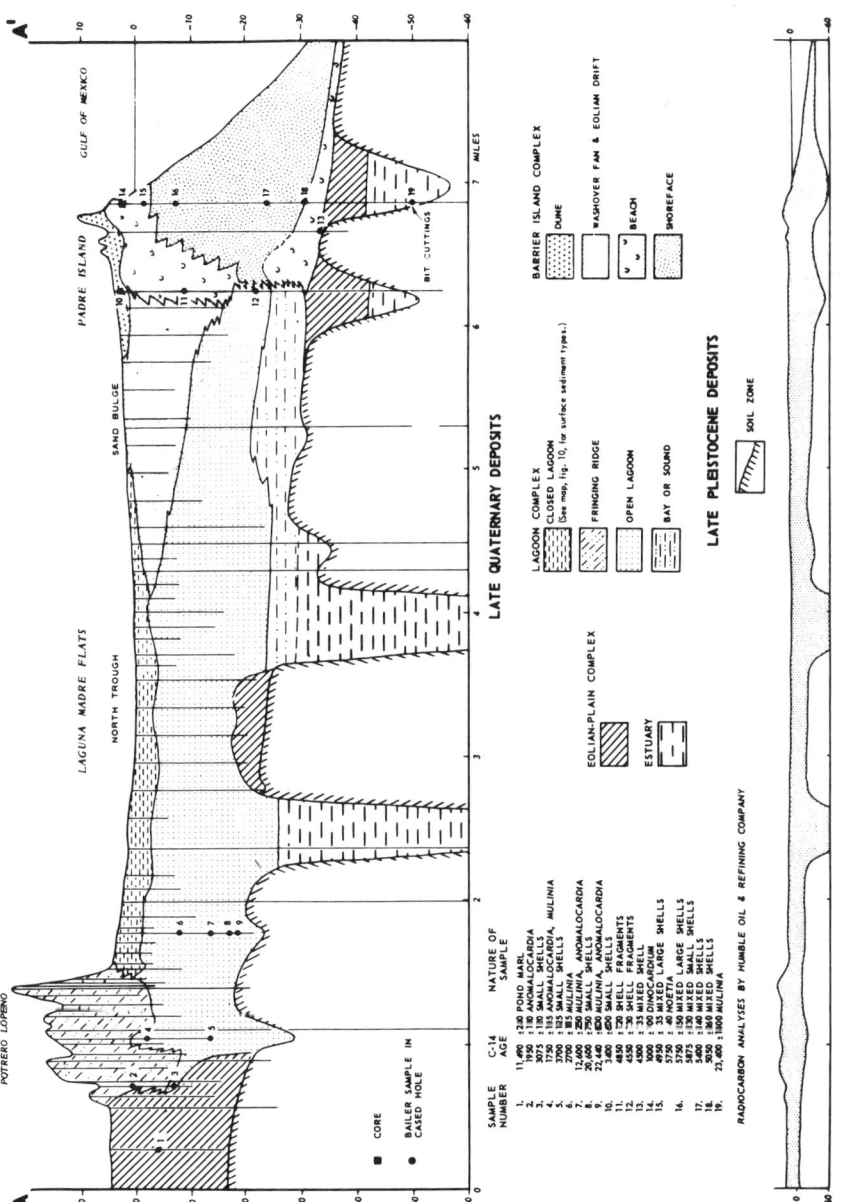

Fig. 8–12 Section showing history of sedimentation in central Laguna Madre, Texas (latitude 27°). From Fisk (1959).

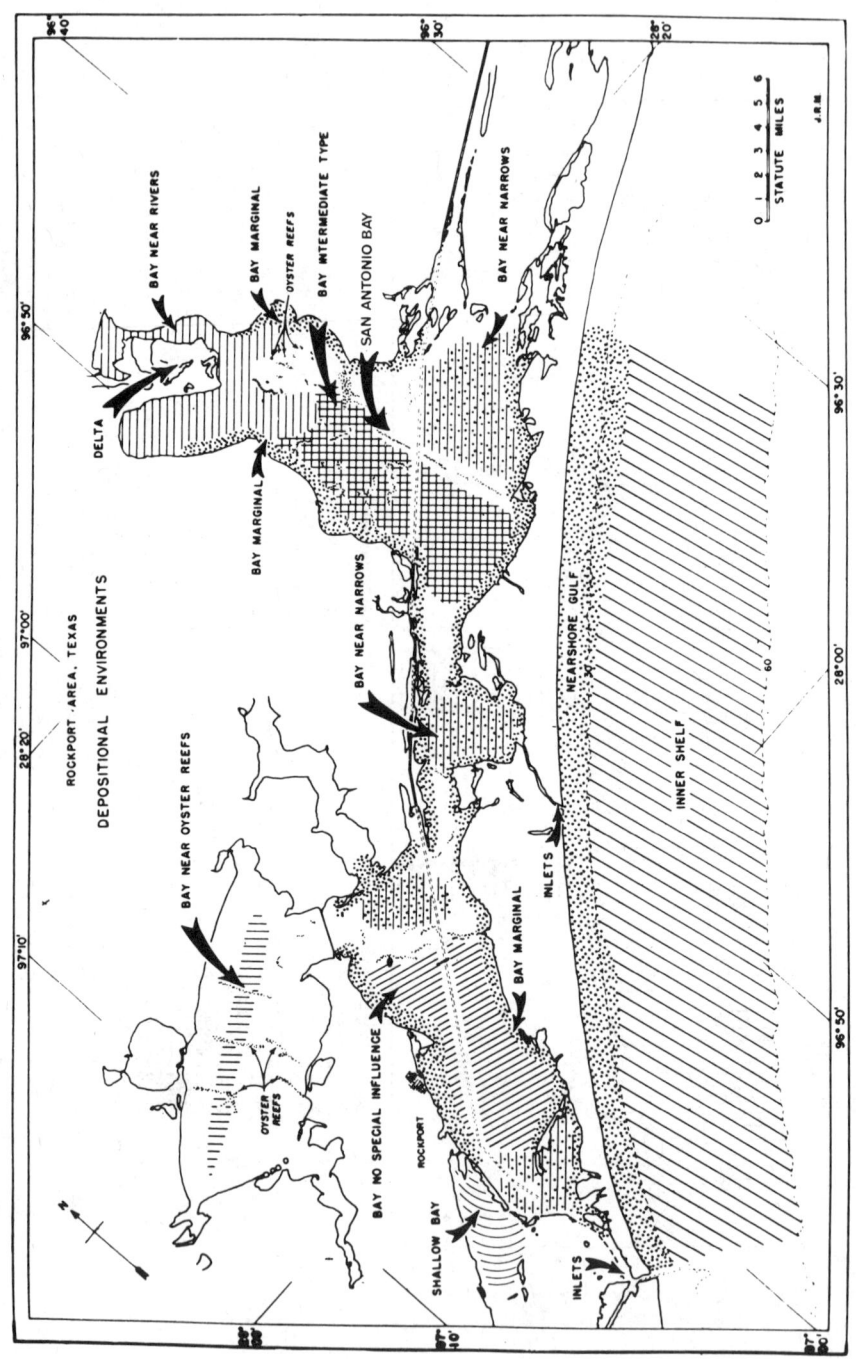

ROCKPORT AREA, TEXAS

DEPOSITIONAL ENVIRONMENTS

BAY NEAR RIVERS

DELTA

BAY MARGINAL

OYSTER REEFS

BAY INTERMEDIATE TYPE

SAN ANTONIO BAY

BAY NEAR NARROWS

BAY MARGINAL

BAY NEAR NARROWS

BAY NEAR OYSTER REEFS

OYSTER REEFS

BAY NO SPECIAL INFLUENCE

ROCKPORT

SHALLOW BAY

BAY MARGINAL

INLETS

INLETS

NEARSHORE GULF

INNER SHELF

N

0 1 2 3 4 5 6
STATUTE MILES

J.R.M.

A

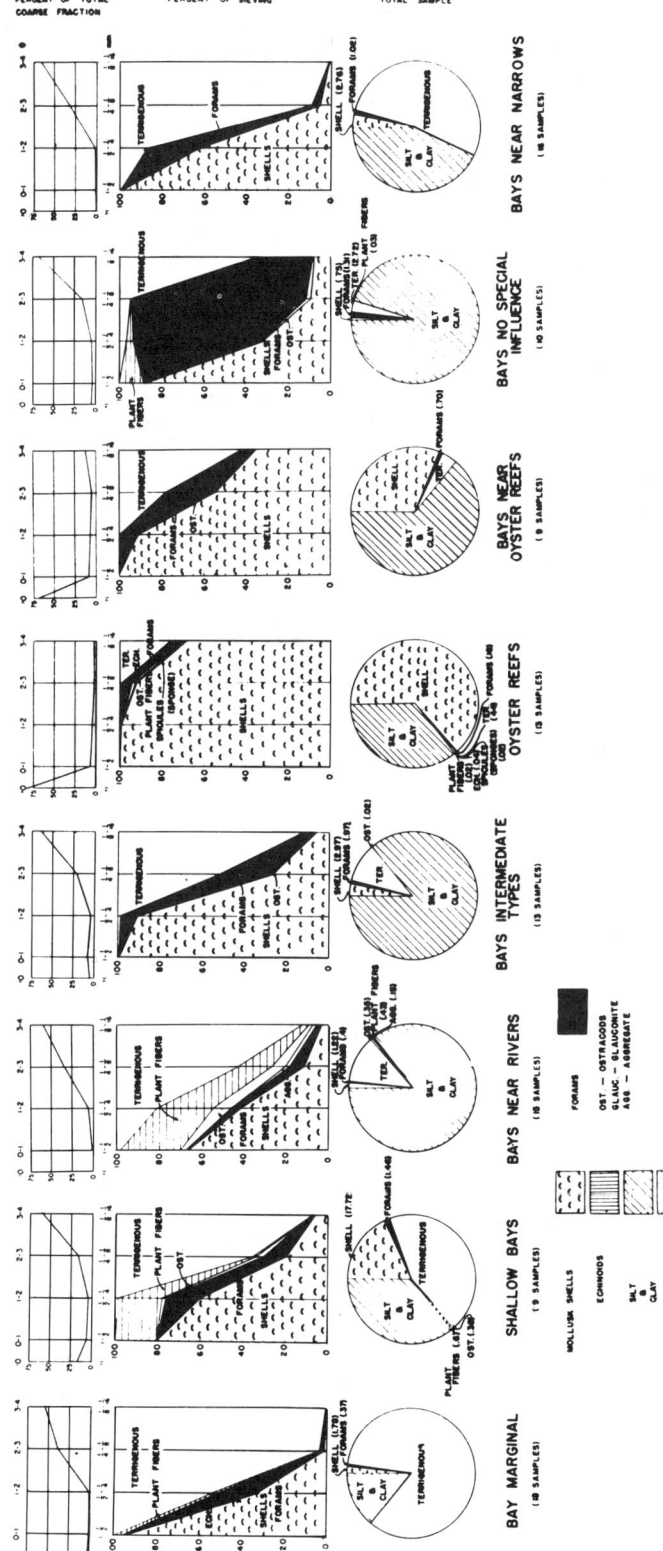

B

Fig. 8–13 A, distinctive environments of deposition in lagoons of central Texas coast. From Shepard and Moore (1955). B, coarse fraction and sand-silt-clay contents of sedimentary environments along the central Texas coast. From Shepard and Moore (1955).

sediment so that it was highest near the marshy islands and lowest in the channels.

Lagoons of Northwest and West Florida A group of lagoons extends along the coast of northwest Florida from Pensacola east to Carrabelle, although some of the lagoons near Panama City, Florida, have been filled. Typical of these are the lagoons inside the barrier islands near the mouth of the Apalachicola Delta (Gorsline 1963A; Kofoed and Gorsline 1963). The depths are all shallow except in the inlets. The floors of the bays slope toward the barrier islands. Like the Rockport bays, these Florida bays have an abundance of oyster reefs. The sediments of Apalachicola Bay and vicinity (Fig. 8–14) show the importance of a river mouth that brings in the fine silt sediment and of the inlet and barrier islands that introduce the sand.

The lagoons along the West Coast of Florida are typified by Charlotte Harbor and vicinity (Huang and Goodell 1967). Here the lagoons are protected by a series of barriers that have grown to the west. The sediments are largely terrigenous sand that decreases in grain size going up the harbor, and their coarsest elements are in the main inlet. Carbonate content, mostly detrital, decreases in an upbay direction, whereas organic carbon increases in the same direction.

Florida Bay Along the south coast of Florida, the sand barriers of the Miami area grade into a coral reef barrier, the Florida Keys (Ginsburg 1956, 1957; Gorsline 1963A; Scholl 1966). Florida Bay, inside the Keys, is essentially a lagoon in its upper end, although it opens out to the west. The floor of the bay is limestone partly covered by anastamosing mud banks consisting of calcareous silts with abundant shell fragments. In between the banks are basins with depths to about 2 m. The banks are exposed at low tide and have a cover of mangroves that locally form marshy islands. Colonizing of the mud banks by marine grasses and mangroves has given them considerable stability. The banks began to grow about 4000 years ago and hence the rate of deposition of this calcareous deposit has been about .04 mm/yr.

Gulf of Batabano, Southwest Cuba Another area of lime deposition in a lagoon is found in the Gulf of Batabano.[1] This has been studied by the Jersey Production Research Company (Daetwyler and Kidwell 1959). The shelf here has a width of 125 km, a length of 270 km, and a depth averaging about 8 m, with many banks and shoals rising to the surface. Along the outside is an almost continuous rim of shoals. These constitute coral reefs to the east, and the large Isla de Pinos lies along the outer shelf in the south center of the bay.

The sea bottom is largely covered with calcareous sediments despite the supply of terrigenous material from the entering streams. According to Daetwyler and Kidwell, the water is carried into the bay from the east by the wind-induced currents and is warmed over the banks, causing precipitation of $CaCO_3$. This produces an ovoid nonskeletal carbonate with a local zone of oölite. These chemical precipitates grade into a zone with relatively high skeletal content and locally into calcareous muds. Along the shores of the mainland there are zones high in molluscan fragments. The grain size in the calcareous sediments show no relation to depth of water nor to distance from the coast. Locally along the north side of Isla de Pinos there are small patches of quartz sand coming from the metamorphic rocks of the island. The calcareous sediments are not very thick, and local borings have shown that

---

[1] This is the "Bay of Pigs" where the unfortunate landing attempt was made in 1961.

SEDIMENT TYPE

SAND
SILTY SAND
SHELLY SAND
SILTY SHELLY SAND
SANDY SHELL GRAVEL
SHELL GRAVEL
SANDY SHELLY SILT
SANDY SILT
SILT

STATUTE MILES
0 1 2 3 4 5

Fig. 8–14 Sediments of Apalachicola Bay and vicinity. From Kofoed and Gorsline (1963).

the underlying platform of limestone contains large cavities, apparently the result of solution during a low sea-level stage of the Pleistocene.

Lagoons of Central East Coast, United States Pamlico Sound is the largest lagoon on the East Coast of the United States. It includes estuary indentations on the landward side but is essentially a lagoon because it is bounded seaward by barriers, including the Cape Hatteras cuspate foreland (Fig. 7–18). A study of 500 samples from the lagoon (Pickett and Ingram 1969) indicates that the sediments are largely fine sand, although silt and clay predominate in the deeper portions. Using coarse-fraction analyses, Pickett and Ingram show the general character of various environments (Fig. 8–15). The sediments of these environments greatly resemble those of central Texas lagoons. Water depths, sediment source, and wave and current action are the controlling factors. The sediments are lenticular and mottled. The importance of organism reworking is evident in most of the bay.

The eastern half of Delmarva Peninsula, which forms the east flank of Chesapeake Bay, is a mass of lagoons with deep tidal channels, marshes, and inlets, all fronted on the Atlantic side by broad sand barriers. Wachapreague Lagoon, in the Virginia area of the Peninsula, has been investigated (Newman and Munsart 1968). The borings in the area show that the lagoons have been in existence for the past 4400 years and the barrier islands for at least a thousand years longer. Boring through the rhizome mat of the salt marsh showed gray shallow-bay sediments beneath. The lagoonal sediments, as elsewhere, increased in thickness to the southeast, but nearing the barrier islands, the sand made drilling and probing difficult. In some places, there was an indication that with depth in the cores, the rhizomes change character to a possible sedge peat, indicative that submergence is more rapid than sedimentation; nearer the surface the condition is reversed. Newman and Munsart believe that the deposits indicate general crustal submergence in addition to eustatic sea-level rise.

Lagoons of Western Baja California The lagoons along the central coast of western Baja California have been studied partly because of their economic importance coming from the mining of salt (Phleger and Ewing 1962; Phleger 1965, 1969A). Guerrero Negro Lagoon and its neighbors are protected from the open Sebastian Vizcaino Bay by sand barriers (Fig. 8–16). The tidal inlets allow a large supply of sand to be introduced, and the intense evaporation in this arid region concentrates the salts in the inner lagoons and produces the wide salt flats at their heads. The bulk of the fine sand in the lagoon is introduced by wind from the barriers. The lagoon is a mass of channels with steep margins separated by extensive tidal flats. The inlet to Guerrero Negro Lagoon has migrated northward 3 km during the past thousand years. As in Texas lagoons, the protective barrier island has grown seaward along with the latest rise in sea level.

In the adjoining Ojo de Liebre Lagoon, marine evaporites are being deposited in diked basins. Aragonite forms in brine pans during the daytime because of the high pH that results from photosynthesis of actively growing algae. At times, wind transports brine into the inner basins, resulting in halite deposition. The amazing feature of the evaporites and brines in this area is that they support large standing crops of organisms.

Lagoons of the California Coast Many of the lagoons of the Southern California coast have been changed rather extensively by man in relation to harbor-building projects, particularly in the Long Beach–San Pedro area. One lagoon that has remained somewhat in its natural condition is at San Diego, where the Coronado barrier has produced an elongate bay. The natural

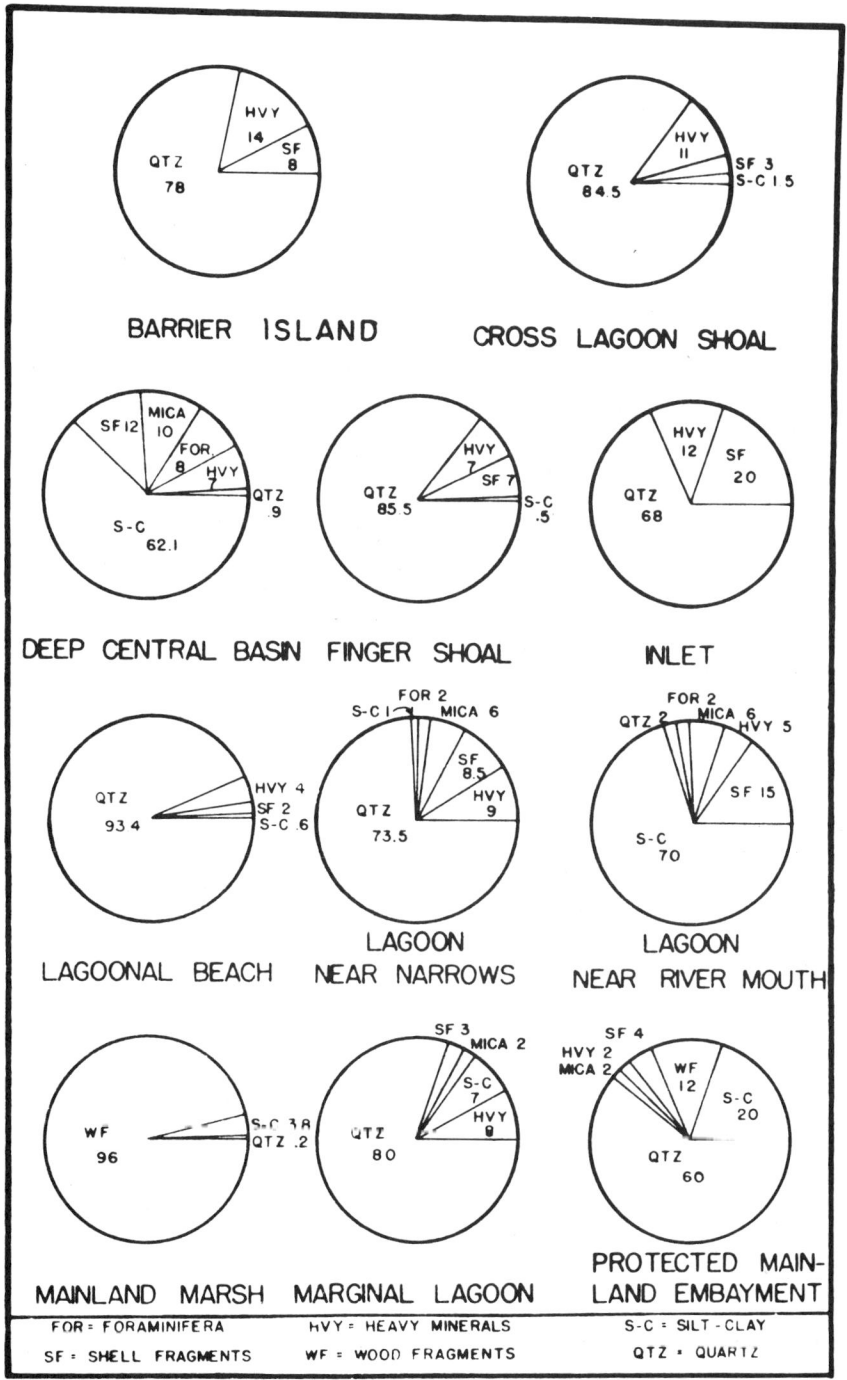

Fig. 8–15 Sediment types of Pamlico Sound, North Carolina, determined by coarse-fraction analyses. From Pickett and Ingram (1969).

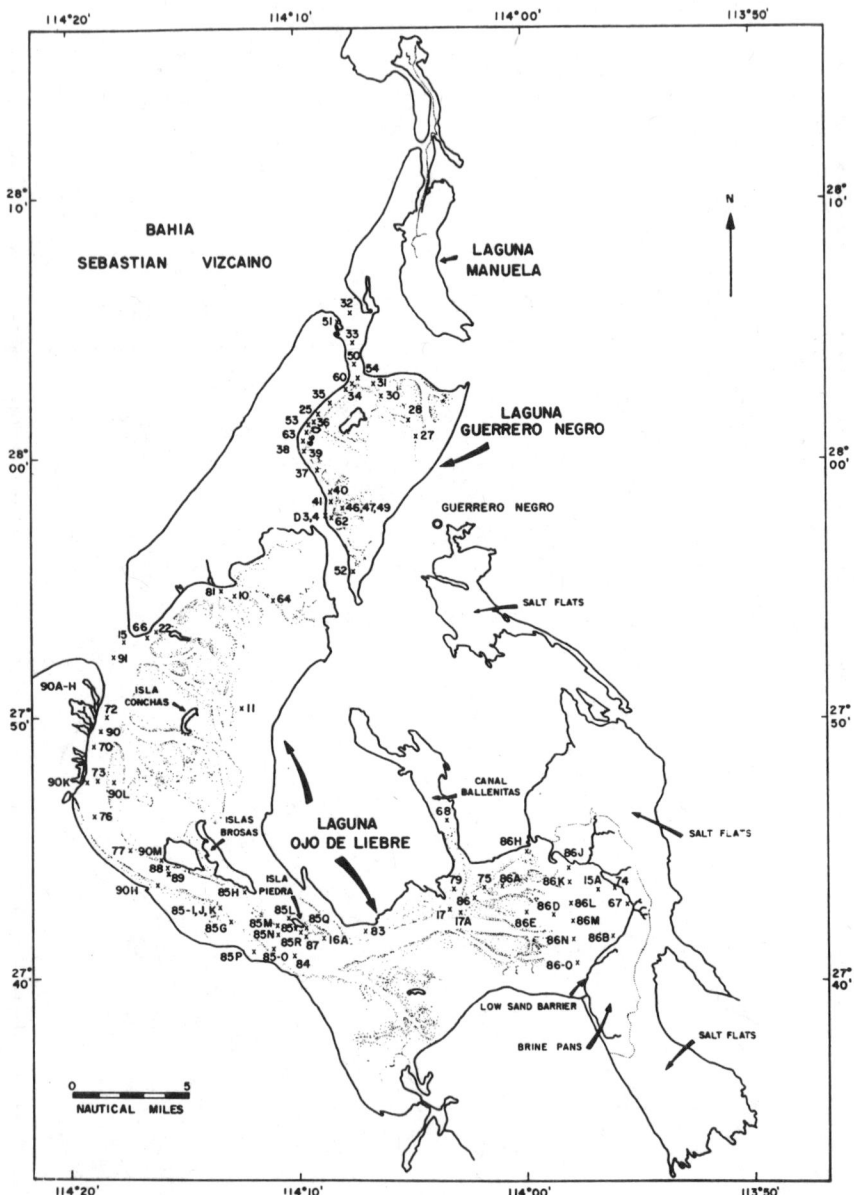

Fig. 8–16   Coastal lagoons of Guerrero Negro Lagoon, Baja California. From Phleger (1965).

opening at the north is maintained largely by the strong tides, but deepening by dredging has cut through the baymouth bar. Also, the marsh areas on the north and east of the bay have been extensively filled by man as the city of San Diego has grown. The bay sediments are sandy near the opening and grow progressively finer toward the head of the bay, where there are broad salt marshes and tidal flats.

Newport Bay, a large yachting center, has been somewhat more changed, but the harbor is essentially a lagoon with broad tidal flats on the inside. According to Stevenson and Emery (1958), the Santa Ana River, which now enters well to the north, formerly flowed through the bay, leaving estuarine deposits in the partly filled channel. The coarsest samples found by Stevenson and Emery were in the entrance channels, and the finest on the lower marshes where flood tides first overflow. Natural levees bordering the marshes have the coarsest fragments and most poorly sorted samples.

West of the Los Angeles area, Mugu Lagoon extends for 7 km inside a sand barrier. Other older lagoons existed inside old barriers farther inland on the delta plain (Warme 1969). Here, also, the sediments become finer away from the inlet, with sand in and near the inlet, and mud forming a veneer in the inner lagoon and on the salt marshes. Silts accumulate on the uppermost intertidal zone. As in many other lagoons, there are large populations of only a few species of invertebrates.

Humboldt Bay, 23 km long and 7 km wide, is the largest lagoon of the California coast. It has been studied by R. W. Thompson (manuscript). It is segmented into three sections: South Bay and Arcata Bay, consisting of marsh lands with many winding tidal channels, and Entrance Bay, having the navigable portion of the lagoon along which the port of Eureka is located. The tidal channels have gravelly and shelly sands, with finer and more muddy material in their upper reaches. The flats have partially organic clayey–silt and silty clay but no lamination because they are stirred by a rich in-bay fauna. According to Thompson, Humboldt Bay has only been filling at the same slow rate as that of the late sea-level rise, causing little recent change in area. Most of the mud comes indirectly from the Eel and Mad Rivers that flow into the open Pacific, but much of the suspended sediment is carried along the coast and swept into the bay through the entrance channel.

Wadden Sea of The Netherlands The Wadden Sea along the Dutch coast consists largely of tidal marshes exposed at low tide and of shallow meandering tidal channels. Extensive areas along the inner margin of the marshland have been reclaimed for agricultural land by dikes but are still partly below sea level. The Wadden Sea is protected by a row of barrier islands, so that it can be classified as lagoonal, although there are many estuaries along the inner margin (Fig. 8–17). The area has been widely studied (Wadden Symposium 1950; van Straaten 1954A, 1954B, 1956; Postma 1957; van Straaten and Kuenen 1957).

The sediments in the Wadden Sea appear to have come principally from the North Sea. Only small streams enter the bays in this area that lies between the mouths of the Rhine and Elbe rivers. Here again, the sediment size in the Wadden Sea depends on the tidal current velocities, and these are highest in the entrance channels. Tidal transportation has been the most important factor in distributing sediment in the channels, but the waves are more important on the flats (Postma 1957). Shell beds in the Wadden Sea are a concentration of mollusks due to upward crawling during slow sedimentation and reworking of the uppermost deposits (van Straaten 1950).

Lagoons of Northern Gulf of Guinea Many lagoons are found along the Gulf of Guinea west of the Niger Delta. Along the low coast of Dahomey (Guilcher 1959), a narrow lagoon consists of the lower ends of two small rivers extending along the barrier. The barrier used to seal off the lagoon during the dry season and the outlets were changeable in position. Recently, the Soviet Union dug a good channel and protected it by jetties. One result was the death of the abundant fish in the lagoon; now native fishermen have

Fig. 8–17 The Wadden Sea of northern Holland. Much more reclamation has been accomplished since this 1943 map. From Wadden Sea Symposium (1950).

to fish in the dangerous surf on the seaward side of the barrier. The lagoon is mostly less than 1-m deep except in the channels. The sediments are mud or sandy mud. Inside the lagoons are broad marshes, and Lake Aheme, an old estuary, trends in from the lagoon for 18 km.

Along the Ivory Coast, near d'Abidjan, the coastal lagoon has a widespread sand layer under a thin mud covering (Debyser 1955). Salinity conditions show a variation from open sea conditions at the inlets to fresh water at the stream mouths. Despite the acid condition of the lagoon, two areas contain sediment high in pyrite.

General Characteristics of Coastal Lagoons The lagoons described can be classified into three groups: (1) those in temperate humid areas with predominantly terrigenous deposits; (2) those in arid regions with terrigenous deposits mixed with euxinic evaporites; and (3) those in humid tropical areas with calcareous sediments, a combination of chemical precipitation and reworking of eroded coral and other types of limestone. The sediments of these lagoons have almost all been formed during the past 6000 to 7000 years, and the offlying barriers have generally prograded seaward during this period (Gorsline 1967; Phleger 1969B).

In humid areas, the lagoons have a very similar pattern with a predominance of silt in the deeper portions, sand along the margins and near the inlets, silty clays at stream mouths, and numerous oyster reefs. Terrigenous sediments come from the entering streams and from the open sea. These sediments differ from those of the adjacent shallow-water marine deposits in having a scarcity of glauconite and echinoids. The sediment cores show a rough stratification, but only those from near river mouths are laminated. Mottling of the sediments is common and is the result of stirring of the sediments by the benthic life. Bottom-dwelling invertebrates are usually very abundant but consist of only a few species; large changes in salinity that occur in these bays tend to eliminate many types. Marshes are common, including both low-marsh and high-marsh vegetation, both with entrapped fine sediments (Phleger 1969B).

In arid regions, the lagoons have a preponderance of sand deposits introduced by wind from the barrier islands and from dunes on the mainland sides. At points well removed from inlets, the salinity becomes very high and salt and other evaporites are formed, particularly on the flats at the inner margins of the bays, where they may come from brine blown in by the wind. Where salinity is very high, as in the Sivash Lagoon on the west side of the Sea of Azov, salt is deposited also on the lagoon floor (Stevenson 1968). Calcareous oölites also form partly along the lagoon margins. Despite the high salinities, productivity is often very high in the lagoons of arid areas. One result is the growth of algal mats that form a covering of the sand areas. Stratification and lamination in lagoonal deposits are more regular than in humid lagoons, because the benthic faunas are poorly developed under the euxinic bottom conditions.

In the lagoons that are protected by limestone reefs, the deposits are mostly fine-grained nonskeletal calcium carbonates due to precipitation of lime from the saturated waters that come in from the open sea and are warmed in passing over the shallow banks. This same process occurs over much of the Bahama Banks (see p. 218). The waves and currents also transport much fine carbonate sediment from the reefs, and the limestone outcrops along the shore add bulk to the deposits. Where grasses are growing on the shallow marine floors, they localize the calcareous deposits that are transported past them and form banks of irregular shape.

ESTUARIES

Although it is difficult to differentiate between many estuaries and lagoons, and small estuaries are present along the inner edges of most of the lagoons described in the preceding section, many embayments are primarily estuaries. The typical long, relatively narrow estuary that tapers inland toward its heads results in an embayment that has a very different effect on sedimentation than the typical lagoon where the elongation is, in general, parallel to the sand barrier. The lagoon is subject to introduction of sand sediments by the overflows from the numerous inlets that are opened during a hurricane, often providing a nearby source of sand to the entire lagoon. The long estuary has an effect that varies with distances from the source of storm sediments. Also, the typical estuary has a continuous decrease in salinity from mouth to head, whereas the lagoon, if provided with many inlets, may have very similar salinity along its length.

Chesapeake and Delaware bays, on the East Coast, are definitely of the estuary type, as are the mouths of the Columbia, and Yaquina rivers and Coos Bay, Oregon, on the West Coast. Perhaps San Francisco Bay should also be included, although it certainly has had tectonic influence in its origin. The numerous indentations along the northwest coast of Spain, referred to as *rias,* are clearly of the estuary type, as are the similar indentations on the coasts of the southern British Isles and of western France. As mentioned previously, fiords are omitted.

Chesapeake Bay The longest estuary in the United States, Chesapeake Bay (Fig. 8–18), is the drowned river valley of the Susquehanna and Potomac rivers. From the mouth of the Susquehanna to the Bay entrance is 300 km. The Bay has a series of branching channels with many small tidally excavated basins along their courses. Depths are as much as 43 m, but the flat floors of most of the Bay are around 5 to 10 m. The shallow waters are well oxygenated, but the deeper basins are periodically low in oxygen (Biggs 1967). The sediments are mud in the deep areas and sand in shoal water, especially near land. Biggs found that cores from along the channels show alternating layers of black and gray mud with interlaced thin sand layers. At the Bay entrance, there are well-developed sand waves related to the tides (J. Ludwick 1970). The currents flow predominantly in on the north side and out on the south.

A surprising feature of Chesapeake Bay is the fast rate of erosion on the eastern side. Singewald and Slaughter (1949) found that the Maryland shore had lost 17,000 acres in an 89-year period and that the islands in the Bay had lost 7600 acres in the same period. The large changes in some of the low islands from 1847 to 1942 are indicated in maps by Jordan (1961). According to Jordan, the sediment eroded from the east shore and the islands has partly filled the depressions and has formed shoals in the lee of the eroded islands.

Estuaries of Southern New England Numerous estuaries extend into the coast of southern New England. Although the area has been glaciated, the bays are more drowned river valleys than of the fiord type, which typifies most bays of Maine, to the north, and Hudson Inlet, to the south. Narragansett and Buzzards bays are the largest of these southern New England embayments, but the former shows considerable effects of glaciation with its numerous islands. The estuary of the Connecticut River and the Thames Estuary are both clearly drowned river valleys.

Narragansett Bay has definite drowned valleys at its head, and a

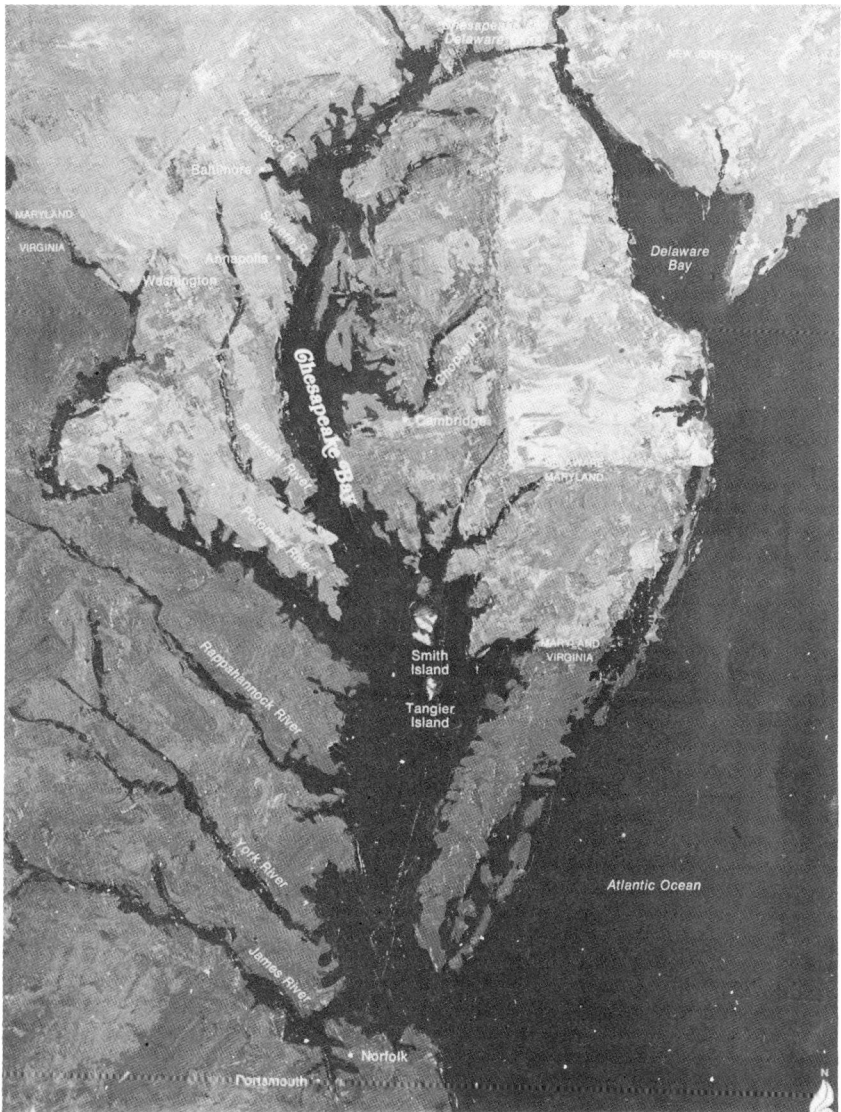

Fig. 8–18 Chesapeake Bay, a typical estuary. Courtesy of *Oceans Magazine.*

drowned valley extends all along the east side of the Bay with a branch entering from the northeast, but the depths of as much as 45 m are no doubt related to glaciation. The Bay sediments (McMaster 1960, 1962) are mostly clayey–silt and sand-silt-clay. The sediments coarsen in general toward the Bay mouths.

Buzzards Bay is more typically an estuary than Narragansett Bay. The sediments of the former have been studied partly because of their proximity to Woods Hole Oceanographic Institution and because of the interesting variety of sediment types (Hough 1940; J. Moore 1963). The sediments show a variation that agrees well with the depths, sand being found along the shore of the mainland and islands, and most of the mud in the deep outer portion. However, mud also occurs locally in the inner bays. The sands are largely related to the strong tides. Gravel is found around rocky points. Protograywackes from some of the deeper areas are described by Moore.

The Merrimack River Estuary in northern Massachusetts penetrates the coast north of Plum Island (Hartwell 1970). The estuary has a tidal delta at its mouth. The coarsest sediments, including sand and gravel, are found in the main channel where point bars are common. Poorly sorted muddy sands comprise the tidal flats along the side of the channels. Yellow-orange feldspathic sands are being carried down the estuary from the river, which has source sediments in glacial outwash. Sand in the marshes of the lower estuary comes from a local source. The marsh deposits show the effect of the most recent gradual rise in sea level and the infilling of what was originally an open bay.

Columbia Estuary   The Columbia, the largest river in the western conterminous United States, has headwaters in the mountains of British Columbia. In its lower course, the river has relatively clean water but transports great quantities of sand along the riverbed in the form of giant sand waves. The length of the estuary is difficult to determine, because all of the lower course of the river has a maze of point bars. However, the tide goes up as far as Bonneville Dam, 230 km from the estuary entrance, and the sandbar channels that appear to represent the estuary extend up at least 125 km. The U.S. Army Corps of Engineers has studied the changes at the mouth and measured the river flow (Hickson and Rodolf 1951), but little study has been made of estuary sediments. Apparently one or more great floods came down the Columbia River during the retreat of the glaciers in Montana which allowed sudden draining of a huge lake (Bretz 1969). Presumably, these floods cut deeply into the lower valley and one can imagine that very coarse sediment, even huge boulders, will be found in the underlying fill. Enormous fans outside Astoria Submarine Canyon (p. 319) may have a definite relation to these catastrophic flows. The postflood fill of the estuary has certainly included much sand, and probably consists of anastomosing point bars with intervening channels.

Yaquina Bay, Oregon   One of several winding estuaries that indent the Oregon coast is Yaquina Bay at Newport. The sediments in the 37-km estuary (Kulm and Byrne 1967) are predominantly sand except in the shoals along a bend and in a tributary where silt and clay are significantly increased. In the lower estuary, a yellow sand is brought in by the strong tides from the open shelf (see p. 64). Toward the head of the estuary, the heavy minerals coming from the Yaquina River show a great increase in mica. Maximum deposition occurs during the winter when the waves introduce large amounts of marine sands. Winds also bring sand from the coastal dunes.

San Francisco Bay   Although San Francisco Bay is to a considerable extent a product of earth movements, it includes definite estuaries (Fig. 8–19). The small Coyote Creek at the south end has a drowned mouth, as do the small Napa and Petaluma creeks to the north. The large flow of fresh water into the Bay comes from the combined Sacramento and San Joaquin rivers, which drain the Great Valley and the Sierra to the east. They have formed a deltaic plain at the head of Suisun Bay to the northeast. Eighty percent of San Francisco Bay is less than 10 m deep, and broad flats exist in San Pablo Bay and along the east side of the large south arm. The great amount of silt

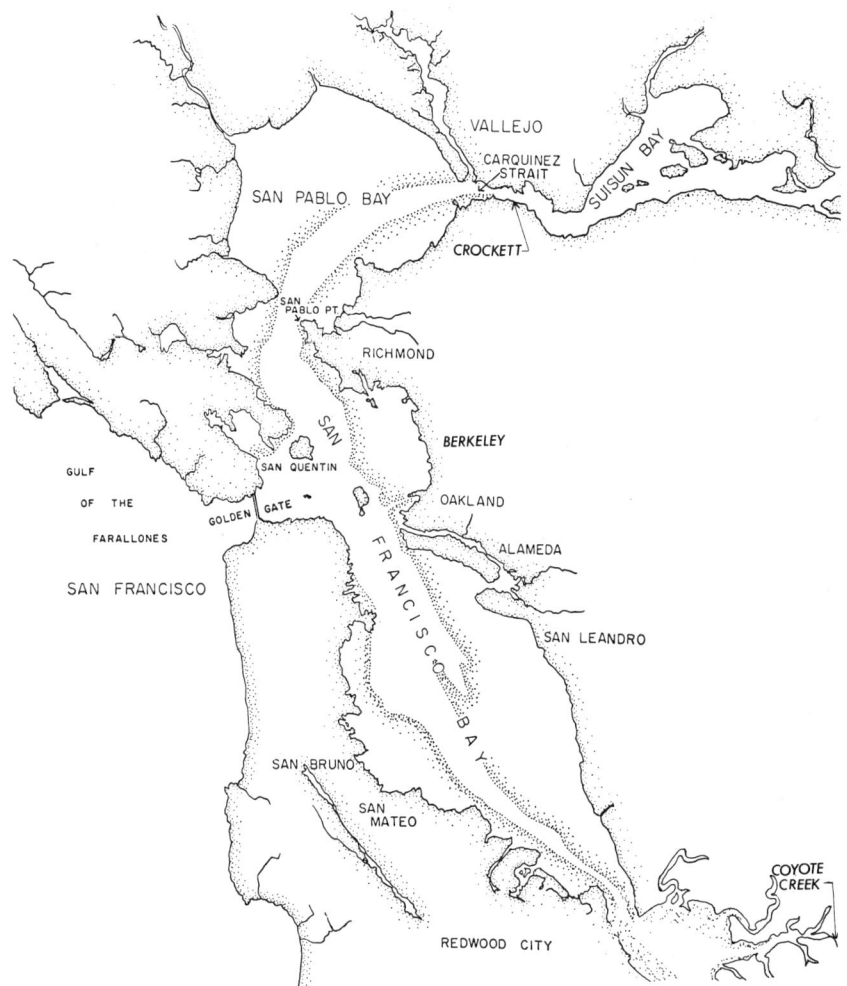

Fig. 8–19 San Francisco Bay with the 6-m contour. From U.S. Coast and Geodetic Survey Chart 5402.

carried in by the large rivers will fill the Bay in a few centuries unless the active subsidence that originally formed the Bay is renewed. Also, the Bay is being filled artificially over rather wide areas; this may be even more of a threat to its future existence.

The San Francisco Bay sediments (Louderback 1939) show a marked relation to depth. The deep channels and even deeper entrance have sand and gravel or rock bottom where the tide is strong in the Golden Gate. The upper Bay flats are almost entirely muds. The finest sediments are found in the tidal marshes, even near some of the cliffed shores. San Pablo Bay cores show alternating sand and mud layers. In the channel along the South Bay, sand layers usually have large numbers of shells and mud layers have scattered shells. The areas of active deposition have a reducing condition, giving the sand a gray color and muds are dark gray to black.

More recent work on San Francisco Bay gives a good picture of the stratigraphy and engineering character of the underlying sediments (Trask and Rolston 1951) and of the basement surface and structure of the sediments (Carlson et al. 1970). Five formations of Quaternary age are recognized. The oldest, Alameda formation, is firm sand with some layers of sandy clay and clay. This thick deposit forms a good foundation for bridge pilings, but the overlying San Antonio formation has more clay and sandy clay and tends to settle under heavy loads. The youngest formation is the Bay mud, which fills old eroded valleys and affords poor support for pilings. Subbottom profiles show a highly irregular basement caused by fluvial and tidal erosion. The thickness of sediments varies from zero over rock knobs to 75 m in parts of the central Bay. Many reflectors are found; one very persistent reflector is thought by Carlson and others to represent an unconformity with deposition above having formed since the last rise in sea level.

Tampa Bay, Florida Several estuaries are located among the bays of western Florida, although in their lower portions they are all essentially lagoons. Tampa Bay has two estuary arms that extend up to the city of Tampa. The sediments (Goodell and Gorsline 1961) are quite uniform in character, consisting of quartz sand and biogenic carbonate detritals. Only in the Bay heads are there mud sediments; these have a high organic content, as do the gravel-shell layers near the Bay mouth. The streams entering the Bay come from a limestone area and carry little sediment, so that the chief source of the Bay sediments is from the erosion of the Pleistocene terrace deposits along the Bay margin and from the shell animals of the estuary. The Bay deposits appear to have had two stages, (1) when the sea was somewhat below present level and the rivers were carrying relatively coarse material from the Florida Peninsula, and (2) since the sea reached the present level there has been little sedimentation, and the sediments are being mostly reworked by tidal currents.

Gulf of Paria The Gulf of Paria (Fig. 9–20), protected by a Venezuela mountainous peninsula on the north, Trinidad Island on the east, and the Orinoco Delta on the south, is only in part an estuary but it has several estuarine arms and will be discussed under that category. The Gulf of Paria has been extensively sampled (van Andel and Postma 1954), and its history is clearly indicated from acoustic reflection profiles (van Andel and Sachs 1964). The principal mouth of the Orinoco, Caño Manamo, has been entering the south side of the Gulf for more than 700 years and is building a large delta into what was a large estuary. The strong current coming into Serpents Mouth on the southeast carries with it sediment from both the Orinoco and the Amazon, but the velocity is high enough to prevent deposition of fine sediments in the entrance. Soldado Bank, just inside the Serpents Mouth, has Pleistocene sediments on its surface. An unusual feature of this embayment is the large area where the sediments have a high content of glauconite, unusual in bays, but suggesting to van Andel and Postma that this sediment is Pleistocene and that open-sea conditions existed at that time in the area.

Estuaries of Western France Innumerable articles have been written about the long winding estuaries of western France, notably the Loire and the Seine (Francis-Boeuf 1947; Berthois 1955, 1960; Guilcher and Berthois 1957; Rajcevic 1957; Bourcart and Boillot 1959; Giresse 1967). Of these the most complete is that of Francis-Boeuf, whose promising career was terminated by an airplane crash shortly after he completed his work on the French estuaries. His observations on tides and tidal bores, on the salinities, and on the chemical content of the waters gives clear insight into the origin of the various types of sediment found along the estuaries. Notable is the *tangue,* a fine-grained calcareous sediment derived from coquina banks and carried into estuaries by tides. The relatively high flat marshes on the sides of the estuaries cut by creeks are referred to as *schorre,* and the sloping estuary sides below the flat upper marshes are called *slikke* (Guilcher 1954, Fig. 12).

The tides coming into some of the French estuaries that extend in from the English Channel are amazingly powerful. At Mont-Saint-Michel (Bourcart and Boillot 1959; Dolet et al. 1965), the tide comes in over 11 km of sand flats and locally reaches speeds in the channels during flood stage too fast for a horse to run. The loosely compacted sediments near the famous castle include many quicksands. Extensive tangue deposits are found, some of them very old.

The studies of the Loire Estuary by Berthois (1960) showed that the sediments are largely from the river. During floods, the river water moving down the Estuary carries not only silt and clay but also sand. Some of the sediment is even carried out to the tidal flats at the Estuary mouth. Sedimentation is more active in summer because of the low viscosity of warmer water. Thus the Loire Estuary has a predominance of fluviatile sediment, unlike most of the estuaries along other coasts. In the Seine Estuary (Larsonneur and Hommeril 1967), the flood tides have a much greater effect than in the Loire, and the sediment is largely carried into the Estuary from the English Channel.

The Wash, Eastern England   The broad-mouthed Wash on the North Sea Coast of England is mostly a filled estuary, although it was definitely a drowned river valley until recent times. Here, the sediment zones have been established by Evans (1965) and include from the sea inland: lower sand flats, lower mud flats, *Arenicola* sand flats, inner sand flats, higher mud flats, and salt marsh. The source of the sediments is thought to be wave-eroded products from the boulder-clay cliffs and from the floors of the North Sea and of the Wash. The material is carried in by waves and flood tides. Each of the zones has distinctive sedimentary structures and organisms. The sediment belts grow in a seaward direction over successive seaward zones. Similar deposits are found also on the coasts of Germany and Holland.

The Rias of West Galicia, Spain   The estuaries of the Spanish coast north of Portugal are the prototype for drowned valleys. According to Pannekoek (1966), the river valleys of the area are in such a drowned state because of continuing subsidence along a rift or fault zone that lies somewhat inland of the coast. The sediments of the Ria de Arosa, one of these estuaries, have been studied by Koldijk (1968). Muds and sandy muds are the predominating type. The muds are high in H_2S in the inner bays but are oxidized in the middle and outer portions. Coarse sediments are mainly found in the coastal zones. Shell debris and lithothamnion rudite occur locally in shallow areas. The sediments have a major marine source, but part of the clays in the fine sediments are introduced by the rivers at the estuary heads. Montmorillonite is carried in from the open sea by bottom currents.

Inland Sea of Japan   The Inland Sea of Japan, a series of oval interconnected basins with shallow depths (Fig. 8–20), is only in part an estuary, but estuarine characteristics are evident and it can be compared with estuaries. The Inland Sea separates the three southern islands and has entrances from the ocean through four straits, all with strong currents and deep scoured depressions in the eastern entrances (Fig. 9–30). The southern sea shows the effect of currents bringing in sand from the open Pacific and continuing into Iya Nada with sufficient velocity to prevent deposition of muddy material. The weaker currents in the shoaler entrance from the Sea of Japan do not continue into Suo Nada, so that mud covers the bottom of the basin. Mud

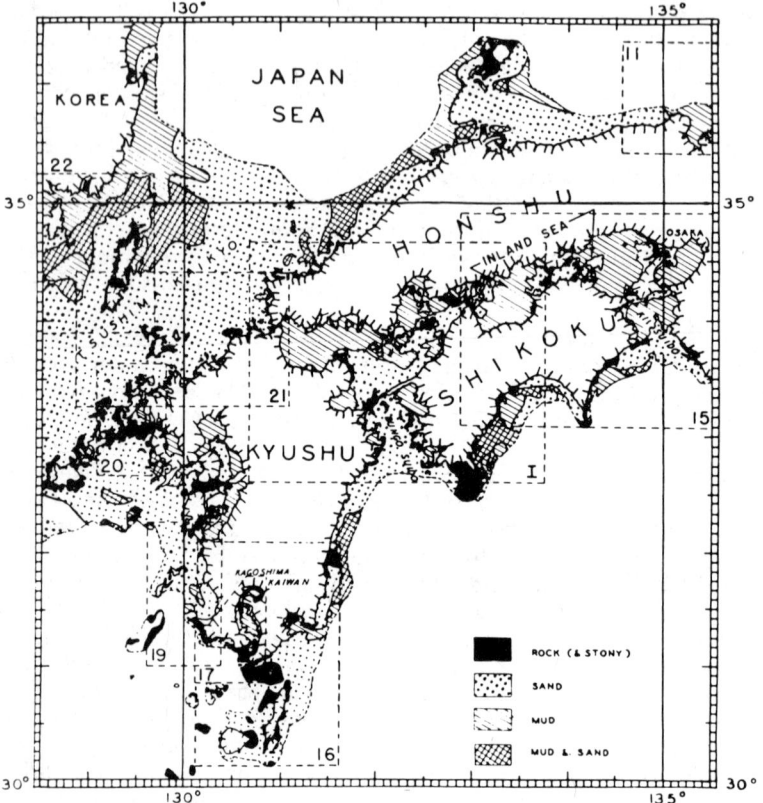

Fig. 8–20 Inland Sea of Japan, showing general character of sediments. From Shepard et al. (1949).

is also found on the floor of the three northern seas, which are also more protected. In the straits that separate the different inland seas, the currents are strong enough to concentrate sandy sediments that have an abundance of shells, and the same is true in the entrance straits except where sediments have been entirely removed leaving rock bottom.

General Characteristics of Estuaries Most typical estuaries that are true drowned valleys have relatively deep winding channels with shallow banks on the sides or in the middle if the channels are anastomosing. In most of these estuaries, the channels have sand bottoms and the banks are mud flats, but usually the mud has a considerable percentage of sand. Marshes are better developed in estuaries than in lagoons. Typical estuaries differ from lagoons in showing increasing fluvial influence toward their heads and increasing marine influences toward their mouths. This is partly indicated by the heavy minerals that are characteristic of the inland areas and of the coast and nearshore shelf at the mouth. Also, the faunas change progressively as the salinity decreases headward in the estuary. Mica and land plant debris are common in the upper estuary sediments. In most estuaries, the sediments become finer with distance from the open sea where sand and other coarse sediments are usually introduced. The trend of the sand bars in the lower estuary is usually along the axis of the estuary and hence at right angles to the coastal trend (Fig. 8–7). However, the bars in lagoons are likely to be parallel to coastal trends. River sediment at the estuary head may also be coarse but usually has a high silt and clay content. Rivers with high gradients and large flow introduce sand. Large rivers of low gradient, like the Amazon, carry great quantities of sediment down their estuaries, consisting mostly of silt and clay but with considerable sand. The Columbia transports great quantities of sand along its estuary and out into the open ocean.

Where the lower ends of the estuaries are largely blocked by barriers, the sediments are usually similar to those of lagoons, with considerable silt and clay, a scarcity of glauconite and echinoids, and many invertebrates but few species.

The estuaries in some arid areas are not so common as in humid regions because of the scarcity of river valleys for drowning by the postglacial rise of sea level.

The other inland seas, which are only in part estuaries, show many of the characteristics of typical estuaries with the strong marine and tidal influences at their mouths, general decrease in grain size in their bulging inner basins, and in their restricted faunal assemblages.

Chapter 9
CONTINENTAL TERRACES

INTRODUCTION

In the earlier editions, the continental shelves and continental slopes were discussed in separate chapters. However, during the past decade with the great increase in knowledge of the two zones, particularly from seismic profiles, we have learned that they are closely interrelated. Therefore, in describing the shelves and slopes of any specific area it will be less confusing to combine them under the term *continental terrace*. This has become evident in the realization that the shelves and slopes are influenced by plate tectonics, as was indicated previously in discussing coastal classifications.

The increased knowledge of the continental terraces since the second edition was written has been truly staggering. I can only attempt to summarize briefly what appears to me to be the most important of these new developments. One achievement stands out above the others. That is the detailed investigation of the shelves and slopes off the East Coast of the United States.

The data from thousands of samples, photographs, and seismic lines from all parts of the terrace from Maine to Florida are based largely on the efforts of the U.S. Geological Survey combined with Woods Hole Oceanographic Institution, a project under the energetic leadership of K. O. Emery and, more recently, Elazar Uchupi.[1] Many other marine institutions have added important information to shelf study. Almost as much attention has been given to the continental terraces along the Gulf Coast of the United States by marine geologists at Texas A & M University and the University of Miami, by the U.S. Geological Survey with headquarters at Corpus Christi, by the ESSA (now NOAA) group of the Department of Commerce, and by petroleum research institutions. Oil companies have become intensely interested in the entire area, although scientific information from their studies is slow to be released.

British marine geologists, despite being handicapped by small budgets, have produced notable results from their investigations of the North Sea and the English Channel. Partly because of the oil and gas boom in the North Sea, marine geologists from England, Holland, Germany, and Norway have had increasing opportunities in recent years to expand their studies. Almost nothing was known of the continental terraces around Australia until the 1960s; here, again, petroleum discoveries have provided incentives so that good progress has now been made. The shelves and slopes off Oregon are being investigated by Oregon State University. The Bering Sea is being studied particularly by the U.S. Geological Survey, the University of Washington, and the Soviet Union. Marine geologists in eastern Canada have been making important progress in studying the local shelves and large embayments. Widespread investigations of the slopes off many parts of the world have been accomplished by D. G. Moore and J. R. Curray, with notable concentration off the California coast. Now there are few areas of the world where the continental terraces have not been studied, at least by preliminary seismic profiling, during the 1960s.

The most significant book (mimeograph edition) to appear covering continental terraces is the result of an American Geophysical Institute course, *The New Concept of Continental Margin Sedimentation,* published under the editorship of Daniel J. Stanley. This gives a good summary of information, particularly of the work off the East Coast of the United States.

DEFINITION OF TERMS

Although definitions occasionally change and have different meanings among scientists, it is advisable to attempt to define the words that will be used in the following discussion of continental terraces. *Continental shelves* are here considered as the shallow platforms or terraces that surround most of the continents and are terminated seaward by a relatively sharp break in slope, called the *shelf edge* or *shelf break.* If the slope has more than one break, the shelf is confined to the zone within that break that is above 550 m (300 fm). The *continental slope* is the relatively steep descent from the shelf break to the deep-sea floor, usually a very irregular area. Off some coasts, the topography seaward of the shelf has basins, and ridges with depths distinctly less than those of the deep ocean. Such a zone is called a *continental border-land* and is typified by the area off Southern California (Fig. 9–22). Outside other shelves are deep terraces, which are referred to as *marginal plateaus,*

[1] Covered in detail in Emery and Uchupi (in press).

typified by the Blake Plateau off southeastern United States (Fig. 9–8). At the base of the continental slope, in some areas, a smooth gently sloping apron of sediments merges into the deep-sea floor. This is called the *continental rise*.

In discussing the continental shelves, sediment terms introduced by Emery (1952) and now well established include: *relict,* meaning deposited at an earlier period and left uncovered by recent sedimentation; and *residual,* meaning weathered products from old rocks underlying the shelf. Referring to relation to the adjacent land, the term *paralic* means coastal environments, such as lagoons or nearshore within the surf zone. *Neritic* is the benthic environment out to the edge of the continental shelf, and *bathyal* is the benthic environment on the continental slope. Terms introduced by Swift (1969) include *sand blanket* or *mud blanket* for a surficial cover of recent sand or mud of a portion of the shelf over older sediment; in *unmixing,* poorly sorted glacial or outwash deposits on the shelf have been worked over by waves and currents or restored to a sorted condition. The most recent addition is *palimpsest* sediments (Swift et al. 1971), meaning the reworked portions of a relict sediment.

EFFECTS OF GLACIATION AND SEA-LEVEL CHANGES

Before describing the character of the continental terraces it may be well to discuss the special effects resulting from continental glaciation and from worldwide sea-level changes that accompanied them. During the maximum glaciations of at least the Illinoian and Wisconsin (called *Riss* and *Würm* in Europe), sea level was lowered enough so that the areas now constituting the continental shelves were mostly laid bare of water. This necessarily produced a very different environment from the present; waves were cutting benches at a much lower level than now, subaerial processes were in operation on the exposed shelves, and rivers were flowing across them and building deltas. As the glaciers melted at the end of the Wisconsin and the sea gradually covered the shelves, benches and barriers were formed at intermediate levels, particularly during pauses in the sea-level rise (Curray 1960). The character of the level changes during the last few thousand years B.P. (Fig. 6–3A) is also highly significant relative to the nature of the continental shelf. If, as claimed by some geologists, the rise terminated about 5000 years ago (Fisk 1959), at least the inner shelf has had an opportunity to become well adjusted to present conditions; if considerable alternation of higher and lower sea levels has occurred during the last millennia (Fairbridge 1961), there should have been very little adjustment. An alternative more in agreement with available carbon-14 dating is that the sea underwent a considerable slowing of the rise rate some 6000 years B.P., and the rise became still slower 3000 years later and virtually ceased in the past millennium.

Another important effect of continental glaciation on the shelves came from the ice tongues, which extended beyond the continental limits, producing much erosion and deposition. This gave the shelves in the glaciated areas very distinctive characteristics, a consideration often overlooked by geologists.

Finally, the effect of glaciation on the continental slopes has been very great because, during lowered sea levels, streams carried their sediment out to the edge of most shelves; the resulting deposition produced very unstable marginal conditions, conducive both to mass movements and to the development of turbidity currents.

TECTONIC TYPES OF CONTINENTAL TERRACES

As important as glaciation to the development of continental terraces is their tectonic history. Now that we appear to have establishd the validity. of plate tectonics and sea-floor spreading, it is necessary to consider the relation of the terraces to this overall scheme (Fig. 6–2). Thus, we have continental terraces in subduction zones where active earth movements are developing mountain ranges on one side and sea-floor trenches on the other. The continental terraces along the coasts with trailing edges are much freer from earth movements, particularly in the case of the Afro and Amero types (see p. 104) of trailing edges where there has been a long period for shelf development. Neo-trailing-edge coasts, such as the Red Sea, however, have had little chance for shelf development and are relatively unstable. Also, terraces in the areas of transform faults along the margins of plates could be expected to have important effects from strike-slip faulting and should be less well developed than in areas of stability. The protective effect of island arcs that considerably reduce wave action on inland seas has also had a marked effect on the development of terraces, particularly along the East Asiatic continent. Furthermore, these marginal seas appear to have undergone extensive subsidence, which, in turn, has influenced the development of the adjacent terraces.

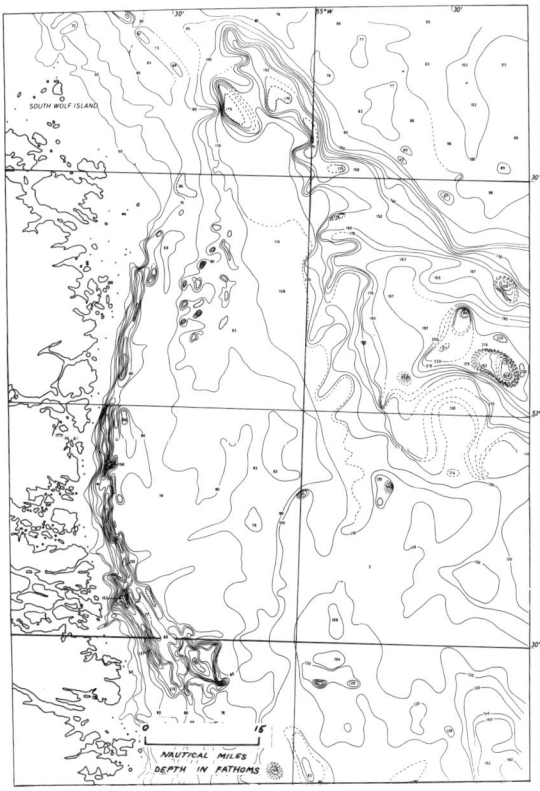

Fig. 9–1 Troughs and depressions along the Labrador shelf north of the Strait of Belle Isle. Recent sounding lines indicate even greater irregularity. From H. Holtedahl (1958).

DESCRIPTION OF CONTINENTAL TERRACES

In describing the continental terraces, a tour will be made around the continents in a clockwise direction, as in earlier editions of this book. Besides considering both the shelves and slopes for each major area, the description that follows will provide available information concerning the structures underlying the terraces. Also, attention will be directed to the relation of the terraces to plate tectonics and, where applicable, to Pleistocene glaciation. Location of place names can be found in the Gazetteer (pp. 483–490).

Eastern North America   The eastern North American coast which, according to the sea-floor spreading hypothesis, is an Amero-trailing-edge type and has been glaciated as far south as New York; south of Cape Hatteras the shelf has been greatly influenced by the powerful Gulf Stream, thus accounting for a tripartite division of the eastern continental terrace.

Labrador   Along the Labrador coast, the soundings are not extensive, but the topography definitely conforms with that off other glaciated coasts. The shelf width is close to 200 km, much greater than the world average of 75 km. Deep troughs and basins are shown by the soundings. These deep

Fig. 9–2   Bathymetric chart of the sea floor off Nova Scotia, Newfoundland, and the Gulf of St. Lawrence. From Canadian Hydrographic Service 801, 802. Contours by Elazar Uchupi.

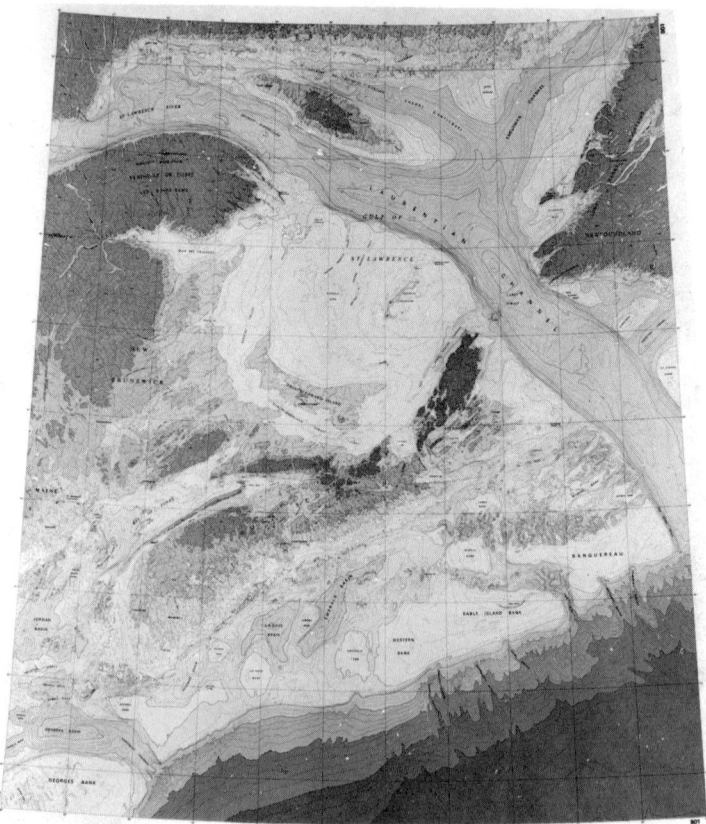

zones extend both longitudinal and transverse to the shelf. The most striking troughs are shown in the figures by H. Holtedahl (1958) and in that of Grant (1966), the former reproduced in Fig. 9–1. To the south, near the Strait of Belle Isle, the principal trough follows the coast for 150 km, but farther north another trough extends for about 400 km roughly parallel to but 70 km seaward of the coast. This trough has depths as great as 800 m. The scarcity of soundings makes it difficult to ascertain if the offshore trough connects with the various fiord inlets along the shore; but clearly there are seaward continuations of the trough across the outer shelf. Holtedahl originally suggested that the longitudinal trough was the result either of faulting or of erosion along old fault lines. However, seismic profiles across these troughs (Grant 1966; Sheridan and Drake 1968) intimate that they are the result of erosion along the contact of the crystalline Precambrian rock on the inside and the softer sedimentary rocks on the outside. The deep-basin depressions as well as the troughs are characteristic of glaciated surfaces. The continental slope is not charted well enough to determine its exact character.

   Newfoundland   The continental shelf widens off Newfoundland (Fig. 9–2), with a pronounced bulge off the southeast corner of the island where

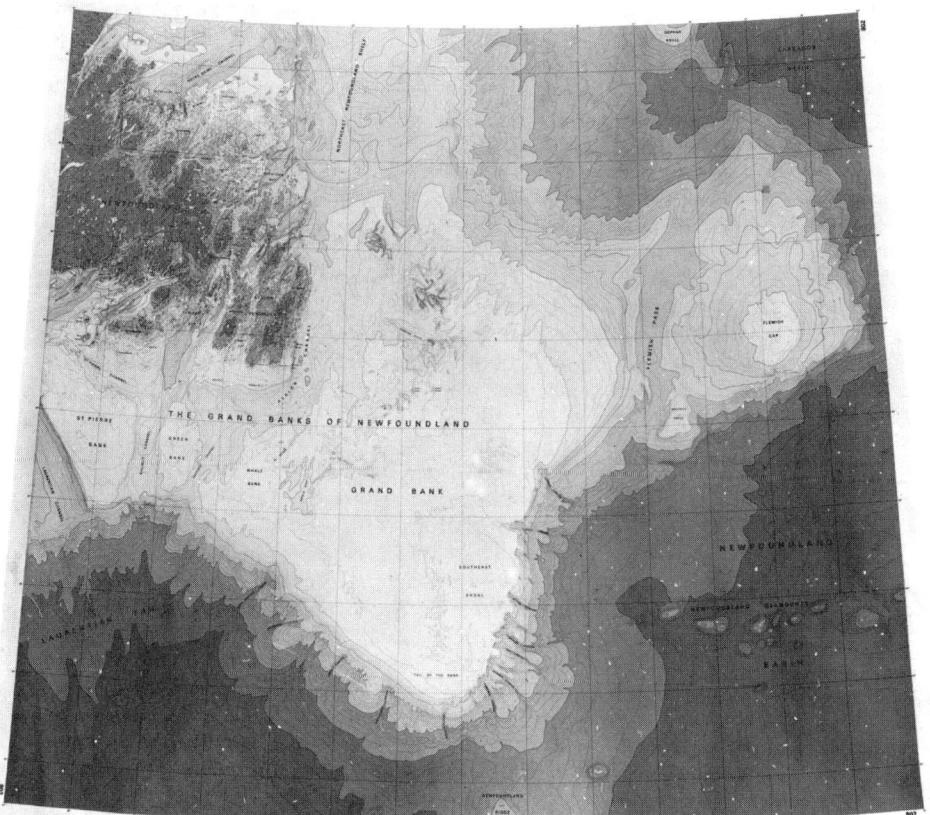

the Grand Banks project seaward for about 500 km. Here, the continental slope bends to the west-northwest, showing the first of several major changes in trend along the North American Continent. There are two distinct parts of the Newfoundland shelf. To the north, most of the water is deeper than 200 m, but it gradually shoals to the south, and portions of it form the broad fishing areas known as the Grand Banks, where the rather irregular bottom is mostly less than 80 m deep. In this southern area, however, the water is distinctly deeper on the inner shelf than outside. This deep-water zone extends up most of the bays of Newfoundland. Again the topography is characteristic of glaciated shelves. Sen Gupta and McMullen (1969), from a study of the southern Grand Banks, conclude that the coarse gravel sediment to the northwest of their studied area probably represents the maximum extent of continental glaciation, and that the finer sediments farther south have been reworked from proglacial material. They note that the trend of individual banks, either northeast-southwest or southeast-northwest, may represent the effects, respectively, of the Labrador Current and the Gulf Stream. According to E. Uchupi (personal communication), the lines may be due to the strike of old formations exposed by glacial erosion. The deeper areas landward of the Grand Banks are probably the result of glacial erosion.

Seismic refraction profile data reported by Sheridan and Drake (1968) suggest that along a north–south line, the Cretaceous to Holocene sediments thicken to the south, as do the Pennsylvanian to Triassic, which underlie them unconformably.

Seaward of the Grand Banks, Flemish Cap rises to within 115 m of the surface and appears to be an outlier of the continent. Also the JOIDES deep-sea drillings cored a deep knoll 560 km northeast of Newfoundland and encountered shallow-water formations, showing that this outlier is probably a mass broken away from the continents (JOIDES 1970).

Gulf of St. Lawrence One of the largest trough reentrants into the continent is connected with the outer shelf through the straits on both sides of Newfoundland. About half of this Gulf has depths in excess of 200 m (Fig. 9–2). A deep trough-shaped channel has been traced from the mouth of the Saguenay River, passing south of Anticosti Island, out through Cabot Strait, and across the entire continental shelf south of Newfoundland. In conformation with recent maps, we shall refer to this as the Laurentian Channel (originally called Laurentian Trough). Another deep branch lies northeast of Anticosti Island and has an arm extending through the Strait of Belle Isle. These troughs are almost entirely deeper than 200 m and have many basins. The principal trough has been attributed to faulting (Gregory 1929), but glacial erosion seemed more reasonable—striations indicate that glaciers had moved out of Cabot Strait across St. Paul Island (Shepard 1931). Recent investigations have apparently confirmed this glacial origin, although they show that erosion has been partly controlled by old fluvial erosion valleys along tectonic lines (Nota and Loring 1964; Conolly et al. 1967). No evidence of faulting was found along the outer channel (Lewis King and MacLean 1970D).

Sediments in the Gulf include poorly sorted coarse-grained deposits nearshore and on the topographic highs, whereas mud covers the deep troughs with a considerable quantity of unsorted coarse material (Nota and Loring 1964). Reddish-brown glacial-marine sediments were deposited in the deep channels during the late Pleistocene, along with brick-red glacial till (Conolly et al. 1967). This is covered with brown silty clay that is rock flour in origin. The thickness of the Holocene sediments, a transparent layer, has been determined largely by echo soundings, the work of Loring (1962) and Conolly and others (1967).

Nova Scotia The continental shelf topography off Nova Scotia (Fig. 9–3) is irregular, as it is farther north. In general, there are basins and troughs on the inner shelf, and relatively shallow banks are widespread on the outer shelf, including Sable Island, a narrow sand strip that extends east and west for about 35 km near the outer shelf edge. Here again, we find most of the deeper portions of the shelf near the land, as in other glaciated areas. Although no trough comparable with that coming out of Cabot Strait is found on the Scotian Shelf, a fairly continuous trough can be traced landward from the gap between Sable Island Bank and Banquereau. This connects with some of the basins on the inner shelf. Also, the 170-km Emerald Basin is comparable in length with some of the troughs off Labrador. The most irregular portion of the shelf is in an east–west belt north of Banquereau, which has small deeps down to 350 m and intervening flat-topped highs. This belt has the appearance of a glacial moraine, but the seismic profiles by King and MacLean (1970C) indicate that the underlying materials are preglacial coastal plain deposits.

Extensive geological studies have been made of portions of the Scotian Shelf, notably the work of L. H. King (1967, 1969, 1970), L. H. King and Mac-Lean (1970A, 1970B, 1970C), and of Stanley and Cok (1967). The map and section by King off the Halifax area show that partly covered glacial till extends virtually across the shelf. According to Stanley and Cok, Sable Island Bank and Banquereau represent outwash areas, accounting for their broad sand plains. Definite end moraines have been mapped by King (1969) in an area 30 to 40 km off the Halifax area. These have only low relief, but the till is as much as 50 m thick above the underlying bedrock. This till is largely masked by ponded silt and clay deposits. Some of the exposed parts of the till have been reworked by marine processes.

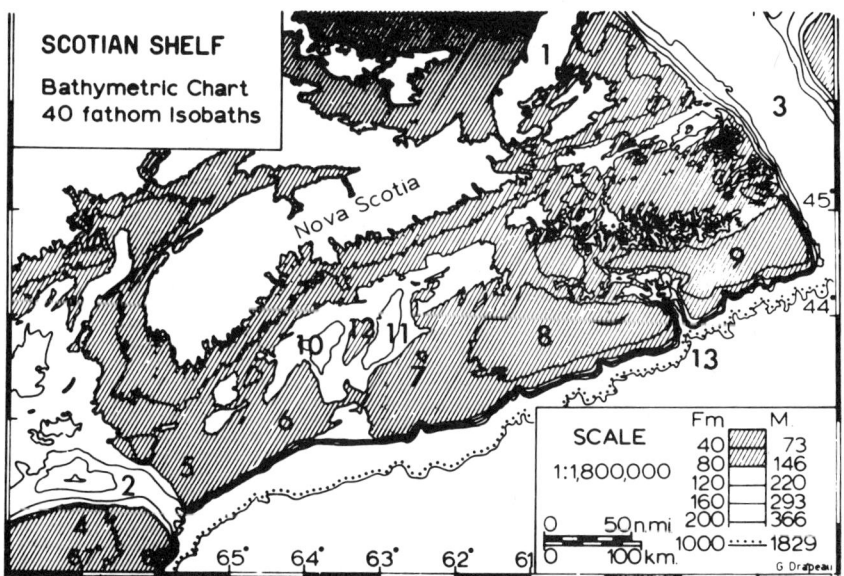

Fig. 9–3 Bathymetric chart of the Nova Scotian Shelf showing: 1, Cape Breton Island; 2, Northeast Channel; 3, Laurentian Channel; 8, Sable Island Bank; 9, Banquereau Bank; 10, La Have Basin; 11, Emerald Basin; 12, Sambro Bank; 13, The Gully Submarine Canyon. From Stanley and Cok (1967).

The sediments of the Scotian Shelf include much evidence of contributions from floating ice, which Stanley and Cok (1967) have shown is now principally derived from ice coming out each spring from the Gulf of St. Lawrence and drifting to the southwest across the shelf. As off Newfoundland, the deeper portions of the shelf are mostly covered with silty clay.

All across its width, the Scotian Shelf is underlain by rock at only moderate depth. Seaward, the rocks are Early Paleozoic overlapped by Cretaceous, and then by Tertiary semicompacted sediments. These deposits are thought to have been bevelled prior to the deposition of the Pleistocene tills and glacial outwash sediments. A diapiric structure was found piercing Cretaceous sediments 45 km north-northeast of Sable Island (L. H. King and McLean 1970A). This was thought to have an evaporitic core.

According to Stanley and others (1968), a series of terraces on the Scotian Shelf appear to conform in depth with terraces found farther north. The deepest, at 146 m, is thought to represent the lowest stand of the sea during glaciation, and was considered to belong to the Illinoian glacial episode.

The continental slope has a southwesterly trend all along the Scotian Shelf as well as across the front of the Laurentian Channel (Fig. 9–2). This trend indicates the second major change in direction of the Canadian continental slope. Apparently an important fault lies near the northern end of this slope, because a huge submarine earthquake occurred here in 1929 (see p. 330). The slope in this area is gentle, about 800 m in 15 km on the upper part and decreasing with depth. The northeast portion of the slope has been sufficiently sounded to show that it has an abundance of submarine canyons, notably The Gully, which extends seaward from the channel that separates the two main shelf edge banks. To the southwest, however, few canyons have been discovered.

Gulf of Maine and Georges Bank   The Gulf of Maine and Georges Bank (Fig. 9–4) show a continuation of the glacially modified topography of the Canadian terrace. Northeast Channel (originally called Northeast Trough) is somewhat intermediate in character between the Laurentian Channel and the channel inside The Gully. Northeast Channel cuts Brown Bank and Georges Bank, terminating seaward at the shelf edge, as does the Laurentian Channel. Apparently the west end of the channel originally was connected with the Wilkinson and Jordan Basins and with the channel coming out of the Bay of Fundy, but these deeps are now separated by several banks and swells, evidently representing glacial moraines. The Gulf of Maine has very irregular topography. Detailed studies by the group at Woods Hole Oceanographic Institution have been made of the Gulf (Uchupi 1965; Emery 1966; Ross 1970A).

Originally, D. Johnson (1925) considered that the basins and channels were the remnants of old river valleys now submerged by sinking. Later, noting the resemblance to other shelf features off glaciated areas, glacial erosion and glacial deposition were suggested for the irregular Gulf floor (Shepard 1931). This origin has now been established by the sampling and seismic reflection profiling of the Woods Hole group. They have found that the central part of the Gulf is underlain by pre-Triassic igneous and metamorphic rocks and Jurassic igneous rocks (Uchupi 1966B). A thin mantle of Pleistocene and Holocene sediments covers much of this material.

The Bay of Fundy was thought by Johnson to be bounded on the northeast by a fault scarp or a fault-line scarp. However, the somewhat steepened margin was found to be typical of glacial troughs (Shepard 1930).

These earlier writings lacked solid evidence and are now superseded by the detailed investigations with seismic profiling. Klein (1961) and Uchupi (1966B) could find little evidence of faulting. Swift and Lyall (1968), however, did find some buried fault scarps, although they believe that the trough along the Bay is largely the result of glacial erosion by an ice mass. Most recently, Ballard and Uchupi (in press) have concluded that the Bay is a rift zone. The sediments of the Bay of Fundy include a large amount of gravel bottom, the result of the strong tidal surge going up the southeast side and down the northwest (Swift 1969). These are relict gravels that can be traced locally into glacial outwash.

The Gulf of Maine sediments (Ross 1970A) are poorly sorted even on the ridges. Reworking has been slight because of the relatively great depth, and there is a high ratio of feldspar to quartz. The sediments are mostly glacially derived.

Georges Bank, a valuable fishing ground since the 1600s, resembles the banks to the northeast along the Canadian shelf. It is bounded on the north by a gentle escarpment, which D. Johnson (1925) correctly interpreted as the rim of a Tertiary cuesta. The northwestern portion of the Bank is very irregular, being covered with elongate sand ridges that trend northwest–southeast. Short-crested sand waves are also found, which trend east–west or northeast–southwest, according to a map by Uchupi (1968A). These were called underwater dunes by Stewart and Jordan (1965). These two groups (Fig. 9–5) were investigated by diving (Stewart et al. 1959) and were found to be related to the tidal currents that sweep across the Bank and to the swell coming from the southwest. According to Stewart and Jordan (1965), the shoaler set of ridges to the north has shifted westward between the surveys of 1930 and 1950 at an average rate of about 12 m a year.

Detailed studies of Georges Bank (Emery and Uchupi 1965) have shown that there is only a small amount of morainic material, and much of the 200 m or more of sediment cover on the bank is outwash from the glaciers, the surface having been greatly reworked by waves and currents. The sediments differ from the muds of the Gulf of Maine in being much coarser and having good sorting and a low feldspar/quartz ratio. According to Ross, the heavy minerals indicate a source partly from the underlying coastal plain sediments and partly from the glaciers that crossed the Gulf of Maine. The glaciers in the Gulf apparently had two main outlets, one through Northeast Channel and the other through Southeast Channel.

The southern portion of the Bank slopes smoothly toward the 110-m shelf edge. Seismic profiles reveal several buried canyon heads (Roberson 1964). Elsewhere, the submarine canyons extend for as much as 25 km into the shelf margin.

The continental slope off Georges Bank is somewhat steeper than off Nova Scotia. Portions of the inner slope are inclined 5° with a decided decrease close to the 2000-m line. The trend of the continental slope seaward of the Gulf of Maine is curving with an indentation in the general vicinity of Northeast Channel and a bulge off the shoal areas of Georges Bank.

Numerous seismic profiles on the slope off Georges Bank (Emery and Uchupi 1965; Hoskins 1967) show that it has, in general, been built seaward with interrupting periods of erosion. Old formations, including Cretaceous, outcrop locally. According to Hoskins, the slope has not been built forward much since the Cretaceous, and most of the sediment that was being carried over the shelf edge has accumulated on the continental rise at the slope base.

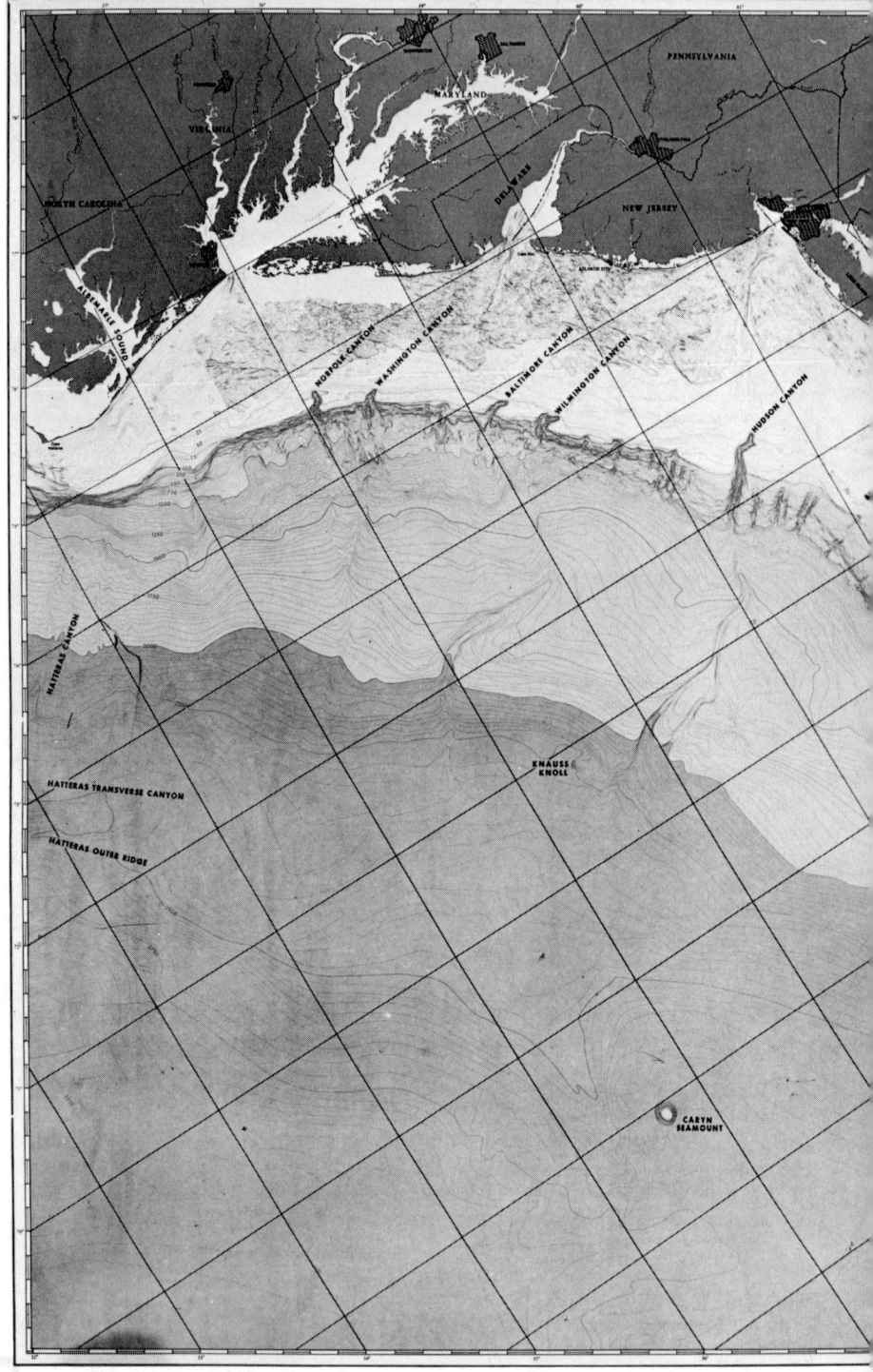

Fig. 9–4 Continental terrace off northeastern United States. From compilation of American Association of Petroleum Geologists Chart 1.

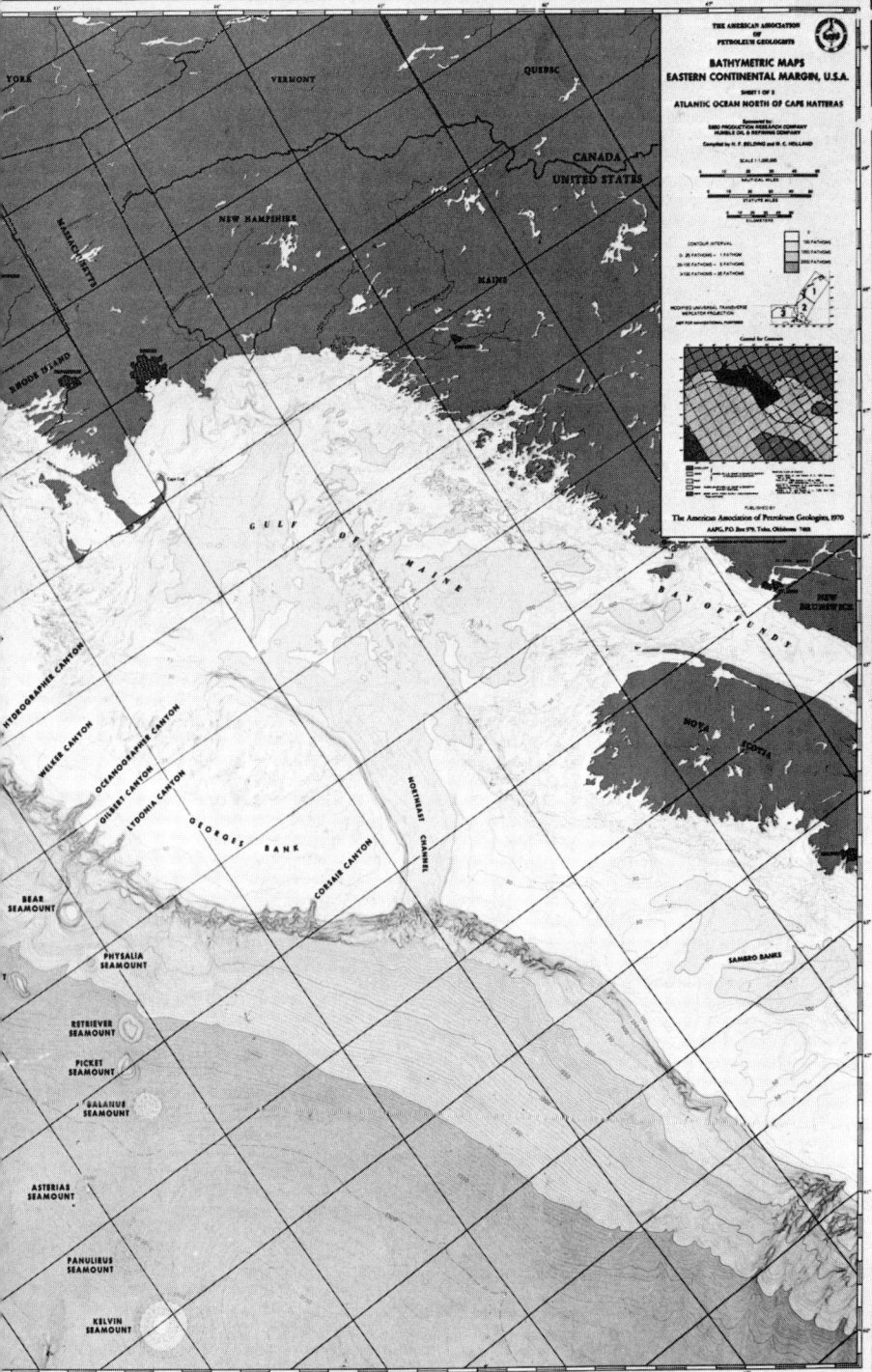

THE AMERICAN ASSOCIATION
OF
PETROLEUM GEOLOGISTS

BATHYMETRIC MAPS
EASTERN CONTINENTAL MARGIN, U.S.A.

SHEET 1 OF 3

ATLANTIC OCEAN NORTH OF CAPE HATTERAS

Sponsored by
ESSO PRODUCTION RESEARCH COMPANY
HUMBLE OIL & REFINING COMPANY

Compiled by H. F. BELDING and R. C. HOLLAND

SCALE 1:1,000,000

PUBLISHED BY
The American Association of Petroleum Geologists, 1970

AAPG, P.O. Box 979, Tulsa, Oklahoma 74101

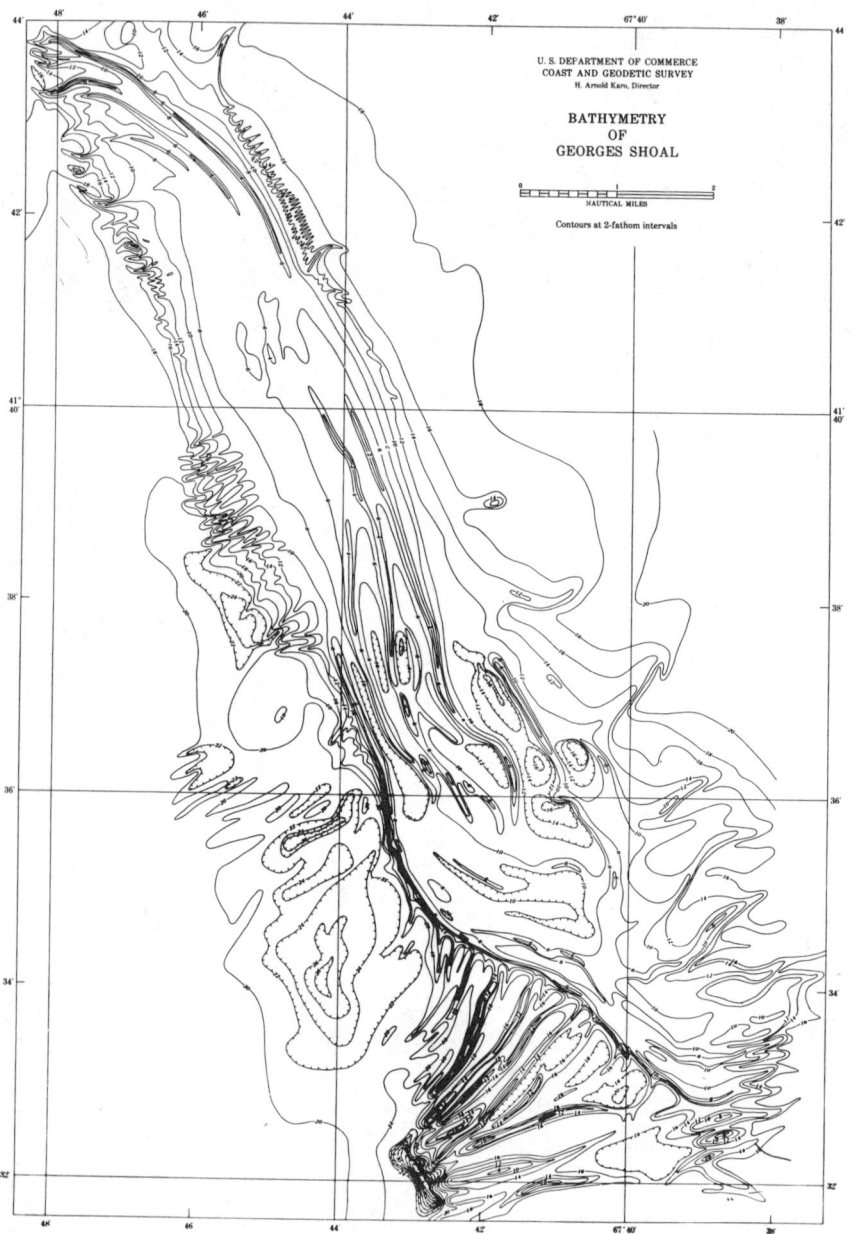

Fig. 9–5 Underwater "dunes" on Georges Bank. Two trends are indicated. From H. B. Stewart, U.S. Coast and Geodetic Survey.

Nantucket Shoals, a westerly continuation of Georges Bank landward of Nantucket Island, has an alternating series of sand ridges and intervening channels that curve around the east end of Nantucket. Like Georges Bank, the Shoals also has a series of sand waves with general east–west trend at right angles to the predominant north–south trend of the longer ridges. These sand waves are constantly shifting. According to Uchupi (1968A), along the eastern edge of the Shoals, the sand appears to have been derived from Cape Cod, but the rest is probably reworked glacial sediment, like that of Georges Bank. It is notable that Nantucket and its neighbor, Martha's Vineyard, are located where the Pleistocene glaciers diverged from the land, and irregular banks begin in the area east of Nantucket. According to Uchupi, the banks and islands are cuestas fluvially carved from shelf strata and later modified by glacial erosion and deposition. The reworking and sorting of the sediments on the Shoals has produced a broad sheet sand comparable to the sandstones of many ancient formations.

At the outer edge of Cape Cod, the slope of the sea floor extends down directly from the coast to at least 120 m with a 1° inclination. This slope is so even that by observing the depth on a fathometer it is possible to make a rough estimate of the distance of a ship from land.

Nantucket Island to Cape Hatteras South and east of Nantucket, the continental shelf undergoes a remarkable transformation (Fig. 9–4). Here, a relatively smooth shelf progressively deepens seaward except where interrupted by long sand ridges of low relief and intervening troughs (Fig. 9–6).

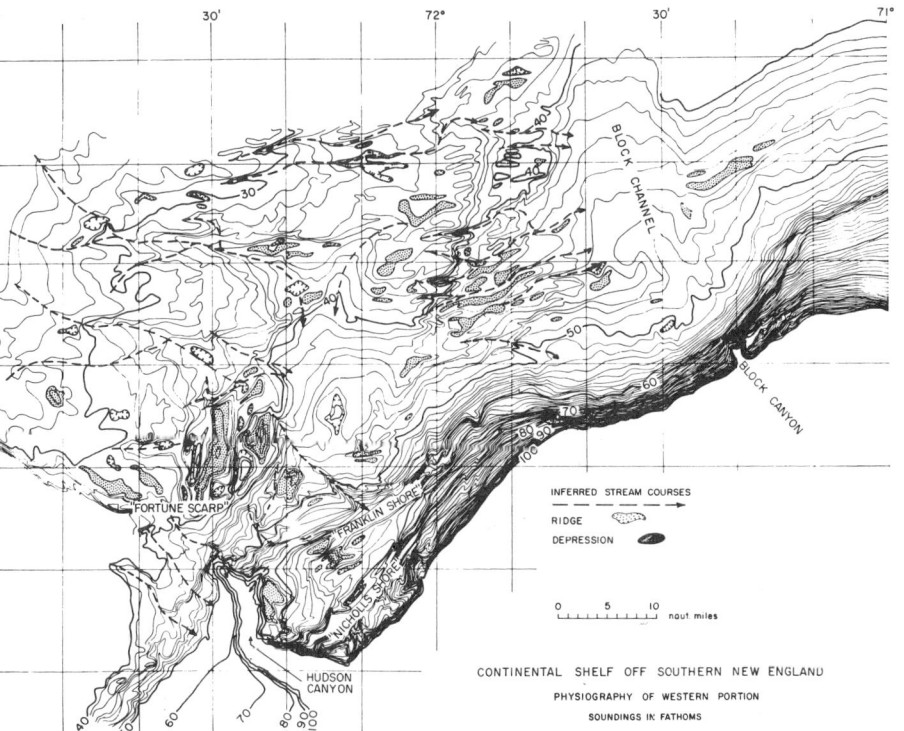

Fig. 9–6 Geomorphology of the shelf off southern New England. From Garrison and Mc-Master (1966).

Shelf widths in this area off the Middle Atlantic states average close to 130 km, and the depth at the shelf margin continues, like that farther north, at about 110 m as far as New Jersey and then gradually shoals as the shelf narrows in approaching Cape Hatteras.

Along the coast from Nantucket to New York, a series of irregular embayments and elongate sounds with basin depressions lie inside a string of islands terminating at Long Island. The depths are mostly less than 40 m except off New London, where some holes are 90 m deep. Seismic profiles of Long Island Sound (Tagg and Uchupi 1967; Grim et al. 1970) show that there is a widespread reflector horizon varying from 16 to 160 m below the bottom. This is thought to represent the top of hard pre-Cretaceous rocks. The irregular upper contact of this older series is said to be the result of both fluvial and glacial erosion. A series of buried valleys were found by seismic reflection profiling across the inner glaciated zone of the shelf between Martha's Vineyard and Long Island (McMaster et al. 1968).

The shelf topography along the Mid-Atlantic states is much smoother than that of the glaciated territory, but detailed soundings show numerous small ridges and sand swales (Figs. 9–6, 9–7). These are illustrated on the 1-fm interval contour charts of the U.S. Coast and Geodetic Survey and the Bureau of Commercial Fisheries. Some of these ridges run parallel to the coast and the edge of the continental shelf, but most of them are diagonal and, in the area north of Cape Hatteras, approach the coast in a direction to the left of the coastal trend. In general, they terminate landward at depths of 10 to 12 m (Uchupi 1968A), but some extend into the coast and may connect with emerged beach ridges. In the area off southern New England, Garrison and McMaster (1966) interpreted a concentration of low ridges at depths between 63 and 71 m and between 77 and 84 m as barriers formed during pauses in the last sea-level rise. However, some of the diagonally approaching sand ridges are known to be moved by storms and hence appear to be related to present-day waves, rather than to old barriers drowned by rising sea level (Swift 1969, DS-4, 28).

Terraces and scarps on the outer shelf off the New York area were first reported by Veatch and Smith (1939), and included their Franklin and Nichols shores (Fig. 9–6). Seismic bottom profiles showed that the Nichols Shore was caused by sediment fill, but the Franklin Shore proved to be a wave-worn platform that cuts across the seaward-sloping sedimentary layers of the outer shelf (M. Ewing et al. 1963; Knott and Hoskins 1968). Many profiles across the East Coast shelf show Pleistocene sediments up to 30 m thick covering older eroded sediments.

Another type of relief on the unglaciated shelves between Rhode Island and Chesapeake Bay consists of transverse but discontinuous valleys. Block Channel crosses most of the shelf, and Hudson Channel extends from the entrance to New York Harbor almost all the way across the shelf. Others emerge from the entrances of Delaware and Chesapeake Bays but cross only a small portion of the shelf. All of these shelf valleys include small basin depressions.

Sediments of the shelf off the Mid-Atlantic states were first described by Shepard and Cohee (1936) and by H. Stetson (1938). These earlier investigations revealed that the sediments are largely terrigenous but do not decrease progressively in grain size in crossing the shelf. The much more detailed studies of the Woods Hole–U.S. Geological Survey group have verified this generalization. They found extensive gravel deposits, especially off New Jersey. Shells indicative of estuarine conditions were obtained in many places

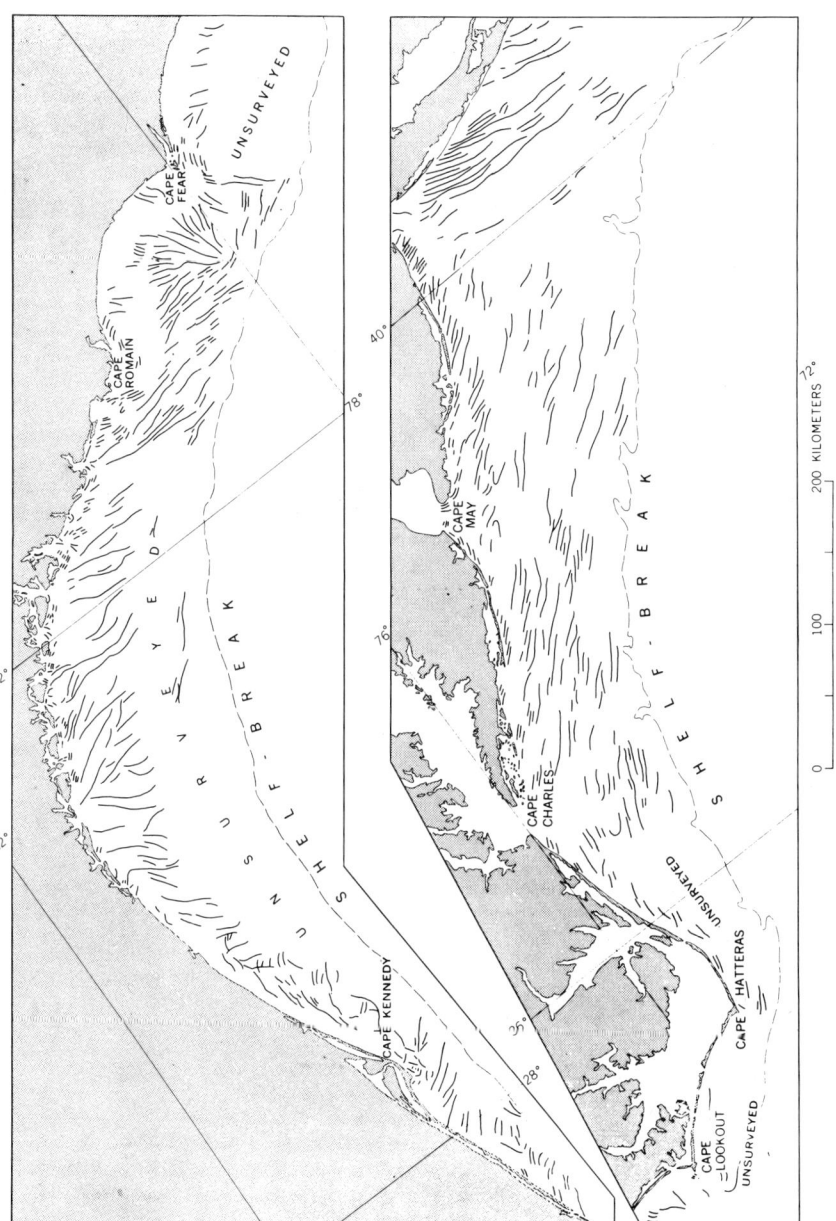

Fig. 9–7 Submerged sand ridges on the shelf from Long Island to Florida. From Uchupi (1968A).

(Emery 1965). Fossil mammoth bones have been discovered, and no doubt a large part of the sediment mantle is relict. A low content of calcium carbonate, largely under 10 percent, characterizes most of the shelf in this area, although some increase is observed in approaching Cape Hatteras.

The continental slope bends to the west in an area off Nantucket, then swings southwest again off Long Island. This local westerly trend is believed to follow a large transform fault that crosses the shelf and extends into the land (Drake et al. 1959, 1963). Seaward, the supposed continuation of this fault connects with a series of seamounts. The fault was first indicted by the offset of magnetic pattern lines.

The continental slope, as off Georges Bank, is cut by many submarine canyons. The slope continues to have approximately the same inclinations as off New England, with some decrease between Block Canyon and Hudson Canyon, a slight increase off Delaware and Chesapeake Bay, and finally a decided increase off Cape Hatteras. Seismic profiles of the slope (Fig. 10–4) show seaward-dipping formations partly eroded and partly truncated by large landslips (Hoskins 1967; Uchupi and Emery 1968). Hummocky disturbed sediments under the sea floor are found, especially at the base of the slope off Block Island (Uchupi 1967A).

Cape Hatteras to Straits of Florida At Cape Hatteras, the continental terrace undergoes a complete change in character (Fig. 9–8). The Cape is a cuspate foreland and has largely overlapped the shelf, reducing its width to 30 km, half of which consists of Diamond Shoals with water depths less than 10 m. The 80-m depth at the shelf break is also considerably lower than farther north. South of Cape Hatteras, the shelf again widens and maintains an average width of about 130 km as far south as Jacksonville, Florida, where it again narrows until it virtually disappears at Palm Beach. The depth at the shelf margin shoals south of Cape Hatteras averages about 55 m off the other Carolina capes. The shelf-break depths then increase to about 70 m until reaching Cape Kennedy, where narrowing, the shelf edge gets continuously shoaler until it is at 25 m near Palm Beach. Beyond, it is difficult to recognize any true shelf.

The sand ridges mapped by Uchupi (1968A) (Fig. 9–7) occur all along the inner shelf from Cape Hatteras to Cape Kennedy. However, they differ from the ridges north of Hatteras in trending at larger angles to the coast. Between Capes Fear and Romain they show a fan pattern with the apex of the fan seaward near the shelf break. In the narrowing shelf south of Cape Kennedy, they also trend at a large angle to the coast. As off Cape Hatteras, there are shoals extending well across the shelf outside of each of the cuspate capes, including Cape Kennedy. The shelf also has occasional hills and depressions (probably sinkholes). Otherwise the shelf slopes seaward rather continuously with a steepening at about 40 m.

The shelf sediments south of Hatteras are very different from those to the north (Goodell 1967). H. Stetson (1938) first called attention to the great increase of shell fragments and calcareous oölites. Off the Carolina capes, terrigenous sediment makes up most of the shelf cover (Goodell 1967). Farther south, the nearshore sediments continue to be predominantly terrigenous, mostly quartz in contrast to the high feldspar contents to the north. According to Gorsline (1963), a few kilometers from the coast the recent detrital materials are replaced by relict sediments that are a mixture of shell, oölite, and quartz, and have an abundance of phosphorite, especially at the shelf edge. The latter is authigenic, indicating chemical action in an area of nondeposition.

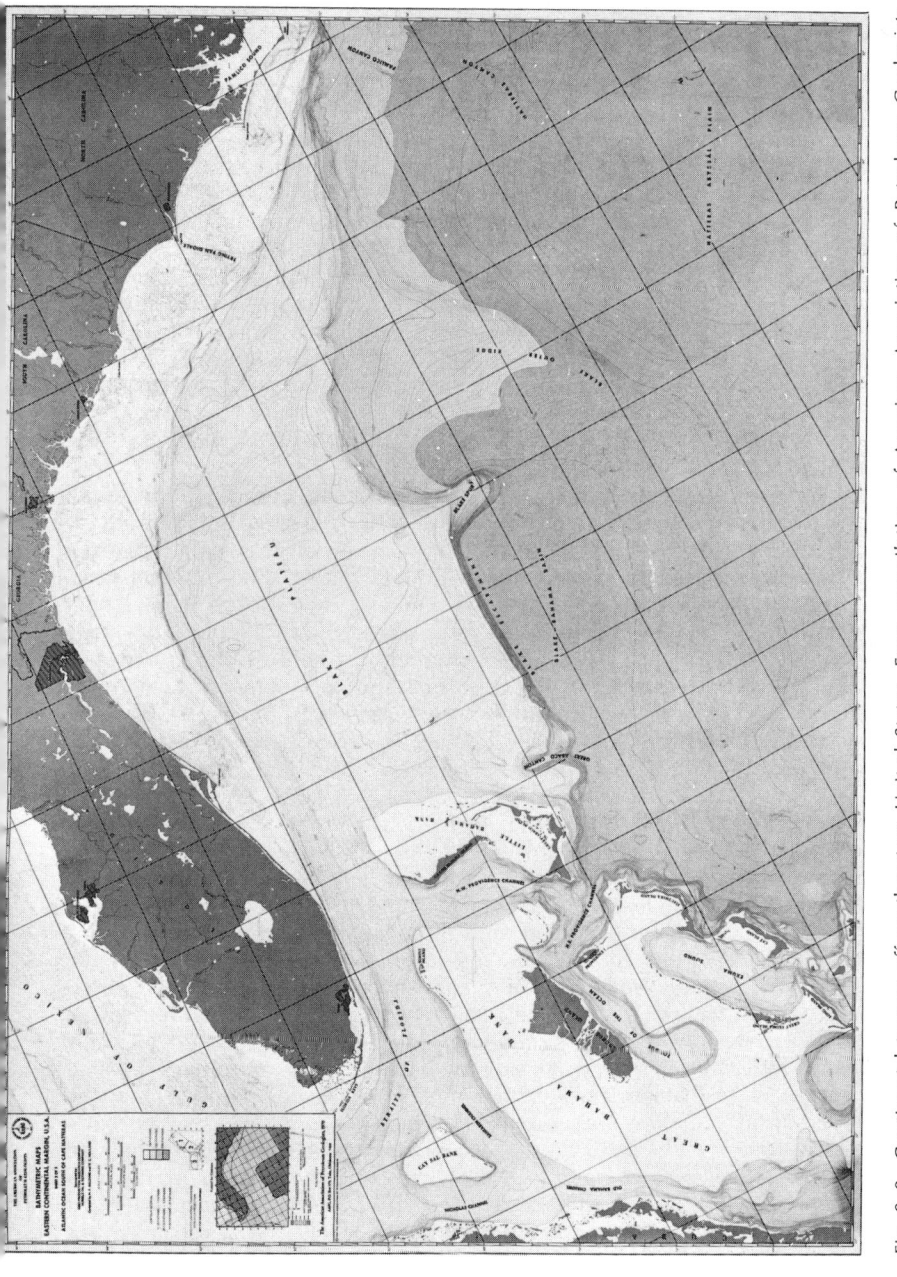

Fig. 9–8 Continental terrace off southeastern United States. From compilation of American Association of Petroleum Geologists Chart 2.

Blake Plateau—Straits of Florida, and the Bahama Banks The bathy-metric map of the area south of Cape Hatteras (Fig. 9–8) shows that the shelf edge turns 30° to the right, but along a line forming the southward continuation of the northern shelf edge there is the outer margin of a broad relatively flat surface, called Blake Plateau. Here, the depths are mostly from 750 to 1000 m, increasing to the south. The Plateau is almost 300 km wide to the south and wedges out to the north. Landward, the Plateau is separated from the continental shelf by a slope that lacks submarine canyons. Far more gentle than the slope to the north, it has local inclinations of up to 2°. The outer edge of the Plateau has a very steep escarpment to the south, locally 20° or more. It has no canyons except near the Bahama Banks, where Great Abaco Canyon incises the edge. The southern border of the Plateau is bounded on the west by the trough that comes from the Straits of Florida and on the east by the Great Bahama Banks, which rise close to sea level over a wide area and have many low islands.

Blake Plateau, extensively investigated by Lamont-Doherty Geological Observatory and Woods Hole Oceanographic Institution, was the first area included in the JOIDES deep-drilling program. To the north, the Plateau has a gentle slope with benches, but to the south it is almost horizontal over wide areas, with relatively shallow basins and low ridges.

Sampling of Blake Plateau has proved difficult because of the hard bottom. However, the combined information from seismic profiling and the JOIDES work across the shelf and Blake Plateau shows that actually there has been very little deposition on the deep Plateau during the 6 million years since the Miocene (Emery and Zarudzki 1967). Locally, Eocene and Upper Cretaceous deposits have been discovered on the Plateau surface, and many indications of bottom currents can be seen in photographs of ripples and rock outcrops. As indicated in Fig. 9–9, the several hundred meters of Quaternary on the northern Florida shelf and on the gentle marginal slope scarcely appear at all on the Plateau, and even local Miocene deposits are very thin. The evidence clearly implies nondeposition and even erosion due to post-Miocene Gulf Stream action. Many seismic profiles of the Plateau have been published (J. Ewing et al. 1966A; Sheridan et al. 1966; Uchupi 1970). Evidence of arches under the Plateau has been discovered in these profiles. Coral mounds and ridges are found over much of this area, particularly beneath the axis of the Gulf stream (T. Stetson et al. 1962). They occur even on the outer edge of the Plateau. These have been sampled and photographed and also seen during dives of *Alvin* (T. Stetson et al. 1962; Squires 1963; Milliman et al. 1967; Zarudzki and Uchupi 1968). The hills consist largely of ahermatypic corals that are now growing in many places. This type of reef, unlike the hermatypic corals of the shallow reef areas, can grow in deep and cold water. Seismic profiles indicate that underlying these growing reefs there are older reefs that grew up as Blake Plateau subsided (J. Ewing et al. 1966A). Dredging on the steep escarpment below these marginal reefs has revealed shallow-water algal limestone of early Cretaceous age (Heezen and Sheridan 1966; Sheridan et al. 1969B). The general opinion is that the outer margin is a barrier reef that grew on top of a Paleozoic structural arch (Zarudzki and Uchupi 1968). Older carbonate ridges have been mostly buried under the advancing slope between the continental shelf and Blake Plateau.

The Straits of Florida, some 85 km across, contain a deep trough or valley separating Florida and the Florida Keys from the Bahama Banks and western Cuba. The valley slopes southward until a 110-km plain breaks the slope at a depth of 850 m. Farther south, the valley continues its slope to a

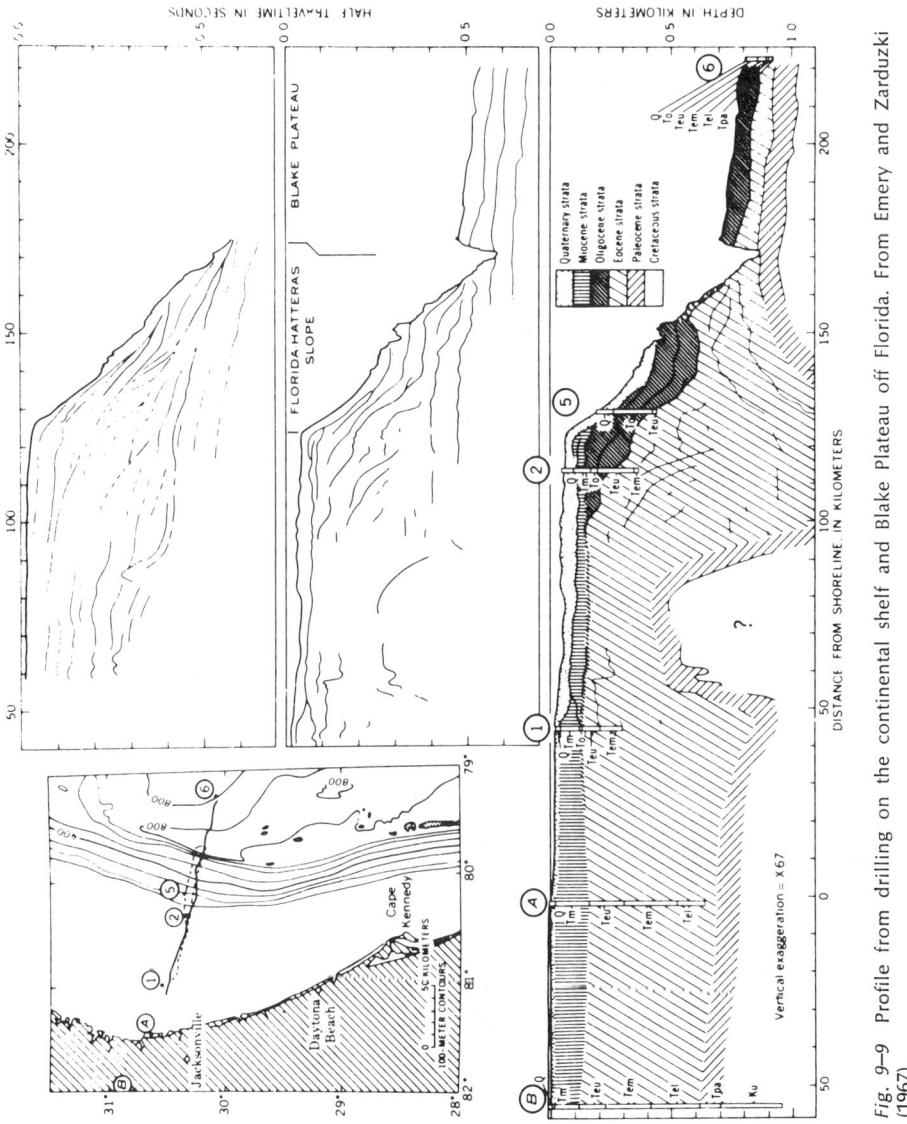

Fig. 9—9 Profile from drilling on the continental shelf and Blake Plateau off Florida. From Emery and Zarduzki (1967).

basin with depths of 1750 m in the Straits south of Key West. Along the Florida side, Miami Terrace and Portales Terrace interrupt the slope down to the floor of the Straits. Miami Terrace (Fig. 9–10) lies mostly between depths of 300 and 400 m and is about 25 km wide. It is bordered on both sides by escarpments—on the west leading up to the narrow shelf off Miami and on the east down to the main valley. Small basins, which may be drowned sinkholes, indent the terrace. Adjacent to the east escarpment, a small ridge has been interpreted as a depositional anticline (Malloy 1968). Portales Terrace, off the western end of the Florida Keys, has depths of 200 to 350 m and a

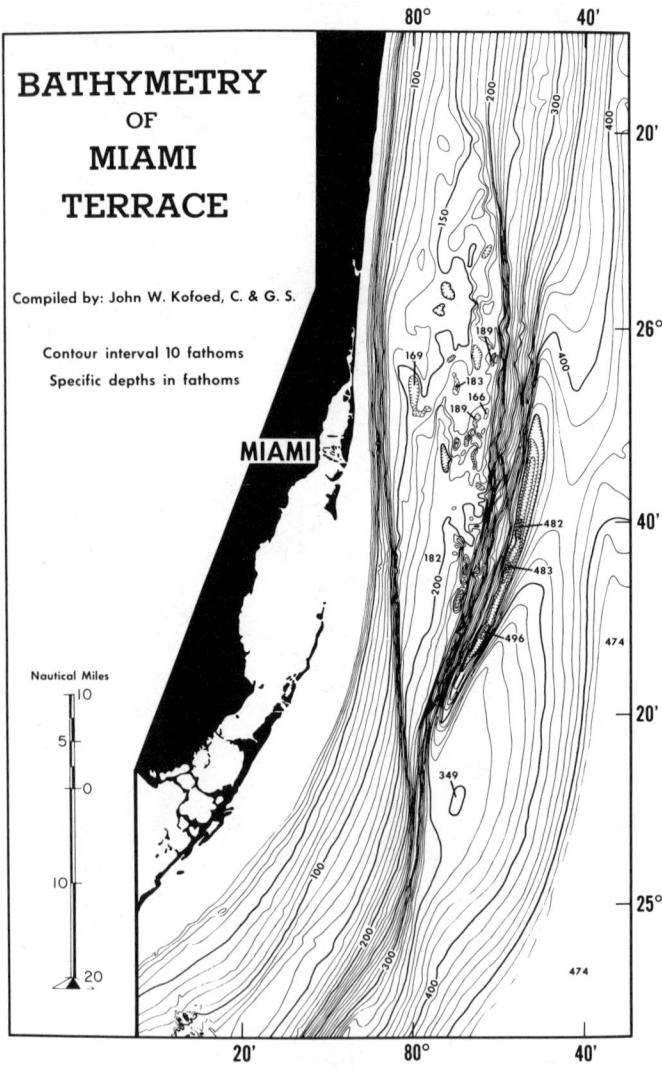

Fig. 9–10 Sea floor off southeastern Florida showing Miami Terrace (100- to 200-fm zone). From Kofoed and Malloy (1965).

maximum width of 35 km. Of special interest in this terrace are the numerous sinkholes, with depths to 160 m below the terrace surface. Jordan and others (1964) explained these as being the result of sufficient subsidence to drown former land areas. The Florida Current has prevented the limestone surface of the terraces from receiving any recent sediments. Farther west, the sloping Tortugas Terrace has escarpments on both sides (Fig. 9–11). After making seismic profiles, Uchupi (1968B) interpreted this terrace as caused by slumping rather than by faulting, as had been suggested by Kofoed and Jordan (1964).

The floor and walls of the deep trough in the Straits of Florida are of particular interest, because they underlie the Florida Current and the main branch of the Gulf Stream. As early as the past century, Agassiz (1888) suggested that the Straits were the result of erosion by the Gulf Stream. More recently, several authors (Talwani ct al. 1960; Sheridan et al. 1966) ascribed this trough, as well as those in the Bahama Banks, to faulting. Jordan and others (1964) and Kofoed and Malloy (1965) considered that the western and northern sides had indications of faulting. Uchupi (1966A), however, on the basis of 1000 km of continuous seismic profiling, found evidence that mostly supported Newell's (1955) contention that the walls of the troughs were the result of upgrowth of carbonate sediments as the basement sank. Uchupi considered that local ancient faults may exist to the south, but that the Straits were mainly produced by currents that prevented deposition in the

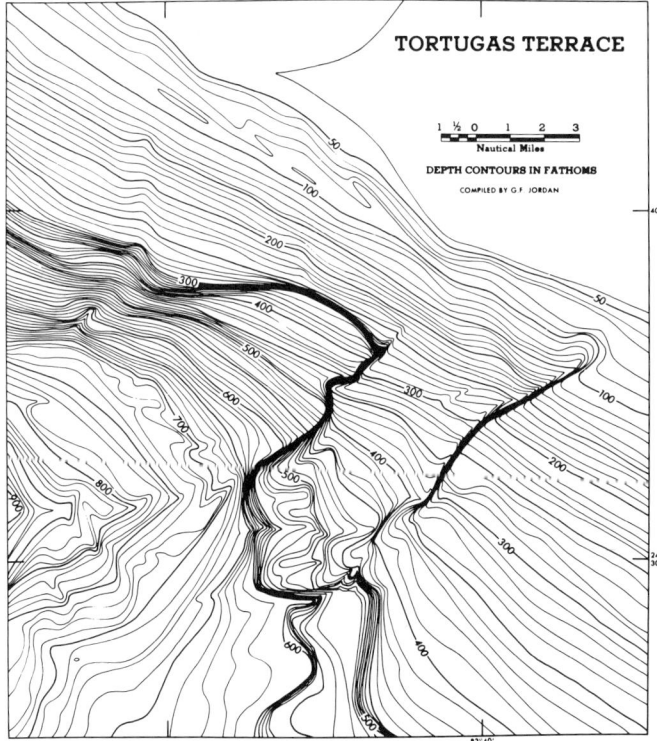

Fig. 9–11 Tortugas Terrace off southwestern Florida interpreted as caused by slumping or faulting. From Kofoed and Jordan (1964).

center, while upgrowth occurred on the sides as the whole block gradually dropped. Evidence of bottom currents has come from the dives of the *Aluminaut* (Neumann and Ball 1970). On the east escarpment of Bimini Island, the currents were found to be flowing north, but on the western side they flow south, counter to the direction of the Gulf Stream. On the west side, the currents may have built a sedimentary anticline. Ripple marks on the floor of the trough also show a south flow in some places (Hurley and Fink 1963).

The Bahama Banks have definitely grown from a subsiding basement, perhaps caused by sea-floor spreading, as suggested by Lynts (1970) and by Dietz and others (1970). This sinking is shown by the shallow-water nature of the formations encountered in drillings made by petroleum companies (Spencer 1967). Apparently, at least some of the deep troughs between the Banks were maintained well below the Bank surfaces during subsidence, because deep-water sediments were found in the limestones on the walls of troughs (Gibson and Schlee 1967). This has recently been supported by data from one of the JOIDES drillings (no. 98). Maintenance of the troughs and their incised canyon valleys is evidently the work of turbidity currents and is not related to the Gulf Stream (Andrews et al. 1970).

The topography of the Bahama Banks includes shelf lagoons, terraced outer platforms, marginal escarpments, and troughs (Newell 1955; Purdy 1963). The lagoons have calcareous muds, pellet muds, and grapestone facies; oölites and coral-algal sediments occur on the outer platforms; coral-algal facies also appear along the top of the marginal escarpments; and the troughs have a mixture of calcareous oozes and coarse sediments, the latter coming by slumps from the escarpments and transported along the floors in part by turbidity currents.

The Bahama Banks are one of the best areas in the world for observing shallow-water sand ridges and sand waves from the air (Fig. 9–12), although from high-flying jets the view is not as good as when flying relatively low in propeller planes. The water on these shoals is mostly from 2 to 10 m deep. The sediment forming the sand ridges and giant ripples consist largely of oölites, which form in the shoal water mainly by chemical precipitation resulting from the warming of cool water that, because of tidal currents, rises from the depths and flows over the heated Banks (Newell et al. 1960). The grains contain considerable organic material that is incorporated into the oölite shells.

Gulf of Mexico   The Gulf of Mexico terrace has had almost as much study as that of the East Coast, although much of the information is retained by petroleum companies. The northwest Gulf was investigated during the 1950s by Scripps Institution scientists for American Petroleum Institute Project 51 (Shepard et al. 1960). Extensive reports have appeared during the past decade from oceanography departments of Lamont-Doherty Geological Observatory, Texas A & M University, the University of Miami, the University of Florida, Woods Hole Oceanographic Institution, Louisiana State University, and the U.S. Geological Survey; petroleum geologists have also contributed much data.

As can be seen from the excellent bathymetric maps of the Gulf of Mexico (Fig. 9–13), the continental terrace has several natural divisions. These include areas off west Florida, Mississippi–Alabama, the Mississippi Delta, western Louisiana–Texas, the east Mexico shelf, Tobasco, and Campeche.

West Florida   Along the relatively straight Florida coast south of Apalachee Bay, the west Florida shelf is about 185 km wide and gradually steepens from 80 m out to about 200 m. The shelf break is indefinite along

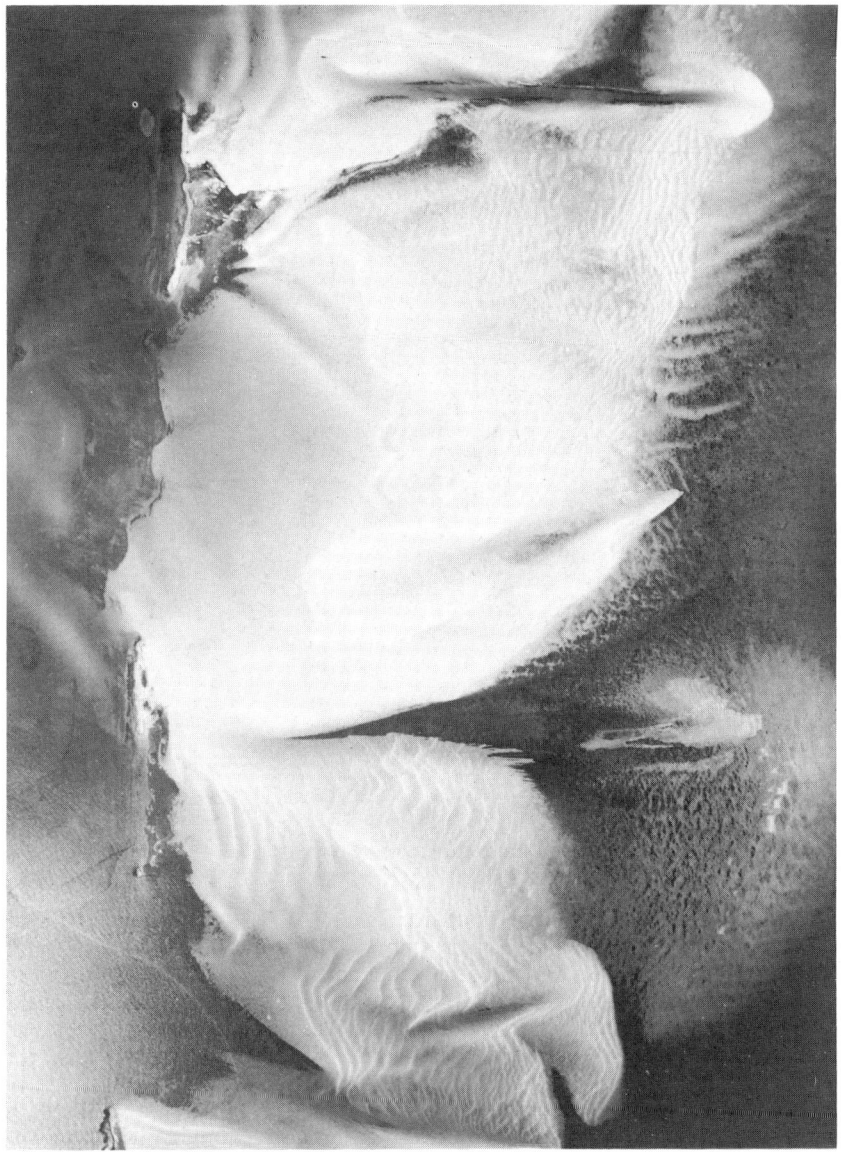

Fig. 9–12 Aerial view of underwater "dunes" on Bahama Banks, northwest of Great Exuma. Courtesy of N. D. Newell.

much of this west Florida shelf. Farther south, a break at about 55 m is followed by a rather gentle slope down to 550 m.

The west Florida shelf's relatively even slope is interrupted by a number of small isolated hills, coral reefs, and other types of bioherms. Along the outer margins there are low sand ridges with depths of about 100 to 160 m (Jordan and Stewart 1959). The shelf sediments studied by Gould and

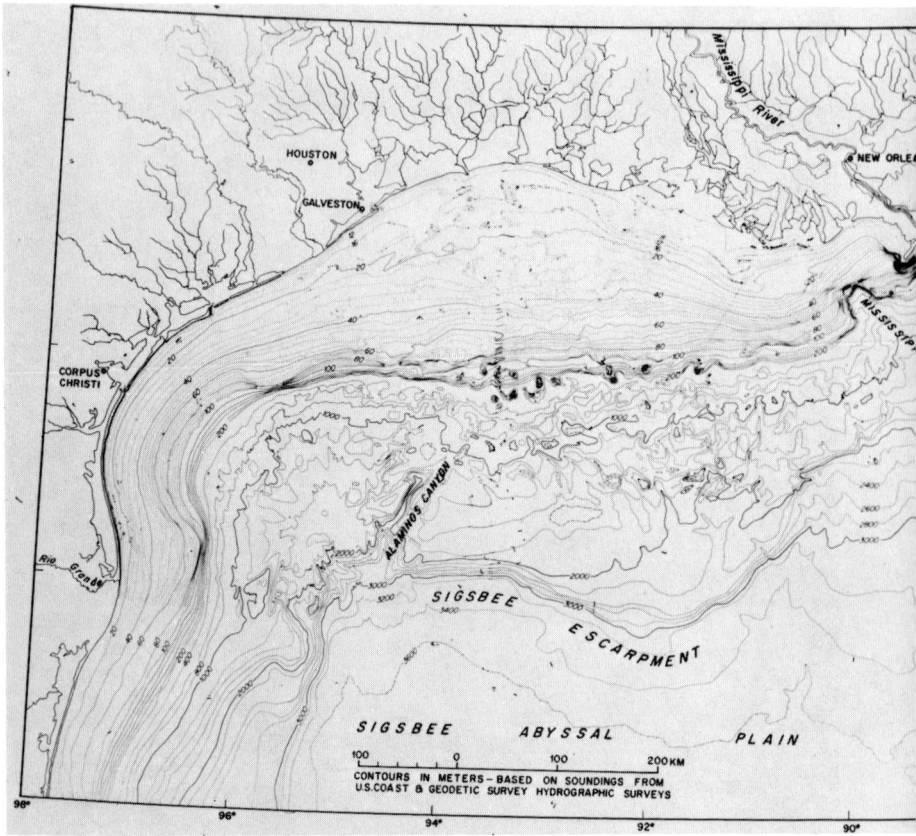

Fig. 9–13 Bathymetry of Gulf of Mexico. Contours by Elazar Uchupi.

Stewart (1955) show a sequence out from shore of (1) quartz sand with shell content increasing with distance from shore, (2) detrital sands, (3) algal sand, (4) oölitic sand, and (5) foraminiferal sand and silt (Fig. 9–14).

To the north where the shelf is narrowed by the outward bulge of the Apalachicola Delta and Cape San Blas, the calcareous outer shelf gives way to quartz sand derived from the weathered rocks of the Appalachians. The sand includes zones with abundant shells.

The escarpment off the west Florida shelf (Fig. 9–13) is particularly interesting in that it is one of the steepest of the entire continental slope. The escarpment is quite straight to the north but has many valleylike indentations to the south. If there were no ocean, this great slope would be comparable in appearance with the fault scarp on the east side of the Sierra Nevada. The natural inference is that this is a fault scarp; however, the area has had no recent earthquakes. It has been suggested that the origin of the escarpment was the upgrowth of a coral reef (Gibson 1962; Uchupi and Emery 1968). Seismic profiling shows a structural anticline in the deep platform just inside the scarp, and dredging on the scarp has revealed shallow-water Cretaceous limestones (Antoine et al. 1967; Antoine and Jones 1967). The anticline appears to be related to coral growth. The most recent results from seismic profiles and magnetic studies, however, have convinced Antoine (1968) that the

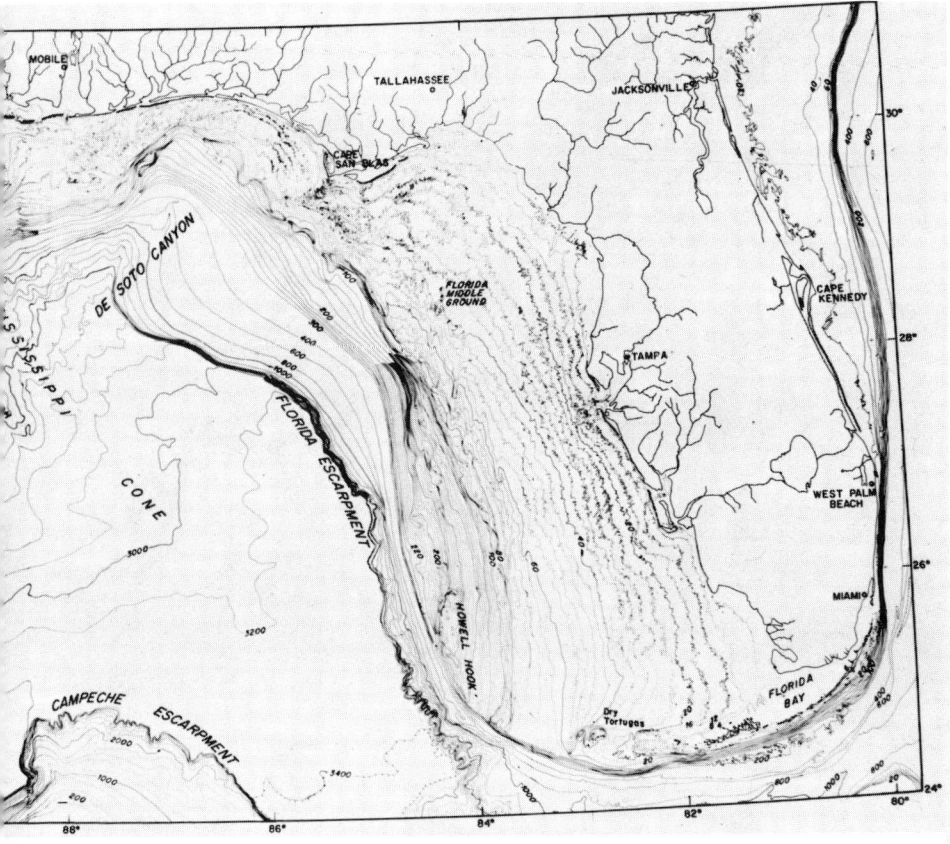

escarpment is more likely due to faulting, but that this has been accompanied by great subsidence on both sides of the fault, much greater on the seaward flank. The base of the escarpment is found to decrease in slope under a cover of abyssal plain sediments.

The plateau landward of the Florida escarpment has sunk at least 2000 m, and little deposition has occurred since the Cretaceous, showing the antiquity of the large loop current in the Gulf of Mexico. This is similar to the nondeposition and erosion that has resulted from the activity of the Gulf Stream on Blake Plateau.

Mississippi–Alabama The relatively narrow shelf between the Apalachicola Delta and Mobile Bay, Alabama, has complex shoals across most of its width (Tanner 1961; Hyne and Goodell 1967; Schnabel and Goodell 1968). Directly seaward of Capes St. George and San Blas, shoals cross the shelf. The highs in these shoals are almost all transverse (parallel to shore), unlike the combined transverse and longitudinal sand waves in the shoals off the Carolina capes. Farther west, a series of shoals trend diagonal to the coast and, like those off Delaware and Virginia, are deflected left from the normal approach to the coast. The shoals off the Florida Panhandle have been interpreted either as relict beach ridges drowned by recent sea-level rise (Hyne and Goodell 1967) or due to occasional storms (Swift 1969, DS 4-28).

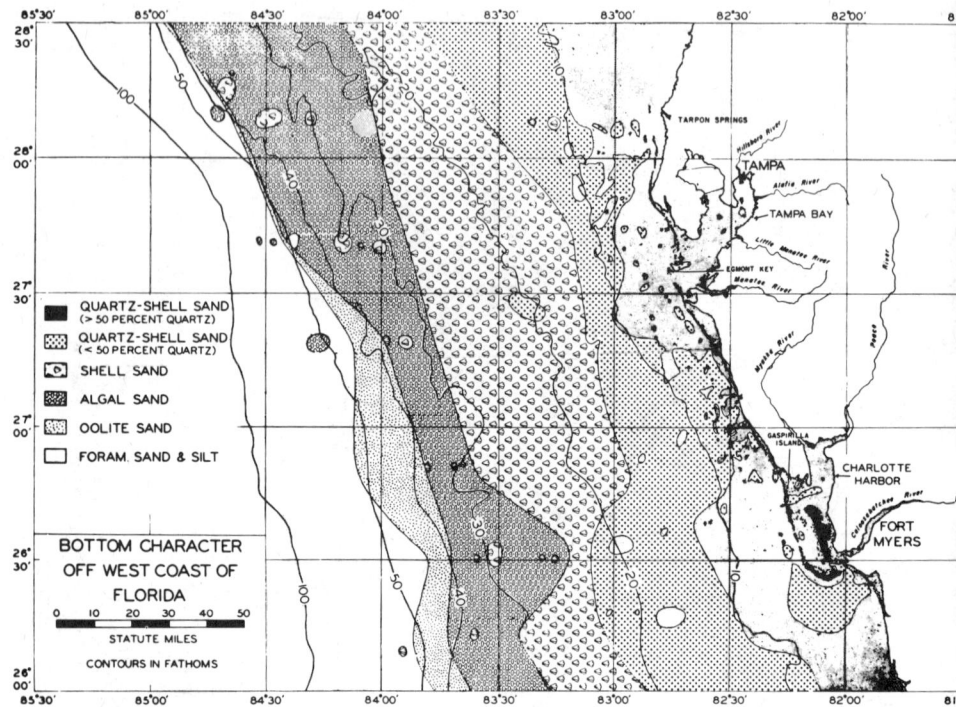

Fig. 9–14 The progressive change in sediment types in crossing the west Florida shelf. From Gould and Stewart (1955).

Another feature of interest is the series of hills, near the edge of the shelf, that become more prominent to the west where the shelf is wider off Mississippi. These hills, according to Ludwick and Walton (1957), were recently flourishing coral reefs that have been killed by the postglacial rise in sea level. Samples from these highs include reef rock and various types of calcareous organisms. Several of the shelf hills are now suspected of having underlying salt domes (Fig. 9–15).

The continental slope off Mississippi and Alabama shows another of the marked changes in trend that characterize the slope off eastern North America. At the northern end of the inner slope off west Florida, the direction changes at right angles, and the same change of trend occurs in much deeper water at the north end of the West Florida Escarpment (Fig. 9–13). De Soto Canyon indents the gentle slope near the change in trend. Several diapiric structures are located along this slope (Fig. 9–15). The continental slope off Mississippi and Alabama is more normal than off west Florida in that it is relatively steep at the edge of the shelf, about 1° as compared with a much gentler slope seaward of about 400 m.

Greater Mississippi Delta On the continental terrace off Louisiana, the shelf has been completely overlapped by the Birdfoot Delta during the past 500 years (Fisk et al. 1954, p. 98). The forward growth of the slope at an angle of 0.5° can be recognized by comparing old charts with new (Shepard 1955).

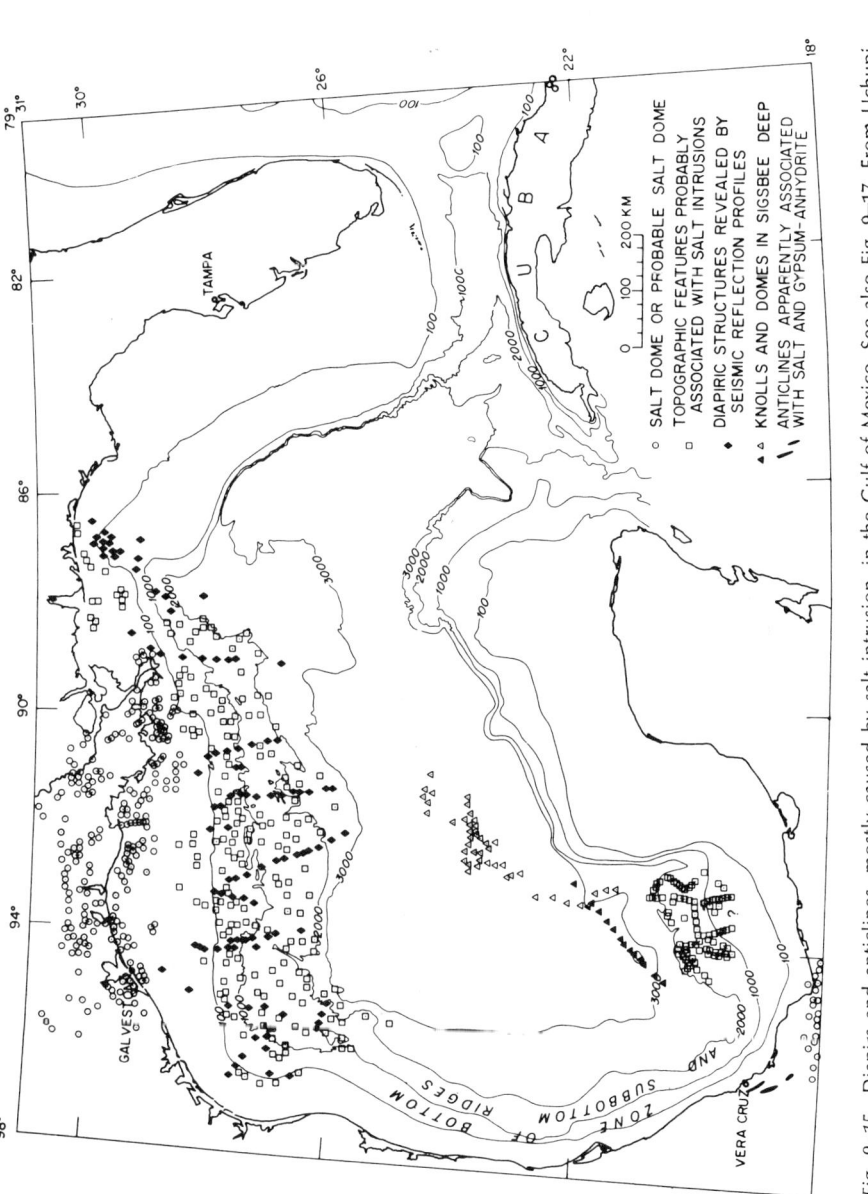

Fig. 9–15 Diapirs and anticlines, mostly caused by salt intrusion, in the Gulf of Mexico. See also Fig. 9–17. From Uchupi (1967A).

Seventy km east of the Mississippi Delta, the sandy shelf sediments change abruptly to silty clays, which represent the bottomset beds of the Mississippi Delta. The sand content of the sediments closer to the river mouths is much lower in quartz than the terrigenous sands to the east because of the intermixture of immature sands high in feldspar and ferromagnesians coming from the glaciated areas of the upper reaches of the Mississippi River system. The coarse fraction of the sediments around the Mississippi Delta has also a high content of mica, wood fibers, and small brown aggregates, all characteristic of shelf sediments bordering many other large rivers.

The continental slope off the Mississippi Delta (Fig. 9–13) has inclinations that are rarely more than 0.5°. West of the Birdfoot Delta, a trough extends about 35 km to the shelf edge and can be traced down the gentle outer slope for more than 100 km, where it debouches into a broad fan, usually called the Mississippi Cone. The cone evidently was developed by deposition of sediment carried out along the Mississippi Trough over a long period of time, especially during periods of Pleistocene lowered sea level.

Texas–Western Louisiana The terrace west of the Mississippi Delta is now one of the best-known areas of the sea floor, largely because of its great economic importance to the petroleum industry. The continental shelf bulges to a maximum width of more than 200 km near the Louisiana–Texas line and then narrows to about 90 km at the Rio Grande Delta. In the most extensive study of this shelf, Curray (1960) noted a number of elongate sand ridges and terraces on the shelf, which he interpreted as drowned barrier islands and beaches. In many places these slightly elevated sand highs contain abundant shells, mostly of shallow-water origin. Off the southwest end of Padre Island, detailed soundings show a series of the same type of low ridges trending diagonal to the coast as described from the northeastern Gulf and East Coasts (Rusnak 1960, Fig. 3). Again, the ridges trend left of normal to the coast and occur in depths of about 10 to 20 m. Rusnak explained them as related to the greater Rio Grande Delta, presumably during a lower sea-level stage. However, their trend is at right angles to that of the barriers around the other Gulf Coast deltas. A reasonable explanation must include the other examples of diagonal ridges where no delta relationship is evident.

A row of hills, mostly along the outer edge of the shelf, was first described and interpreted as salt domes (Shepard 1937). These hills have been extensively studied in more recent years (Parker and Curray 1956; Lankford and Curray 1957; Neumann 1958). Now they are known to be largely salt domes and partly mud diapirs (Lehner 1969). Calcareous growths have formed on most of them, because they were elevated by diapirism above their surroundings, allowing good circulation that favored coral and algal growth.

The sediments of the Texas–Louisiana shelf are mostly a mixture of silty clays and silty sands (Fig. 9–16). The sand content is particularly high off the Galveston–Port Arthur area, which is farthest from the large deltas. Studies of the foraminiferal and molluscan content of the sediments have shown that the mollusks are often relict of lower sea-level stages, whereas the foraminifera are mostly related to present depths of water, suggesting that the latter have sifted down into the older sediments. Foraminifera usually show a change from benthic to planktonic in crossing the shelf, so that a comparison of the ratio of the two gives an approximate distance of a sample from the present shore.

The continental slope topography off the Texas–West Louisiana area is strikingly different from any other known slope (Uchupi and Emery 1968;

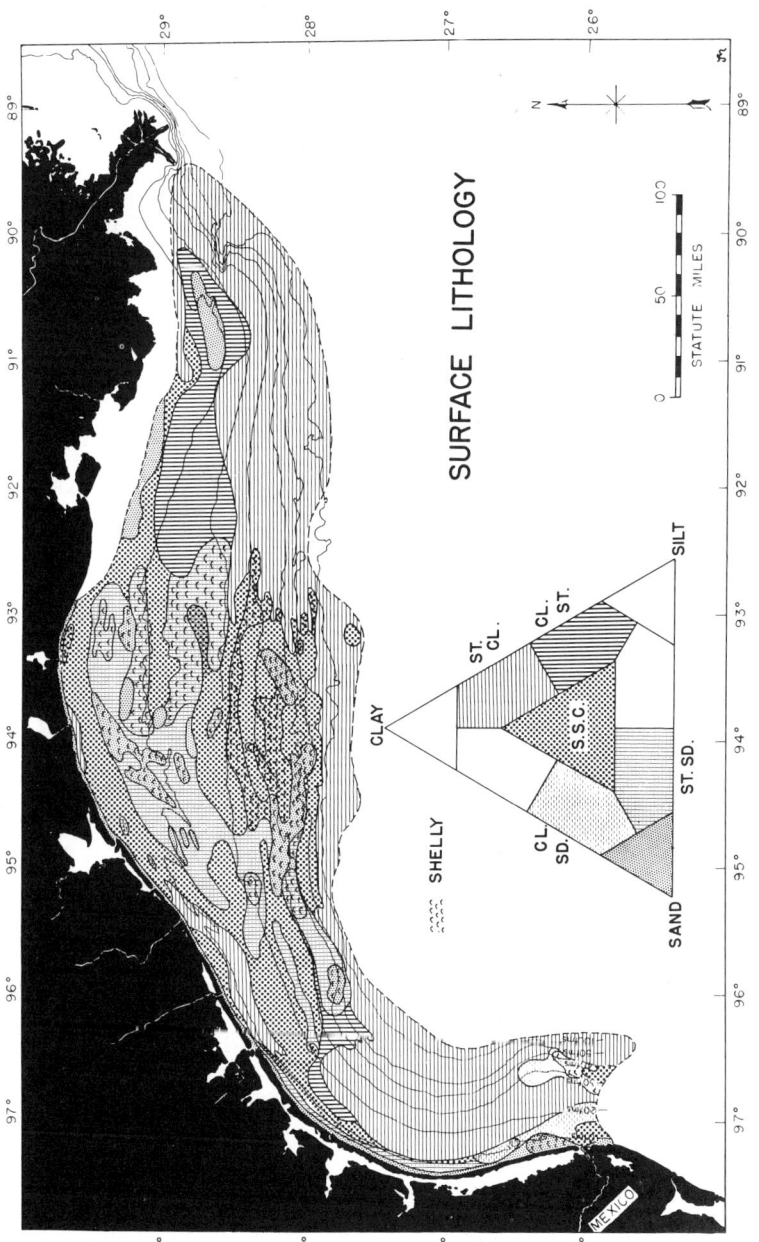

SURFACE LITHOLOGY

Fig. 9–16 Sediment types on the continental shelf west of the Mississippi Delta. For nature of sediment, see inset triangle diagram. From Curray (1960).

Lehner 1969). This area first became known from a map by Gealy (1955). The characteristics include an upper portion with oval hills, basins, and discontinuous valleys (Figs. 9–13, 9–17). The average slope of 0.5° is similar to that off the Mississippi Delta; otherwise the topography is entirely different. The oval hills and depressions are suggestive of a seaward continuation of the salt domes and diapirs of the shelf edge and are now generally so interpreted, although Gealy originally thought the topography was the result of mass movements. More than 50 of the hills are diapiric intrusions (Fig. 9–15). The hill and basin topography terminates at a depth of about 1800 to 2300 m at Sigsbee Scarp, which is attributed either to faulting or being the surface expression of a salt wall (Lehner 1969). This scarp has heights of about 500 m and is far gentler in slope than the precipitous West Florida Escarpment.

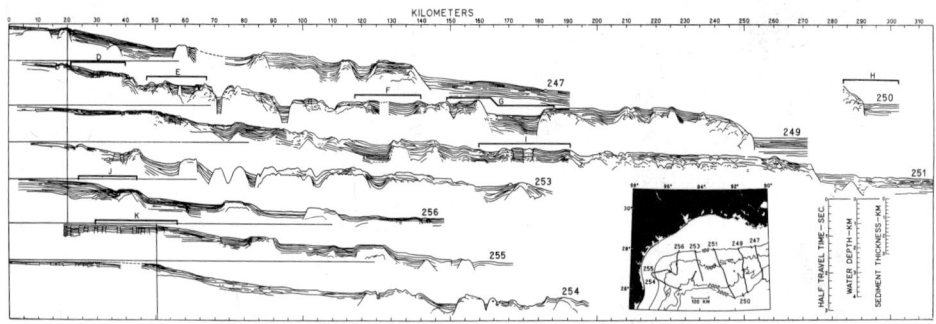

Fig. 9–17 Seismic reflection profiles of northwestern Gulf of Mexico showing importance of salt intrusions. From Uchupi and Emery (1968).

Eastern Mexico South of the Rio Grande, the continental shelf narrows to approximately 35 km and continues at about that width to the head of the Gulf of Campeche, except where Cabo Rojo, a large cuspate foreland, has covered half of the shelf. The continental slope off eastern Mexico is almost unbelievable. Fathograms and seismic profiles (Bryant et al. 1968) show that the slope consists of parallel anticlinal ridges with a relief of about 500 m and an average width of 10 km (Fig. 9–18). These extend roughly parallel to the shelf edge and cover the entire continental slope. They even penetrate the shelf margin. Between a northern and southern zone of these anticlinal ridges there is a zone with east-west trending escarpments having maximum relief of more than 1800 m. The escarpments are interpreted as a possible seaward continuation of the Zacatecas Fracture Zone (Murray 1961, p. 11). The longitudinal ridges north and south of the escarpments are an example of a feature almost unknown on the continents—folds that produce the present-day relief.

The Gulf of Campeche has a trough trending north and south, bounded on the east by the Campeche Escarpment. This trough, first described by Creager (1953) and later studied by Worzel and others (1968), contains a zone where numerous diapirs can be traced seaward to the center of the Gulf of Mexico and landward to a group of salt domes in Campeche.

A shelf with widths up to 200 km flanks the west and north sides of the Yucatan Peninsula. Water depths are mostly less than 50 m. Various oval coral banks rise above this shelf and one of them, Aricife Alcaran (Fig. 12–17),

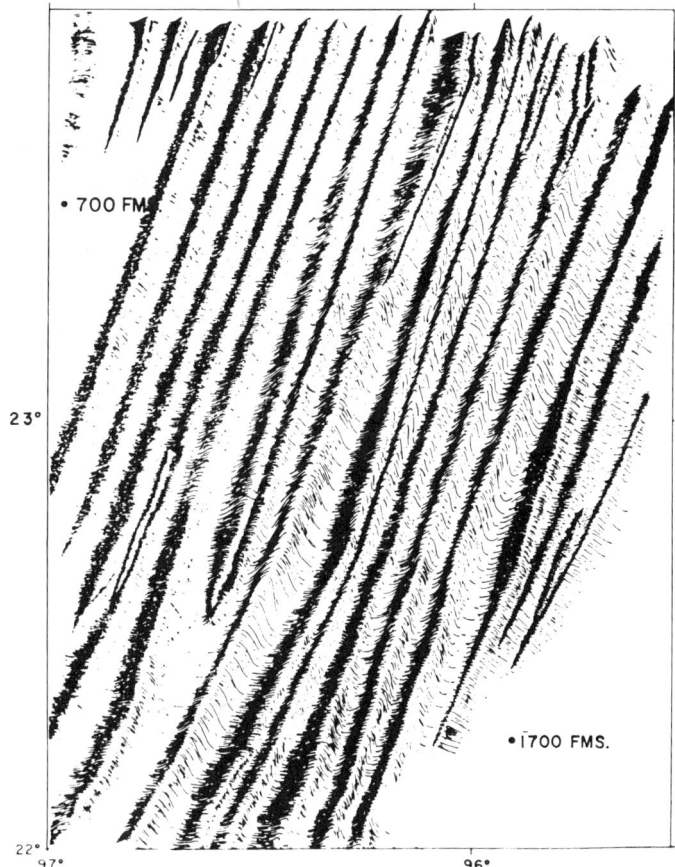

Fig. 9–18 Diagram showing anticlinal ridges off eastern Mexico. From Bryant et al. (1968).

resembles an atoll in having a central lagoon (Kornicker and Boyd 1962). Most of the sediment on this shelf is calcareous. The Campeche Escarpment that borders the broad shelf has an irregular margin similar to the southern portion of the West Florida Escarpment. One large basin depression in the margin is suggestive of the effect of solution on the limestone within the escarpment. On the east side of Yucatan, as on the east side of southern Florida, the·shelf virtually disappears, and here also the Gulf Stream may have played an important part in producing this asymmetry. Two narrow ridges off this coast may be remnants of a Paleozoic fold belt (Baie 1970). Farther south, a wide zone off eastern Honduras and Nicaragua has extensive coral reefs and other calcareous shoals.

    Northern South America The east–west coast along the north side of South America (see Fig. 13–17) is bordered mostly by narrow continental terraces with offlying rows of islands. Along much of this coast there appear to be large strike-slip or transform faults related to independent movement of the South American and Caribbean plates (Ball and Harrison 1969). The long straight coast of Venezuela that almost lacks a shelf can be explained in

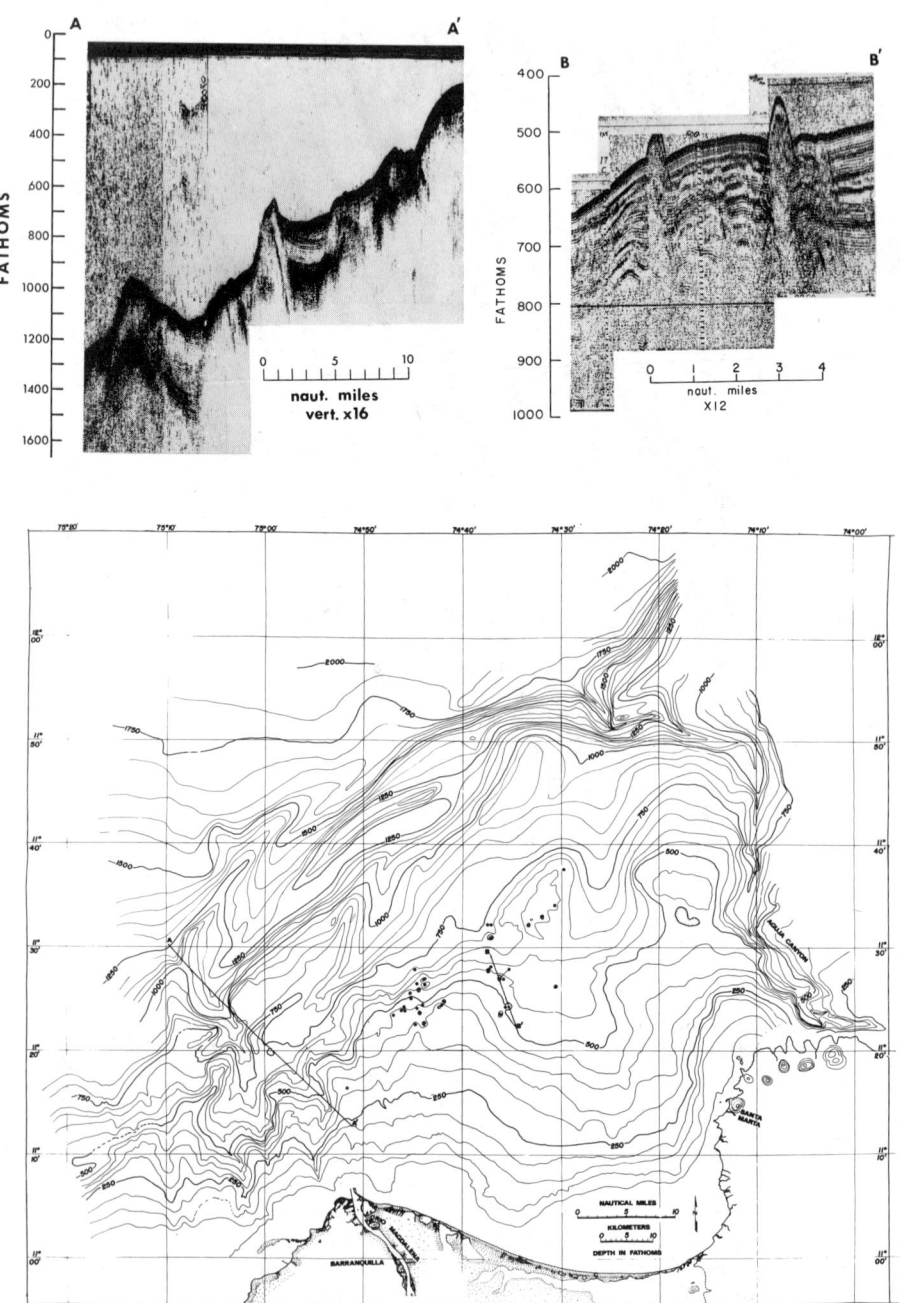

Fig. 9–19 Sea-floor character off Magdalena Delta and Santa Marta, Colombia. Note how the head of Aguja Canyon extends parallel to the mountainous coast. Mud diapirs that have pushed up the sea floor shown by solid circles, and two are indicated in seismic profile B–B'. Profile A–A' shows the fault troughs and horsts to the northwest of the Delta. Contour interval 50 fm (92 m). Survey and contours by the author.

this way: To the west, the Gulf of Darien, separating Panama from Colombia, has some indications of being a rift valley, although it has been sufficiently filled so that its depths are now less than 50 m. Deposition of mud sediments is very rapid, as much as 1 m in 7 years (Zeigler and Athearn 1968).

The Magdalena River of Colombia, which drains most of the northern Andes, has built a great delta into the Caribbean and, like the Mississippi, has entirely eliminated the continental shelf over a wide area (Fig. 9–19). The foreset slope, about 3°, is much steeper than off the Mississippi. The slope has complex topography, being cut by valleys and intruded by numerous mud diapirs, some rising as much as 200 m above their surroundings (Shepard et al. 1968). Along the outer slope north of the river mouth are northeast–southwest trending ridges that are apparently related to faulting; the slope to the west has fold ridges and depressions that conform with the underlying structure, like those off eastern Mexico. A fault scarp near the base of the slope flanks much of the deep Caribbean Basin.

The largest shelf sea along the north coast of South America is the Gulf of Venezuela and its connecting inland sea, the Gulf of Maracaibo. The sediments of the Gulf of Venezuela (Zeigler 1959) are largely silty clay with terrigenous sand around the margin; the former include several zones high in fecal pellets. Indicating the recency of tectonic movements, a submerged arch crosses the Gulf of Venezuela and persists as a high despite heavy sedimentation. The Gulf of Maracaibo has very fine sediments with low oxygen content (Redfield 1958).

The Cariaco Trench, off the north coast of Venezuela south of Tortuga Island, extends west from the narrow Gulf of Cariaco and has depths of 1400 m. Athearn (1968) found that the recent sediments of much of the Trench were deposited under euxinic conditions and, as a result, are well laminated and low in benthic fauna. However, piston cores showed that normal marine conditions existed during the Pleistocene. The JOIDES drilling at Site 147 in the Trench penetrated anaerobically deposited sediments.

Eastern South America Off the main mouths of the Orinoco, both Nota (1958) and van Andel (1967) found conditions that are in many ways similar to those off the east side of the Mississippi Delta. A broad shallow platform off the river (Fig. 9–20) has laminated silty sediments that become clayey offshore and lose their lamination. Halfway across the shelf on the outside, sand increases and calcareous sediments become important. The calcareous sands of the outer shelf represent relict deposits, like those of the northern Gulf Coast. Near the outer edge, a series of calcareous-covered banks, said to be bioherms, rise above the general level. Unlike the Mississippi, the Orinoco has not built appreciably across the shelf, probably because of the North Equatorial Current.

Farther south, the Amazon, despite its enormous sediment transport, has failed as yet to fill the last remnant of a large estuary. However, a submarine delta is being built on the inner continental shelf outside the river mouth (Russell 1958). According to Ottmann (1968), the shelf off the Amazon has shelly sand sediments on the outside and mud on the inner shelf. The continental slopes off Brazil are still little known, but apparently are low off the large rivers, comparable to that off the Mississippi. The slope has a broad outbend off the Amazon (Hayes and Ewing 1970). A valley, possibly similar to that off the Mississippi, extends down the slope and is a path for transporting sediment to the deep ocean floor. A large slide block was found near the base of the slope (T. C. Moore et al. 1970). South of Cape San Roque, along the narrow shelf zone, the slopes are steep for about 2500 km, having

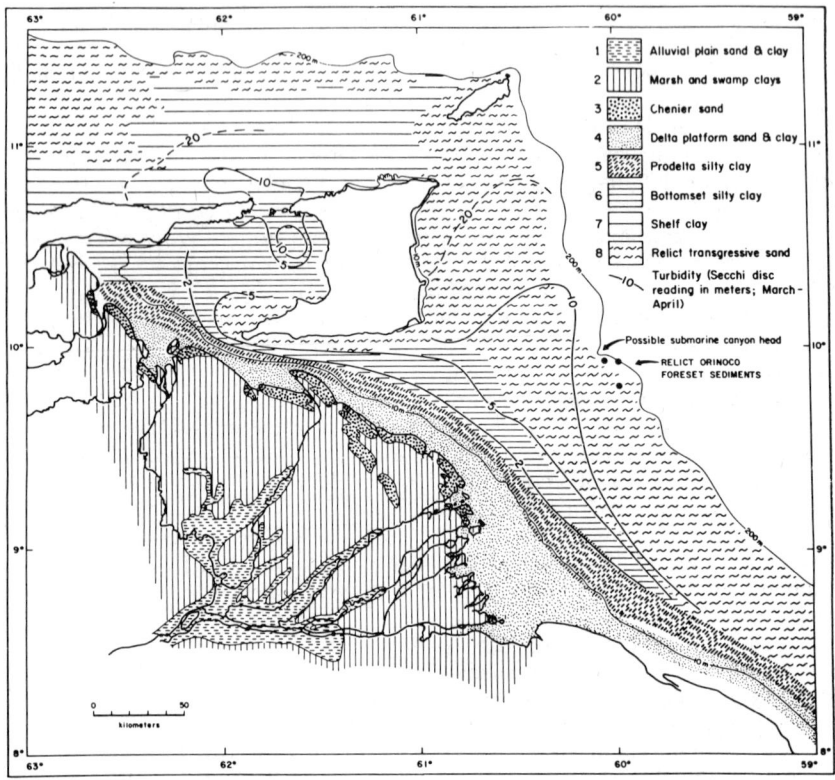

Fig. 9–20 Deltaic and marine sediments of the Orinoco Delta and adjacent shelves. From van Andel (1967).

inclinations from 4 to 20°. This is perhaps the longest stretch of continental slope in the world with almost consistently steep inclinations.

South and east of the Amazon, the generally wide shelves of the northern lowlands shrink and become largely calcareous, with many coral banks and other types of shoal-water reefs. South of Cape San Roque, a shelf only 20 km wide extends along the southeastern length of Brazil as far as Queen Charlotte Bank, where a large coral reef with abundant shoals increases the shelf width to 120 km. Farther south, the shelf is irregular both in width and depth. Off the large Rio São Francisco, there are again a mud zone near shore and sand on the outer shelf. South of Rio de Janeiro, the shelf widens to 180 km off the large estuary, Rio de la Plata. The sediments in the estuary (Ottmann and Urien 1965) are a mixture of sand and mud, with clean sand on the open shelf and shelly sand near the outer margin.

The continental slope off southern Brazil and Uruguay has been investigated (Butler 1970). Here submarine canyons are lacking, although they occur both to the north and south. The slope south of Rio de Janeiro has diapirs and associated grabens that result in topographic depressions. A series of prograded sedimentary wedges have built the continental slope forward, accompanying the subsidence that began in the Late Cretaceous. The history shows a relationship to that of Angola in West Africa from which it is thought to have been rifted. From latitudes 39 to 49° S. the shelf break averages 140 m but commonly reaches 180 m; to the north, it is rarely more than 90 m. The Golfo San Matias, south of the Rio de la Plata, was investigated by the *Vema* (Granelli 1959). Ripples 7 m high were found at the entrance, and giant ripples of 15 m were discovered in the gulf with numerous smaller ripples in between. Strong tidal currents up to 8 knots are indicated on the charts. These are related to the large tidal range, as much as 7 m. Despite the currents and ripple marks, fine sediment is reported in the gulf. According to Granelli, a core had 9 m of clay overlying 3 m of sand.

Farther south, the 200-km-wide Golfo San Jorge, like the Golfo San Matias, is remarkable in having water more than 50 m deep along the entire coast and a large basin with gentle slopes in the center of the bay. The strong tidal currents along this coast may explain these unusual shelf features; more detailed investigations of this portion of the shelf are needed. Cores obtained from the Argentine shelf by Lamont-Doherty Geological Observatory (Fray and Ewing 1963) show a homogenous sand layer underlain by sand with abundant shells. The latter are indicative of shallow cold-water conditions and out to depths of 120 m prove to have carbon-14 ages from 11,000 to 16,000 years. Evidently these sands and shells were deposited during low stands of sea level, like similar deposits off the East and Gulf Coasts of the United States. Seismic refraction shows two elongate sedimentary basins extending across the shelf (M. Ewing et al. 1963). These have sedimentary rocks with a total thickness of 8 km (Fig. 9–21). The continental slope off Argentina has a series of submarine canyons, apparently similar to the slope off most of the East Coast of the United States.

The southern tip of South America was covered by glaciers during the Pleistocene, which influenced the topography. Apparently there are troughs off Patagonia, and heterogeneous types of sediments are indicated. A trough with depths between 200 and 700 m extends up along the south and west sides of the Falkland Islands. This is bordered to the south by Burwood Bank. The Strait of Magellan is similar to the deep inlets in other glaciated zones.

Western South America and North America to Gulf of California Most of the West Coast of South America and North America up to the Gulf of California is interpreted as in the underthrusting zone between a west-moving South American plate and an east-moving plate coming from the East Pacific Rise. As would be expected in an underthrusting zone, large earthquakes are common. The continental terrace is predominantly very narrow, and a deep trench extends along most of the outer slope. The exceptions are at the southern end of South America and off Panama , where the trench is missing and few earthquakes occur.

Northwest of Cape Horn, the irregularity of the shelf increases considerably, and a series of troughs cross it with depths comparable to those off Labrador and Newfoundland. The deepest fiords known (maximum depth 1650 m) indent the coast, and these inlets occur as far north as 42°S latitude, the terminus of glaciation. To the north, the shelf becomes relatively flat but quite deep, mostly more than 100 m, and very narrow. The slopes of the inner shelf are generally quite steep as far north as 7°S latitude. As off Argentina, the shelf edge is deep, usually close to 200 m. From 14 to 6°S latitude, the shelf is wider, about 55 km across, and at the Golfo de Guayaquil it reaches 75 km.

Off the West Coast of South America, the continental slopes are notable for their great vertical range (Zeigler et al. 1957; Scholl et al. 1970B).

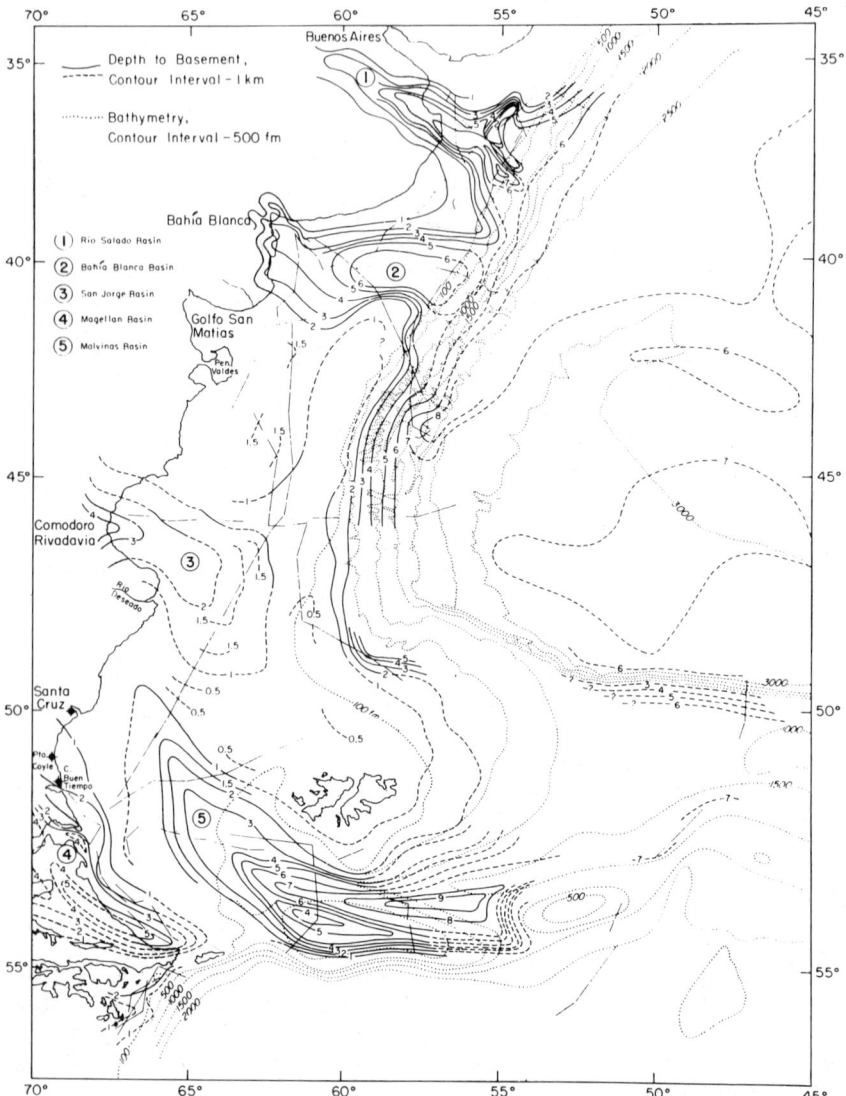

Fig. 9–21 Subsurface structural map of area off Argentina shows approximate depth to basement (in kilometers) and the bathymetry. Note the two elongate sedimentary basins. From M. Ewing et al. (1963).

Many of them go down to 5500 and even 8000 m, terminating in the Peru-Chile Trench. Including the adjacent slopes of the Andes, these escarpments have the greatest vertical extent of any in the world, amounting to about 13,000 m in one place. However, despite these great vertical ranges, the slopes are for the most part not very steep, averaging about 5°. Seismic profiles (Scholl et al. 1970B) indicate that the slope terraces with their structural basins have Cenozoic sedimentary fill, and the gentler portions of the slope have a thin sedimentary cover of disturbed Cenozoics. The lower slopes, mostly 10 to 15°, have basement rocks, and the basal escarpment is partly buried by the Late Cenozoic trench deposits. Lines run along the continental slopes show canyons off some of the main rivers, but no detailed surveys exist.

North of Peru, little information is available concerning the South American slope. In Central America from Costa Rica north to the Gulf of California, steep slopes descend to the Middle America Trench (Fisher 1961) and are comparable to those off most of western South America. Ross and Shor (1965) found a bench along the slope underlain by a basin mostly filled with sediments deformed by folding , faulting, and slumping. In the San Blas area, south of Mazatlan, investigations have shown that sediments from the land have built the terrace forward for many kilometers (Curray and Moore 1964).

At the Gulf of Panama, the shelf widens to about 140 km. Flat bottom with depths of about 90 m is conspicuous in this gulf. The shelf terminates near the 200-m contour. The Gulf of Panama has been well charted for bottom character (Golik 1965). The sediments become increasingly coarse away from shore. Mud occurs in the inner Gulf, and sand and shells become more abundant away from shore and out beyond the lee of the Perlas Islands. Rock is reported in several places near the shelf edge. To the north of Panama, the shelf is narrow or nonexistent up to 11°15′N latitude. Beyond this point, the shelf widens to 55 km and continues wide on the west side of the Gulf of Tehuantepec, where the maximum width is 100 km. Soundings are scarce, but apparently this is a deep shelf with some zones more than 200 m. From Puerto Angelo to Puerto Vallarta, the shelf is virtually nonexistent. At Puerto Vallarta, Banderas Bay has a deep fault trough reentrant into the coast. Just south of San Blas, the shelf starts again to have appreciable width; this widened zone extends north of Mazatlan to the Gulf of California.

The Gulf of California is now recognized as being a rift valley, and is thought to be an initial stage of sea-floor spreading and to represent the northward continuation of the spreading East Pacific Rise (Rusnak and Fisher 1964). The shelves are narrow, particularly on the west side of the Gulf. The slopes have virtually no submarine canyons except at the mouth of the Gulf (Rusnak et al. 1964). The eastern shelf has apparently had a depositional origin, as has the shelf at the head of the Gulf where sediments come from the Colorado River (van Andel 1964).

Western Baja California and California North of where the East Pacific Rise enters the Gulf of California, the continent appears to be overlapped onto the deep-ocean floor. A series of ranges and fault blocks extend diagonal to the continental margin. The same characteristics are found in the continental borderland of this area as on the lands (Fig. 9–22). Much of the faulting is strike-slip, and some of it is clearly related to the transform faulting of plate tectonics (see p. 92).

The shelves off Baja California are mostly narrow except for the 80-km shelf in San Cristobal Bay, north of Magdalena Bay, and the 100-km shelf in Sebastian Vizcaino Bay, north of Cedros Island. The shelf along the west coast of Baja California was studied rather extensively during and since World War II. The sediments of San Cristobal, Sebastian Vizcaíno, and Todos Santos bays are described by Emery and others (1957). The wide-open San Cristobal Bay has a sand shelf with little or no relation of grain size to distance from shore. Phosphorite is found in most samples. D'Anglejan (1967) investigated the phosphate nodules and grains in San Cristobal Bay and discovered that a large deposit of phosphorite occurs on the margin of the shelf adjacent to

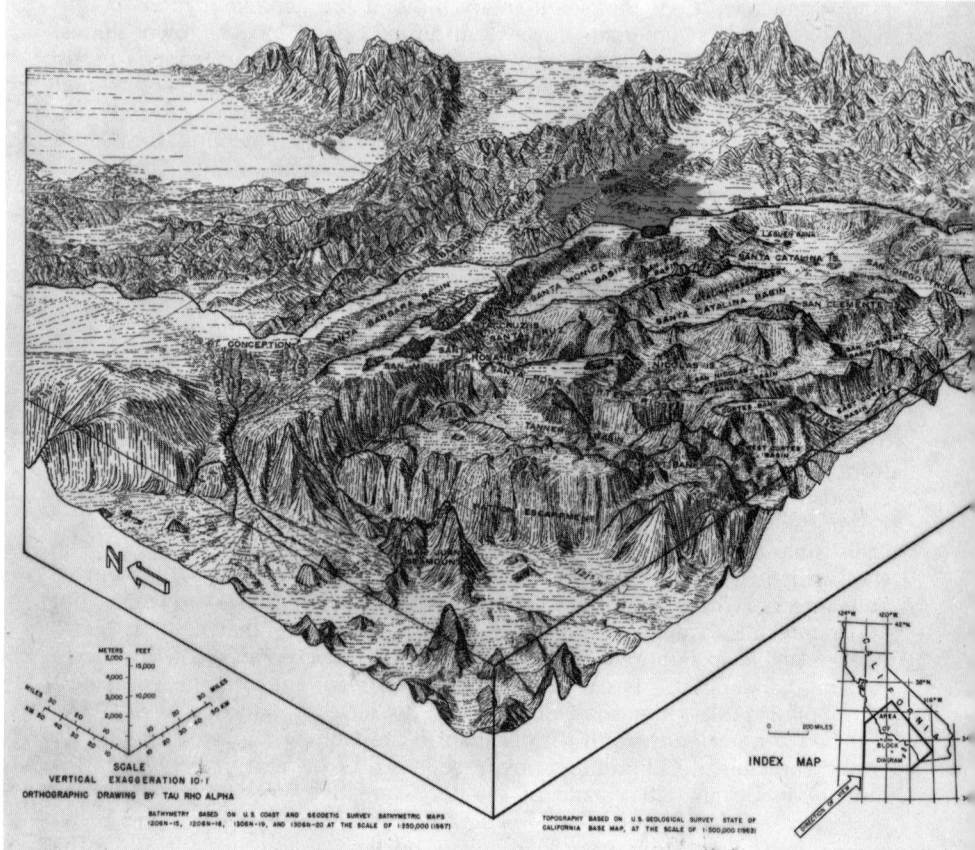

Fig. 9–22 Continental borderland off Southern California shows ridges and basins, largely the result of faulting. Physiographic diagram by Tau Rho Alpha, U.S. Geological Survey.

a shallow trough. Vizcaíno Bay, partially protected by Cedros Island, has much more sediment with some decrease in grain size outward. The coarsest sediment is found in the strait at the south end of Cedros Island where the currents are strong. Todos Santos Bay, at Ensenada, is also protected by several islands and by Punta Banda to the south. The shelf sediments landward of the islands grade out from clean, fine-grained sand along the shore to micaceous silt in the deeper water; to the north of the islands, the outer shelf is rocky with sand and gravel deposits. The strength of the currents is indicated by the giant ripple marks seen in bottom photographs. Some of the deep outer sand is definitely shown to be relict by the presence of extinct shallow-water foraminifera (Walton 1955).

North, along the west coast of Baja California, only miscellaneous lines of Scripps Institution, Hancock Foundation, and U.S. Navy soundings are available as far as Cedros Island. These slopes are cut in some places by submarine canyons. Elsewhere they are remarkably even. South of Cedros Island, an escarpment leads down to a small trench. These slopes are said to contain metamorphic rocks (Uchupi and Emery 1963). Somewhat north of Cedros Island, the trench terminates and a continental borderland comparable to the basin-and-range topography off Southern California begins (Uchupi and Emery 1963; Krause 1964A; D. G. Moore 1969), continuing to Point Conception, west of Santa Barbara, California.

The shelf off San Diego was studied in detail during World War II (Emery et al. 1952). This shelf is only 20 km wide, but it has many features of interest (Figs. 9–23, 9–24). On the outside it is flanked by a rocky submarine bank that continues north as a submarine ridge. To the south, the bank is cut by a canyon, but beyond the canyon it rises to form the Mexican Coronado Islands. Inside the bank, a shallow longitudinal valley borders the coastal shelf. Sediments off the Tia Juana River, near the Mexican border, alternate in a seaward direction between coarse and fine. The sediments now introduced from the land locally grade outward from coarse to fine; but relict, and hence presumably Pleistocene, sediments are represented over wide areas (Fig. 9–24). Thus, an extensive cobble and boulder patch is found 1 km outside the Tia Juana River, and apparently the same deposit underlies a much larger area to the west and north. Farther seaward, a large deposit of brown semioxidized, medium-grained sand is also clearly relict and, like the boulders, was probably subaerial in origin. During World War II, a large amount of sediment was dredged out of San Diego Harbor and dumped onto the inner-shelf sediments. This formed an oval patch of sandy mud that was later removed by currents until now there is a fine sand in the area

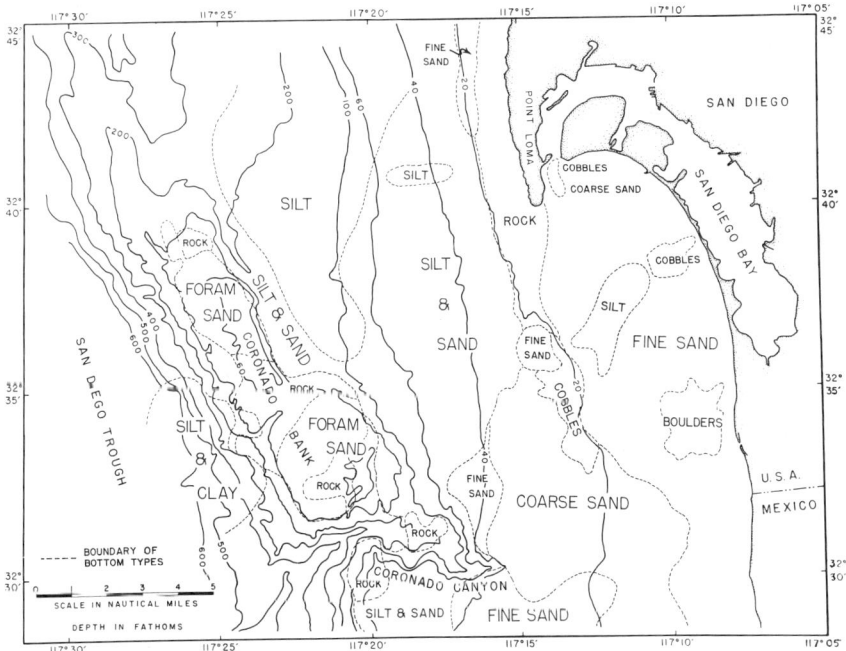

Fig. 9–23 Bottom sediment types and topography on the shelf off San Diego. The silt zone near shore has now disappeared, having been a temporary result from dredging of San Diego Bay. See also Fig. 9–24.

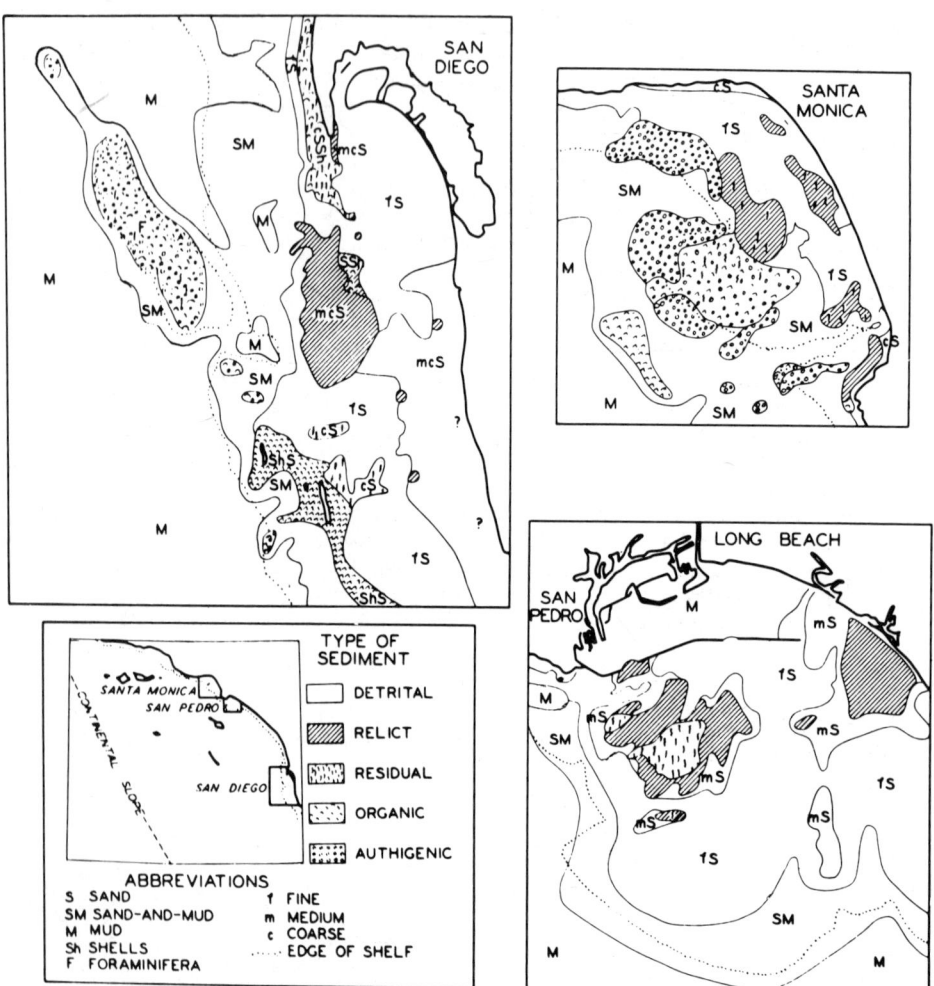

Fig. 9–24 Types of sediments of three areas in Southern California, classified as to origin. Dotted line is shelf break. After Emery (1960, Fig. 180).

comparable to that of the surrounding deposits. Similarly, after rainstorms, muddy sediments settle temporarily on the coarser sediments outside but are soon carried away. The outer bank is not receiving any appreciable amount of sediment from the continent at present. Between outcrops of rock on this outer shelf, there are sediments consisting of foraminifera along with some relict terrigenous material, partly silt. Many of the foraminiferal tests are filled with glauconite. Phosphorite nodules are also found, especially on the deeper parts of the bank beyond depths of about 140 m.

Inman (1953) investigated the nearshore sediments north of Point La Jolla where the narrow shelf is cut by two canyons. Most of the shelf sediments are fine sands like the local beach, except south of La Jolla Canyon where medium-grained sand comes from the pocket beaches. Wimberley

(1955) studied the sediments directly north of Scripps Canyon and found that the texture is in general related to water depth with a decided increase of mud at about 55 m. However, a sand suggesting a former beach was located at 75 m at the shelf margin.

Farther north, a narrow shelf extends to the Long Beach area where it again widens. Here, rocky zones were indicated along the central and outer portions (D. Moore 1954), and Pleistocene sediments surround the central rock area (Fig. 9–24). Both relict and residual sediments are found. The central shelf has Miocene rock flanked by residual coarse brown sand. To the southwest is a large expanse of relict sands that has a Pleistocene fauna. This relict sediment is recognized also by a lithological resemblance to Pleistocene dunes bordering Santa Monica Bay and by brown iron-stained sand like that off San Diego. The shelf is very narrow off the Palos Verdes Hills, as one would expect because of the Pleistocene uplifts (Woodring et al. 1946). It consists of a wave-cut platform overlain by 15 m of layered coarse sediments (Uchupi and Gaal 1963) that are partially capped by 12 m of homogeneous silty sands. In Santa Monica Bay, a well-studied shelf (Terry et al. 1956; Emery 1960, p. 200) has rock and coarse relict sediment on the outside (Fig. 9–24). Some of this relict sand is also brown-stained. Terry and others (1956) compared the sediment of Santa Monica Bay reported by Shepard and Macdonald (1938) with that collected 20 years later and concluded that the grain size had definitely decreased. They attributed this to the diversion of the mouth of the Los Angeles River to Long Beach and to other artificial changes, thereby reducing the supply of sand sediment introduced into the Bay. Natural oil and gas seepages occur in this area. The shelf edge averages about 80 m, typical of the entire Southern California coast.

West and north of Santa Monica, the shelf is mostly narrow, much of it either rocky or covered only with a thin veneer of sediments. Oil exploration, particularly in the vicinity of Santa Barbara, has shown that rock structures underlie much of this shelf, and drilling is taking place here when permitted. From Ventura to Point Conception, much oil and bitumen are seeping naturally out of the underlying rocks, so the water surface over wide areas is often oil-covered, especially near Coal Oil Point, west of Santa Barbara. Other oil spills come from drilling operations. In places, scuba divers reported streams of oil rising 10 m high, and asphalt domes 3 m high and 15 to 35 m across (R. F. Dill, personal communication, 1960). These tar mounds were subsequently described by Vernon and Slater (1963) off Carpinteria, Goleta, and Point Conception. Smaller seeps occur in Santa Monica Bay and in other places along this part of the coast (Emery 1960, Fig. 244). Near Point Conception, the bottom currents are occasionally so strong that little sediment is found on the shelf; it is often difficult for scuba divers to stem these currents.

Using a bottom-penetrating echo sounder, studies of sediment thickness and shallow structure on the shelves off Southern California (D. Moore 1960) showed that relatively thin lenticular masses cover smooth sloping rock platforms. In general, the sediment thickens landward of the shelf margin as well as down the slope beyond. Great local variation in sediment thickness on these shelves is caused by topographical control and proximity to land sources. Most of the mid-portion of the shelves has 10 to 15 m of sediment, except off La Jolla where the shelf is mostly bare rock. Rock structure is evident in seismic profiles, showing anticlines and synclines in some places.

The Southern California continental borderland province is about 280 km wide (Fig. 9–22), terminating westward with an escarpment comparable in height and declivity to that of the eastern Sierra Nevada, but differs in the absence of the great canyons that cut the land escarpment. This area was discussed by Shepard and Emery (1941), and more recently by Emery (1960) and D. Moore (1969), the last based on seismic reflection. All borderland basins are shoaler than those typical of the deep ocean. To the south off San Diego, San Diego Trough, a former basin, has been filled to its sill depth, but the other basins are only partly filled (Emery 1960, p. 52). Some of the intervening fault blocks rise above the surface as high islands, such as San Clemente and Santa Catalina. Others form shallow banks, such as Cortes and Tanner. Near the outer escarpment, the tops of the banks are several hundred meters deep. Studies of the Tanner Bank (D. Moore 1969) indicate an anticline with the topography a reflection of the structure. Others of the basin-range highs owe their topographic expression to faulting, and their internal structure is discordant with their topography. On the sides of the basins there are relatively straight escarpments in most places. Near the continent and around some of the islands and banks, these escarpments are cut by valleys.

Sediments in the borderland basins and in San Diego Trough are mostly a silty clay but also contain many layers of sand, especially around the fans outside submarine canyons. The sand layers in the outer basins are largely calcareous, as are the sands on the outer banks (Emery 1960, pp. 210–227; Gorsline et al. 1968). These calcareous sands are a mixture of foraminifera and shells of other types. In many places, the foraminiferal sands are associated with glauconite and phosphorite (Dietz et al. 1942). Rather extensive areas on the banks have rock bottom and well-rounded gravel up to cobble size.

To the north, the outer escarpment of the continental borderland is interrupted west of the east–west ranges of the Santa Barbara area. However, a northward continuation of the scarp shows an offset to the east. Farther north, there is no borderland inside the scarp, but instead a 1° slope extends to 1500 m as a marginal plateau with a slope of 12° beyond. This seaward change of slope is suggestive of the gentle slope and steep outer escarpment off west Florida and western South America.

The shelves are narrow or even missing along the mountainous coast from Point Conception to Monterey Bay. Near San Francisco, the shelf again widens and maintains its width off most of northern United States and British Columbia. Off Pigeon Point (about 75 km south of San Francisco), D. Moore and Shumway (1959) studied the 28-km-wide shelf. Samples and bottom-penetrating echo-sounding data convinced them that the broad flat outer shelf was cut into Pleistocene sediments, and that the inner steep shelf has a rock floor with little sediment cover. The Golden Gate, the entrance to San Francisco Bay, is a locus of strong currents with resulting rock and gravel bottom. Outside the Gate, a sandbar some 9 m deep forms an arc with natural openings only at the south and north ends. This bar is constantly shifting in position and depth. The open shelf is mostly sand-covered, but near the outer edge a rocky rim rises to the surface to form the Farallon Islands. Cordell Bank to the north has littoral zone shells at 62 m, and Miocene rocks have been dredged from the area (Hanna 1952). North of Drakes Bay, the shelf shows an alternation seaward of sand, then mud, and then sand, as do so many other shelves.

In the San Francisco area and to the north, the shelf edge is deeper than off Southern California, mostly 130 to 145 m. North of San Francisco, the shelf locally narrows to 18 km at Point Arena, and the inner part is steep. Evidence of greater marginal depths of the rock floor comes from the bottom-

penetrating echo-sounding studies of D. Moore and Shumway (1959). Off
Pigeon Point, they found that the sloping rock terrace terminates seaward at
about 165 m, whereas the Holocene sediment terrace overlying the rock has
a margin at about 110 m. Another rock terrace near shore terminates at about
90 m where the rock outcrops. These records show the difficulty of estimating
the depth of terrace cutting from topography alone.

Along the Santa Lucia Mountain coast, north of Cape San Martin, the
steep portion of the slope is near land, in contrast to the gentle slope farther
south. Approaching Point Sur, a series of large canyons with numerous tributaries cut into the slope. These culminate in the great Monterey and Carmel
Canyons but continue intermittently almost to the northern border of California. The continental slope off the San Francisco area (Curray 1965B) has
basement rock at or near the surface. These rocks outcrop on the outer shelf
near the Farallon Islands. This basement ridge has been traced by Curray both
north and south of the Farallons (Fig. 9–25). Sediment basins are found landward of the basement high and seaward at the base of the slope. This is
similar to the two basins underlying the outer shelf and slope base off eastern
United States (Drake et al. 1959). A profile across the slope shows contorted
sedimentary rock, indicating large landslides seaward of the Farallon Islands.
At Punta Gorda, just south of Cape Mendocino, the entire continental slope
is offset and an east–west escarpment extends seaward for more than 75 km
(Fig. 9–26). Many earthquakes are known to have originated along this escarpment (Tocher 1956; Shepard 1957). Furthermore, the inner end of the scarp
bends to the southeast and points toward Shelter Cove, where the northernmost observed displacement occurred at the time of the San Francisco earthquake. South of the escarpment, a submarine canyon terminates headward
directly against a smooth mountain wall rather than opposite the land valleys.
This appears to give evidence of a large horizontal movement similar to that

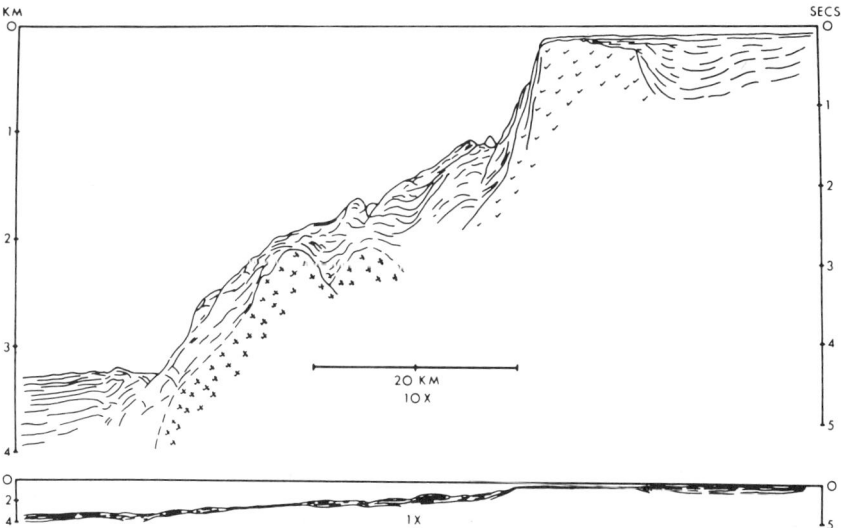

Fig. 9–25 The basement ridge along the outer continental shelf north of San Francisco. Note
landslide blocks along the upper slope and the sedimentary basins inside the shelf edge and
seaward of the slope base. Compare with Fig. 10–3B. From Curray (1965B).

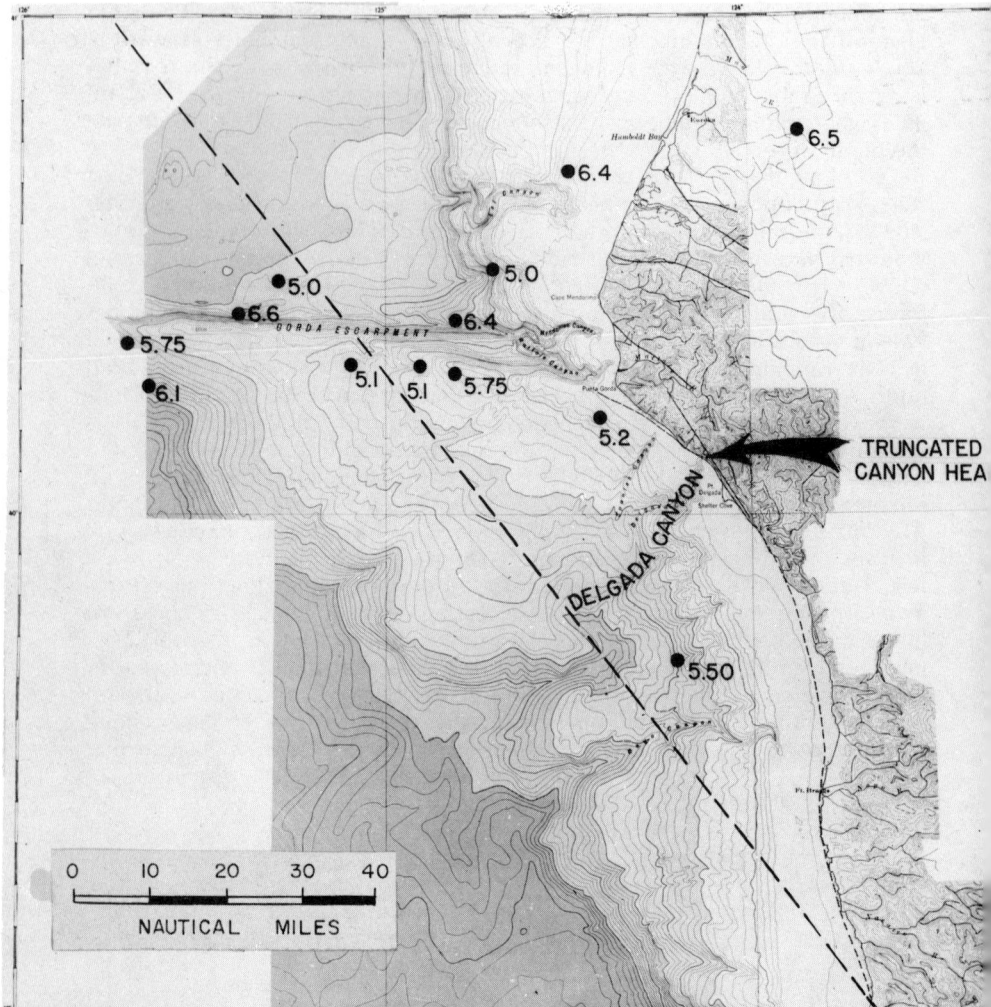

Fig. 9–26 Gorda Escarpment with large offset of the continental slope where the San Andreas Fault apparently turns seaward directly inside the Mendocino Fracture Zone. The old interpretation of the San Andreas Fault is given by the dashed line, but this line shows no offset of the Gorda Escarpment and hence is discredited. Exact continuation of the fault is not known definitely, but the head of Delgada Canyon appears to have been truncated, indicating displacement near the coast. Major earthquakes follow the Gorda Escarpment seaward.

of the San Andreas Fault to the south (see also p. 380). The area was investigated more recently by Nason (1968). However, the research by Curray and Nason (1966) shows that the main San Andreas Fault may lie several kilometers seaward of this truncated canyon head. The whole region is complicated by the Mendocino Fracture Zone that lies seaward of the Gorda Escarpment (see p. 382).

Northernmost California, Oregon, and Washington   North of the Mendocino–Gorda Escarpment, the continental terrace again comes under the influence of a subduction zone between a narrow block east of the reappearing East Pacific Rise and the west-moving block of the American plate (see Fig. 5–4). Despite this change in influences, the general appearance of the shelf and slope is remarkably similar north and south of the great escarpment. A narrow shelf on both sides of the juncture is deeply indented by submarine canyons. However, a decided change does take place near latitude 41°20′. To the north, there are no large canyons, and the slope (well illustrated on Coast and Geodetic Survey charts 1308N-12, 17, and 22) has a series of basins and ridges trending north and south and cut by an escarpment that is roughly between depths of 2000 and 3000 m, the latter being the slope base. Byrne (1966) interprets this basin and ridge topography as due to compression caused by sea-floor spreading from the East Pacific Rise. Beginning off the mouth of the Columbia, submarine canyons again penetrate the outer continental shelf all the way to the Strait of Juan de Fuca. However, the elongate basin and ridge topography on the outer slope has somewhat less relief than farther south (Nelson et al. 1970). The canyons and fan valleys off this area are described in many papers and some doctoral theses by Oceanography Department students of Oregon State University (see also p. 318).

Off the Klamath River in Northern California, G. Moore and Silver (1968) discovered deltaic sediments 60 m thick overlying folded Pliocene rocks. The shelf off Northern California is only about 15 km wide, but it widens irregularly to the north and is mostly about 50 km across along the coast of Oregon and Washington, the greatest width for the West Coast of the conterminous United States. South of the Columbia, the outer shelf has many small ridges and depressions but these are rare to the north. The shelf break off Oregon and Washington is close to 180 m, unusually deep for an unglaciated shelf. South of the Columbia River, the inner slope of the shelf is distinctly greater than the outer, but farther north the slope is quite uniform. The shelf sediments consist largely of sand on the inner shelf and coarse silt along the outside (Gross et al. 1967). Modern sediment is found on the inner shelf, much of it derived from the Columbia River estuary (White 1970). Currents move the sediment in both directions but predominantly north. Relict sediments of various types have been sampled on the outermost shelf. The sediments seaward of about 150 m contain abundant glauconite. Sedimentary rocks have been collected from both the shelf and slope off Oregon (Byrne et al. 1966). These rocks are of Miocene and Pliocene age, some containing foraminifera that indicate deposition in water as much as 1000 m deeper than where they were recovered. It is rare to find that a continental terrace has been uplifted rather than submerged. However, geophysical data (Byrne et al. 1966) indicates as much as 6000 m of sediments have accumulated in the area of the continental shelf along the central Oregon coast. This, of course, implies that large subsidence has also occurred.

British Columbia and Southern Alaska At the northwest corner of Washington we see the same abrupt change in continental shelf topography that characterizes most other boundaries between glaciated and unglaciated shelves. Deep troughs and shoal outer banks are found all the way north and west to the Aleutian Peninsula. The southernmost of the deep troughs comes out of the Strait of Juan de Fuca, which separates Washington from Vancouver Island. Landward, the Strait connects with a network of deep fiords and troughs, including Puget Sound to the south and the Strait of Georgia to the north. Off the coast of southwestern Vancouver Island, La Perouse and Swiftsure Banks are similar to the many banks off New England and southeastern Canada. The 75-km shelf off the Strait of Juan de Fuca narrows to the north along Vancouver Island until it is almost eliminated by the protruding Brooks

Peninsula; north of Vancouver Island, Queen Charlotte Sound has a deep shelf 130 km wide. Still farther north, Queen Charlotte Islands are located near or at the outer edge of the shelf. On the landward side of these islands there is a broad shallow shelf, Hecate Strait. Petroleum companies have been investigating the shelf and islands in the vicinity (Shouldice 1970). Seismic profiles show a considerable amount of folding, large unconformities, and faulting. The Tertiary strata are up to 4500 m thick. At Dixon Entrance, there is another wide shelf where glaciers have cut a wide trough across the marginal islands of the Inside Passage. These fiords continue as far as the turn in the coast, where a broad coastal plain lies at the foot of the St. Elias Range. Off the central part of the south-facing Alaskan coast, the shelf widens to as much as 170 km. Its topography is complicated by glacial troughs and by still active diastrophic movements. One of the world's greatest earthquakes occurred along this coast in 1964 and was accompanied by extensive changes of level on the land and sea floor (Plafker 1969). The shelf changes included an uplift of 15 m southeast of Kodiak Island (Malloy 1964), which is the world's largest authenticated movement during an earthquake. Pamplona Searidge, with minimum depths of 130 m near the shelf margin off Cape St. Elias, may represent the submerged rock reef on which vessels were wrecked in the eighteenth century (Jordan 1958). If true, this submergence was probably due to repeated displacements.

The shelf sediments in this glaciated area contain the usual mixture of coarse and fine material with many pieces of gravel among muddy sediments. In some of the fiords with deep basins, stagnant conditions allowed preservation of varved layers because they inhibited the existence of the benthic organisms that normally burrow into the bottom (Gross et al. 1963). Sediment deposition in the Alaskan bays with glaciers at their heads is often very rapid (Jordan 1962). Most of the sediment comes from outwash but is partly attributable to landslides from the steep walls.

The continental slope off British Columbia and southeastern Alaska is relatively narrow and terminates mostly at depths of 2000 and 2500 m; far less than the average for the world. Beyond the turn in the coast, the slope changes drastically. The Aleutian Trench, which has an eastward terminus at or near Yakutat Bay, deepens to the west until it is about 7000 m off the Aleutian Islands. The continental slope descent to the Trench is comparatively gentle in the upper part but steepens near the base with local inclinations of as much as 30°. This slope and trench are probably related to sea-floor spreading and the general northwesterly movement of the Pacific plate. Drilling into a ridge on the slope revealed deformed Pleistocene formations, according to a JOIDES report of 26 July 1971. Off the south-facing Alaskan coast, the unusual feature is a shelf that is remarkably wide for a subduction coast.

Bering Sea and East Asia Beginning with the Alaska Peninsula and the Aleutian Islands, we encounter coastal types different from those discussed previously. A series of arcuate islands extend along the east and southeast Asiatic coasts with partly filled deep basins on the mainland side. On the seaward side, we have the arc type of subduction coast and a marginal coast on the inside. The continental terraces are greatly influenced by this relationship, being very narrow along the outside of the arcs and having a series of trenches at the base of the continental slopes.

The 600-km continental shelf along the Alaskan coast of the Bering Sea, one of the widest in the world, has been studied in detail by the U.S. Geological Survey, various U.S. Navy laboratories, and by Soviet Union geologists. Numerous sounding profiles across the shelf indicate an amazing lack of relief (Buffington et al. 1950). This Bering shelf was predominantly free from glaciation during the Pleistocene and therefore lacks the glacial troughs that characterize the shelf off Southern Alaska. The volcanic Pribilof Islands rise above the outer shelf to the south and, to the north, St. Matthew and St. Lawrence Islands are nearer the mainland. The latter are also largely volcanic but include some sedimentary formations. Despite the flatness of most of the shelf, seismic reflection profiles show that the underlying basement is irregular, and the basement rocks are overlain by layered Cenozoic sediments with an average thickness of 1 km (Scholl and Hopkins 1969). Figure 9–27, based

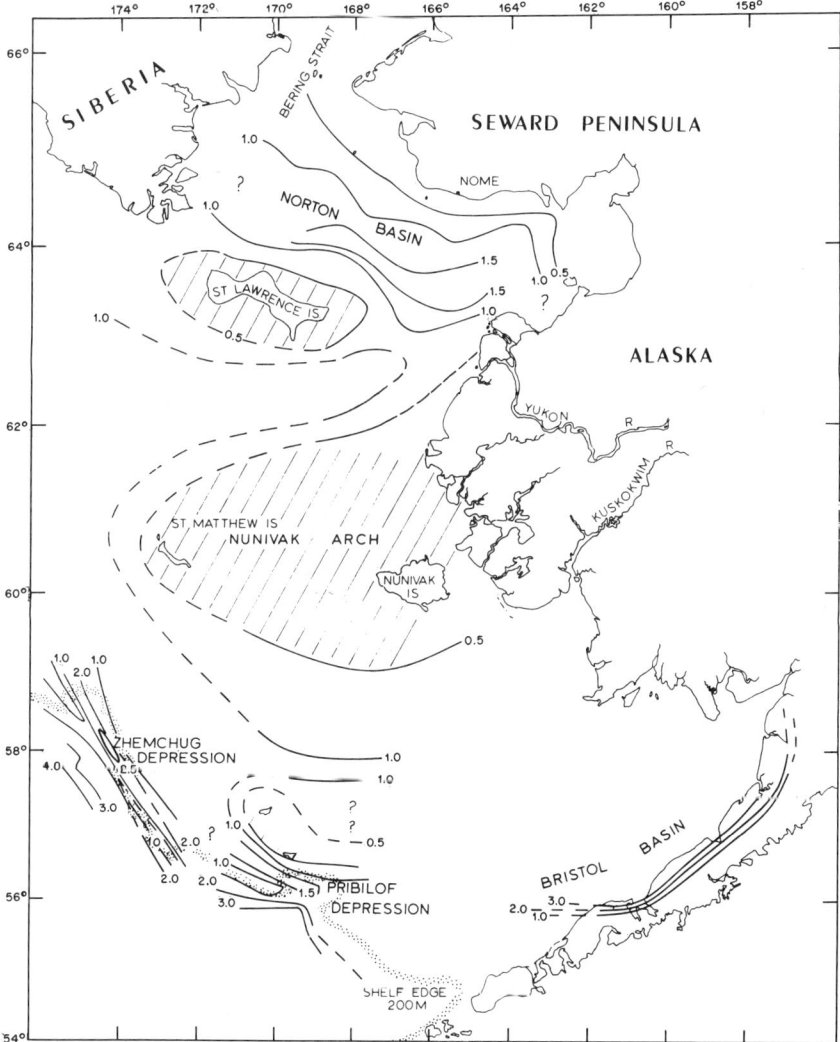

Fig. 9–27 Structure contours (0.5-km interval) on acoustic basement of Bering shelf. From Scholl and Hopkins (1969).

on seismic reflection profiles, shows several Cenozoic basins with intervening arches.

The Bering shelf sediments have been little studied except in and near Bering Strait. Norton Sound, to the southeast, has largely a silty-sand bottom (Creager and McManus 1967). The shelves on the Siberian side of Bering Sea have been investigated by Lisitzin (1966, 1969).

The slope along the southwest margin of the Bering shelf, with relief of more than 3000 m, has very rugged topography (Scholl et al. 1968). The largest and longest submarine canyons in the world are cut into this slope, partly along faults (see p. 319). Rocks as old as Cretaceous outcrop along the canyon walls and even on the slope. Locally, the layered Cenozoics overlap the slope, but prograding has been minimal. At the southeast end of this slope, there is a plateau or borderland. Along the south side of the Aleutian arc (Fig. 9–28), the slopes have complex topography due to the combined results of faulting, marine erosion, and volcanic activity (Nichols and Perry 1966; Marlow et al. 1970). On the north side of the arc, the slopes are also very complex, including precipitous fault scarps and large submarine canyons with many dendritic tributaries.

The Sea of Okhotsk has been intensively explored by Soviet Union scientists (Bezrukov 1960). A wide shelf on the north side is said to lack the typical glacial troughs (Fig. 9–29). In the Sea, a very uneven slope was discovered with continental-borderland topography, including various basins and ridges, the former having depths of more than 3000 m. Bezrukov (1960) has mapped the sediment distribution in this borderland (Fig. 9–29). Much of the area is covered with mud high in diatoms. Sand and gravel deposits occur on some of the highs.

The elongate Gulf of Tatary, south of Tatar Strait, is largely a shelf sea covered with mud, probably supplied by the Amur River. The shelf terminates to the south at a depth of more than 200 m. Off Vladivostok, in the Sea of Japan, the shelf is about 55 km wide and has irregular patches of sand, mud, and rock. Mud occurs in the bays, sand on the open shelf, and rock bottom off the projecting points of land. South of Vladivostok, a narrow shelf extends down the Korean (Tsushima) Strait. This Korean shelf is sand covered to the north, but mud increases and prodominates in the approach to the Strait. This Strait is all part of the shelf, although there is a relatively deep zone along the west side of Tsushima Island. The sediment in the Strait is principally sand, with mud on the Korean side and rock bottom skirting the islands and reported from the banks. Tsugaru Strait, between Honshu and Hokkaido islands, has been investigated with considerable care because of plans for tunneling and because of petroleum possibilities. Seismic profiles show that the shelf in the Strait and to the west has a thin sediment cover overlying folded and faulted rock formations (Sasa and Izaki 1962).

The nature of the sediment on the shelf and slopes around Japan is indicated on the Japanese bottom sediment charts published in 1949 by the Maritime Safety Board. Work during World War II resulted in more detailed sediment charts of the same area (Shepard et al. 1949). In both cases the basis for the charts is largely the bottom-character notations made in great detail by Japanese surveyors and included on the Japanese navigational charts. Except off Hokkaido, sediments show little indication of decreasing grain size in a seaward direction. The predominant sediments are sand on the open shelf with mud in the bays. Bungo Strait, between Kyushu and Shikoku, has a depression 294 m deep (Fig. 9–30), the deepest known cut in the world that can be attributed to tidal action in a narrow bay entrance. In the large Kago-

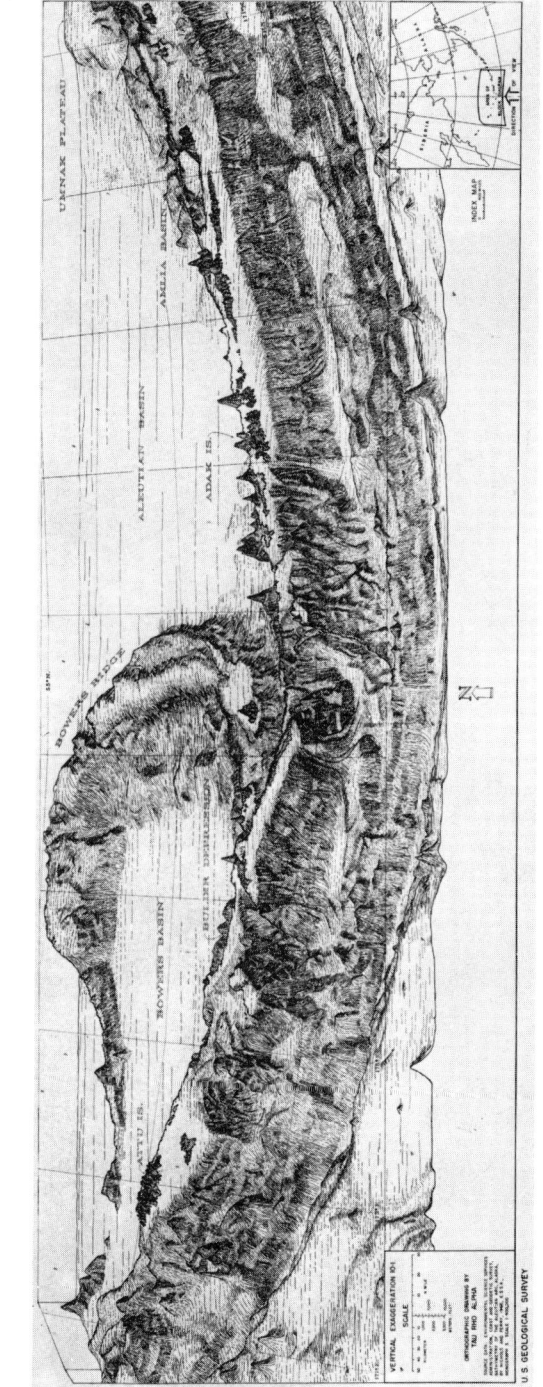

Fig. 9–28 Western Aleutian Islands and Bowers Ridge. Slope south of Aleutians, leading to Aleutian Trench, is cut by many faults and includes a plateau and ridge. The relief in the islands is largely volcanic. Physiographic diagram by Tau Rho Alpha, U.S. Geological Survey.

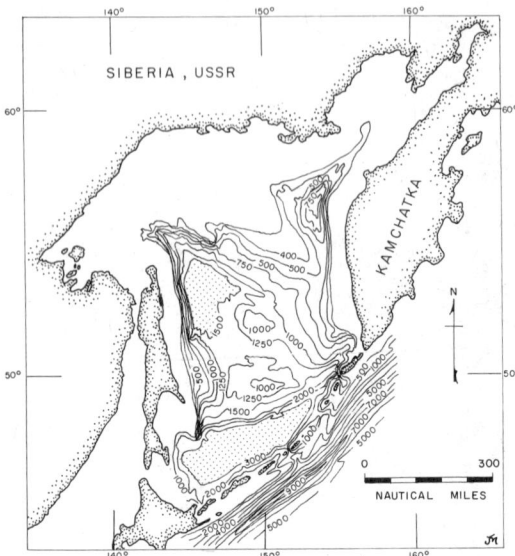

Fig. 9–29A Contours of the Sea of Okhotsk. Depths in meters. After Bezrukov (1960).

shima Bay, at the southern end of Kyushu, an active volcano has grown up across the northern end of the Bay and has apparently contributed much of the bottom sediment.

The continental slopes east of the Japanese islands have been subjected to great diastrophic and volcanic activity throughout the Tertiary and Quaternary. North of the Tokyo area, the continental slope terminates in the Japan Trench. However, seaward of the Tokyo area, a volcanic arc of islands extends south as the South Honshu Ridge. Nankai Trough, another deep trench, borders the slope off the southern islands of Japan, and this in turn is interrupted by a transverse ridge. The slopes of the Trench floors are mostly steeper near the base. A seismic profile into the Nankai Trough (Hilde et al. 1969) shows several step faults and a crumpling of the sediments on the Trough floor, suggesting subduction. Valleys on the slopes penetrate landward into most of the coastal indentations and appear to be related to the tectonic development.

To the southwest of Japan, Ryukyu or the Nansei Shoto Island arc extending through Okinawa to Taiwan, forms a barrier that encloses one of the largest shelf seas in the world. This includes the East China Sea, the Yellow Sea, and the Gulf of Po Hai (Fig. 9–31). Using seismic reflection and geomagnetic methods, exploration of all but the inner reaches of this vast sea (Emery et al. 1969; Wageman et al. 1970) has shown that there are actually five barriers that close off the various basins (Fig. 9–31), most of the latter having sedimentary fill of 1 km or more. The inner basins are considered to be zeugogeosynclines, following Kay's classification (1951). Here, as in the Bering Sea, a wide shelf owes its breadth to deposition in subsiding basins inside natural barriers. The source of the sediments has probably been the great rivers of China, which are still major contributors to recent shelf sedimentation.

Depths of these huge shelf seas show a progressive increase from the shallow Gulf of Po Hai, mostly less than 20 m, to the Yellow Sea, with a belt

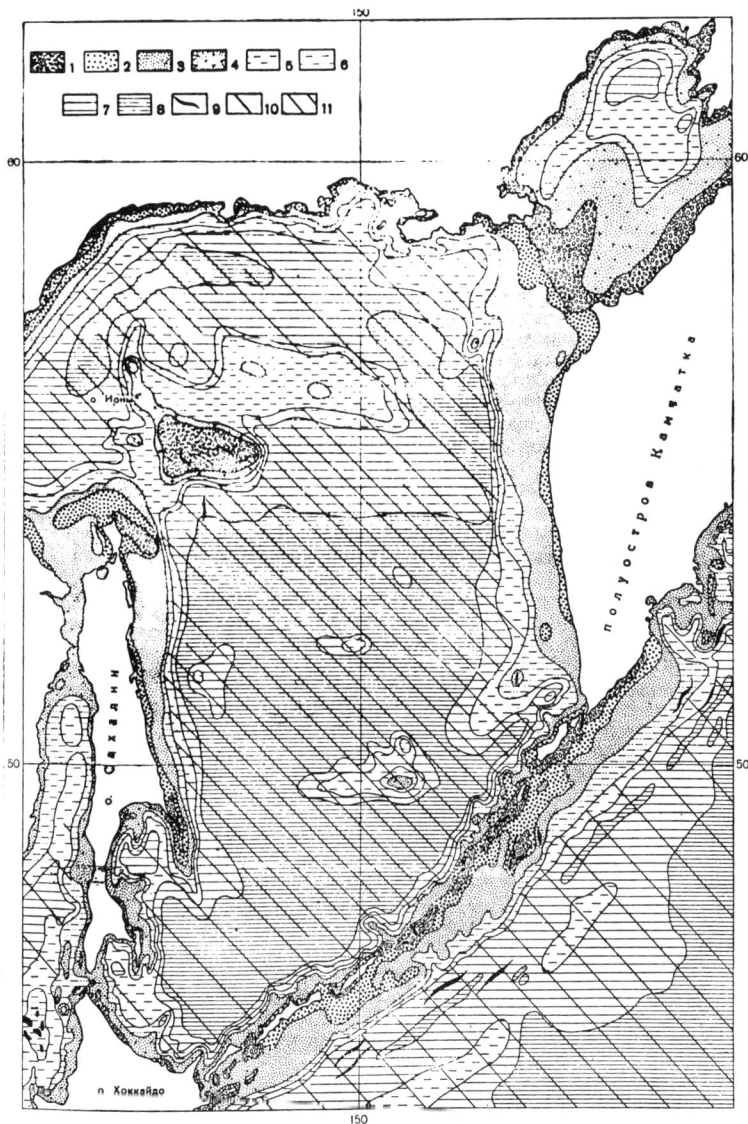

Fig. 9–29B The sediments of the Sea of Okhotsk. Interpretation of pattern in upper left-hand corner: 1, shingle-gravel sediments; 2, coarse sands; 3, fine sands; 4, unanalyzed sands; 5, coarse silt; 6, fine-silt oozes; 7, silty-clayey mud; 8, clayey mud; 9, area of rock outcrop; 10, slightly siliceous-diatomaceous oozes; 11, siliceous-diatomaceous oozes. From Bezrukov (1960).

toward the Korean side that is in excess of 80 m, to marginal depths of 150 m at the break in slope in the East China Sea. Broad shallow areas occur along the China coast off the old mouth of the Hwang Ho and the present Yangtze Kiang River. On the Korean side are broad shoals where the Yalu River enters the sea from North Korea and off Seoul, South Korea. Off most of these coasts

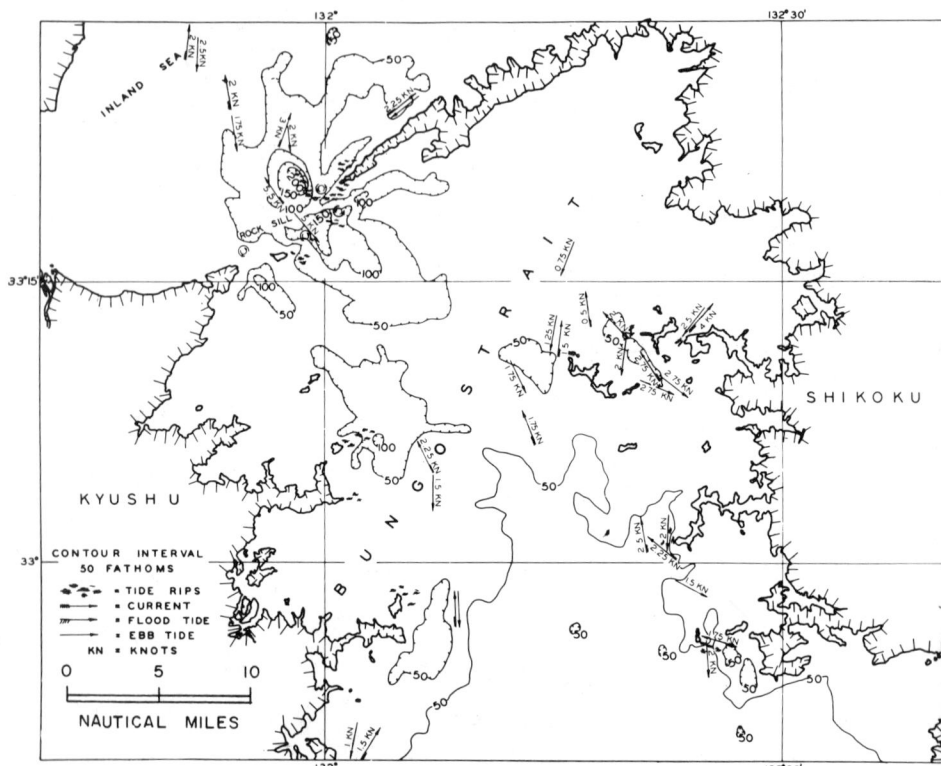

Fig. 9–30 The unusually deep hole resulting from tidal action at the southeastern entrance to the Inland Sea of Japan. Based on Japanese Navy charts.

where there are large tidal ranges and heavy sedimentation from the rivers, a series of shoals and troughs extend normal ·to the land. The bars are tens of kilometers in length and several kilometers in width. Off the old Hwang Ho, bars are parallel to the coast, possibly because there is no longer any entering river at this place.

Sediments of these broad shelves have been studied by Niino and Emery (1961). Their work has verified the earlier indications (Shepard et al. 1949) that the sediments are generally finer near the China coast and coarser at the shelf margins and along the Korean coast (Fig. 9–32A). The inner shelf muds are contributed by the great rivers of China, and the sand zones are predominantly relict with a considerable amount of authigenic and organic material. The north flow of the Kuroshio with branches extending into the Korea Strait and along the east side of the Yellow Sea and the south flow of a countercurrent on the west side contribute to this sediment distribution (Emery et al. 1969).

Along the east coast of Taiwan, a very narrow shelf is bordered by a steep irregular slope. In Formosa Strait, the sediment is largely relict sand (Fig. 9–32B). Off southeast Taiwan, a large submarine canyon penetrates deeply into the shelf, and this is bordered on the west by a series of banks and small islands, including many coral reefs. The shelf sediments as far south

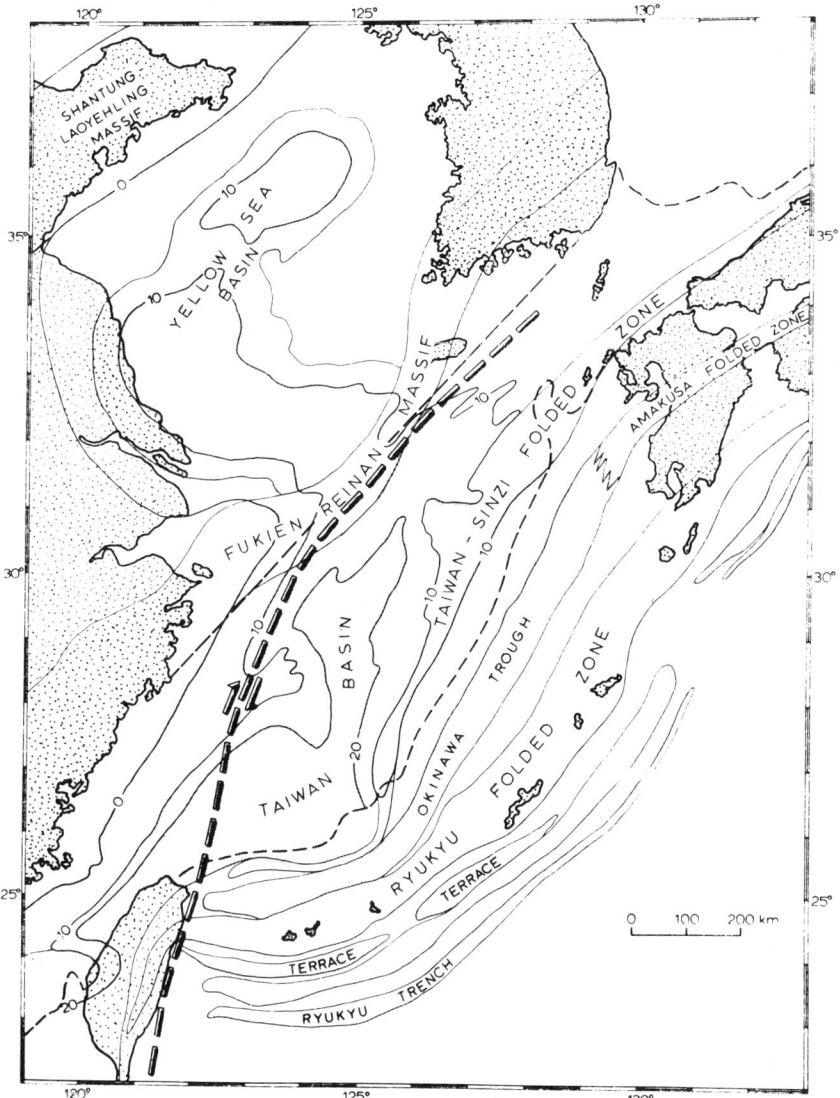

Fig. 9–31 Patterns of ridges and structural depressions in the East China and Yellow seas. Contours show approximate thickness of sediment in hundreds of meters. Note the series of dams that enclose these structural depressions. From Emery et al. (1969).

as Hong Kong are mostly relict. Beyond this, large rivers contribute recent muds, which form the major sediment cover of the shelf except for narrow sand belts where the Kuroshio has prevented recent deposition. Beyond the wide shelf that includes the Gulf of Tonkin, the shelf narrows and becomes very irregular along the coast of South Vietnam. The margin of the shelf southwest of Taiwan is not clearly defined. There appear to be relatively deep

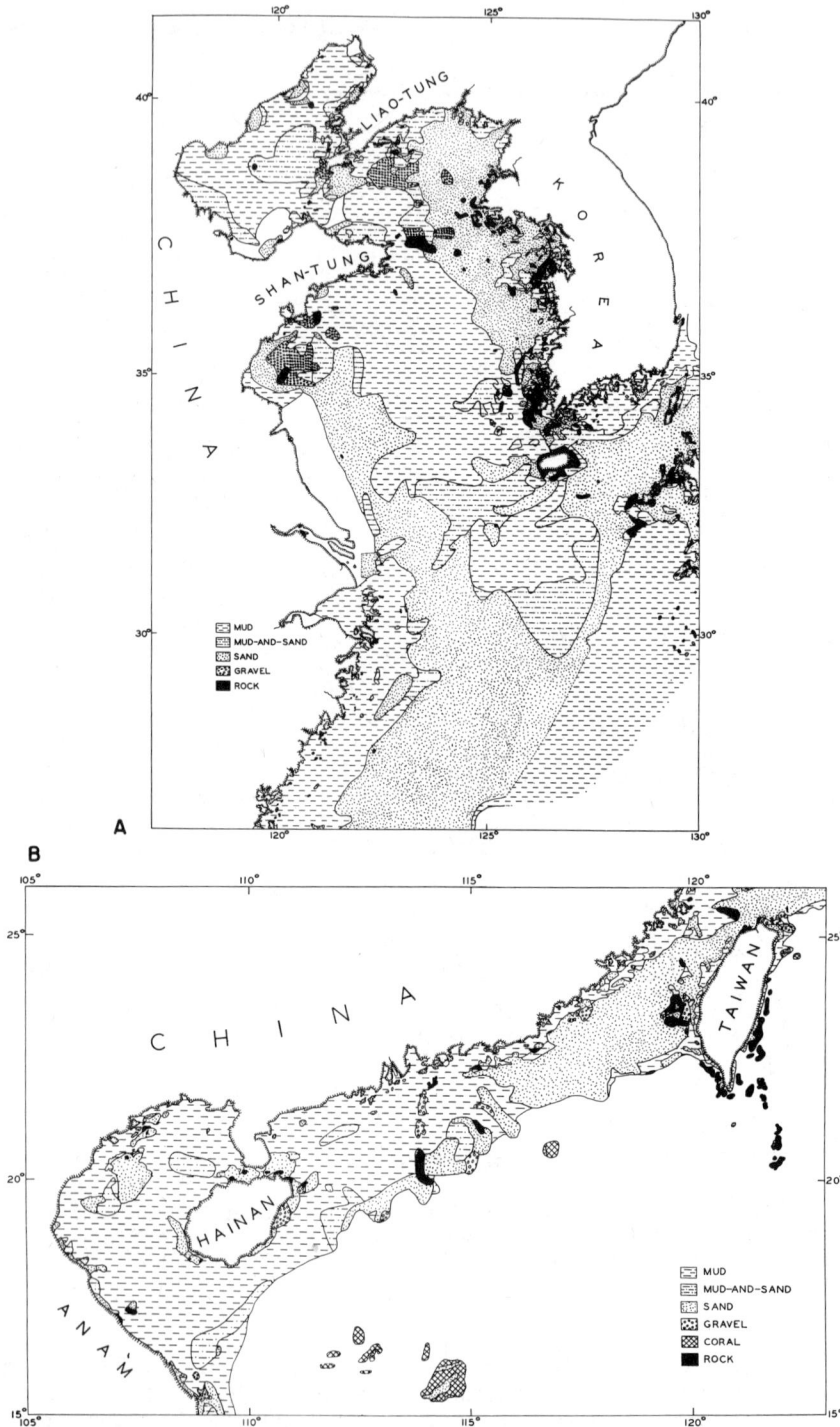

Fig. 9–32 Bottom sediment charts from Niino and Emery (1961) showing the extensive sand zones along the China coast north (A) and southwest (B) of Taiwan. Based on 53,000 bottom notations, partly from Shepard et al. (1949).

terraces above which small coral reefs and other shoals rise almost to the surface. Southwest of Hainan, a large group of coral banks lies seaward of a marginal trough, with depths of about 1500 m. The gentle continental slopes contrast greatly with the steep escarpments that border the shelves farther north.

Another of the world's widest continental shelves is in the Gulf of Siam. This broad shallow sea has also been investigated by Emery and Niino (1963), who report that the northern part of the Gulf is a structural trough that was formed during the late Tertiary and contains thick sediments, also constituting a zeugogeosyncline. The shelf sediments are mostly mud, although on the outside is a broad band of relict sand with reef patches. The north-flowing surface currents here are also important, and the Kuroshio sends an arm up the east side of the Gulf with a return south flow to the west. The inner Gulf has a broad basin depression with depths of over 80 m. According to seismic profiles of Parke and others (1971), the sill of the depression has a buried rock ridge. Two other sediment-filled basins were found in the South China Sea by the same group. One basin extends into southern South Vietnam and along the Vietnam coast to the northeast and is separated by a sill from another larger basin extending east to northwest of Borneo. Sediments in all the basins are at least 2 km thick, are folded, and include diapiric intrusions.

The southward continuation of the broad shelf off Thailand and Malaysia has an arm that curves gradually to the east and maintains a width of 400 km or more through the South China and Java seas. Emery (1969A) has constructed a sediment map of the area. The partly enclosed shelf sea has a mass of reefs and islands in its northwestern portion. Farther south, Bangka, Belitung, and the Karimata Islands partly fill the strait between Sumatra and Bornea. In this constricted area the sediments are mostly relict sand and the bottom is part rock and coral. Both north and southeast of this passage are broad mud-covered shelves, the mud coming from the numerous rivers of adjacent tropical islands with their high equatorial rainfall. The eastern border of the Java shelf sea has a dam produced by extensive coral growth, where many reefs rise to or near the surface. The Java Sea sediments (van Baren and Kiel 1950) have been found to have several petrological provinces including: (1) a Java group which borders the north coast of Java, except to the west where the Krakatau group is influenced by the volcano of that name; (2) an extensive Bornea group covering almost half of the Java Sea; (3) a Malaccan group; and (4) a South China Sea group.

Shelves of the South China and Java seas are said to have many branching valleys with tributaries coming from the main rivers of the adjacent islands (Umbgrove 1949). However, the soundings now available do not support this contention (Shepard and Dill 1966, p. 285). Such channels as exist are very discontinuous. The best developed extend northeast across a portion of the South China Sea between Great Natuna and South Natuna islands. Unless future surveys with high-frequency seismic profilers verify the existence of partly filled channels, it seems quite unwarranted to use this area as an example of valley cutting during low sea-level changes. The Proto-Lupar Channel off Sarawak, northwest of Borneo, is said to have been partly filled since Pleistocene low sea levels (Pimm 1964).

Tracing the South China Sea shelf along the northeast coast of Borneo, we find that it narrows to less than 100 km and is still narrower along the Philippine island of Palawan. These narrow shelves are partly covered with a mass of shallow coral reefs. A few samples off Sarawak have been studied

by Pimm (1964). In some places, as much as 90 m of sedimentation has occurred during the Holocene.

Around the Philippine Islands, the shelves are mostly very narrow, although indentations provide some appreciable shelf widths. The shelves have
many coral shoals. On all sides of the Philippines the slopes are unusually
steep, averaging 11° down to 2000 m. The slopes descend to great depths
around Samar and eastern Mindanao but do not appear to steepen at depth.
Off Mindanao, the slopes have a vertical range of 9200 m, which added to
the land height of 600 m, makes a total of 9800 m, the highest of the Eastern
Hemisphere. The slopes around eastern and northern Luzon are cut by
numerous canyons, although valleys are not shown elsewhere. Along the west
coast of Bougainville Island, the slopes descend to 9000 m, but the few soundings suggest that these slopes are not steeper than about 12°.

Australia   East of the Java shelf sea, two deep basins, Flores Sea and
Banda Sea, intervene before we encounter the tremendously broad shelves of
the Timor and Arafura seas that include the broad Gulf of Carpentaria. This
shallow platform extends 1300 km from the Aru Islands southeast to Kimberley
at the head of the Gulf of Carpentaria, and about the same distance from the
shelf edge in the Timor Sea east to Torres Strait, similar to the distance across
the combined East China and Yellow seas. At Torres Strait we complete covering the group of wide shelves that lie in the partial protection of the offlying
islands from the Bering Sea to northern Australia.

Although Australia is a relatively stable area and has had no Mesozoic
or Cenozoic mountain building, the continental shelf, according to Phipps
(1967), has been extensively warped and faulted, even as recently as in the
Holocene (Fig. 9–33). The shelf structures can be traced landward into basins
and ridges. In general, the wide shelves occur where basins have bowed down
the land, as at the Bonaparte Depression west of Darwin (van Andel and
Veevers 1967), the adjacent Arafura Depression, and the Gulf of Carpentaria
of North Australia, which Phipps describes as a graben. The Queensland shelf
of northeastern Australia, with its vast coral reefs, has topographic depressions
including the Capricorn Channel inside the Swain Reefs. Bass Strait, separating
Tasmania from southeastern Australia, is a structural basin and important in
petroleum operations. Along the west coast, a geosyncline borders the shelf
and is found also as far north as the Timor Sea (van Andel and Veevers 1967).
The narrow shelf off Sydney has a downwarped terrace seaward of the
Sydney structural basin on the bordering land.

The wide Sahul shelf and the adjacent Timor Sea (Fig. 9–34) have been
investigated (van Andel and Veevers 1967; van Andel et al. 1967). The striking
feature here is the Bonaparte Depression that lies near the entrance to Bonaparte Bay. This was a lagoon during stages of low sea level, and cores reveal
that it maintained its salinity during this time, indicating that the climate on
the lands was drier than at present. The recent sediments of this area are
generally thin and predominantly calcarenites and calcilutites. Sand-sized
materials cover about 85 percent of the shelf, including the bank areas on
the outside. The deeper portions of the Bonaparte Depression have silt-sized
sediments. Farther east along the north coast, terrigenous green silts cover
most of the broad flat areas of the Gulf of Carpentaria (Phipps 1966).

The wide Queensland shelf has recently become economically important to petroleum companies but has long been of scientific interest because of having the largest coral reefs in the world. A comprehensive report
on this region is by Maxwell (1968). About half of the shelf is very narrow,
being covered with patches of reef. Some of the reef zones are as wide as

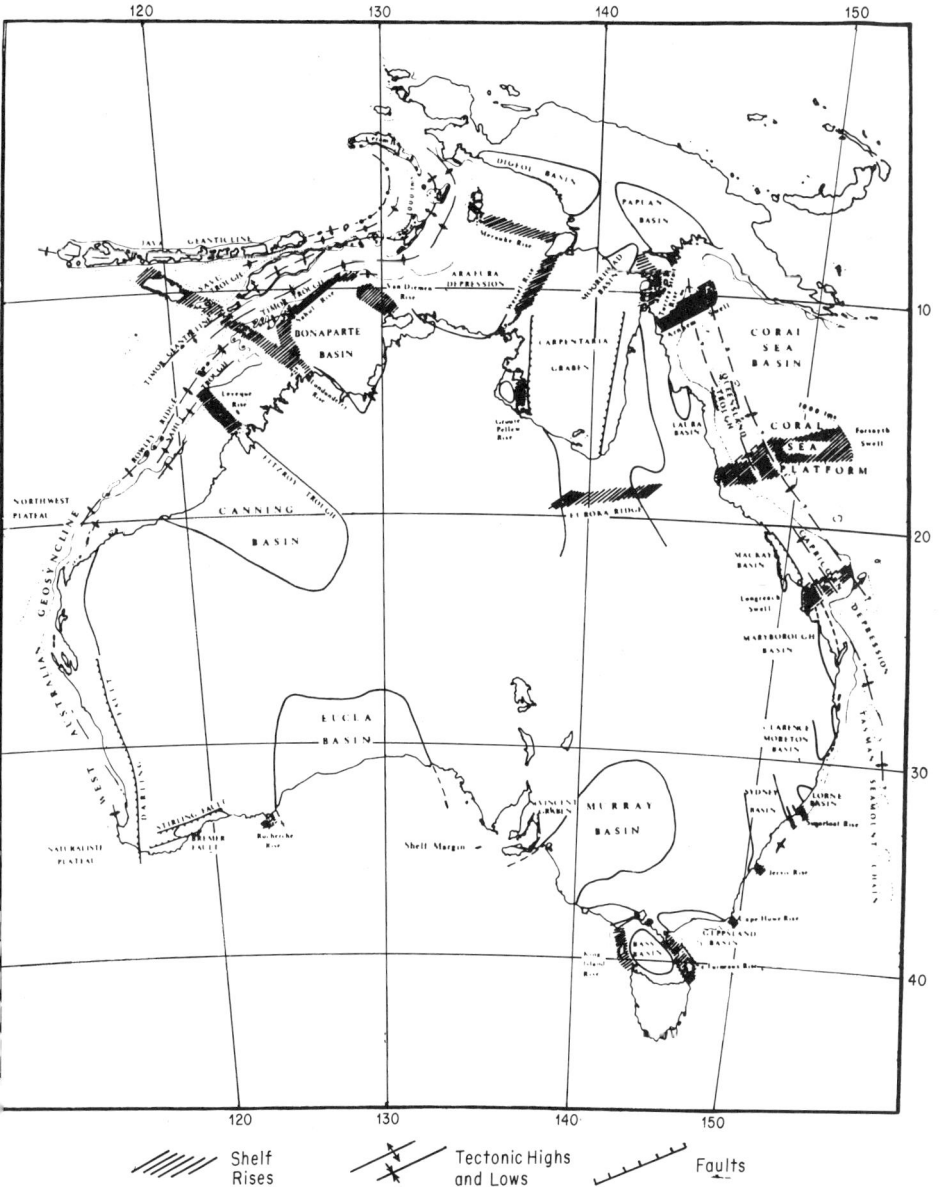

///// Shelf Rises	Tectonic Highs and Lows	Faults

Fig. 9–33 A structural interpretation of the continental shelves around Australia. From Phipps (1967).

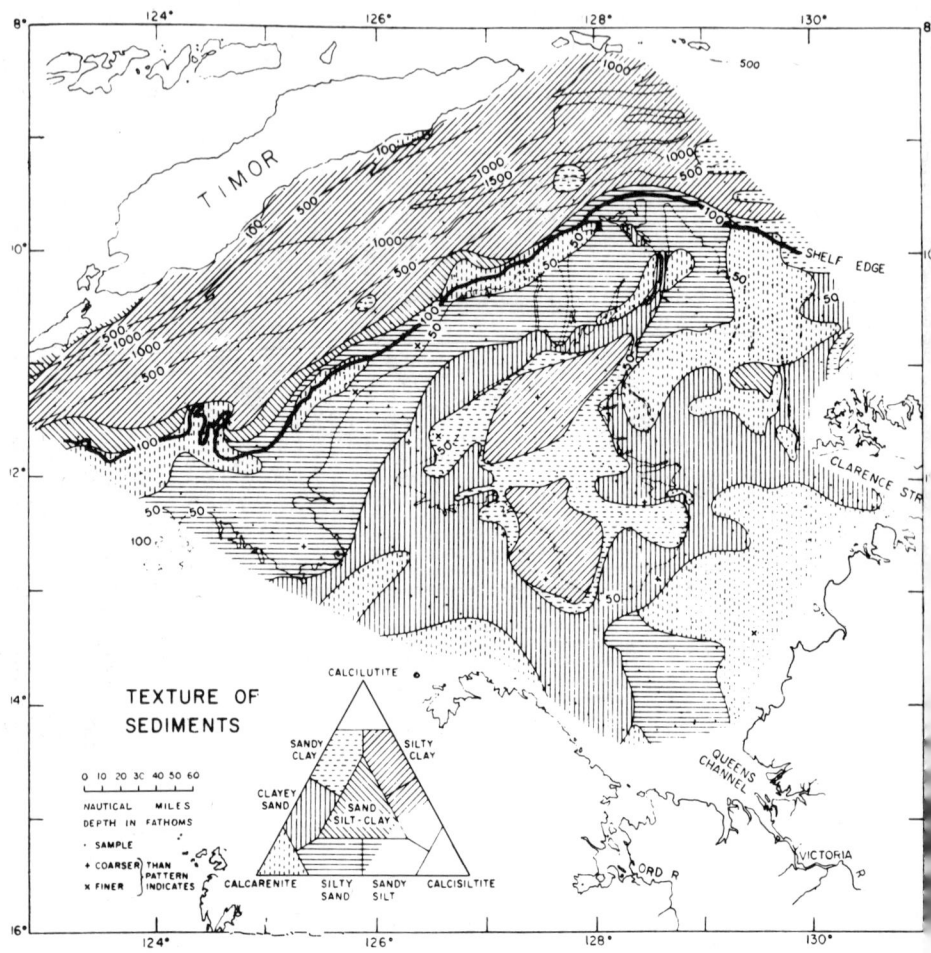

Fig. 9–34   Sediment textures in the Timor Sea off northwestern Australia. Type of sediment indicated by triangular diagram. From van Andel and Veevers (1967).

130 km. Maxwell's book shows the topography of the ship channel that follows the coast inside the main reef tract in the area south of latitude 15°45′ and swings seaward, passing inside the Swain Reefs where the depth is 150 m. The channel has a mud bottom along most of its length. Farther north, the reefs have grown across almost the entire shelf. Borings into the reef have penetrated thick Tertiary limestones, although directly below the present reef, terrigenous sands and gravels were found (Richards 1940).

South of the coral reef area, the shelf narrows to about 40 km, and is largely covered with detrital and detrital-organic sediments (Maxwell and Maiklem 1964). Off southwest Australia, sand is the most common sized material but gravel and rock are found locally, and a few rocks rise above the surface on the outer shelf. The break in slope is at about 110 m.

The shelf off South Australia is mostly wider than 100 km and attains 200 km in Bass Strait on the east side. Aside from the mud zones of Bass

Strait in the lee of Tasmania, the south shelf is largely covered with calcareous sands (Wass et al. 1970). These consist principally of bryozoans, mollusks, algae, and foraminifera (Conolly and von der Borch 1967). Many of these sediments have reworked Tertiary and Pleistocene calcareous detritus.

The Rowley shelf of western Australia, as described by Carrigy and Fairbridge (1954), is largely covered with very much the same type of calcareous sand as off the south coast. The terrigenous constituents of insoluble residues consist mostly of quartz. The arid conditions on land result in little runoff of muddy material. Off southwest Australia, the shelf break is at 150 m (Phipps 1963), but north of 20°, extensive areas of the shelf are deeper than 200 m, and in some places the shelf break is at about 450 m, with a marginal plateau like that off part of Queensland. Coral reefs here also rise above the deep outer shelf.

The continental slopes off Australia are surprisingly steep, particularly off the southwest where inclinations are as much as 27°, and off the southeast where they reach 20°. Off Victoria, the slope generally steepens below 2000 m. The inner slope was found to be a prograded wedge of sediments ranging from Tertiary to Quaternary (Conolly 1969). Off South Australia, the slopes are somewhat more gentle and include offlying plateaus (Conolly and von der Borch 1967; Conolly et al. 1970; von der Borch et al. 1970). Rocks interpreted as old as Precambrian outcrop on the slope, and seismic profiles indicate little outward growth since Australia was broken away from Antarctica by sea-floor spreading. Canyons indent the South Australian slope and broad fans lie at the slope base, apparently fed by sediment moving down the canyons.

Southern Asia Malacca Strait, separating Sumatra and Malaya, is shallow, mostly less than 75 m. The bottom is largely muddy sand (Keller and Richards 1967). Sand associated with gravel and shells is found along the axis of the Strait and near the shelf break. Toward the outer margin, sand, presumably calcareous, becomes predominant.

The enclosed Andaman Sea (Fig. 9–35) is protected from the open Indian Ocean by the Andaman and Nicobar Islands (Weeks et al. 1967), another of the arcs like those along the eastern margin of the Asiatic continent. According to Rodolfo (1969), the Andaman Sea has rhombe-shaped basins due to extension caused by the southward movement of southeastern Asia. As one consequence of this spreading, faulting has produced deep terraces adjacent to the shelf.

The shelf north of Malacca Strait is relatively narrow off Phuket but broadens farther north in the Gulf of Martaban to 280 km. To the west, the Irrawaddy Delta partly overlaps this broad shelf. In the Gulf of Martaban a broad flat area with depths of 18 to 30 m is probably a submerged delta. In this area, the inner shelf is almost entirely mud covered, but on the outer shelf, sand predominates, as off so many other large rivers. The shelf margin is from 110 to 130 m. Currents sweeping along the margin prevent deposition of mud. Along the coast of Malay, the inner shelf has numerous islands, the Mergui Archipelago. On the outer shelf margin, many shoals are reported with coral and rock bottom.

Along the Burma coast north of the Irrawaddy Delta, the shelf narrows to 40 km but continues to be mud-covered. The east–west shelf at the head of the Bay of Bengal skirts the Ganges–Brahmaputra Delta. To the east, an indentation into the land is comparable to the marginal sags off other deltas. This embayment contains low ridges and shallow troughs characteristic of other areas with strong tides. Off the Ganges, the shelf is largely covered with mud, but fine-sand bottom is reported along much of the coast and extends

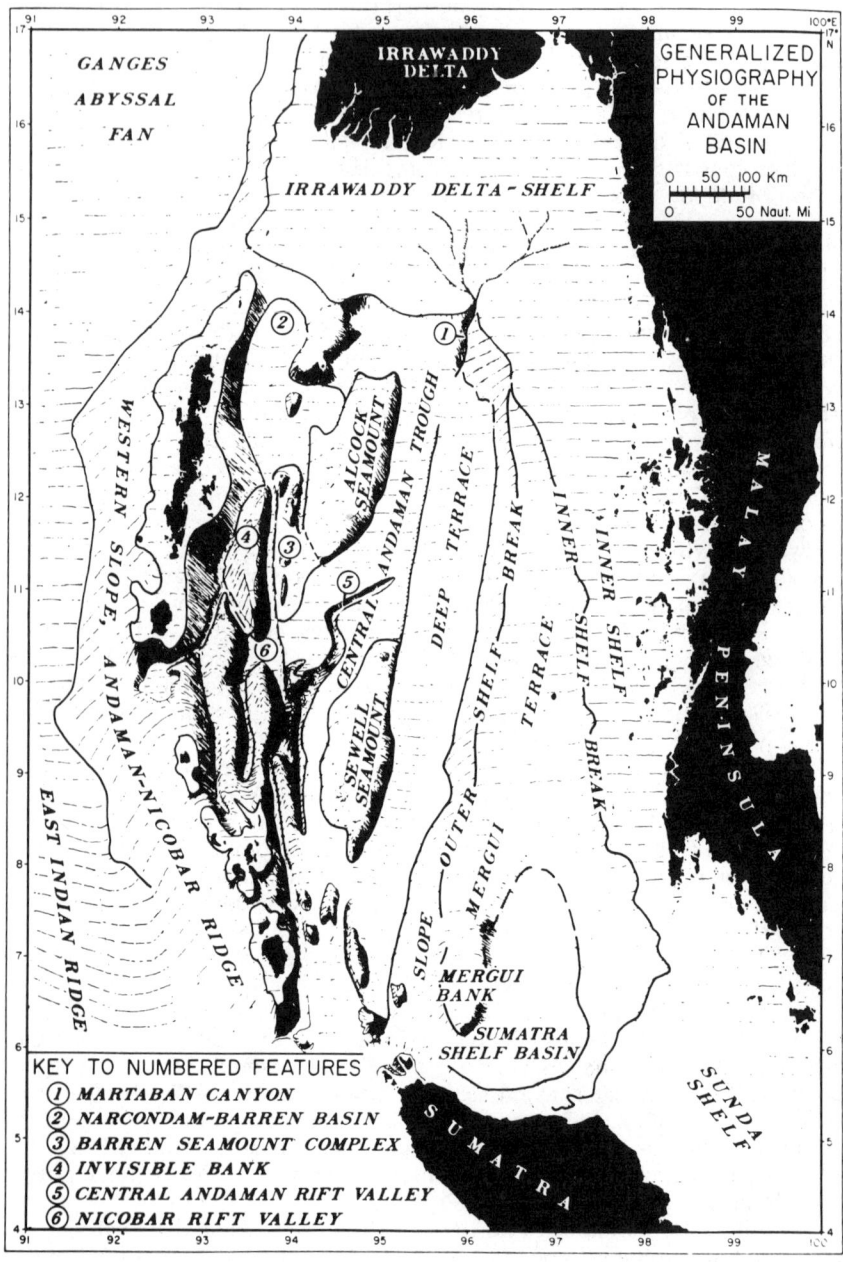

Fig. 9–35 Physiographic diagram of the Andaman Sea, Irrawaddy Delta, and the Andaman–Nicobar Islands and ridge. From Rodolfo (1969).

as a sand tongue outside the main river mouth. Sand is also found among the elongate shoals on the west side of the Delta where subsidence, due to faulting and tilting, appears to have occurred (Morgan and McIntire 1959). Off the middle of the Delta, a troughlike valley cuts across most of the shelf (see p. 334). The continental slope is relatively gentle as off most large deltas. East of the Ganges Trough, the upper slope is as much as 3° down to 1000 m but not over 2° in the area west of the Trough.

On the narrow shelf along the east coast of the India Peninsula, the typical sediment distribution (Mahadevan and Rao 1954; Rao and Mahadevan 1959) is terrigenous sand near shore with black-sand concentrates grading into mud at about 55 m, and then into calcareous sand along the outer shelf beyond 75 m. The outer sand includes oölites similar to the oölitic sand off both sides of Florida. The foraminifera in these oölitic sands, according to F. B Phleger and R. R. Lankford (personal communication), have an abundance of shallow-water types even at depths of about 180 m, that suggest deposition under low sea-level conditions and nondeposition since the sea rose. The narrow shelf is almost entirely overlapped by the Godavari and Kistna (Krishna) deltas, and the margin runs comparatively straight despite these deltas. The shelf edge is around 90 to 130 m except off the deltas where it shoals to 40 m. Most of the shelf is 40 km wide, but near Pondicherry, at the south end of the India Peninsula, the 200-m curve bends near the coast.

The slope off the eastern India Peninsula is steeper than that off the Ganges, having inclinations of 4 to 6°. Palk Strait, separating India and Ceylon, is 75 km in width with water largely 9 to 13 m in depth, except where local coral reefs rise above this shallow floor, particularly to the northwest near Adams Bridge. The latter, judging from its appearance from the air, is a coral reef virtually blocking the southern end of Palk Strait.

Ceylon has a fairly consistent 20-km-wide shelf cut in many places by submarine canyon heads. An average depth at the margin is 65 m. The canyon at Trincomalee Bay comes into the estuary and crosses quartzite ridges. The Ceylon shelf is largely covered with sand and coral. The continental slopes off Ceylon are mostly steep, averaging about 10°, and in one place near Trincomalee, a slope of 43°32' was recorded (Stewart et al. 1964). This is the steepest slope discovered to date other than those on walls of submarine canyons. Elsewhere on the Ceylon slopes are many of these canyons.

Off the southwest tip of India, the shelf widens to 110 km and deepens to as much as 220 m. To the north, the shelf narrows to about 55 km at 11°N latitude but again widens off the Gulf of Cambay to as much as 350 km. Much of the western Indian shelf has a margin between 90 and 110 m, and a broad flat terrace at 90 m extends from 17 to 21°N latitude. Schott and von Stackelberg (1965) show a zonation of sediments along the west coast with coastal sands bordered seaward by muds and then limesands. Mud occurs along parts of the coast and, during the southwest monsoons, this mud is stirred up into the water to such an extent that it actually disrupts the wave trains sufficiently so that small boats can land behind a "breakwater" of soupy mud called a *mudbank* (Hiranandani and Gole 1959). This phenomenon is apparently unique. The continental slopes off western India are gentle, mostly 2 to 3°, but with coral reefs rising above them.

In northwestern India, there are two large indentations, the Gulf of Cambay and the Gulf of Kutch. It is difficult to determine if these are akin to the delta marginal depressions in many other areas. The Gulf of Kutch and its easterly continuation of vast salt flats, the Rann, are probably related to the Indus Delta. The Gulf of Cambay has the same type of sand banks converging toward the head of the bay as in the indentation east of the Ganges. The Gulf of Kutch and the Rann experienced a large earthquake in 1819 that changed the area tremendously, because sinking allowed the sea to spread widely over the low country.

Seaward of the Gulf of Cambay on the broad shelf is a large flat surface with depths close to 90 m (Fig. 10–5). Its dimensions of 70 km across and 130 km along the shelf trend makes this the largest submerged terrace on the entire continental shelves of the world. Just north of the Gulf of Kutch, another terrace, Kori Great Bank, has depths of 27 m and appears to be a submerged extension of the Indus Delta. The Bank is terminated to the north by Indus Submarine Trough that cuts almost entirely across the shelf. To the northwest of the canyon, the 27-m terrace is missing. The shelf averages 140 km in width from the Gulf of Cambay to Karachi, where it narrows abruptly with the change in trend.

Off Baluchistan, the width averages 37 km with gradual decrease to the west. The shelf continues to be mud-covered, according to the sparse bottom notations. The edge is much shoaler along this east–west coast and averages about 37 m. Along the coast of Iran, the shelf is still narrower, about 18 to 30 km wide, but the edge appears to have deepened to 110 m.

The Gulf of Oman has a broad shelf at its head and this leads into the shallow water of the Persian Gulf. The Gulf of Oman is largely mud-covered, as is also the northern and northeastern portion of the Persian Gulf (Emery 1956C). The Tigris and Euphrates rivers, the source of these muddy sediments, are said to have built their combined delta forward into the Gulf 305 km since 325 B.C., but warping has played an important role and the delta area is subject to periodic marine incursions (Berry et al. 1970). The submarine slope of the delta at the head varies from 00′ 50″ to 01′ 40″. The southern and western portions of the Persian Gulf have sandy sediment, much of it calcarenite (Houbolt 1957); but along the Arabian Desert coast to the west, most of the sand is terrigenous and of eolian derivation. Small coral reefs grow along the Qatar Peninsula. Emery (1956C) found relatively high organic contents in the Gulf sediments, especially near the southern end where mixing of the Arabian Sea water with the underlying saline water from the Persian Gulf has encouraged planktonic growth in the Strait of Hormuz. This mixed water is carried into the Gulf, and the plankton sinks to the bottom. Precipitation of calcium carbonate occurs frequently, producing patches of milky water (Wells and Illing 1964). Many hills and islands in the southern Gulf have proved to be salt domes.

Along the coast of Arabia, the mud bottom disappears, and sand along with rock is indicated on the generally narrow shelf. The shelf margin is fairly straight, but the land has a number of projections and islands which virtually or entirely eliminate the shelf because there is no appreciable marginal outbend around them. Approaching the entrance to the Red Sea, the shelf is more regular in width, averaging about 38 km, and terminating at 75 m. At the Red Sea entrance, there is a long central channel with axial depths of 230 to 340 m. This is one of the rift valleys.

East Africa   The East Africa coast is an Afro-trailing-edge type (Inman and Nordstrom 1971), but the Gulf of Aden at the north is a neo-trailing-edge type because of the recency of the splitting of Africa and Arabia.

The shelf on the east coast of Africa is narrow and poorly surveyed. In places there appears to be no shelf along the straight coast, suggestive of a fault origin (p. 119). As along the Arabian coast, the indentations are accompanied by a widening of the shelf, but narrow zones occur off the points and islands. A coral-covered shelf surrounds Zanzibar and Mafia islands. The shelf widens south of the Zambezi River to 130 km off Beria at 20°S latitude. Most of the shelf is shallow with marginal depths around 55 m. A narrow stretch follows, and then the shelf again widens to about 130 km between 25 and 26°S latitude, but here the increase is related largely to a marginal plateau with depths of the outer portion about 380 to 550 m. The shelf is narrow again to the south except for a bulge north of Natal, where an unusually even bench terminates at 90 m.

The sediment of the East Africa shelf has been little studied. A few samples taken along the shore off Somali and Kenya consist of calcareous sands (Schott and von Stackelberg 1965; Müller 1966). Shells and coral sand are indicated on the charts farther south. The continental slope off East Africa is not well sounded but appears to be mostly less than 3°, with the steeper areas off Somali Republic and a portion of South Africa south of Durban. Very gentle slopes are found off Kenya and Tanzania. The slope base is rarely deeper than 3000 m.

Off South Africa, the shelf bulges considerably with a maximum width of 240 km. The slope is steep along the coast out to about 75 m. The outer shelf soundings are scarce, but it extends out to well over 200 m and includes Agulhas Bank with its many hills and depressions. Most of this outer portion is between 130 and 180 m deep. Irregularities occur in an area where a strong westerly current flows. The shelf sediment is mostly sand with abundant shells. Small areas to the west are mud-covered, and rock is also reported in a number of places, especially on the outer shelf. Investigations of the Agulhas Bank were made by Dingle (1970), who found basement rock outcropping to the west and overlapped on the east by Cretaceous sediments in the Alphard Banks and then by Tertiary limestones, the latter mostly covered by recent unconsolidated sediments.

West Africa Off West Africa, the terrace is also related to an Afro-trailing-edge coast. Currents run north along this coast as far as Cape Lopez at the equator.

The shelf narrows temporarily west of the Cape of Good Hope, but its edge deepens considerably. Between 32 and 28°S latitude the shelf is 180 km wide, with the outer margin close to 400 m, according to van Andel and Calvert (1971). The slope is steep near shore in most places. Sparse bottom notations show mud on the inside with sand and rock on the deep outer shelf. Aside from Antarctica, this shelf is perhaps the deepest in the world, if one rules out the Blake Plateau that is entirely deeper than the arbitrary 550-m limit set for shelves. North of 28°S latitude, the shelf narrows to 90 km and continues to the north at about that width for more than 1000 km. All along this area the inner shelf is steep, whereas terraces occur at about 130 m. The exact depth of the shelf break is difficult to determine.

Farther north, the shelf narrows locally and disappears from 16 to 13°S latitude, but approaching the equator it widens again off the mouth of the Congo to 90 km. Here, Congo Submarine Canyon crosses the entire shelf and penetrates a deep estuary on the inside. The terraces on this shelf are more than 200 m deep in some places, but the margin adjacent to Congo Submarine Canyon is at about 180 m. The sediments are muddy around the Congo mouth, but sand predominates to the south. To the north, the shelf shoals and is 55 to 70 km in width as far as the Niger Delta, except where it is overlapped by Cape Lopez (1°S latitude). Muddy sediments are reported from most of the shelf between the Congo and the Niger.

Coastal currents run east all along the bulge of the African coast as far as the Niger Delta. The shelf off this Delta has an unusual broad outbend that conforms with the arcuate Delta margin (Fig. 9–36). The Delta and continental terrace have been investigated by Allen (1964, 1965B). He considers

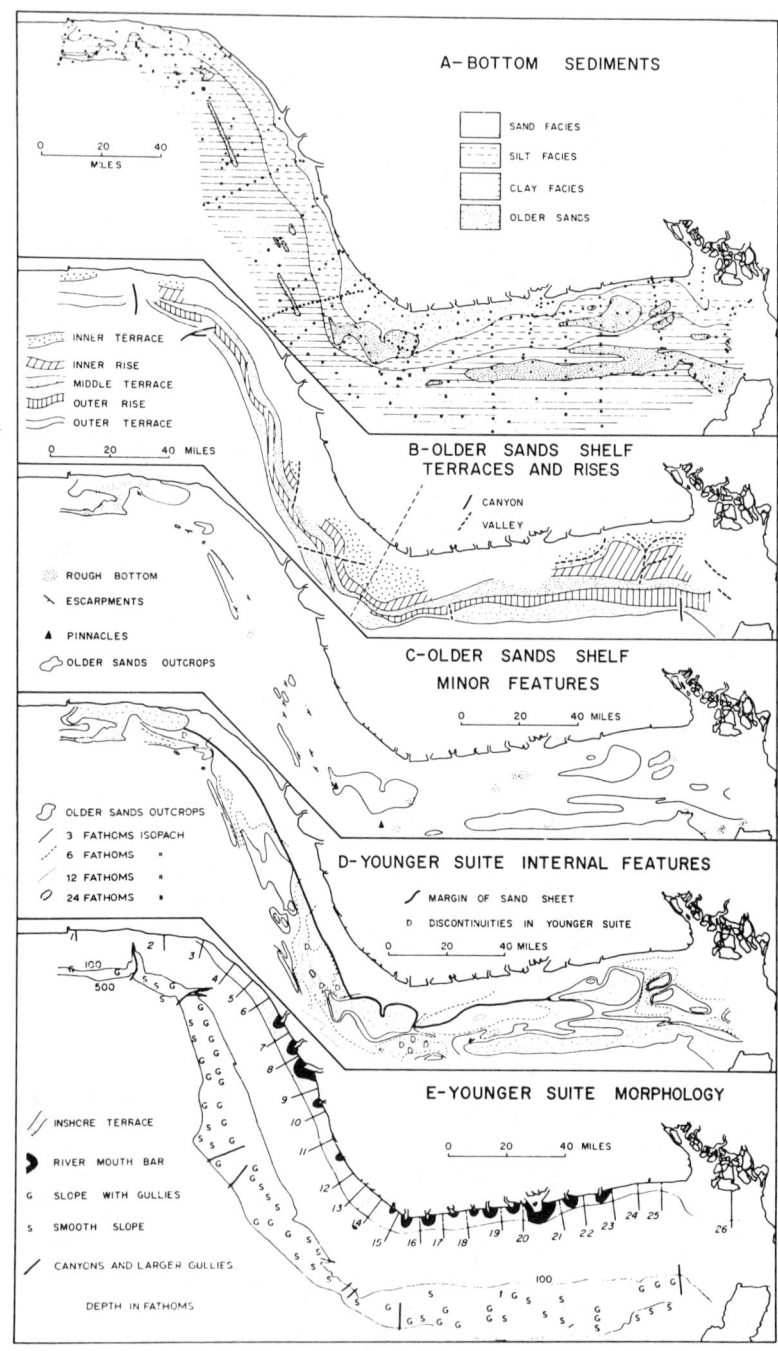

A– BOTTOM SEDIMENTS

- ☐ SAND FACIES
- ☐ SILT FACIES
- ☐ CLAY FACIES
- ☐ OLDER SANDS

0   20   40
MILES

⬚ INNER TERRACE
⬚ INNER RISE
— MIDDLE TERRACE
⬚ OUTER RISE
— OUTER TERRACE

0   20   40 MILES

B–OLDER SANDS SHELF
TERRACES AND RISES

/ CANYON
// VALLEY

⬚ ROUGH BOTTOM
⬚ ESCARPMENTS
▲ PINNACLES
⬚ OLDER SANDS OUTCROPS

C–OLDER SANDS SHELF
MINOR FEATURES

0   20   40 MILES

⬚ OLDER SANDS OUTCROPS
/ 3 FATHOMS ISOPACH
⬚ 6 FATHOMS "
⬚ 12 FATHOMS "
⬚ 24 FATHOMS "

D–YOUNGER SUITE INTERNAL FEATURES

/ MARGIN OF SAND SHEET
D DISCONTINUITIES IN YOUNGER SUITE

0   20   40 MILES

⬚ INSHORE TERRACE
▶ RIVER MOUTH BAR
G SLOPE WITH GULLIES
S SMOOTH SLOPE
/ CANYONS AND LARGER GULLIES

DEPTH IN FATHOMS

E–YOUNGER SUITE MORPHOLOGY

0   20   40 MILES

Fig. 9–36 Interpretation of the sediment facies and morphological features around the Niger Delta. From Allen (1965A). See also Fig. 8–5.

the shelf as a depositional feature in spite of small erosional valleys that truncate part of the slope. The building out of this shelf after South America was separated from Africa would help account for the slight overlap that is indicated where an attempt is made to fit the two continental margins together (Bullard 1969).

Allen found concentric terraces and sediment zones conforming to the subaerial delta front and apparently related to the Holocene sea-level rise. The outer shelf has outcrops of an older sand series, which is found also nearer shore on the distal ends of the arc. Coral banks were discovered along the outer margin, partly buried by sediment (Allen and Wells 1962). Carbon-14 dates have shown that the Delta is sinking, like the Mississippi and other large deltaic masses. Various distributaries are contributing a much larger proportion of sand to the inner shelf than does the Mississippi. The Niger is a classical example of the concentric advance of sediment zones associated with a prograding delta.

The east–west trend of the African coast beyond the Niger Delta is said to be related to the Romanche Fracture Zone that crosses the Mid-Atlantic Ridge (Fail et al. 1970) (see p. 374). West of the Niger Delta, a narrow mud-covered shelf continues with a bulge west of the Volta Delta, where it is 90 km wide off the south side of the African bulge. The wider continental terrace of Sierra Leone and the Guineas has been investigated by two groups (Sheridan et al. 1969A; McMaster et al. 1970A, 1970B, 1971). The geophysical work of both groups has indicated that here the shelf, with a maximum width of 250 km, has been built forward by deltas, particularly during stages of low sea level. The slightly submerged Bissagos Delta has built across three-fourths of the shelf (Fig. 9–37) but without producing any outward-curving shelf margin. Two submerged remnants of Pleistocene deltas have been found along the shelf margin southwest of the Bissagos Delta (McMaster et al. 1970A). The same investigators also located various other indications of the low sea-level stand, including relict shell-sand, reef corals dating as old as 19,000 years B.P., a diagonal shoal which they interpreted as a submerged spit, and various drowned river channels. Their seismic profiles show three superimposed sets of imbricating deltas formed on a subsiding shelf. At 12.5°N latitude along the shelf edge, an arc of 12 salt domes was found under the sediment, producing local upbowing of the Tertiary formations (Aymé 1965).

The shelf farther north is covered with biogenous sand (McMaster and LaChance 1969). It narrows to only 8 km at Cap Vert, an old volcano attached to the coast by a tombolo. The continental terrace of northwestern Africa is narrow, mostly less than 65 km, and the mean depth averages 105 m near 28°N latitude and about 150 m northward to the Strait of Gibraltar. The coastal currents flow south all along this coast except for a countercurrent in the embayment north of Cap Vert. The shelf has many outcrops but is largely sand-covered with small quantities of silt that increase on the outer shelf. Much of the sand is carbonate. Glauconite is common, especially on the upper slopes. Seismic profiles show that the northern shelf has underlying folds and faults with decreasing deformation to the south (McMaster and LaChance 1968). In general, the shelf has prograded. Rona (1970A) has found comparable symmetry between the continental margins off Cap Blanc and Cape Hatteras, another indication of sea-floor spreading because these two areas were presumably once joined. The slopes off northwest Africa are generally low, about 2°. The topography of the slope landward of the Canary Islands might be classified as a continental borderland, although unlike that off Southern California, the relief is controlled principally by volcanism.

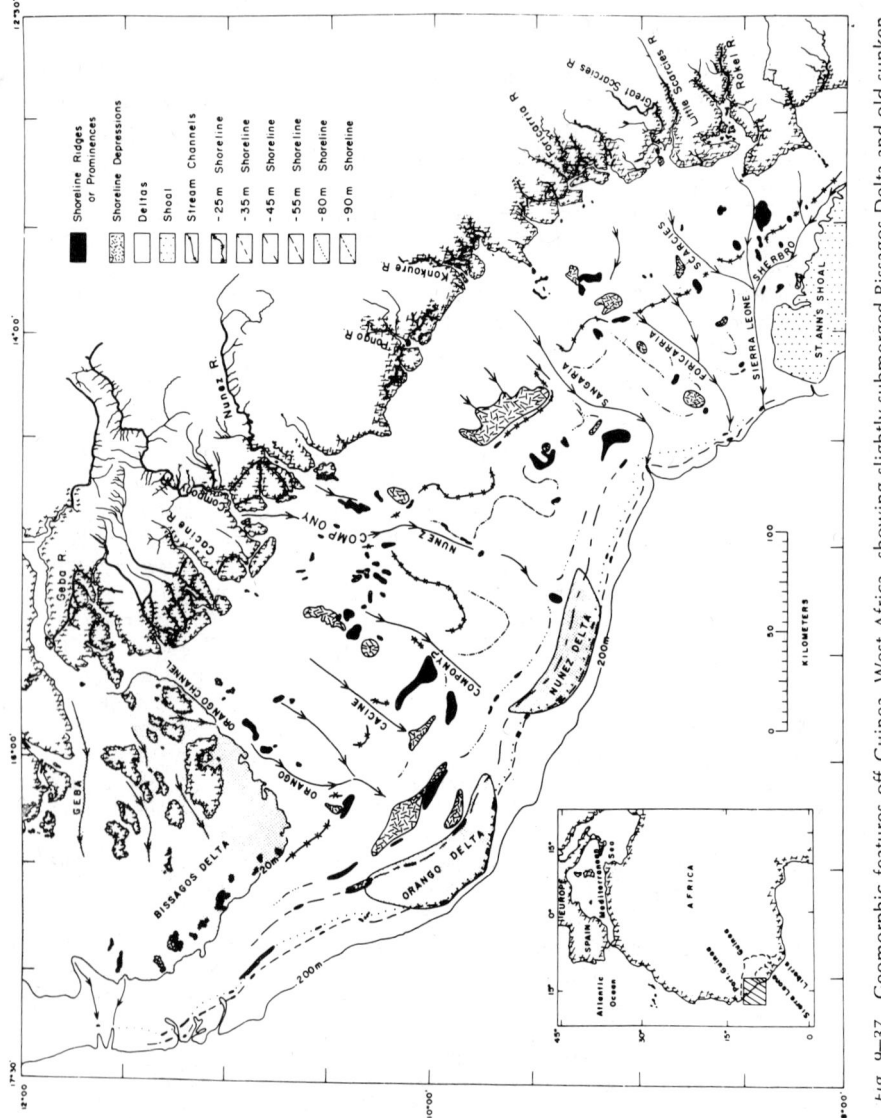

Fig. 9–37 Geomorphic features off Guinea, West Africa, showing slightly submerged Bissagos Delta and old sunken channels. From McMaster et al. (1970B).

The Mediterranean According to the latest information (March 1972) from the JOIDES drillings, the Mediterranean appears to be a rather special type of collision zone, resulting from intermittent northward movement of Africa toward Europe (see p. 425). The Strait of Gibraltar has a sill depth of 286 m traversed by strong bottom currents. Narrow shelves are found on either side and irregular troughs and ridges in the middle (Giermann 1961). Along the Atlas Mountain coast of Africa east of Gibraltar, a narrow shelf and a slope averaging about 6° is cut by numerous small canyons (Rosfelder 1955). The shelf sediments have been studied by LeClaire and others (1965). As far east as Algiers they found many patches of rock and gravel, usually off projecting points. The embayments have mud bottom with bioclastic material. At the turn of the coast near Tunis, the shelf widens to about 150 km, and the Strait of Sicily, approximately as shallow as that of Gibraltar, cuts the Mediterranean into two main basins. The Gulf of Tunis is largely mud-covered but wide sand areas lie off the points on both sides. Blanc (1954) found that the sand in the Gulf of Tunis and the Strait of Sicily is mostly calcareous with many shells, bryozoans, and foraminifera. South of the Strait, the shelf widens to as much as 300 km in the Gulf of Gabès. The shelf edge is rather indefinite but is apparently around 370 m in much of this area. The floor of the Gulf of Gabès is largely sand-covered, but mud and clay are reported for the deeper water.

East of the Gulf of Gabès, the shelf continues as a 20- to 40-km band along the desert coast of Libya. The sediments, principally sand, have a desert source. The first mud appears on the outer shelf at 28°E longitude and becomes the dominant sediment approaching the Nile, although off Alexandria, the inner shelf is still sandy but changes to a sandy mud farther east with mud outside. Only mud is deposited directly off the Rosetta and Damietta River mouths, but relict sand zones are found at various depths on the outer shelf (d'Arrigo 1936). The shelf edge curves out around the Nile Delta, as off the Niger Delta. The shelf is relatively steep from the shore to 10 m, followed by a gentle slope to 30 m, then somewhat steeper to about 95 m, the average shelf break. The nature of the continental slope off the Nile Delta is indicated in a series of profiles (Emery et al. 1966). The typical inclination is only 0°20′, somewhat less than off most other large deltas.

The shelf along the coast of Israel has been studied (Emery and Bentor 1960). Along the shore there is a relatively steep descent to about 20 m. The shelf edge shoals to the north from 120 m to about 75 m, and the flat areas also shoal in that direction. This suggests the influence of the weight of the Nile Delta in depressing the southern end. The sediments are derived mainly from the Nile, although they are patchy, as on many other shelves. In general, the sediment is sand near the shore and mud outside, but rock bottom with shells and coral often occurs between the two. Some rock is found also along the outer shelf. The same type of narrow and mostly mud-covered shelf extends north around Lebanon and then along the south coast of Turkey. The continental slopes along the Israeli coast are gentle to the south, 1.5°, but increase to 5° farther north. They are cut by many valleys. Off Lebanon and Syria the declivity averages 10°, and this steepness continues along the coast of Turkey with many irregularities, probably fault troughs.

Most of the northern Mediterranean continental terraces are narrow. An exception is the rectangular 900-km Adriatic Basin where the northern 550 km has been built up by sediments to form a shelf. The Adriatic has been studied by van Straaten (1965B) and by Pigorini (1968). The shelf edge is somewhat indefinite but appears to lie close to the 200-m line off the Gargano promontory on the Italian side. Several small islands rise above the sediment fill near the shelf edge. Well inside the edge, a transverse basin has depths to 270 m and is apparently a graben. The sediment map (Fig. 9–38), based mainly on Pigorini's studies, shows mud off the Po Delta, the chief recent sediment source for the area. This mud extends down the Italian side of the sea, whereas there are sands and muddy sands in a pocket northeast of the Po Delta, along the Yugoslavia side of the shelf, and in the outer basin beyond the shelf margin. An old Po Delta was recognized by the minerals composing the sands of the outer shelf. Much of the sand, according to van Straaten (1965B), has been considerably reworked. He explained that some of the irregularities at the shelf edge are attributable to creep, but part of the outer slope represents foreset beds of the old delta.

The remainder of the Italian peninsula has narrow shelves. The south portion has been mapped by d'Arrigo (1959) and the sea floor off Naples has been described by Müller (1958). The narrow shelves continue along the French Riviera and shelves are lacking in some of this area (Bourcart 1954A, p. 31). From the Italian border to Marseille, four sediment zones extend out from shore (Nesteroff 1959) consisting of: (1) *terrigenous sand and gravel* along the shore; (2) *Poseidonia beds (les herbiers)* (Fig. 9–39) between 6 and 45 m wherever terrigenous sediments are scarce; (3) *shelly sands* between 28 and 55 m; and (4) *mud* covering most of the sea floor below 55 m except for gravel and sand along the floors of canyons. West of Marseille, the shelf widens off the Rhone Delta area to 75 km. However, the shelf edge does not bend out here as off the Nile Delta on the other side of the Mediterranean. The laminations in the foreset beds of the Rhone Delta are due to variations in discharge and descend to a depth of 75 m, much deeper than in the Mississippi foresets. The outer shelf, which generally lacks laminated sediments, is largely mud-covered off the Rhone Delta, but the sediments analyzed by Kruit (1955), van Andel (1955), and Duboul-Razavet (1956) are much sandier than those off the Mississippi.

The continental slopes around the Italian peninsula are mostly quite low but steepen off the Italian and French rivieras (Bourcart 1959), where they are cut by many submarine canyons and troughs. These slopes are mostly mud-covered (Bourcart 1954B). The slopes are somewhat gentler off the relatively broad shelf of the Gulf of Lyons.

Along the east coast of Spain, the shelf is mostly narrow, but off the Ebro Delta and for 110 km to the south, it is about 85 km across, much wider than to the north and south. Muddy zones occur off the Ebro, but gravel is also found out near the shelf edge, partly coming from a group of small

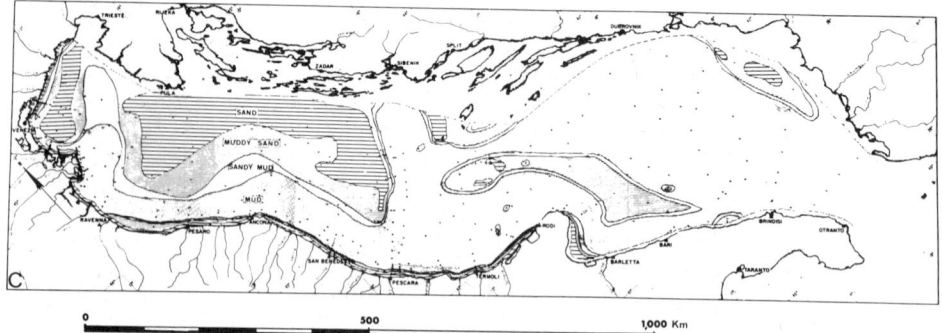

Fig. 9–38   Sediment map of the Adriatic Sea. From Pigorini (1968).

Fig. 9–39 *Poseidonia* beds along the French coast near Nice. The plants have a height of about 2 m. Underwater photo by R. F. Dill.

islands, Islotes Columbretes. Along the coast of eastern and southern Spain, some slopes are as steep as 10° out to the 2000-m line. They also have many canyon indentations.

Western Europe The continental shelves along the Atlantic side of the Iberian Peninsula continue narrow, rarely more than 50 km. The nature of the bottom is well shown along the Portuguese coast by the fishery Carta Lithologica Submarina. The notable feature on these charts is the large but irregular areas that have rock or gravel bottom, and bottom notations indicate that rocks are also common off northern Spain. The depths of the shelf edge in the Gulf of Cadiz, west of Gibraltar, are difficult to determine because the extremely gentle slope is cut by numerous hills and ridges that continue seaward to depths of at least 1000 m. Farther west, off southern Portugal, an almost equally gentle slope is underlain by well-bedded sediments that are folded, the ridges conforming to the anticlines and the troughs to the synclines (Roberts and Stride 1968). These authors interpreted the folds as due to slumping of the slope sediments against an older mass, called the Betic Block (see also Roberts 1970). Along the west coast of the Iberian Peninsula, profiles (Berthois et al. 1965B) show that the slopes are steeper. The shelf edge deepens from about 200 m to the south to almost 400 m to the north. Seismic profiles (Stride et al. 1969) off western Iberia (Fig. 9–40) indicate

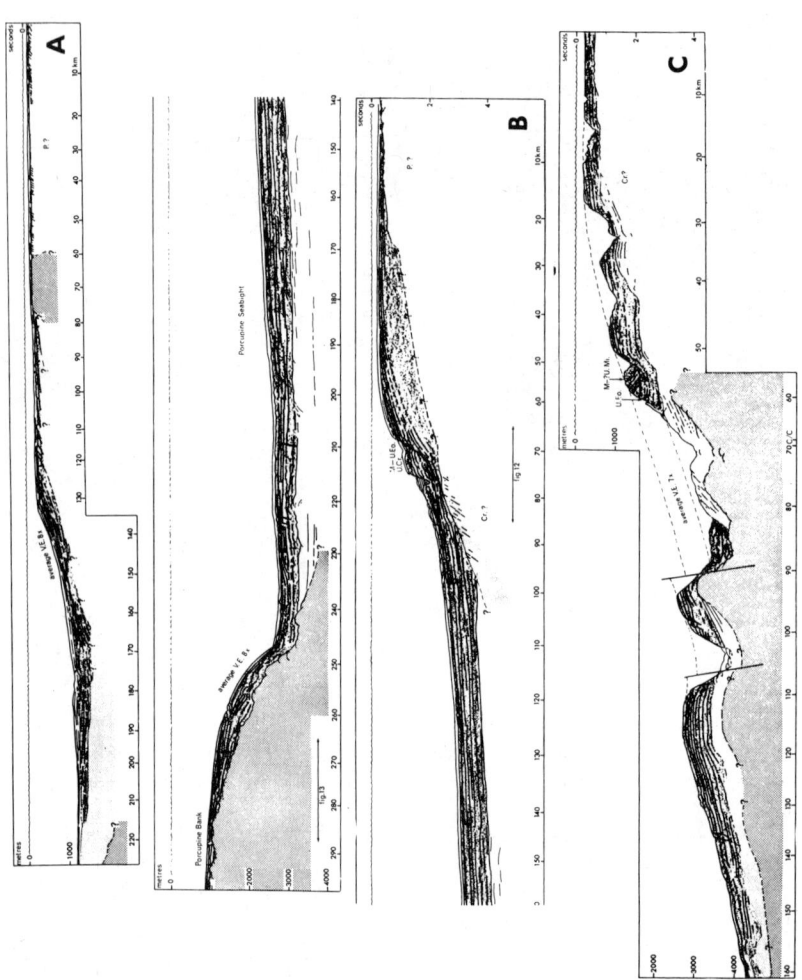

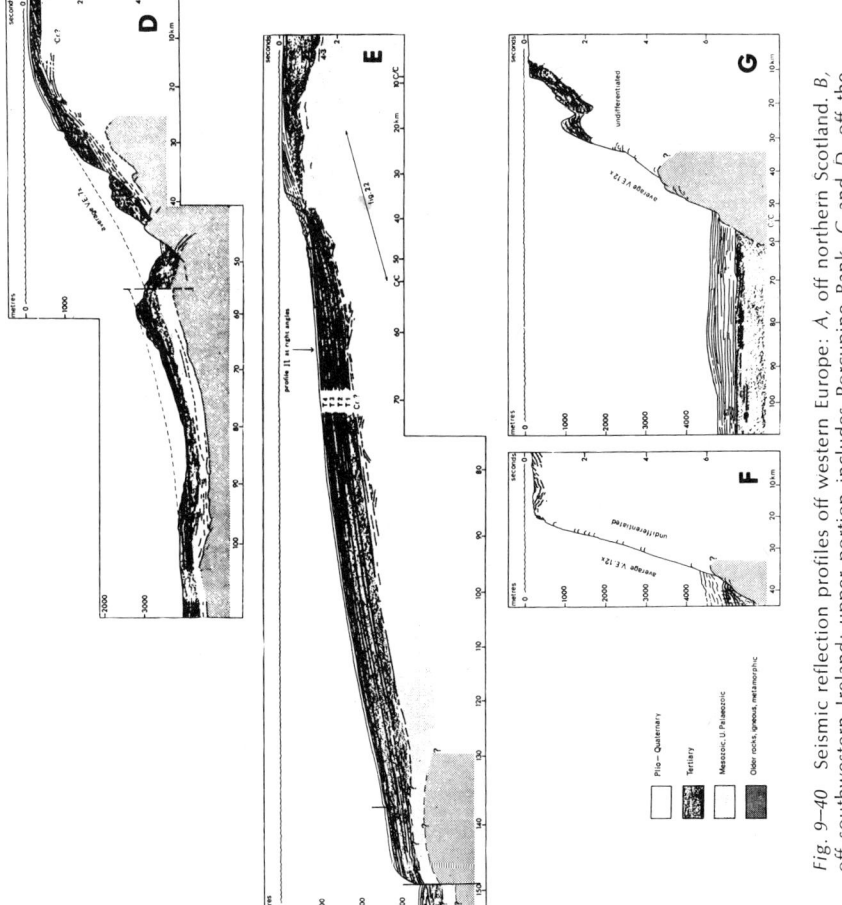

Fig. 9–40 Seismic reflection profiles off western Europe: *A*, off northern Scotland. *B*, off southwestern Ireland; upper portion includes Porcupine Bank. *C* and *D*, off the English Channel. *E*, off southwestern France. *F*, off northcentral Spain. *G*, off west Portugal, north of Lisbon. See also Fig. 10–3A. From Stride et al. (1969).

that this continental terrace has not been prograded appreciably, although sediments have accumulated at the base of the inner slope in a trough that forms part of what could be classified as a continental borderland. Galicia Bank, seaward of the trough, has outcrops of Cretaceous rock (Black et al. 1964). The western three of the four seismic profiles off northern Spain (Stride et al. 1969) indicate a similar absence of Tertiary and Quaternary outbuilding, but one profile of Boillot and d'Ozouville (1970) shows some prograding (Fig. 10–3A). Old rock masses appear near the base of an escarpment. Thin Cretaceous (?) and Tertiary beds outcrop on the outer shelf, according to these profiles. The eastern profile off Santander shows forward-building Tertiary on the slope cut by slumping and faulting, the latter where the slope terminates in Cap Breton Canyon.

The continental shelf off western France (Fig. 9–41) widens to the north from 55 km at Biarritz to 180 km off Brest. The greater width of this shelf than off the Iberian Peninsula is in keeping with the broad coastal plains that characterize most of the French coast, in contrast to the mountainous nature of most Iberian coasts. To the south, the shelf terminates close to 120 m, but to the north, the edge deepens to nearly 200 m. In the northern part, the 100-m contour is closer to the coast than to the 200-m contour, an unusual shelf feature. Sediments of the French shelf are a mixture of sand, mud, shells, and gravel, with coarser sediment more common on the outer shelf than inside (Berthois and Le Calvez 1959). Mud deposits locally overlie sand, the latter partly relict from what may have been a wide sand-dune belt, like that of the Gascony coast. The French shelf is said to have received little sedimentation since the last rise in sea level.

The inclination of the continental slope off the French coast averages about 5°, decreasing to the north to 2°. The slope is cut by many canyons, but only Cap Breton Canyon penetrates the shelf (Berthois and Brenot 1960; Berthois et al. 1965A). Seismic profiles (Fig. 9–40) on the slopes off western France and the British Isles (Stride et al. 1969) indicate a complicated history. Cretaceous outbuilding was followed by erosion that removed much of the continental shelf; then, after downwarping, there was extensive outbuilding of Tertiary sediments, and finally the slopes were modified by submarine erosion, faulting, and slumping.

The widening shelf off the French coast reaches its apex off the English Channel and the Irish Sea. This bulging platform, known as the Celtic Sea, is 300 km wide off Lands End, England, but including the adjacent channels the shelf is more than 600 km across. The marginal depths are close to 200 m and, as off France, much more than half of the Celtic Sea is deeper than 100 m.

The English Channel and St. Georges Channel both have elongate deeps; of these, the Hurd Deep, near Cherbourg Peninsula, is cut 50 m below its rock rim. In the English Channel, rock outcrops were first reported by Dangeard (1928) and later confirmed by others (Hill and King 1954; Curry et al. 1965; A. Smith et al. 1965). Rocks underlying the Channel are Paleozoic, Mesozoic, and Tertiary in age. They have been studied partly because of interest in constructing a submarine tunnel under Dover Strait. Sediments of the English Channel (Fig. 9–42) are largely gravel, sand, and shells, with little mud. The strong tides are responsible for the thin sediment cover and the coarseness of the sediments. The tides apparently have less influence along the south coast of Ireland and in the western part of the Irish Sea, because much more muddy material is found there than in the English Channel (Stride 1963B; Belderson 1964).

The most remarkable feature of the Celtic Sea is the many elongate

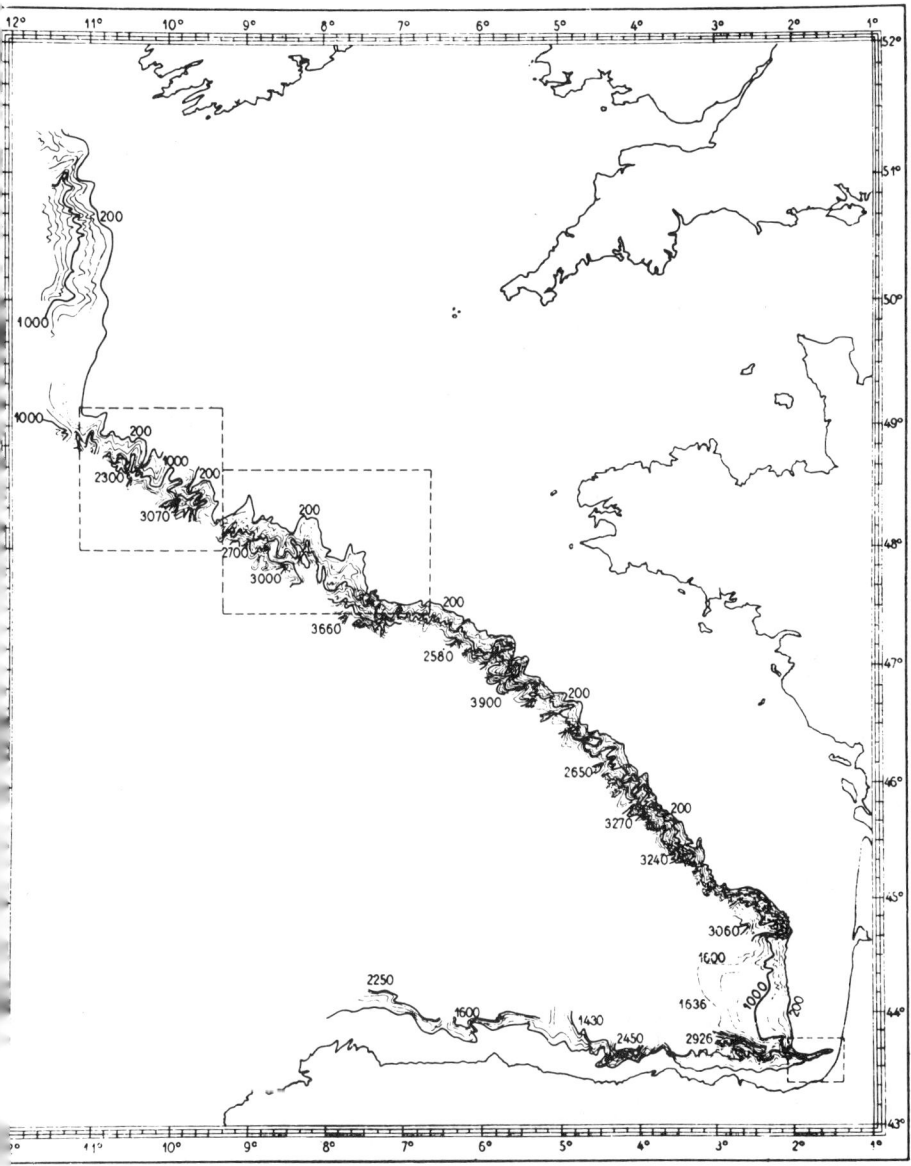

Fig. 9–41 General contours of the continental slope off northern Spain, western France, and the southern British Isles. The large Cap Breton Canyon is to the southwest. From Berthois and Brenot (1960).

sand ridges trending northeast–southwest, which are especially concentrated on the outer shelf off the Irish Sea beyond the 100-m contour. Among these, Great Sole Bank is about 150 km in length and 10 km in width. The ridges rise 30 m or more above their surroundings. Stride (1963A) compares these with the various tidal-induced ridges in the entrances to estuaries and

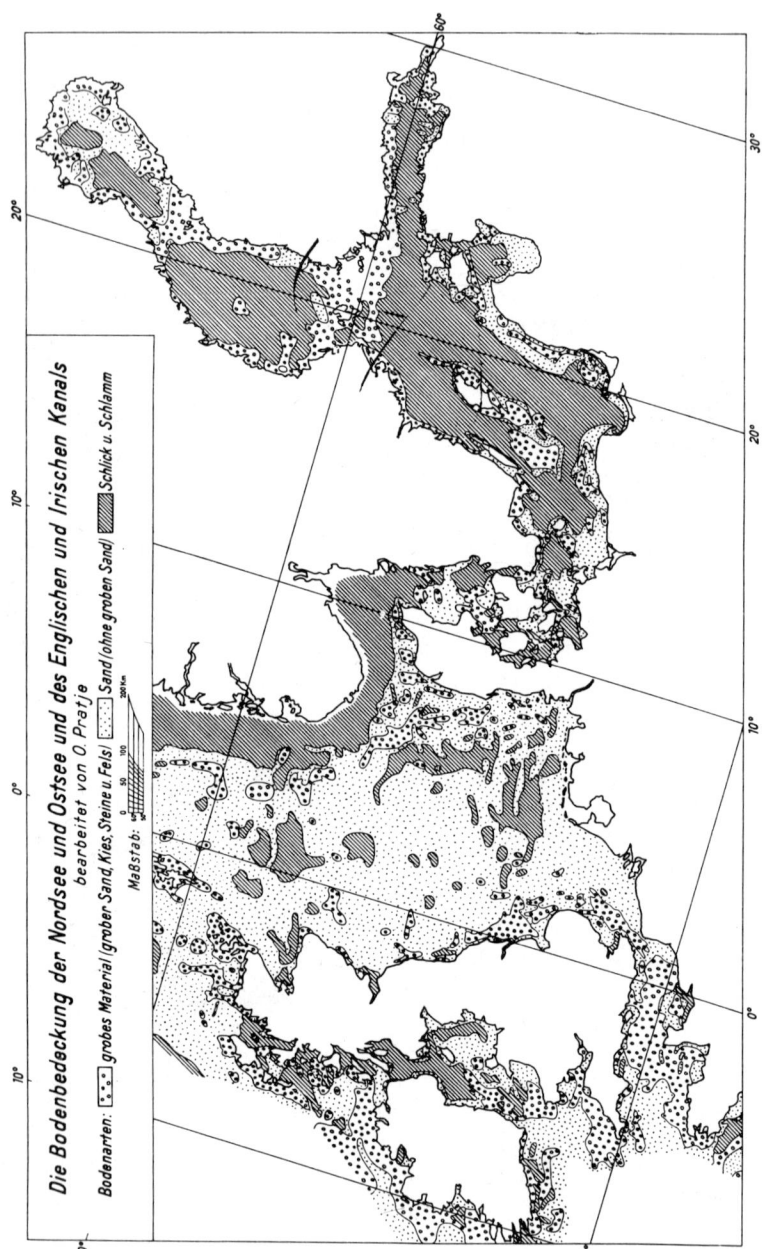

Die Bodenbedeckung der Nordsee und Ostsee und des Englischen und Irischen Kanals
bearbeitet von O. Pratje

Bodenarten: [∴∵∴] grobes Material (grober Sand, Kies, Steine u. Fels) [:::] Sand (ohne groben Sand) [////] Schlick u. Schlamm

Maßstab: 0 50 100 200 km

Fig. 9–42   Bottom sediment types of the North Sea and Baltic. From Pratje (1949).

adjacent to great deltas. The trends of these ridges suggest that they are related to the tides coming up into the Irish Sea and the English Channel, but they differ from other tidal sand ridges in their much greater depth, greater length, and greater relief, and, according to Stride, in having tidal currents of less than 1 kn. Currents associated with other sand ridges of this type (Off 1963) are mostly from 1 to 5 kn. Berthois and Le Calvez (1959) believe that the Celtic Sea ridges are due to stronger tidal currents during stages of lowered sea level.

The Celtic Sea and its estuaries have numerous sand waves, which were first described by van Veen (1936), and have been investigated particularly by English geologists (Stride 1963B; Kenyon and Stride 1968). These sand waves, like those of Georges Bank and the Bahamas, form at right angles to the sand ridges and are caused by currents, as shown by their steep lee sides and gentle stoss (on current) sides. Sand waves off Great Britain have been located in part by side scanning and sonar (Asdic), but are also clearly shown in echo-sounding profiles (Fig. 9–43). Their charting determined the predominant currents for various parts of the Celtic Sea (Stride 1963B).

Sand ribbons, also observed in this area, are shown in sonar (acoustic) records where sand has been spread along the bottom by currents (Stride 1963B). Other linear patterns in the English Channel are also described by Stride (1959B).

The continental slope off Ireland could be included among the continental borderlands. The shelf edge is indefinite but mostly deeper than 200 m. The offlying Porcupine Bank, explored by Berthois and Guilcher (1961) and by Stride and others (1969), apparently has a thin cover of Tertiary sediments but may also have older igneous rocks in the center. It is evidently a part of the continental block. Farther north, a 2000-m channel separates Rockall Bank from the mainland. This Bank is now considered to be a block that has broken off from the continent.

The North Sea, with its valuable petroleum and natural gas resources, is being explored by many marine geologists. The topography to the south has been mapped in great detail (Stocks 1956; Houbolt 1968), including many sediment maps (Pratje 1949; Jarke 1956; Houbolt 1968). The sediments are mostly sand but mud is found to the north, especially in the deep Skagerrak Channel around the southern end of Norway (Fig. 9–42). The North Sea has been subjected to various influences. During glacial episodes, ice caps coming from Norway and the British Isles spread out onto the North Sea shelf and flowed as far south as 53°N latitude. One result was the forming of Dogger Bank, a glacial moraine (Valentin 1955; Stride 1959A). After partial retreat of the glaciers, Dogger Bank was left above the sea as an island and was inhabited by ancient man, whose artifacts along with mammal bones are now being dredged from the Bank. During glaciation, the outwash was carried south of the ice margin, and vast quantities of sand were deposited in the southern part of the North Sea and even down into the English Channel, covering areas which were then above sea level. After the glaciers retreated, the tides swept sand back through Dover Strait and up along both sides of the North Sea (Houbolt 1968). The ridges and sand waves of the southern North Sea (Fig. 10–6) have been studied by many geologists (Guilcher 1951; Stride 1963B; Robinson 1966; Houbolt 1968). These highs are moved during great storms (Stride 1965A; Robinson 1965). Houbolt found that the ridges consist of cross-bedded sands overlying a flat surface. They are derived from Pleistocene sediments of the sea floor rather than from Holocene sediments of entering rivers.

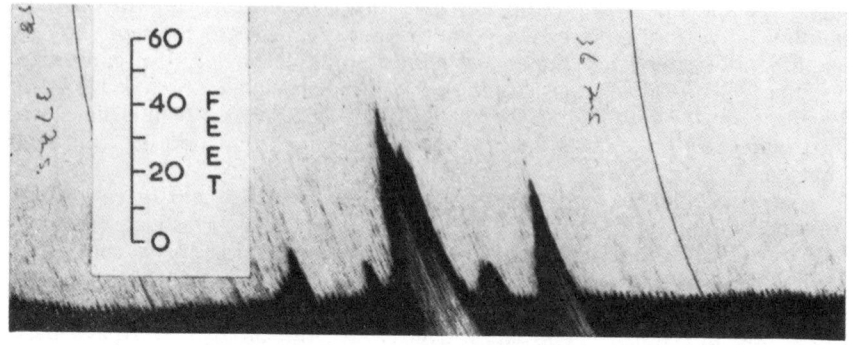

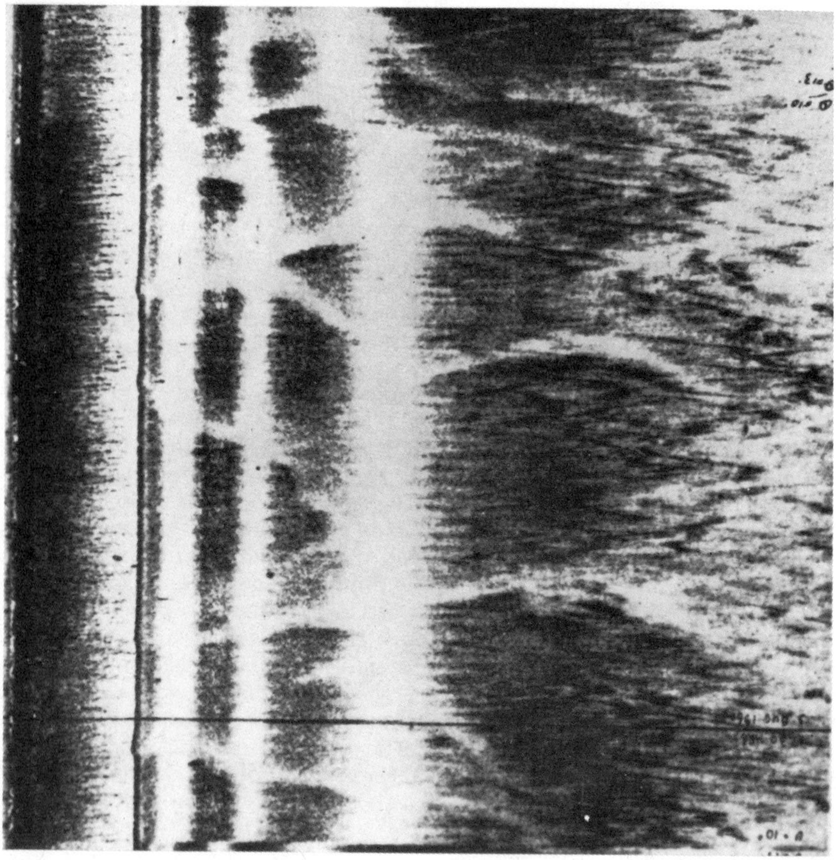

Fig. 9–43 Sand waves in the North Sea shown in echo-sounding profile (above) and side scanning (below). From Stride (1963B).

The effects of glaciation on the continental shelf are much more evident around Scandinavia than in the North Sea. The deep Skagerrak Trough, which swings around the south coast of Norway and terminates at the shelf edge despite its trend at right angles to the Scandinavian ice cap, is clearly the same type of feature described off other glaciated coasts. It was at first interpreted as a result of faulting (O. Holtedahl 1940, 1964; H. Holtedahl 1955), like so many other troughs in glaciated areas, but because of the use of seismic profilers it is now thought to be erosional in origin. It is related to the contact of old rocks of Norway and younger sediments of the sea floor (O. Holtedahl 1970): Farther north, the continental shelf is crossed by many other glacial troughs, as shown so well on the Olaf Holtedahl maps (1940). The shelf off Norway averages about 160 km in width and has extensive areas more than 200 m in depth. The offlying banks also resemble those outside other glaciated coasts, although they are somewhat deeper. The shelf sediments are poorly sorted (H. Holtedahl 1955) and contain many rock fragments indicative of glacial marine origin.

The continental slopes off northern Scotland and the Shetland Islands have deep channels, like those off Ireland, and banks and islands, such as the Faeroes, on the outside. There is no evidence that erosion produced the channels (Stride et al. 1967). Presumably this is a part of the European continent that has broken away from the main mass (see p. 422). Off Norway, the continental slope is very gentle, averaging about 1° out to the 2000-m contour, but north of latitude 67° it steepens and 4° slopes occur off the Lofoten Islands.

The Arctic   The Barents Sea, along the north coast of Europe, has a very wide and deep shelf (Fig. 9–44). The 1300-km width compares closely with the shelf seas of China and Australia, but it is far deeper than the others, ranging mainly between 200 and 400 m. It includes many deep troughs. Two large groups of islands, Spitsbergen and Franz Joseph, rise near the northern shelf border. This sea has been thoroughly explored by Russian marine geologists, notably Klenova (1940, 1960). The sediment is largely mud but contains an abundance of stones (Bissett 1930), as in other deep areas off glaciated coasts. The Kara Sea, on the east side of Novaya Zemlya, has a deep trough following the coast of the island.

East of the Kara Sea, the shelf undergoes a marked change. Aside from some deep troughs around the presently glaciated outlying islands of Severnaya Zemlya (100°E longitude), the Siberian and adjacent Alaskan shelves are very flat and usually less than 75 m in depth. Very shallow zones extend out beyond most of the large rivers of northern Siberia. The shelf is decidedly narrower where it is overlapped by the Severnaya Zemlya Islands, but widens again to about 550 km off northeastern Siberia, then narrows in the Chukchi Sea off Alaska to about 100 km, still maintaining its general flatness, except where cut by Barrow Sea Valley, which penetrates the shelf west of Point Barrow, Alaska. Otherwise, the flatness of the shelf north of Bering Strait corresponds with that of the Bering shelf on the south side. According to the fathograms in both places, the shelf appears to be as flat as any from which echo soundings are available. Locally, however, small hills about 2 to 4 m high are located near shore in depths out to 28 m (Rex 1955). Other hills were discovered at greater depths on the side of the shelf sea valleys (Carsola 1954B).

Sediments of the Chukchi Sea (Carsola 1954A) are predominantly mud, apparently from the Mackenzie and other rivers, along with a scattering of sand and gravel that is rafted into the sea by drift ice. This mixing produces

Fig. 9–44   Glacial erosion topography of the Barents Sea north of Norway and Russia. From Klenova (1960B).

a bimodal sediment characteristic of all high-latitude shelves off both glaciated and unglaciated areas. The brown color of the Chukchi Sea sediments suggests oxidizing conditions of deposition. Decomposable organic material comprises about 1.5 to 2.0 percent. These sediments appear to be similar to those of the northeast Siberian shelf (Böggild 1916).

At the eastern end of the Beaufort Sea are a number of deep troughs, and farther east, all the glaciated lands are cut up into islands and deep troughs, similar to those off virtually all other glaciated territory. As far as is known, the depths of the passageways between the islands are no greater than those around Norway and Newfoundland. Hudson Bay is another relatively deep basin. Not much is known about the sediments of these embayments nor of the irregular continental shelves on the outside.

Exploration of the continental slopes in the Arctic Ocean has been difficult because of ice conditions. The area off Point Barrow has been found to have offlying ridges suggestive of a continental borderland (Fisher et al.

1958). Farther east, the slope to 2000 m is only 1.5° and is cut by a series of submarine canyons (Carsola et al. 1961).

The Antarctic   Despite extensive exploration of the continental terraces of Antarctica during the past two decades, considerable uncertainty remains as to their nature. The best summaries that I have found of the relief of the sea floor around Antarctica are by Zhivago (1962, 1964), but these have little concerning the continental terraces. Apparently the shelf is very deep, much of it more than 200 m. Sounding lines show that it is not so irregular as are many shelves off other glaciated areas, but there are undoubtedly many deep troughs and basins. The continental slope generally is gently inclined, much of it less than 0.5°. No one has determined if it is cut by submarine canyons. Most voyages to the Antarctic continent cross the slope at right angles to its trend; thus the nature of the slope relief is still not indicated. Off the Davis Sea sector, Soviet geologists have noted a 5° slope.

Chapter 10
ORIGIN AND HISTORY
OF CONTINENTAL TERRACES

INTRODUCTION

In Chapter 9, the continental terraces of the world were described with only a brief reference to their origin and history. The present chapter, after summarizing the characteristics of the shelves and slopes, will attempt to explain the various types that border the continents of the world. It will show how processes of erosion and deposition have interacted with diastrophism and sea-level changes to develop present-day conditions. Because we have not yet given proper documentation to the principal erosional feature of the slopes—submarine canyons—this phase of development of continental terraces will be considered in Chapter 11, in which canyons are described.

Since writing the second edition, publication of results from important investigations from all parts of the world have led to many new ideas concerning the origin of the continental shelves and slopes. Many interpretations that seemed well-founded 10 years ago must now be modified. In reaching the tentative conclusions that will be offered here, I have drawn heavily on the publications of J. R. Curray, K. O. Emery, H. D. Hedberg, D. G. Moore, A. H. Stride, D. J. P. Swift, and E. Uchupi.

CONTINENTAL SHELF CHARACTERISTICS

Shelf Topography In Chapter 9, the description of the shelves will have made it evident that the typical shelf profile is far different from the smoothly graded curve of the old concept of a wave-cut terrace bordered seaward by a wave-built terrace. The so-called profile of equilibrium does exist in shallow water along many beaches but has little application to the shelves as a whole. Where closely spaced soundings are available, the typical shelf is seen to include many terraces, hills, and depressions.

Unfortunately, we do not have any recent compilation that would allow us to obtain a good statistical analysis of shelf depths, widths, slopes, and relief. Lacking such material, it is necessary to use the old statistics, which I obtained with the help of students more than 30 years ago. The figures are derived from contoured charts and profiles of all parts of the world and from measurements taken at intervals of approximately 18 km along all the continental coasts.

1. The continental shelf has an average width of 75 km.
2. The average depth at which the greatest change of slope occurs at the shelf margin is 130 m.
3. The average depth of the flattest portion of the shelves is about 60 m.
4. Hills with a relief of 20 m or more were found in 60 percent of profiles crossing the shelves.
5. Depressions 20 m or more in depth were indicated in 35 percent of the same profiles. Many of these are basins, but others represent longitudinal valleys.
6. The average slope is 0°07′, being somewhat steeper in the inner than in the outer half. Hayes (1964) found that out to 60-m depths average slopes were 0°12.4′.

The same statistics show little relation between marginal depths and shelf widths. Although shelves narrower than 40 km have slightly shoaler margins than the rest, there is no other progressive deepening of the margins with increasing width. Furthermore, the narrow shelves with shallow margins usually occur in areas of active coral growth. Elsewhere, the shoaling of the narrow shelves is related to the overlapping of the shelf by deltas or cuspate forelands. It is surprising that the shelves along the East Coast of the United States have shoaler marginal depths than those of the West Coast north of San Francisco. This seems contrary to the evidence from various sources that the East Coast is sinking and the West Coast is rising. The greater depths on the West Coast may be related to greater wave attack during the last glacial stage of lowered sea level. However, an examination of the marginal depths off all coasts with prevailing offshore winds and those with onshore winds (omitting the deep glaciated shelves) showed that those exposed to onshore winds average 109 m and the leeward coasts 132 m. This and other comparisons indicate that other factors are more important than exposure of coasts to wave erosion. Obviously, the depth of shelf margins is related to the amount of deposition since sea level rose after exposing the outer shelf to wave erosion during the Wisconsin glacial stage, and is dependent on the effect of ocean currents, like the Gulf Stream, in either preventing deposition or eroding the bottom. Furthermore, there has been much warping and faulting of continental margins since the last period of low sea level.

Shelf Sediments   Prior to the 1930s when geological investigations of the shelves were initiated, it was virtually geological dogma that the shelf sediments grade outward from coarse near shore to fine on the outer margins. Now, however, the vast collections of sediments obtained in recent years have confirmed an opinion based only on chart-bottom character notations (Shepard 1932) that the outward gradation of sediments is exceptional and confined mainly to narrow zones near the beach. We now know that the size grades of the outer shelf sediments show little relation to distance from shore. Coarse sediment is common on the outer shelves, and tracts of sand border many areas covered by mud on the inner shelves. Rock and gravel are found in abundance on the outer shelves. In fact, the typical shelf bottom-sediment chart looks like a patchwork quilt.

Despite the lack of any general rule of decreasing sediment-grain size seaward across the shelf, some definite trends are related to character of land area, coastal configuration, strength of coastal currents, and topography of the shelf. All of these have a marked influence on the bottom character. These relationships can be summarized as follows:

Sand, which is the most common of the shelf types, is generally found in the following environments: (1) on open shelves with small relief, (2) off long sand beaches, (3) on elongated shoals, (4) off sandy points, (5) outside narrow bay entrances, and (6) on banks off glaciated areas (often mixed with gravel and stones).

Mud bottom is the dominant type in the following situations: (1) off large river mouths and downcurrent from them, (2) in sheltered bays and gulfs, and (3) in depressions of the open shelf.

Rock bottom, often associated with gravel, pebbles, or stones, is found commonly: (1) in straits, (2) off rocky points, (3) along coasts having rocky sea cliffs, (4) on an open shelf with strong currents, and (5) on shelf hills, except off glaciated areas.

Hayes (1967), using bottom-sediment notations on the inner shelves (0 to 60 m) of the world exclusive of the Arctic, correlated sediment types with climate. He found the following relations:

1. *Mud* is most abundant off areas with high temperature and high rainfall (humid tropics).
2. *Sand* is abundant everywhere and increases to a maximum in intermediate zones of moderate temperature and rainfall in *arid* areas of all except extremely cold temperatures.
3. *Coral* is most common in areas with high temperatures.
4. *Gravel* is most common in areas of low temperatures (subpolar and polar).
5. *Rock* is generally more abundant in cold areas, but its distribution correlates strongly with inner shelf slope (more abundant on steep slopes).
6. *Shell* distribution is not diagnostic with regard to climate (except more abundant in hyperarid).

The use of notations from only the inner shelf provides a good idea of the nature of present-day sedimentation, because it eliminates a large proportion of the relict sediments, but relict sediments are found also on portions of the inner shelf.

CONTINENTAL SLOPE CHARACTERISTICS

The continental slopes are still far less well known than the continental shelves, partly because echo soundings fail to indicate the most precipitous slopes. However, the numerous sounding profiles along with seismic profiles show the underlying structure and give us a far better picture of these slopes than was available in the past decade. No one has attempted to obtain new statistical averages, so it is necessary to repeat those presented in earlier editions. From the profiles then available, it was determined that the slopes averaged 4°17′ down to the 1800-m depth. In general, these slopes decrease at greater depth, but in the Pacific Ocean most slopes increase with depth. This is true not only where the slopes terminate in trenches, but also off some areas where trenches are missing, as off much of the West Coast of the United States and southeastern Australia. On both sides of Florida, the steepest inclines are on the outer continental slope beyond a plateau on the east and beyond a gentle inner slope on the west.

The depths of the base of the continental slopes vary extensively. In general, they are deeper in the Pacific, averaging greater than 4000 m and as much as 10,000 m where the slopes terminate in the deepest trenches of the western Pacific. On the other hand, the base of the slopes are generally less than 3000 m along the West Coast of North America and north from Northern California to the change in trend near Yakutat Bay, Alaska. In this zone there are no present-day trenches. In the Atlantic, many of the continental slopes terminate at less than 3000 m, but are bordered by a continental rise that continues the descent as a gentle slope to more than 4000 m. In most of the inland seas, the slopes terminate at less than 3000 m, and the same is true of the large embayments, such as the Bay of Bengal and the Arabian Sea.

The typical sediments of the continental slopes are finer than those of the shelves. A count of bottom notations on published charts indicated that about 60 percent are mud, 25 percent sand, 10 percent rock and gravel, and the remaining 5 percent shells and ooze. In general, sampling across the shelf break shows a change from sand to mud, but on the slopes are patches of coarse sediment, many of them related to hills and ridges. Rock bottom is found along most steep slopes and in localities with strong slope currents. The rock is partly slide debris but includes outcropping layers truncated by the slopes.

BASIC TYPES OF CONTINENTAL TERRACES

Seismic profiles have recently provided us with a good basis for classifying the structural frames of the continental terraces of the world. Fig. 10–1 is a compilation partly redrawn from D. Moore and Curray (1963), Emery (1968C, 1969B), and Hedberg (1970). In this classification there is no attempt to show the special results of glaciation of the shelves, the special effects of migrating sea levels during the Pleistocene, the changes produced by shelf currents, or the modification of the slopes by slumping or by erosion of various types of currents. These will be considered later.

The first two types with faulted margins are probably very common around the slopes of the world, although the fault scarps have generally been greatly modified by slumping and erosion and by deposition in the depressions left by slumping or in the depressed zones between step faults. The continental terraces that are bordered by trenches are presumably of this

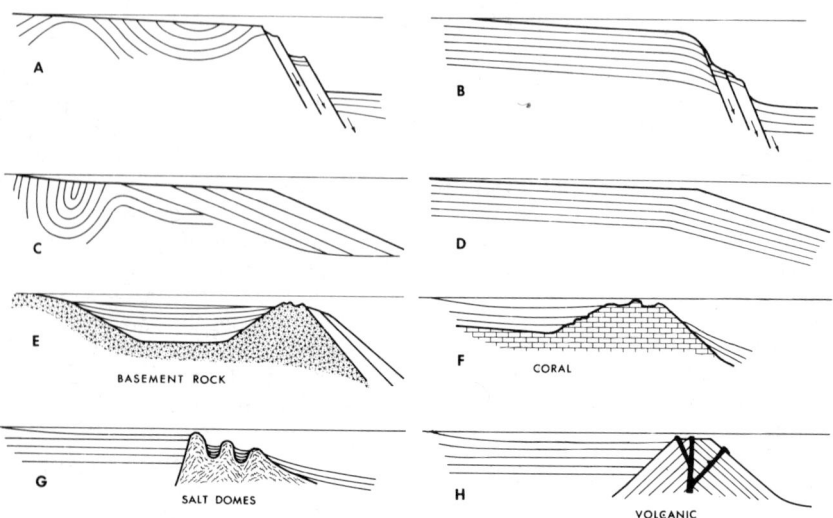

Fig. 10–1   Principal types of continental terraces.

type. Emery (1969B) classifies all of the West Coast shelves of the Americas south of Canada as having tectonic dams (his figure on p. 44). However, his example (Fig. 9–25) shows what is probably a faulted margin with slumps below the upper scarp. It could also be type E in the classification of Fig. 10–1. It is to be expected that faulting would be common at the juncture between the heavy oceanic blocks and the light continents. Evidence that faulting has been important even along continental slopes where there are no trenches at the base comes from the occurrence of earthquakes on many of these slopes, on a steepening of the basal portion of slopes in many localities, on the numerous places where the structure of the land comes to the sea diagonally and is truncated by relatively straight continental slopes (Fig. 10–2), and on the numerous abrupt changes in trend of the slopes.

The shelf inside the fault scarp margins may have several origins. In profile A (Fig. 10–1), it is assumed that the shelf has been eroded, developing a type of wave-cut terrace related only in part to present sea level. An example of such a combination can be seen in Fig. 9–40, sect. 7, and in Fig. 10–3A. The material eroded from the shelf could, of course, accumulate on the slope. However, if the slope were sufficiently steep to cause sediments to slump to the base, the fault scarp might be essentially preserved. In profile B, the shelf is indicated as having been built upward by deposition as the underlying crust sank. Here, we are confronted with the problem of how the edge could grow without slumping over the adjoining scarp. Probably there would be extensive slumping around the shelf margin unless the marginal sediments were coral or another type of reef, or if elevated fault blocks along the margin formed a dam of hard rock, as in Fig. 10–3B, or alternatively, by landward tilting of a hard layer (as in Hedberg 1970, Fig. 18C).

Type C in Fig. 10–1 shows a forward-built slope, usually called a *wave-built terrace*, and previously considered to be the normal origin of the continental slopes. In the earlier editions, I argued against such an origin because of (1) the presence of extensive rock formations on the outer shelves and even

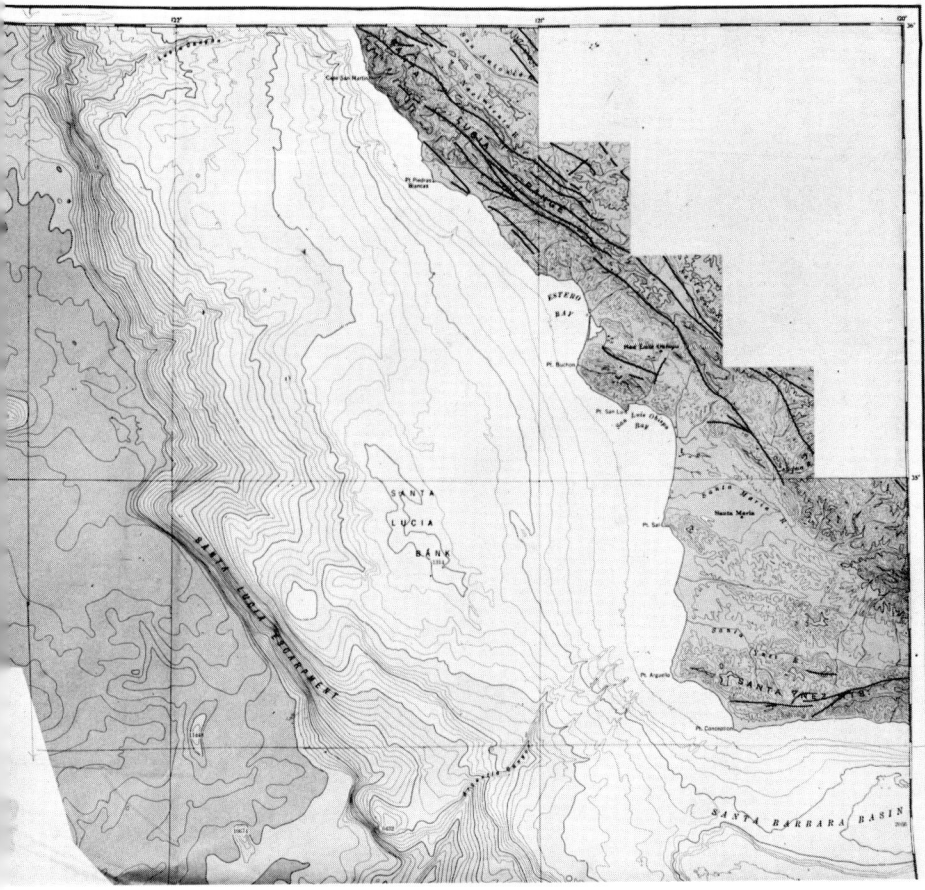

Fig. 10–2 Truncation of coastal ranges of California by the continental slope south of Monterey. From Shepard and Emery (1941, Chart II).

on the slopes, (2) the great irregularity of the slopes, and (3) the frequent increase in angle near the slope base. These considerations, however, merely show that prograded continental terraces are not the only explanation. Numerous seismic profiles now indicate that prograding has been very important (see Fig. 10–4) and that it has definitely widened many shelves, notably along the Atlantic coasts of the continents. On the other hand, even the prograded Atlantic shelves have outcropping rock or slightly buried old formations on the outer shelves, as, for example, on Georges Bank and off the English Channel. In most places, prograding has simply added to the outer margin. Around many great deltas, however, the margin has advanced over the deep ocean for many miles. This is almost certainly true off the Niger Delta (Fig. 9–36) and the Nile. Very likely the forward growth here has been as foreset beds of the delta rather than as an actual wave-built terrace. Without long cores from these prograded slopes, it is difficult to distinguish between the two.

Type D, a combination of deposition on a subsiding basement with

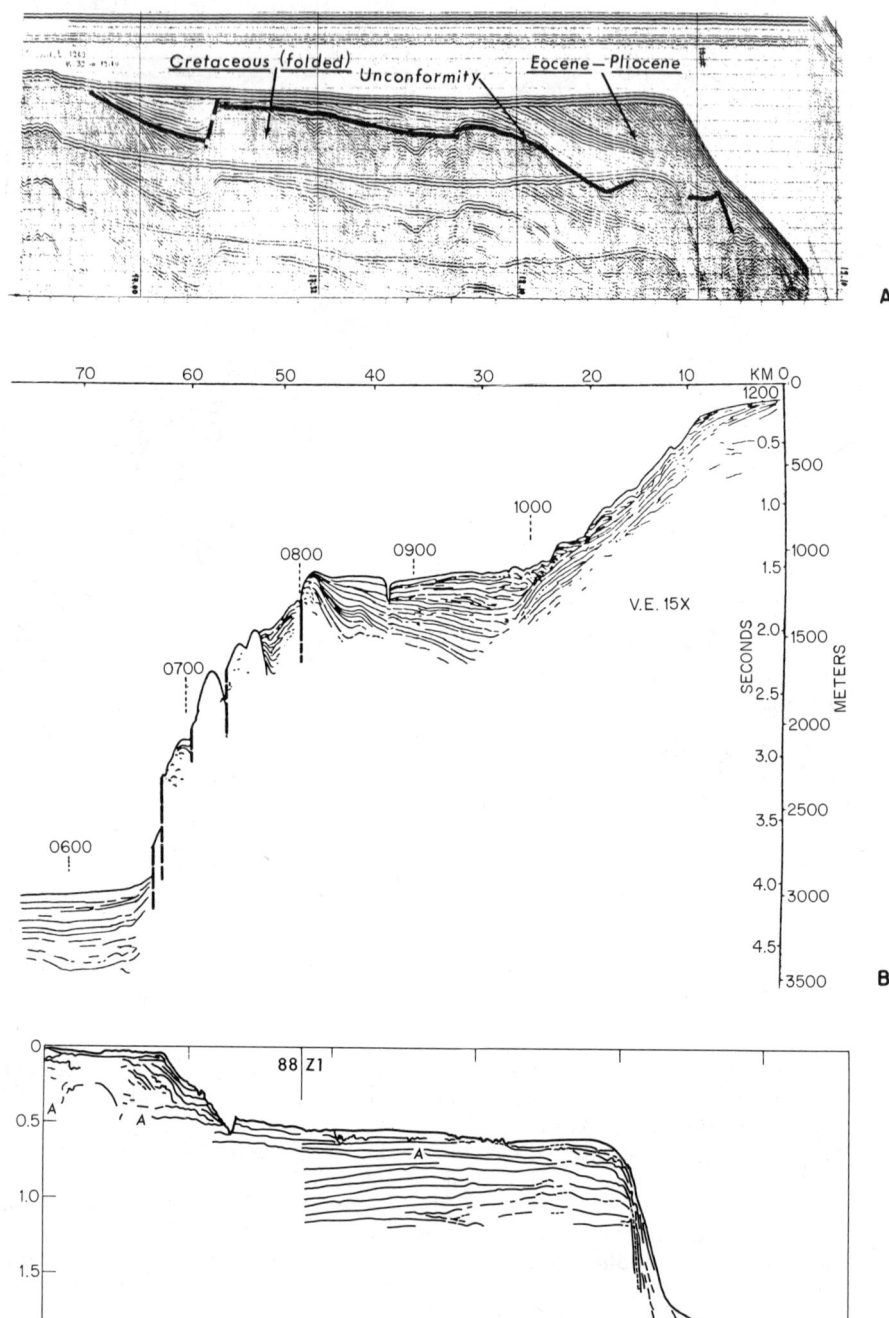

Fig. 10–3   A, seismic profile off northern Spain, an example of a wave-cut terrace on a continental shelf and a fault-scarp margin with small outbuilding. From Boillot and d'Ozouville
(1970). B, seismic profile off Northern California shows how a fault block has formed a dam
at the shelf margin. From Silver (1971). C, seismic profile off Blake Plateau shows how a coral
reef has formed a dam at the margin of the Plateau. From Uchupi (1970).

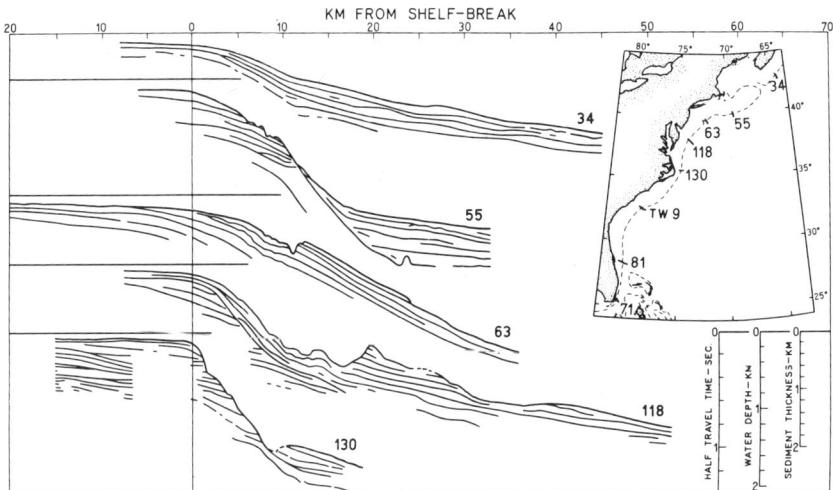

Fig. 10–4 Seismic profiles from East Coast of the United States indicate a prograded continental terrace combined with local slope erosion and slumping. From Emery (1968A).

prograding of the slope beyond the shelf, has proved to be relatively common. The narrow inner shelf off Florida (Fig. 9–9) is now definitely demonstrated to have this type of origin. Another good example has been found off the Nayarit coast of Mexico south of the Gulf of California (Curray and Moore 1964).

Several authors have stressed the importance of dams along the outer edge of continental shelves as a primary cause for the shelves. The most common of these is indicated in Fig. 10–1E. In the second edition, such an explanation was suggested in Fig. 125B, taken from the San Francisco area where basement rocks have been eroded on the outer shelf, leaving the Farallons as erosion remnants. Deposition has occurred on the inside of this hard rock barrier, filling the former basins (Fig. 9–25). Such situations are referred to as *tectonic dams* by Emery (1969B) and as *basement ridges* by Hedberg (1970). Emery indicates that shelves of this type extend along the margins of the Americas from British Columbia south to Chile. This, of course, is a broad generalization; some seismic profiles fail to substantiate such an origin for the terraces in this long tract.

Another type of shelf-margin dam is shown in Fig. 10–1F. Here, coral growth has built up the margin. The slope beyond may be entirely a reef slope, but is more commonly tectonic because of the rare event that shallow-water reefs have sunk to the base of the entire slope. Examples of coral-reef dams are found off the East Coast of Florida, probably at the margin of Blake Plateau (Fig. 10–3C), and also in the Java Sea. Dredging and profiling of the West Florida Escarpment shows that at least the outer portion of the shelf grew upward as a coral reef while the slope was sinking. This, in turn, allowed deposition of other noncoral sediments on the landward side.

Type G, having a dam the result of a diapiric intrusion, has proved in the past decade to be very important. The type example is the northwest Gulf of Mexico (Fig. 9–17). Here, the diapirs are believed to be mostly salt domes (Lehner 1969), but many other hills on outer shelves and possibly ridges or anticlines may have been formed by mud diapirs. Unless these hills have been studied by geophysical methods, they usually are interpreted as bioherms because of the surficial covering of coral or algae. Even with seismic profiles and gravity methods, interpretation may be difficult.

Volcanic dams (Fig. 10–1H) are included by Hedberg as among the barriers on the continental margins, giving the Ryukyu Islands as an example, although actually these islands form an arc marginal to the relatively deep Okhotsk Basin. Probably a better illustration is the Pribilof Islands, along the southern margin of the Aleutian shelf.

There seem to be numerous examples of what may become future continental terraces. Thus, each of the narrow seas along the east and south-east Asiatic coast has a framework of islands that is allowing sedimentation to create a shelf on the inside. Wave erosion may bevel the islands to produce a submerged dam.

PROCESSES INFLUENCING SHELF TOPOGRAPHY

The nature of shelf topography has been influenced in many ways by glaciation of the continents. Direct effects have come from glaciers spreading out over the continental shelves beyond most parts of the glaciated continents, and indirect effects have developed from changing sea levels that accompanied the waxing and waning of the great glaciers. Additionally, shelf topography has been influenced by wave and current erosion and by the growth of coral reefs. Shelves have been narrowed by overlap of deltas and cuspate forelands and, to a smaller degree, by prograding barriers. Locally, topographical features may be influenced by faulting and folding, as well as by diapiric intrusions.

Shelf Glaciation The topography of most shelves that have been covered by glaciers is strikingly different from those that have not. Deep troughs cross or partly cross the shelves, and others follow zones of weak rock parallel to the shelf trend. Basins with depths commonly exceeding 200 m are found, especially near land. The outer shelves have many banks that rise above the levels of the inner shelves. These are a combination of morainic ridges due to the dumping of sediments at the margins of glaciers and to outwash fans built along the glacial termini. The effect of outwash has in many cases masked the direct glacial deposition and accounts for the smooth surfaces on the outside of Georges Bank, Banquereau, and Sable Island Bank. In the past, failure to appreciate that glaciers have produced these various features has led to many misinterpretations of the sea-floor topography in glaciated areas. The most common difficulty has come from mistaking the relatively straight margins of glacial troughs for fault scarps and interpreting the troughs as grabens. Fortunately, this can now be avoided by seismic profiles, although, of course, it will be found locally that some of the troughs follow old faults, which can be detected in the profiles.

The landward margins of glaciated shelves are also distinctive. The coast is penetrated by deep reentrant bays, like the Gulf of St. Lawrence, Hudson Bay, Dixon Entrance, the Strait of Juan de Fuca, and the Strait of Magellan. Even larger bights have been eaten into Antarctica, but most of these are still partially or largely buried in ice.

Effects of Changing Sea Level The lowering of sea levels during glacial stages allowed rivers to flow out over the shelves and cut shallow channels into the exposed surfaces (see Fig. 9–6), although most of these

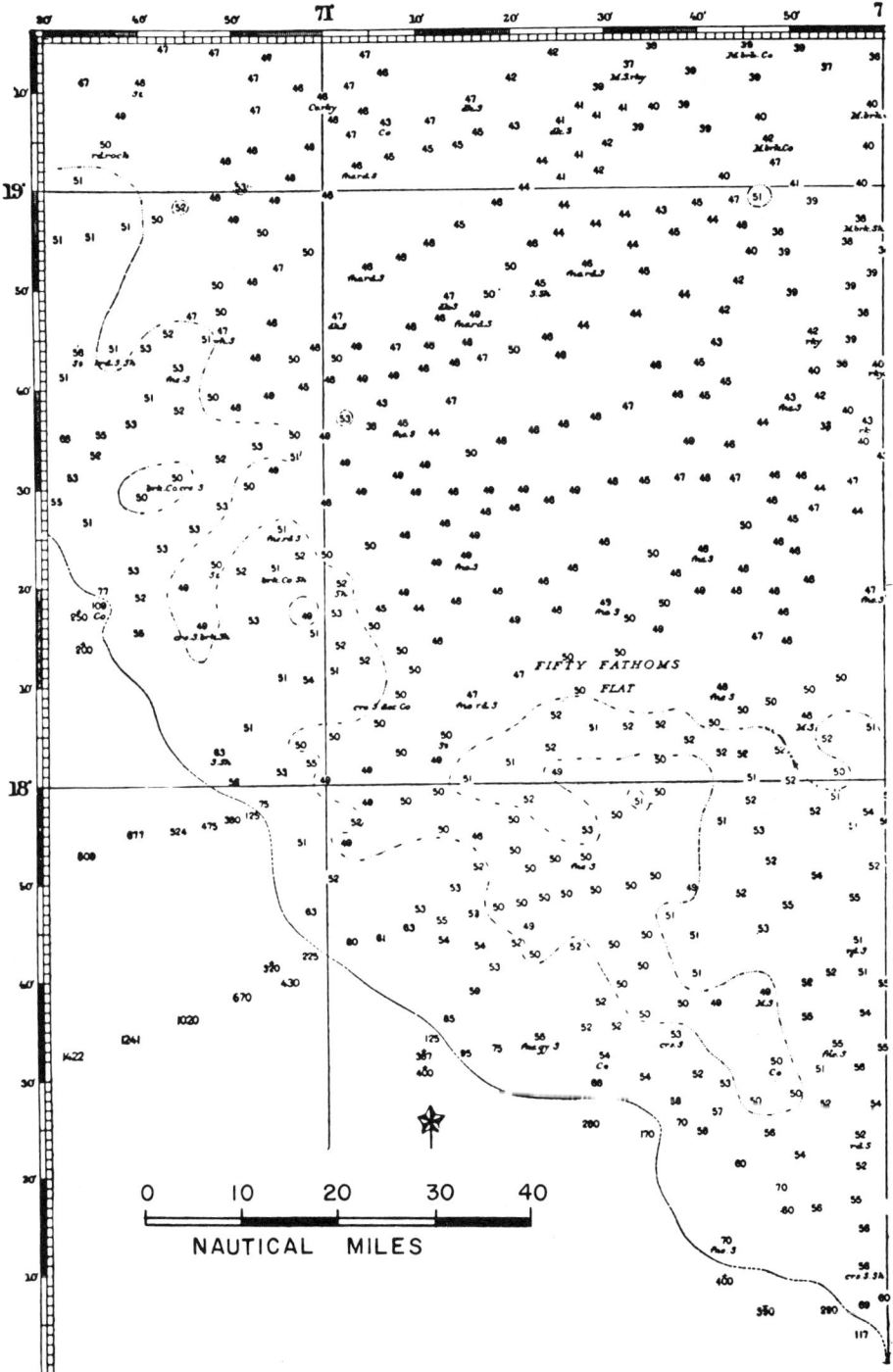

Fig. 10–5 The broad flat terrace on the outer shelf of western India just south of the Indus Delta. From U.S. Navy Hydrographic Office Chart No. 1590.

were filled or partly filled as the sea again rose. The best-preserved example is the valley that crosses the shelf off New York. Terraces have been cut into the outer shelves when the sea level stood still for a sufficient period. These are illustrated by the Franklin Shore of Fig. 9–6, which also shows ridges on the outer shelf that have been interpreted as barrier islands formed during low sea-level stages. Rivers were also forming their deltas at low levels, and these no doubt account for many of the broad flat areas on the shelves off the present deltas (Fig. 10–5). Off limestone areas like Florida, the lowered sea levels produced sinkholes now covered by the sea.

Permanent ocean currents can both prevent deposition and produce erosion. Undoubtedly, the depth of Blake Plateau is far greater than that of the shelf to the north, because the Gulf Stream prevented deposition as the Plateau sank. In contrast to the shallow inner shelf, there has been little deposition since the Oligocene on Blake Plateau and, in some places, the Eocene outcrops on the surface suggest that even the Oligocene has been eroded. To the south where the Gulf Stream runs close to the coast, the shelf is virtually eliminated. Less-drastic effects are those of the North Equatorial Current, seen along the front of the Orinoco Delta where advance is inhibited. The absence of an appreciable shelf along the east coast of Yucatan may be explained in part by the powerful northwest-flowing current. It is suggested that the deep shelf off southwest Africa (van Andel and Calvert 1971) and the lack of a shelf off southern Angola may also be the result of the Benguela Current, although little is known about this area.

Another effect of currents is observed near the mouths of bays where the tides are concentrated. The resulting holes in the bottom are surprisingly deep. The entrances to the Inland Sea of Japan have the deepest (Fig. 9–30), but basins even exist in such virtually tideless seas as the Gulf of Mexico, where the water is 50 m deep at the entrance of Barataria Bay. The concentration of ocean currents in straits is even more effective. Probably the 1750-m depth of Florida Strait is largely the result of the Gulf Stream concentration between Cuba and Florida.

Coral Growth on Shelves   In tropical regions away from large rivers, the profuse growth of coral on the shelves accounts for the numerous banks that rise to or nearly to the surface over wide areas. The Queensland shelf of Australia is the best example, and many other tropical areas on the east side of the continents have extensive shelf reefs.

Sand Ridges on Shelves   One of the most amazing features of the continental shelves is the variety of sand ridges, usually rising only a few meters above their surroundings. They are particularly common in the shallow water near the coast but are found also on the middle and outer shelves (Fig. 9–7). In areas with high tidal range, these ridges ordinarily converge toward the entrances of bays and straits, notably in the English Channel and the south end of the North Sea (Fig. 10–6). Off cuspate forelands, such as Cape Hatteras on the East Coast and Cape San Blas on the Gulf Coast, zones of sand ridges extend seaward with elongations both parallel and at right angles to the trend (Fig. 10–7). Off the straighter parts of low barrier coasts, the ridges show a strong tendency to approach the coast diagonally to the left (Fig. 10–8). This is true off the Mid-Atlantic East Coast, off the Panhandle of Florida, and off Padre Island, South Texas. Some ridges shown in Fig. 9–7 appear to trend nearly normal to the shore; others have a relation to bay entrances, as in Figs. 8–7 and 10–9.

The nearshore ridges that are roughly parallel to the coast and those on the outer shelf parallel to the shelf margin (Fig. 9–7) appear to be drowned

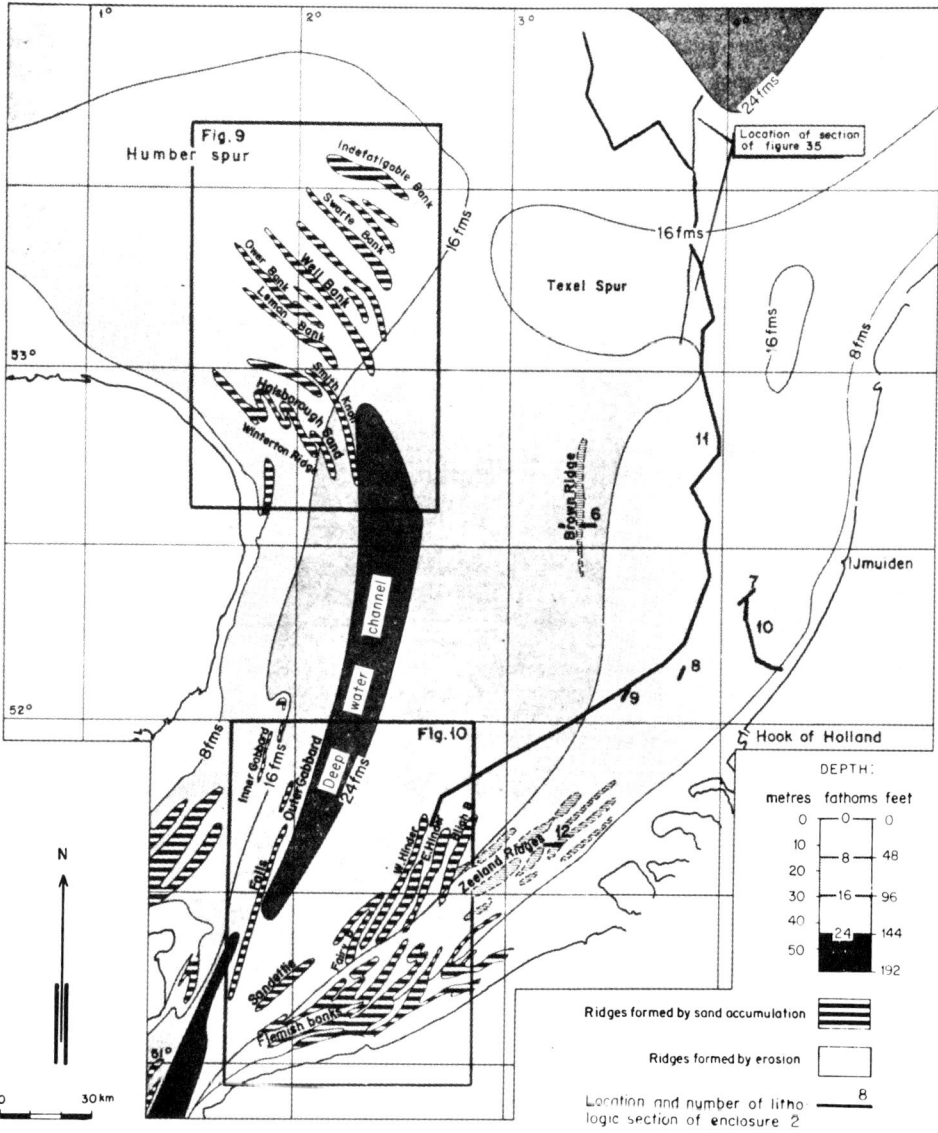

Fig. 10–6   Sand ridges of the English Channel and south end of the North Sea. From Houbolt (1968).

barrier islands. Some of the diagonal shoals have also been interpreted as drowned beach-dune ridges (Sanders 1963). However, Swift (1969, DS-4), referring to comparative surveys by D. Moody, has noted that some of the ridges have moved southeastward during recent storms, and he believes that they are related to modern processes. The possibility that originally they may have been the old beach-dune ridges of submerged cuspate forelands and still subject to movement should be considered. Examples like those in Fig.

Fig. 10–7 Cape Lookout Shoals, North Carolina. Typical buildup of sediment on the shelf outside of cuspate foreland. From U.S. Coast and Geodetic Survey, 1959.

N

10–8 could be explained as having originated at slightly lower sea level. Perhaps the predominant current in all three areas of left-trending ridges has been to the left along the shore, tending to eliminate the ridges on the right side of the forelands and to preserve only those on the left. The ridges that converge toward embayments (Figs. 8–7, 10–9) are tidally induced (Off 1963). Some of the transverse sand ridges of the outer shelf are also the result of tidal currents, which were directed toward bay mouths during periods of lower sea level. It will be recalled that those off the English and Irish channels have been explained in this way.

Nondepositional and Relict Sediment Zones on the Continental Shelves
In many areas on the continental shelves, the character of the bottom indicates that deposition has been negligible or entirely lacking since the major rise in sea level at the close of the last glacial stage. Thus, rock platforms indicative of wave erosion during low sea levels are found at depths too great to suggest present-day activities. Far more widespread than rock are types of sediment that are clearly not being introduced under present conditions. Vast tracts of sand bottom lying outside a mud-covered shelf suggest that the sand is relict. Extensive sampling, especially of the shelf off the eastern United States, has verified the relict nature of the outer sand areas by showing that associ-

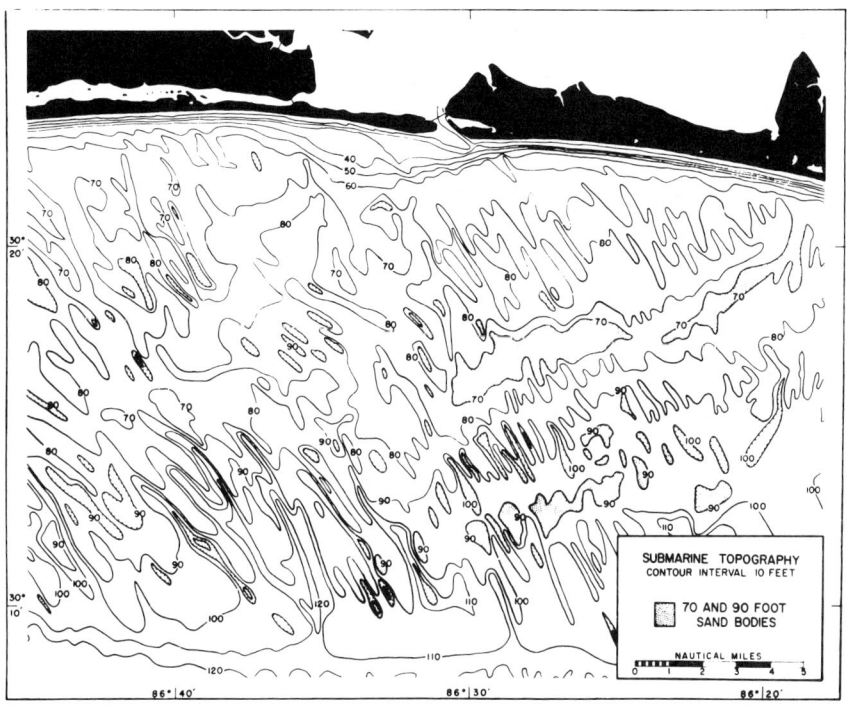

Fig. 10–8 Sand ridges diagonal to coast off northwest Florida. From Hyne and Goodell (1967).

ated with the sand are numerous zones with shells of distinctly coastal character, partly estuarine. Also, freshwater peat is widespread, and even fossil land mammals have been collected. In glaciated areas, the till and outwash deposits of former glaciers are still uncovered. Relict sediments have, of course, been subject to changing conditions as the water first covered the shelves and then grew progressively deeper. Thus, while they were in the surf zone, the sediments may have been worked over by the waves, removing some of the fine sediments and concentrating the coarse and heavy grains. Also, as the water deepened, planktonic foraminifera and other microorganisms may have sifted down into these sediments. One effect of the encroaching seas on the relict sediments has been the production of polymodal mixtures (Curray 1960).

Approximately 70 percent of the sediment cover of the continental shelves of the world have been classified as relict (Emery 1968B). These include most of the sediments off the East Coast (Fig. 10–10). Other examples are the relict sediment in the zone marked sand or gravel in Fig. 9–32 and the outer-shelf sand and shell zones in Fig. 9–16. As Emery has indicated for the East Coast, the generally landward-moving bottom currents from the middle shelf and the seaward-moving currents on the outer shelf are an important factor in maintaining this relict sediment. Also, the recent rise of sea level has drowned so many river mouths that most deposition takes place in estuaries. As a result, the open shelf currents do not have to compete with great quantities of sediments and are capable of keeping the occasional contributions

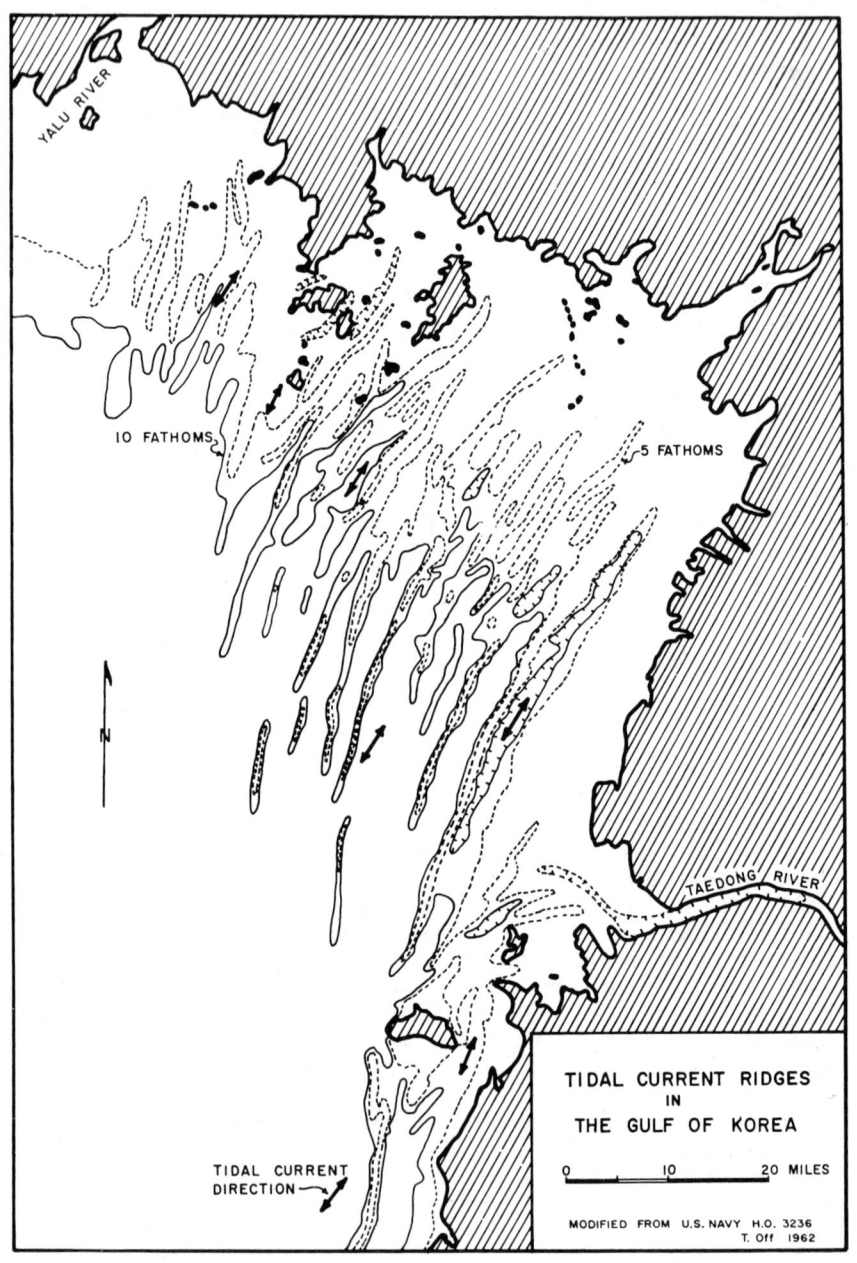

Fig. 10–9   Sand ridges extending into an embayment in the Gulf of Korea. From Off (1963).

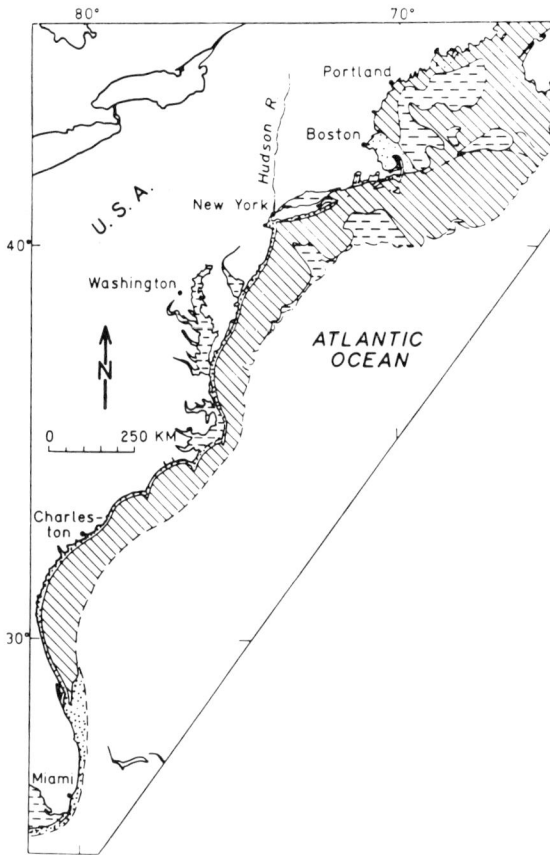

Fig. 10–10 Relict sediment zones (diagonal lines) off the East Coast of the United States. From Emery (1968B).

of fine sediment moving across the shelf. The strong permanent currents, such as the Kuroshio off East Asia and the North Equatorial Current off the Amazon and Orinoco, contribute to an important degree to this effect.

In some areas, the shelf sediments are the result of weathering of rocks in place (see Swift 1969, DS-5). Thus, the phosphatic sediments on the shelf south of Cape Hatteras have been shown to be weathered from Miocene rocks, and the glauconite off the Carolina coasts has been traced to underlying strata. Swift has noted the importance of what he calls "unmixing" of sediments on the continental shelves. As a clear-cut example, waves and currents have worked on the poorly sorted glacial till off Nova Scotia while it was being covered by the sea and going through the surf zone. This process removed the fines and concentrated the coarser elements, producing good sorting. Sorting action may also take place in relatively deep water during hurricanes, as emphasized by Curray (1960). Using cores from the shelf off New Jersey, examples of well-sorted fine sediments were found at the surface overlying very mixed grain sizes (Donahue et al. 1966). In the North Sea, the strong currents coming both from the north and the south have reworked poorly sorted Pleistocene morainic deposits and glacial outwash.

EXPLANATION OF WIDE SHELVES

The widest shelves in the world are found inside the island arcs along the East Asiatic coast and in the Arctic. Explanation for the former seems fairly clear. Dams, in part volcanic and in part diastrophic, have allowed many of the marginal basins from the Bering Sea to North Australia to fill with sediments to the shelf level inside the barriers. The fill has been provided in most cases by large rivers. Why more of these basins have not been filled up to present sea level is not easily explained. However, the growth of deltas during low sea-level stages and subsequent submergence is probably at least part of the answer. Perhaps some of these shelves were built up well above their present level during preglacial periods but were eroded during glacial stages and have not grown back to sea level in the short period that has followed.

The broad Arctic shelves north of Siberia are probably also related to deposition of large rivers. Although these coasts are bordered by frozen seas during most of the year, some of the rivers discharge sediment underneath the ice pack, as is the case at the mouths of the north-flowing rivers of Alaska and the Mackenzie in nearby northern Canada (Leffingwell 1919). This sediment may be circulated by currents flowing under the ice. The broad shelf in the Barents Sea, north of Russia, has been severely glaciated, and some of its width may be due to erosion of broad tracts that had previously been above sea level. The broad shelf in the Gulf of Maine has almost certainly grown wider as a result of ice excavation during the glacial period.

The preceding explanations of most of the broad shelves do not include outbuilding, as illustrated in Fig. 10–1C, D. Growth by progradation probably accounts for narrow additions to the outer edge of some wide shelves, but seismic profiles have made it clear that the widest shelves, like those of the Yellow and Bering seas, are attributable to fill behind barriers with little external widening. The same is probably true of most of the others.

MAXIMUM SEA-LEVEL LOWERING DUE TO GLACIATION

An interesting question, still unanswered, is the maximum extent of sea-level lowering during various episodes of continental glaciation. One method to estimate this is to look for the average level of the landward edge of the deepest terraces on the continental shelves. For such an undertaking to be meaningful we need to make use of seismic profiles; otherwise we may be dealing with a depositional terrace that would not have been at sea level during the time of deposition. The elevation of the inside of a wave-cut bench is close to mean sea level where the bench is cut in rock (Fig. 10–11), although terraces cut in unconsolidated material may be 2 m or more below mean tide level (Jordan 1961); retreating cliffs may not have any terrace beneath them, as, for example, along the outer shore of Cape Cod (Fig. 10–12). Similarly, deltas formed at lowest sea levels may be close to mean sea level at their outer edge but may later sink due to compaction of sediment or, in some cases, may develop several meters below the low-tide level (Thompson 1955).

Despite these difficulties, it is reasonable to expect that the depth of eroded surfaces near or at the outer edge of the shelves may be a good approximation of sea level during the stage when it was at the lowest point at which it stabilized over a period of a few decades or perhaps a few centuries. In most well-surveyed areas, the shelf break lies between 130 and 150 m, and the 130-m world average is in line. Therefore, it seems likely that

Fig. 10–11   A wave-cut bench cut across inclined layers of Eocene rock at La Jolla, California. Photo taken during a low spring tide.

the sea stood for an appreciable period at this level. Thus, the terrace at 145 m off the Mid-Atlantic coast of the United States (J. Ewing et al. 1960) provides partial confirmation. The shoaler marginal depths for the narrowest shelves could easily be interpreted as the result of their emergence, particularly because most of these shelves are in unstable areas.

It is possible that very short periods occurred when the sea stood lower than the 130- to 150-m level. We do have such indications. Using narrow-beam echo profilers, Dill (1969A) found an abundance of very narrow terraces and coral reefs around Australia at depths between 175 and 230 m. These would not appear on ordinary echo-sounding profiles. During descents in deep-diving vehicles off San Diego, California, he also observed narrow benches with rounded gravel and shallow-water shells in a zone between 180 and 230 m along the upper slope, and he saw similar features off Baja California. Dill obtained a date of 14,380± years for some shells from such a terrace. Also, Veeh and Veevers (1970) dated an Australian terrace at 175 m as 17,000 years, and Koldewijn (1958) dated algal material samples from 187 m off the Orinoco Delta as 14,200 B.P. These scattered stations may not mean sea-level changes, but most of them are from relatively stable areas, and they certainly suggest short episodes of very low sea level. A lowering of about 200 m would mean that the usual estimates of volumes of ice during the Wisconsin stage have been underestimated, because insufficient attention was given to ice growth in Antarctica; nor was the possibility considered that most of the dry basins may have been filled and ground water greatly increased in arid areas (Dill 1969A). On the other hand, it should be

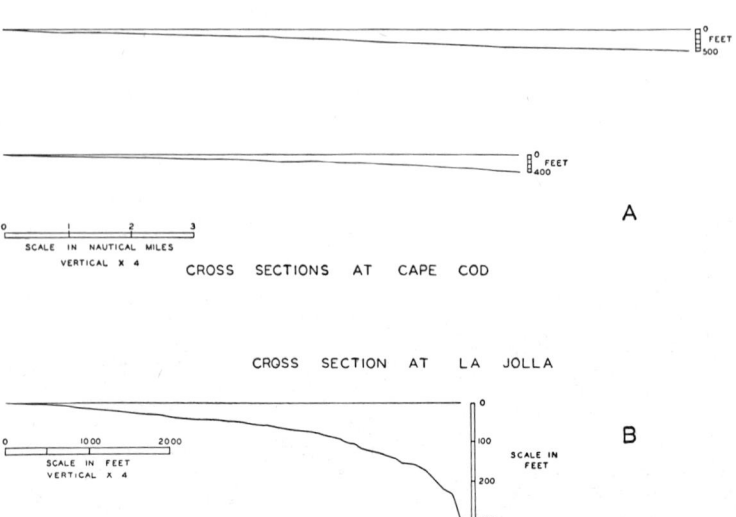

Fig. 10–12 *A*, profiles off outer Cape Cod where wave erosion has cut into the fluvio-glacial cliffs at about 1 m per year for the past century. Note absence of wave-cut terrace. *B*, profile off La Jolla, California, just south of Scripps Institution, where the waves have been cutting back the alluvial cliffs at about 25 cm per year except for a recent cessation. The gentle slope terminates outwardly with a steep descent into a submarine canyon.

remembered that evidence favoring this rather considerable lowering of sea level is still meager and requires much more investigation, especially from deep-diving vehicles.

CAUSE OF CONTINENTAL SLOPE IRREGULARITIES

Examination of continental slope profiles shows them to be mostly irregular. There appear to be three principal causes of the relief: (1) mass movements resulting in tilted slump blocks, which also cause unevenness of slopes on land; (2) step faulting; and (3) erosion, predominantly submarine but including subaerial where slopes have been temporarily above the sea during their formation. In addition, the same factors that produce dams at the shelf margin may produce irregularities on the slopes, that is, diapirism, volcanic activity, and coral growth on submerged platforms. Slope irregularities in a few areas appear to be caused by folding, partly the result of slumping.

Mass Movements on the Slopes   We seem to be confronted with some conflict between those who determine slope stability by measuring such properties as shear strength, Atterberg limits, and consolidation characteristics of slope sediment (D. Moore 1961A; Inderbitzen 1964; Morelock 1969), and those who look at seismic profiles of slopes to determine if the structures are indicative of slumping and other types of mass movements. Soil-mechanics measurements indicate that many of the slopes even up to 15° are stable, but seismic profiles include structures that look like mass movements even where the slopes are much less than 1°. Seismic profiles off Long Island and Nova Scotia (Fig. 10–13) indicate slumping on very gentle

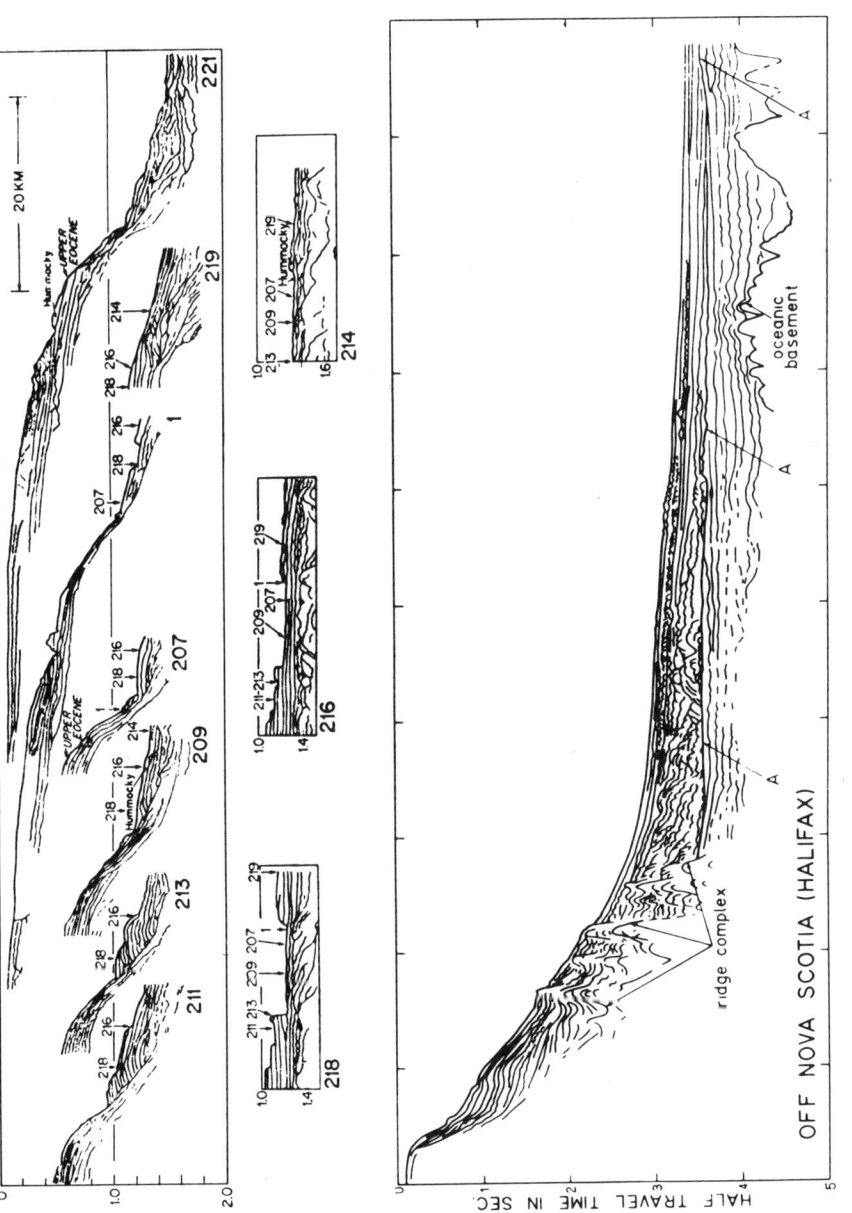

Fig. 10–13 A, seismic profiles seaward of Long Island, New York, indicate major slumping on gentle slopes. From Uchupi (1967B). B, seismic profile off Nova Scotia indicates the importance of mass movements on slope and continental rise. From Emery et al. (1970).

slopes. On the other hand, gentle slopes off rapidly advancing deltas have metastable sediments, and sliding should be expected (Terzaghi 1956). Gullies with hummocks at their base that are found on the foreset slopes of some deltas (Mathews and Shepard 1962) are almost certainly the result of mass movements. However, the irregular profiles of the slope off Texas and western Louisiana (Fig. 9–17), formerly interpreted as slide blocks (Gealy 1955), have now been determined to be largely the result of diapirism. The measurement of shear strength of sediments on these slopes (Morelock 1969) established that stability was such that slides would be unlikely. Yet, during the Pleistocene low sea-level stages, rivers building deltas on the outer shelf caused overloading and led to local slides that left records on the upper slopes (Lehner 1969, Fig. 41).

Faulting and Folding on the Slopes   Many irregularities of the continental slopes are probably due to faulting, although the recognition of faults is often difficult, even in seismic profiles. Undoubtedly, the rugged

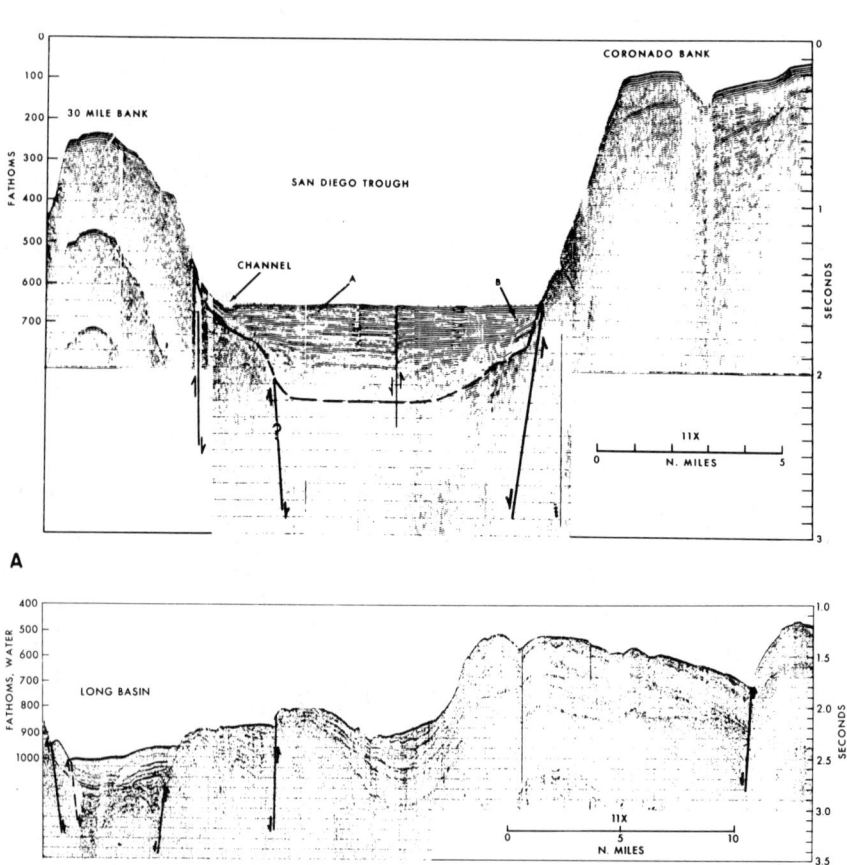

Fig. 10–14   Faulting and folding in the continental borderland off Southern California. From D. G. Moore (1969). *A*, profile across San Diego Trough shows faulting and distributary channels. *A* and *B* are buried channels. *B*, profile across central borderland off Southern California. Folds are largely transformed to faults in this area.

topography of the continental borderland off Southern California is produced in a large part by faulting (Fig. 10–14). Another example is indicated in a profile from Northern California (Fig. 10–3B). Other slope-disruptive faults are suggested by seismic profiles seaward of the English Channel (Fig. 9–40C, D).

In a few places, folds on the slopes appear to conform with the topography, indicating their recency (Fig. 9–18; 10–15). It is difficult to determine the extent to which these folds are the result of gravitative creep on the slopes.

Slope Erosion The significance of submarine erosion in developing slope irregularities has become evident from the exploration of submarine canyons. Detailed studies of the slopes along the northeast coasts of the United States have shown that erosion has been important, especially during episodes of low sea level. Off western Europe, two episodes of erosion were believed to have occurred; first, during the late Cretaceous and Early Tertiary, and second, in the Late Teritiary and Quaternary (Stride et al. 1969).

Another type of slope erosion is caused by contour currents (see p. 63), the return underflows from major ocean currents. The work of Heezen and others (1966) has shown how important these are along the East

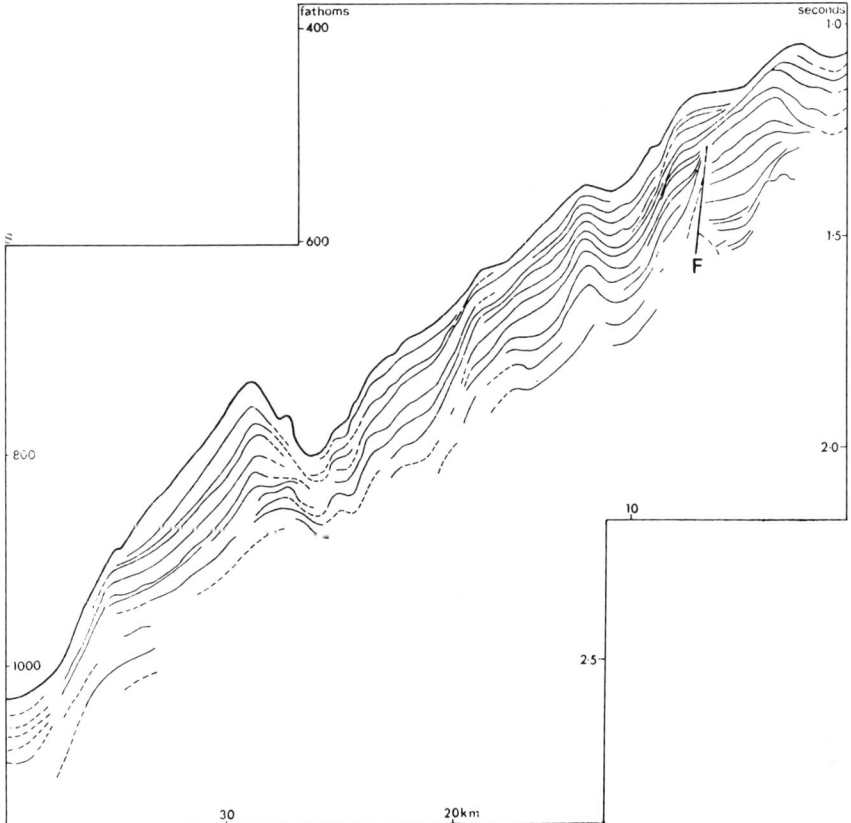

Fig. 10–15 Slump folds on the slope off southern Portugal. From Roberts and Stride (1968).

Coast and in the Scotia Sea. Slope erosion by these currents may well be the explanation for the outcrop of rock formations along the continental slopes in many places. Rona and others (1967) have explained the cutting of slope valleys near Cape Hatteras by this process.

UNDERLYING CAUSES OF CONTINENTAL SLOPES

Until we have solved the problem of why the light continental crust covers only about 35 percent of the globe, we cannot entirely explain the continental slopes. Meanwhile, it seems reasonable that very appreciable slopes should separate the light continents and the heavy ocean basins. We can assume that the continental slopes on the two margins of the Atlantic were initiated during the Jurassic or Triassic, when the east and west hemispheres were separated and heavy sima rose in the crack that developed between the two masses. According to evidence from JOIDES drilling, the deep-sea floor sank as the continents drifted apart and thus increased the height of the escarpment separating continents and ocean floor. Also, masses from the edge of the continents could be expected to break away, sinking part way down the slope. Such blocks have been found off Newfoundland, Portugal, and the British Isles.

Erosion of the continents would lead to sediments pouring down the marginal escarpment, and this sediment would, in turn, produce the great fans, which we call the *continental rises*. These are much better developed along the margins of the Atlantic coasts (Fig. 10–16) than in the Pacific. The growth of these fans must have produced some adjustments between the blocks, leading to some of the faulting. Possibly the development of the eugeosyncline found by refraction profiling under the upper continental rise off the East Coast (Fig. 10–17) is at least in part the result of the deposition at the base of the slope.

Locally, the Atlantic margins have been built out over the ocean basins.

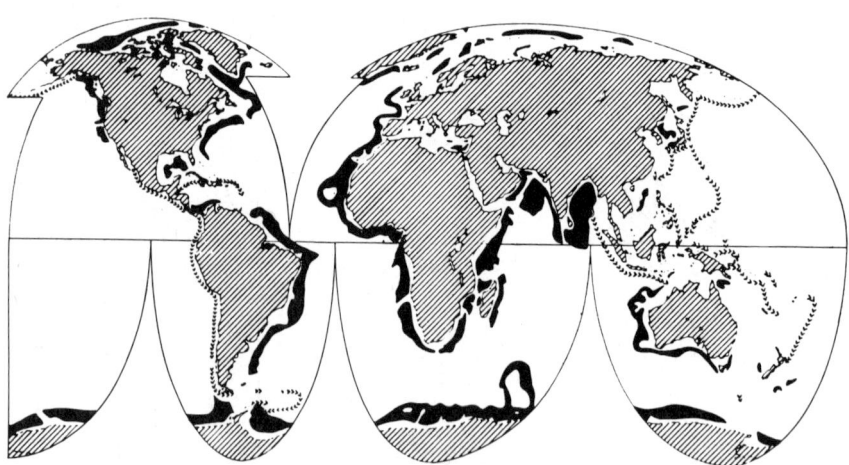

Fig. 10–16   Distribution of continental rises. Note they are well developed around the Atlantic and Indian Oceans and rare around the Pacific. From Emery (1969C).

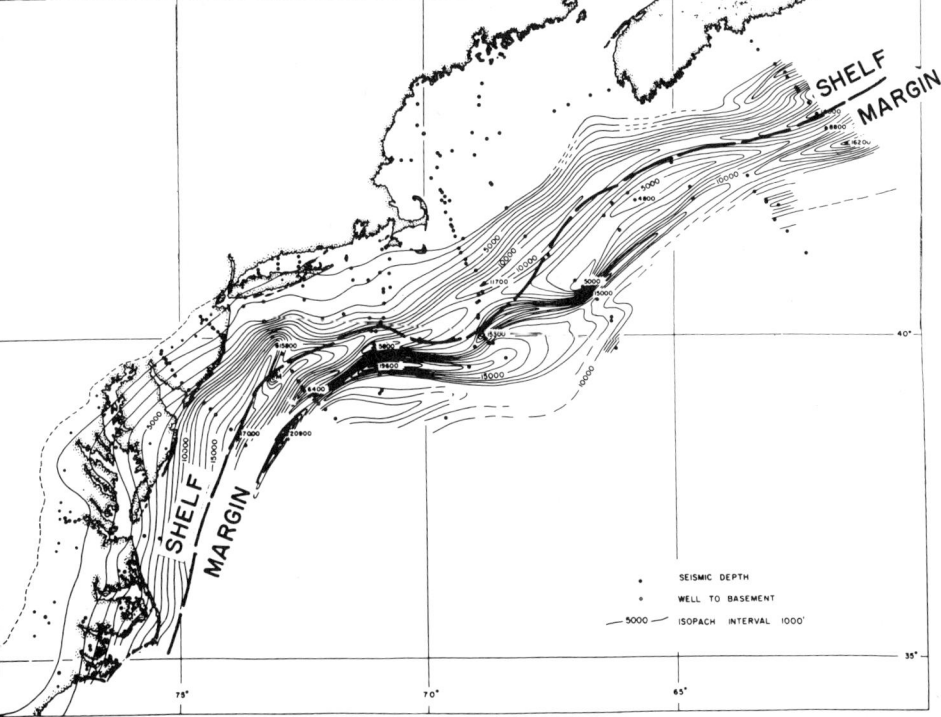

Fig. 10–17 Buried eugeosyncline off the northeast coast of the United States. The thickness of sediments (given in feet) indicates two synclines with a ridge between them. From Drake et al. (1959).

Thus, the bulging shelf margin off the Niger Delta conforms with the subaerial front of the delta. All such slopes are gentle. Dietz and others (1970) consider that the Bahama block is another mass built onto oceanic crust. They suggest that when the Atlantic rift first opened the relatively narrow intervening basin that formed around the Bahamas, it became filled to sea level with turbidites coming from both continents. Subsequently, coral was established and grew over the turbidites, keeping pace with the sinking block. This could explain the continued existence of shallow seas during the deposition of at least 4400 m since Early Cretaceous. This is not a geosyncline and requires another explanation for the continuous sinking. A base consisting of oceanic crust could account for it.

Most of the Pacific continental slopes have had another origin, according to the sea-floor-spreading plate-mechanics hypothesis. They are largely associated with collision zones and downbending of the great ocean trenches, along with the subduction of the margin of ocean plates under the continents (see Chap. 5). As would be expected, the base of these slopes is commonly steeper than the upper portions because of the sinking of the trenches. The upper slope appears to represent step faults that have been partly covered by deposition from the lands.

EVOLUTION OF CONTINENTAL RISES AND THEIR FUTURE

If most continental slopes have developed as the result of sea-floor spreading, then the slopes formed by the separation of the two sides of the Atlantic should have originated in the Mesozoic and only slightly steepened because of (1) very gradual sinking of the ocean crust as it moved farther from the active rift of the Mid-Atlantic Ridge, and (2) isostatic adjustment due to accretion of the continental rises. However, the continental slopes along most of the Pacific margin should be in a development stage, being constantly renewed by collision of the continental and oceanic plates and the subduction of the ocean crust beneath the continents or island arcs. The seismic profiles certainly show a great contrast between these two types, especially in the general absence of continental rises along the Pacific margins and their good development in the Atlantic. One can imagine that if the major rifts were shifted to new centers of activity, like the Red Sea and the rift valleys of Africa, the subduction coasts would stabilize, and sediment would begin to fill the moribund Pacific trenches and then build continental rises. Apparently this happened in the remote past along the margins of the Atlantic (Fig. 10–17), as shown by the buried eugeosynclines discovered by refraction profiling (Drake et al. 1959).

If the somewhat moribund slopes of the Atlantic are left undisturbed for further long periods, one can suppose that, as suggested by Dietz (1952, 1963), they will be overlapped by sediment deposition until the rise builds up over the slope and overlaps the shelf. Actually, the seismic profiles indicate that the diagrams by Dietz do not give enough credit to the buildout of the shelf margins. Dietz considered that outbuilding could be due only to parallic deposition, largely in the form of deltas that develop only in very shallow water. Some of the seismic profiles suggest an actual growth because of sediment carried across the shelves and deposited on the slopes. Long cores from slopes indicate that deposition has occurred on the slopes since sea level had risen sufficiently to drown the outer shelf deeply. Probably much more deposition on the slopes will take place after the filling of present-day estuaries, which are now receiving most sediment from rivers. Fine sediments can easily be carried across wide shelves. Great storms stir bottom sediments into suspension to depths of 100 m or more (Curray 1960; Lyall et al. 1971), and the muddy sediments introduced by floods settle toward the bottom and develop outflows in the basal water layers because of greater density and general seaward slope. Therefore, deposits from the upgrowth of the continental terrace might interfinger with those caused by outgrowth of the continental slope. Whenever the rivers had built deltas to the shelf margin, the slope deposition would be accentuated as it was during low sea-level stages of the Pleistocene.

ECONOMIC RESOURCES OF THE CONTINENTAL TERRACES

The past decade has shown a tremendous increase in the development of resources of the continental shelves. By the end of 1969, petroleum companies were producing oil from the margins of the Gulf of Mexico, the Persian Gulf, the North Sea, Cook Inlet, and from the shelves off Nigeria, Australia and New Guinea, Venezuela, Trinidad, and California (*Ocean Industry*, Jan. 1970). Approximately 700,000 sq. km were under production. Wildcat drilling at this time was much more widespread, covering many places on the shelf off West Africa, the west Japanese shelf, the Java Sea, the South China Sea, various areas off eastern Central America, and scattered areas off eastern South America. Most of the continental shelves and continental slopes of the world have underlying sediment basins that are considered favorable for future development, whereas almost half of the continents are unfavorable because of igneous and metamorphic rocks at or near the surface. It is now believed that about 20 percent of future oil reserves for the world are under the sea.

Virtually all oil and gas produced from the sea floor has come from the shelves, but drilling is now being attempted on the upper slopes in various areas, particularly along the Gulf Coast of Texas. Both the shelves and slopes are largely underlain by Tertiary sediments with many structures that are potential oil producers. Salt domes have proven important along the Gulf Coast and in the Persian Gulf and are being drilled along the bulge of West Africa.

Placer deposits are also being investigated on some of the shelves (Fig. 10–18). These include the very heavy minerals, gold, tin, and platinum; and relatively heavy ilmenite, rutile, zircon, and monazite. Most of these placers are found along the shelf continuations of river valleys and submerged beaches (Emery and Noakes 1968). Thus, the tin deposits along the coast of Thailand, Malaysia, and Indonesia come from submerged alluvium. The submerged beaches off Nome, Alaska, are being prospected (Tagg and Greene 1971). Off South Africa, diamonds are being mined from old submerged stream deposits.

Intensive studies of the continental terrace off the eastern United States have indicated that future exploitation of several resources seems probable (Fig. 10–19). Thus, there are possibilities of obtaining petroleum along the outer shelf off New England, Newfoundland, Nova Scotia, and Cape Fear, North Carolina (Emery 1968A). Phosphorite is abundant along the slope that lies between the East Florida shelf and Blake Plateau. Manganese nodules are common on Blake Plateau. The need for sand and gravel is making the large areas of coarse sediment off the East Coast good possibilities for future devel-

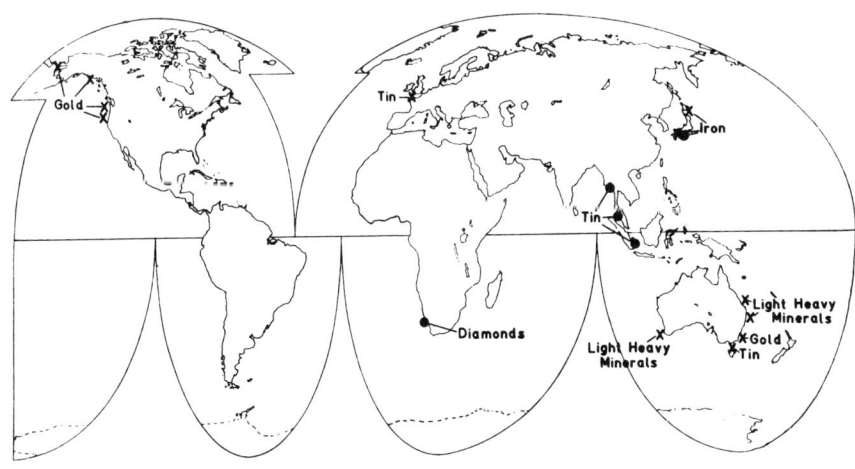

Fig. 10–18 Areas on continental shelf where minerals are mined (●) and actively prospected (x). From Emery and Noakes (1968).

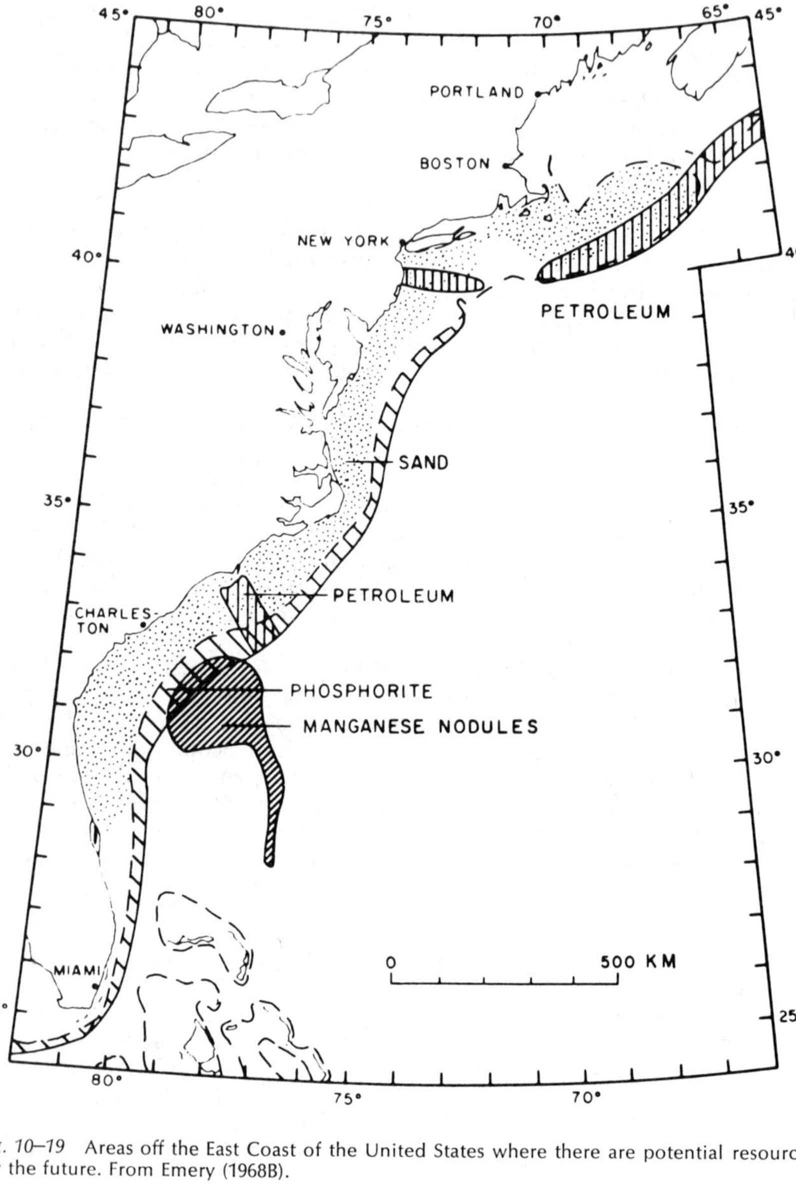

Fig. 10–19 Areas off the East Coast of the United States where there are potential resources for the future. From Emery (1968B).

opment. In fact, nearshore dredging is already producing a considerable supply, particularly near New York. The shell sediments off Florida are being dredged for use on small secondary roads.

On the relatively shallow banks off Southern California are extensive deposits of phosphorite nodules. Several companies have been considering mining these but have encountered difficulty, partly because many of the deposits are in water more than 200 m in depth. Also, investigation of the deposits on Fortymile Bank, off San Diego, showed that explosives had been dumped there during or after World War II, making the area hazardous for development.

Mining of iron ore on the continental shelves is very limited. Small production comes from Bell Island in Conception Bay near northeastern Newfoundland. Great thicknesses of oölitic hematite occur in this area (Trumbull et al. 1958, p. 52).

Chapter 11
SUBMARINE CANYONS
AND OTHER MARINE VALLEYS

INTRODUCTION

For more than a century, geologists have known of the existence of valleys on the sea floor, but only recently has their importance as a conduit in transporting sediments to deep basins been appreciated. This role becomes especially significant now that we also realize that many old formations, notably turbidites and graywackes, were probably deposited in deep basins. In the past decade, much more has been learned about submarine valleys; numerous descents in deep-diving vehicles have made possible their exploration below scuba limits by visual methods. Bottom photographs taken within the canyons from sleds pulled along the canyon axes and from cameras that were kept at proper elevations above the canyon floors and walls by electronic methods have covered some of the deeper portions where descents in vehicles have not yet been feasible. New echo-sounding surveys, including those made by the deep-tow method, have also been an important factor in enlightening us on the nature of these remarkable relief features of the sea floor. Seismic profiles have given evidence of the outcrops on canyon walls. All this new information has helped explain the origin for the various types of marine valleys but has not yet solved all the problems.

TYPES OF MARINE VALLEYS

As land valleys are formed by various processes, the valleys of the sea floor appear to have had several origins, producing distinctive types. Therefore, it seems unfortunate that some marine geologists still refer to all marine valleys as "submarine canyons." The following types, illustrated in Fig. 11–1, have been discovered partly as the result of extensive surveys.

1. *Submarine canyons* Here the term "submarine canyons" is restricted to the steep-walled, sinuous valleys, with V-shaped cross sections, axes sloping outward as continuously as river-cut land canyons, and relief comparable even to the largest of land canyons.

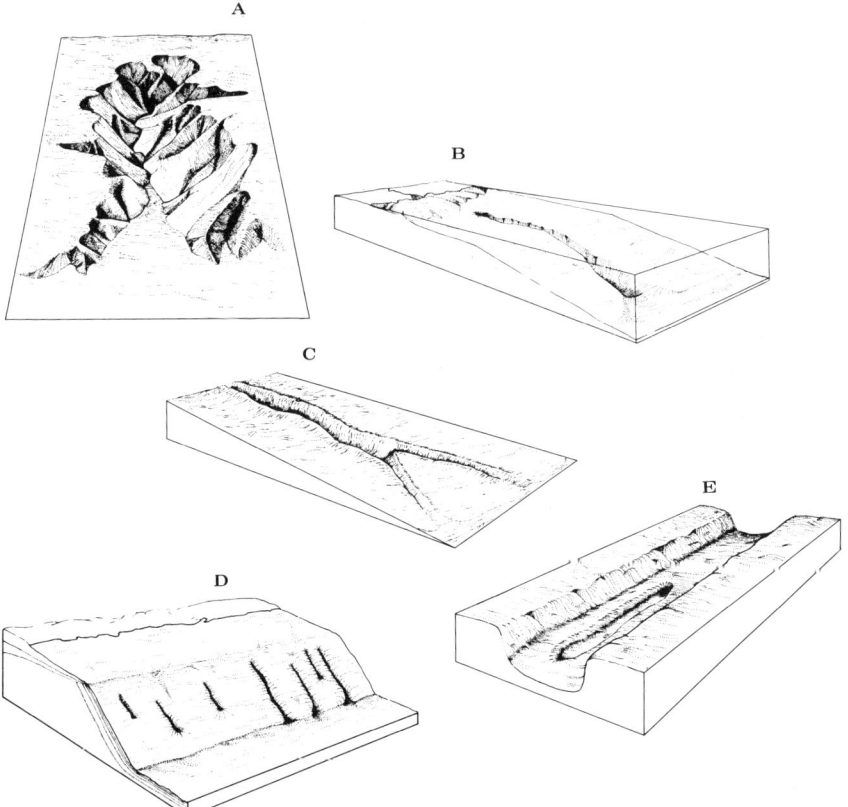

Fig. 11–1 Various types of sea-floor valleys: *A*, submarine canyon. *B*, delta-front trough. *C*, fan valley. *D*, slope gully. *E*, fault valley.

Tributaries are found in most of the canyons and rock outcrops abound on their walls.

2. *Delta-front troughs* These may be closely related in origin to submarine canyons, but have U-shaped transverse sections, comparatively straight courses, few if any tributaries, and are located exclusively on the fronts of a few large deltas. No rock outcrops have been found on their walls.

3. *Fan valleys* The seaward continuation of submarine canyons and delta-front troughs across the sediment fans at the base of the continental slope. Many have bordering levees that are built above the fans, most have distributaries, but very few have tributaries, and the walls are locally steep and are cut into unconsolidated fan sediments.

4. *Slope gullies* Discontinuous valleys of small relief with few tributaries, found on prograding slopes, such as delta fronts.

5. *Fault valleys* Trough-shaped, broad-floored valleys that follow structural trends, have few if any tributaries, but have a few deep basin depressions.

6. *Shelf valleys* These have been described in Chapter 9.

7. *Deep-sea channels* These will be described in Chapter 13.

It is not possible to classify all marine valleys into one of these categories. There are many borderline cases, and possibly such types as slope gullies and fault valleys may gradually evolve into submarine canyons, so we are confronted with transitional stages.

DESCRIPTION OF SUBMARINE CANYONS AND THEIR FAN VALLEYS

The most completely studied canyons may be atypical so that their description may lead to false premises concerning canyon origin. However, the detailed investigations provide our best sources of information, and those described include contrasting types, such as the valleys off California, off the northeastern United States, and off the French Riviera. The descriptions of a few others less well investigated will be included because of their distinctive features.

La Jolla and Scripps Canyons Scripps and La Jolla canyons and the La Jolla Fan Valley are the most completely studied of any of the marine valleys in the world. In addition to the observations and photographs by scuba divers (Dill 1964), the records obtained by 30 dives in the bathyscaph *Trieste*, Cousteau's *Diving Saucer*, and the Westinghouse *Deepstar* have given us an intimate knowledge of most of the length of these canyons (Shepard et al. 1969). The bathymetry shown in the detailed maps is largely based on electronic positioning (Shepard and Buffington 1968). Out to depths of 300 m, the steep-walled canyons (Fig. 11–2) were surveyed with wire soundings, because echo soundings fail to show the steepness. The vertical walls indicated by wire-sounding surveys have now been completely confirmed by diving, using both scuba and deep-diving vehicles. In fact, Scripps Canyon is found to have overhanging cliffs in many places. (In one descent, the *Diving Saucer*, 2.7 m in diameter, could not reach the bottom because of the narrow space between the walls.) The cliffs are covered largely with organisms, many of them acting as protective cover but others destroying the rock by boring

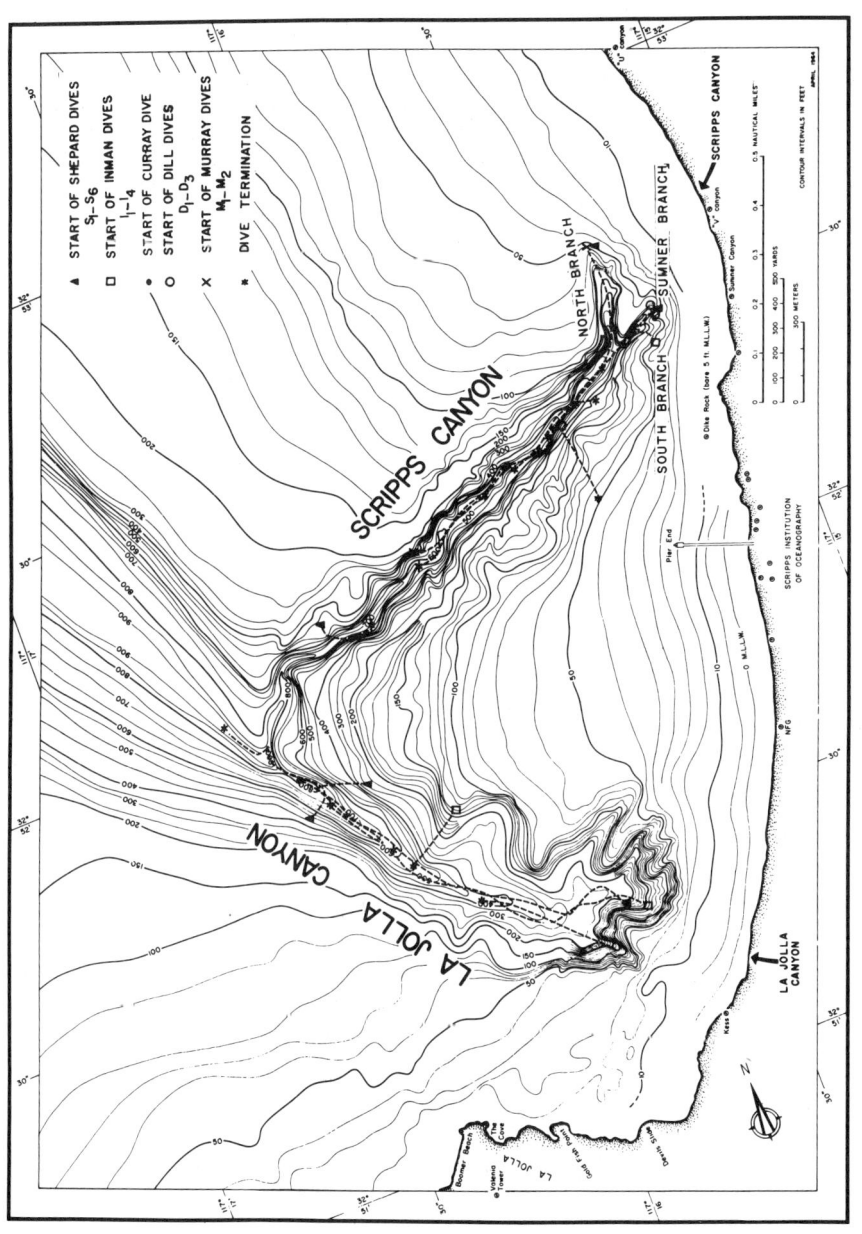

Fig. 11–2 Head of La Jolla Canyon with Scripps tributary and routes followed by Cousteau's *Diving Saucer.*

(Warme and Marshall 1969; Warme et al. 1971). Gaps in the organic cover were seen where rocks had broken off recently and fallen to the canyon floors. Also along the base of the cliffs, the walls were often bare of organisms, caused by a scouring action that left grooves and other signs of erosion on rock that formerly had been covered with sediment.

Two of the heads of Scripps Canyon are essentially sand chutes, although by using a water jet to penetrate the sand, Chamberlain (1960) showed that a rock-walled gully below the sand extends to within 60 m of the low-tide shoreline. Following the sand chutes seaward for about 100 m, one finds narrow gorges with precipitous rock walls. These gorges continue along the entire length of the canyon, 1.5 km, to the junction of Scripps and La Jolla canyons. The floor of Scripps Canyon has sand intermittently covered with kelp, surf grass, and other plant material. A few terraces, were seen with a marked increase in gradient due to boulders that formed a temporary dam, causing sediment to accumulate on the upcanyon side. Locally, clean rock was found exposed on the floor.

Investigations, primarily by Dill (1964), have shown that blocks of cement and even an old auto body placed on the canyon floor had disappeared, partly by creep and slumping, but perhaps also by sand flows and turbidity currents. Repeated soundings along profiles across the canyon heads (Shepard 1951) have determined that sediment accumulates rapidly. Periodic seaward flushings probably occur, usually during storm conditions (Dill 1969B), but such flushings occasionally may be induced by earthquakes of unusual magnitude.

Compared with Scripps Canyon, the head of La Jolla Canyon has less precipitous walls and a broader floor, locally 70 m wide. The canyon head has a series of small tributaries where the walls are of flat-lying unconsolidated lagoonal clays. Comparisons with old surveys show that the head has retreated at an average rate of 0.7 m/yr. Beginning at depths of about 40 m, the walls are of rock but largely covered with organisms, so it is difficult to determine their nature. Locally, rock is exposed on the canyon floor. In part, the canyon is cut along a fault between Eocene rocks to the north and Cretaceous to the south (Emery and Shepard 1945). During *Diving Saucer* descents, we saw a number of steps or terraces suggestive of slump blocks in the canyon fill. These steps change periodically, as we saw from our repeat dives made after a half-year interval (Shepard 1965). Large blocks had fallen from the walls along much of the La Jolla Canyon floor, some having formed grooves where they crossed the talus slopes. In a *Deepstar* dive at 316 m, an interesting overhanging wall was observed by Harold Palmer (personal communication) (Fig. 11–3). Great numbers of boring organisms had apparently caused the spalling off of a column from the siltstone cliff, lending evidence to the importance of organisms as an erosion agent in the canyons.

La Jolla Canyon gradually changes in character seaward (Fig. 11–4). At approximately 365 m of axial depth, the floor widens, and a little farther seaward there appears to be a marginal terrace near the bottom with a winding channel cutting through it. At 550 m, the walls, although still steep, consist of only slightly consolidated fan material. On the north side, a gradual development of levees rises somewhat above the general slope. At a depth of 658 m, levees are found on both sides, and the seaward slope outside the valley has evolved into a fan shape. Thus, in its outer part it can be considered a true fan valley, although it is not clear where the break occurs between the two types. The outer valley decreases in depth below surroundings until it merges with the floor of San Diego Trough at a depth of 1100 m.

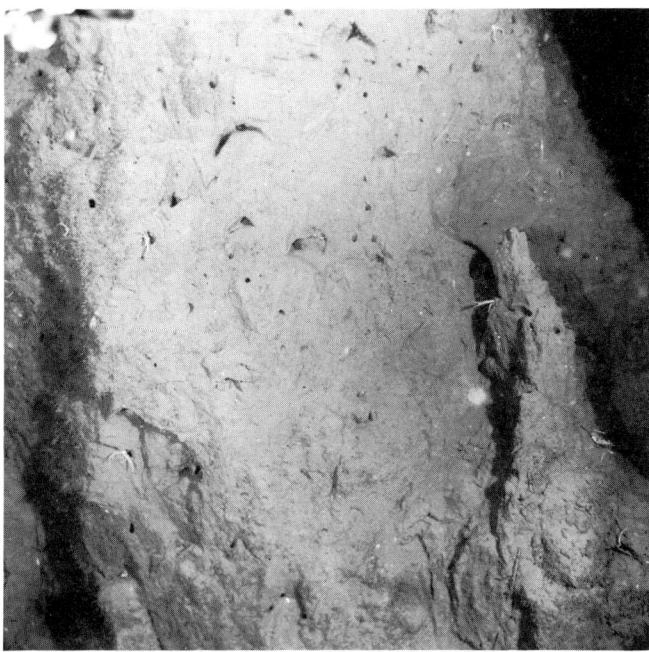

Fig. 11–3 A pinnacle separated from the vertical wall of La Jolla Canyon by combination of erosion by organisms and gravity creep. Note shrimp and crabs protruding from excavated holes. Photo courtesy of Harold D. Palmer, taken in *Deepstar* dive.

Geologists at Scripps Institution and the San Diego Naval Undersea Research Center have taken several hundred samples along the course of La Jolla Canyon and Fan Valley (Fig. 11–5). These include 98 box cores that show undisturbed sedimentary structures and stratigraphy of the fill on the valley floor and on the lateral terraces and levees (Shepard et al. 1969). Block diagrams of Fig. 11–5 indicate how the sediments vary seaward along the valley. Most cores show a covering of silty clay over sand and gravel layers. Of the sand layers, 26 percent are graded, 59 percent have parallel laminations, and 41 percent have ripple structures or cross laminations. There are indications of deposition from traction currents of a pulsating character, such as distinct separation or unconformity between the covering mud layers and the underlying sand and laminae of heavy minerals; but the unconformity may be due to bioturbation, and the laminae could be related to turbidity currents. In general, the sediments are coarser in approaching San Diego Trough; the opposite might be expected if they were deposited by turbidity currents. However, Piper (1970) explains this coarsening as the result of a powerful turbidity current that did not allow deposition until it had reached the lower end of the fan valley where it began to spread out. This may account for some coarse sediment on the fan outside the lower end of the valley. Also, the levees have many thin sand layers, indicating that sediment-laden waters with considerable velocity were filling the fan valley, even where the floor is as much as 91 m below the crest of the levees. Furthermore, Piper found that the top sediments on the fan showed a gradation toward fine material

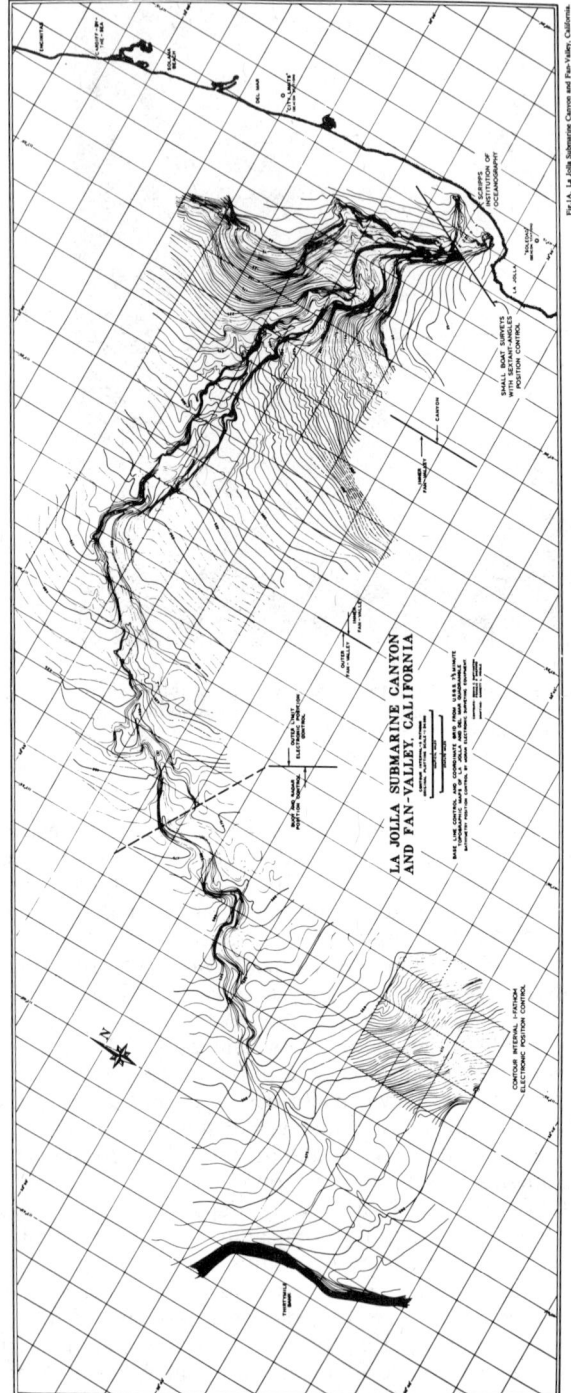

Fig. 11–4 Complete contour map of La Jolla Canyon and Fan Valley shows its termination in San Diego Trough. Nature of surveys indicated. For details of canyon head, see Fig. 11–2. From Shepard and Buffington (1968).

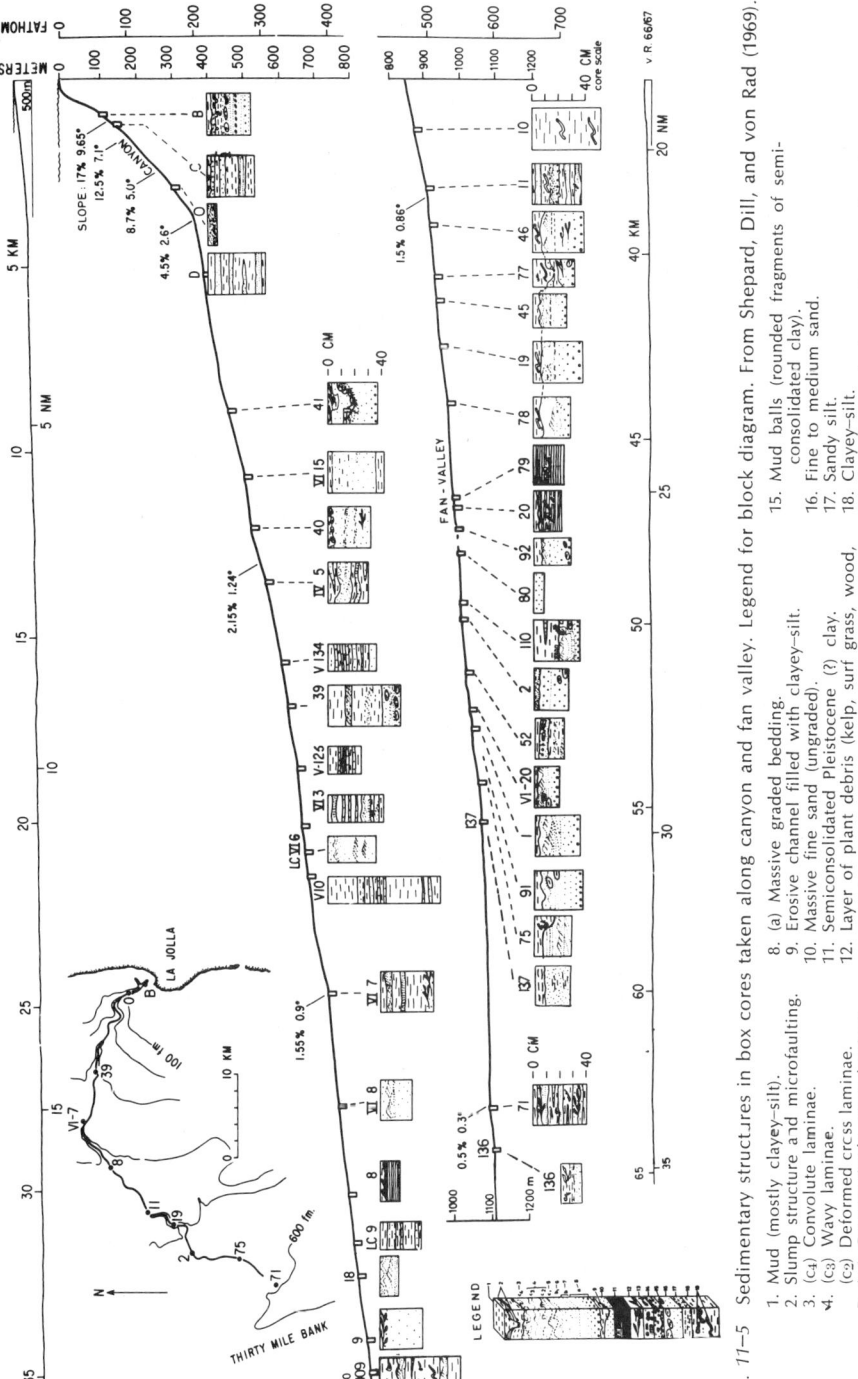

Fig. 11–5 Sedimentary structures in box cores taken along canyon and fan valley. Legend for block diagram. From Shepard, Dill, and von Rad (1969).

1. Mud (mostly clayey-silt).
2. Slump structure and microfaulting.
3. (c₃) Convolute laminae.
4. (c₂) Wavy laminae.
   (c₃) Deformed cross laminae.
5. (c₁) Current-ripple cross laminae.
6. (b₁) Inclined laminae.
7. (b) Parallel laminae.

8. (a) Massive graded bedding.
9. Erosive channel filled with clayey-silt.
10. Massive fine sand (ungraded).
11. Semiconsolidated Pleistocene (?) clay.
12. Layer of plant debris (kelp, surf grass, wood, etc.)
13. Layers of coarse sand, gravel, and pebbles.
14. Layers with gravel and shells.

15. Mud balls (rounded fragments of semi-consolidated clay).
16. Fine to medium sand.
17. Sandy silt.
18. Clayey-silt.
19. Bioturbation by burrowing organisms (worms, mollusks, echinoids, etc.)

away from the levees, as would be expected if turbidity currents in the valleys are thick enough to pour over the levees. On the valley floor, sand layers generally have a thin covering of mud. Piper investigated the sand content of some of these covering muds that were free of bioturbation and found that they showed an increasing sand content downward. This suggests that the muds represent the tail of a turbidity current of decreasing velocity. Also, the sand fraction in many of the mud layers contains shallow-water foraminifera, indicative of turbidity-current emplacement or other downcanyon currents.

An acoustic transponder towed at considerable depth directly over the terraces in the fan valley produced records indicating that the terraces were the result of relatively hard layers of sediment outcropping on the walls of the fan valley (Normark 1969, 1970). These deep-tow records also show that the fan valley is incised into the fan sediments, rather than being the result of nondeposition as the fan levees were built upward.

San Lucas Canyon, Baja California   At the southern end of Baja California, a maze of bordering canyons extend out from the coast to depths of 2400 m (Fig. 11–6). These have been investigated for many years by scientists from Scripps Institution (Shepard 1964) and from the Naval Undersea Research and Development Center, San Diego (Dill 1966; Shepard and Dill 1966, pp. 101–130). Scuba dives and descents in deep-diving vehicles have been concentrated in this area, particularly in San Lucas Canyon.

San Lucas Canyon heads almost at the shore with one branch directly off the pier of a fish cannery. When observed from deep-diving vehicles, the sediments in the upper part consisted of soft mud and showed no signs of the current activity found in the La Jolla canyons, but Dill found that stakes driven in here became tilted seaward by creep. At a depth of 100 m, sand appears in San Lucas Canyon; a little farther seaward the sand is highly rippled and partly covered with angular blocks of granite and other igneous rocks that have fallen from the south wall. This wall is well known for its sand flows and sand falls (Fig. 11–7), which have been seen after periods when large waves swept sand over a tombolo beach that connects the peninsula to a large rock stack. The walls on both sides of the canyon consist of intrusive crystalline rocks, but the falling rock and sand flows testify to active erosion, at least on the south wall.

During dives in the *Deepstar*, the canyon has been examined intermittently down to depths of 1220 m. At this depth we found one vertical wall of granite and the other sloping about 45°. The 30-m-wide floor consisted partly of mud but mostly of coarse sand, the latter varying from smooth to rippled, the ripples due to downcanyon current. Locally, large granite boulders on the floor were bordered upcanyon with depressions and downcanyon with lee dunes. Dives higher up the axis also showed variability in bottom character, alternating between coarse sand and mud, as if the floor had been hit by discontinuous flows resembling the path of a tornado that at times touches the earth and then jumps over a wide area only to descend again along its path.

The floor of San Lucas Canyon has been sampled out to its greatest depths. Coarse sand and gravel are as common at the canyon mouth as in its upper portion. Piston cores were difficult to obtain, probably because of the sand and gravel bottom. Box cores (Fig. 2–14) showed that here, as in La Jolla Canyon, there is no consistent decrease in sediment size in a downcanyon direction.

Fan valleys are poorly developed around Cape San Lucas at the canyon mouths. According to the seismic profiles of Normark and Curray (1968),

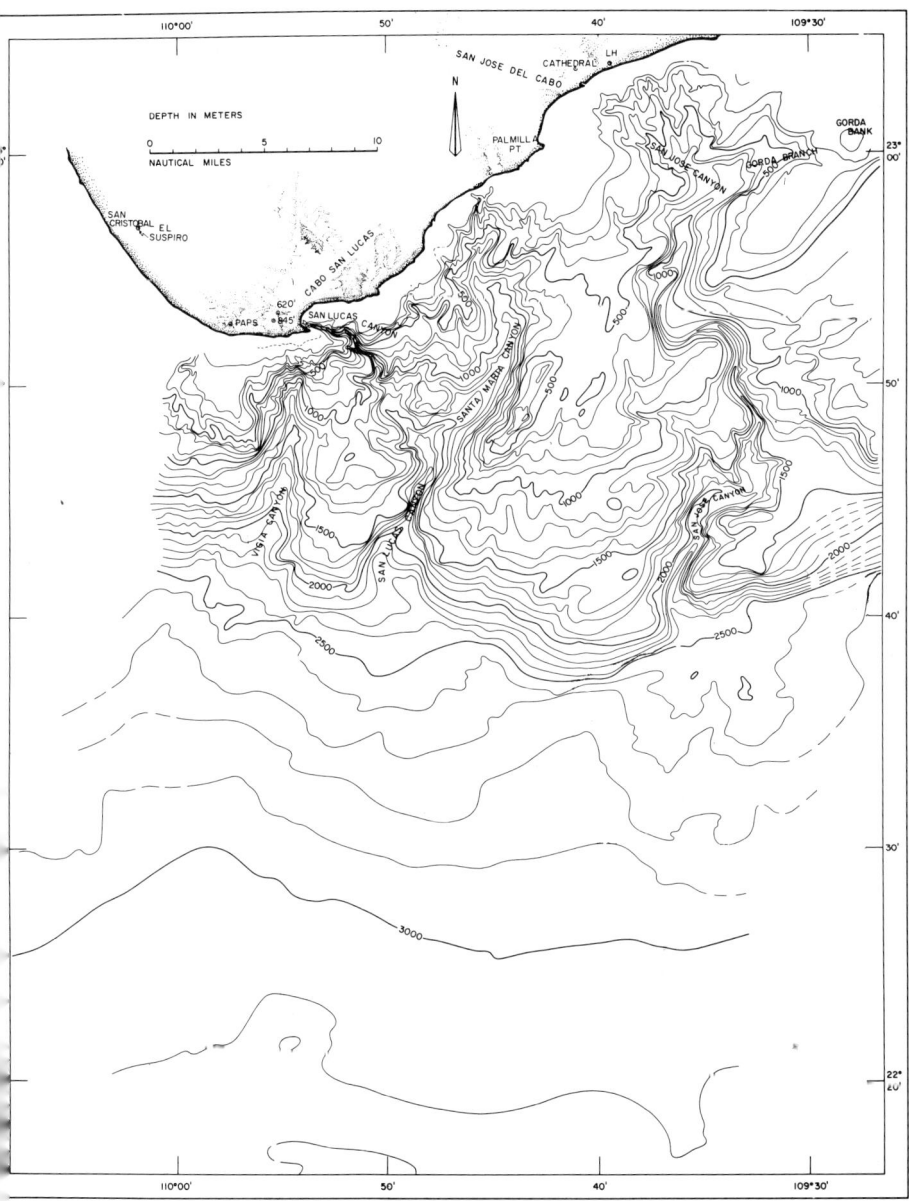

Fig. 11–6 The canyons off the southern end of Baja California are partly cut in granite and in some places follow contacts between granite and sedimentary rocks. Vigia Canyon may have been pirated by the head of San Lucas Canyon. Note fans at seaward termini of canyons and a small fan valley off San Jose Canyon. From Shepard and Dill (1966).

Fig. 11–7   A, underwater sandfall at depth of 40 m in a gully leading into San Lucas Canyon, Baja California. Naval Undersea Research and Development Center photo by R. F. Dill. From Shepard and Dill (1966). B, granite blocks on the floor of San Lucas Canyon at a depth of 270 m are located below the sandfall. Photo by R. F. Dill from *Diving Saucer*. C, ripple-marked sand at depth of 290 m in San Lucas Canyon. Current was flowing downcanyon to produce ripples. Photo by R. F. Dill from *Diving Saucer*.

they probably have been filled. A short fan valley is located seaward of but separated from San Jose Canyon.

Monterey Canyon The most impressive canyon along the California coast heads in Monterey Bay and has a prominent tributary that enters Carmel Bay (Fig. 11–8). Monterey Canyon, which has at one place a profile comparable to the Grand Canyon, was formerly considered the largest in the world, but recent work has shown that it is surpassed considerably in wall height by a canyon in the Bahama Island area and in length by two Bering Sea canyons. In addition to its high walls, Monterey Canyon is impressive because its dendritic pattern makes it closely resemble many land canyons. Actually, its drainage pattern is partly related to structures, and locally the main canyon follows a fault contact between granite on its south wall and Tertiary sediments on the north (Martin and Emery 1967; Greene 1970B). Although Monterey Canyon is not located off any present-day river, it is generally believed to be related to the former drainage of the Great Valley of California, and, according to Starke and Howard (1968), an old buried canyon of Miocene age on land is directly in line with the submarine canyon head. Starke and Howard suggest that the present canyon may represent the reexcavation of an old gorge that locally has a fill as thick as 1500 m.

The head of Monterey Canyon is cut into unconsolidated formations, and the first rock walls are located where the axis has reached a depth of 500 m. The fill in the head is very unstable. Slumping has twice wrecked the Moss Landing pier built into a southern tributary. Also, part of one jetty at Moss Landing Harbor has slid into another tributary of this canyon.

Traced seaward, the rocky walls reach heights of almost 1500 m. The floor has been extensively cored and most cores contain sand layers, but some have gravel. At a depth of 1900 m, the narrow floor of the inner canyon broadens to 5 km, and at 2700 m, the floor has widened to about 6 km. Here, the valley extends southeast and is bordered on the northeast side by a ridge that rises as much as 400 m above the floor. This appears to be a levee. Dredging on the levee wall revealed some gravel but no outcrops. The wide straight course continues for 20 km to the southeast where a narrow fan valley is bordered by levees on both sides, forming a large meander (Fig. 11–9). This had formerly been interpreted as three distributaries, but detailed surveys showed clearly that the meander interpretation is correct. Farther seaward, the fan valley is less incised but continues for at least another 400 km. After exploring the Monterey Fan Valley with reflection profiling and a pneumatic sound source, Normark (1970) found that renewed erosion had increased the depth of the fan valley as much as 200 m after it had pirated the neighboring Ascension Fan Valley. The latter is left as a hanging valley. Sediments in the outer fan valley have sand layers with gravel up to 10 cm in diameter. At one point, Menard (1964) found flat pebbles in a muddy matrix, suggesting a possible mud flow.

Astoria Canyon[1] One of the two major canyons off Oregon and Washington heads at 100 m, 17 km seaward of the Columbia River mouth (Fig. 11–10). This canyon has been studied by marine geologists at Oregon State University (Carlson 1967; Nelson 1968; Carlson and Nelson 1969; Nelson et al. 1969, 1970). The canyon winds seaward for 120 km where, at a depth

---

[1] Called Columbia Canyon in earlier editions. This appropriate name was changed by the Board of Geographic Names after having been approved by the U.S. Coast and Geodetic Survey and used in a publication (Shepard and Beard 1938). However, since Astoria Canyon has been used extensively in the literature and on charts, it will be adopted here.

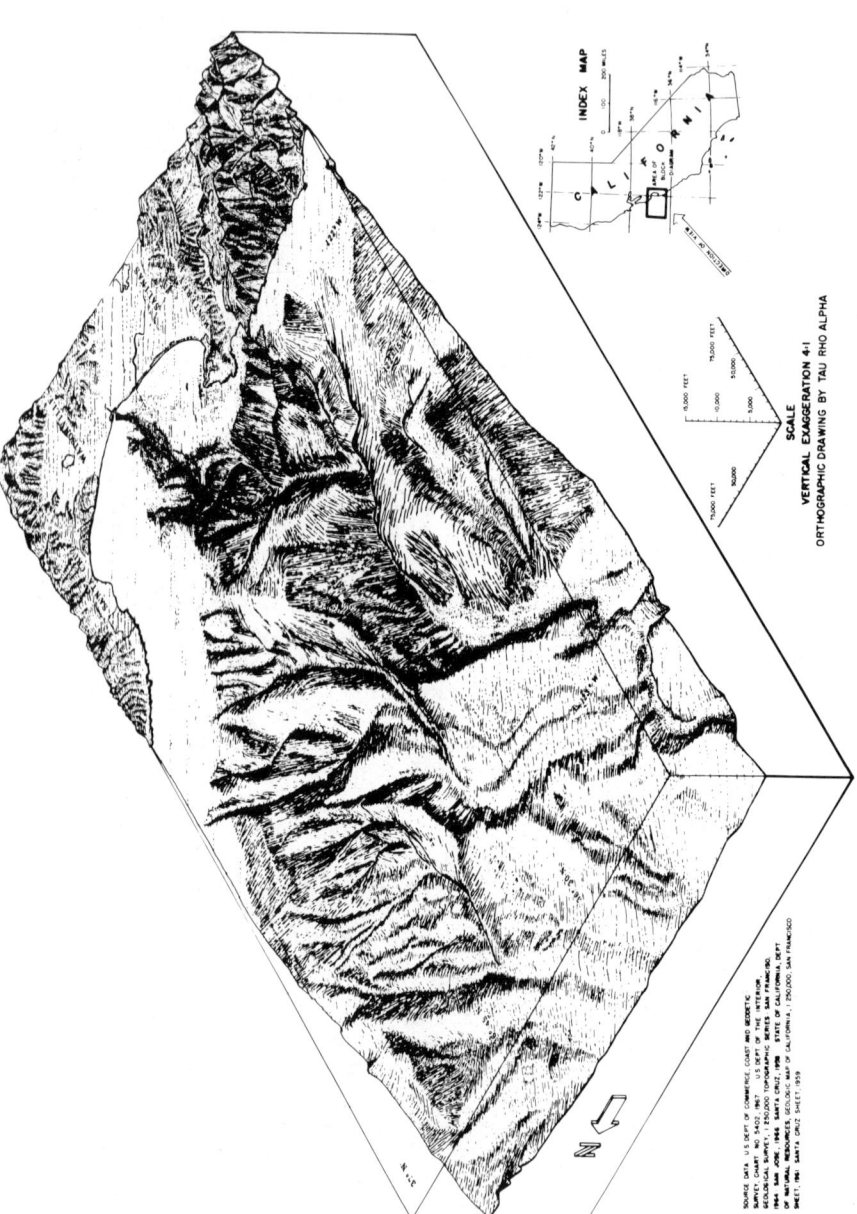

Fig. 11–8   Monterey and Carmel Canyons off central California. For detail of southern continuation, see Fig. 11–9. Physiographic diagram by Tau Rho Alpha, U.S. Geological Survey.

Fig. 11–9 Meander in Monterey Fan Valley. Note gap in the levee where a channel appears to have been partially excavated because of an overflow of a turbidity current at this point. Contours by the author.

of about 2000 m, it emerges onto a fan. From here it can be traced seaward as Astoria Fan Valley for another 170 km. The break between canyon and fan valley is distinct. Shoreward of the canyon head, seismic profiles have indicated buried channels heading toward the estuary at the mouth of the Columbia (Nelson et al. 1970). The canyon is somewhat U-shaped but it changes to a V-shape at greater depth. Numerous small tributaries enter it from both sides. The concave longitudinal profile has an average gradient of about 1°. The outer canyon crosses several rock ridges of the continental slope. Walls of the canyon have heights up to 900 m and slopes as steep as 30°.

Astoria Fan Valley is the principal of several distributaries that cross the fan beyond the lower canyon. The youngest fan valleys definitely follow the south side of the fan, as pointed out by Menard (1955), who explained the "left hook" as due to Coriolis force in the Northern Hemisphere. The higher levees on the right-hand side support this idea, as do a sequence of ancient valleys that appear to have migrated leftward throughout fan history (Nelson et al. 1970).

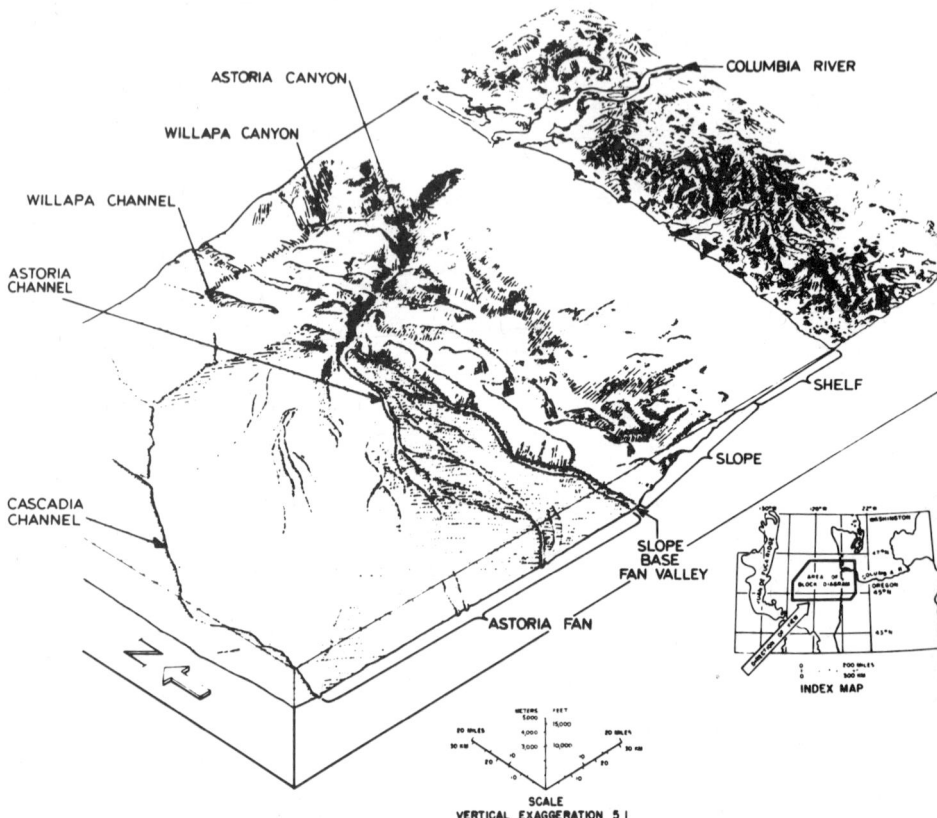

Fig. 11–10   The fan valleys and ridges seaward of Astoria Canyon, Oregon Physiographic diagram by Tau Rho Alpha, from Nelson et al. (1970).

Cores taken along the floor of the canyon and fan valley (Carlson and Nelson 1969) commonly show both late glacial and postglacial sediments. Most of the postglacial deposits are hemipelagic silty clays, although in the canyon and in Astoria Channel some coarser turbidity-current deposits are interbedded in the early Holocene sediments. The last major turbidity-current events in the canyon-fan system were related to the cataclysmic eruption of Mt. Mazama (6600 B.P.) (Nelson et al. 1968). The glacial age sediments of both canyon and fan include thin silty clay interbedded with turbidites that consist of silts, sands, and gravels. The turbidites of the fan valley have thick coarse sediment beds containing gravels and some pebbly mudstones, along with current-rippled laminae and other features similar to those in cores from the La Jolla Fan Valley. The Astoria Canyon cores have more organic material, more glauconite, and more evidence of slumping from the walls than those of the Astoria Fan Valley. The lower fan has better sorting, more grading, and greater abundance of turbidites than the upper fan, where structureless turbidites with unsorted materials of all size grades are most common.

Nelson and others (1970) suggest that the canyon cutting and the seaward building of the fan could all have been accomplished during the

Pleistocene. They believe that the sediment load introduced by the Columbia River during low sea-level stages of the Pleistocene would account for the fan morphology and for all of the unconsolidated sediment of the fan, as computed from seismic surveys (Shor et al. 1968).

Bering Sea Canyons The continental slope along the east side of the wide Bering Sea shelf contains what are probably the largest and longest submarine canyons in the world (Fig. 11–11). All main canyons head many kilometers from shore. Bering Canyon, the longest in the world, extends for at least 1100 km, and Zhemchug Canyon, according to a Russian survey (Kotenev 1965), appears to have the largest volume in the world with 8500 cu km (Scholl et al. 1970A). Bering Canyon, heading off Akutan Island in the Aleutians, has a large number of tributaries on the south side that come down from the eastern end of the Aleutian Island arc, but on the north, only one tributary has been found. Seismic profiles (Scholl et al. 1968B, 1970A) show that these huge canyons are partly related to the faulting and warping that has taken place at the continental margin (Fig. 11–12). Basinward sliding is thought to have been responsible for much of the cutting of the canyons. Bering Canyon is interpreted as having undergone alternate cutting and filling during its development. The huge size of Zhemchug Canyon is explained by breaching of the seaward wall of an outer shelf basement depression. Scholl

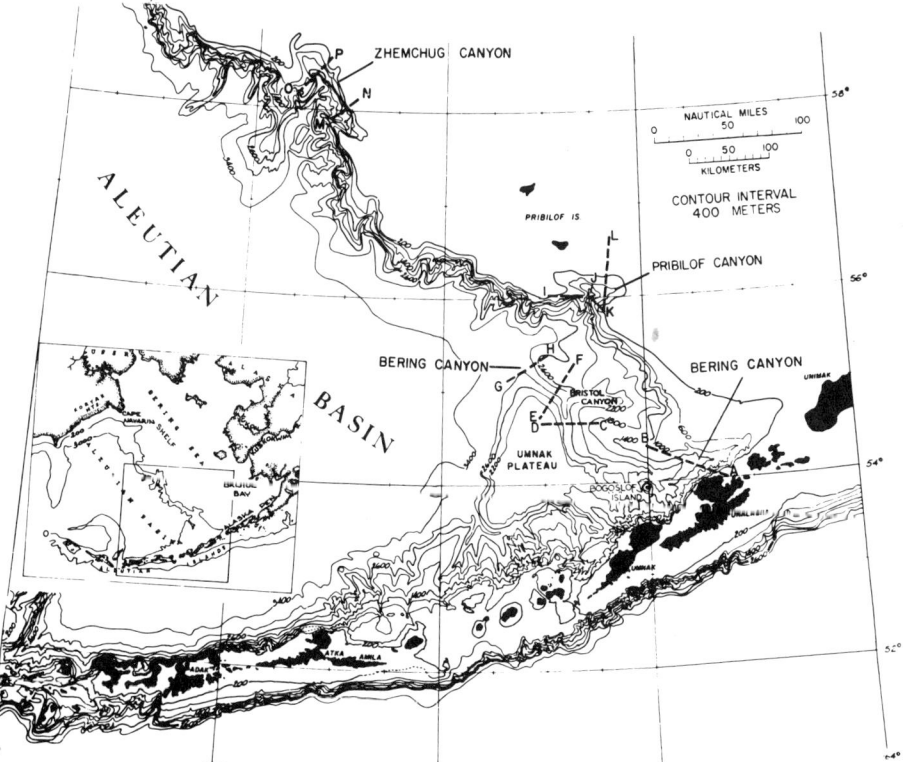

Fig. 11–11 Canyons of the Bering Sea, the largest and longest in the world. Dashed lines show position of two seismic profiles in Fig. 11–12. From Scholl et al. (1970A).

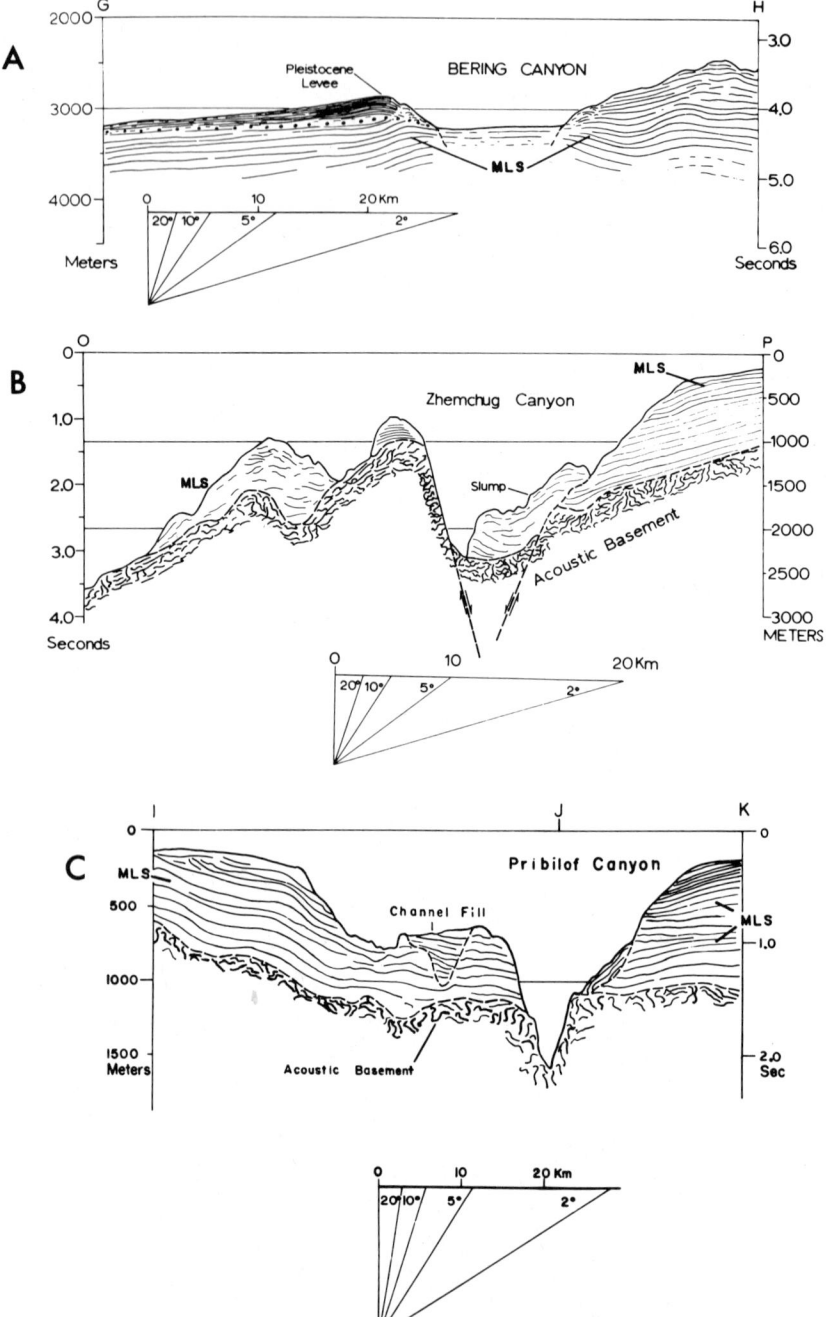

Fig. 11–12 Geological interpretations of seismic reflection profiles in Bering Canyon. MLS means main layered sequence. From Scholl et al. (1970A). *A,* fan valley at the terminus of Bering Canyon with Pleistocene levee. For location, see G–H in Fig. 11–11. *B,* faults and slumps in Zhemchug Canyon. Acoustic basement interpreted as Cretaceous. From Scholl et al. (1968B). *C,* unconformity and old channel fill in profile of Pribilof Canyon. For location, see I, J, K, in Fig. 11–11.

and others (1970A) estimated that 4500 cu km of Tertiary sediment have been removed from the depression. The outer portions of the canyons appear to continue as fan valleys. A Pleistocene levee rises 350 m above one side of the Bering Fan Valley (Fig. 11–12A).

Probably the large Yukon and Kuskokwim rivers flowed into the heads of these canyons during the times of Pleistocene sea-level lowering. Hopkins (1967) suggested that a supply of sediment across the elevated shelf probably existed also during much of the Tertiary, allowing a long period for the formation of these huge canyons, although Scholl and others (1970A) think that the cutting could have occurred during the Pleistocene.

Tokyo Canyon At the mouth of Tokyo Bay, a sinuous canyon with dendritic tributaries extends down into the Sagami Bay Fault Trough where it terminates at 1460 m. This is the largest of a series of canyons along the northeast side of Sagami Bay. Tokyo Canyon has a mud bottom in its inner course, but sand layers are found in cores of the outer portion. The walls are largely soft sedimentary rocks with a few outcrops of lava. Bottom photos of the narrow, steep-walled Mero Canyon, which parallels Tokyo Canyon to the south, show that currents are important along the canyon floor and that rock bottom occurs locally. An area where current ripples were seen is marked on the published chart H. O. 2724 as having "overfalls," caused by the eddies at the contact between the Kuroshio Current and the tide coming out of Tokyo Bay.

Congo Canyon At the mouth of the Congo, the largest river in Africa, a large submarine canyon (Fig. 11–13) extends 25 km into an estuary and is about 450 m deep at the bay mouth. This is significant because the sediment load of the Congo, although small for a large river, could fill the estuary in a short period (Veatch and Smith 1939, p. 27). Elmendorf and Heezen (1957) and Heezen and others (1964B) have shown that the depths in the canyon change frequently, and it has been impossible to maintain submarine cables across the canyon. Slumps or turbidity currents occur about 50 times a century, especially during rainy seasons when large loads of sediment are carried into the canyon head. The north longshore drift along the coast also introduces abundant sediment (Veatch and Smith 1939).

Congo Canyon crosses the 90-km-wide shelf and continues down the slope, where it makes a sharp bend to the left at an axial depth of 2560 m, and then swings to the right at 3100 m. Here, 190 km from the head, natural levees are detected on both sides; therefore the valley can be interpreted as a fan valley. In this area, there are many distributaries and some of them have been traced seaward for 500 km. Cores from the floor of the canyon and from the fan valley have layers of silty sand alternating with mud and containing large quantities of plant fragments. Again, little change in sediment size is found with distance from the canyon head.

Canyons of West Corsica and the French Rivera Extensive French surveys and sampling of the canyons along the Riviera and West Corsican coasts have made these among the best explored marine areas in the world. The Corsican canyons (Fig. 11–14) are located exclusively on the west side of the island where each extends into an estuary. This is the only place in the world where the submarine canyons look like recently submerged continuations of the land canyons or, conversely, as if the inner portion of a group of submarine canyons had been elevated to become land canyons. Corsican submarine canyons have slightly steeper gradients than those of land canyons (Shepard and Dill 1966, p. 186), but the difference is much less than in most other places. At the mouths of the land canyons are small deltas,

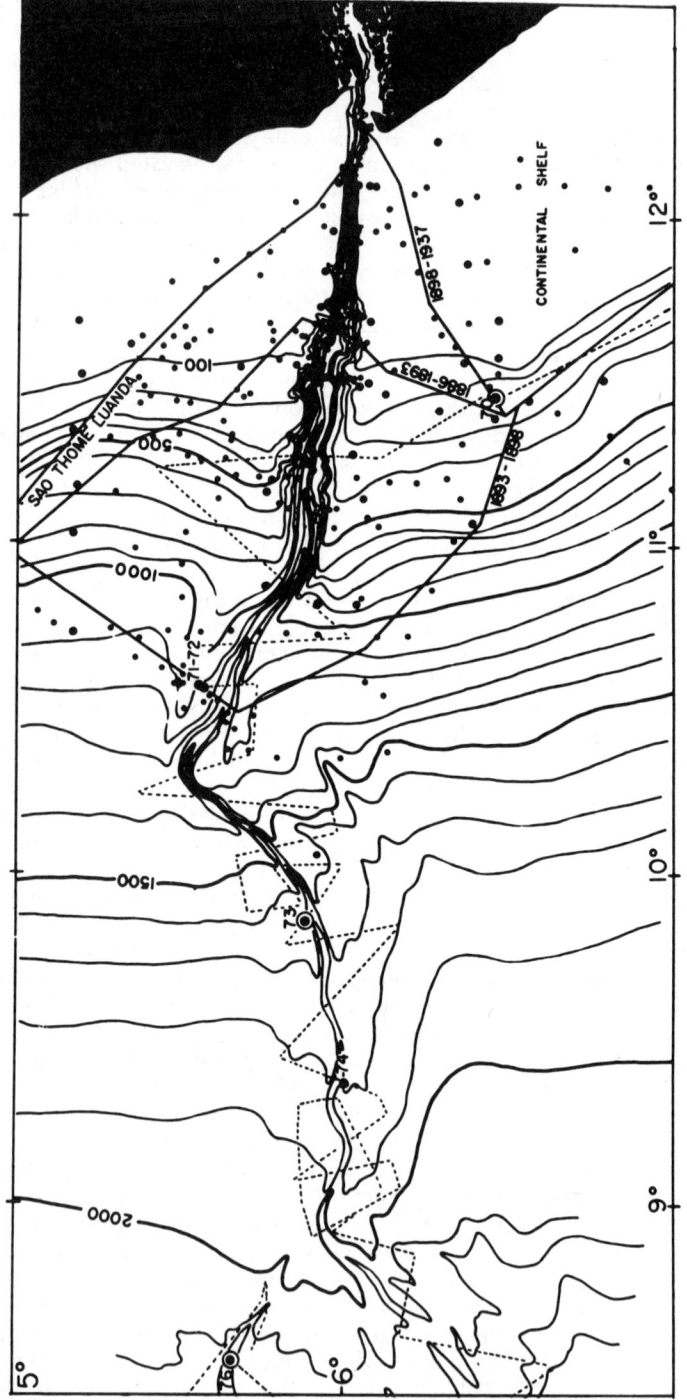

Fig. 11–13 Congo Canyon and Fan Valley showing where submarine cables were laid and later broken by turbidity currents or slides. Note the canyon head enters the Congo Estuary. From Heezen et al. (1964B).

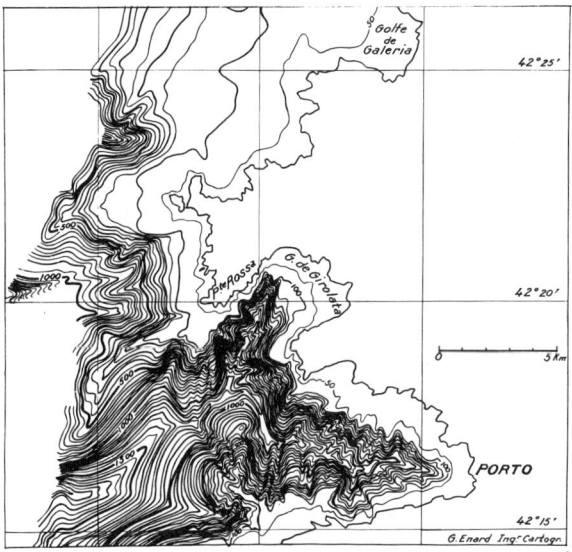

Fig. 11–14 Typical submarine canyons off western Corsica, showing the close relationship to the embayments. From Bourcart (1959).

as would be expected had there been an interval since the land and sea canyons became separated.

Tracing the sea-floor topography north and west from Corsica, one finds that the canyons continue and are well developed all along the French Riviera, and are somewhat less distinct along the lowland coast of France farther west. However, along the French coast, most of the canyons do not extend deeply into embayments (Fig. 11–15), having instead a rough relationship to the broad indentations of the coast. The French Mediterranean canyons have been sampled and photographed in detail. Abundant ripple marks and cobbles are seen in photographs taken from Cousteau's *troika* (sled) (Gennesseaux 1966), which was pulled up the axes of several canyons. In a western canyon near the Spanish border, grooves and notches in the canyon walls, like those of Scripps Canyon, and many boulders were seen on the canyon floor during dives in Cousteau's *Soucoupe* (*Diving Saucer*) (Dangeard 1962; Dangeard et al. 1968). Using seismic profiles, Dangeard also found evidence of earlier canyon excavation to greater depths than now exist. This was followed by filling and then reexcavation.

Canyons off the Northeastern United States and Nova Scotia The continental slope of eastern North America north of Cape Hatteras is cut by many canyons at least as far north as the limits of detailed soundings off the Grand Banks of Newfoundland (Figs. 9–2, 9–4). None of these canyons extends more than a few kilometers into the wide continental shelf, but they all run down the slope to the well-developed continental rise, and many of them continue beyond as fan valleys. The new U.S. Geological Survey and Canadian Hydrographic Service charts compiled with the help of Woods Hole Oceanographic Institution, under the direction of Elazar Uchupi, give a clear picture of the situation. Examination of these maps shows a shelf too wide to corre-

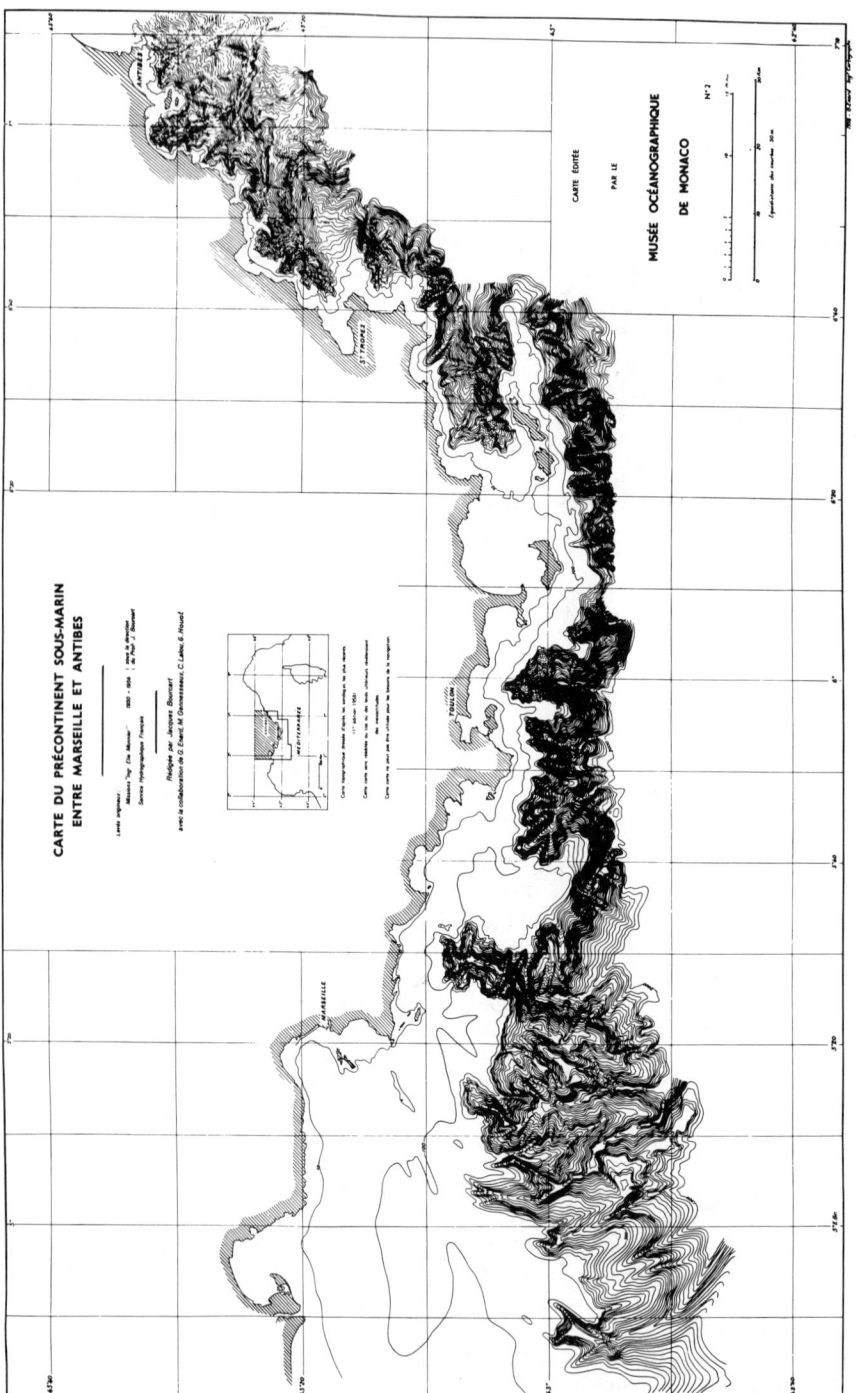

Fig. 11–15 Submarine canyons from the Rhone Delta (37 km west of Marseilles) to Antibes in southern France. Note change in character of canyons off the hard-rock coast east of Marseilles. From Musée Océanographique de Monaco. Courtesy of J.-Y. Cousteau.

late definitely all the canyons with rivers and embayments on land. However, Chesapeake Bay and Delaware Bay each have two clearly related canyons, and Hudson Canyon is adjoined by a shallow shelf valley that extends into New York Bay and hence into the Hudson. Block Canyon is less definitely connected with a shelf valley coming out from the gap between Block Island and eastern Long Island (Fig. 9–6). Three major canyons off Nantucket Shoals and the southwest part of Georges Bank have no apparent relation to the coastal area. Two small canyons are located off the entrance to Northeast Channel, but the seven canyons off Nova Scotia's Sable Island Bank and Banquereau have little relation to anything except that The Gulley, the largest of the group, is just north of Sable Island (Stanley 1967).

The East Coast canyons as a group differ from most others in having only gentle curves along their axes and few tributaries, although at the head of The Gulley three tributaries join to form the main canyon. Also, a number of small valleys enter the sides of Hudson Canyon. Seismic profiles across the eastern canyons show that they are cut through sedimentary layers (Uchupi 1968C).

The work of H. Stetson (1936, 1949) established that most of the East Coast canyons were cut into sedimentary rock ranging in age from Cretaceous to Late Tertiary. Some of the dives in Woods Hole's *Alvin* and in *Deepstar* (Dillon and Zimmerman 1970) have proved Stetson's contention that the Georges Bank canyon walls are locally vertical. Another dive showed the importance of mass movements in transporting blocks down the canyon axes (Trumbull and McCamis 1967). Cores indicate that, despite the distance of the canyon heads from shore, sand layers are found under a mud covering. Some of the sands were probably introduced during glacial stages of low sea level; others appear to be very recent. Ross (1968) found ripple marks in Corsair Canyon. Biological activity is important in eroding the canyon walls.

The fan valley off the Hudson has been traced seaward for 280 km. It has a distributary at a depth of 3000 m, and this side valley apparently rejoins the main valley 110 km farther out. Seventy-four km from the head of the fan valley, transverse profiles show a decided deepening that forms an outer gorge, which differs from most other fan valleys in its Tertiary material instead of recent sediments on the walls (Ericson et al. 1961). Wilmington Canyon, the subject of much recent study (Stanley and Kelling 1967, 1970; Stanley 1971), and its neighbors are active despite their great distance from shore. In the vicinity of the canyon, Lyall and others (1971) found that the sediments are resuspended during large storms at the shelf edge. This resuspension may account for the bottom currents observed in the canyon. They are funneling great quantities of sediment to the deep-sea floor.

Great Bahama Canyon It is indeed puzzling that the submarine canyon with the highest walls in the world is found incised between two of the low islands of the Bahamas. Its maximum wall height of 4347 m is perhaps greater than any land canyon. Great Bahama Canyon (Fig. 11–16) is also one of the longest, with a total length of 278 km including Northwest Branch and Northeast Continuation. Both Northwest Branch and Tongue Branch are V-shaped cuts below two of the trough-shaped depressions that separate several of the Bahama Islands. In fact, the broad-floored trough into which the head of the V-shaped valley in Tongue of the Ocean is cut can be traced southward for 100 km, with depths of 1460 to 1280 m.

Dives into Tongue Branch (Gibson and Schlee 1967) have yielded Tertiary rock with deep-water foraminifera from the canyon walls between 914 and 1576 m, in contrast to the shallow-water fossils from well cuttings at

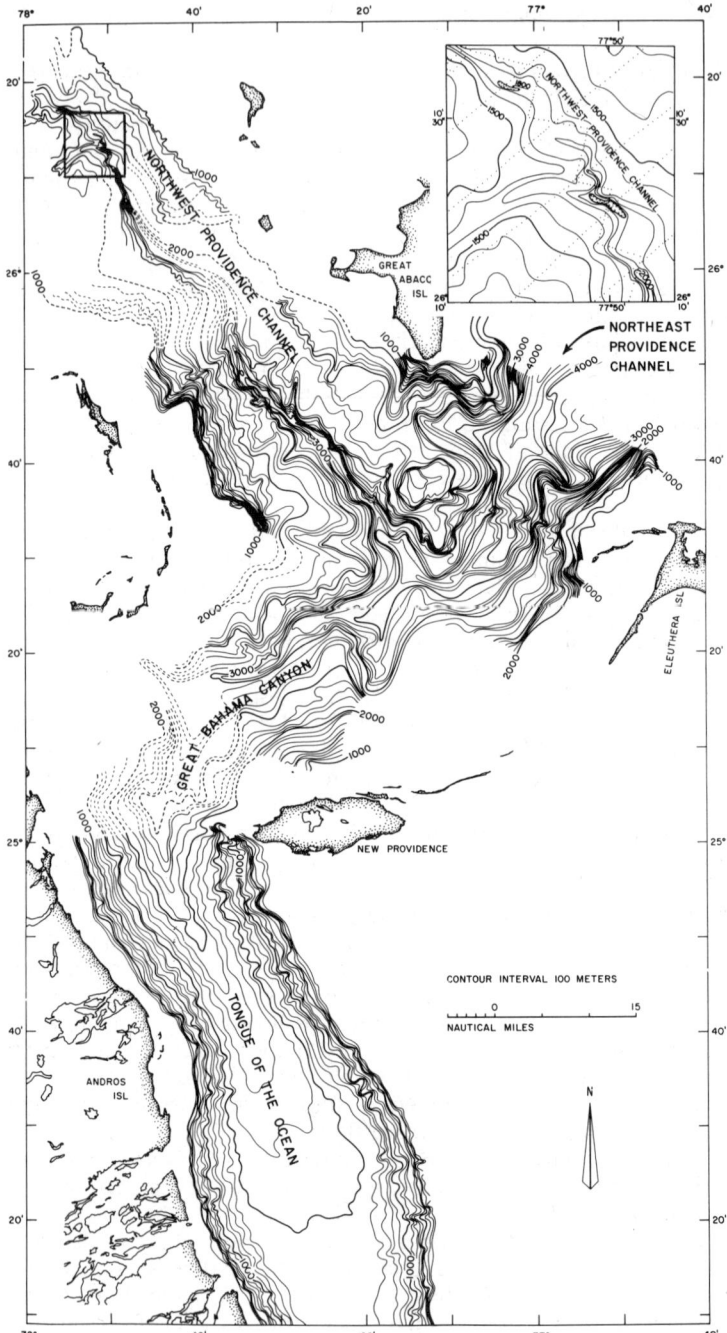

Fig. 11–16   Great Bahama Canyon. Northeast Providence Channel has the highest canyon walls of any in the world. Northwest Providence Channel has basin depressions along its course (inset). Note that the head of the main canyon is bowl-shaped in the deep water of the Tongue of the Ocean. Contours by the author from soundings of the U.S. Oceanographic Office, partly unpublished, and surveys by Shepard and Hurley (Andrews et al., 1970).

Fig. 11–17 Photo taken in or near axis of Great Bahama Canyon at depth of 3500 m. Note steep rock walls and rounded cobbles on the bottom. Photo by Hurley and Shepard (Andrews et al., 1970).

similar depths in the island borings. This indicates that valleys existed during at least part of the long period when the Bahama Banks were growing upward apace with the sinking of the entire block.

The inclination of the canyon walls do not average more than a few degrees, but observations from dives in submersibles and suspended-camera photographs (Fig. 11–17) confirm local precipitous slopes, at least of moderate height. Camera shots also revealed numerous cobbles and ripple marks along the canyon axes (Andrews et al. 1970). The compass in the photographs indicated that the currents producing the ripples were traveling downcanyon. Box cores contained layers of sorted calcarenites, one with a ripple-marked surface. Limestone walls contain local caves, seen from submersibles, and soundings suggest basin depressions along the canyon axes at various depths (Fig. 11–16, inset). Both of these findings indicate the strong possibility that solution has been important in developing this canyon.

Great Bahama Canyon loses its canyon character just seaward of where it emerges from the strait between Great Abaco and Eleuthera islands. Beyond this, it appears to be a fan valley. The high levee on the east side showed evidence of having migrated to the west and north during growth, which pushed the fan valley against the steep slope that leads down from Great Abaco Island (Andrews 1970).

ORIGIN OF SUBMARINE CANYONS

The question of submarine canyon formation has caused much disagreement. To explain successfully these sea canyons it is important to consider some of their significant features (see Shepard and Dill 1966, pp. 223–231 and Appendix):

1. Typical submarine canyons have V-shaped profiles, like river canyons, winding courses, dendritic tributaries, and almost continuous seaward-sloping floors.
2. Most canyons can be traced to the base of the continental slope, although they may change into leveed fan valleys where they emerge below the relatively steep rocky continental slopes and are cut into the basal fans.
3. Canyons are essentially worldwide on all continental slopes, except they are rare on (a) those with less than 1° of inclination, (b) those separated from continental shelves by continental borderlands or plateaus, and (c) those seaward of shelves with barrier coral reefs.
4. Rocks on canyon walls are mostly soft shales but include hard rocks, such as granite and even quartzite.
5. The majority of the canyons are directly off river valleys, but the latter usually have much wider floors and much gentler gradients than the adjoining sea-floor canyons.
6. Sediments on the canyon floors include an abundance of sand and even gravel, although coarse sediments are often covered with mud.
7. Current ripple marks are common at all depths on the canyon floors and, where known, the steep side of the ripples is predominantly downcanyon.
8. The organisms contained in sand layers, both on the canyon floors and on the adjoining fan valleys, include many shallow-water forms and land plants that apparently have been transported downvalley.
9. Heads of active canyons are known to have access to rapid influx of sediment and to have extremely unstable fill, which is carried seaward periodically by flows, slides, or unusual currents.
10. Canyons are found off stable and unstable coasts, but many of the largest appear to be related to shelves that have a long history of submergence and upbuilding of sediments, such as the shelves of the Bering Sea, the Bahamas, and the west coast of Europe.

Discredited Hypotheses When the first edition of this book was written, geologists were still using a number of hypotheses as canyon explanations. As explained in the second edition, these hypotheses have been mostly discarded. A commonly accepted idea is that the canyons are the result of turbidity currents, but some of us who have been investigating the canyons feel that this explanation is still far from proven. Alternatives that should be considered are (1) that canyons were initiated by drowning of land valleys and have been maintained and enlarged by various marine processes, such as mass movement and turbidity currents; and (2) that the principal cause of the canyons has been the marine processes, but that the great heights of the walls are due to upward growth by sedimentation on the surrounding marginal areas.

The idea that the canyons are the product of artesian springs arising along the slopes (Johnson 1939) was based on the East Coast canyons, which appear to be related to the seaward-dipping coastal plain formations; these were thought to act as aquifers. The hypothesis does not explain the canyons cut in granite and other resistant rocks, nor is there even any likelihood of there being sufficient underground circulation to produce the deep canyons beyond the wide shelf off the East Coast of the United States. The absence of sinkholes or other basins along the length of the canyons, except perhaps in the Bahamas, adds another serious difficulty.

Tsunamis were suggested by Bucher (1940) because these long-period waves transmit some of their energy to the deep-ocean floor. These currents, however, are significant only in shallow water and on the slopes would be concentrated only on ridges. Finally, the canyons are definitely not confined to areas where tsunamis are common.

Canyons have been explained by diastrophic movements (Wegener 1924, p. 177). No doubt both faulting and folding can produce valleylike depressions on the continental slopes. The winding patterns and the V shape of the cross section of submarine canyons, however, show clearly that most of them are not diastrophic. Fault valleys would not show as much sinuosity nor would they show the dendritic tributaries that enter many canyons of the sea floor. Furthermore, there is little indication that canyons are more common in unstable areas, although they often follow old fault lines.

Prior to the postwar period of intensive investigations of the sea floor, the idea of drowned river valleys as a sole or principal cause of submarine canyons was held by many geologists, including myself (Shepard 1948, pp. 241–248). This supposition has virtually disappeared as we have learned more about the active erosion presently taking place in the canyons. Also, the great contrast between most submarine canyons and the adjacent land valleys has discouraged the idea, and the failure to find any acceptable means for exposing the continental slopes of the world to subaerial erosion has furnished the knockout blow.

Subaerial Erosion as a Contributing Cause The resemblance of submarine canyons to river-cut canyons on land (although not in their vicinity except along West Corsica) and the juxtaposition of many land and sea canyons have always provided a strong argument that at least the heads of the latter were originally cut by rivers. The well-established lowering of sea level during glacial stages of the Pleistocene must have allowed the rivers to cut 100 m or even 200 m below the present datum. Furthermore, various coasts are known to have undergone subsidence, which should have submerged many land canyons. Marine processes are clearly capable of preventing a number of these land-cut canyons from being filled with sediments. The deep drillings (JOIDES) in the Mediterranean have yielded convincing evidence that the Mediterranean had partially evaporated during portions of the Late Miocene (Ryan, Hsu, et al. 1970). Whenever the sill at Gibraltar closed and the Mediterranean shrank, the potential existed for rivers from mountains, such as the Alps, to cut canyons into the exposed slopes. The study of cores taken in the submarine canyon off Toulon, southern France (Mascle 1968), may provide evidence of a sea-level lowering or subsidence. Mascle discovered littoral formations of an age between Pliocene and Quaternary at the base of the cores underlying deep-water beds. These littoral beds do not appear to be turbidites and thus may show the result of the lowering of the Mediterranean. The import of this factor in the production of the Mediterranean submarine canyons awaits further study of JOIDES results.

On the other hand, there are formidable obstacles against the subaerial erosion origin for the deep submarine canyons of the continental slopes. Thus, if subaerial origin was the principal cause, the canyons should show a direct relationship to coastal history; that is, they should be well developed where coasts have undergone large submergence and rare or nonexistent along coasts that have a history of emergence. Examination of world maps shows no such relationship. Canyons are as common off the rising West Coast of the United States as along the sinking East Coast. Nor are the margins of shelves any deeper where canyons abound than where they are relatively scarce. Furthermore, if subaerial origin were important we should find distinct breaks in the continuity of the canyons at the level where the subaerial erosion ceased. No such breaks exist. They not only continue without modification past the 100- to 200-m level of the maximum glacial stage but continue with little change to the base of the continental slopes. Finally, if submarine canyons were produced largely as the result of submergence of land canyons, they should show a more striking resemblance than they do to adjacent land canyons.

Turbidity-Current Erosion of Canyons   Turbidity currents (see p. 66) are widely accredited with transporting great quantities of sediment down submarine canyons to form the great fans at the base of the continental slopes. These fans contain far more volume than would come solely from canyon excavation. These currents apparently account for the formation of levees in the valleys that cross many of the fans. Because coarse sediments are common in the fan valleys and even on the levees, the latter often 100 m or more above the channel floors, it would appear that the currents moving down the canyons are sufficiently powerful to produce erosion on the continental slope. Sediment-laden water is heavier than adjacent clear water and hence is capable of descending to the bottom of slopes, whereas other currents have to contend with a general but slight increase in density at greater depths. Thus, clear-water currents tend to flow above the bottom after reaching a water layer with higher density.

Although it is likely that turbidity currents are at least an active contributing cause of submarine canyons and certainly keep them open, we still lack substantial proof that they are of major importance. It has often been assumed that the sequence of cable breaks that occurred after the Grand Banks earthquake in 1929 was caused by a turbidity current that had a velocity up to 50 knots (92 km/hr) (Heezen and Ewing 1952). However, as explained elsewhere (*Submarine Geology*, 2nd ed., pp. 339–343), the cable company data suggesting the high velocity are far from conclusive and are actually more indicative of a 15-knot (28 km/hr) current (Fig. 11–18). The samples collected from the area clearly do not support any indication of a powerful flow sweeping down the slope. Furthermore, the Grand Banks earthquake appears to have produced a large landslide (Heezen and Drake 1964). As this developed, successive slumps may have broken cables in the observed sequence.

Other sediment characteristics of submarine canyons are perhaps hard to reconcile with claims of rapid erosion by turbidity currents. The sand and gravel layers found in the canyons and fan valleys often contain abundant delicate foraminifera, including arenaceous types which should have been broken by high-speed currents. The nature of layering and unconformities on the valley floors is often more indicative of traction currents than turbidity currents (Shepard et al. 1969). The frequent absence of any graded fine material at the top of the sand layers is also hard to visualize had turbidity currents been responsible for emplacement of the sand. Finally, if due to turbidity currents, it may be strange that the sediment so often coarsens in a downvalley direction. An exception is the Mazama Ash in Astoria Canyon (Nelson et al. 1968). Other objections are cited by van der Lingen (1969).

These objections may not be valid. Most theoretical discussions of turbidity currents based on hydraulic laboratory experiments favor high

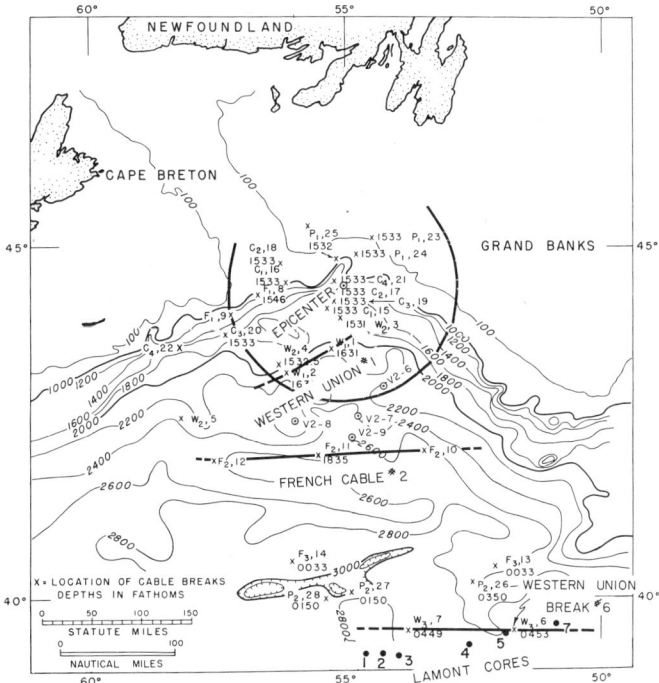

Fig. 11–18 The times and locations of all cable breaks in the Grand Banks area after and during the earthquake in 1929. The arc of a circle (100 nautical miles from epicenter) circumscribes the area in which most cable breaks occurred simultaneously with the earthquake. The station numbers represent cores taken by Lamont-Doherty Geological Observatory. Numbers 1 to 7 were all on cruise A-180 but numbers 3 to 6 obtained no core.

velocities (Inman, *Submarine Geology*, 2nd ed., pp. 132–140; Kuenen 1965; Middleton 1966; Komar 1969). If confined to valleys rather than to the open slope, as originally suggested, the 15-knot velocity for the Grand Banks earthquake cable breaks (Fig. 11–18) may be rather ordinary. Komar calculated that the velocities due to the filling of the channels in the Monterey Fan Valley would be of the order of 16 to 40 knots. Unfortunately, we still have no clear evidence that will support or contradict these theoretical considerations.

We still know very little about turbidity currents except those induced in the laboratory (Kuenen 1965, 1967). Perhaps the only appreciable turbidity current observed in the sea was that reported by Reimnitz (1971) in the Rio Balsas Submarine Canyon off western Mexico. During a period of large surf, Reimnitz observed current pulses with estimated velocities of 4 km/hr flowing down a tributary and transporting large quantities of sand. It was producing erosion of the tributary walls. Otherwise, only very slow turbidity currents have been observed due to suspension of muddy sediments (D. Moore 1969). Attempts to start turbidity currents, even by setting off explosives buried in canyon fills, so far have proven unsuccessful (Buffington 1961; Inman 1968; Dill 1969B).

Combination of Submarine Processes as Canyon Origin   As submarine canyon studies have continued, results appear to favor a combination of processes as the cause. Thus we are finding much support for mass movements, especially in the canyon heads (Shepard and Dill 1966, pp. 332–334). Perhaps gravity-induced slides and slumps are not important except in the steeper heads, but seismic profiling is showing us that slides take place even on gentle continental rises (Emery et al. 1970). However, typical mass movements on land slopes leave many undrained depressions and, as far as we know, basins of this type are very rare in submarine canyons. Also, the general shape of the canyons, with their dendritic tributaries, is much more indicative of a type of erosion that is comparable with that of rivers; it could also be produced by descending sea-floor currents (Reimnitz and Gutierrez-Estrada 1970).

In addition to turbidity currents, there is growing evidence that other types of currents are important. The records of up- and downcanyon flows (Fig. 11–19) that have been measured, especially in La Jolla Canyon, are of low

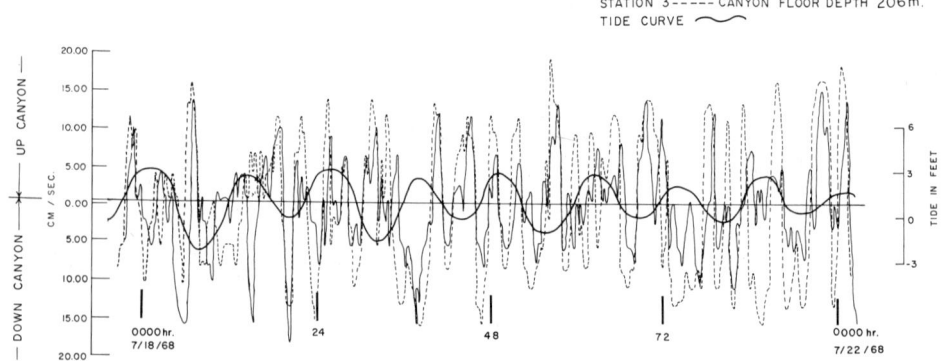

Fig. 11–19 Four days of simultaneous records from two current meters located along the axis of La Jolla Submarine Canyon at depths of 167 and 206 m. Note the frequent reversals between up- and downcanyon direction. The downcanyon flows usually last longer and the net flow is downcanyon in most of these records. The relation to the tides will be seen to be small. From Shepard and Marshall (1969).

velocities, but downcanyon velocities may be increased as the result of surf beat brought about by strong onshore winds (Inman 1970), and a 160-cm/sec (3.6-knot) current was measured under strong onshore wind conditions by Inman during one episode. Probably it reached a higher velocity, because the cable holding the meter in place broke at that time. Inman explained this current as the piling up of water along the shore and downcanyon pulses occurring as a result of surf beat. Such downcanyon flows may continue past the denser lower temperature water layers if the currents stir up sufficient sediment to cause a turbidity current. The alternating canyon-floor currents are partly related to tides and probably also involve internal waves. Submersible dives to deep-canyon floors have provided us with a considerable number of observations of down-flowing currents of velocities up to at least 15 cm/sec. These do not appear to be related to sediment-laden water, suggesting some driving mechanism other than that of turbidity currents. They may account for the numerous observations of asymmetrical current ripples on the canyon floors with their steep side downcanyon.

Another type of current that flows along bottom contours and may produce valleys has been described as occurring on the East Coast continental slope (Heezen et al. 1966; Rona et al. 1967). These recent discoveries certainly suggest that it is inadvisable to attribute all indications of current erosion to turbidity currents. Probably various types of flows, still not well understood, may be equally important. In some areas, cascading currents due to cooling of water along the coast may be significant (see p. 64).

Upbuilding of Canyon Walls   If various processes on the sea floor are both excavating and maintaining previously excavated canyons, undoubtedly deposition either on the shelves or on the continental slopes is taking place and increasing the total wall heights of the canyons. Such a process seems particularly probable in Great Bahama Canyon where evidence has been found that Middle or Upper Tertiary formations within the canyon were deposited in deep water (Gibson and Schlee 1967), whereas the formations of the same age underlying the Bahama Islands apparently were all deposited in shallow water (Spencer 1967). Kuenen (1953) thought this idea was not applicable to the East Coast canyons, because there should be a change in slope between the old formations into which the canyons were cut and more recent formations that could have been deposited subsequently on the shelves and slopes along the sides of the canyons. However, profiles show slope changes on the sides of the Hudson (Fig. 11–20). Also, where the recent formations bordering canyons are of coral that grew upward apace with the sinking, the upper slopes might be steeper than the cuts made into the older formations. This is obviously true of the slopes in the Bahamas where the reef fronts have precipitous slopes.

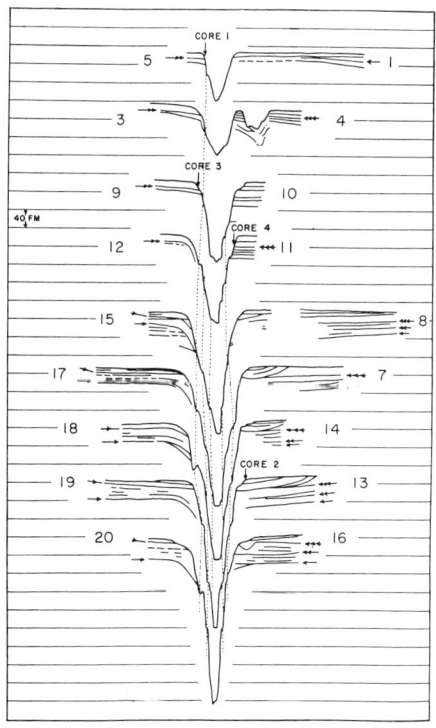

Fig. 11–20   Profiles of the Hudson Canyon by bottom-penetrating echo soundings. Showing the inclined layers and terraces covered by recent sediments along the walls. From J. Ewing et al. (1960).

The advantage of the combined downcutting and upbuilding hypothesis is that it provides a long period of time for the formation of huge canyons such as those in the Bering Sea and in the Bahamas. This hypothesis seems decidedly preferable to the idea so frequently expressed that all submarine canyons have been excavated by turbidity currents during the low sea-level stages of the Pleistocene (Kuenen 1950, p. 503). Discovery of old filled canyons lying below the floor of present canyons adds impetus to the newer hypothesis and suggests that canyon cutting may have alternated with fill at different episodes. The numerous examples now reported of filled marine canyons in the stratigraphic section are further indications of such alternations.

DELTA-FRONT TROUGHS

A few trough valleys extend seaward from several of the great deltas. It may be a mistake to separate them from submarine canyons. However, they are distinctive in character, as will be shown from the following descriptions.

Ganges Trough   Outside the 450-km-wide Ganges–Brahmaputra Delta, a trough-shaped marine valley, called the Swatch of No Ground, crosses the shelf diagonally in a southwesterly direction (Fig. 11–21A). As indicated by Coleman (1969), the bars and channels off the mouth of the combined rivers point toward this trough and undoubtedly allow sediment to be carried into the head of the trough during floods. This, in turn, helps explain why the delta and the bars are not being built seaward, despite the enormous load of sediment coming to the sea at this locality. The Ganges Trough has a comparatively flat floor 5 to 7 km wide and walls of about 12° inclination. As far as could be determined, the floor slopes outward continuously. At the edge of the shelf, depths in the Trough are about 1200 m. The walls are relatively straight and only minor tributaries are indicated by the detailed Pakistan Navy survey. A curious phenomenon is often observed in crossing the boundary between the shelf and the Trough. The color of the water changes from brown to a pale blue, and slicks are seen parallel to the sides of the Trough. These either indicate upwelling or that currents are deflected by the underlying topography. Seismic profiles (Fig. 11–21B) across the Trough show that it has considerable fill and that either faulting or slumping occurred on one margin.

The Ganges Trough has a seaward continuation for almost 2000 km down the Bay of Bengal in the form of fan valleys with levees (Curray et al. 1971). These can be detected even south of Ceylon (Fig. 13–15A). The fan valleys wind across the great fan with many distributaries, and in some places have a braided pattern. The huge fill in the Bay of Bengal is apparently derived primarily from the Ganges and Brahmaputra.

Indus Trough   A valley similar to Ganges Trough is located off the main mouth of the Indus River (Fig. 11–22). Trough depths at the shelf edge are 1130 m. A continuation seaward as a fan valley has also been discovered, but it is not as well mapped as off the Ganges. Again, according to the Pakistan Navy, current lines occur at the margin of the Trough. The floor is not quite as wide but the profiles indicate flattening, in contrast to the V shape of typical canyons.

Mississippi Trough   Off the west side of the Mississipi Delta, a trough has been located that heads 55 km from shore (Fig. 11–23). This is definitely the same type as those troughs described above, but here the inner portion of the trough has been filled. Drillings and seismic profiles show that the

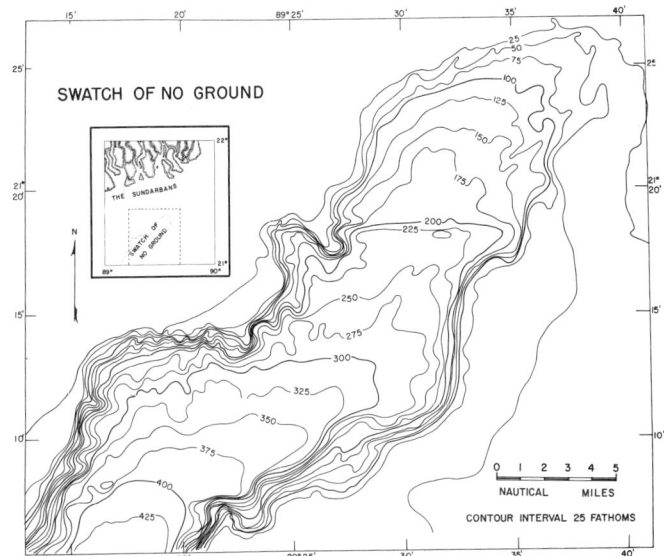

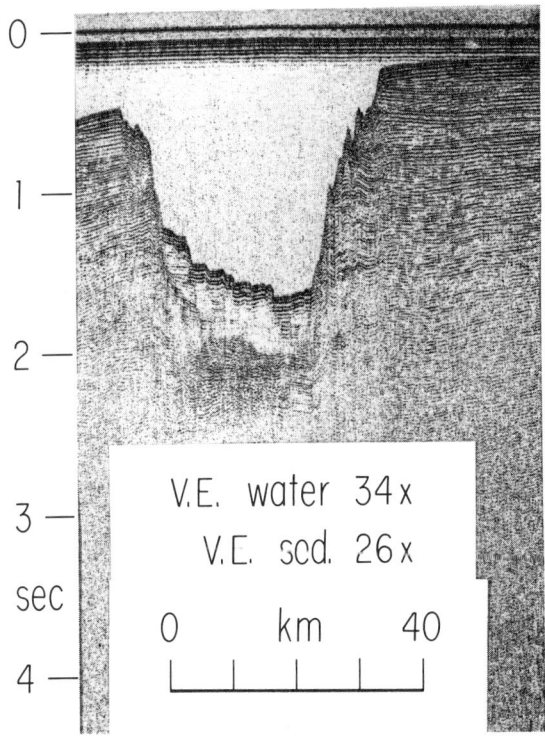

Fig. 11–21 A, the Swatch of No Ground, a delta-front trough that extends diagonally seaward across the outer shelf off the Ganges Delta. Contours from soundings by Pakistan Navy. Local steepening of floor has been disproved. See also Fig. 13–15. B, seismic profile across the Swatch of No Ground. Note thick fill with minor faults and/or slump scarps. Courtesy of J. R. Curray and D. G. Moore.

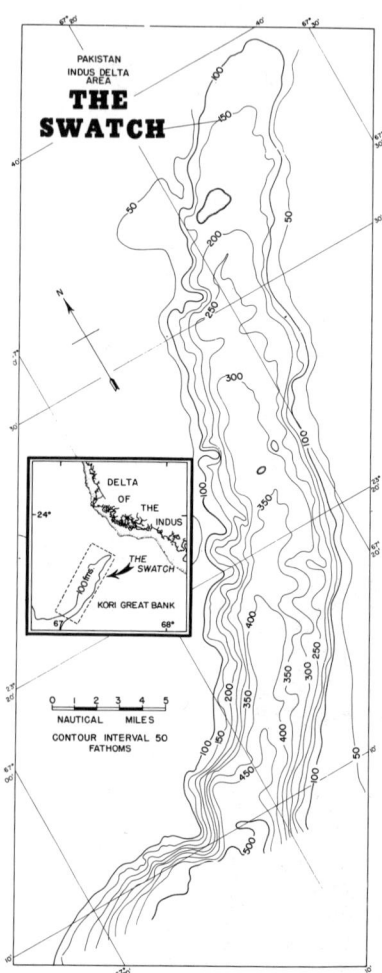

Fig. 11–22 Delta-front trough off the Indus Delta. Contours from soundings of the Pakistan Navy.

filled Mississippi Trough extends into shore. The trough floor on the outer shelf is 19 km wide, but the buried portion inshore narrows near the sub-aerial delta front. The walls of the outer trough are more irregular than those off India, probably as a result of salt dome or mud diapir intrusions. The Mississippi Trough has been traced down the continental slope to about 1460 m, where it apparently disappears on a large fan that has built forward into the Gulf.

Perhaps it is a coincidence that where all these troughs cross the continental shelf, the floors have inclinations of 8 to 10 m/km. This is also the typical frontal slope of several large deltas built out beyond the shelf.

Origin of Delta-Front Troughs   We have much less knowledge of delta-front troughs than of submarine canyons. Therefore, conclusions concerning the troughs must be even more tentative. The three described are comparable in length with the longer of the canyons, have floor depths hundreds of meters below their surroundings, and are important corridors for sediment transportation. Transverse seismic profiles (Fig. 11–21B) suggest,

Fig. 11–23 Submarine trough and fan off the western portion of Mississippi Delta. Depths in fathoms. Contour interval 100 fathoms to 1400, then 10 fathoms. From M. Ewing et al. (1958).

but do not prove, that the troughs were cut through shelf sediments. However, they may be simply nondepositional, being maintained by currents as the shelves were built up. Like some of the canyons, they have indications of thick deposits on their floors. Deposition may alternate with cutting. Despite their contrasting character with submarine canyons, they probably have undergone a similar history.

SLOPE GULLIES AND THEIR ORIGIN

Many submarine slopes of relatively uniform gradient are cut by small discontinuous gullies. These are found on the forward-building slopes of deltas where the deltas are advancing into comparatively deep water, as in fiords (Mathews and Shepard 1962), or where the delta has built across the continental shelf. The best-known slope of this type is off three of the distributaries of the Mississippi Delta: Southwest Pass, South Pass, and Pass a Loutre (Fig. 11–24). The contrast between these gullies and typical submarine canyons is usually great. Slope gullies begin and end at various depths along the slope, and they have few if any tributaries. At their lower limit they usually have small hills. Resurveying of the gullies off the Mississippi Delta (Shepard

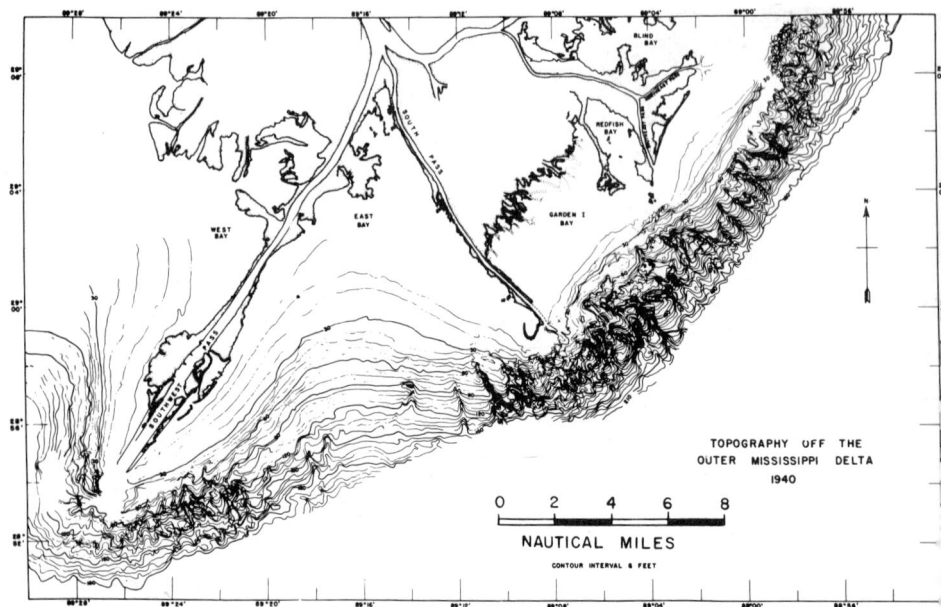

Fig. 11–24 Gullies in the foreset slope of the Mississippi Delta that are concentrated off the most advanced distributaries. These gullies are apparently the result of slides in the advancing slope. They terminate at depths of 60 m (200 feet).

1955) has shown that they are unstable in position. Old gullies have been filled and new ones developed.

Slope gullies are located also along submarine fault scarps. Examples are found on the northeast slope of San Clemente Island, off California, and in the Gulf of California along the faulted margin of Ceralbo Island. Apparently these slopes were built by sediments from the island after most of the faulting was completed. Another example (Emery and Terry 1956) lies along the fault scarp off the Palos Verdes Hills, Southern California.

Slope gullies are evidently due primarily to slumping or other mass movements. Despite the very gentle slope off the Mississippi Delta, according to Terzaghi (1956), rapid sedimentation makes the material metastable and subject to sudden slumps that open small gullies. Similar gullies on land are due to mudflows.

INTERMEDIATE TYPES OF VALLEYS

Some sea-floor valleys appear to be intermediate in character between slope gullies and submarine canyons. A particularly good example is located on a deltaic coast of northwestern Luzon, Philippine Islands (Fig. 11–25). Here, the entire slope is cut by valleys of moderate depth. Unlike slope gullies, these have tributaries, and the valleys may even extend to the base of the slope. The valleys off the Rio Balsas Delta, along the west coast of Mexico, are also good examples of this intermediate type. There is no continental shelf and the upper slopes are entirely deltaic sediment. The valleys

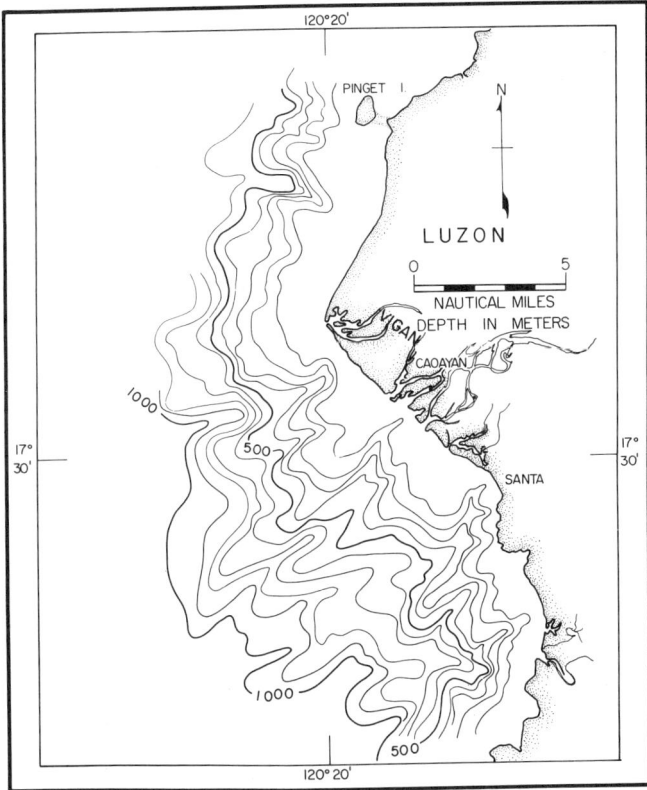

Fig. 11–25 The slope valleys off the deltaic coast of northwestern Luzon, Philippine Islands. From Shepard and Dill (1966).

extend in close to the shoreline off the present active mouth of the Rio Balsas and off recently abandoned mouths. They have many tributaries. Reimnitz and Gutierrez-Estrada (1970) show conclusively that these have undergone very recent cutting.

The intermediate valleys, all on sediment slopes, may formerly have been old slope gullies caused originally by slides but later modified by erosion from submarine currents, which gave them some of the erosional characteristics of submarine canyons. They may be better classified with ordinary canyons, although they are related in most cases to deltaic foreset slopes, which are unstable, and the excavation of the various tributaries can be very rapid because of the unconsolidated character of the formations.

FAULT VALLEYS OF THE SEA FLOOR

Many land valleys are primarily the result of faulting. Other subaerial valleys are caused by a combination of faulting and erosion. Because earthquakes are common on the sea floor near many continental margins and since land faults can be traced seaward in many places, it can be assumed that many of the valleys of the sea floor are also the result of faulting. The characteristics of land fault valleys, like Death Valley, include straight walls, absence of accordant tributaries, trough-shaped floors, and termination in basin depressions. Their fault origin is confirmed in some cases by movements along the fault scarps at the times of earthquakes. On the sea floor the evidence is not so easily obtained. Some of the valleys on the south side of the Aleutian Islands (Fig. 11–26) appear to fit the above criteria because they lack tributaries, have straight walls, trough-shaped transverse profiles, basin depressions at their lower ends, and extend parallel to the fault system on the adjacent lands. The longitudinal valley in Sagami Bay, Japan, which curves to the east to form a large arc, has a floor that slopes continuously seaward and a transverse profile that is more V-shaped than trough-shaped, but it appears to be connected with a fault valley at the bay head where there was considerable displacement during the 1923 earthquake (Yamasaki 1926; Shepard et al. 1964). This valley is probably a structural feature that has been modified

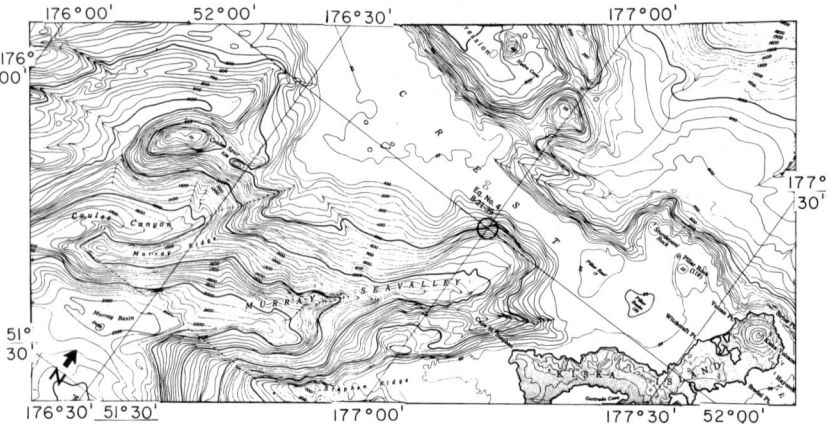

Fig. 11–26 Trough-shaped valleys near Kiska, Aleutian Islands, in an area of active tectonism. Note basin depressions in Murray Sea Valley. Contour interval is 50 fathoms (92 m). From Gibson and Nichols (1953).

by marine erosion, which developed its present characteristics. Similarly, a valley off the southwest end of Baja California lies almost parallel to the coast and terminates to the north where a major land fault has been discovered. In this case, faulting evidently offsets the head from the lower end of a submarine canyon that was originally cut athwart the fault valley (Shepard 1964). No basin depressions exist in the fault valley, and it has a V-shaped transverse profile, suggesting again the importance of marine erosion in its development. The great trenches of the sea floor are a type of valley and undoubtedly are of tectonic origin. However, even these trenches show some indication that they have been modified by marine erosion.

FOLD VALLEYS

Presumably, fold valleys exist on land, although most geologists have hesitated to describe them as such. The recent wide coverage of the sea floor by seismic profiling has revealed many places where valleys and ridges are underlain by folds essentially parallel to the surface. That these profiles indicate fold valleys and ridges on the sea floor is uncertain because seismic profiles can often be deceptive. Some of the apparent folds may actually be faults. Features of this sort seem to be well illustrated in the Caribbean to the east of the Magdalena Delta of Colombia, off various portions of the California coast (Curray 1966), and off western Europe (Stride et al. 1969). Because these features are known only from single profiles, it is not possible to describe their continuity. Probably they contain basin depressions, like fault valleys.

UPLIFTED SUBMARINE CANYONS

If some submarine canyons are downwarped river valleys, it should be equally probable that some land valleys were excavated by marine processes and then elevated to their present position. This idea was developed by Winslow (1966) and critically discussed by Soons (1968), with a reply by Winslow (1968). Soons did not object to Winslow's idea but took exception to many of his examples, particularly to those from New Zealand where she was familiar with the situation. Winslow unfortunately used as one of his principal arguments that fiords are largely uplifted submarine canyons that have been slightly modified by glacial erosion and fill at their lower ends. As we have seen (p. 112), fiords and glacial troughs on the shelf are unique to glaciated areas, and have the same characteristics as the basins and troughs found well inland from the sea where marine erosion could not be considered. Fiords and troughs are as well developed off relatively stable coasts, such as eastern Canada and Norway, where the only diastrophism is elastic rebound due to deglaciation, as along young mountain range coasts where elevation of submarine canyons might be expected.

On the other hand, Winslow's discussion of the short dry valleys cut into the edge of elevated terraces as examples of elevated canyon heads may have some merit. The somewhat anomalous valley heads in Point Reyes Peninsula, north of San Francisco, may represent elevated canyons, but no evidence is now available that these valleys continue on the sea floor. The small submarine canyons off La Jolla, California (Fig. 11–2), have several land valley continuations that could also be elevated marine canyons. However, many land canyons of the same character farther north have no connecting submarine canyons, nor is there any evidence that there are buried canyons on the shelf and slope outside these land canyons. Perhaps a better illustration is the series of land canyons that appear to connect with sea canyons on the north side of Molokai, Hawaii (Shepard and Dill 1966, Fig. 104). In any case, before any faith can be put into considering a land valley as an uplifted submarine canyon, field examination is necessary for confirmation.

Chapter 12
CORAL REEFS

INTRODUCTION

The past decade has seen progress in the study of coral reefs. Notable among the published results are a compendium on the Great Barrier Reef (Maxwell 1968), a report on the drilling into the lagoon of Midway Atoll (Ladd et al. 1970), a series of reports on the reefs of French Oceania (Guilcher, 1965B; Guilcher et al. 1965, 1969), a book on atoll environment and ecology (Wiens 1962), discussion of the Florida Keys (Hoffmeister and Multer 1968), a study of Ifaluk Atoll (Tracey et al. 1961), a good summary on the history of reefs (Newell 1971), and a symposium on the results of an expedition to the Caroline and Marshall islands (Curray et al. 1970).

The first important study of coral reefs was by Darwin (1842) as a result of his observations during the world encircling cruise of the *Beagle*. For half a century Darwin's book and his ideas of reef upgrowth during submergence of volcanoes were supported and strengthened by many other scientists, including Dana (1885) and Davis (1928). They were seriously challenged by Agassiz (1903) and Daly (1910, 1915) but were established as fundamentally correct by the deep borings into Eniwetok and Bikini atolls (Emery et al. 1954). The debates between Davis and Daly make interesting reading, but neither of these famous geologists made the careful field observations that characterize the work of Ladd and Hoffmeister (1936), Yonge (1940, 1963), Wells (1954), Newell (1956, 1959), and Goreau (1961). Laboratory studies of coral growth by Vaughan (1916) and Mayor (1924) were also valuable in setting the stage for the solution of many coral-reef problems. The investigations that accompanied and followed the drillings in the Marshall Islands are still the most important geological development in the field of organic reefs.

BACKGROUND INFORMATION

Organic Reefs Defined   Organic reefs, usually referred to collectively as coral reefs, are structures built by organisms with a framework strong enough to withstand the attacks of ordinary wind waves by a baffle effect (Ladd 1961). Unusually large waves break fragments off the front of the reef, and often pile coral masses, some weighing tons, above sea level to produce islands. An actively growing reef will soon recover from the storm erosion and rebuild its damaged frame.

The most common frame builders of present-day reefs are the hermatypic corals, the coralline algae, and the hydrocorals, along with foraminifera. During the remote past, the reef framework included sponges and bryozoans in addition to corals and algae. Emery and others (1954) estimated that in the Marshall Islands 30 percent of the total reef consists of the framework. The frame and binding material together make a rigid stable structure but leave large cavities. This produces a mass high in pore space; for example, 25 to 50 percent was estimated by Newell (1956) for Raroia Island. As the reef grows upward, the cavities generally have communicating passages with the sea, and sediment from the reef is carried into them. Cullis (1904, p. 396) suggested that after being buried by approximately 6 m of coral, the cavities are filled by chemical precipitation from the circulating sea water and that, deeper in the upgrowing reef, solution may take place. This may attack the original coral framework, leaving the filling of the old cavities as the new framework. In this way the coral reefs may remain almost as porous after deep burial as when they were first formed. Alternatively, the solution and reprecipitation may occur as the result of intermittent uplift of the reefs during the long periods of gradual submergence of the reef platform. Such alternations have been found by Ladd and Schlanger (1960) in the study of the Eniwetok borings. This, of course, is important in providing the space for oil accumulation in ancient reefs.

Types of Reefs   The most common types of reefs in the present seas are: (1) fringing reefs, which grow out from a land mass but are connected with it; (2) barrier reefs, which are separated from land by a lagoon; (3) atolls, oval-shaped reefs rising from deep water and surrounding a lagoon in which there is little or no land; (4) faros, ring-shaped reefs located on banks or shelves; (5) table reefs or coral banks that rise above the adjacent sea floor as plateaus without any appreciable rim; and (6) coral knolls, also called patch reefs or pinnacles, small masses that rise above the lagoon floor inside barrier reefs or atolls. Most of these are illustrated in Figs. 12–1, 12–2, and 12–8.

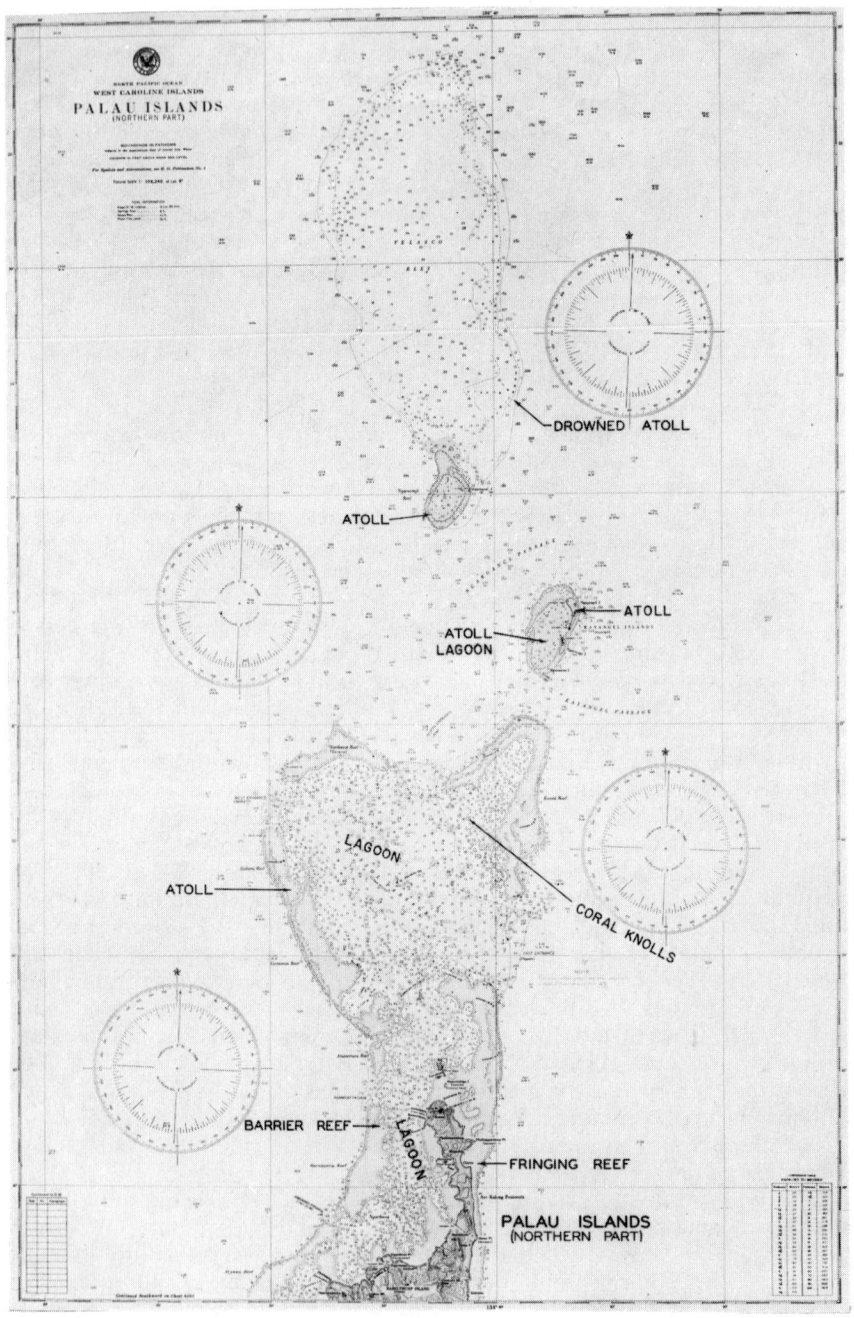

Fig. 12–1 Various types of coral reefs. From U.S. Oceanographic Office Chart 6074. See also Figs. 12–2 and 12–18.

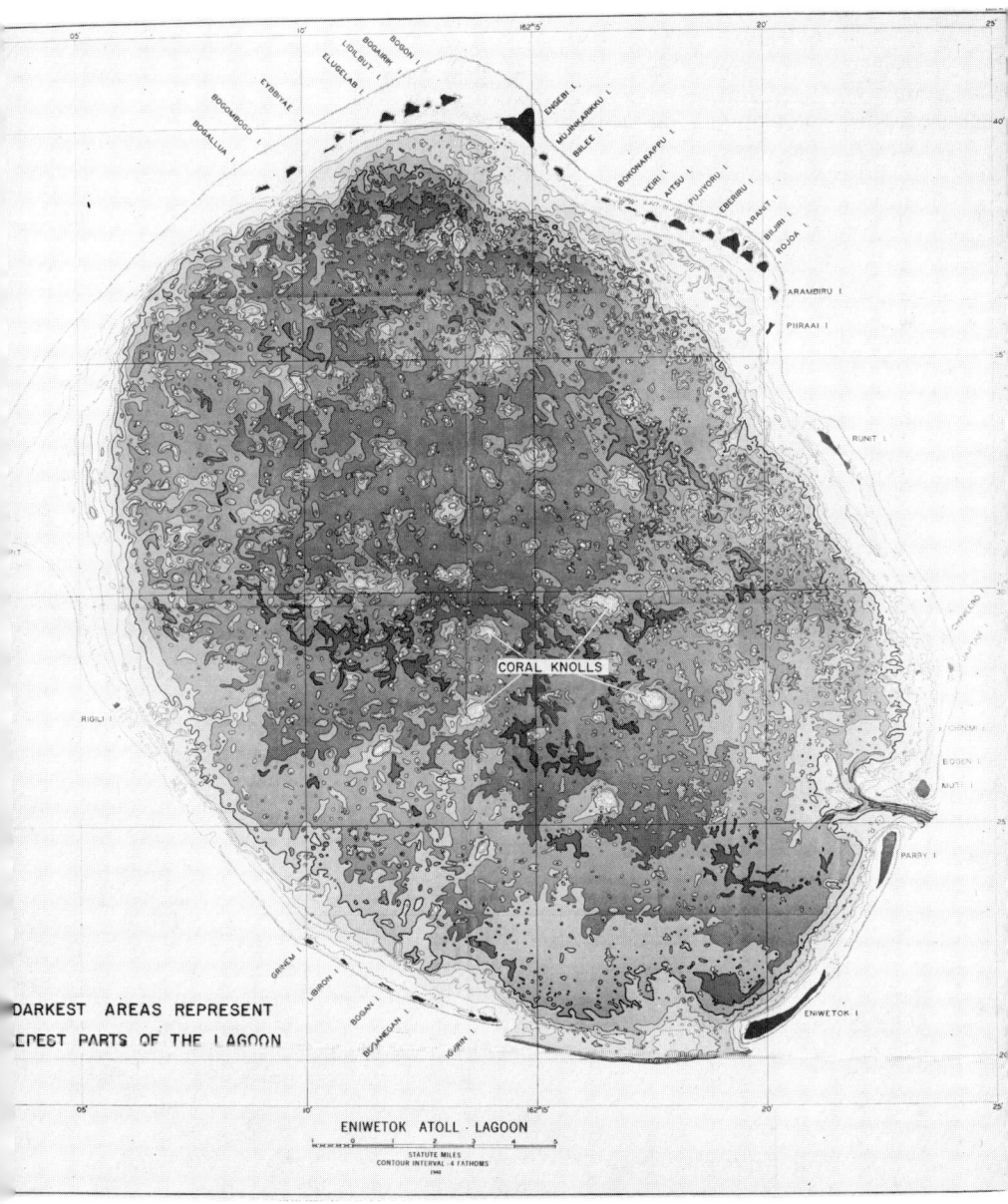

DARKEST AREAS REPRESENT
DEEPEST PARTS OF THE LAGOON

CORAL KNOLLS

ENIWETOK ATOLL · LAGOON

STATUTE MILES
CONTOUR INTERVAL 4 FATHOMS

Fig. 12–2 The relief in the lagoon on Eniwetok Atoll, Marshall Islands. The darkest shades have depths of 32–36 fathoms (59 to 66 m). Note the large entrance channel on the east side of the island. From Emery et al. (1954).

Reef Ecology   The common impression that coral reefs grow only in shallow and warm water, and hence ancient reefs are indicative of warm climatic conditions, is not actually correct. The modern corals (Scleractinia) include hermatypic and ahermatypic varieties. The former contain symbiotic flagellates (Zooxanthellae), which can grow only in warm shallow environments; but ahermatypic corals that lack the symbiotic flagellates occur in all depths and at temperatures as low as $-1.1°C$ (Vaughan and Wells 1943). Ahermatypic reefs do not flourish as well as the hermatypic, but they are found forming reefs in many areas, notably the deep shelf and slope off Norway (Teichert 1958), on the deep Blake Plateau (Squires, 1963), and along the continental slope off western France.

By far the most common reef formers of the present are the hermatypic corals with symbiotic flagellates and the coralline algae, both largely confined in their growth to tropical belts, extending a little north or south of the tropics where conditions are favorable. The vigorous growth of these reef formers takes place only to depths of 45 m or less, although some hermatypic reef corals have been found living as deep as 165 m. Winter temperatures in reef areas are rarely below 18°C. In the Persian Gulf, however, where the summer temperatures are well above 30°C, small reefs are growing where winter temperatures may reach 13 to 14°C.[1] Maximum temperatures of above 35°C will kill most corals, but such temperatures are rarely, if ever, attained in present seas, except in extremely shallow water where there is negligible circulation.

The relationship of symbiotic flagellates with corals is understandable; the flagellates obtain phosphorous and nitrogen from the corals, and are protected by them. The advantages to the corals are not so evident. The flagellates probably do not provide food. Early investigations established that the reef corals are carnivorous, preying exclusively on zooplankton, whereas the flagellates are plants. Goreau (1961) found that removal of the flagellates from the corals greatly reduced the rates of calcification. Flagellates may also provide corals with a product of photosynthesis, which is used to form the chitinous organic component of the coral skeleton.

Reef corals are also dependent on the proper salinity conditions, living within margins of about 27 to 40⁰/oo. As a result, a reef may be killed by a great flood of fresh water that sweeps out over it from the land. According to Mayor (1924), this happened in Pago Pago Harbor of American Samoa after a 37-in. rain in four days had deluged the area.

The studies of coral reef platforms by Vaughan (1916) and Hoffmeister and Ladd (1935) have shown that reefs develop on a great variety of surfaces, although Vaughan claimed they cannot grow on a soft mud bottom. Their absence on muddy deposits is explained by the deleterious effect of mud in the water. Actually, borings in west Sumatra revealed coral reefs, overlying mud deposits (Umbgrove 1947). Studies of the reef at Waikiki, Oahu, suggested to Edmondson (1928) that the mud from a stream which had formerly entered here had prevented growth in the immediate vicinity. His experiments showed that most reef-forming corals could not withstand burial in silt except for a short period. The absence of reefs along the coast at the southern end of the Queensland Great Barrier Reef is apparently related to the muddy water coming out from the land in this zone. However, the effect of the mud in the water may not be very important by itself. Ladd and Hoffmeister (1936, p. 82) have called attention to the existence of reefs off the Rewa River in the Fiji Islands ". . . in spite of the tremendous amounts of silt brought to the

---

[1] From records of the Arabian-American Oil Company of Dhahran, Saudi Arabia.

sea in this area." Kuenen (1933, p. 65) also refers to floods covering a reef with silt without killing it. Reefs exist off the south coast of the Hawaiian Island of Molokai despite the muddy water. Maxwell (1970) found that the reef-covered surfaces off Queensland, Australia, are located mainly on what was previously a bathymetric elevation. He believed that this relationship was due to the greater effect of waves and currents on the shoaler areas, resulting in removal of mud and concentration of sand.

The general impression has been that corals and algae grow somewhat more actively on the outer edge of a reef, particularly on the windward side because of the large supply of food and the clear water. However, little evidence exists to show either that there is more food or that corals are growing more actively at this place than elsewhere. The studies of Odum and Odum (1955) on Eniwetok suggest that this outer reef may not obtain any net gain from the larger plankton that are swept across it. Zones of different types of corals extend in roughly parallel bands from the outer reef to the inner lagoon (Wells 1954). Mayor (1924) found that *Acropora* thrive at the reef margin, whereas *Porites* and other genera thrive along shores in rather silty water away from the breakers. The plankton tows in coral-reef areas indicate that copepods, a particularly important source of food for corals, are more abundant in lagoons than on the marginal reefs. Also, some of the most thriving reefs are found in lagoons. Where there is a good food supply in the lagoons inside barrier reefs, fringing reefs are developing actively and filling the lagoons so that the entire reef is evolving into a fringing reef. Furthermore, the actively growing coral knolls within the lagoons of many atolls (Fig. 12–2) indicate that these lagoons will be completely filled in a few thousand years unless conditions change. Figure 12–3 illustrates an atoll with a largely filled lagoon.

A helpful study of calcium-carbonate production in coral reefs is that of Chave, Smith, and Roy (1970). A few definitions are required to discuss this. In relation to reef growth, potential production is the amount of $CaCO_3$ produced by an individual organism per unit area of reef surface that is covered by that organism. Gross production is the amount of $CaCO_3$ produced per unit area of sea floor; the sum of the potential product of each organism times the proportion of sea floor covered by it. The net production is the rate of growth compared with $CaCO_3$ retention by the reef.

It is important to be able to estimate the gross production for the various environments of the reef. Chave, Smith, and Roy found that "Sand flats with green algae had—4×10^2 (g/m²)/yr; coral mounds on sand—10^4 (g/m²)/yr; complete coral coverage—6×10^4 (g/m²)/yr; algal ridges—9×10^3 (g/m²)/yr." From estimates of the relative areas covered by each of these associations, they suggest the following gross production figures: "Lagoon—7×10^3 (g/m²)/yr; reef flat—4×10^3 (g/m²)/yr; algal ridge—9×10^3 (g/m²)/yr; upper slope—6×10^4 (g/m²)/yr; and lower slope—8×10^3 (g/m²)/yr." They concluded that coral reefs "commonly have gross production rates in the neighborhood of 1×10^3 (g/m²)/yr."

If sea level is now rising about 1 mm/yr, as is indicated from tide-gauge records, the production is sufficient so that they can overproduce and hence expand the reefs aerially. Chave and others consider that during the maximum rise of sea level in the Holocene (about 1 cm/yr), the reef community structure must have been altered but probably could have kept pace with the rise and thus have prevented the reef from drowning.

Vaughan (1916) and Mayor (1924) estimated coral growth from laboratory studies that indicated they might grow upward as much as 5 to 10 cm per year. The work of Chave, Smith, and Roy shows that this is probably

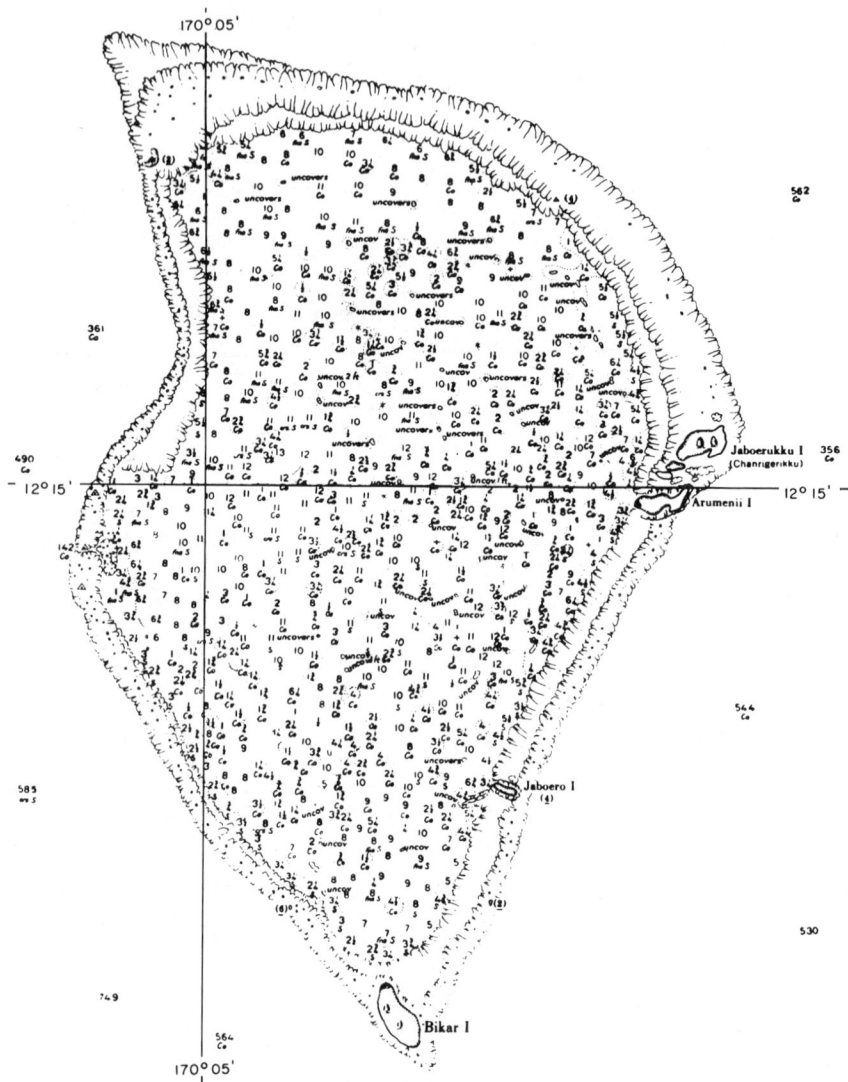

Fig. 12–3 Bikar Atoll of the Marshall Islands has been largely filled by coral growth. The shape of this atoll with its projecting corners is more typical than that of Eniwetok (Fig. 12–2).

excessive, but a rate of 1 cm/year seems probable where a well-covered coral surface exists with adequate food supply.

ATOLLS

General Character   Atolls are the most common type of coral reef. Most table reefs and coral banks are closely related to atolls because they usually have raised rims, although they do not extend to the surface. Including

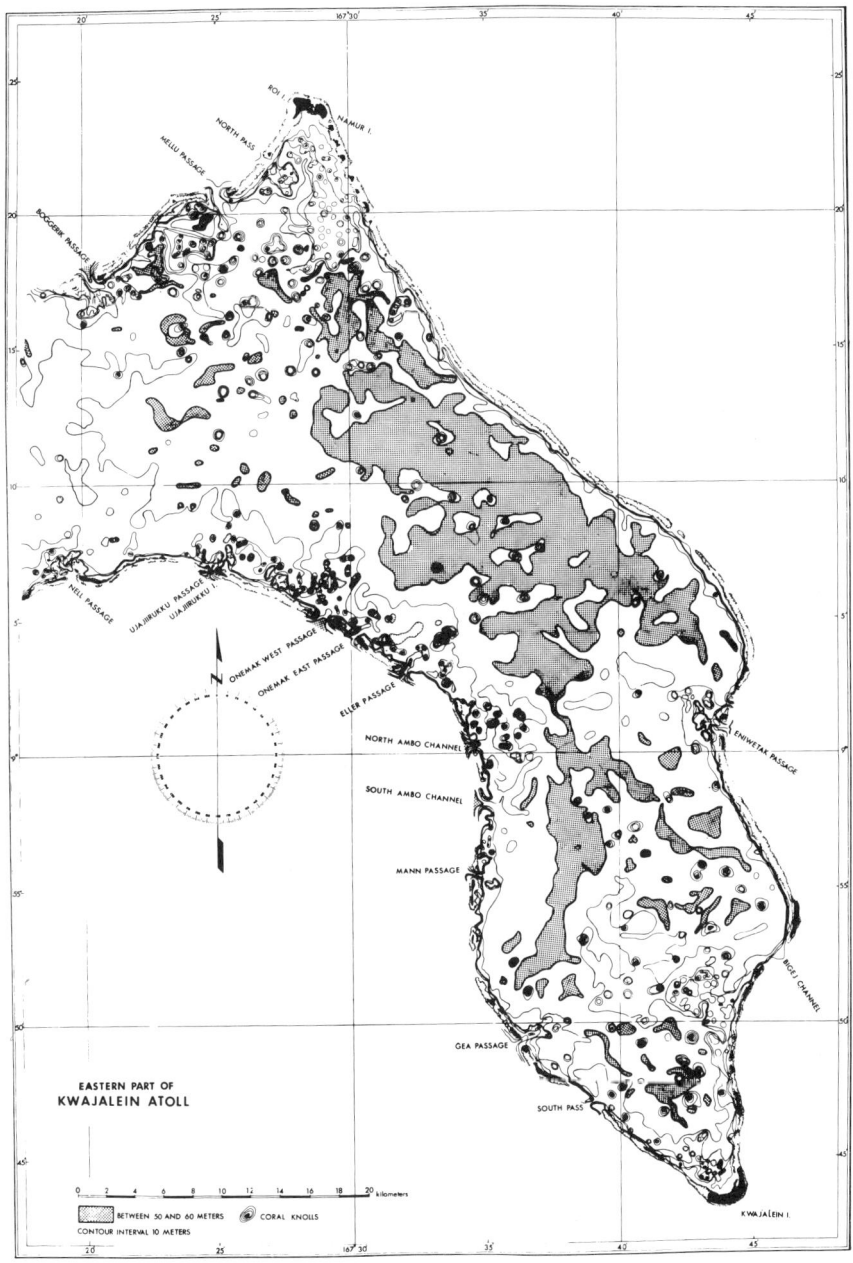

Fig. 12–4 Kwajalein Atoll of the Marshall Islands, the largest atoll in the world, has typical southeast spur. Most inhabitants of atolls live on islets in this position. From U.S. Oceanographic Office Chart 6027.

reefs with submerged rims, Bryan (1953) counted 400 atolls. Cloud (1958) counted 330 atolls exclusive of the banks and table reefs. More than half of the latter are in the Tuamotus (62), eastern Indonesia (37), the Carolines (32), the Marshalls (29), and the Fiji Islands (25). Only about 10 modern atolls lie outside the Indo-Pacific tropical area. These exceptions include Midway and Kure in the central Pacific, two atolls in the Red Sea, the Dry Tortugas (west of Key West, Florida), and Hogsty Reef in the Bahamas, and possibly Bermuda, a borderline case.

An impression prevailed for many years that atolls are circular platforms fringed by island-studded reefs with a flat-floored lagoon in the center. This description scarcely fits any of the hundreds of atolls. Actually, most atolls have surrounding reefs that are either elongate ellipses, rectangles, triangles, or irregular-shaped masses with several protruding corners. Numerous charts show that a southeast spur is particularly common (Fig. 12–4), and spurs are also usual on the northeast. Where both are present, a relatively straight reef usually lies between the spurs facing the trade winds. Marginal reefs are mostly awash with the tops exposed only at lowest tides, but low islands rise above much of the encircling reefs. These islands are concentrated in the southeast quadrant where many of the settlements are located, as at Kwajalein (Fig. 12–4). There are fewer islands on the leeward side of atolls, and reefs are somewhat wider to windward (Fig. 12–5). Most atolls that have

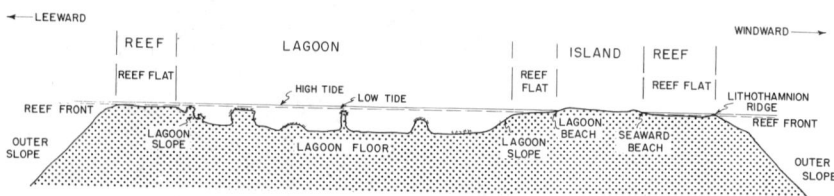

Fig. 12–5 Diagrammatic cross section of an atoll. After J. I. Tracey, Jr., U.S. Geological Survey; P. E. Cloud, University of California, Santa Barbara; and K. O. Emery, Woods Hole Oceanographic Institution. Vertical scale exaggerated.

been well surveyed have irregular floors (Fig. 12–2) with numerous coral knolls, which form shoals, and many depressions. The average lagoon depth, as far as determined, is 45 m, and it is unusual to find any lagoonal depths of more than 100 m. Figure 12–6 shows the average depth (in fathoms) of a group of lagoons in the Southwest Pacific. Two of them are only 3 fm (5.5 m) deep.

A cross section of a typical atoll is shown in Fig. 12–5. Bordering the steep talus slope leading to deep water is a growing reef often fringed by an algae ridge (called the *Lithothamnion ridge*, not a correct generic name) emerging above sea level (Fig. 12–7). The outer growing reef has an abundance of *Acropora*. A partly dead reef flat lies farther in, followed by a zone near the island where *Porites* is particularly common. A calcareous sand and gravel beach flanking the island consists largely of rubble but in some localities contains old slightly elevated reefs. A study of an inter-island reef on the windward side of Eniwetok (Odum and Odum 1955) showed that there are six zones represented in a traverse from the open ocean to the lagoon. These consist of (1) a windward buttress about half coral, (2) a coral-alga ridge, (3) an encrusting coral zone, (4) a zone of small coral heads, (5) a zone of large coral heads, and (6) a zone of sand and shingle carried in

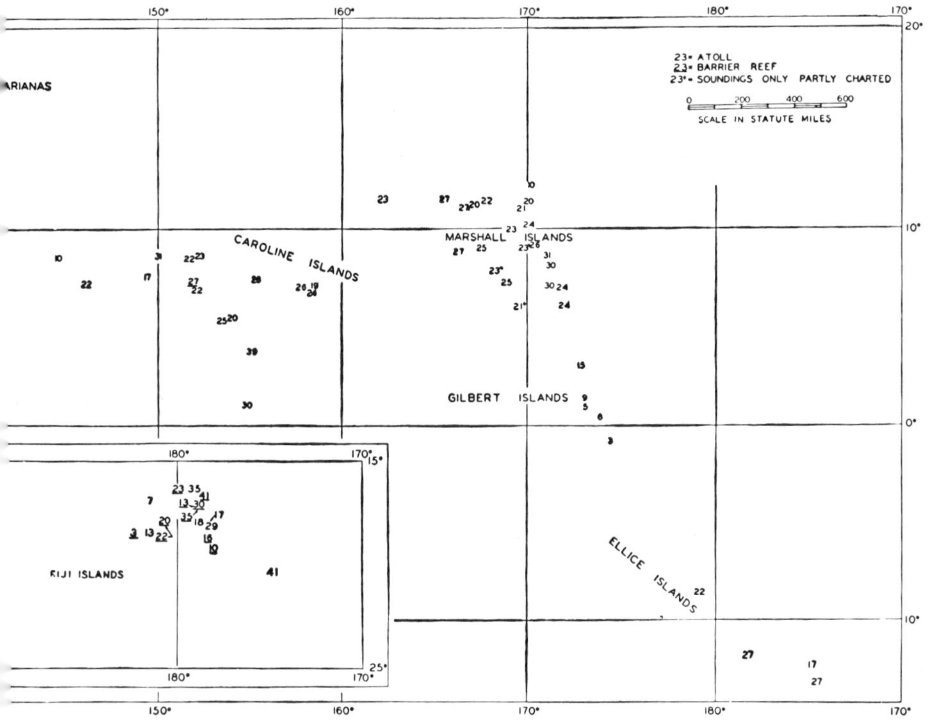

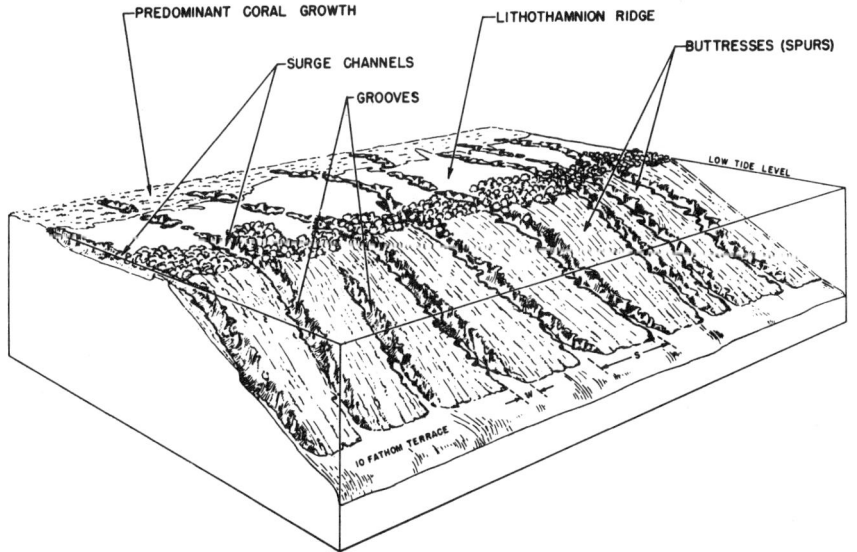

Fig. 12–6   Chart showing the average depths of the lagoons in the principal atoll area of the Southwest Pacific.

Fig. 12–7   Typical growth and erosional features of the reef front. From Munk and Sargent (1948).

by waves. On the lagoon side are another reef flat, a slope of live coral, a shallow terrace, and then a further descent to the sediment-covered lagoon floor.

The atolls in the Maldive Islands of the Indian Ocean differ from most of the others in having numerous faros, both in the lagoons and on the encircling reefs (Fig. 12–8). The Maldives are also somewhat rounder in outline than most other atolls. Canton Island, in the Phoenix group of the South Pacific, has another unusual feature, a series of ridges extending across the lagoon at right angles to the east–west elongation of the atoll. This lagoon is very shallow, mostly less than 5 m, and the western end is largely filled with coral.

An interesting characteristic of the outer margin of atolls was first reported by Hanzawa (1940) and has been seen in recent years by numerous geologists flying over them. The margin has numerous grooves, particularly on the windward side (Fig. 12–7). These may be related to wave action and have been explained by Munk and Sargent (1948) as representing the form with the greatest natural breakwater effect. Some of them, however, are apparently being covered over by coral growth, and, off Jamaica, their origin by ridge growth has been clearly demonstrated by Goreau (1959). In many places it is possible for a scuba diver to swim from a surge channel on the reef flat through a tunnel and come out on the reef face in one of the grooves. Although less common, grooves also occur on the front of other types of reefs (Shinn 1963).

The lagoon floors have been extensively sampled (Emery et al. 1954; McKee 1958). Adjacent to the inner margin of the reef platform, these authors found a belt with a predominance of either coral debris or foraminiferal sands, and next toward the center a belt with the debris of the alga *Halimeda,* then a second band of foraminifera; and in the deeper lagoons, coral mud. This order is greatly complicated by the irregularity of the floors. At Kaneohe Bay, Oahu, it is thought that no debris is carried into the lagoon from the marginal reef (K. E. Chave, personal communication).

Unusual Atolls   Two ocean island platforms, Bermuda and Johnston islands, greatly resemble atolls, although not usually so classified. The Bermuda platform, rising from the deep Atlantic, has a shallow lagoon with a submerged rim covered with calcareous sand on three sides and a group of sandy islands on the southeast. Johnston Island, southwest of the Hawaiian group, has a triangular platform with a raised rim that is well developed only on the northwest side, and has only isolated shoals along the other sides (Emery 1956B). The lack of a continuous marginal reef around the Johnston Island platform is explained by Emery as either caused by erosion of part of the rim during glacially lowered sea levels or by tilting that has submerged the south and northeast portions. It is unusual for an island and reef to be confined to the leeward side of a platform.

Bermuda's islands are entirely on the windward side. The land geology of Bermuda has been studied in detail (Sayles 1931; Land et al. 1967; Mackenzie 1969). Because there is no evidence of earth movements during the Pleistocene (Land et al. 1967), the islands appear to be an excellent place to determine eustatic sea-level changes, and the available information shows a close agreement with the center of the sea-level curves in Fig. 6–3A. The Bermuda hills, rising to 78 m, consist of eolianites and were formed, according to recent data, during high sea-level stands of interglacial episodes. This contradicts an earlier view (Sayles 1931) that the eolianites were formed during glacial low sea levels. However, Stanley and Swift (1967) found a sub-

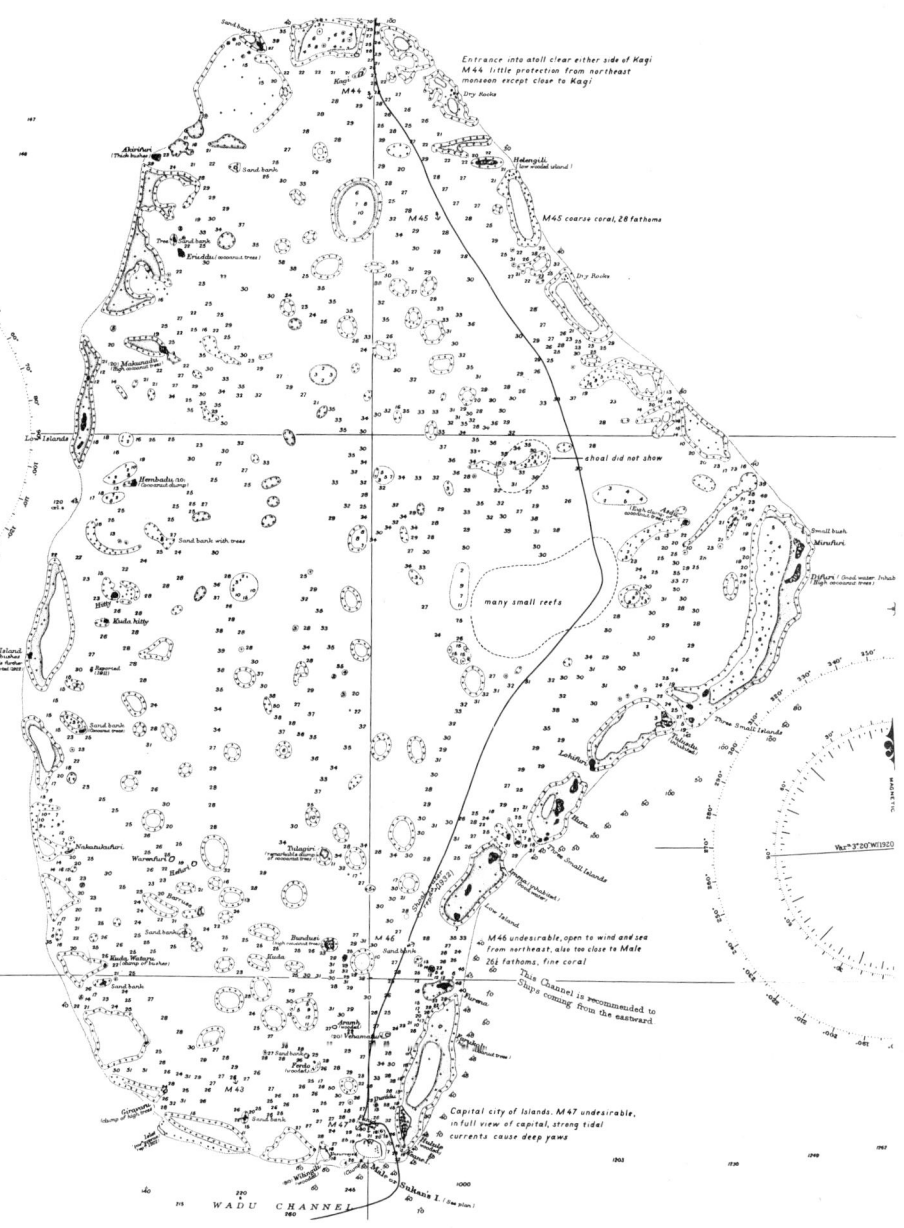

Fig. 12–8 Male Atoll in the Maldive Islands of the Indian Ocean. Numerous faros character-ize the lagoon and marginal reef. From U.S. Oceanographic Office Chart 5604.

merged dune at a depth of 18 m on the rim. Red soils on all the islands (Bricker and Mackenzie 1970) were produced by good water circulation during low stands.

All of the Bermuda Lagoon is shallow, less than 15 m, and partly filled with calcareous sand flats and coral knolls. The marginal reef, according to Stanley and Swift, has zones of calcareous sandstone, and they are inclined to believe that the reefs and the 18-m reef-front terrace are submerged ridges of windblown sand, rather than true coral reefs (Stanley 1970).

Drowned and Emerged Atolls   In unstable areas, as expected, some of the atolls have been uplifted, bringing the lagoon out of water as a bowl with a raised rim. Examples are Tinian, in the Mariana Islands, and some of the Lau group of southeast Fiji (Davis 1917). Drowned atolls are said to include some of the deep guyots (see p. 363), although there are few, if any, of these platforms with raised rims. Macclesfield Bank, in the South China Sea, has a completely submerged rim with depth of 13 to 25 m and a central area with average depths of 75 m, decidedly deeper than typical atoll lagoons. Therefore, it seems reasonable to assume that this is a submerged atoll. However, some of the coral banks with entirely submerged rims may not represent drowned atolls, because their central portions have typical atoll lagoon depths, suggesting that, rather than submergence, the margins have failed to grow upward apace with the rising postglacial sea levels, as appears to have happened among the atolls. Examples are Mogami and Gray Feather banks in the Caroline Islands (Shepard 1970). Another example referred to as a drowned atoll (Fairbridge and Stewart 1960) is Alexa Bank, northwest Fiji Islands, but the lagoon floor is shoaler than the average for atoll lagoons.

Slopes to Sea Floor Outside Atolls   Seaward of the atolls are generally narrow terraces, especially on the windward side. Locally, vertical slopes occur on the leeward sides. The slopes down to 450 m in the Marshall Islands average about 35° with gradual decrease at greater depths (Emery et al. 1954). In the Marshalls, the steepest slopes are around the projecting spurs. Kuenen (1933, p. 96) found the slopes around the East Indian (Indonesian) atolls are mostly more than 45° down to 200 m. Profiles in Micronesia taken in 1967 on the Scripps Institution ship *Horizon* showed slopes of about 30°. Almost all slopes around atolls are far steeper than the continental slopes. In fact, virtually the only continental slopes that can be compared with seaward margins of atolls are those off West Florida and east of Blake Plateau, where it is probable that upgrowing coral has been largely responsible.

Many of the atoll slopes extend precipitously down to the deep-ocean floor, but many others terminate on ridges or plateaus of intermediate depth. Notable examples of the latter are found off the northeast coast of Australia and in the atoll belts shown in Fig. 12–6.

The slopes generally have actively growing coral down to about 20 m, and sporadic growth below that. Talus material on the deeper slopes consists of coral blocks, *Halimeda* fragments much larger than those of the lagoons, and foraminifera with abundant shallow-water types that have slid down the steep slopes. At depths of about 1800 m, the debris grades into *Globigerina* ooze. Large blocks of coral are found mostly on the upper slopes and are rare below 180 m (Emery et al. 1954, pp. 70–73).

In a few places, the slopes have outcrops of lava and old sedimentary rocks. Probably most of these are exposed as the result of mass movements of sediments sliding from the steep slopes. Slides have also been suggested as an important cause of the indentations into the margins of atoll reefs (Fairbridge 1950A). Although this may be partially true, it was discovered that many projecting spurs extend as ridges to great depths, which could scarcely be explained as landslides. More likely the upward growth of the atoll followed a pattern established by the nature of the underlying base.

Geophysical Prospecting Seismic-refraction methods have been used to determine the character of the basement under the atolls at Bikini, Eniwetok, Kwajalein, and Midway (Raitt 1952, 1957; Shor 1964). Some of the results are indicated in Fig. 12–9. Rock with velocity of 4.15 km/sec occurs at depths of 1.6 to 1.8 km under a north–south section of Eniwetok, which is somewhat too deep for the 1267 m and 1405 m at which basalt was found in the two borings. Raitt explains this either by the assumption that the velocity of the second layer may have been increased by the presence of thin layers of consolidated material, or that the upper part of the volcanic material may have been highly fractured and hence showed no greater velocity than the overlying reef material.

Raitt's section across Sylvania Seamount and the adjacent Bikini Atoll shows an irregular surface of the volcanic rock (3.68–4.13 km/sec). Somewhat more than 1 km of supposed coral was found, enough to account for the failure of the Bikini drilling to penetrate to the basalt. In a northeast–southwest section across Bikini Atoll, velocities showed that volcanic rock comes to the sea floor on the side of the island; this was confirmed by dredging.

The geophysical surveys at Midway Lagoon (Shor 1964) indicated that the underlying reef was thicker on the north side than to the south. This was found to be the case in the two drillings, but volcanic rock was penetrated at a much shoaler depth than had been anticipated by the survey. This is an example of the great difficulty in obtaining good seismic results in coral.

Borings into Atolls Charles Darwin was the first to urge that a well be drilled into an atoll to determine the origin of the great masses of coral that lie more or less isolated and rise above the deep ocean with no direct evidence of the basement on which they must have been built. He had hoped that some millionaire would provide the funds; actually, various governmental agencies, following recommendations of scientific societies, have supported most of the drillings. To date, seven atolls have been drilled, including two slightly elevated (Fig. 12–10). Volcanic basement was encountered in two holes in Eniwetok and Midway, one in Mururoa, and several holes in Bermuda. In 1897 and 1898, borings were made into Funafuti Atoll (Fig. 12–10) in the Ellice Islands by the Royal Society and the government of New South Wales. The main hole passed through 335 m of coral reef material, which is described by Hinde (1904) as follows:

1. 0–45 m, corals growing in place and surrounded by foraminifera and other organisms.
2. 45–228 m, largely fragmental material with a small percent of coral and consisting largely of foraminifera and organic debris.
3. 228–340 m, a mass similar to that from 0–45 m except that it was dolomitized.

Hinde considered that the entire section represented continuous deposition of reef material. Subsequent study of the cores (Grimsdale 1952) has shown that the 171- to 235-m zone consists of submarine talus. Hinde found that a 21-m boring in the lagoon of Funafuti Atoll had cores consisting entirely of uncemented joints of *Halimeda* akin to the present lagoonal sediment; but a short section below showed rubble limestone with coral and foraminifera similar to the upper part of the main boring. There is no certainty that the

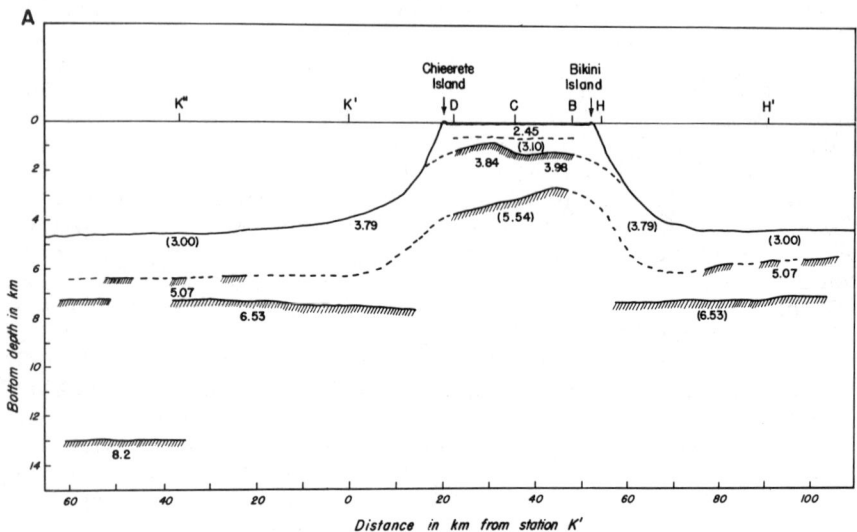

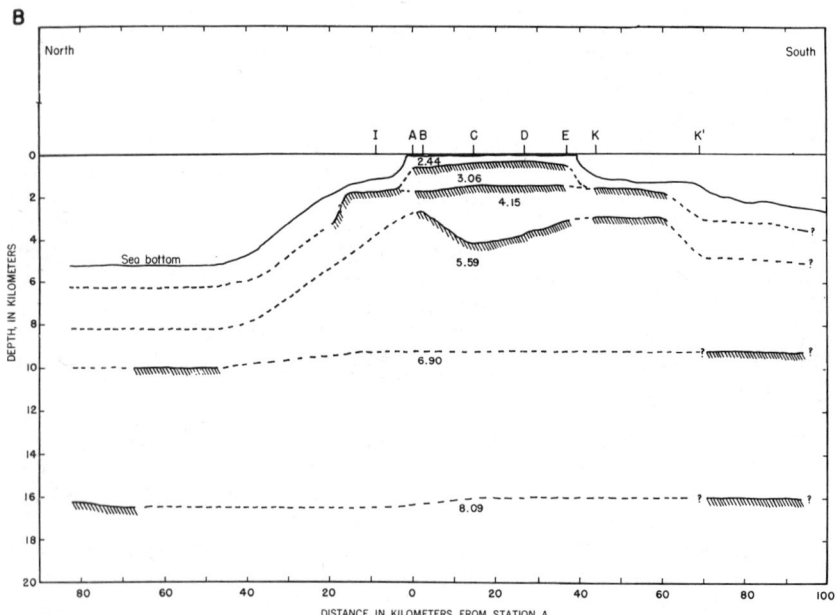

Fig. 12–9 A, velocity of sound in the substructure of Bikini Atoll. Velocities given in kilometers per second. From Raitt (1952). B, cross section beneath Eniwetok Atoll. Sound velocities in kilometers per second. After Raitt (1957).

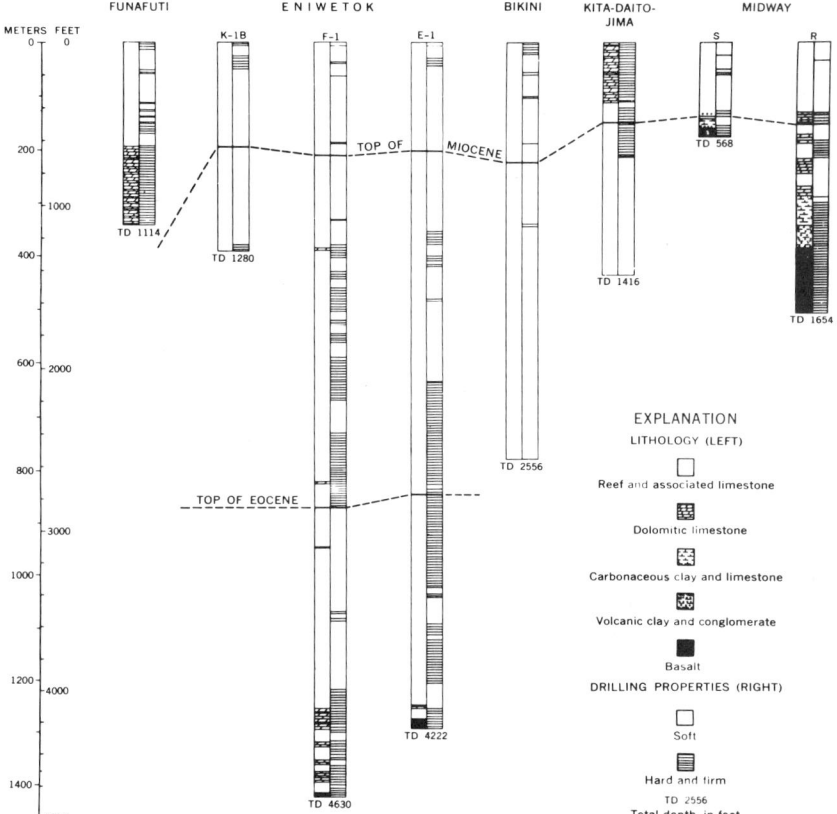

Fig. 12–10 Results of drillings into various atolls of the Pacific. From Ladd et al. (1970).

main boring penetrated beyond Pleistocene, but Pliocene and Pleistocene are difficult to differentiate by the foraminifera, the chief index fossils.

A Bermuda boring (Pirsson 1914) penetrated through the coral cap at 75 m, where it entered a weathered volcanic rock extending to 139 m, below which there is a volcanic sand and gravel zone traced to 171 m, where basaltic lava was encountered. Borings penetrated to 432 m in the elevated atoll, Borodino Island (Kito–Daito–Jima), 330 km east of Okinawa (Hanzawa 1940). This well was drilled through 213 m of hard limestone and encountered Miocene at 152 m. A boring into Maratoa, a small elevated atoll island on the shelf northeast of Borneo, extended 427 m, all in reef material suggestive of lagoonal fill (Kuenen 1947).

In 1947, four holes were drilled in Bikini Island by the Navy in co-operation with the U.S. Geological Survey (Emery et al. 1954, pp. 80–85). One of these reached a depth of 780 m (Fig. 12–10), and all of this appeared to represent a shallow-water lagoonal or near-reef environment. Some of the sections at various depths probably consist of what was a growing near-surface reef. According to Todd and Post (1954), shallow-water foraminifera were found throughout the section. No *Globigerina* oozes or exclusively deep-water organisms were found to indicate that an environment such as the deep outer slope was involved. Miocene rock was reached at about 259 m, and, in the deepest hole, Oligocene was encountered at 610 m.

Eniwetok, also in the Marshall Islands, is the most drilled of any atoll. From 1950 to 1952, a total of 17 shallow holes were drilled in six of the islands and three deep holes. In two of the latter, from islands on the north and southeast sides of the rim, basalt was encountered at 1267 m (in E 1) and 1405 m (in F 1). Most of the sedimentary sections were of soft lime material or weakly consolidated rock and probably represent largely shallow-water lagoonal material. However, in hole F 1 below 884 m, 330 m was drilled in *Globigerina*-rich limestone, indicating relatively deep-water conditions. One of the shallow holes was mostly through solid rock to 40 m and appears to have penetrated reef framework. In the deep holes, Miocene was recovered at 200 m, and Eocene in the two deepest at about 850 m, somewhat deeper in F 1. Thus there are three indications that the basement at hole F 1, on the north side of the atoll, was sinking faster than at the other deep hole on the southeast. Dolomite was cored at various depths. Weathered limestone at several depths indicates temporary emergence of the reefs; this is supported by finding fossil land snails (Ladd 1958).

According to Ladd and others (1970), in 1965, two holes drilled in Midway Atoll penetrated the reef to the underlying basalt. One in Sand Island, well inside the marginal reef, reached an alkalitic basalt at 157 m, and the other just inside the marginal reef reached basalt at 384 m. The drillings show that the old underlying volcano was weathered and partly truncated by wave erosion. After the volcano sank, it was first covered with reworked clays and volcanic conglomerates, and these in turn were covered by reef limestones. The sinking of the island was interrupted by three periods of emergence, and apparently some of the uplifts were sufficient to produce relatively high islands. As a result of these emergences, the limestone was leached and re-crystallized. Dolomitic sections occur at irregular intervals.

Mururoa Atoll, in the Tuamotu group, was drilled at about the same time as Midway (Chauveau et al. 1967). Here, basalt was recovered at about 400 m. The overlying limestone has been dated as more than 100,000 B.P., from a depth of 7 m, and the dates below 19 m were found to range from 200,000 to 500,000 B.P.

In the Bikini, Eniwetok, and Midway drillings, large amounts of soft rock were encountered, as was also the case at Funafuti. Much of it was poorly cemented coral sand. Large caverns caused difficulties with loss of tools. Dolomites at different depths in all the holes (Fig. 12–10) indicated no regularity in the limestone-to-dolomite sequence.

ISLAND BARRIER REEFS

Many tropical islands in the Southwest Pacific are surrounded or partly encircled by barrier reefs separated from the volcanic islands by a lagoon. Among these the best known are Tahiti and Moorea (Crossland 1928), Tutuila, Samoa (Chamberlin 1924; Bramlette 1926), Truk, in the Caroline Islands (Curray et al. 1970; Shepard 1970), and Bora Bora (Davis 1925; Guilcher et al. 1969). The continental island, New Caledonia, is also almost completely surrounded by a barrier reef (Guilcher 1965B).

The barrier lagoon at Truk is the most completely surveyed (Fig. 12–11). Wartime activities of the Japanese are responsible for most of the soundings, and the Scripps Institution CARMARSEL Expedition added many sounding

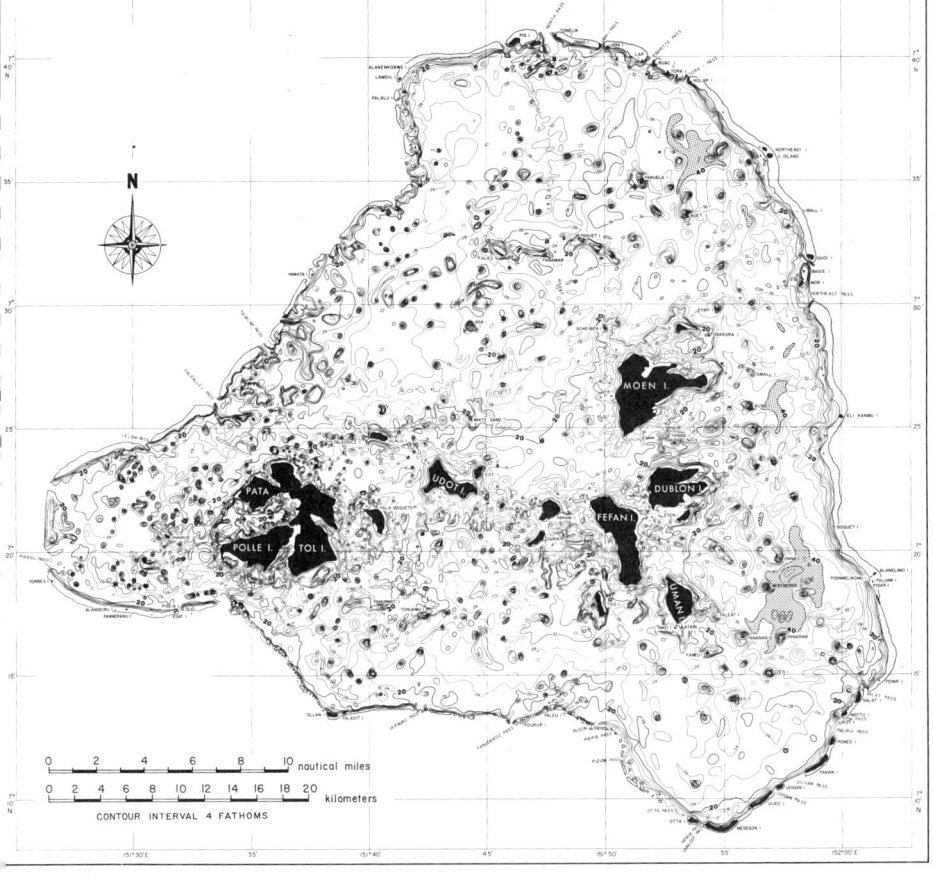

Fig. 12–11   Truk Lagoon and its high islands (black). Based on Japanese cartography and surveys by Scripps Institution. Contours by author.

lines in critical areas (Shepard 1970). The Truk Islands are old, extensively eroded volcanoes, and the lagoon shows great similarity to those inside typical atolls. The depths are close to the 45-m average for atolls, and the topography has the same abundance of knolls and basin depressions. As far as is known, the sediments are largely *Halimeda* debris with fragments of coral surrounding the knolls and adjacent to the marginal reef. Scattered reef islands rise above the reef flats and prevail on the eastern half of the barrier, as is also true of atoll reefs.

The reef of New Caledonia surrounds a high and continuous continental island, and the lagoon could therefore be expected to receive much more terrigenous sediment. However, Guilcher (1965B) found that the lagoonal sediments were at least 87 percent calcium carbonate. The average width of the lagoon is 15 km. On the northeast side, lagoonal depths are similar to those of atolls. On the southwest, the depths are generally shoaler, and in many areas the lagoon has been largely filled, leaving a broad reef flat.

The lagoons inside the barrier reefs of Tahiti and Moorea are mostly less than 1 km in width and very shallow, except on the east side of Tahiti where the depths are typical of atolls.

The lagoon inside the barrier reef of the volcanic island of Bora Bora averages about 2 km in width and has typical atoll depths, but the floor is smoother and has fewer irregularities. The reef islands are more common on the northeast side, but the reef flat is wider on the south and southwest. The lagoonal sediments have only very small percentages of terrigenous material. However, the sediments in the Tahiti Lagoon have high percentages of volcanic debris (Guilcher et al. 1969).

TERRACES AND BASINS OF ATOLL AND ISLAND BARRIER LAGOONS

Scuba divers have been impressed with the terraces both inside and outside the reefs of atolls and barriers. A general impression is that these

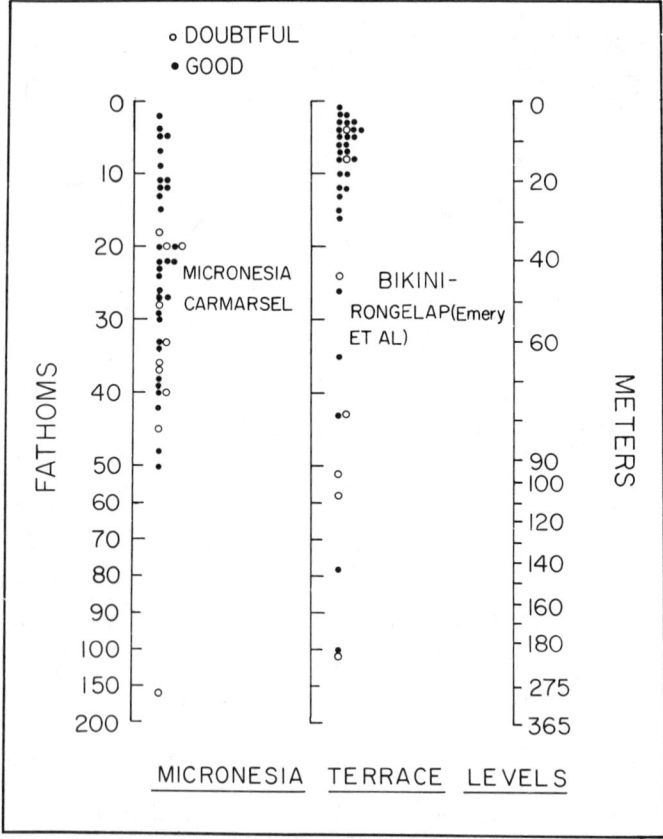

Fig. 12–12 Terrace levels in coral lagoons of Micronesia. Note absence of supposed 18-m terrace. Left, from Shepard (1970); right, from Emery et al. (1954).

terraces are from 15 to 18 m in depth (see, for example, Ladd et al. 1950). Terraces at this depth are common on continental shelves. During the CARMARSEL Expedition, an attempt was made to look for such terraces in the lagoons. Echo sounding profiles failed to confirm any concentration of terraces at this depth (Shepard 1970). Study of profiles from Bikini and Rongelap lagoons (Emery et al. 1954) showed the same result (Fig. 12–12). At least from these fathograms, one would conclude that the so-called 10-fathom terrace is quite rare. Scientists probably have been referring to terraces of variable but shallow depths as the "10-fathom terrace." Actually, it would be surprising if terraces that developed when the sea level stood for some time at approximately 15 to 18 m below the present should have remained uncovered by coral growth in the 7,000± years since the sea stood at that depth (Fig. 6–3). It is more likely that the present-day concentration of coral growth near 15 to 18 m has produced some flattening of the seaward slope at about these depths. More careful study should be made of reef terraces before generalizations are warranted. It is important to determine if the terraces are due to reef growth, talus accumulation, or wave erosion. Observations during the CARMARSEL Expedition suggested that all three types of terraces exist and that the levels change radically when a terrace is followed along its strike.

The abundant oval or round depressions found in most coral lagoons may have more than one origin. It seems improbable that most of them are the result of upgrowth of coral above a former flat platform. If this were true, the bottom of the depressions should be at the same level, and no such level has been found in the numerous sonic profiles that cross the lagoons. The irregular depths and the round or oval shape of the depressions are more reasonably explained as the result of sinkholes that developed during stages of low sea level. The importance of solution has been suggested by various scientists (see, for example, MacNeil 1954).

ORIGIN OF ATOLLS AND ISLAND BARRIER REEFS

Since the time of Darwin, the origin of atolls and barrier reefs has been of great interest to geologists and biologists. The knowledge that, aside from the unusual ahermatypic corals, reef growth is limited to shallow water indicates that virtually all reefs originated on shallow platforms, presumably in the case of oceanic atolls on the sides or tops of volcanoes. As far as is known, all oceanic islands are volcanic. (New Zealand, Fiji Islands, New Caledonia, and others in the vicinity are considered continental.) The existence of more than a hundred atolls in several adjoining groups of the Southwest Pacific with no exposed volcanic rock is particularly puzzling.

Darwin (1842) proposed his hypothesis as an outgrowth of his extended voyage on H.M.S. *Beagle*. As illustrated in Fig. 12–13, the hypothesis is remarkably simple, requiring first the formation of fringing reefs on the sides of volcanic islands or along the margins of other islands and, second, the gradual subsidence of the land with the upgrowth of corals along the outer margin keeping pace with subsidence. As the island sank, the outer margin of the fringing reef would outgrow the inner reef, producing a barrier and a lagoon. Then eventually, if the entire island sank, the reef would form an atoll. It has often been overlooked that Darwin postulated other ways in which atolls might form. He was the first to suggest that some atolls could have formed on flat, subsiding platforms and have grown upward into atolls without passing through the barrier-reef stage. He further suggested that a

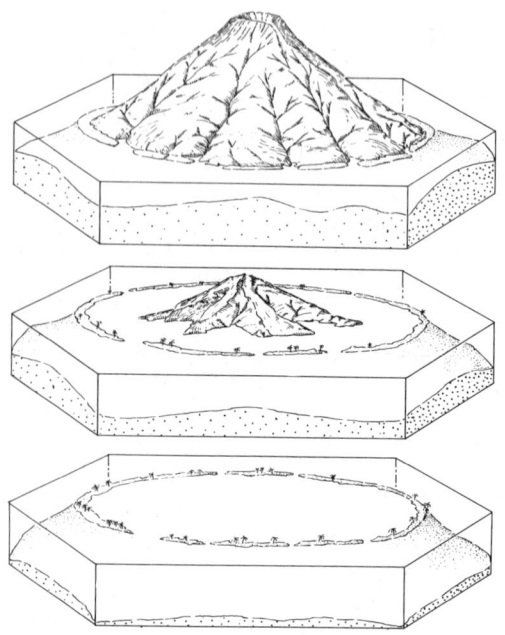

Fig. 12–13 The Darwin hypothesis of the submergence of a volcano and the upgrowth of the fringing reefs to form successively a barrier reef and then an atoll. Drawing by D. B. Sayner.

shallow bank might, by simple growth of coral and without any subsidence, "produce a structure scarcely to be distinguished from a true atoll." It is now generally recognized that the atolls of the northern Marshalls originated in part on relatively flat-topped banks and were converted into atolls by subsidence without going through a barrier-reef stage; thus, Darwin anticipated the findings of modern studies. The Darwin hypothesis was supported among others by Dana (1885) and Davis (1928), who both called attention to the numerous embayments into volcanic islands inside barrier reefs as evidence for submergence. Some of these embayments, however, are probably the result of volcanic collapse; and others could be due simply to submergence of river valleys resulting from the rise in sea level at the end of the last glacial stage.

Daly (1910, 1915) first emphasized the importance of glacial control of sea level in the evolution of coral reefs, although the idea of glacial control was first suggested by Penck (1894). Daly considered that the lowering of sea level was accompanied by colder water that killed many of the reefs and hence allowed erosion of the volcanic islands previously protected by their coral fringes. He also thought that the lowered sea level led to the development of muddy water, because it exposed the muds of the outer shelf and lower slope to wave turbulence. This also would tend to eliminate the corals because they live better in clear water. During these epochs of low sea level, the reef-free volcanic shores would have been eroded by the waves; and wherever the islands consisted of soft volcanic ejecta, the erosion might have been quite extensive. Daly thought that in this way one could account much better for the flatness and the shallow nature of the lagoons than by submergence. It is now clear that he greatly overestimated the lagoonal flatness, but the hypothesis might help explain the relatively shallow nature of many lagoons, which has been an embarrassment for the advocates of the Darwin hypothesis.

Davis believed that some of the islands in the marginal coral belt probably had their reefs killed during glacial stages, but he argued that most of the island shores inside lagoons were not cliffed as they would have been if the reefs were killed during low sea-level stages. This appears to be a good observation and tends to negate the glacial-control hypothesis.

Neither Daly nor Davis seemed to have realized fully the importance of reef building and growth of *Halimeda* within the lagoons or the importance of wave action in providing sediment from the growing rim to help fill the lagoons. It is, however, rather surprising that submergence has not resulted in more deep lagoons because of the more rapid growth around the margins.

Daly also failed to appreciate the vast size of many of the atolls. It would not take long for open-ocean waves acting on a cinder cone to produce a wave-cut platform a kilometer or more in width, but it is scarcely credible that volcanoes with diameters of 40 to 55 km could have been truncated completely during glacial stages. It is as if the island of Hawaii were truncated by waves during a fraction of a million years. Surely no one familiar with the relatively slow attack of waves on most rock coasts would expect anything of the sort to happen.

The final test of the two hypotheses has been the borings. Daly (1948) believed that the Funafuti borings and even the Bikini borings had penetrated only the coral talus slopes on the sides of the volcanoes. This position was not well supported because the detailed study of the cores in both cases revealed mostly shallow-water organisms and almost none of the typical talus aspects; the core material was much more suggestive of lagoonal deposits than of the outer reefs, although both were found. However, the borings at Eniwetok seem to have proven definitely that there is no shallow wave-beveled platform under that island, and the geophysical measurements on several of the islands have shown clearly that coral lies far too deep below the surface to fit the Daly hypothesis.

Thus, the evidence favors the ideas proposed by Darwin. It is remarkable that the actual sequence of events that led to the development of atolls such as Eniwetok and Bikini were anticipated by this great scientist. Probably many of the volcanoes had been considerably truncated by erosion before extensive submergence set in. Guyots (flat-topped seamounts described in Chap. 13) form the base of several of the atolls, and the coral has grown up either on a portion of the flat surface or on what appears to be an eroded volcanic peak rising somewhat above the guyot (Fig. 12–14).

Effects of Glacially Controlled Sea Levels Even if Daly's hypothesis of wave-beveled platforms formed during glacial stages of low sea level is not the principal explanation of atolls and barrier reefs, undoubtedly the glacially controlled sea levels have had important effects on reef development. Some wave erosion may have occurred as the result of coral reefs being killed in the marginal zone by cooler water temperatures, but there is little evidence that the reefs in the Pacific were appreciably affected. The apparent unconformity in the Eniwetok borings at about 12 m (Thurber et al. 1965) between postglacial and interglacial coral suggests an erosion interval, although it may be only a time of nondeposition. Solution of the elevated reefs during times of low sea level is probably the greatest effect. This was first suggested by Hoffmeister and Ladd (1935), although Murray (1880) had suggested that solution was responsible for the lagoons without realizing that glacially lowered sea levels would have made this possible. MacNeil (1954) called

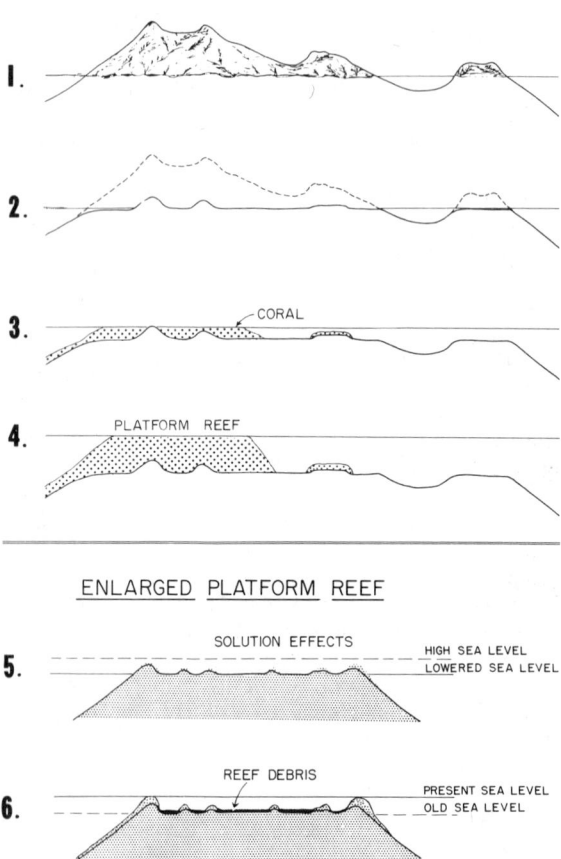

Fig. 12–14   Suggested stages of development of an atoll such as Eniwetok.

attention to the solution in limestone islands that has often produced "basin-shaped platforms with prominent raised rims," and thought this could account for the morphology of atolls.

A controversy has occurred concerning the existence of a postglacial high stand of sea level, said to have risen as much as 3 m above the present levels, as for example, 1 m at Midway (Ladd et al. 1970). The CARMARSEL Expedition to Micronesia (Curray et al. 1970; Newell and Bloom 1970) was conducted largely to test the possible existence of raised reefs in this supposedly stable area of the Pacific. Visits to 33 islands failed to show examples of reef rock that grew in place and is now above low-tide level. Instead, we found flat-topped, somewhat cemented rubble masses at approximately high-tide level. It seems unlikely that any of these 33 islands had experienced positive stands of the sea in postglacial time. It is possible for corals such as the blue *Heliopora* and *Acropora* to live in tide pools slightly above low-tide level. Present evidence appears to be against high postglacial sea levels, but some indications, notably in Midway and in Brazil (van Andel and Laborel 1964), leave the question open for further study, particularly by blasting into reef platforms in stable areas.

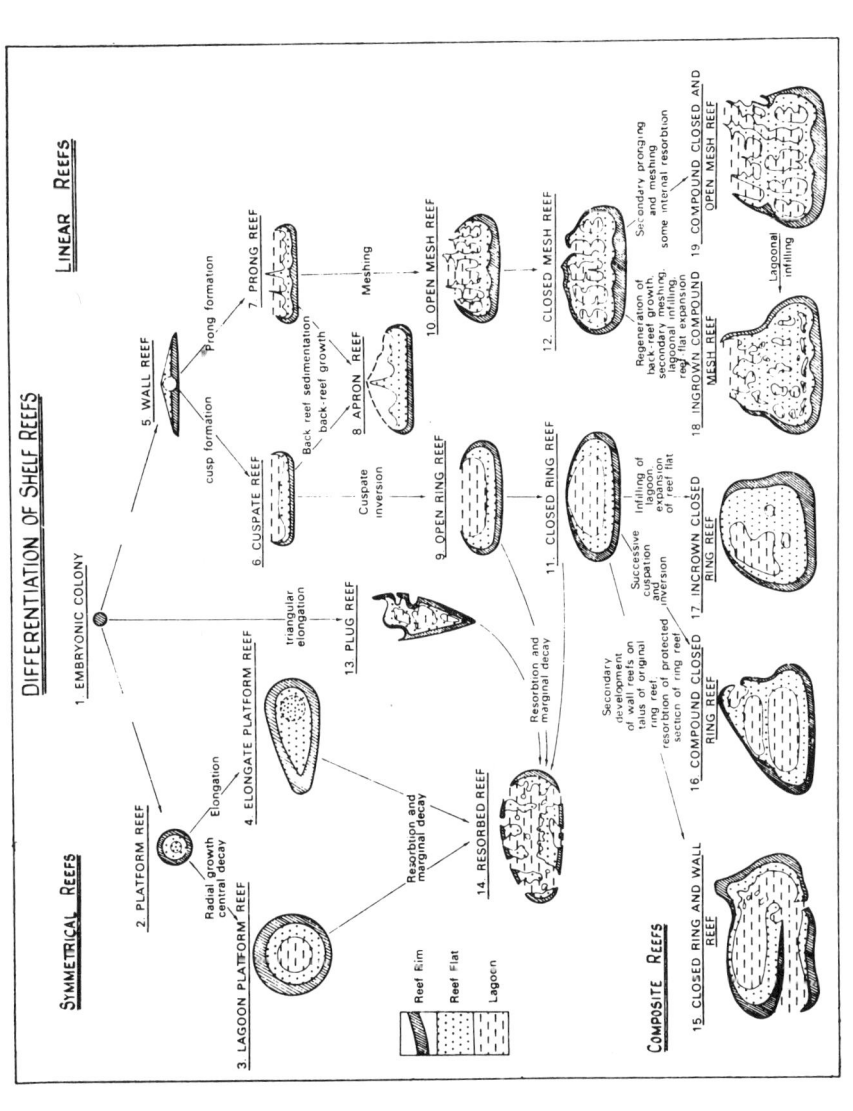

Fig. 12—5 · Numerous shapes of reef patches on the Great Barrier Reef of Australia. From W. Maxwell (1970).

QUEENSLAND GREAT BARRIER REEF AND OTHER SHELF REEFS

The shelves and platforms on the east sides of continents in tropical areas have an abundance of reefs. The best known of these are the Great Barrier Reef of Queensland, Australia, and the Bahama reefs off the Florida coast. An extensive barrier reef follows almost the entire coast of British Honduras, and a scattering of submerged patch reefs are growing on the shelf north of Rio de Janeiro.

The Great Barrier Reef is the largest in the world and has the finest coral growths, although large areas are now threatened by the crown-of-thorns starfish. This region, most completely described by Maxwell (1968), extends 1600 km along the northeast side of Australia from 10 to 24°S latitude. In the southern portion, the reef growth is largely confined to the outer shelf, but north of 16°, reefs are scattered across the shelf and form an almost continuous barrier along the outer edge. Similar continuous outer barriers are found in some of the southern portions, notably at Hardline Reefs 20 to 21°S latitude. The reefs inside these barriers are mostly ring-shaped faros, oval-shaped coral banks, and fringing reefs around small coral rubble islands. The large variety of shapes among these reefs is illustrated in Fig. 12–15. According to Maxwell (1970), the composite forms result "from changing growth patterns caused by subsequent eustatic fluctuations, hydrological variation, or bathymetric change." He also discovered that tidal deltas develop among the reefs. The large tidal range in the southern portion of the reef has an important effect, and the unusually strong currents transport great quantities of coral debris. He suggests that some channels inside the southern reef zones are the result of old delta distributaries (Maxwell 1968, Fig. 22B).

Borings into the Great Barrier Reef were made at Michaelmas Cay, Heron Island, and Wreck Island (Jones 1966). At Michaelmas Cay, the drill passed through 157 m of reef material and then penetrated quartz sand with glauconite, shell fragments, and foraminifera. At Heron Island, coralline material extended to 166 m, where quartz sand was obtained. In Wreck Island, reef material was cored down to 130 m. This is underlain by marine Pleisto-

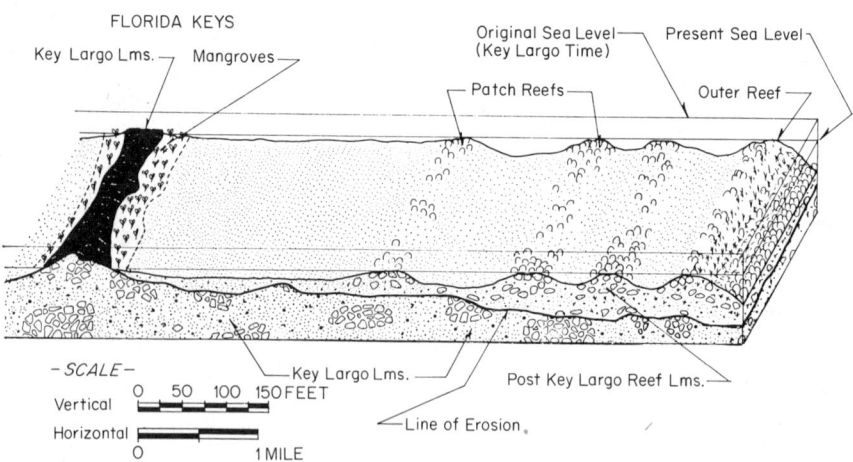

Fig. 12–16 The relation of reef growth to sea-level changes in the Florida Keys. From Hoffmeister and Multer (1968).

cene and Tertiary, which is in turn underlain by volcanic tuffs at 547 m. The entire coral-bearing layer is generally believed to be of Holocene age, although Jones thought this indicated that the shelf had subsided because active coral growth has depth limits less than the thickness of coralline sediments. This is not necessarily true, because sea level was rising rapidly during most of the Holocene.

The Bahama reefs have been studied principally by Newell (1955),

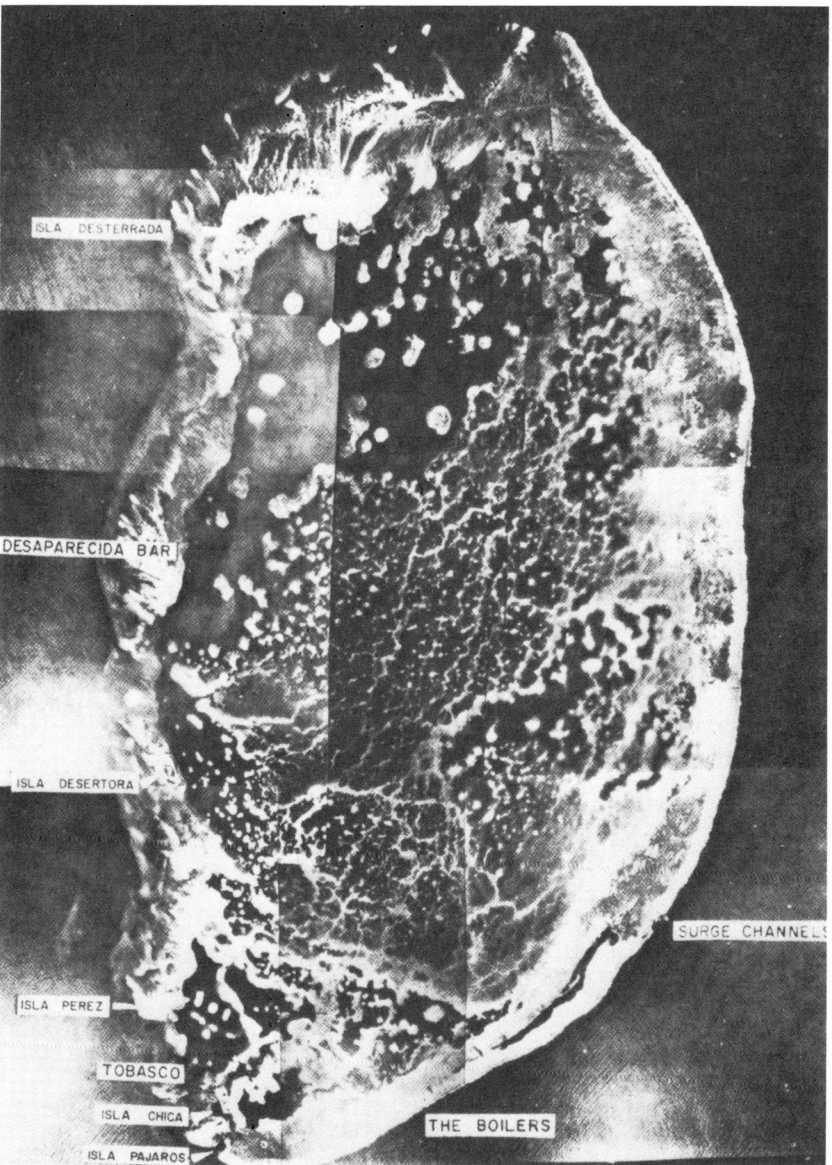

Fig. 12–17 Aerial view of Alacran Reef north of Yucatan. From Kornicker and Boyd (1962).

Newell and others (1951, 1959), and Storr (1964). The somewhat detached Bahama Banks resemble the rectangular type of atoll. Coral growth has concentrated primarily around the rims of these banks, leaving depressed lagoons in the center. Newell and other scientists believe that the almost vertical escarpments at the margins are the result of upgrowth of corals that continued at the edge of the platforms as they subsided through long periods. However, those reefs seem to have been largely exterminated, but new reefs have formed around the bank margins. The sediments of the interior of the banks are mosty calcareous oölites and aragonite-pellet mud with very little reef material. Deep borings into Andros Island apparently have encountered little, if any, coral-reef material (Goodell and Garman 1969).

The Florida Keys, located several kilometers seaward of the Florida coral-reef platform, include islands that are either elevated Pleistocene coral reefs or oölitic limestones. According to Hoffmeister and Multer (1968), the Keys were formed as patch reefs in a back-reef area with an outer reef that has not been eroded and covered by more recent reef material. The present reefs seaward of the islands consist of rows of patch reefs (Fig. 12–16). Shinn (1963) studied the spurs and grooves on the outer slope of the growing reef. He found that the great fronds of *Acropora palmata* on the spurs have grown toward the direction facing the prevailing seas. This greatly strengthened the colony against wave attack, even during hurricanes. Growth here continues until the corals reach the surface and become overcrowded and partly covered with calcareous algae and *Millepora*.

Platform reefs grow locally on the shelf off eastern Mexico, as at Isla de Lobos off Veracruz (Rigby and McIntire 1967). The Alacran Reef, a shallow lagoon faro (Fig. 12–17) on the Campeche shelf north of Yucatan, is elliptical with a 22-km-long axis (Kornicker and Boyd 1962). It rises above a 50-m-deep limestone shelf. The lagoon has abundant small patch reefs. A few small sand islands on the marginal reef are destroyed and rebuilt annually. Reef corals are flourishing, and frame-building coral is actively building around the margins. Encrusting algae are contributing to the stability of the windward reef.

Chapter 13
DEEP-OCEAN FLOOR
TOPOGRAPHY

INTRODUCTION

The deep oceans are separated from the continental terraces at the approximate boundary between the area of the thick continental crust (avg. 35 km) and the thin oceanic crust (avg. 6.5 km). Actually there are relatively small zones seaward of this boundary which have continental crust or crust that is intermediate between continental and oceanic, but here these are included with the oceanic provinces.

During the 1960s, extensive exploration of the deep oceans continued. The topographic mapping was concentrated in the Indian Ocean, bringing that area somewhat on a par with the well-charted Atlantic and Pacific. One of the great accomplishments of the 1960s has been the geophysical investigations that have now led to a reasonable explanation of the major relief features of the oceans (see Chap. 5). Of equal importance has been the series of drillings into the deep-ocean floor (JOIDES, discussed in Chap. 14). These

are extending down for as much as 1000 m and yielding stratigraphic information that is helping to revolutionize the entire science of geology and is
apparently establishing the sea-floor-spreading plate-tectonics hypothesis.

Relative to sea-floor topography, the greatest advances of the past
decade have been in locating the numerous fracture zones crossing the
oceanic ridges and rises of the Atlantic and Indian oceans. These are similar
to those of the northeast Pacific, which had been located previously by H. W.
Menard and his colleagues but were not at first known to be related to spreading ridges. These features are graphically indicated on the Heezen and Tharp
relief maps, published by the Geological Society of America and in small
revised form by the *National Geographic* (Oct. 1967, June 1968, Oct. 1969). In
using these maps, it should be noted that they are generalized and the actual
number of transverse fracture zones are actually not known and may not be
so numerous as indicated. A more accurate representation of the fracture
zones is given for the North Pacific in the 1970 maps of Chase, Menard, and
Mammerickx. However, these maps also generalize somewhat, especially for
the smaller hills shown as patterns. The detailed charting of portions of the
Indian Ocean ridges (Laughton et al. 1970B; Fisher et al. 1971) has added
much helpful new information. Additional valuable material was published in
The Sea, Volume 4 (Maxwell 1970).

Among other important developments of the 1960s have been the
great increase in deep-ocean bottom photography (Hersey 1967; Heezen and
Hollister 1971). These photographs include an amazing number of ripple-
marked surfaces, especially on the seamounts, but also in deep passageways
and in fault valleys transverse to the deep ridges.

TOPOGRAPHIC FEATURES DEFINED

The results of deliberations by an international committee (Wiseman
and Ovey 1953) led to a considerable degree of uniformity in the nomenclature of the topographic features of the ocean floor. In a few cases the rulings
of the committee appear to have been subsequently disregarded by marine
geologists and certain terms have become rather well established, although
they were not approved by the committee. The following terms and definitions given alphabetically are, therefore, not entirely in accord with the
committee report:

> *Abyssal plain*   A very flat surface found in many of the deep-ocean basins or
> adjacent seas.
> *Archipelagic apron*   A fan-shaped slope around an ocean island, differing from
> deep-sea fans in having little, if any, sediment cover.
> *Basin*   A depression on the sea floor that is more or less equidimensional.
> *Continental rise*   A gentle even slope at the base of the steeper continental
> slope.
> *Deep*   The well-defined deepest part of a depression, not used unless depths
> are in excess of 5500 m.
> *Deep-sea channel*   An elongate valley that cuts slightly below the surface of
> many of the deep-sea fans and may extend out to the basin floor.
> *Deep-sea fan*   A gently sloping sediment-covered, fan-shaped plain that bor-
> ders the continental slopes in many places. The fans also occur in basins
> of intermediate depth on continental borderlands.
> *Deep-sea terrace*   A benchlike feature bordering an elevation of the deep-sea
> floor at depths generally greater than 550 m.
> *Fracture zones*   Ridges and narrow troughs extending at right angles to the

mid-ocean ridge system and displacing their crests. These fracture zones also displace the magnetic bands that are parallel to the mid-ocean ridges.

Gap   A steep-sided furrow that cuts transversely across a ridge or rise.

Guyot   A flat-topped seamount, also called a *tablemount*.

Mid-ocean ridge   The ridge and rise system related to the axes of sea-floor spreading. This runs down the center of the Atlantic Ocean and continues around into the Indian and Pacific oceans where it has branches and comes into the coast in two places, so that the name is not entirely applicable.

Plateau   An ill-defined extensive elevation of the deep-sea floor.

Ridge   A long elevation of the deep-sea floor with relatively steep sides and irregular topography.

Rise   An elongate broad elevation of the ocean floor with gentler and smoother sides than those of a ridge.

Seascarp   An elongate and comparatively steep slope of the sea floor.

Sill and *sill depth*   A submarine ridge or rise that separates two partially closed basins. The greatest depth over the sill is called the sill depth.

Seaknoll   A submarine hill or elevation of the deep-sea floor less prominent than a seamount.

Seamount   An isolated or comparatively isolated elevation of the deep-sea floor with a relief of 900 m or more and with comparatively steep slopes and relatively small summit area.

Tablemount   See Guyot.

Trench   A deep, long, and narrow depression of the deep-sea floor having at least relatively steep sides.

Trough   An elongate but relatively broad depression on the sea floor.

GENERAL SHAPE OF OCEAN BASINS

That the three major oceans of the world each have distinctive outlines has long been of interest. The Pacific margin is essentially along a great circle so that the basin covers almost half of the globe. The Atlantic is elongate with curving but roughly parallel margins. The Indian Ocean has an isosceles-triangular shape with irregular boundaries; the Indian peninsula truncates one point of the triangle.

All the major oceans are interconnected but a ridge almost severs the connection between the Atlantic and Pacific. More pronounced barriers partially separate the major oceans from the series of deep marginal seas. The latter include the Gulf of Mexico, the Caribbean Sea, the Mediterranean Sea, the Black Sea, the Arctic Ocean, and the group of relatively deep marginal basins landward of the festooned arcs along the east Asian continent and eastern Australia. Most of the marginal basins are circular or oval in shape; the chief exception is the Mediterranean with its three irregular basins.

THE ATLANTIC

The parallel-sided Atlantic has an area of 86 million sq km (Menard and Smith 1966), which increases to 94 million including its marginal seas. The mean depth is 3736 m, reduced to 3575 m when including its marginal seas. The latter is slightly less than the average for the world oceans, 3729 m.

The Atlantic Ocean shows remarkable symmetry, with the large Mid-Atlantic Ridge following down the center and, including the Bermuda Rise, occupying 32 percent of the total area (Menard and Smith 1966). This is flanked on either side by deep-ocean basins that cover 39 percent, and in turn by continental rises covering 8 percent. Most of the rest is the relatively wide continental shelf and slope with 17.7 percent. The mean depth in the basins is 4670 m and on the Ridge 4008 m. However, in many places, the Ridge has a relief of as much as 3000 m above the basins. These subdivisions are illustrated in Fig. 13–1.

Mid-Atlantic Ridge and the Rift Valley   The Mid-Atlantic Ridge has been known to exist in the North Atlantic since the *Challenger* Expedition, and its continuation into the South Atlantic was well documented by the pioneering echo soundings of the *Meteor* Expedition after World War I (Stocks 1933). The real significance of the centralized position of the Ridge was first pointed out by Heezen (1960), who determined that the Ridge probably continued south of Africa into the Indian Ocean and south of Australia into the Pacific. In the Atlantic, the Ridge extends north into the Arctic. Heezen suggested that the central rift valley of the Mid-Atlantic Ridge, which had been first observed by the British (Hill 1956), continued around the world along the crest of the Mid-Ocean Ridges. At the time of writing the second edition, it seemed quite unwarranted to claim such a continuation for the supposed "rift valley," because many profiles crossing the Ridge failed to show such a valley. Although this is still true in many areas, the suggestion of the continuous valley was prophetic, because we now are convinced that this world-encircling Ridge has at its center active tensile forces that are pulling the ocean crust apart to allow magmas to enter from below. Probably the resulting crack is filled with magma, particularly where spreading is rapid (Menard 1967A) and local cracks develop on the sides of the Ridge to relieve the tension.

Along the crest of the Ridge or over the rift valley, which follows much of the crest, are very distinctive magnetic anomalies. These were first described by M. Ewing and others (1959B) for the Atlantic Rift Valley, where they found high positive anomalies. Further studies (Vine and Matthews 1963; Vacquier and Von Herzen 1964) have shown that large anomalies at the crest of the rift valley are commonplace, but they may be a combination of large negative and positive anomalies and do not always coincide with the crest. The parallel bands of anomalies on either side of the Ridge crest indicate a spreading rate of about 3 cm/yr. Heat-flow measurements along the crest also show high values (Vacquier and Von Herzen 1964).

A portion of the rift valley was given detailed study by van Andel and Bowin (1968). They found various fairly parallel valleys, but only the rift valley showed good linearity. Their relief map and a generalized profile give a clear impression of the valley in low northern latitudes (Fig. 13–2). They found metamorphic and deep intrusive rocks on the walls of the median valley. The most common rock on the Mid-Atlantic Ridge is basalt of the tholeiite variety (Engel et al. 1965). There is very little or no sediment cover over the rock on the central part of the ridge. Rock is shown in many photographs.

That the Mid-Atlantic Ridge might represent the sunken continent of Atlantis, described by Plato, has been a popular belief. Actually, the only possible evidence of such a sunken continent is the fresh-water diatoms found in a core from the tropical Mid-Atlantic Ridge (Kolbe 1957). Malaise (1969, p. 169) used this as evidence for his idea of an emergent ridge, but it seems more likely that the diatoms were blown out from the Sahara by the trade winds and concentrated on the ridge by currents.

Atlantic Fracture Zones   The location of the fracture zones cutting the Mid-Atlantic Ridge have been largely established during the past decade. It is interesting to compare the 1968 Heezen and Tharp physiographic diagram

Fig. 13–1   Physiographic provinces of the North Atlantic. From Heezen et al. (1959).

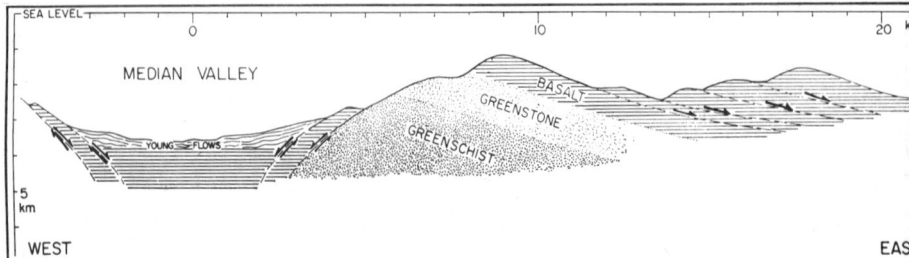

Fig. 13–2   Schematic structural diagram of Mid-Atlantic rift valley and eastern crestal range near 22° 30'N Lat. From van Andel and Bowin (1968).

with their 1959 edition. Their 1961 South Atlantic diagram shows that the concept had developed at that time and a few of the fracture zones are indicated in the equatorial area. The actual number is still unknown; probably many of these transform faults offset the continuity of the great Ridge along most of its length.

Selected fracture zones have been studied by Heezen and others (1964A, 1964C), Krause (1965), and van Andel and others (1967, 1969). Large earthquake epicenters are located along these transform faults in the zone between the offset ridge crest (Sykes 1967). The offsets at these epicenters are, as would be expected from the sea-floor-spreading hypothesis, opposite to the direction in which the Ridge appears to be displaced (Fig. 5–3).

In addition to ridges, the fracture zones have deep elongate valleys, including the Romanche Deep, 7856 m, one of the deepest holes in the Atlantic. The general relationship is shown in Fig. 13–3. It is significant that the large east–west sections of South America and Africa near the equator are related to two of the greatest fracture zones, the Vema and Romanche. The eastern continuation of the Vema Fracture Zone, off Sierra Leone, is called the Guinea Fracture Zone (Krause 1964B). It shows 230 km of displacement and extends into the African continent. Although the fracture zones are rarely seismically active except in the gaps between parts of the Ridge, the topographic relief can be traced into the continental margin in some places (Fail et al. 1970). The New England Seamount chain, which has the finest seamounts in the Atlantic, is said to lie along a fracture zone (Uchupi et al. 1970). These seamounts are a continuation of an east–west transcurrent fault that extends ashore at 40°N latitude and can be traced inland for 740 km (Drake and Woodward 1963). However, this fault is believed to be Devonian or late Paleozoic; therefore the relationship does not indicate contemporaneity— the seamounts are clearly of Mesozoic age. In the South Atlantic, somewhat similar volcanic chains are located off southeast Brazil, including the Rio Grande Rise and the Trinidad Island group, and, on the African side, the Walvis Ridge and the Guinea Ridge. These are presumably other examples of volcano growth along fracture zones or old tension cracks. The Walvis Ridge has outcropping Eocene sedimentary rocks of shallow-water origin, showing subsidence of more than 1 km (M. Ewing et al. 1966A). The North Brazil Ridge (Hayes and Ewing 1970) flanks the slope and is thought to represent oceanic volcanism following the early splitting of the Atlantic.

The Azores Islands lie along a fracture zone east of the Mid-Atlantic

Fig. 13–3 Bathymetric sketch of portions of Chain and Romanche fracture zones. Arrows indicate probable direction of bottom flow. Contour interval 200 fathoms (366 m) except where too steep for contouring. From Heezen et al. (1964A).

Ridge. According to Krause (1965), this zone can be traced east to Gibraltar, and a near continuation on the west side of the Ridge may continue as far west as the New England Seamounts, crossing the entire ocean. Large displacements apparently occurred along these fracture zones but the evidence is not clearly established.

Bottom photographs show ripple-marked surfaces in the fracture-zone valleys, indicating local relatively strong currents (Heezen and Hollister 1971). McGeary (1969) found plant-bearing sand layers in the Vema Fracture Zone, showing that turbidity currents had flowed as far seaward as the Ridge, probably from the Amazon during the last glacial stage of lowered sea level.

Deep-Sea Channels of the Atlantic We have already described the fan valleys that cut sediments built along the base of the continental slopes (p. 298). We have ample evidence that in some localities either these fan valleys continue across portions of the abyssal plains or that some form of erosion has cut the gentle slopes of these plains, developing winding channels that have a general trend which is more parallel to the distant continental coasts than at right angles to them. Mid-Atlantic Channel,[1] one of the best known, extends south down Baffin Bay, around the toe of the Newfoundland Rise, and then down the Atlantic Basin into the Nares Deep (Fig. 13–4) with a gradient of 1/2900. This has been explored by Heezen and others (1959) and has been found to slope continuously to the south. Apparently it has relatively steep walls and a flat floor at least 2 km wide. More than one core from the channel floor had sand, presumably a turbidite. One seismic profile showed a reflector that outcrops on the wall of the channel (Embley et al. 1970).

On the other side of the Atlantic, another deep-sea channel system has been found in a gap connecting the Biscay and Iberian Plains. Laughton (1960, 1968) discovered that this channel has benches along the side that are correlated with the sub-bottom strata found in the seismic profiles of the area. These benches are comparable with some of those in La Jolla Fan Valley (p. 312). This suggests erosion, and bottom photographs show slumping and eroded cliffs that seem to strengthen this opinion.

Off the northeast coast of Brazil, Vidal Channel is located between Barracuda and Demerara abyssal plains (Embley et al. 1970). This channel is 800 km in length and the gradient is 1/1100. Apparently it has tributaries to the south and is believed to show evidence of headward erosion. Thus, despite the very gentle gradients and great distances from land, currents must be cutting or maintaining channels into the deep-ocean floor. Whether these are turbidity currents or some deep-flowing sort of high-salinity current, like the contour currents along the continental rise off the East Coast of the United States (p. 411), is still not clear.

Trenches of the Atlantic The Atlantic Ocean has two small trenches, both insignificant compared with those of the Pacific. The greatest depth in the Atlantic is in the Puerto Rico Trench, where it reaches 9200 m north of Puerto Rico. It has been traced around most of the volcanic and seismic arc of the West Indies but apparently dies out topographically east of Dominica. Seismic profiles suggest that the trench may be due to underthrusting of the North Atlantic plate (Chase and Bunce 1969). The steep north wall of this trench has been interpreted as a fault scarp (J. Ewing and M. Ewing 1962). Serpentinized basalt has been found outcropping on the lower slope (Bowin

---

[1] Called Mid-Ocean Canyon, but clearly not of the character to which the name *canyon* applies.

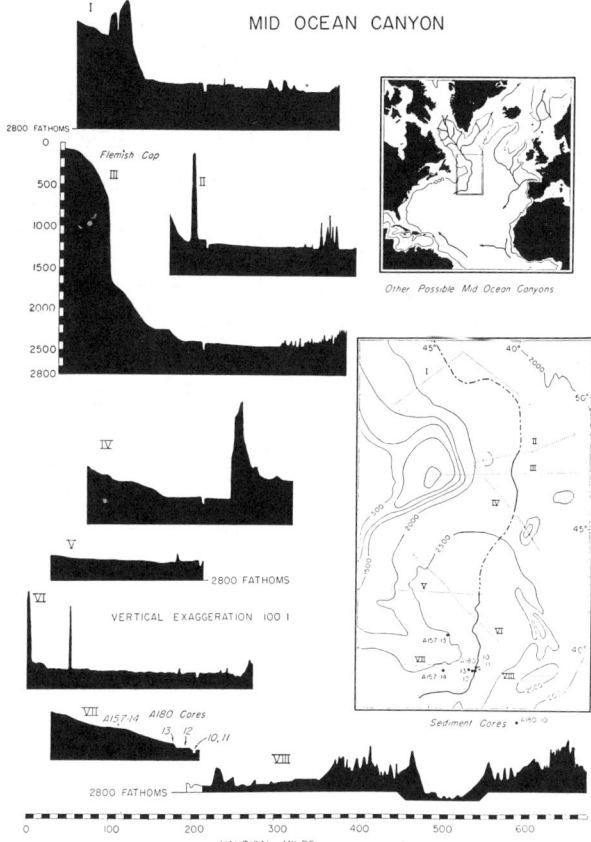

MID OCEAN CANYON

Fig. 13–4 The deep ocean channel extending down from Baffin Bay to the Nares Deep. From Heezen et al. (1959).

et al. 1966). The floor of the trench is thought to be flat and to have a thick sediment cover.

The South Sandwich Trench lies around the arc of the Sandwich Islands that connects South America with Antarctica. The arc is the locus of an earthquake belt and has active volcanoes. Recent information on the trench was compiled by Heezen and Johnson (1965). The trench walls are steep and the floor up to 7 km in width. The greatest floor depth is 8264 m.

Effects of Concentrated Bottom Currents The bottom currents associated with oceanic circulation (see p. 63) are concentrated in certain localities and are capable both of transporting sediments and producing bottom relief, such as ripple marks and even wavelike hills of sediment. The western boundary current (Fig. 3–16) flows with special force along the lower continental rise off the Carolinas (Heezen et al. 1966A). Rona (1969) found elongate hills at about 5000 m depth with crestal lengths of tens of kilometers and relief up to 100 m, which are oriented with the south-moving current and have their steep faces on the southwest side. These are obviously the result of the current.

Farther south, the Blake–Bahama outer ridge (Fig. 9–8) deflects south-eastward from the base of the Blake Escarpment at 32°N latitude and forms a hook to the south, separating the Blake–Bahama Basin from the Hatteras Abyssal Plain. This ridge, first described by Pratt and Heezen (1964), has been investigated in detail by Markl and others (1970). The seismic profiles suggest that this is a depositional ridge with sediment 1 km or more in thickness (see p. 424). It is believed that it was derived from the western boundary current, but, as Markl and others have shown, the history or the ridge building has been complicated. Andrews (1967) suggested that gravity tectonics produced some of the structure of the ridge, which might be indicated from some of the disturbed strata. Giant ripples with 2.6-km wavelength were found on the western flank by Markl and others (1970). Whatever the final explanation, it is evident that deep currents can be very effective in transporting large quantities of sediment.

The effect of strong bottom currents in Drake Passage and the Scotia Sea, south of South America, is evident in many bottom photographs (Heezen and Hollister 1964, 1971). Although no clear proof of erosion is visible, the ripples and other current indications suggest the probability that erosion and large transfer of sediment is taking place in this area. The Antarctic bottom current is also important farther north along the slope base off Argentina and Brazil. In fact, erosive processes are probably operating in many areas of the Atlantic.

Fig. 13–5   Bathymetry of the North Pacific indicating the three distinctive divisions. From Chase, Menard, and Mammerickx (1971) with shading added to bring out features.

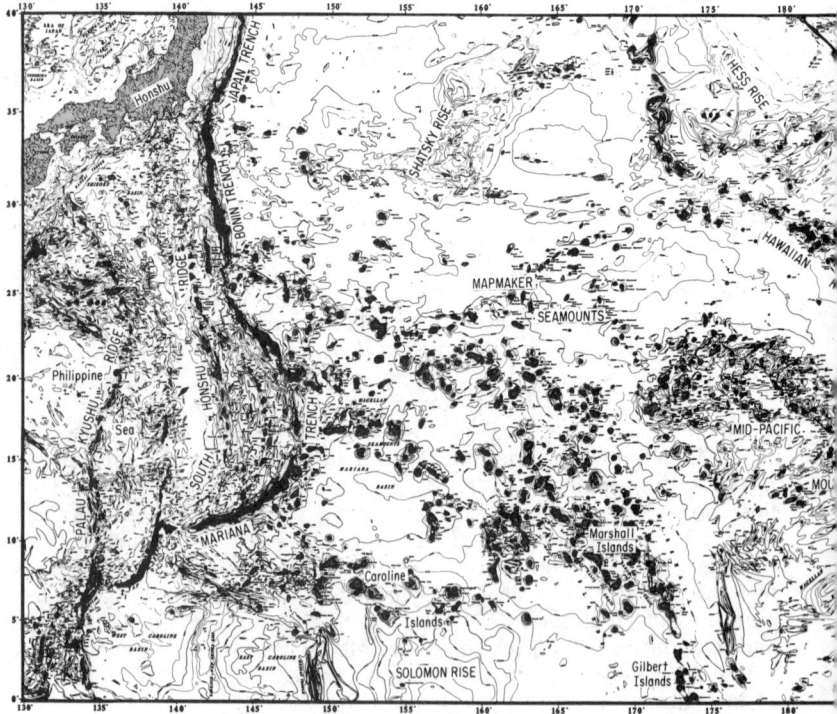

THE PACIFIC

The Pacific, the largest and deepest of oceans, covers 166 million sq km, or 181 million including the marginal seas (Menard and Smith 1966). The mean depth is 4188 m or, including the marginal seas, 3940 m.

An examination of various bathymetric maps of the Pacific shows clearly that the topography lacks the symmetry of the Atlantic and appears to have three major divisions (Fig. 13–5). The eastern sector, north of the Gulf of California, is dominated by the east–west fracture zones; and south of the Gulf of California, by the East Pacific Rise and its associated fracture zones. The East Pacific Rise curves gradually to the west, crossing the South Pacific. Farther west in the Pacific, the topography of the second division is dominated by the northwest- or north-trending ridges, which are crowned by many islands, including the Hawaiian chain, the coral atoll belts, and the coral islands of the Southwest Pacific. Tectonically, this is a relatively quiet area, but farther west the third belt with its great island arcs and their trenches is highly unstable. The arcs merge locally with the continents and larger islands. The other major feature of the Pacific is the great trenches that circumscribe most of the ocean boundary. The largest gap in the trenches extends from British Columbia to the Gulf of California.

During the past decade, exploration of the Pacific has become more localized with special emphasis on (1) the arcs and basins of the southwest,

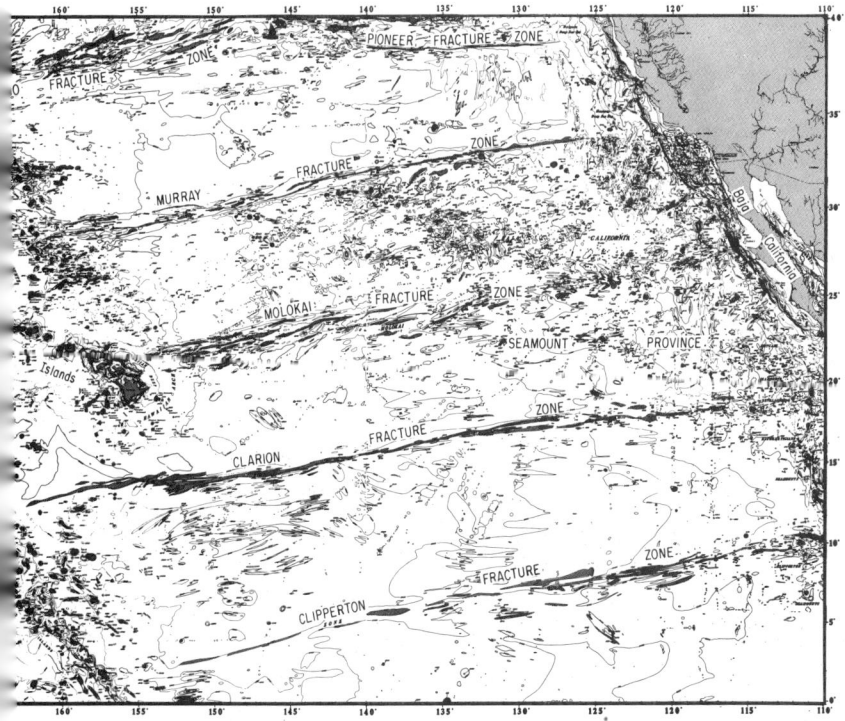

(2) the Gulf of Alaska, (3) the area off Oregon and Washington, (4) the north end of the East Pacific Rise, and (5) the Hawaiian Islands area.

East Pacific Rise   The East Pacific Rise, apparently a continuation of the system of mid-ocean ridges including the Mid-Atlantic Ridge, has a crest 2 to 3 km above the deep-ocean floor and is thousands of kilometers wide. The side slopes are gentler than those of the Mid-Atlantic Ridge, only 0.001 to 0.002 percent (Menard 1964, p. 121). The Rise is interpreted as having the same origin as the Ridge. However, it is notable that Menard's extensive explorations failed to find the rifts in the crest that are so significant in the Mid-Atlantic Ridge. On the other hand, there are many low ridges and troughs parallel to the crest, as are the magnetic anomalies. Many volcanoes rise above the flanks of the Rise, and the crest has little or no sediment cover over the volcanic rocks.

To the north, the East Pacific Rise definitely enters the Gulf of California (Larson 1971). Apparently, Baja California was rifted away from the mainland about 6 million years ago. At the Gulf entrance, according to Larson, the new sea floor is being added at the rate of 6 cm/yr, about average for the Rise farther south and considerably faster than in the Mid-Atlantic Ridge. The Rise crest near the Gulf of California is almost barren of sediment. Both volcanic and tectonic processes are in operation along the crest.

A presumed continuation of the East Pacific Rise under the North American continent is not definitely known, and the idea was questioned by Blackwell and Roy (1971). However, where the San Andreas Fault finally leaves the continent at Punta Gorda, the spreading zone again appears in the Gorda and the Juan de Fuca ridges (Menard 1960). In the intermediate zone, from the head of the Gulf to Punta Gorda, the San Andreas Rift is known to have a displacement rate of 6 cm/yr, which is comparable to the Ridge spreading south of the mouth of the Gulf. The significance of this relationship is discussed by Atwater (1970). Exploration of the Gorda Ridge (Atwater and Mudie 1968) determined that it has a trough between two ridges and that the trough is a graben with many faults on both sides. Active spreading of the Gorda Ridge during the past 700,000 years is indicated (G. W. Moore 1970).

Pacific Fracture Zones   Early bathymetric maps of the Pacific had no indication of the east–west topographic ridges and troughs, now shown to be such an important element in the relief of the eastern Pacific. Menard and Dietz (1952) first found that the north-sloping Gorda Escarpment, off Northern California, had a westward continuation, the Mendocino Ridge, although its steep side is on the south. This was the first discovery of a fracture zone in any ocean. Later, the investigations of Menard and his associates led to the finding of many other fracture zones; these were traced westward beyond the base of the East Pacific Rise and even beyond the Hawaiian Islands and the Phoenix Islands, the latter in the South Pacific (Fig. 13–6). Although it was thought at first that the fracture zones are generally very straight (Menard 1964, p. 43), further study showed that they are often offset due to change of spreading direction (Menard and Atwater 1968). Also, they are ordinarily parallel to each other and the larger ones have rather even spacing. The North Pacific bathymetric maps show many lengthy depressions as well as ridges extending along these fracture zones (Fig. 13–5).

The relationship of fracture zones to the magnetic anomalies on the west side of the East Pacific Rise has now been rather well established (Atwater 1970). The displacement of the anomalies along the fracture zone by many hundreds of kilometers (Fig. 5–6) has long been a source of surprise to structural geologists, but it now appears to be explainable by the hypothesis

Fig. 13–6 The fracture zones of the eastern and central Pacific and a portion of the East Pacific Rise (● ● ● ●). From Menard (1967A).

of sea-floor spreading and transform faulting. The displacement of the 3500-m-depth curve on the two sides of the Mendocino and Murray fracture zones is about equal to the displacement of the magnetic anomalies. This appears to require more explanation than is now possible. The complicated history of motion of the plates along western North America has been discussed by Atwater (1970) and off Northern California by Silver (1971). They show how the right lateral motion along the San Andreas Fault, supposed to have been initiated 40 million years ago, is related to these fracture zones with a major change in plate motions, having taken place during the Late Mesozoic. During mid-Tertiary, a trench is thought to have existed off western North America with a subduction zone under the continental margin. This is indicated by the nature of volcanism in Southern California and Baja California (Hawkins and Atwater 1971). They consider that subduction continued through Middle and Late Mesozoic and resulted in the generation of andesite and dacite rocks. Subduction terminated when the ridge and continent converged. Later, the present type of tectonics with regional dilation and right lateral strike-slip faulting has been due to differential movement between the Pacific and American plates.

The fracture zones are seismically active only in the portions between displaced segments of the East Pacific Rise and of the Gorda and Juan de Fuca ridges (Tobin and Sykes 1968). A notable example is at 55°S latitude where the East Pacific Rise is displaced 1100 km (Sykes 1963), about the same distance as the anomaly belt displacement at the Mendocino Ridge.

The Mendocino Fracture Zone has been extensively explored (Krause et al. 1964). Topographically, the most impressive feature is the 1200-m-depth increase of the ocean floor on the south side of the Mendocino Ridge. The ridge has as much relief as 3000 m and slopes up to at least 30° on the south side. Dredge hauls have yielded pyroxene-rich basalts, glassy basalts, olivine basalts, and palagonite tuff breccias (Nayudu 1965). Many pebbles and cobbles from the ridge are well rounded. The lithology indicates that they were eroded from the ridge rather than being rafted from another area (Krause et al. 1964).

Channels and Abyssal Plains of the Northeast Pacific   The greatest known concentration of deep-sea channels of any ocean is in the northeast Pacific (Fig. 13–7). Some of the channels extend from glacial troughs of the shelves along the coast of British Columbia and southeast Alaska. Others seem to have been beheaded by the development of the Aleutian Trench (Hurley 1960; Hamilton 1967; Grim 1969; Mammerickx 1970; Horn et al. 1971). According to Mammerickx, the Aleutian Abyssal Plain is a complex of fans formed from the Alaska Range, presumably from the Susitna River. The channels have shifted successively from west to east. These channels have well-developed levees that are larger and higher on the west (right) side and are apparently of depositional origin (Hamilton 1967). The sediment of the western part of the abyssal plain, largely turbidites, is thought to be fossil, because the source of sediments is now cut off by the trench and the turbidites of the plains are covered by a layer of pelagic material about 270 m thick (JOIDES press release, 13 July 1971).

Cascadia Channel (Fig. 13–8) is apparently connected with the submarine canyons coming from the continental slope off Washington. This deep-sea channel extends south for 550 km, then passes through a gap in the Blanco Fracture Zone, and from there continues west-northwest across much of the Tufts Plain. Cores from the channel floor show that continent-derived gravel has been transported for great distances, and the steep slopes, particularly in the gap, suggest that erosion has been important (Hurley 1960).

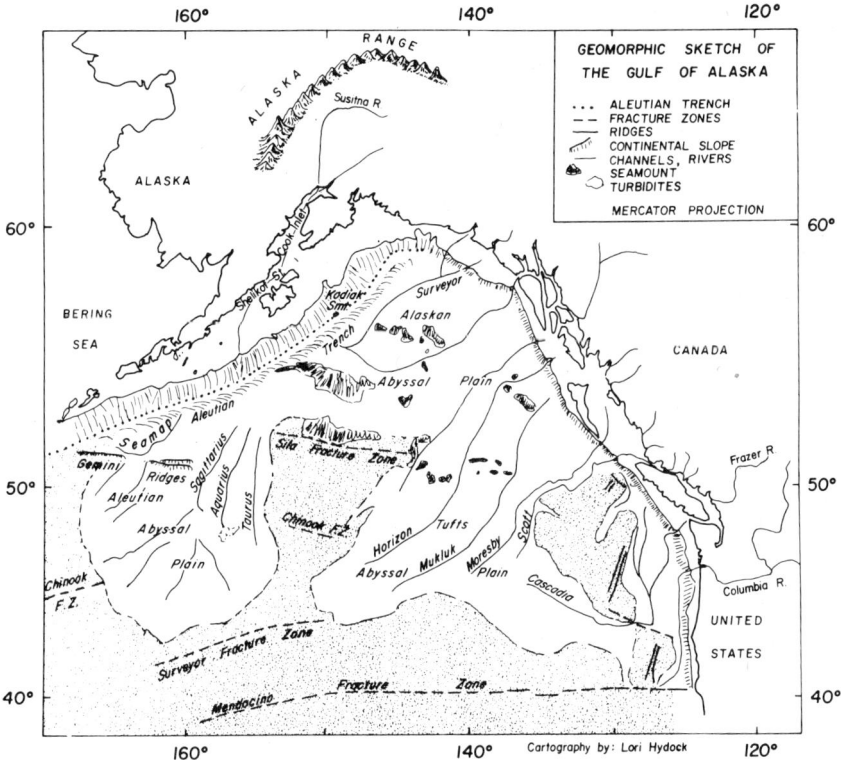

Fig. 13–7 Deep-sea channels of the northeastern Pacific and other geomorphic features. From Mammerickx (1970).

A few other deep-sea channels have been found in the Pacific. Wilde (1966) described channels on the Cocos Ridge in the east-central Pacific. Because continental material is not available here for cutting these channels, he thought that turbidity currents laden with volcanic ash from the submarine volcanoes may have caused them.

Volcanic Ridges and Guyots of the Central Pacific Although seamounts and guyots are scattered over much of the Pacific Ocean, their concentration is associated with the general northwesterly trending ridges of the Central Pacific. Of these, the most important is the Hawaiian Ridge, which extends northwest of Hawaii for 4500 km, where it connects with the north-trending Emperor Seamounts (see p. 98). South of the Hawaiian chain are the Mid-Pacific Mountains, which, combined with the Line Island Ridge, have a general northwesterly trend although connected with the Hawaiian group by a northeasterly trending Necker Ridge. Farther south and west are the great lines of coral reefs, including the Marshalls, the Carolines, the Gilberts, and the Ellice Islands. Then to the southeast, the Tuamotus, Samoa, the Society Islands, and the Austral Ridge. The high islands of these ridges are volcanic, and the borings in the atolls confirm that volcanic material underlies the coral island ridges.

A new bathymetric chart of the South Pacific (Mammerickx et al., in preparation) shows that these northwest-trending volcanic ridges are cut by a

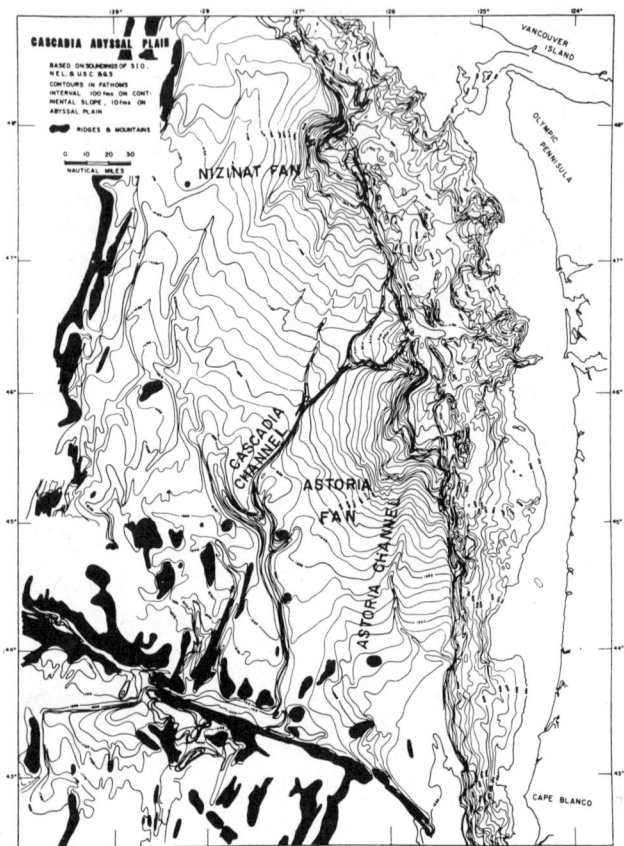

Fig. 13–8 Cascadia abyssal plain and the Cascadia and Astoria channels. Note that the connections between the channels and the canyons near shore are not very definite. The left hook of the channels emphasized by Menard (1956) is well illustrated. From Hurley (1960).

depression, Nova-Canton Trough, elongate in a WSW–ENE direction. It extends from the equator at 165°W to 5°S latitude, 177°E. Apparently this is a rift valley like those of the mid-ocean ridges, as it has bands of magnetic anomalies parallel to it on both sides (E. L. Winterer, personal communication).

To the north, the volcanic ridge zone is represented by the Emperor Seamount chain, which, unlike the others to the south, does not break the surface. Diverging from the north end of this ridge is the Emperor Trough or Emperor Fracture Zone (B. Erickson et al. 1970). This has more of the northwest trend characteristic of the ridges to the south, and clearly cuts across the trends of the various fracture zones farther east. The magnetic anomalies to the east and west of the Emperor Seamounts and the Emperor Trough trend more at right angles to the seamount ridge than to the trough, which might suggest that the seamounts rather than the trough represent a fracture zone. It is also possible that despite their divergence both are fracture zones.

The Hawaiian Ridge is well known because of the high islands to the southeast. The study of volcanism of the Hawaiian Ridge has shown that activity has progressed from northwest to southeast at a rate of 14.8 cm/yr from Kauai to Midway (Jackson et al. 1972). This is interpreted as indicating that the Pacific plate in this area has been moving northwest at this rate over a hot spot in the underlying crust or mantle, and that new volcanic activity is started when new sections of the plate cross over a hot spot. Once started, the volcanism could continue after the crust had passed beyond the source. It is possible that the same explanation applies to other volcanic ridges. The Hawaiian Islands are believed to have undergone crustal subsidence of 2 to 6 km (Moberly and McCoy 1966). This sinking resulted in the forming of an adjacent deep and the upwarping of the Hawaiian arch, some 200 km northeast of the islands.

The area south and west of the Hawaiian Ridge is of particular interest because of the large number of seamounts, guyots, and coral reefs (discussed in Chap. 12). The flat-topped guyots have been of interest since Hess (1946) called attention to them after studying his vast number of trans-Pacific soundings taken during World War II. They have generally been considered to represent wave beveling of old volcanoes during former emergence, and therefore should show the amount of subsidence. The map by Menard (Fig. 13–9) shows that the depths of the shelf break at the edge of the guyots is quite irregular, but average depths of more than 1200 m in the Mid-Pacific area appear to exist. This would imply considerable subsidence. Menard had suggested that the guyot depths along with other information from dredging and seismic profiles can be used to indicate the existence of what he called the "Darwin Rise," which he thought trended northwest during the Mesozoic and was an old spreading ridge. Drillings in the deep central basin between the Line Islands and the Gilberts (Winterer, Riedel, et al. 1969) showed that these areas on the hypothetical Darwin Rise have not sunk, as would have been the case if there had been a former rise in this area. Also, there is no evidence of the great amount of lava and sediment that would have come from this elevated tract. There is even some doubt that the flat tops of the guyots all represent sea-level erosion. Some flat tops develop in volcanic peaks by filling calderas with lava.

According to Menard (1956), many smooth portions of the Pacific Ocean are found as aprons around "groups of existing or ancient islands," notably in the ridge area of the Central Pacific. These archipelagic aprons slope evenly away from the high areas resembling fans, but through seismic-refraction measurements (Shor 1960) they are found to consist largely of lava flows with a rather thin covering of sediments, sufficient only to develop extreme smoothness. These sediments, like those of the continental rise and the abyssal plains, were presumably introduced to a considerable extent by turbidity currents.

Three elevated tracts all along a northwest line add a special character to the ridge and seamount province of the Pacific. From north to south these include the Shatsky Rise, the Magellan Rise, and the Manihiki Plateau. Each of these has a broad summit thousands of meters above the surrounding deep ocean floor, and seismic profiles show that each has a sediment cover at least 1000 m thick. The Shatsky Rise and Magellan Rise have already been drilled and the sediment has proved to be all deep pelagic (Fischer, Heezen, et al. 1970; Winterer, J. Ewing, et al. 1971) with no suggestion that these large areas were ever above sea level. However, M. Ewing and others (1966B) dredged some shallow-water Cretaceous molluscs from the Shatsky Rise, which they believed to have come from one of the narrow peaks rising above

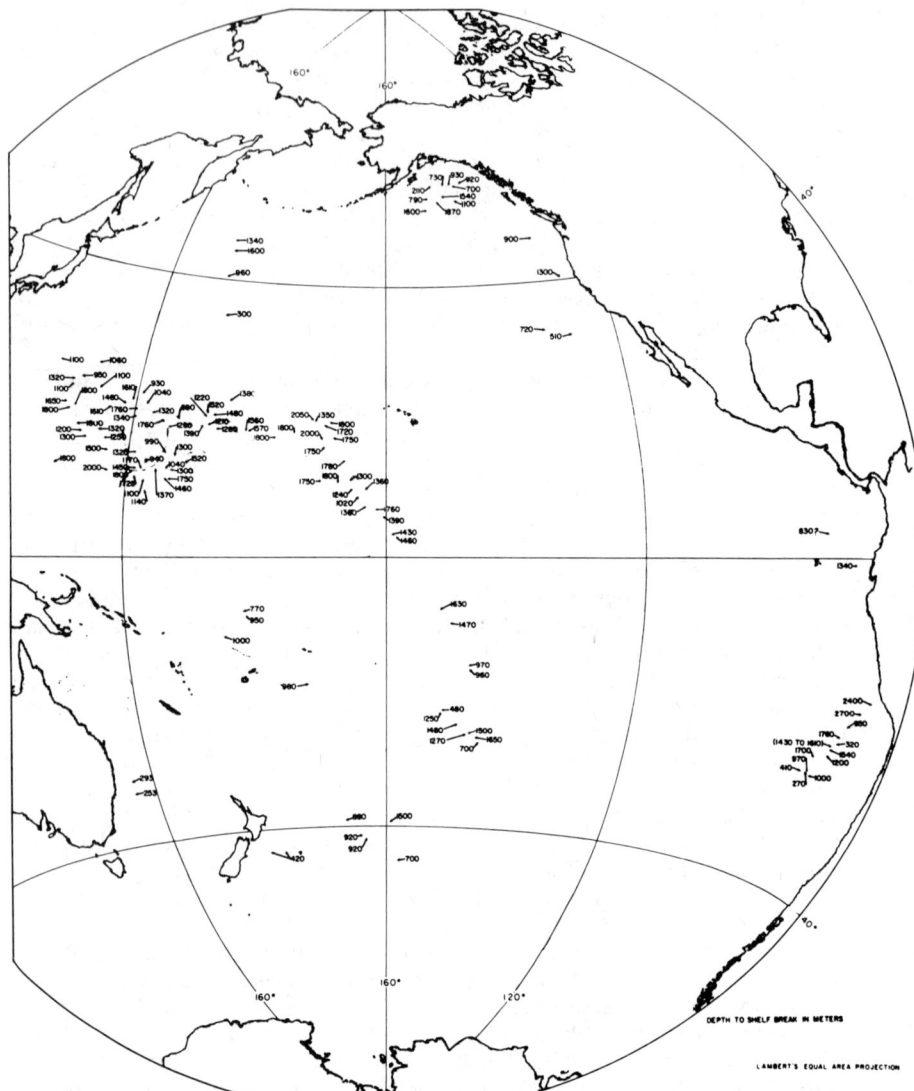

Fig. 13–9 Depth of shelf break in guyots of Pacific. From Menard (1964, Fig. 6–14).

this broad feature. Shatsky Rise profiles from Johnson and others (unpublished manuscript) are reproduced in Fig. 13–10. It will be noted that the basement under the sediment layers is very irregular with sharp peaks that could provide sediment to the lower areas on the broad summit.

The Manahiki Plateau south of the equator has an area of about 600,-000 sq km (Heezen et al. 1966B). Bottom photographs show ripple marks and scour zones around the numerous manganese nodules on the Plateau summit. Dredging yielded Cretaceous fossils suggesting that the beveling of the Plateau had occurred before that time. Cretaceous coral reefs had been dis-

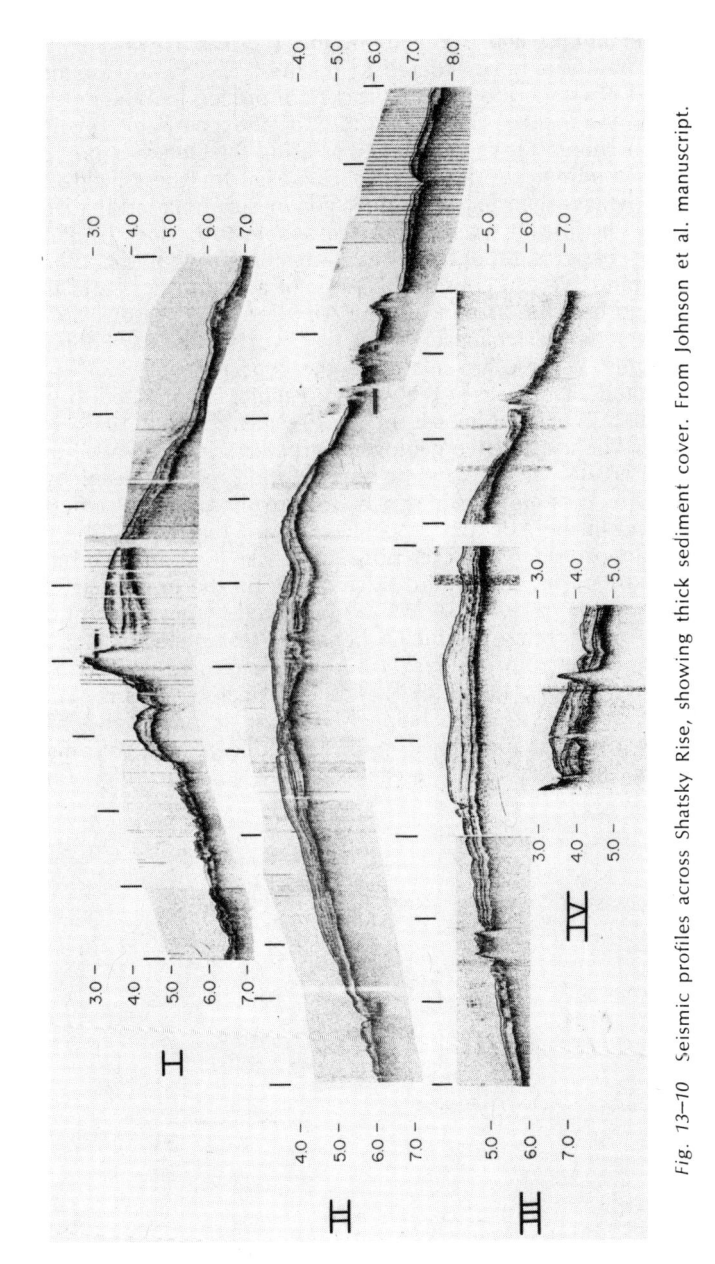

Fig. 13–10 Seismic profiles across Shatsky Rise, showing thick sediment cover. From Johnson et al. manuscript.

covered in earlier work in the Mid-Pacific Mountains (E. Hamilton 1956), and there are other indications that the Cretaceous was a period when the guyots may have been close to sea level.

Arcuate Ridges and Trenches of the Western Pacific   Formerly, the Pacific Basin was usually considered to be the area seaward of the andesite line, landward of which the volcanic rocks are predominantly andesitic (Macdonald 1949) and hence similar to those of the continent. Menard (1964, pp. 27–31) has shown the inconsistency of using this line as a boundary, calling attention to numerous localities where crustal thickness inside the line is oceanic, that is, less than 6.5 km. Other geophysical data add to the confusion of using this line. It seems more reasonable to bound the Pacific Basin largely on the basis of topography, as is the practice for the other oceans. With this in mind, we can consider the area of arcuate ridges and islands and their associated trenches as part of the Basin. Borderline cases exist inside the inner arcs, such as the Japanese islands, Ryukyu Islands, the Philippines, and Indonesia. Some of these are related to the continents.

The island-arc system of the western Pacific, investigated in the past by many geologists (Vening Meinesz et al. 1943; Umbgrove 1945; Hess 1938, 1948), has had the most active geological structures in the world, the deepest trenches, the greatest gravity anomalies, and the greatest seismicity (Menard 1964, p. 97). The arcs are mostly convex toward the Central Pacific Basin, just as the two arcs in the Atlantic are convex toward that basin. The typical arc includes, on the outside: an outer ridge, a trench, then an island arc (frontal arc), an inter-arc basin, and a third ridge or arc on the inside (Fig. 13–11). As indicated previously (p. 93), the arcs are bordered by a collision zone where the oceanic crust is being subducted beneath the arc system, and a thermal diapir may be rising from the deeply buried oceanic crust to form the inter-arc basin. Karig (1970A, 1970B, 1971) has investigated the arcs with modern methods and appears to have helped clarify their origin and history. His work suggests that first the trench is formed, followed by volcanism on the ridge, and by extension of the area inside the volcanic ridge forming the

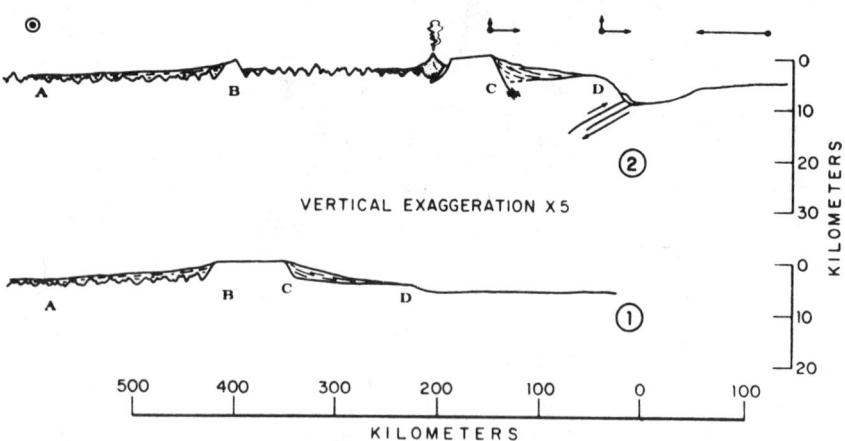

Fig. 13–11   Diagrammatic section across Tonga-Kermadec arc in the early Pliocene (1), and the present (2). Shows development of interarc basin. Crustal displacement to the east given by vectors along the top of figure. Note extension between *B* and *C* forming basin and major crustal shortening in trench to right. From Karig (1970A).

inter-arc basin, and finally by large-scale sinking of the third ridge. The basin fills are, as far as known, all Quaternary. On the west of the third ridge in the Mariana Islands area, a chain of andesite volcanoes produced a thick sediment apron. The idea of the inter-arc basin being extensional was first suggested by Dietz (1954). It is now well confirmed by Karig and partly from the work by Sclater and others (1972), who investigated the Lau Basin between the Tonga and Lau ridges, north of New Zealand. Here, the extension is apparently related to transform fault motion between the Indian plate and another smaller plate to the east. Intrusion of hot material accounts for the Lau Basin being 3000 m above the deep Pacific floor, despite its oceanic crust, shown by refraction shooting and rocks obtained in many dredge hauls (Sclater et al. 1972). The extension of the basin and the underlying intrusion continued during the past 10 million years.

Using earthquake epicenter data from ESSA (now NOAA) and from Katsumata and Sykes (1969), Karig (1970A) mapped the subduction (Benioff) zones under the Pacific arcs. He concluded that active extension of the inter-arc basins is confined to arcs where earthquakes on the Benioff zone go deeper than 350 km. Basins inside the arcs farther west are also very likely due to extension. According to Gardner (1970), the Coral Sea Basin was the result of Early Tertiary rotational spreading and large subsidence at a rate of 17 to 24 cm/1000 yrs since the Miocene. The Coral Sea Basin is underlain by normal oceanic crust (J. Ewing et al. 1970), but is shoaler than most of the oceanic basins because of its 2.5 km of sediment fill. The South China Sea Basin has crustal thicknesses even lower than the average for ocean basins (Ludwig 1970). The Celebes and Sulu basins between Borneo and the Philippines have been studied by Krause (1966), who found evidence of recent large-scale faulting, producing great subsidence. He also found that strike-slip faulting was an important cause of the basins.

Inside the Aleutian arc, the Bering Sea has a most unusual feature, Bowers Ridge (Ludwig et al. 1971) (see Fig. 9–28). This bank is an arc that rises from the floor of the Bering Sea and intersects the Aleutian arc at right angles off the Rat Islands. Unlike the Aleutian arc, it is not seismically or volcanically active at present. According to Nichols and others (1964), it is a major reverse fault block. They compare it to the arcuate structures such as the Big Horn Mountains in Montana and Wyoming.

Trenches of the Pacific The relation of the Pacific trenches to the continental slopes and sea-floor spreading has been discussed, but their significance relative to the Pacific Basin needs further emphasis because they virtually surround it and are an important factor in the development of this ocean.

The trenches along North and South America closely approach land but have several gaps in their continuity, especially the one along the West Coast of the United States and Canada. This particular gap is indeed fortunate for the numerous west coastal cities of the United States that are located on low ground, because the trenches are the world's greatest source of tsunamis. The gap between the Middle America Trench and the Peru–Chile Trench is apparently related to the Caribbean block; and this in turn is bordered by a trench that curves around the West Indian arc (see Fig. 5–4). Also, the southward termination of the Peru–Chile Trench is perhaps related to the only other trench in the Atlantic that borders the South Sandwich arc.

The East Pacific trenches are only slightly deeper than those of the Atlantic and are not nearly so deep as those of the western Pacific. The latter, starting off Kamchatka, form a series of somewhat disconnected deeps extending in a general southerly direction as far as the Kermadec Islands, north of New Zealand. The greatest depths of these trenches are given in Table 4. The similarity of many of these depths is striking. The maximum soundings of the Mariana Trench and Tonga Trench are so close that we are not yet certain which is deeper; some evidence favors the former. The Challenger (Trieste) Deep in the Mariana Trench is now well sounded and the depth of 10,915 m is based on corrected echo soundings, on a wire sounding, and on corrected pressure readings taken on the trench floor in the bathyscaph *Trieste* by Jacques Piccard and Lt. Don Walsh in January 1960.

Table 4   Depths of Pacific Trenches

Trenches	Meters
Aleutian	7,679 (not accurate)
Kuril	9,750 ± 100
Japan (Idzu–Bonin)	9,810
Mariana	10,915 ± 20
Mindanao (Philippines)	10,030 ± 10
New Britain	8,320
New Hebrides	9,165 ± 20
Tonga	10,800 ± 2
Kermadec	10,047
Peru–Chile	8,055 ± 10

Source: Fisher and Hess (1963) and Udintsev (1959).

Some of the trenches have V-shaped bottoms, but more commonly the profile indicates a relatively narrow flat floor several kilometers wide. Where the bathyscaph landed on the bottom of the Mariana Trench, a diatomaceous ooze was encountered. The same type of bottom was reported by Petelin (1960) from both the Yap and Mariana trenches. Sand and gravel were found in the sample taken by the Galathea in the Mindanao Trench (Bruun and Kiilerich 1955). Pillow lava was dredged from the walls of the Mariana and Bougainville trenches (Petelin 1960). Extensive dredging on the walls of the Tonga Trench (Fisher and Engel 1969) and subsequent work in 1970 shows that the offshore wall is consistently basalt, but the nearshore wall has coarsely granulated serpentinized peridotite and dunite. These ultramafics are probably exposed by faulting. Hess (1964) considered such rocks as derived from the mantle.

The floor of the Aleutian Trench slopes quite evenly toward the west as far as longitude 180°. Farther west, the slope is reversed. The even slope in the eastern portion is a strong indication that turbidity currents have been flowing along this trench. Profiles of the Mid-America Trench, off western Central America and Mexico, are V-shaped to the south and generally flat-floored profiles to the north (Fisher 1961). The flat floor was shown by reflection shooting to have a roughly V-shaped floor under the sediment. As discussed previously, the sea-floor-spreading hypothesis explains the trenches as due to subduction of the oceanic crust where two plates are in collision. The finding of undisturbed flat-lying sediments in the base of the trench off Chile (Scholl et al. 1968A) and under the eastern Aleutian Trench (von Huene and Shor 1969) is somewhat disturbing to this concept but may simply indicate that subduction has not been active in Late Cenozoic or that thrusting goes on beneath the trench fill (Elsasser 1968).

The trenches are bordered in many places by rows of volcanoes and arc-shaped island chains, like the Aleutians, Kurils, and Marianas. Virtually all of these occur on the concave side of arcuate trenches. Earthquake epicenters are near the surface in the vicinity of the trench but at progressively greater depths in a direction toward the continents (Fig. 13–12).

Gravity measurements have been taken at many sites over trenches in the Atlantic, the Pacific, and the Indian Ocean (Vening Meinesz 1929, 1955; Worzel and Shurbet 1955). These trenches have some of the greatest negative anomalies in the world. There is evidently a deficiency of mass at these places, such as would be produced if the crust had been downbuckled and was being held down against isostasy by pressure from the sides, as suggested by Vening Meinesz (1929). Downbuckling is also indicated by the greater depth of the Mohorovičić discontinuity (see p. 89) under the Tonga Trench (Raitt et al. 1955), under the Mid-America Trench (Shor and Fisher 1961), under the Kuril Trench (Sisoev et al. 1959, Fig. 2), and under the Puerto Rico Trench (Talwani et al. 1959).

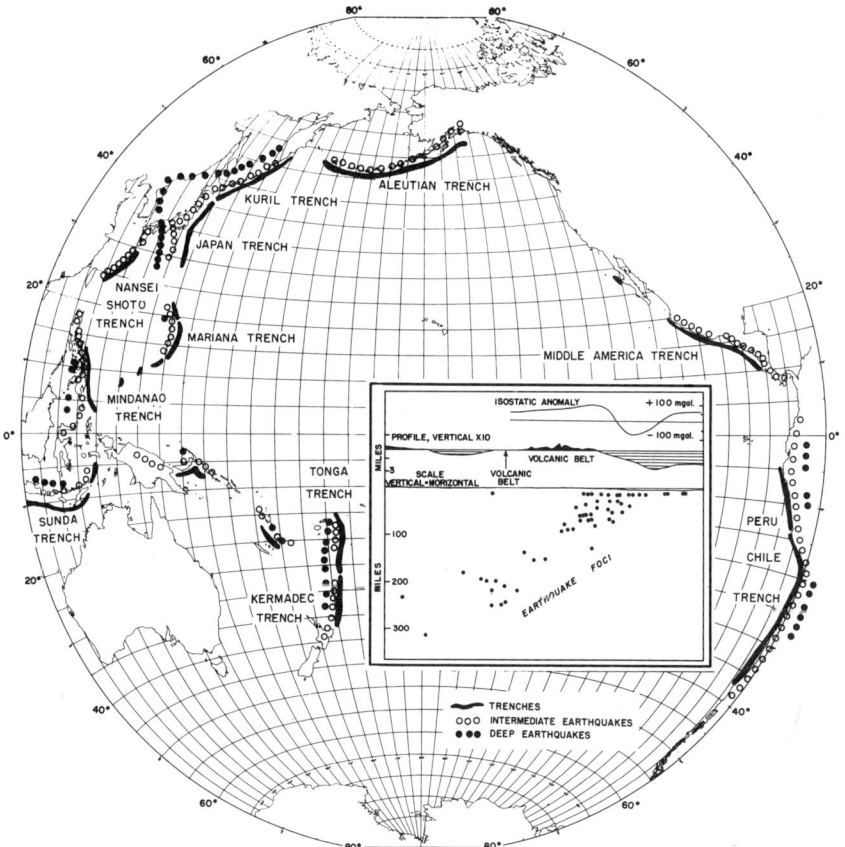

Fig. 13–12 The relation of trenches to the Pacific margin and the depths of earthquake foci in relation to the continental margins. After Gutenberg and Richter (1954) and Fisher and Revelle (1955). See also Fig. 5–5.

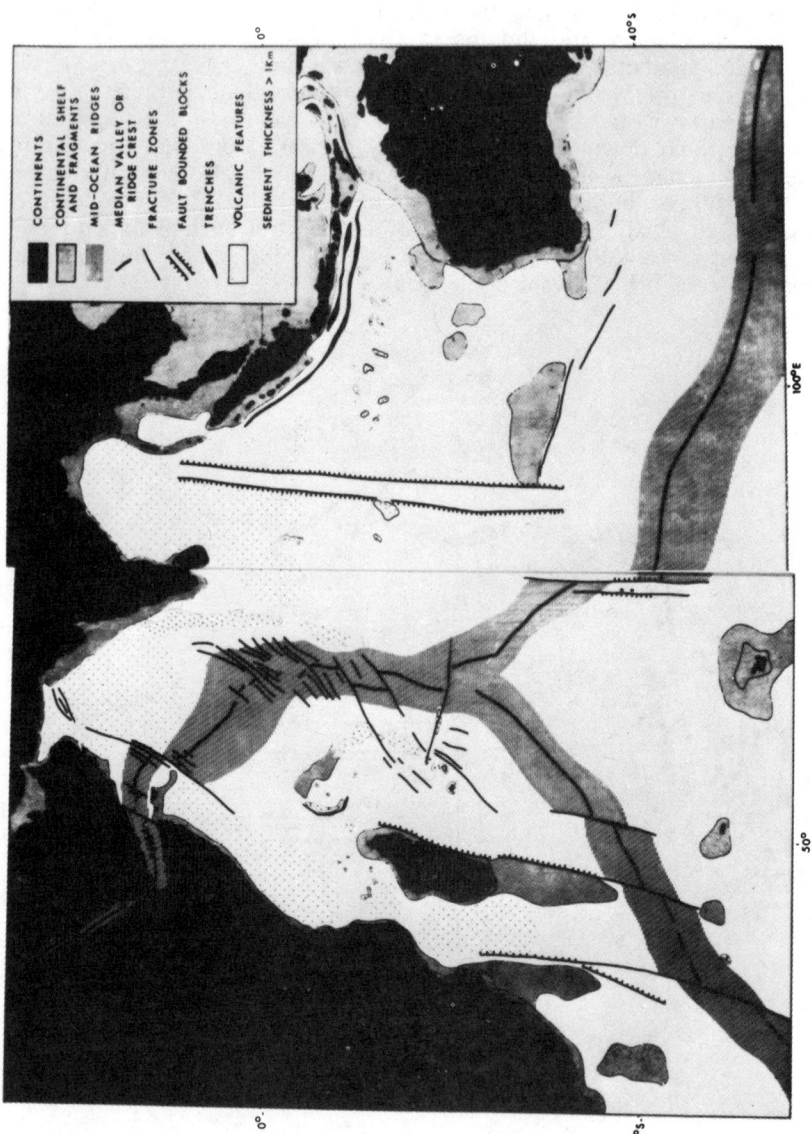

Fig. 13–13 Structural and morphological map of Indian Ocean. From Laughton et al. (1970).

INDIAN OCEAN

The Indian Ocean was intensively studied during the past decade. The results are still being compiled, although much information concerning the topography and geophysics has appeared (see, particularly, Laughton et al. 1970; Fisher et al. 1971). The Indian Ocean covers an area of 73 million sq km and has a mean depth of 3872 m. The general character, shown in Heezen and Tharp's diagammatic sketch, is updated in Fig. 13–13. This ocean has its own peculiarities, as do the Atlantic and Pacific. The most striking feature of the Indian Ocean is the inverted Y ridge that resembles the Mid-Atlantic Ridge and the East Pacific Rise of the other oceans. One arm of the inverted Y, the Southwest Indian Ocean Ridge, extends around Africa and connects with the Mid-Atlantic Ridge; while the other arm, the Southeast Indian Ridge, bends around Australia and joins the East Pacific Rise. North of the Y junction, the Central Indian Ridge runs north toward the Gulf of Aden and bends northwest as the Carlsberg Ridge, where it connects with the rifts of Africa and the Red Sea.

Another distinctive feature of the Indian Ocean is the north–south ridges, including Ninetyeast that runs south from the east side of the Bay of Bengal, and the Chagos-Laccadive Ridge that skirts the west side of India. Many other lineations in a north–south direction are indicated on the Heezen and Tharp diagram, but few of them are well established. The Indian Ocean has one arcuate deep, the Java (Sunda) Trench, that appears to be an adjunct from the Pacific arc zone. Another notable topographic feature of the Indian Ocean is the somewhat linear plateaus capped with reefs that have continental rocks, such as the Mascarene Plateau with the granitic Seychelles Islands. The largest fan in the world is located in the Indian Ocean off the Ganges Delta and a second, less well known, off the Indus.

Indian Mid-Ocean Ridges The inverted Y-shaped ridge that dominates the Indian Ocean is part of a major seismic zone. Just as the more detailed study of the Mid-Atlantic Ridge has shown that it is far more complex than was originally supposed, so the extensive surveys of the Central Indian Ridge have indicated far greater complexity than was shown, for example, in the Heezen–Tharp diagrams. The portion of the Ridge between the equator and 30°S has been intensively surveyed (Fisher et al. 1971). The results, partly indicated in Fig. 13–14, show numerous parallel ridges and troughs but, at least in this zone, there is no continuous central rift valley, and the earlier indications of such a valley apparently were the result of crossing the small trenches of the fracture zones that cut the ridge here in a northeasterly direction. The fracture zones were in part discovered by magnetometer profiles. The east–west Rodriguez Ridge, which is out of line with the other fracture zones, has proved to be a very recent feature. Fisher and others (1971) have attempted to explain the development of the Carlsberg and Central Indian Ridges in terms of plate tectonics. They consider that prior to 50 million years ago (about the time of anomaly 21), the Carlsberg Ridge extended east and west and was cut by a large north-south transform fault, the Chagos Fracture Zone. Between 50 and 10 million years ago, spreading was very slow and volcanics were extruded. Subsequently, about 10 million years ago (anomaly 6), this north–south fracture zone broke up into north-trending crestal segments separated by northeast-trending transform faults that now dominate. By Miocene, the ridge spreading had changed from north–south to northeast–southwest, both in the Central Indian Ridge and along the Carlsberg Ridge. The northern end of the Carlsberg Ridge is offset by the Owen Frac-

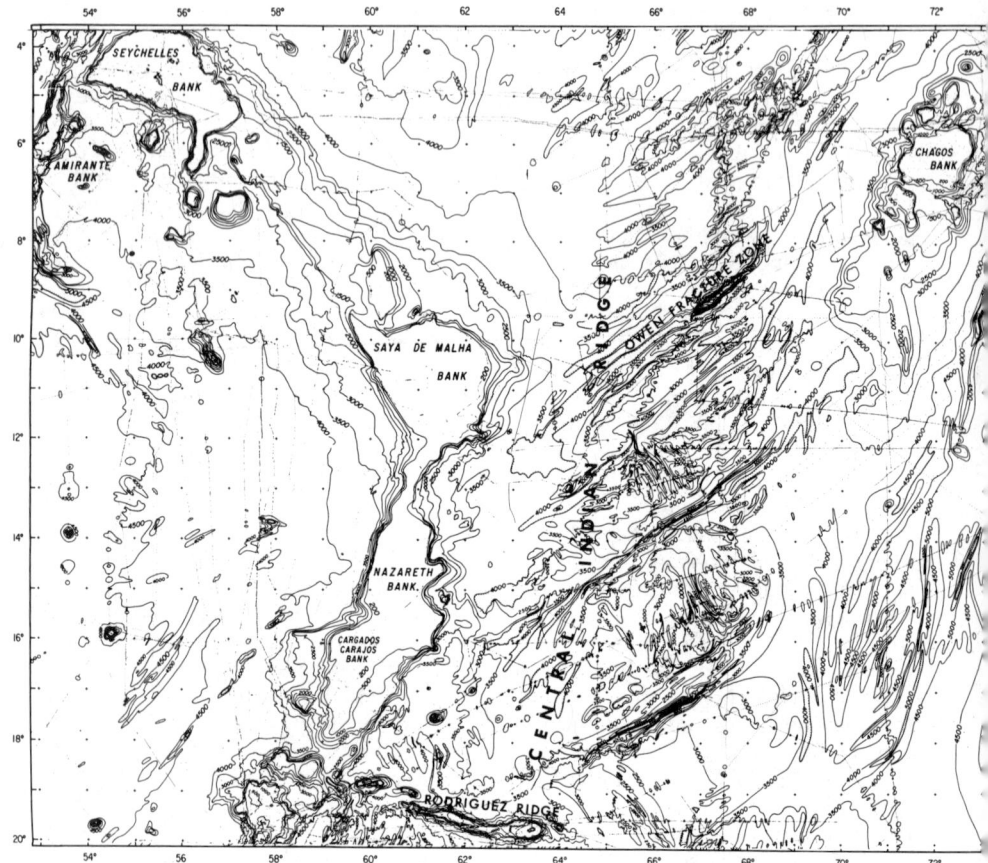

Fig. 13–14 Bathymetry of Central Indian Ridge and vicinity. Contour interval 500 m (and 200-m contour). Dots are ship tracks. From Fisher et al. (1971).

ture Zone from the Sheba Ridge, which enters the Gulf of Aden (Laughton and Tramontini 1969). The Owen Fracture Zone extends for hundreds of kilometers north and south and may connect with some large right-lateral wrench faults along the west side of the Himalayas (Abdel-Gawad 1971). Here again, we find a mid-ocean ridge cut by numerous fracture zones (Fig. 13–13), but, unlike the Central Indian Ridge, a rift valley is well developed in portions of the Sheba Ridge (Laughton et al. 1970). There is also a rift valley in the Red Sea (Laughton 1970), but this valley is not cut by transform faults. The Red Sea reflection profiles (Phillips and Ross 1970) show highly disturbed materials along the axis of the trough, suggesting recent rifting action. Much attention has been given to the hot brines found here and their associated metallic deposits (Degens and Ross 1969). The 1971 expedition on the R/V *Chain* (Ross et al. 1971) cored into heavy-metal-rich sediments in three of the deeps of the Red Sea. Iron and copper deposits of great potential value exist in the area and are thought to be still forming. According to Ross, the temperature of 56.2°C recorded in 1966 had increased to 59.2°C and actually showed an increase during the three weeks of the expedition. The hot brines are considered to be a manifestation of recent sea-floor spreading.

The Southwest Indian Ridge is apparently one of the least well developed of the spreading mid-ocean ridges, and the usual magnetic anomalies associated with these ridges are not found (Vine 1966). The Southeast Indian Ridge has a central magnetic anomaly, and the crest of the Ridge is bare of sediments (Fisher 1966). There is no evidence of a rift valley (Udintsev 1966), but epicenters of earthquakes follow close to the crest of the Ridge. Rapid spreading rates, about 6 cm/yr, are reported from the Southeast Indian Ridge (Le Pichon and Heirtzler 1968), in contrast with the low rates for the Southwest Ridge.

Rocks dredged and cores taken from the Indian Mid-Ocean Ridge (summarized in Laughton et al. 1971) include many basalts of the tholeiite type common to deep oceans. Carlsberg Ridge contains breccia of ultramafic rocks partly replaced by quartz. The Owen Fracture Zone has gabbro and, farther south, the Ridge has low-grade metamorphic rocks. Serpentinized rocks of possible mantle origin come from near the center of the Y in the Ridge, and lherzolite, thought to be upper mantle, is found as bedrock in the center of the Ridge (C. Engel and Fisher 1969).

Aseismic Ridges and Plateaus The shoal portions of the Indian Ocean not associated with earthquakes include several distinct structural types (Laughton et al. 1971). The most unusual is represented by the Mascarene Plateau (Fisher et al. 1967), extending 2300 km from the Seychelles Islands south to Mauritius Island. Structurally, it is characterized by block faulting. The Precambrian granite of the Seychelles is clearly continental, as in the bank directly to the south. However, farther south, the Saya da Malha Bank has a coral mass overlying oceanic crust. Other nearby banks and islands are underlain by volcanic crusts, and recent extrusions have alkali olivine basalts. The usual interpretation is that at least the northern part of the plateau is a continental fragment related to the general disruption of Pangaea. Other similar plateaus are found farther south, including Agulhas Plateau off South Africa.

The north–south Chagos-Laccadive Ridge is probably volcanic but partly covered with coral. It may possibly be an example of migration over a hot spot, as suggested for the Hawaiian Island chain (p. 98). Ninetyeast Ridge has been interpreted as a horst where ocean crust has been upthrust, pehaps due to convergence of crustal plates (Le Pichon and Heirtzler 1968).

Bengal Fan There appears to be no place in the deep ocean where the topography has been so influenced by deposition coming from the adjacent land as in the Bay of Bengal. The gently sloping plain, extending for 2000 km from the slope base off the great Ganges–Brahmaputra Delta to the 5000-m contour far south of Ceylon, is unique (Fig. 13–15). The slope has an almost even gradient of 1.5 m/km, and the fan is crossed by many anastomosing channels with low levees on the side. The most complete study of the fan has been that of Curray and Moore (1971). Their profiles show that whatever type of current is producing the channels must continue with little change in character down across the entire length of the plain.

MEDITERRANEANS

The relatively small, deep basins that are largely surrounded by continents or island arcs are called *mediterraneans*. Their depths are less than those of the large oceans, although most of the basins have small zones in

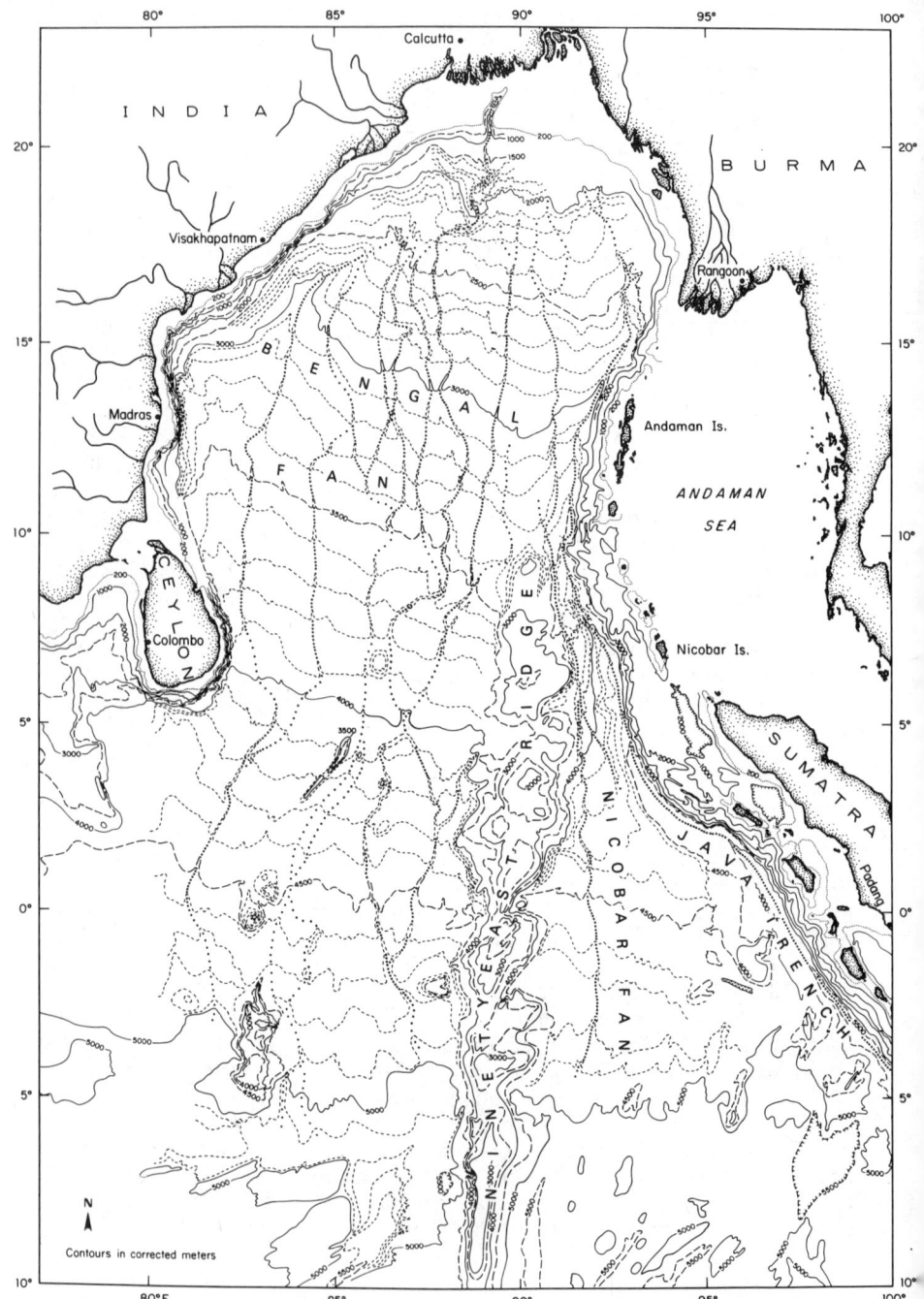

Fig. 13–15A Bengal Fan, channels, and Ninetyeast Ridge in the Indian Ocean. From Curray (1971).

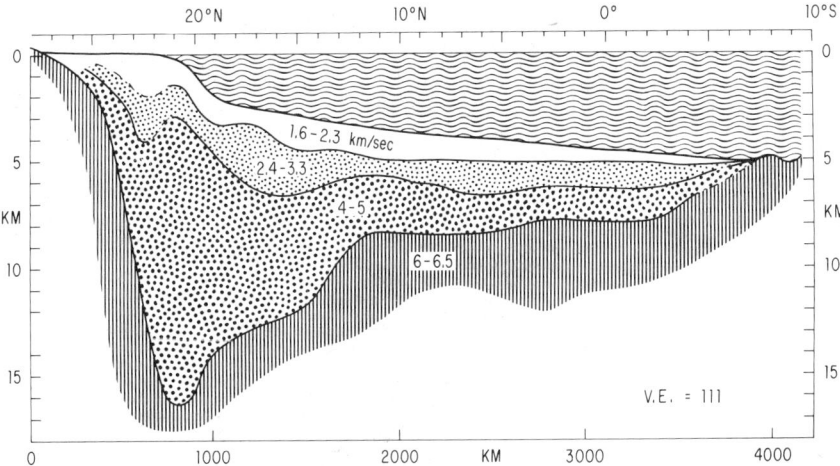

Fig. 13–15B   Diagrammatic profile down the Bengal Fan. Courtesy of J. R. Curray and D. G. Moore, based on recent Scripps Institution expedition.

excess of 4000 m. The crustal thickness beneath the mediterraneans, as far as is known, is intermediate between that of continents and ocean basins but generally somewhat closer to oceans (Menard 1967B). The mediterraneans include the Arctic with an area of 9.5 million sq km, the Caribbean and Gulf of Mexico with 4.4 million, the Mediterranean and Black Sea with 3 million, and the group of basins, already discussed, from Bering Sea to the South China Sea with about 6 million.[2]

Arctic Ocean   The largest of the deep landlocked basins is in the Arctic. It has two relatively shallow connections with the Atlantic and a very shallow connection with the Pacific. Its outline is elliptical but the narrow axis of the ellipse is crossed by a series of elongate ridges and basins (Fig. 13–16). The relief of the Arctic Ocean has been determined through a combination of soundings from submarines, drifting ice islands, and from icebreakers. The results are summarized in Dietz and Shumway (1961), Gakkel and Dibner (1967), Hunkins (1968), and Demenitskaya and Hunkins (1970). There are two principal deeps—the Canadian Basin with depths to 3900 m and the Eurasian Basin with depths of 4090 to 4500 m. The ridges include the aseismic Lomonosov, which rises to 1290 m and is parallel to the seismically active Arctic Mid-Ocean Ridge. The latter, according to Johnson and Heezen (1967), is a continuation into the Arctic of the Mid-Atlantic Ridge and is cut by many fracture zones. It has axial-oriented magnetic anomalies, although they are of small magnitude. Alpha Cordillera, an elevated tract between Lomonosov Ridge and the Canadian Basin, is thought to be an extinct mid-ocean ridge (Hunkins 1969).

Gulf of Mexico and the Caribbean   The two mediterraneans that intervene between the main masses of North and South America are quite different in character (J. Ewing et al. 1970C). The Gulf, except for a group of diapric hills, has an extraordinarily flat floor (Fig. 9–15) (M. Ewing et al. 1958). The slope toward the center of the oval Sigsbee Abyssal Plain is about 20 m

---

[2] The depth averages Menard and Smith (1966) used for the oceans include so much continental shelf area in the mediterraneans as to be unrealistic so are not included.

Fig. 13–16   Physiographic provinces of the Arctic Ocean. From Hunkins (1969).

in 130 km. One branch of a great fan, the Mississippi Cone (Wilhelm and Ewing 1972), leads from the Mississippi Delta to the central plain, and another branch extends along the base of the Florida Escarpment. These fans are exceeded in size only by the great Bengal Fan in the Indian Ocean. The Sigsbee Knolls, rising as much as 330 m above the abyssal plain, were explained by M. Ewing and others (1962) as salt domes, but considerable skepticism about this origin existed until one of the first of the JOIDES drillings penetated salt in a Sigsbee Knoll and even yielded hydrocarbons (Burk et al. 1969). The Knolls are now known to connect with a group of diapirs in the Gulf of Campeche (Worzel et al. 1968). The crust of the Gulf is about 19 km thick (J. Ewing et al. 1970C). The history of the Gulf of Mexico (p. 218) has been well documented (Wilhelm and Ewing 1972).

The Caribbean (Fig. 13–17) is far more active tectonically than the

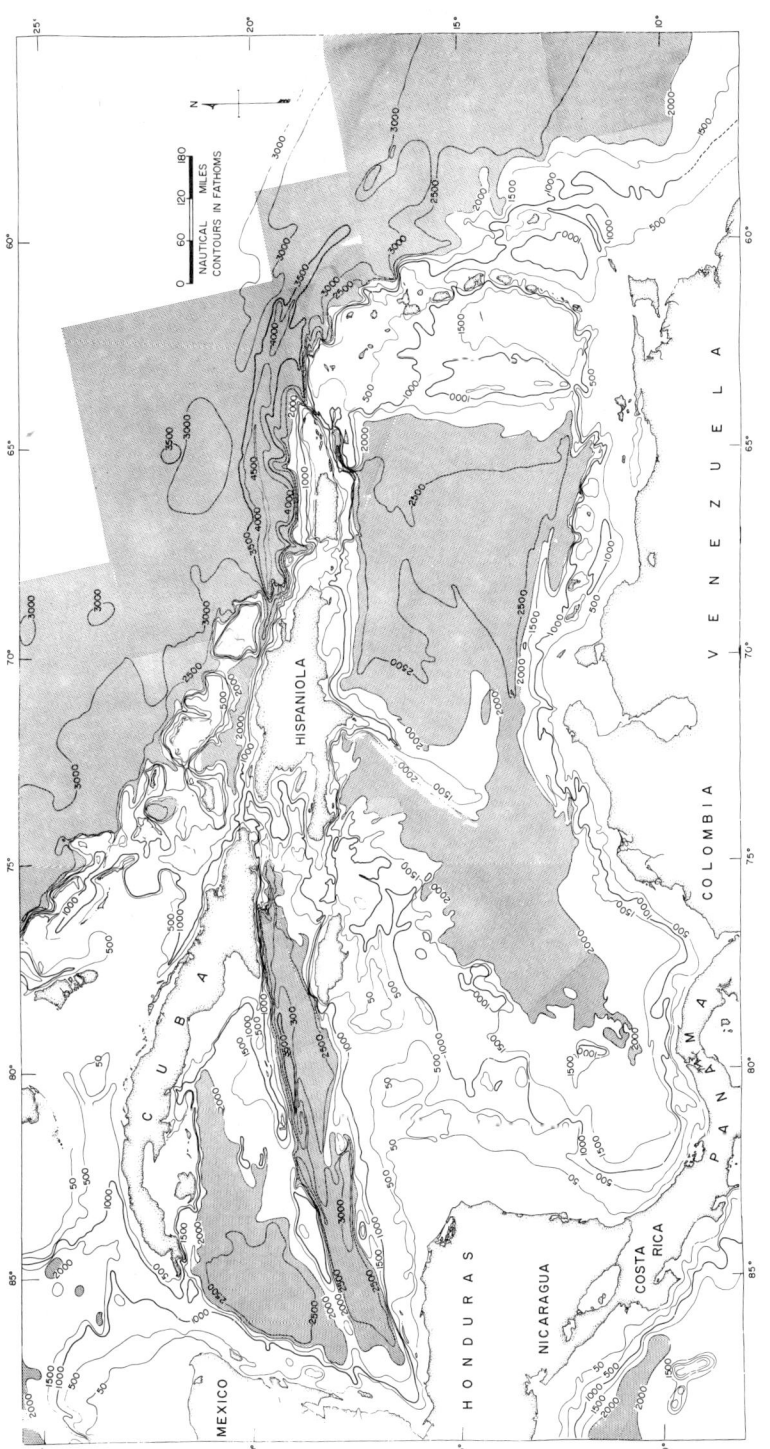

Fig. 13–17 The relief of the Caribbean and the series of basins and ridges that characterize this sea. Shaded areas represent depths in excess of 2000 fathoms (3658 m). From U.S. Hydrographic Office charts.

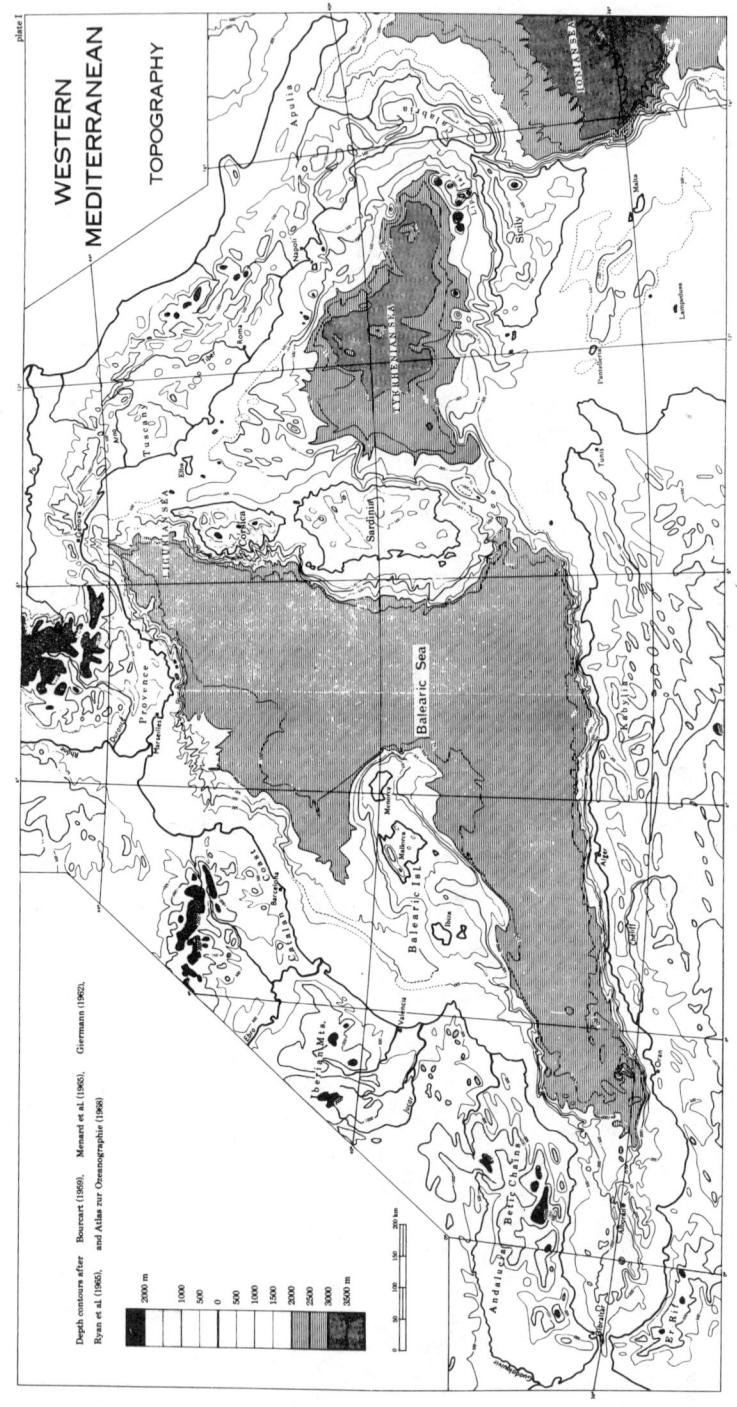

Fig. 13–18A Sea-floor relief of western Mediterranean. From Pannekoek (1969).

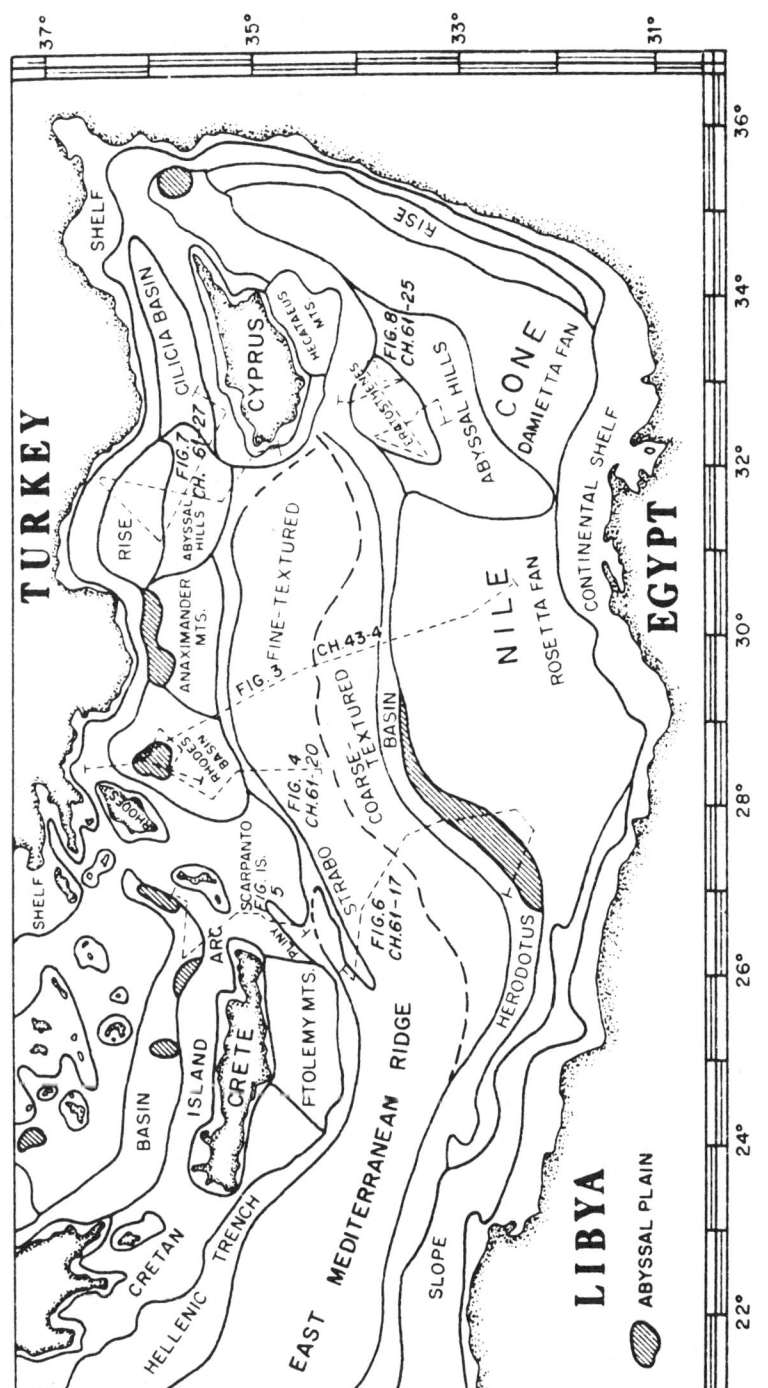

Fig. 13–18B Physiographic provinces of eastern Mediterranean. From Wong et al. (1971).

Gulf of Mexico. Most of the Caribbean is believed to consist of a plate that has been thrust eastward and is bounded on the north by the Cayman Trough, on the northeast and east by the arc of the West Indies, and on the south by the seismic zone of northern South America (Molnar and Sykes 1969; Malfait and Dinkleman 1972).

The Caribbean contains a series of trough-shaped elongate basins (Fig. 13–17). From west to east, these include the east–west Yucatan Basin with a maximum depth of 4848 m; the east–west Cayman Trough with Bartlett Deep having a sounding of 6945 m; and then west and south of Jamaica, a series of north–south basins and ridges including the Colombian Basin with depths to 4169 m; the Venezuelan Basin 5462 m deep; and the Grenada Trough with depths to 4120 m. These basins have crustal thicknesses varying from 13 to 20 km (J. Ewing et al. 1970C). The Beata Ridge (Fox et al. 1970), south of Haiti, has rocks indicating great submergence since the Eocene.

On the east side of the Lesser Antilles, the Tobago Basin separates the Grenadine Islands from Barbados. The latter island has often been referred to as containing elevated deep-sea deposits, which may be true, although the foraminifera do not appear to have originated in depths much in excess of 1000 m (Beckmann 1954), which is comparable to the Tobago Channel. The West Indies are separated from Venezuela by only a shallow channel, about 180 m deep, and from Nicaragua by a channel about 1170 m deep, which compares with about 750 m separating the Bahamas from Florida.

The Mediterranean and the Black Seas   The Mediterranean has various basins and ridges, and in its active relation to tectonism in southern Europe and northern Africa resembles the Caribbean; whereas the Black Sea is less complex in its bathymetry and has features in common with the Gulf of Mexico. The bathymetry of the Mediterranean has been delineated by Pfannenstiel (1960), Giermann (1962), and Ryan and others (1970B). The physiographic divisions of the West Mediterranean are depicted by Pannekoek (1969), and the East Mediterranean by Emery and others (1966) and by Wong and others (1971) (Fig. 13–18). The Mediterranean is separated from the Atlantic by a narrow, 284-m sill at Gibraltar, and the Mediterranean has another sill at 294 m in the Strait of Sicily. There are three major basins in this sea (Fig. 13–18)—the Ionian between southern Italy and southwest Greece, with a maximum depth of 5045 m; the Tyrrhenian west of Italy, reaching 3970 m; and the Balearic, west of Corsica and Sardinia and south of the Balearic Islands, with a 2850-m maximum. Diverse opinions have been expressed concerning the origin and history of the Mediterranean (see Glangeaud 1968; de Roever 1969; Caputo et al. 1970; McKenzie 1970; Lort 1971). The JOIDES drillings provide extensive factual information that will, of course, have a great influence on hypotheses of the future. At present, all one can say is that there are strong indications of both compressional and extensional tectonics in the Mediterranean, and the history has been very complex (see also p. 263). A feature of special interest in the Mediterranean is the many diapirs, probably salt domes, in the northern part of the Balearic Basin (Hersey 1965; Menard et al. 1965; Alinat et al. 1970). The JOIDES drillings have shown ample Miocene source beds for the salt and anhydrite in the domes.

The Eastern Mediterranean Ridge, 1600 km in length (Fig. 12–18B), is extensively deformed (Giermann 1969; Wong et al. 1971), and seismic profiles show that sedimentation in the eastern part of the sea has not kept pace with the tectonic activities except off the Nile Delta. North of the Ridge, two arcuate trenches have the greatest depths of the Mediterranean. Here, the JOIDES drillings have produced evidence of overthrusting.

Using the seismicity and fault-plane solutions, McKenzie (1970) has interpreted the plate tectonics movements for the Mediterranean. He finds that there are two small plates, the Aegean plate moving southwestward over the Mediterranean and the Turkish plate moving westward. Also, the large African plate is moving north against the Eurasian plate. Further discussion of these plates by Lort (1971) gives evidence of crustal destruction caused by the northward-moving African plate, and of intense compression in the eastern Mediterranean.

The Black Sea is almost landlocked, having only the shallow Dardanelles and Bosporus straits connecting it to the Mediterranean. The Black Sea has been studied mostly by the Soviet Union scientists (Goncharov and Neprochnov 1960) and most recently by Woods Hole Oceanographic Institution (Degens and Ross 1970; Ross 1971). The two relatively smooth-floored basins of the Black Sea are approximately 2000 m deep, with thick and largely undeformed sediments, whereas the slopes are intensely folded and faulted.

Chapter 14
DEEP-SEA DEPOSITS
AND STRATIGRAPHY

INTRODUCTION

Among the outworn ideas in geology was the belief that deep-sea sedimentation was monotonously uniform. This purely theoretical concept was opened to question as soon as deep-sea cores began to be taken. Many of these showed a change between Holocene and glacial Pleistocene (Philippi 1912; Correns 1939). Not only did the glacial-age sediments show a change in type of sediment; they also had striking differences in the planktonic foraminifera (Schott 1935; Phleger 1939; Bramlette and Bradley 1940). As coring continued and methods for increasing their length developed, complete sequences of the glacial and interglacial stages of the Pleistocene were discovered (Ericson et al. 1961). Also, much sediment was obtained of Tertiary age, either beneath a thin cover of Quaternary or directly below the sea floor where sedimentation had been prevented or erosion had occurred because of bottom currents (Riedel 1952).

The development of the JOIDES (Joint Oceanographic Institutions for Deep Earth Sampling) program (p. 29) in 1968 provided an opportunity to study the stratigraphy of all ages of deep-sea sedimentation, and truly remarkable results have been pouring into the scientific literature since that time. The borings have penetrated as much as 1 km into the bottom and have extracted formations as old as Jurassic and revealed the nature of underlying igneous rocks constituting the basement. It is still too early to obtain more than a brief synopsis of the vast fund of information that is being promulgated, but this text would be sadly incomplete without including a modest attempt to interpret the amazing results of this large and continuing project.

During the past decade, another important source of information has come from the thousands of kilometers of reflection and refraction profiles that have been run across the deep oceans. These provide us with a much clearer idea of the good sound reflectors that underlie the ocean floor and of the layers that make up the sub-oceanic crust. With the help of the deep drillings, it is now possible to interpret many of these reflectors.

SURFACE SEDIMENTS OF THE DEEP SEAS[1]

In this chapter, we shall discuss first the types of sediments found on the deep-ocean floors as well as the sources from which these sediments have been derived. The first clear indication that deep-sea sediments are distinctive in character from those of the continental terrace came from the many samples obtained during the voyage of the H.M.S. *Challenger*, 1872–1876, painstakingly described by Murray and Renard (1891), and still housed in the British Museum of Natural History in London. Later expeditions gradually supplemented the knowledge of these deep sediments, but the ground work of the *Challenger* reports still gives a remarkably good picture of the material that forms the top layer of the deep-ocean basins.

Sources of Deep-Sea Sediments The sediment sources of the deep sea are somewhat similar to those of shallow water, except that the sediments washed into the sea from the lands are largely deposited relatively near shore. As a result, the deep sea has a much greater concentration of products precipitated from the dissolved material in the ocean, either by activity of organisms or by chemical reactions, and also of the colloidal material brought in from the lands by runoff and which, having entered the sea, settles at a very slow rate. Silt-sized material is introduced by the wind. Eruptions from suboceanic volcanoes and from volcanic island craters provide a source that appears to be appreciable. Another source of deep-sea sediments is from continental ice sheets, which enter the sea as icebergs that contain an abundance of sediment of all sizes. When the ice melts the sediment is dropped to the sea floor. According to Lisitzen (1970, p. 90), most of the terrigenous sediment in the South Pacific has come from the ice of Antarctica.

The migration and source of the common deep-sea deposits is illustrated in Fig. 14–1. The products can be classified as: *lithogenous*, if they come from rock weathering or volcanic sources and have undergone little or no alteration during their slow passage through the ocean; *biogenous*, if they have been taken from the dissolved material of sea water by animals or plants; *hydrogenous*, if they have been precipitated inorganically from sea water; and *cosmogenous*, if they came from outer space as meteorites.

[1] Adapted in part from Shepard, *Submarine Geology*, 2nd ed., Chap. XVI by E. D. Goldberg.

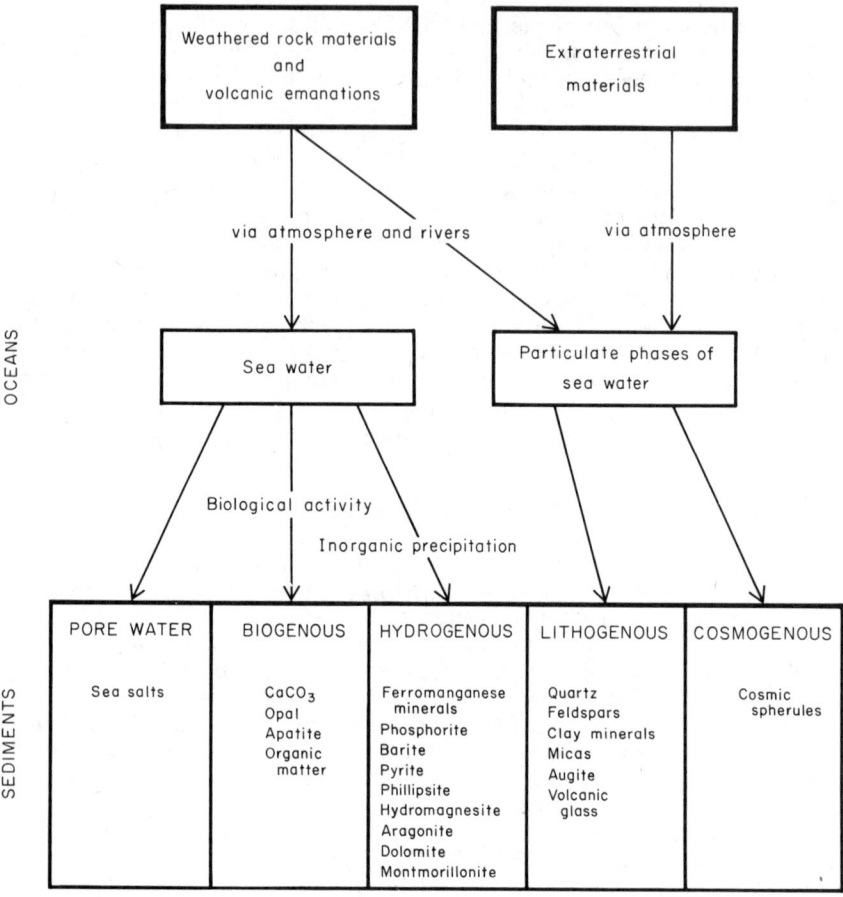

Fig. 14–1   The migration of dissolved and particulate materials through the ocean.

Classification of Deep-Sea Deposits   The classification of deep-sea deposits by Murray and Renard (1891) is still used widely but is subject to revision, partly because of recent discoveries. The older classifications included principally *red clay* and various types of organic ooze along with terrigenous muds of assorted colors. A much-needed revision has been either to abolish the term red clay or use it in quotes. There are a few localities in which deep-sea clays low in carbonates are red, but these are certainly rare in the surface sediments, and, aside from some reddish-brown silty layers said to be deposited by turbidity currents (Ericson et al. 1961, p. 203), reddish clays are apparently confined to portions of the South Indian Ocean. The typical deposit to which the name red clay has been applied has a color that is mostly a shade of brown, particularly a chocolate-brown indicative of an oxidizing environment. Therefore, it is now common practice to call this *brown clay*. *Brown lutite* was suggested by Ericson and others (1961), but the word clay seems more appropriate. Lutite is generally defined as consisting of silt and clay, whereas very little silt is present in the material in question.

Some of the deposits formerly called red clays are not even clay, because they have been largely altered to zeolites (Arrhenius 1963). Another difficulty with the older classifications is the failure to include the sandy sediments usually called *turbidites*. This oversight resulted from the short length of the early cores and from the failure to recognize fine-grained turbidites until recent years. Similarly, until the time of the Piggot North Atlantic cores (Bramlette and Bradley 1940), the importance of ice-rafted sediments, called *glacial marine*, had not been realized. These are abundant in high latitudes, especially around Antarctica. Also, the definition of *pelagic sediments* is still not settled. Therefore, without hope that any broad agreement can be reached on the definition, pelagic sediments will be considered here as those sediments of the deep, open ocean that settled out of the overlying water at a considerable distance from land and in the absence of appreciable currents, so that the particles are either predominantly clays or their alteration products, or consist of some type of skeletal material from open ocean animals or plants. The clays may be derived from the land, either by water or by wind, or may come from volcanic dust or meteorites.

Classification of Deep-sea Sediments
Pelagic
Brown clay (lithogenous material having less than 30 percent biogenous).
Authigenic (halmeric) deposits (consisting dominantly of minerals crystallized in sea water, such as phillipsite and manganese nodules).
Pyroclastic (derived from the ash of volcanic eruptions).
Biogenous deposits (having more than 30 percent material derived from organisms).
  Foraminiferal ooze (having more than 30 percent calcareous biogenous, largely foraminifera, usually called *Globigerina* ooze).
  Chalk (nannoplankton) ooze.
  Diatom ooze (having more than 30 percent siliceous biogenous, largely diatoms).
  Radiolarian ooze (having more than 30 percent siliceous biogenous, largely radiolarians).
  Coral reef debris (derived from slumping to the deep-sea floor from coral reefs).
  Coral sands.
  Coral muds (white).
Terrigenous
Terrigenous muds (having more than 30 percent of silt and sand of definite terrigenous origin).
Turbidites (derived by turbidity currents from the lands or from submarine highs).
Slide deposits (carried to deep water by slumping).
Glacial marine (having a considerable percent of allochthonous material derived from iceberg transportation).

Abyssal Clays Because of the low influx rate of particles in the deep ocean, coagulation of suspended colloids, averaging 1 micron in diameter, takes place very slowly, causing deposition of rarely more than a few millimeters per thousand years (Arrhenius 1963). For the development of clay deposits on the ocean floor it is necessary to eliminate most of the plankton, which are falling like snow through the water column and would normally constitute more than 30 percent of the sediment were it not for solution effects, particularly of the calcium carbonate below the *compensation depth* where solution exceeds input (Parker and Berger 1971). The level at which the solution rate of calcium carbonate shows a rapid increase is called the *lysocline*. Below this depth, the foraminifera in the samples are noticeably dissolved and only the least soluble are well preserved. The compensation depth and the lysocline are generally much deeper in the Atlantic than in the Pacific, so that clay deposits occur at shoaler depths in the Pacific and hence over much wider areas than in the Atlantic. On the average, clays predominate below 4400 m in the Pacific and 5300 m in the Atlantic (Arrhenius 1963). In parts of the Pacific, the lysocline has been mapped (Parker and Berger 1971) and found to be considerably deeper, about 4400 m around 20°S latitude of the western Pacific, than near the South American continent and Antarctica, where it rises to less than 3000 m. The slow deposition of biogenous material combined with the high hydrostatic pressure and high partial pressure of CO_2, due to oxidation of organic matter, favor solution of calcite and aragonite of marine organisms, and even siliceous skeletons are partially dissolved. This leaves the clay and zeolite particles as the dominant materials on the ocean floor. Because of the retarded deposition, the clays become highly oxidized, but rarely does the oxidation produce a red color in the surface sediments.

Clay minerals of the oceans and their distribution are clearly depicted in an article by Griffin and others (1968). These minerals include: chlorite, particularly abundant in the higher latitudes; montmorillonite, indicative of an oceanic volcanic source but may come directly from the continents; illite, coming largely from argillaceous sediments of the mica group and associated with the large river mouths; and kaolinite, coming from intensely weathered continental soils, notably from the tropics.

Quartz and Feldspar   As a major constituent of crustal rocks and a mineral very resistant to weathering, it is not surprising to find quartz of fine-silt size well represented among the minerals of deep-sea sediments. Rex and Goldberg (1958) found that the quartz content of pelagic sediments had a latitudinal dependence in the Pacific, being concentrated at 30 to 40° north and south. Thus, there is a relationship to the desert areas. Also, quartz is common off the Sahara Desert in the deep Cape Verde Basin of the Atlantic. Here, many of the quartz grains are coated with red-brown hematite, indicative of desert origin. Hence, the inferences are clear that much of the quartz of deep-sea deposits is introduced by the wind.

Feldspars are somewhat less resistant to weathering but are also well represented in deep-sea deposits, occurring as fine silt. In some regions, feldspars are of volcanic origin, and their mineralogy is indicative of the nature of volcanism in the vicinity (Peterson and Goldberg 1962). A distinct relationship of alkali feldspars to the area around the East Pacific Rise contrasts with the calcic feldspars from areas with major volcanic ridges and islands of the Pacific. However, the dissemination of feldspars, rather than concentration in thin beds, indicates that the origin of most feldspars has not been a single volcanic explosion.

Zeolites (Phillipsite)   The authigenic or hydrogenous products found abundantly in the deep oceans and represented principally by zeolites were first recognized by Murray and Renard (1891). The zeolite phillipsite is particularly abundant in the Pacific (Bonatti 1963). In fact, it is the principal sediment of so much of the ocean floor that it can be considered as among the most abundant minerals in the surface deposits of the world. According to Arrhenius (1963), zeolites were probably formed at the sediment-water interface, coming from elements concentrated in the interstitial water by decomposition of volcanic glass. Bonatti considers that the size and state of preservation of the zeolite crystals are such as to indicate that they have not been transported far from where they were deposited.

Although minerals formed in situ are abundant in much of the Pacific, they are apparently rare in the Atlantic. Investigation of 500 cores by Biscaye (1965) showed that almost all of the clay material is detritus from the continents.

Manganese Nodules and Crusts The increasing number of photographs of the deep-ocean floor have shown that manganese nodules consisting of hydrous manganese and iron oxides are extremely common in areas of low accumulation rates. In many photographs, the bottom is almost covered by these nodules (Fig. 14–2). Menard and Shipek (1958) estimated that between 20 and 50 percent of the sea floor in the Southwest Pacific consists of these accretions. Manganese is found also in the adjacent sediments, indicating that it is accumulated around nuclei and by direct rain of materials coming out of solution (Bender et al. 1970). The growth rate of the nodules is of the order of 110 mm/million yrs (Goldberg 1961; Bender et al. 1966). Because the clays are often accumulating at a faster rate, some process must keep the nodules at the surface and hence allow them to grow. Moore and Heath (1966) investigated in detail a small central Pacific area and found that the nodules were largely near the sediment surface and that they are much more common on steep slopes. Menard (1964, Chap. 2) discovered that the nodules in the Pacific are more common in areas with abundant benthic organisms. These facts suggest that the nodules are being kept near the surface by the organisms, particularly where there are steep slopes. The

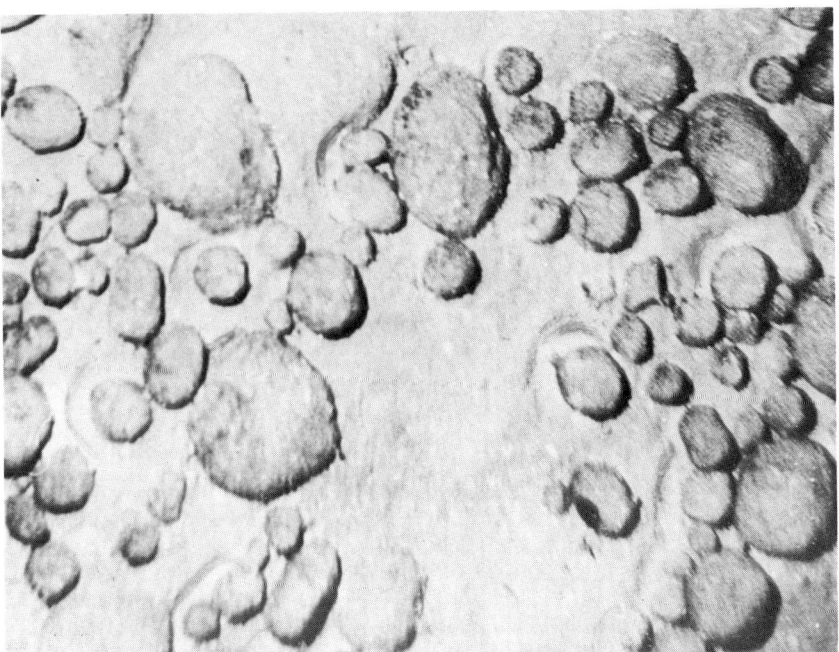

Fig. 14–2 Photo at 5500-m depth, 500 km east-southeast of Bermuda, showing nodules with current scour marks around their margins. Courtesy of David Owen.

presence of currents, as on Blake Plateau where there is a thick crust over large areas and on top of seamounts, favors the development of manganese deposits. Metallic objects on the sea floor, like old Navy shell fragments, accumulate manganese very rapidly, as much as 3 cm/50 yrs (Goldberg and Arrhenius 1958).

The composition of manganese nodules is quite variable. Distinctive types occur near the continents where the nodules are low in cobalt, indicating a derivation from continental manganese, whereas around the areas of large volcanic activity, as in the central Pacific, the nodules with a high cobalt and nickel content are believed to have come from submarine volcanism (Arrhenius and Bonatti 1965). The volcanic origin of many nodules has been established by various authorities (Pettersson 1956; Bonatti and Nayudu 1965).

The feasibility of mining manganese nodules is discussed by Mero (1965), who considers that it is quite possible to undertake such operations on a large scale. He points out that if the entire consumption of nickel in the United States were obtained from the nodules, as could be expected, the by-products would include about 300 percent of the annual consumption of manganese, about 200 percent of that of cobalt, and many times that of titanium, vanadium, zirconium, and other rare metals.

Cosmogenous Components of Sediments The first indications that meteorites have contributed to deep-sea deposits came from Sir John Murray's description of highly magnetic spherical particles from samples of the *Challenger* Expedition. Later studies by Pettersson and Fredriksson (1958) and Smales and others (1958) helped establish the meteorite origin of these spherules, and the rate of accrument to the earth surface has been estimated as 2500 to 5000 metric tons annually. Stony meteorites may add considerably to this amount. They are difficult to recognize in deep-sea deposits, but stony spherules suggestive of meteorites have been described by Murray and Renard (1891) and Hunter and Parkin (1960).

Biogenous Deep-Sea Sediments Biogenous deep-sea deposits, usually referred to as *oozes*, contain more than 30 percent skeletal material. Most have a large percentage of clay and other lithogenous and hydrogenous minerals. Plants and animals contribute three types of skeletal material: calcium carbonate, silicon dioxide, and calcium phosphate. Most of this material is taken out of solution by plankton in the upper waters of the oceans. After death the plankton sinks slowly toward the bottom, but some of it is dissolved en route and more as it lies on the bottom prior to coverage by additional sediment. Contributors to the calcium carbonate sediment include the abundant foraminifera and the less common coccoliths and pteropods. The smaller organisms, less than 60 microns, are generally referred to as *nannoplankton*.

Berger (1970B) has compared the Atlantic and Pacific, respectively, to marginal basins with deep-water outflow and to basins with deep-water inflow. Where there is outflow, as is true of the Atlantic, conditions are favorable for preservation of calcareous sediments, and inflow favors silica-rich sediments. As a result, the Atlantic with its outflowing bottom water is decidedly richer in calcareous oozes than the Pacific, which is higher in CO_2, increasing the solution capacity. According to Pytkowicz (1965), the Pacific water is undersaturated in all except the uppermost few hundred meters. This leads to the destruction of most calcareous sediment on the Pacific floor, especially below 3500 m (Berger 1970A).

The siliceous skeletal remains consist mostly of diatoms and radiolarians in opaline form (Riedel 1959). Diatoms are found in abundance in the deposits of high latitudes in both hemispheres (Riedel 1971). They live in the upper 200 m of ocean water and are particularly abundant in colder water where there is a large upwelling belt in the convergence areas near 60°S latitude and in a narrower belt in the northern North Pacific. Radiolarians, unlike diatoms, are most abundant in warm water, and radiolarian oozes cover an area in the Pacific just north of the equator. Because of solution at and near the sediment surface, large areas of radiolarians occur only within a few centimeters of the sea floor (Arrhenius 1952; Riedel and Funnell 1964, p. 361). The presence of pyroclastic material in the water helps preserve siliceous skeletons from solution on the bottom because of the liberation of silica from weathering pyroclastics.

The calcium phosphates among the deep-sea sediments consist mostly of fish teeth and ear bones of mammals, such as whales. This debris often forms the nucleus of the abundant ferromanganese nodules. Considerable quantities of heavy metals are often associated with these phosphates.

Volcanic Sediments Volcanic sediments in the deep ocean may have been derived by submarine volcanism or by great ash falls coming from land volcanoes. In several areas, it is possible to identify historical eruptions from the ash layers in the cores. In general, a concentration of shards of volcanic glass in a layer is an indication of a volcanic eruption. The huge 1883 eruption of Krakatau in Indonesia caused a worldwide ash fall that is recognizable in many cores. The Katmai eruptions of 1912 in Alaska formed a layer of ash for hundreds of miles to the southeast of the volcano. Ninkovich and Heezen (1967) found an ash layer widely over the eastern Mediterranean that represents an eruption of Santorini Volcano north of Crete. In the North Pacific, ash layers are largely confined to a zone 1000 to 1300 km wide seaward of the island arcs and continents (Horn et al. 1969).

Worzel (1959) reported a widespread ash layer in the Pacific off Central and South America. Maurice Ewing and others (1959A) suggested that this might represent a worldwide ash fall. However, as a result of extensive dredging, Menard (1960) recovered many angular consolidated slabs of ash partially covered with manganese. He suggested that the ash deposits off the eastern Pacific came from various sources, and his echo soundings showed discontinuous layers at varying depths. The ash discovered by Nayudu (1964) in cores of the Gulf of Alaska are apparently older than the ash found by Worzel. Thus, the evidence for worldwide ash falls is not convincing.

Deposits of Turbidity Currents and Other Ocean-Floor Currents (Contourites) The widespread sands and coarse silts in deep-ocean deposits have been given the general name *turbidites*. It is quite probable that most of them are introduced by turbidity currents, because many contain shallow-water fossils and even land-plant debris, indicating their continental derivation. However, we now know that other currents traverse much of the ocean floor with maximum velocities capable of transporting silt and even sand sediments. The *contourites* of the East Coast of the United States are an example (Heezen et al. 1966A). They are largely calcareous silts and have been carried south along the contours by geostrophic currents. Locally, ripple marks and streamers behind mounds are developed. It seems reasonable to expect that some of the so-called turbidites were transported by the geostrophic currents.

Turbidites (using the common name for the coarse sorted sediments of the deep-ocean floor) are found in many cores from continental rises and the adjacent abyssal plains wherever these plains are not separated from the continents by ridges or depressions. Their widespread distribution in the western Atlantic has been documented by Ericson and others (1961) and by Hubert (1964), and in the North Pacific by Horn and others (1969, 1971). The same authors show the distribution around the Hawaiian Island area. In most cores, the turbidites are covered by pelagic sediments or terrigenous muds but they also occur as surface sediments, indicating recent turbidity currents.

The existence of turbidite-sand layers can be inferred from the sound velocity in the surface and near-surface sediments (Frye and Raitt 1961). The typical deep-sea sediments have a slightly lower velocity than the adjoining sea water, whereas the sand layers have a slightly higher velocity. In seismic profiles, some reflecting layers are interpreted as turbidites but they may be other types of sediments, for example chert, as determined by the JOIDES operations.

Apparently some turbidity currents transport only silt-sized material. These can be recognized by indications of rapid deposition, such as the absence of foraminifera, in contrast to their abundance in adjacent layers. Also these silt layers may lack the mottling caused by bioturbation; or this mottling may be only in the top of the rapidly deposited turbidite bed. In tropical areas, the turbidites may be dominantly $CaCO_3$, coming from the calcareous shallow-water formations. Such calcareous layers alternate with brown clay in cores from the Puerto Rico Trench (Ericson et al. 1961, p. 252).

Glacial Marine Sediments   In both the Antarctic and the Arctic, deep-water sediments contain an abundance of coarse material in which mud is predominantly silt rather than clay. These sediments have been termed *glacial marine*, and they are thought to have been ice-rafted from glaciers. The first described (Philippi 1912) were from the Antarctic, where they extend to the margin of the ice pack, but they have been found farther north under the diatom oozes in the South Atlantic and the South Indian oceans (Hough 1956). In high latitudes of the North Atlantic, glacial marine deposits underlie the *Globigerina* ooze in many cores, notably those taken by C. S. Piggot (Bramlette and Bradley 1940). Glacially transported material was also dredged by Menard (1953) from seamounts in the Gulf of Alaska. As expected, cores from the Norwegian Sea show glacial marine sediments that are covered by more recent sediments in the deeper areas (H. Holtedahl 1959).

Sands on the Sea-Floor Highs   Another type of sand is common on the highs of the sea bottom. Fox and Heezen (1965) reported sand in cores from 24 locations along the Mid-Atlantic Ridge. These sands are mostly volcanic, but on seamounts, sands are usually a concentration of foraminifera of sand size. Currents are clearly more powerful in these shoaler areas, and the sands are concentrations due to the sorting action of currents removing the finer pelagics or volcanics. As mentioned previously, the currents on the highs are powerful enough to produce ripple marks.

Erosional and Nondepositional Areas   Rates of accumulation of sediment on the deep-sea floor are extremely variable. There are large areas where scant deposition seems to have occurred since the Tertiary. This was first suggested, although on doubtful grounds, by the discovery of manganese nodules during the *Challenger* Expedition (Murray and Renard 1891). The most extensive areas of nondeposition known at present are in the Pacific. The Swedish Deep-Sea Expedition provided the first evidence that Tertiary formations are encountered at or near the surface over wide areas in the low latitudes of the Pacific (Riedel 1952). Subsequent study of the 900 short cores (not more than 3 m) taken by Scripps Institution from 1950 to 1961 showed Tertiary sediments in 85 of them (Riedel and Funnell 1964; Saito and Funnell

1970). In general, the cores contain only Quaternary near the equator, but to the north or south the cores contain successively older material usually mixed with the Quaternary (Riedel 1959; Heath 1969). Thus, Eocene is found at about 12° north and south latitudes. The large number of cores obtained by Lamont-Doherty Geological Observatory from the Indian Ocean have shown formations as old as Cretaceous with only a small covering of more recent sediments (Herman 1963). Many dredgings by H.M.S. Egeria on the higher areas contained Tertiary rocks as old as Eocene (Wiseman and Riedel 1960).

Evidence of erosion of the deep-sea floor of the equatorial Pacific has been obtained by D. A. Johnson (1971) using deep-towed instruments. Angular unconformities are evident in the seismic profiles of these near-bottom projectors. The work led to the conclusion that hundreds of meters of sediment have been eroded by bottom currents; this is confirmed by cores taken in the area. Much of the eroded material has been redeposited in a nearby basin where the sediments are considerably thicker than elsewhere. The erosion appears to have begun after Late Tertiary sediments were deposited, and is thought by Johnson to have occurred almost entirely during glacial episodes when the bottom currents were stronger than at present. A thin layer of Holocene covering the unconformity shows that erosion has ceased since the last glacial episode. Erosion on the sides of the Bahama Outer Ridge is also indicated by the JOIDES operations.

Distribution of Deep-Sea Sediments   The distribution of the principal types of sediments in the ocean is given in Fig. 14–3.[2] A vast amount of speculation is still involved in any map of this type. Many changes have been made from the widely reproduced map that appeared in The Oceans (Sverdrup et al. 1942). Now the diatom ooze extent is greatly restricted in the North Pacific, and the radiolarian ooze in the high-productivity zone just north of the equator in the eastern Pacific is considerably narrowed. A sharp drop in the $CaCO_3$ content occurs at 5°N latitude and is almost lacking in the sediments north of 9°N latitude (Arrhenius 1963). Furthermore, extensive areas of terrigenous sediment are now included in both the Atlantic and the Pacific, and areas of glacial marine sediment are shown bordering the Antarctic continent.

The sediments formerly called "red clay" in the South Pacific are found by Arrhenius (1963) to consist dominantly of the alteration product phillipsite, making it an authigenic deposit in contrast to the more extensive brown clay area in the North Pacific, which was also included in the "red clay" zonation.

A map of the sediment in the northeast Pacific has been supplied by Nayudu (1959). Based on 150 cores, his map shows that there are seven well-defined zones (Fig. 14–4), including terrigenous sediments around the margins, a large zone with Katmai volcanic ash from the 1912 eruption, glacial marine sediments with diatoms, clays with radiolarians, Globigerina-rich silts, and diatom-rich sediments.

The Soviet Union has made extensive studies in the ocean around Antarctica, which lend considerable authority to the sediment map of the Pacific by Lisitzin (1970, Fig. 9; 1972, Fig. 87). These maps give more detail concerning the content of organisms in the biogenic material, and are probably much more accurate in the high latitudes of the South Pacific than the maps included here.

[2] A large new source of information is available in Lisitzein (1972).

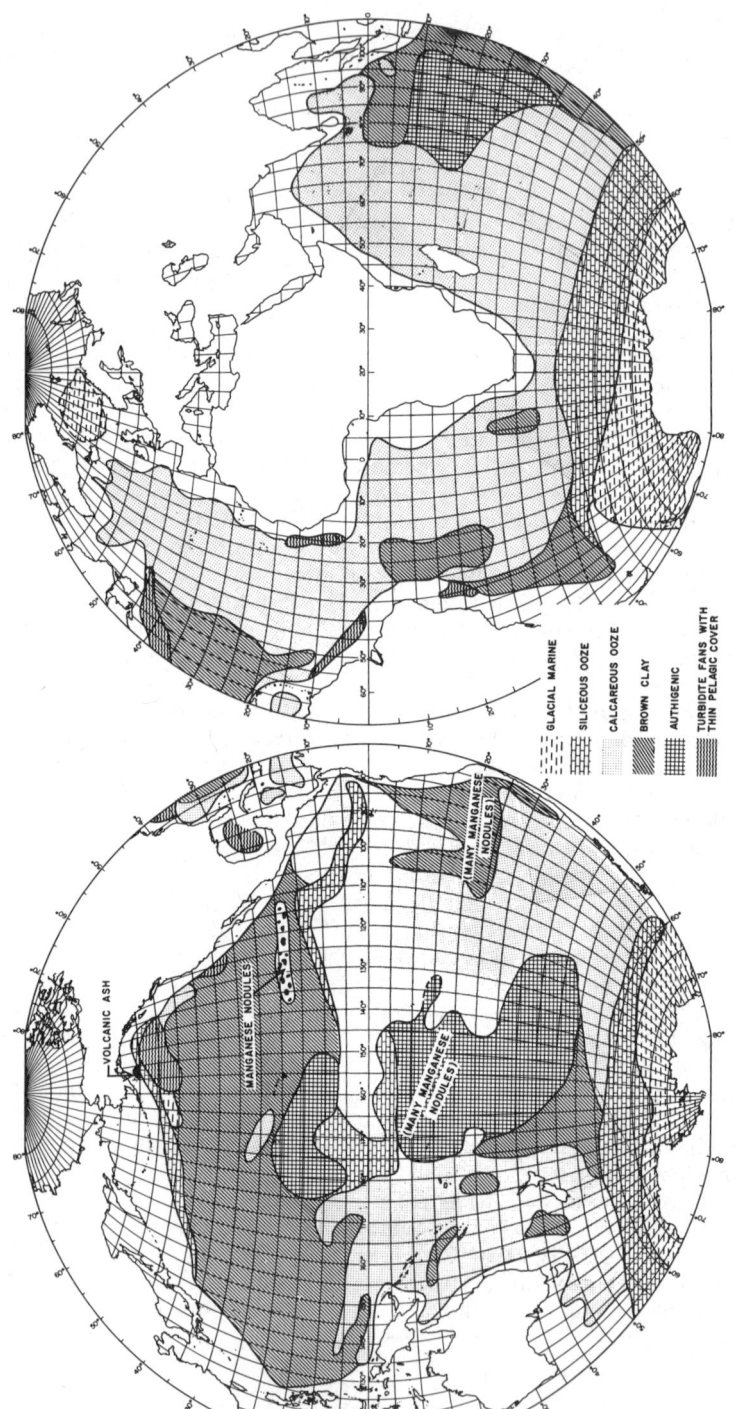

GLACIAL MARINE
SILICEOUS OOZE
CALCAREOUS OOZE
BROWN CLAY
AUTHIGENIC
TURBIDITE FANS WITH
THIN PELAGIC COVER

VOLCANIC ASH

MANGANESE NODULES

(MANY MANGANESE NODULES)

(MANY MANGANESE NODULES)

(MANY MANGANESE NODULES)

Fig. 14–3 Distribution of deep-sea sediments. Compiled from information by Arrhenius (1961) and Nayudu (1959) and from suggestions by D. B. Ericson, H. W. Menard, and W. R. Riedel. It is quite likely that the areas with abundant manganese nodules cover more territory than shown and that there are many more areas where turbidite layers are interbedded with normal pelagic deposits. All boundaries should be considered as subject to extensive changes as information becomes more abundant.

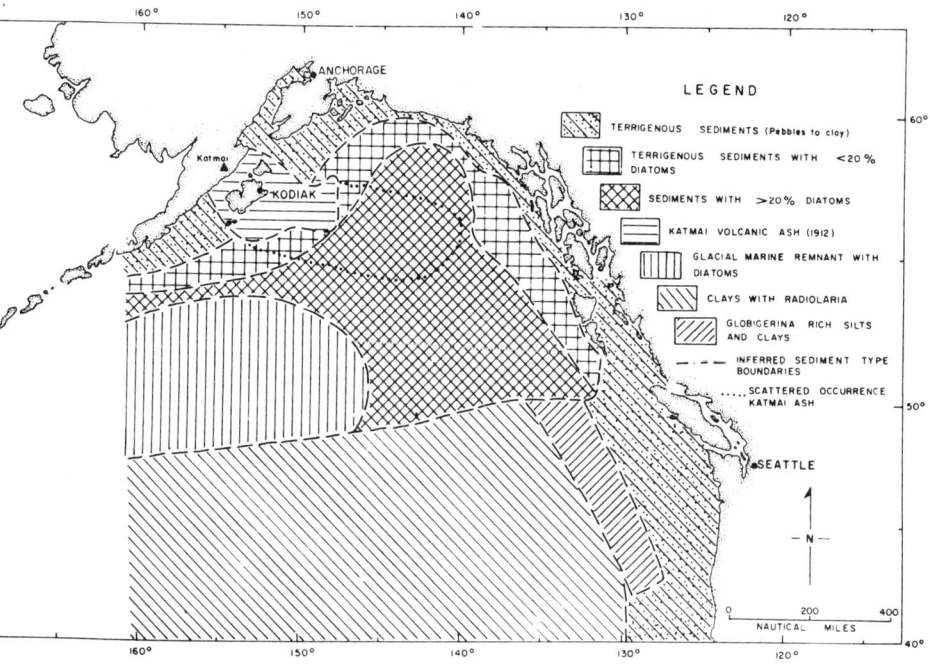

Fig. 14–4   Lithologic distribution of surface sediments in northeast Pacific, based on upper 5 to 10 cm of the cores taken by Nayudu (1959).

QUATERNARY SEQUENCE FROM LONG CORES

Even before JOIDES probed deep into the ocean floor, the ordinary piston corers of 10 to 30 m were sufficient in most places to penetrate well into the Pleistocene or even to the underlying Pliocene. The glacial stages of the Pleistocene can be recognized in much of the Atlantic by the changes in the foraminiferal assemblage. Cooler-water plankton in the layers represent the glacial stages. In the higher latitudes, these faunal changes are combined with scattered pebbles in the glacial-stage layers, which were transported by icebergs. These sequences were first well documented by studying Piggot transatlantic cores (Bramlette and Bradley 1940). Coiling direction changes of *Globorotalia truncatulinoides* are also useful. Temperature at the time of deposition can also be estimated by O^{18}/O^{16} measurements, a system developed by Urey (1947) and applied to deep-sea stratigraphy, mostly by Emiliani (1955, 1966). The foraminifera and O^{18}/O^{16} methods have been applied equally well to the Indian Ocean (Oba 1969). Many cores from the lower latitudes of the Pacific do not show the faunal changes, the glacial marine sediments, or the O^{18}/O^{16} changes that serve so well in the Atlantic and in the high latitudes of the Pacific (Wollin et al. 1971). However, in two low-latitude Pacific cores, the same faunal changes were found as in those of the Atlantic (Blackman and Somayajulu 1966). Arrhenius (1952) discovered that the alternating cold and warm periods are well represented by variations in the carbonate content. The higher carbonate content in glacial stages is the result of the intensification of wind-driven surface circulation leading to increased vertical transport of nutrients to the surface waters in the upwelling areas. Emiliani (1955) found small temperature drops in a few of the layers with high carbonate content.

The dating of cores has been made by a combination of methods. The Holocene and latest Pleistocene have been extensively dated by carbon-14, but this becomes unreliable beyond about 30,000 years B.P. At first, the greater core ages were estimated by assuming a continued even rate of deposition of the organic components of the sediment. Uranium and thorium have been used by several investigators to about half a million years (Sackett 1965; Ku and Broecker 1966). A particularly useful tool has been the magnetic reversals (the periodic change of the earth's magnetic poles) that have been detected in many of the long cores (Opdyke et al. 1966; Opdyke and Glass 1969; Hays et al. 1969). These magnetic reversals total 130 and have been dated as far back as about 79 million years (Heirtzler et al. 1968), so that they extend into the Cretaceous.

The Pleistocene was considered for many years to have had a length of about 300,000 to 1 million years. Now, with the knowledge of magnetic reversals in deep-ocean cores, it has become evident that glacial marine layers extend back at least 2 to 3 million years (Conolly and Ewing 1965; Hays 1965; Donn and Ewing 1968; Ericson and Wollin 1968). There is uncertainty whether the base of the Pleistocene should be considered as the time when glaciation was initiated. It probably existed throughout the Pliocene in Antarctica (Margolis and Kennett 1970). However, the faunal changes, such as the appearance of *Globorotalia truncatulinoides* and the extinction of the Discoasteridae

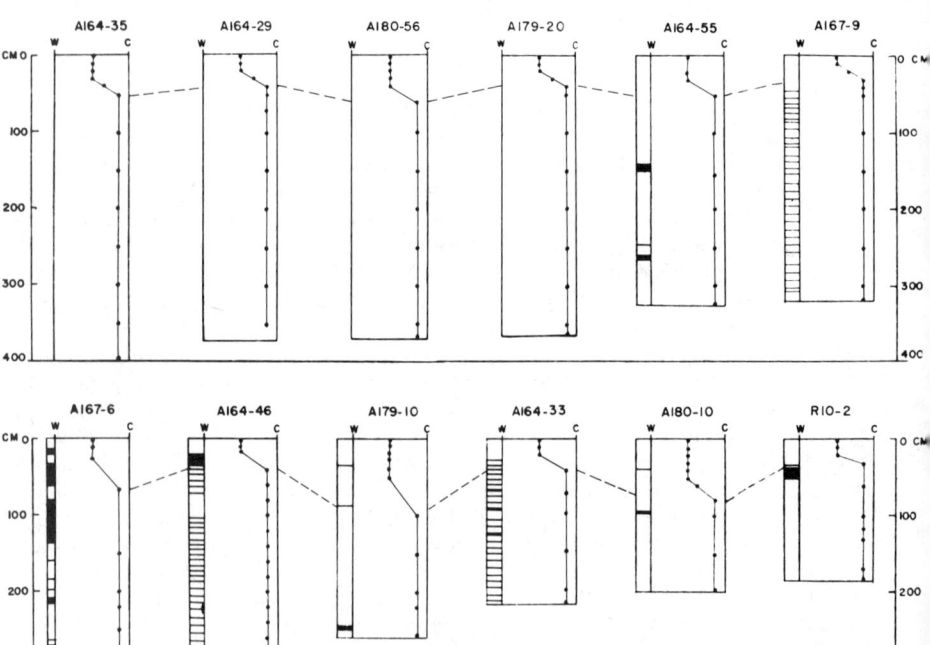

Fig. 14–5 Attempted correlation of horizons in cores from the deep Atlantic, using relative numbers of warm- and cold-water foraminifera. From Ericson et al. (1961).

(Ericson and Wollin 1968), which are said to represent the boundary, appear to agree rather well with the beginning of major glaciation. An attempt to correlate results from a group of cores is illustrated in Fig. 14–5. The extensive explorations of the oceans around Antarctica has produced much new evidence of the boundary between the Pleistocene and Pliocene (Hays 1965; Opdyke et al. 1966).

The dividing line between the Holocene and Pleistocene is also rather indefinite; a common suggestion places it at about 11,000 B.P. This is approximately halfway between the maximum glaciation of the last phase of the Wisconsin (Würm) and the attainment of the near-present sea level, 2000 to 3000 B.P. (Fig. 6–3). Also, at approximately 11,000 B.P., the sea had risen sufficiently so that the Bering Strait was flooded, allowing water to come into the Arctic from the Pacific (Olausson and Jonasson 1969). The foraminifera in North Atlantic cores appear to have changed from cold to warm types also at about this time.

The dating of interglacial stages is not well established by cores, as will be seen by examining Fig. 14–6. However, aside from the possible relatively high sea level of the Mid-Wisconsin, the Sangamon (Riss–Würm), the last of these episodes, is rather well documented as having taken place between 80,000 and 150,000 B.P. (Veeh 1966; Ku 1968; Broecker and Ku 1969). The record of coral reef development on Barbados Island shows a close agreement with three warm episodes during the Sangamon in a core analyzed

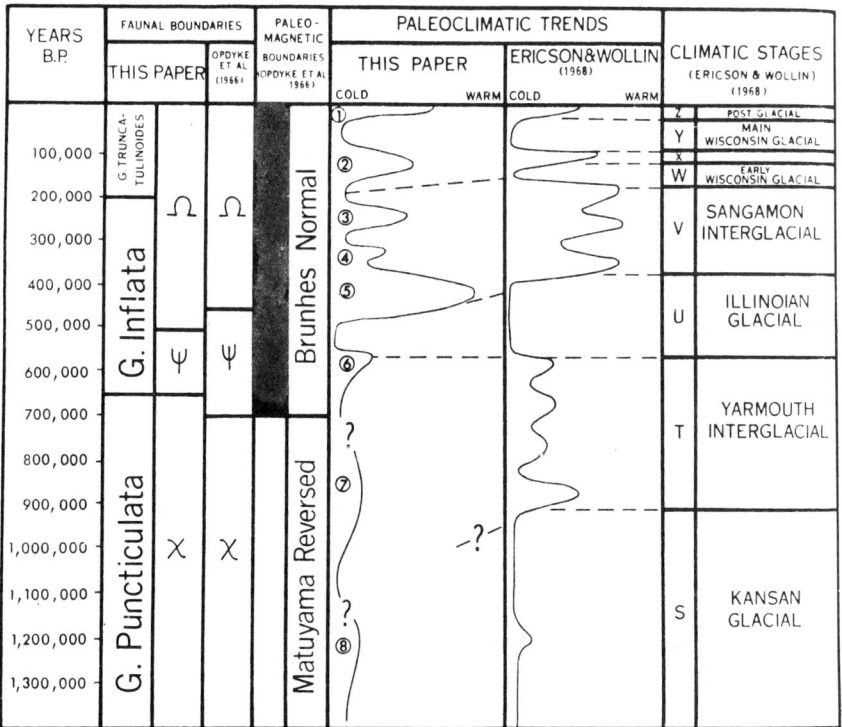

Fig. 14–6 Dating of paleoclimatic events during the upper and middle Pleistocene, including relation to the last paleomagnetic reversal. From Kennett (1970).

by Broecker and Ku. Also, various dates of elevated reefs in Florida and in the western Pacific come into the same interval.

The much-discussed hypothesis of an ice-free Arctic during portions of the Pleistocene (M. Ewing and Donn 1956; Donn and Ewing 1968) was not well received in its original form, which called for ice-free stages during each glacial stage and an ice-covered sea during interglacial periods. Now, however, Donn and Ewing call for freezing of the Arctic as a glacial stage advances and then gradual melting of the ice pack in the last stages of glaciation, causing growth of new ice sheets due to local sources of moisture from the open Arctic Ocean. The most recent contribution to this discussion (Clark 1971) is based on studies of more than 100 cores. Clark concludes that the evidence, particularly from the foraminifera, indicates that the Arctic Ocean may have been continuously covered with ice since Middle Pliocene. He suggests that layers with scarcity of foraminifera developed when the ice cover thickened independently of continental glaciation.

TRANSOCEANIC SEISMIC PROFILES

With the discovery that seismic reflection profiles could be run at speeds up to 12 knots, it has become standard practice for oceanographers to run the reflection devices during most crossings of the deep ocean. Lamont-Doherty ships have obtained a vast collection of these records from all over the world, and their scientists have been leaders in interpreting the results (M. Ewing 1964; M. Ewing et al. 1964, 1969B; J. Ewing et al. 1965, 1966B; Windisch et al. 1968). Profiles by Scripps Institution scientists have been run especially in areas where deep drilling was anticipated in the JOIDES program and in small areas where the deep-tow method was put into operation (Heath and Moore 1965). Much of the information from seismic profiles has not been published, but the reports, particularly from Lamont-Doherty scientists, have shown that there are good reflectors that can be traced for hundreds or thousands of kilometers across the oceans. These reflectors have generally been referred to as Horizon A, B, etc. (Fig. 14–7). The age of some horizons was tentatively established (J. Ewing et al. 1966A; Windisch et al. 1968) by dredging at points where the layer apparently outcrops. Drillings by JOIDES have now shown that many of the good reflectors are from chert formations of Eocene or Cretaceous age (J. Ewing et al. 1970B). The seismic reflection profiles across the ocean show contrast between "transparent" and "opaque" layers. The former are clear, showing little reflection; and the latter have closely spaced or continuous reflectors.

Another feature of interest from the reflection profiles is the rough idea they provide of the thickness of the sediments in the ocean basins. Thus the sediments are found to thin over the mid-ocean ridges and in general to thicken with distance from the ridges, deriving their greatest thicknesses at the base of the slopes on the west side of the Atlantic (M. Ewing et al. 1964). The refraction method of shooting has been more successful than the reflection profiles in measuring the very thick sediment masses. The two methods supplement each other. A recent example of the use of these cooperative methods was from surveys of the Bay of Bengal (Fig. 13–15B), where a sediment thickness of 16.5 km was obtained, the thickest of any place in the world (D. Moore et al. 1971).

Eventually the JOIDES operations will provide much better information concerning the significance of the reflectors. As explained in the next section,

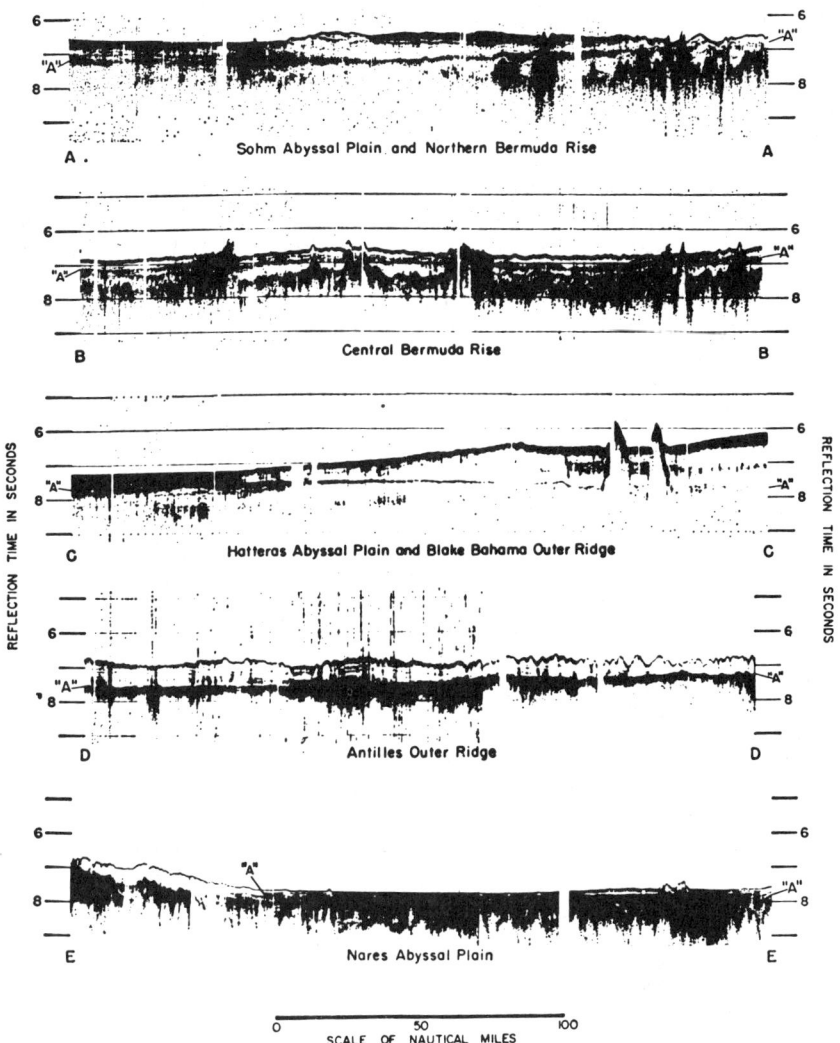

Fig. 14–7 Horizon "A" and other good reflecting horizons in transoceanic seismic profiles. From J. Ewing et al. (1966B).

the deep drillings have confirmed some of the assumptions from seismic reflection and refraction but have shown that others were misleading.

DEEP-SEA DRILLING (JOIDES)

The great revolution in geological thought initiated by the sea-floor spreading or plate tectonics hypothesis (Chap. 5) is often said to have been verified by deep-sea drilling (see p. 429). It is interesting that two such seemingly impossible feats as landing on the moon and drilling for a thousand meters into the deep floor of the ocean should have been accomplished in the same decade.

JOIDES is in a sense an aftermath of the temporarily abandoned Mohole Project, which was intended as a one-hole attempt to penetrate the earth's crust to find what constituted the underlying mantle. This project would have meant drilling at least 5 km into a part of the ocean floor where the crust appears to be particularly thin. The expense estimates grew to staggering proportions, and much skepticism was expressed by engineers as to the possibility of success for Mohole. With its abandonment, after having accomplished some relatively small deep-sea drillings (AMSOC 1961), drilling of shallower holes, using similar principles, was begun (see p. 29). At present, the JOIDES program has been continuing for three years (six including the trial holes off Florida) and the results have been outstanding (Peterson 1971). It is now possible to report some of the principal discoveries from 170 sites in the Atlantic, Pacific, the Gulf of Mexico, the Caribbean, and the Mediterranean (Fig. 14–8). Most sites have had more than one hole. The information quoted below comes largely from the bimonthly summaries that have appeared in *Geotimes*. Other data are from Volumes I to VI of *Initial Reports of the Deep Sea Drilling Project* and from several papers published in journals, as well as discussions with my colleagues at Scripps Institution who have participated in various legs of the project.

The Atlantic   Included in Legs 1 to 4, 11, 12, and 14, the Atlantic has been extensively explored with holes at 48 sites. The drillings in this ocean confirmed many of the conclusions that had come from previous geophysical prospecting. Thus, the sediments were found to be thin over the Mid-Atlantic Ridge and, in general, to increase in thickness away from the Ridge. The fossils, mostly plankton, showed that the oldest sediment overlying igneous basement increases in age with distance from the Ridge (Peterson, Edgar et al. 1969; Maxwell, Von Herzen et al. 1970). The estimated ages of magnetic anomalies agree quite well with radiometric dating of oldest sediments, although the basement rocks often were found to be sills rather than lava flows and may have had thick underlying sediments. Spreading half rates are estimated for the Atlantic as from 1 to 4 cm/1000 yrs and average 2 cm.

Some idea of the nature of formations encountered is shown in Fig. 14–9. Section A, across the North Atlantic, indicates the depth graphically; Section B, across the South Atlantic, gives the distance from the Mid-Atlantic Ridge, which can be compared with basement age. On the left of Section B are two drillings from the Rio Grande Rise off southern Brazil. This Rise had shallow water in Early Cretaceous, representing either a foundered continent or a subsided group of guyots. In Section A, Eocene cherts are shown; these caused termination in drilling by wearing out the bits but provided the good reflectors of Horizon A in the seismic profiles (J. Ewing et al. 1970B). Deep-sea clays are shown in two of the deep holes in Section A, and in both cases these clays underlie calcareous oozes. They have a reddish-brown color and are called "red clay." The other formations are chalky or marly oozes, as might be expected considering their position on topographic highs. As shown in Section B, the Oligocene contains a remarkably persistent white chalk formation in all drillings except on the ridge where the basalt basement is Miocene. This chalk with *Braarudosphaera* (a genus of coccolith) was formerly thought to be indigenous to shallow water but here obviously was found in a deep-ocean deposit. It is considered a time marker for a broad area. Apparently the bed indicates a widespread condition of rapid deposition, so

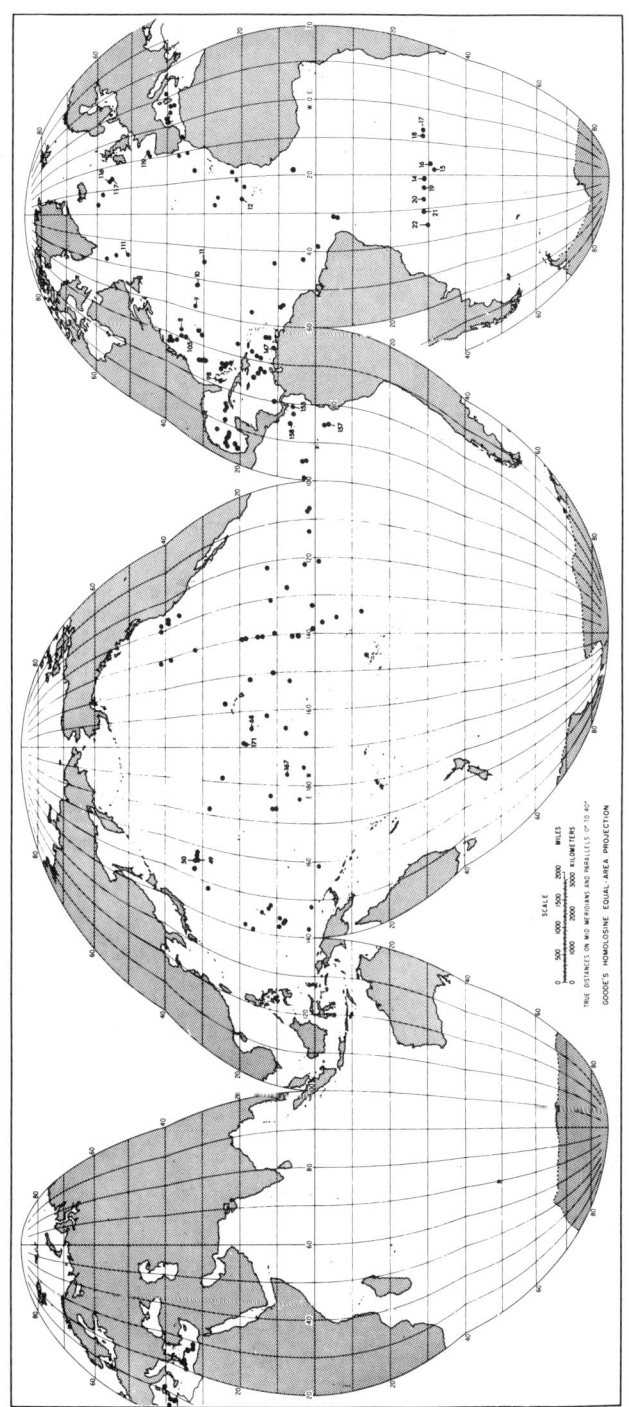

Fig. 14–8 Map showing location of JOIDES drilling holes up to July 1971. Locations discussed in text are numbered.

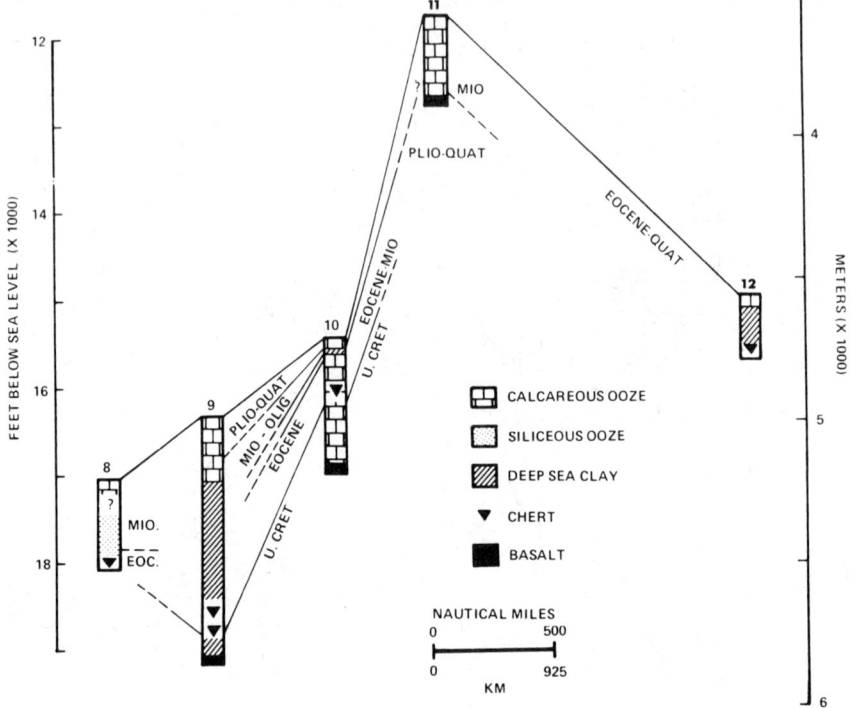

Fig. 14–9A Section showing correlations of five drilling holes across the North Atlantic. Note increase in age away from the Mid-Atlantic Ridge. Hole 11 was near Ridge. From Peterson, Edgar et al. (1970).

that the calcium carbonate was not dissolved, even in such deep water that the bed lies between "red" clays. The rate of deposition for foraminiferal ooze in this South Atlantic section averages 1.8 cm/1000 yrs, and for "red" clay 0.16 cm/1000 yrs.

The age of the North Atlantic is tentatively established as about 155 million years, but apparently it began to form around 175 million years ago. On the east side of the mid-Atlantic, the oldest formations are only about 110 million years old, so there may have been a small Proto-Atlantic on the west that preceded the ocean development on the east side of the Ridge.

The series of drillings well north in the Atlantic on Leg 12 (Laughton, Berggren et al. 1970) provide information from continental masses that have broken away from Newfoundland (Site 111), Great Britain (Sites 116, 117), and northwest Spain (Site 119). These drillings all show large unconformities and large changes in depth. Other cores on this leg indicate the importance of strong bottom currents that evidently are related to the deep circulation of the Labrador Sea and over the sills into the Norwegian Basin. Very high rates of deposition are found here. The Quaternary cores from the Labrador Sea contain thick glacial marine deposits with pebbles from icebergs.

Drillings of the continental rise off the southeastern United States and northwestern Africa produced some evidence of turbidite deposits, but most of the cores are clay. Bahama Outer Ridge was shown by cores from several

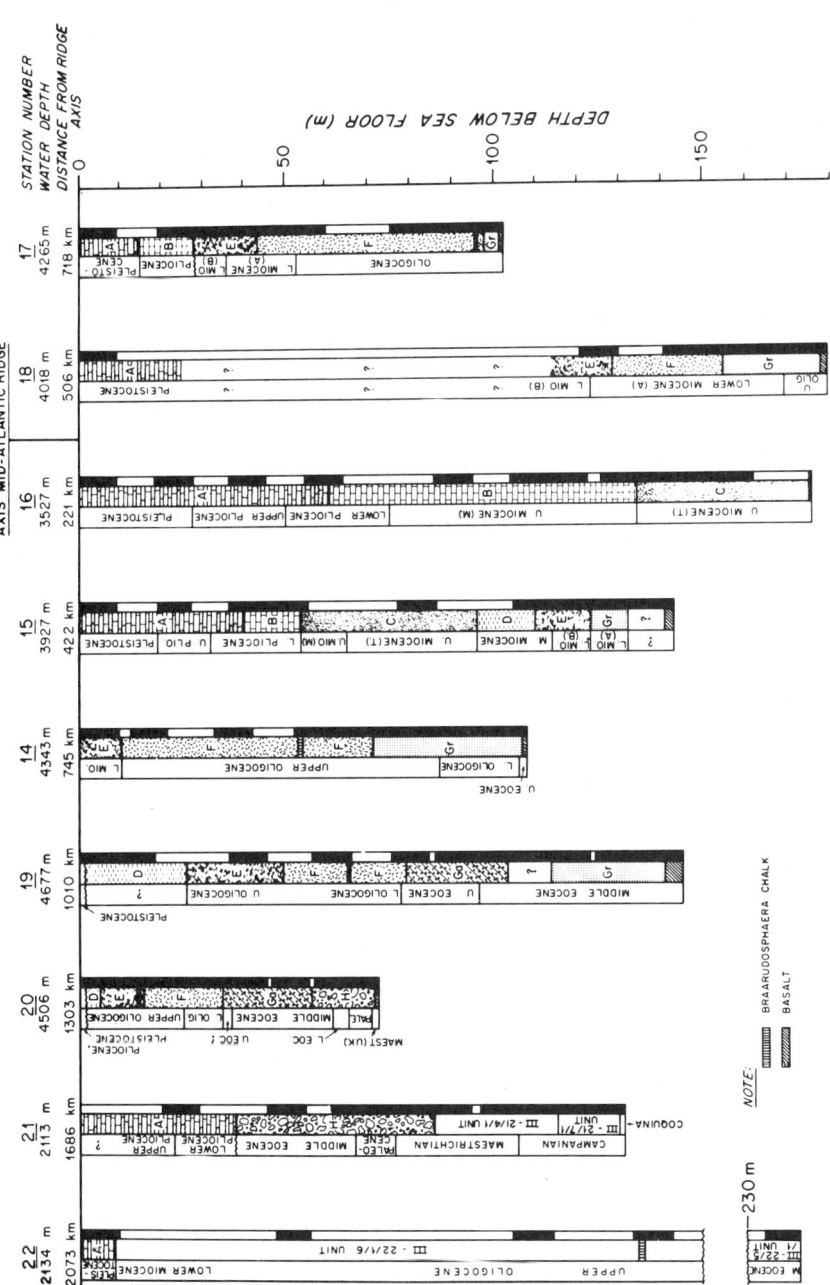

Fig. 14-9B Section of drillings across the South Atlantic, showing relation of age of basement to distance from Mid-Atlantic Ridge and general character of the formations: A to C are nannofossil chalk oozes, D is "red" clay, E and F are nannofossil chalk and marl oozes, Ca nannofossil ooze and clay, Gr and H nannofossil chalk ooze. From A. E. Maxwell, Von Herzen et al. (1970)

holes to be an erosion remnant with beds outcropping on both sides (J. Ewing et al. 1970B). The apparent anticline shown in seismic profiles probably represents a diagenetic change in depth. Sediments from which the Ridge was built are said to be derived from the north and were deposited rapidly, especially during Middle Miocene when a rate of 19 cm/1000 yrs existed. Continued deposition on the Ridge up to Pleistocene was at a slower rate. Whether erosion on the sides was contemporaneous with deposition or subsequent is not known. Rona (1970A) had interpreted a group of diapirs near the base of the continental rise north of Cape Verde Islands as possible salt diapirs. Drilling into one of these (Site 141) encountered basalt (Hayes, Pimm et al. 1971). However, as Rona (1971B) has emphasized, the nearby drillings at Sites 139 and 140, also on the lower continental rise, in the midst of the other diapirs showed marked salinity gradients rising to 75 $^o/oo$ in one place at a 665-m drill penetration. This appears to be evidence that at least some of the diapirs may have salt cores.

Site 142 on the Caera Rise, 700 km northeast of the Amazon mouth, provided evidence of 1200 m of uplift prior to the Pliocene. Thin sandy lenses in the cores at this site may be turbidites. Site 26, in the mid-ocean Vema Fracture Zone, has 483 m of Mid- to Late-Pleistocene that are definitely turbidites from the distant Amazon. Some of these beds are largely land plants.

Several bizarre features, such as native copper and sphalerite crystals, in a hole at the northern end of the Hatteras Plain (Site 105) were found just below 240 m and near the basalt base. In this same hole, an interval of 50 m has an age change from Oligo-Miocene to Early Cretaceous.

The formations have undergone considerable diagenesis with burial, notably the cherts in the North Atlantic. Also, the clays have undergone changes to zeolites in depth. Near the contact with basalt, many vein formations were found and iron deposits were encountered.

Gulf of Mexico The drillings into 16 sites in the Gulf of Mexico on Legs 1 and 10 have produced some of the most interesting results of all the JOIDES operations to date. It had been predicted that the Sigsbee Knolls would prove to be salt domes, but it was still surprising to many to find salt doming in Site 2 under the deep central portion of the Gulf (M. Ewing et al. 1969C). The discovery of gas and small amounts of petroleum caused excitement, but, because of the oil leaks and pollution scare at Santa Barbara, stringent restrictions were imposed on future drillings in all areas where oil might be encountered. The caprock in Site 2 proved to be typical of that of the diapirs throughout the large salt-dome area to the north. It consists of calcite and gypsum with abundant sulphur.

Two holes near the Sigsbee Scarp showed the importance of salt tectonics in the development of the escarpment. Chevron folds in the thick Pleistocene of the cores at Site 1 may be explicable by the salt. Site 92, just north of the escarpment, showed clearly the effects of the underlying salt intrusion, and the salinity increased with depth in the hole, adding some confirmation to seismic profiles that indicated the presence of salt a short distance below the hole base.

The thickness of the Pleistocene in the central Gulf is much greater than in other deep-water areas drilled to date. At Site 1, drilling down to 770 m did not get below the Pleistocene. Most of the Pleistocene sediment was terrigenous silty clay, but some pelagic material was found in the upper portions. Borings from the central area of the Sigsbee Deep traversed very thick Pleistocene underlain by thinner Pliocene and Miocene. The Pleistocene is predominantly turbidites, probably coming from the Mississippi Fan. The Pliocene, however, is mostly pelagic.

In the western part of the Sigsbee Basin, Miocene below the pelagic Pliocene contains thick turbidite sands that stopped the drilling. This coarse sediment was thought to come from a western source. Here, the Middle Miocene depositional rate appears to be equal to that of the Pleistocene.

A hole drilled in the lower part of the Campeche Scarp reached Lower Cretaceous, and the sediments of the entire penetration were apparently deposited in deep water, indicating that the deep Gulf of Mexico dates back at least to Late Cretaceous. Further evidence of Cretaceous deep-sea conditions was found in the western approaches to the Florida Strait, with pebbly mudstones and black cherts having a deep-water fauna. Shallow-water dolomites discovered in Lower Cretaceous on the Campeche Scarp, however, were covered by Late Cretaceous bathyal chalks and oozes.

Caribbean Sea The drillings into 11 sites in the Caribbean were conducted on Legs 4 and 15. During the latter, the first hole reentry was accomplished. This allowed drilling to continue through the Eocene chert formations, which had stopped several earlier operations by wearing out the bit. Drillings of several holes in the basins show that thicknesses of the Pleistocene are approximately as great as those of the Pliocene, although the sediments of these ages are primarily pelagic lacking the abundant terrigenous turbidites of the cores from the Gulf of Mexico.

The oldest formations above the dolerite basement are Coniacian of Late Cretaceous age, about 80 million years ago, and indicate that the deep Caribbean dates back to this time but probably not much farther. Therefore it seems probable that the Caribbean is not nearly as old as the Atlantic. In the early stages of the opening up of the Caribbean, volcanism was a very important factor, as shown by the high content of volcanic products. This decreased gradually and was not important after the Cretaceous except in the easternmost sectors near the volcanic islands of the Antilles. The cores from Site 154, in the western Caribbean, are relatively free of volcanic terrigenous material down to near the base of the Pliocene, where pelagic material is largely replaced by volcanic sands, silts, and clays.

The cores from several of the sites show the significance of diastrophism in the area. On the Aves Ridge at 1232-m depth, a subaerially weathered layer was penetrated but this is overlain and underlain by deposits suggestive of bathyal conditions.

At Site 147 in the Cariaco Trench off Venezuela, where a basin is surrounded by land or by water not deeper than 200 m, a hole was drilled in the floor to 892 m, which did not go below the Pleistocene. As expected, it penetrated an anaerobically deposited sediment series lacking benthic foraminifera. A high rate of sedimentation has taken place here, about 50 cm/1000 yrs.

Mediterranean Sea The Mediterranean was drilled at 15 sites during Leg 13 (Ryan, Hsu, et al. 1970A). The results of this leg were particularly exciting, because they yielded evidence of drying up of the deep basins, followed by sudden incursions of water, presumably from the Atlantic. Large deformations including overthrust faulting were indicated, and new information on plate tectonics was also obtained in the Mediterranean.

In the Hellenic Trough (see Fig. 13–18B for locations), Lower Cretaceous limestones were found overlying flat-bedded middle Pliocene pelagic ooze. In the Levantine Basin, it was revealed that the Mediterranean Ridge was part of the Nile Abyssal Plain as late as middle Pleistocene. These Nile sediments were upheaved onto the ridge. In the Alboran Basin, east of Gibraltar, oceanic basalt underlying Upper Miocene suggests that Africa was rifted from the Iberian Peninsula.

The intrusive material of the domes in the Balearic Basin is apparently derived from Late Miocene salt deposits. Halite was penetrated here with intercalated layers of Late Miocene age. This is part of an evaporite series, which forms a good seismic reflector widely over the area (called the M-reflector).[3] Gypsum and anhydrite were also found in the Valencia Trough, the Tyrrhenian Basin, and the Ionian Basin. Prior to the period of evaporation, normal marine limestones were deposited in the eastern Mediterranean.

At or near the beginning of the Pliocene, a sudden incursion of ocean water drowned the evaporite basins of much of the Mediterranean. At Site 132, in the Tyrrhenian Sea, a flat-pebble Pliocene conglomerate overlies a gypsiferous current-bedded marl of Miocene age. In the eastern Mediterranean, some sapropel layers in the Pleistocene suggest temporary stagnation at this late date.

The very good sections of Pleistocene age in the Mediterranean cores should prove useful for comparison with the much-studied Pleistocene stratigraphy of Italy. The entire Quaternary has many tephra layers indicating intermittent volcanism. A hiatus between the Pliocene and Pleistocene appears in some cores. The seismic profiles show that submarine channels were cut prior to this break in sedimentation. Some of these channels cut into evaporite beds.

The preliminary examination of the cores from the Mediterranean suggests that they have much in common with the rocks of the Alpine chains.

Pacific Ocean   In the Pacific, Legs 5 to 9, 16, and 17 have been completed by June 1971 and holes had been drilled at 69 sites. As this is written, Leg 18 is being run and plans are being completed for three more legs in the Pacific to follow. To date, most of the drilling has been concentrated between the equator and 25°N latitude (Fig. 14–8).

The primary purpose of the program in the Pacific has been to investigate the nature of sea-floor spreading and plate tectonics. With the improvement in drilling methods, particularly in longer-lasting bits and hole reentry possibility, the basement has been reached more successfully than in the Atlantic, and about 30 holes have extended into the underlying lava flows. The westward spreading of the basement from the East Pacific Rise was roughly mapped after Leg 6 (Fischer, Heezen et al. 1970). Using the information acquired up through Leg 17, Winterer, Ewing and others (1971) have somewhat improved the earlier map, as indicated in Fig. 14–10. This map shows increasing age of the basement away from the East Pacific Rise, and spreading rates indicated in Fig. 14–11 are definitely faster than in the Atlantic. The spreading rates to the south near the Clipperton Fracture Zone are the highest, averaging about 7 cm/yr, and between two individual stations as high as 13 cm/yr. To the north, the rates decrease to about 5 cm/yr near the Mendocino Fracture Zone, and 3.5 cm/yr near the Surveyor Fracture Zone. Some increase to the south would be expected because of the pole rotation to the north, but some other factor is involved. In view of the displacement of the East Pacific Rise by the fracture zones, as shown by studies of the magnetic-anomaly belts, the ages of the basement are largely in agreement with distance from the Rise.

In the western Pacific, Cretaceous lava flows have evidently covered an older basement, so that the present basement is younger than it is to the east. Possible Jurassic and earliest Cretaceous were found above the basement

---

[3] In the Nile Cone area, one of the M-reflectors proved to be cemented sandstone of late Quaternary age rather than evaporite.

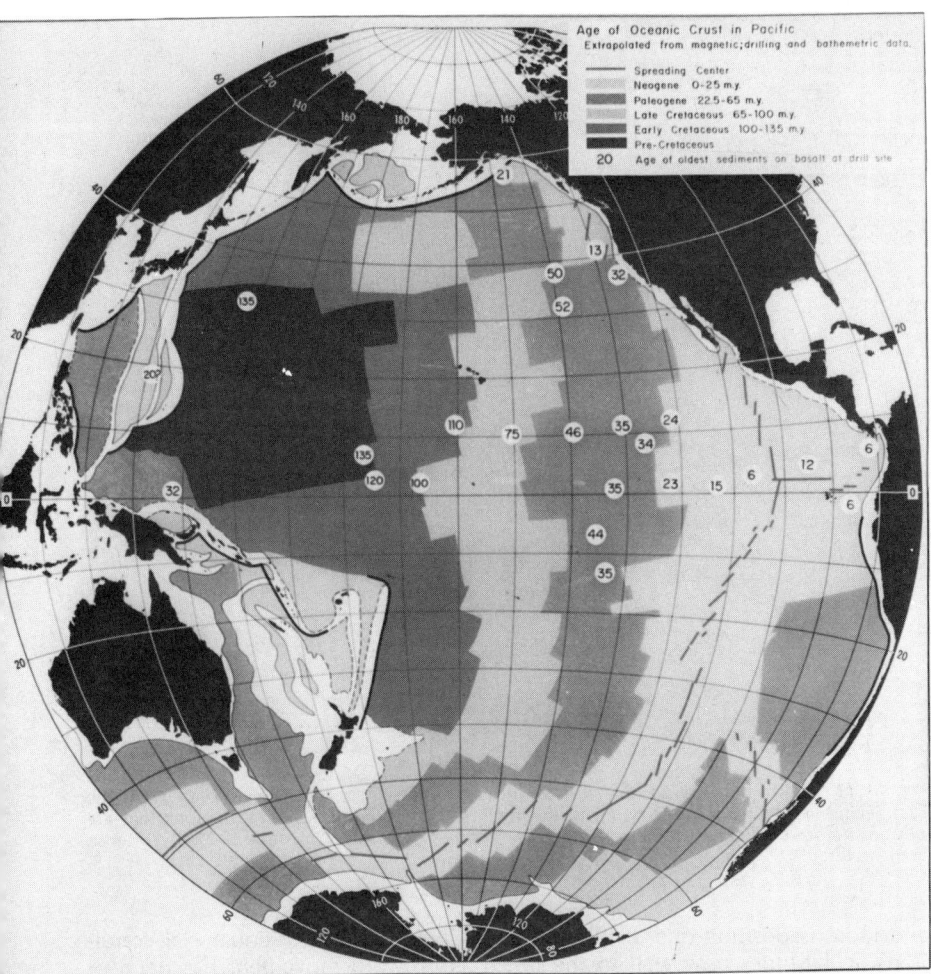

Fig. 14–10   Generalized map of basement age in the North and Central Pacific. Based on latest information (as of this writing) from JOIDES and dredging operations of Scripps Institution. Courtesy of E. L. Winterer, manuscript.

on the Shatsky Rise (Sites 49, 50) to the northwest, and on the Magellan Rise (Site 167) northeast of the Marshall Islands, apparently representing an elevated tract where the old basement was not covered by Tertiary flows. As shown in seismic profiles, the Tertiary lavas have a flat basement, in contrast to the rough basement where the flows have not covered the old lavas that came from the East Pacific Rise. Oligocene lavas underlie the sediments in the East Caroline Basin and on the Caroline Ridge.

In addition to the sea-floor spreading, the drillings show that the entire Pacific block is moving north. The preliminary magnetic determinations from some of the basement suggest a migration of as much as 30° (3200 km) to the north. This is partly borne out by finding calcium carbonate sediments

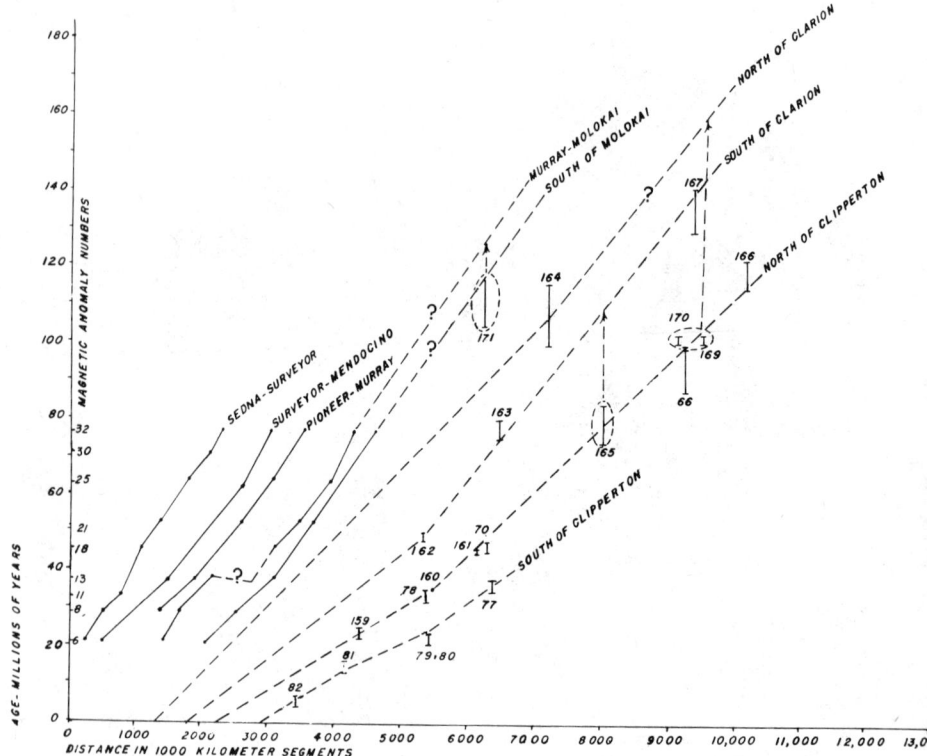

Fig. 14—11 Graph showing age of crust (millions of years) in relation to distance from the East Pacific Rise. Grouped according to the fracture-zone-bounded blocks. Stations that are out of line are connected to the related fracture zones by arrow. Courtesy of E. L. Winterer, manuscript.

and rapid deposition due to high productivity, indicative of equatorial conditions, in latitudes now well to the north. Northward movement has been of the order of 1° in 4 million years.

In Leg 16, two drillings near Central America indicated that an old ridge, which formerly extended east from the Galapagos Islands, had partly moved north (van Andel, Heath et al. 1971). This was suggested after discovering that cores from Site 157 near the equator on the Carnegie Ridge to the west contained nannofossil chalk of equatorial type all the way to the Miocene basement; at Site 158 on Cocos Ridge and Site 155 on Coiba Ridge to the north and northeast, the Miocene chalk grades up into calcareous ooze and finally clay, indicating a northward migration into less productive areas north of the equator. Also, sedimentation rates do not change on the Carnegie Ridge but show a decided decrease in the northern cores followed by an increase, the latter apparently when the area approached the continent. The group on Leg 16 suggests that the eastern portion of the Carnegie Ridge moved northward and filled a former easterly continuation of the Middle American Trench, which in turn caused an isostatic rise of the Isthmus of Panama. This Isthmus had been submerged until the Miocene, allowing a connection between the Pacific and Caribbean.

The drillings off northwestern United States on Leg 5 have given some indications of the basement ages in the more northerly fracture zones (Mc-Manus, Burns, et al. 1969). Where comparisons between magnetic-anomaly ages and the oldest paleontological ages were available, the best information appeared to infer that the latter were about 6 million years younger than the magnetic-anomaly ages.

The siliceous oozes and brown clays of the sediments in the Pacific drillings in the deep western basins indicate that deposition was continuously abyssal back to at least the Cretaceous. East of the Hawaiian Islands, however, the deep basin cores have siliceous oozes underlying the clays, signifying a gradual deepening of the water in the Tertiary. Turbidites were encountered in drillings near the West Coast of the United States, where the drillings penetrated portions of the great fans built out from the land by turbidity currents. Site 35 in the median valley of the Gorda Ridge is significant, because it is the only place where drilling was attempted into a rift valley of a spreading ridge. A 390-m hole did not penetrate below the Pleistocene, indicating rapid deposition of turbidites with many sand layers. Apparently, turbidity currents from a channel have brought sediment from the land and this has been filling the valley rapidly. Also, turbidites with some reef material were found in cores from the basins near the seamounts of the western Pacific.

The Pacific cores have quite consistent Eocene cherts with a zone beneath where sedimentation appears to have been largely missing over a period from 50 to 65 million years ago, between the Lower Eocene and horizons well down in the Upper Cretaceous (Winterer, Ewing et al. 1971). This is essentially a repeat of what was found in the Atlantic. Winterer and others suggest that during this time something may have happened to the oceanic circulation that virtually stopped deposition. It will be interesting to see if the same event is recorded in the Indian Ocean drillings.

The Pacific drillings have penetrated below areas where the manganese-iron nodules are common at the surface. The nodules appear to be rare at depth.

Cores from the holes on the flank of the Shatsky Rise indicate that deepening was taking place, because Jurassic or earliest Cretaceous calcareous ooze is covered with brown clay of Cretaceous and younger age (Fischer, Heezen et al. 1969). Other cores in the western Pacific also suggest former shoal-water conditions, as in the saddle of Horizon Guyot (Site 171), where sandstones and conglomerate with shallow-water fossils are overlain by an unconformity and then calcareous ooze (Winterer, Ewing et al. 1971).

Summary of JOIDES Results   Probably the most important result of the deep-ocean drillings during the first three years (1968–1971) has been the confirmation of the concept of sea-floor spreading. This is clearly indicated by the increasing age of the sediments overlying the volcanic basement with distance from the great ocean ridges in both the Atlantic and Pacific oceans. Furthermore, the sediments overlying the basement appear to have moved into progressively deeper water as they migrated away from the ridges. No sediments appear to be older than Jurassic underlying the ocean floors, as would be expected from the plate tectonics theory. The results of the drillings in the Pacific suggest that since Cretaceous a northward movement of about 30° has occurred at least in some areas. Thus, there is abundant new evidence of the migration of the crust over the asthenosphere.

One of the important results of the drilling has been the discovery that there are lengthy diastems, breaks in the continuity of deposition. These indicate the importance of both erosion and nondeposition over large areas of the ocean floor. The Eocene epoch seems to have been a time when deposition was very slow and widespread chert formations developed. This is found in the Pacific and in the Atlantic.

Although the drillings show in general that deep-water conditions were continuous over most of the column penetrated, deep-sea deposits are occasionally underlain by shallow-water formations. Drillings near the ocean margins show indications of large slides or faulting. A large overthrust was discovered in the Mediterranean. In some localities, thick masses of turbidites were penetrated, showing good evidence that turbidity currents can transport relatively coarse sediment for great distances out into the ocean basins.

The discovery that salt, and even petroleum, underlies the deep Gulf of Mexico has caused great interest. Salt beds were penetrated in the Mediterranean. Finding a few economic deposits in the Atlantic drillings are further indications that the deep-ocean floor may yield resources some time in the future. For paleontologists, the cores provide a remarkable opportunity for tracing the evolution of various forms of life. Nowhere else has such a stratigraphic record been found.

REFERENCES

Note: Single- and double-author references are arranged alphabetically, but with three or more authors, given in the text as "et al.," the references are listed by date sequence after the single- or double-author references.

Abdel-Gawad, Monem, 1971. Wrench movements in the Baluchistan Arc and relation to Himalayan-Indian Ocean tectonics. *Geol. Soc. Amer. Bull.*, v. 82, pp. 1235–1250.

Adams, K. T., 1939. On soundings. *U.S. Naval Inst. Proc.*, v. 65, no. 8, whole no. 438, pp. 1121–1127.

Agassiz, Alexander, 1888. Three cruises of the *Blake. Bull. Museum Comp. Zool.*, Harvard College, Cambridge, Mass., v. 14, 15.

Agassiz, Alexander, 1903. *The Coral Reefs of the Tropical Pacific.* Mem. Museum Comp. Zool., Harvard College, Cambridge, Mass., 28, 409 pp.

Alinat, Jean, Gilbert Bellaiche, Günter Giermann, Olivier Leenhardt, and Guy Pautot,

1970. Morphologie et sédimentologie d'un dome de la plaine abyssale ligure. *Bull. Inst. Oceanog. Monaco*, v. 69, no. 1400, 22 pp.

Allen, J. R. L., 1964. The Nigerian continental margin: bottom sediments, submarine morphology and geological evaluation. *Marine Geol.*, v. 1, pp. 289–332.

Allen, J. R. L., 1965A. Coastal geomorphology of eastern Nigeria: beach-ridge barrier islands and vegetated tidal flats. *Geol. en Mijnb.*, v. 44, no. 1, pp. 1–21.

Allen J. R. L., 1965B. Late Quaternary Niger Delta, and adjacent areas: sedimentary environments and lithofacies. *Amer. Assoc. Petrol. Geol. Bull.*, v. 49, no. 5, pp. 547–600.

Allen, J. R. L., 1970. Sediments of the modern Niger Delta. In *Deltaic Sedimentation, Modern and Ancient*, J. P. Morgan, ed., Soc. Econ. Paleontol. and Mineralog., spec. publ. no. 15, Tulsa, Okla., pp. 138–151.

Allen J. R. L., and J. W. Wells, 1962. Holocene coral banks and subsidence in the Niger Delta. *Jour. Geol.*, v. 70, no. 4, pp. 381–397.

Altemüller, H. J., 1962. Verbesserung der Einbettungs-und Schleiftechnik bei der Herstellung von Bodendünnschliffen mit Vestopal. *Z. Pflanzenern., Düngung u. Bodenk.*, v. 99, pp. 164–177.

AMSOC Committee, 1961. *Experimental Drilling in Deep Water at La Jolla and Guadalupe Sites.* AMSOC Comm. Div. Earth Sciences, Nat. Acad. Sci., Nat. Res. Council, Washington, D.C., 183 pp.

Andrée, K., 1920. *Geologie des Meeresbodens.* Borntraeger (Publ.), Leipzig, v. 2, 689 pp.

Andresen, A., S. Sollie, and A. F. Richards, 1965. N.G.I. gas-operated, sea-floor sampler. *Proc. 6th Intl. Conf. Soil Mech.*, pp. 8–11.

Andrews, J. E., 1967. Blake outer ridge: development by gravity tectonics. *Science*, v. 156, pp. 642–654.

Andrews, J. E., 1970. Structure and sedimentary development of the outer channel of the Great Bahama Canyon. *Geol. Soc. Amer. Bull.*, v. 81, pp. 217–226.

Andrews, J. E., F. P. Shepard, and R. J. Hurley, 1970. Great Bahama Canyon. *Geol. Soc. Amer. Bull.*, v. 81, pp. 1061–1078.

Antoine, J. W., 1968. A study of the West Florida Escarpment. *Trans. Gulf Coast Assn. Geol. Soc.*, 18th Ann. Meeting, pp. 297–303.

Antoine, J. W., and B. R. Jones, 1967. Geophysical studies of the continental slope, scarp, and basin, eastern Gulf of Mexico. *Trans. Gulf Coast Assn. Geol. Soc.*, 17th Ann. Meeting, pp. 268–277.

Antoine, John, William Bryant, and Bill Jones, 1967. Structural features of continental shelf, slope, and scarp, northeastern Gulf of Mexico. *Amer. Assoc. Petrol. Geol. Bull.*, v. 51/2, pp. 257–262.

Arnborg, Lennart, H. J. Walker, and John Peippo, 1967. Suspended load in the Colville River, Alaska. *Geografiska Annaler*, v. 49, ser. A, pp. 131–144.

Arrhenius, G., 1950. The Swedish Deep Sea Expedition. The geological material and its treatment with special regard to the eastern Pacific. *Geol. Fören. i Stockholm Förh.*, v. 72, no. 2, pp. 185–191.

Arrhenius, G., 1952. Sediment cores from the East Pacific. *Swedish Deep Sea Expedition Rept.*, no. 5, Göteborg, 186 pp.

Arrhenius, G., 1961. Geological record on the ocean floor. In *Oceanography*, M. Sears, ed., Amer. Assoc. Adv. Sci., Washington, D.C., pp. 129–148.

Arrhenius, G., 1963. Pelagic sediments. In *The Sea*, Vol. 3, M. N. Hill, ed., Interscience, New York, pp. 655–727.

Arrhenius, G., 1967. Deep sea sedimentation, a critical review of U.S. work 1963–1967. Rept. for 14th Gen. Assembly Intl. Union Geodesy and Geophysics, *Scripps Inst. Oceanog.*, Ref. 67–14, 25 pp.

Arrhenius, G., and E. Bonatti, 1965. Neptunism and vulcanism in the ocean. In *Progress in Oceanography*, Vol. 3, M. Sears, ed., Pergamon, New York, pp. 7–22.

Arrhenius, G., J. Mero, and J. Korkisch, 1964. Origin of oceanic manganese minerals. *Science*, v. 144, no. 3615, pp. 170–172.

Arthur, R. S., 1951. The effect of islands on surface waves. *Bull. Scripps Inst. Oceanog.*, Univ. California, v. 6, no. 1, Univ. Calif. Press, Los Angeles, Calif., pp. 1–26.

Athearn, W. D., 1965. Sediment cores from the Cariaco Trench, Venezuela. *Fourth Caribbean Geol. Conf.*, Trinidad, pp. 343–352.

Atwater, Tanya, 1970. Implications of plate tectonics for the Cenozoic tectonic evolution of western North America. *Geol. Soc. Amer. Bull.*, v. 81, pp. 3513–3536.

Atwater, Tanya, and J. D. Mudie, 1968. Block faulting on the Gorda Rise. *Science*, v. 159, no. 3816, pp. 729–731.

Axelsson, Vlater, 1967. The Laitaure Delta, a study of deltaic morphology and processes. *Geografiska Annaler*, v. 49, ser. A, pp. 1–127.

Aymé, J. M., 1965. The Senegal salt basin. In *Salt Basins Around Africa*, Elsevier, New York, pp. 83–90.

Bagnold, R. A., 1954. Experiments on a gravity-free dispersion of large solid spheres in a Newtonian fluid under shear. *Proc. Roy. Soc. London*, A, v. 225, pp. 49–63.

Bagnold, R. A., 1963. Mechanics of marine sedimentation. In *The Sea: Ideas and Observations*, Vol. 3, M. N. Hill, ed., Interscience, New York, pp. 507–526.

Baie, Lyle, 1970. Possible structural link between Yucatan and Cuba. *Amer. Assoc. Petrol. Geol. Bull.*, v. 54, no. 11, pp. 2204–2207.

Ball, M. M., and C. G. A. Harrison, 1969. Origin of the Gulf and Caribbean and implications regarding ocean ridge extension, migration, and shear. *Gulf Coast Assn. Geol. Soc. Trans.*, 19th Ann. Meeting, Miami, Fla., pp. 287–294.

Ballard, R. D., and Elazar Uchupi (in press). Rift structures in the Gulf of Maine. *Symposium Soc. Econ. Paleontol. and Mineralog.*, R. W. Faas, ed.

Barr, K. W., 1951. The jet sampler—a method of obtaining submarine samples for geological purposes. *Jour. Inst. Petroleum*, v. 37, no. 334, pp. 658–661.

Bartlett, G. A., 1968. Mid-Tertiary stratigraphy of the continental slope off Nova Scotia. *Maritime Sediments*, v. 4, no. 1, pp. 22–31.

Bascom, Willard, 1964. *Waves and Beaches*. Anchor Books, Doubleday, New York, 267 pp.

Beach Erosion Board, 1941. A study of progress in oscillatory waves in water. *Beach Erosion Board Tech. Rept. No. 1*, Office Chief of Engineers, U.S. Army.

Beach Erosion Board, 1954. Shore protection planning and design. *Beach Erosion Board Tech. Rept. No. 4*, 242 pp. Also, 1961, Corrections, revisions, and addenda for *Technical Report No. 4*.

Beal, M. A., and F. P. Shepard, 1956. A use of roundness to determine depositional environments. *Jour. Sed. Petrology*, v. 26, no. 1, pp. 49–60.

Beaumont, Elie de, 1845. (See de Beaumont.)

Beckmann, J. P., 1954. Foraminiferen der oceanic formation, Barbadoes. *Ecologae Geol. Helv.*, v. 46, p. 303.

Belderson, R. H., 1964. Holocene sedimentation in the western half of the Irish sea. *Marine Geol.*, v. 2, pp. 147–163.

Bender, M. L., T.-L. Ku, and W. S. Broecker, 1966. Manganese nodules: their evolution. *Science*, v. 151, no. 3708, pp. 325–328.

Bender, M. L., T.-L. Ku, and W. S. Broecker, 1970. Accumulation rates of manganese in pelagic sediments and nodules. *Earth and Planetary Sci. Letters*, 8, pp. 143–148.

Berger, W. H., 1968. Planktonic Foraminifera: selective solution and paleoclimatic interpretation. *Deep-Sea Res.*, v. 15, pp. 31–43.

Berger, W. H., 1970A. Planktonic Foraminifera: selective solution and the lysocline. *Marine Geol.*, v. 8, pp. 111–138.

Berger, W. H., 1970B. Biogenous deep-sea sediments: fractionation by deep-sea circulation. *Geol. Soc. Amer. Bull.*, v. 81, pp. 1385–1402.

Bernard, H. A., and R. J. LeBlanc, 1965. Resume of Quaternary geology of the north-western Gulf of Mexico province. *Shell Development Co.*, Houston, Texas, 68 pp.

Berry R. W., G. P. Brothy, and Adnan Haqash, 1970. Mineralogy of the suspended sediment in the Tigris, Euphrates, and Shatt-al-Arab Rivers of Iraq, and recent history of the Mesopotamian Plain. *Jour. Sed. Petrology*, v. 40, no. 1, pp. 131–139.

Berthois, Leopold, 1955. La sédimentation en Loire en grande marée de vive eau. *Compt. Rend. Acad. Sci.*, 240, pp. 106–108.

Berthois, Leopold, 1960. Étude dynamique de la sédimentation dans la loire. *Cah. Océanog.*, 12, pp. 631–657.

Berthois, Leopold, and C. Berthois, 1954, 1955. Étude de la sédimentation dans l'estuaire de la Rance. *Bull Lab. Dinard*, 40, p. 15 and 41, p. 18.

Berthois, Leopold, and R. Brenot, 1960. La morphologie sous-marine du talus du plateau continental entre le sud de l'Ireland et le Cap Ortegal (Espagne). *Jour. Conseil Permanent Intern. Exploration Mer*, v. 25, no. 2, pp. 111–114.

Berthois, Leopold, and Andre Guilcher, 1961. Étude de sédiments et fragments de roches dragues sur le banc Porcupine et a ses abords (Atlantique du nord-est). *Rev. Trav. Inst. Pêches marit.*, v. 25, no. 3, pp. 355–385.

Berthois, Leopold, and Yolande Le Calvez, 1959. Deuxième contribution a l'étude de la sédimentation dans le Golfe de Gascogne. *Rev. Trav. Inst. Pêches marit.*, v. 23, no. 3, pp. 323–376.

Berthois, Leopold, R. Brenot, and P. Ailloud, 1965A. Essai d'interpretation morphologique et tectonique des levés bathymétriques exécutés dans le partie sud-est du Golfe de Gascogne. *Rev. Trav. Inst. Pêches marit.*, v. 29, no. 3, pp. 321–342.

Berthois, Leopold, R. Brenot, and P. Ailloud, 1965B. Essai d'interpretation morphologique et géologique de la pente continentale a l'ouest de la Peninsula Iberique. *Rev. Trav. Inst. Pêches marit.*, v. 29, no. 3, pp. 343–350.

Bezrukov, P. L., 1960. Bottom sediments of the Okhotsk Sea. *Repts. Inst. Oceanology*, v. 32, Acad. Sci. USSR, pp. 15–95.

Bezrukov, P. L., and V. P. Petelin, 1960. Outline on collection and simple working instruments for marine sediments. *Trudi Inst. Okeanol. Akad. Nauk. USSR*, 44, pp. 81–112 (in Russian).

Biggs, R. B., 1967. The sediments of Chesapeake Bay. In *Estuaries*, G. H. Lauff, ed., Publ. 83, Amer. Assoc. Adv. Sci., Washington, D.C., pp. 239–260.

Bird, E. C. F., 1965. A geomorphological study of the Gippsland Lakes. *Res. School of Pacific Studies*, Dept. Geogr. Publ. G/1, Australian Nat. University, Canberra, 101 pp.

Bird, E. C. F., 1969. *Coasts, An Introduction to Systematic Geomorphology*, Vol. 4, Massachusetts Institute of Technology Press, Cambridge, Mass., London England, 246 pp.

Biscaye, P. E., 1965. Mineralogy and sedimentation of Recent deep-sea clay in the Atlantic Ocean and adjacent seas and oceans. *Geol. Soc. Amer. Bull.*, v. 76, pp. 803–832.

Bissett, L. B., 1930. Sediments from the Barents Sea and the Arctic. *Trans. Edinburgh Geol. Soc.*, v. 12, p. 207.

Black, M., M. N. Hill, A. S. Laughton, and D. H. Matthews, 1964. Three non-magnetic seamounts off the Iberian coast. *Quart. Jour. Geol. Soc. London*, v. 120, pp. 477–517.

Blackett, P. M. S., Sir Edward Bullard, and S. K. Runcorn, eds., 1965. *A Symposium on Continental Drift*. The Royal Society, London, 330 pp.

Blackman, Abner, and R. S. Anderson, 1969. Description of a dual corer for obtaining paired piston or gravity cores. *Marine Technol. Soc. Jour.*, v. 3, no. 2, pp. 73–74.

Blackman, Abner, and B. L. K. Somayajulu, 1966. Pacific Pleistocene cores: faunal analyses and geochronology. *Science*, v. 154, no. 3751, pp. 886–889.

Blackwell, D. D., and R. F. Roy, 1971. Geotectonics and Cenozoic history of the western United States. *Geol. Soc. Amer. Abstracts with Programs*, v. 3, no. 2, Cordilleran Sec. 67th Ann. meeting, p. 84.

Blanc, J. J., 1954. Sédimentologie sous-marine du Détroit Siculo-Tunisien. *Études sur le seuil Siculo-Tunisien* (suite), 7, pp. 92–126. Voir Ann. Inst. Océanog., v. 32, p. 233.

Bloom, A. L., 1965. The explanatory description of coasts (with German and French abs.), *Zeitschr. Geomorphologie*, v. 9, no. 4, pp. 422–436.

Böggild, O. B., 1916. Meeresgrundproben der *Siboga* expedition. *Siboga-Expeditie Monographie*, v. 65, pp. 1–50.

Boillot, Gilbert, and Laurent d'Ozouville, 1970. Étude structurale du continental nord-espagnol entre Aviles et Llanes. *C. R. Acad. Sci. Paris*, v. 270, pp. 1865–1868.

Bonatti, Enrico, 1963. Zeolites in Pacific pelagic sediments. *Trans. N. Y. Acad. Sci.*, ser. 2, v. 25, no. 8, pp. 938–948.

Bonatti, Enrico, and Y. R. Nayudu, 1965. The origin of manganese nodules on the ocean floor. *Amer. Jour. Sci.*, v. 263, pp. 17–39.

Bougis, Paul, and Mario Ruivo, 1954. Sur une descente d'eaux superficielles en profondeur (cascading) dans le sud du Golfe du Lion. *Bull. d'Info. Comité Central Océanog. et Étude Côtes,* v. 6, no. 4, pp. 147–154.

Bouma, A. H., 1969. *Methods for the Study of Sedimentary Structures.* Wiley–Interscience, New York, 458 pp.

Bouma, A. H., and J. A. K. Boerma, 1968. Vertical disturbances in piston cores. *Marine Geol.,* v. 6, pp. 231–241.

Bouma, A. H., and W. H. Bryant, 1969. Rapid delta growth in Matagorda Bay, Texas. In *Coastal Lagoons,* A. A. Castañares and F. B Phleger, eds., Univ. Nacional Autonoma Mexico, pp. 171–190.

Bouma, A. H., and F. P. Shepard, 1964. Large rectangular cores from submarine canyons and fan-valleys. *Amer. Assoc. Petrol. Geol. Bull.,* v. 48, no. 2, pp. 225–231.

Bourcart, Jacques, 1954A. *Le Fond des Océans.* Presses Université de France, Paris, 108 pp.

Bourcart, Jacques, 1954B. Les vases de la Méditerranée et leur mécanisme de dépôt. *Deep-Sea Res.,* v. 1, pp. 126–130.

Bourcart, Jacques, 1959. Morphologie du précontinent des Pyrénées a la Sardaigne. *Colloq. Intern. Centre Natl. Recherche Sci.* (Paris), v. 83, pp. 33–50.

Bourcart, Jacques, and Gilbert Boillot, 1959. Les sédiments de la Baie du Mont-Saint-Michel. *Intern. Oceanog. Congr. Preprints,* Amer. Assoc. Adv. Sci., Washington, D.C., pp. 602–605.

Bowen, A. J., and D. L. Inman, 1969. Rip currents, laboratory and field observations. *Jour. Geophys. Res.,* v. 74, no. 23, pp. 5479–5490.

Bowin, C. O., A. J. Nalwalk, and J. B. Hersey, 1966. Serpentinized peridotite from the north wall of the Puerto Rico Trench. *Geol. Soc. Amer. Bull.,* v. 77, pp. 257–270.

Bradley, W. H., C. S. Piggot, et al., 1942. Geology and biology of the North Atlantic deep-sea cores between Newfoundland and Ireland. *U.S. Geol. Survey Prof. Paper* 196, 163 pp.

Bramlette, M. N., 1926. Some marine bottom samples from Pago Pago Harbor, Samoa. *Carnegie Inst. Publ.* no. 344, Washington, D.C., pp. 1–35.

Bramlette, M. N., and W. H. Bradley, 1940. Lithology and geologic interpretations. Pt. 1, in Geology and Biology of North Atlantic Deep-Sea Cores. *U.S. Geol. Survey Prof. Paper* 196, pp. 1–24.

Bretz, J. H., 1969. The Lake Missoula floods and the channeled scabland. *Jour. Geol.,* v. 77, no. 5, pp. 505–543.

Bricker, O. P., and F. T. Mackenzie, 1970. Limestones and red soils of Bermuda: discussion. *Geol. Soc. Amer. Bull.,* v. 81, pp. 2523–2524.

Broecker, W. S., and T.-L. Ku, 1969. Caribbean cores P6304-8 and P6304-9: new analysis of absolute chronology. *Science,* v. 166, pp. 404–406.

Bruun, A. F., and A. Kiilerich, 1955. Characteristics of the water-masses of the Philippine, Kermadec, and Tonga Trenches. In *Papers in Marine Biology and Oceanography,* Bigelow Vol., Deep-Sea Res. suppl. to Vol. 3, Pergamon, New York, pp. 418–425.

Bruun, P. M., 1969. Tidal inlets on alluvial shores. In *Coastal Lagoons, A Symposium,* A. A. Castañares and F. B Phleger, eds., Universidad Nacional Autonoma de Mexico, pp. 349–365.

Bryan, E. H., Jr., 1953. Check list of atolls. *Pacific Sci. Bd. Atoll Res. Bull.,* no. 19, pp. 1–38.

Bryant, W. R., John Antoine, Maurice Ewing, and Bill Jones, 1968. Structure of Mexican continental shelf and slope, Gulf of Mexico. *Amer. Assoc. Petrol. Geol. Bull.,* v. 52, no. 7, pp. 1204–1228.

Bucher, W. H., 1940. Submarine valleys and related geologic problems of the North Atlantic. *Geol. Soc. Amer. Bull.,* v. 51, pp. 489–512.

Buffington, E. C., 1961. Experimental turbidity currents on the sea floor. *Amer. Assoc. Petrol. Geol. Bull.,* v. 45, no. 8, pp. 1392–1400.

Buffington, E. C., A. J. Carsola, and R. S. Dietz, 1950. Oceanographic cruise to the Bering and Chukchi Seas, summer 1949. Pt. 1, Sea floor studies. *Navy Electronics Lab. rept.* no. 204, 26 pp.

Bullard, E. C., 1954. The flow of heat through the floor of the Atlantic Ocean. *Proc. Roy. Soc. London*, A222, pp. 408–429.

Bullard, E. C., 1963. Heat flow through the floor of the ocean. In *The Sea*, Vol. 3, M. N. Hill, ed., Interscience, New York, pp. 218–232.

Bullard, E. C., 1969. The origin of the oceans. *Scientific Amer.*, v. 221, no. 3, pp. 16–25.

Bullard, E. C., and R. G. Mason, 1963. The magnetic field over the oceans. In *The Sea*, Vol. 3, M. N. Hill, ed., Interscience, New York, pp. 175–217.

Bullard, E. C., J. E. Everett, and A. G. Smith, 1965. The fit of the continents around the Atlantic. *Philos. Trans. Roy. Soc. London*, A 258, pp. 41–51.

Burk, C. A., M. Ewing, et al., 1969. Deep-sea drilling into the Challenger Knoll, central Gulf of Mexico. *Amer. Assoc. Petrol. Geol. Bull.*, v. 53, no. 7, pp. 1338–1347.

Burke, J. C., 1968. A sediment coring device of 21-cm diameter with sphincter core retainer. *Limnol. and Oceanog.*, v. 13, no. 4, pp. 714–718.

Butler, L. W., 1970. Shallow structure of the continental margin, southern Brazil and Uruguay. *Geol. Soc. Amer. Bull.*, v. 81, pp. 1079–1096.

Byrne, J. V., 1966. Effect of the East Pacific Rise on the geomorphology of the continental margin off Oregon. *Abs. Program Geol. Soc. Amer.*, Nov. 1966, San Francisco, pp. 33–34.

Byrne, J. V., D. O. LeRoy, and C. M. Riley, 1959. The chenier plain and its stratigraphy, southwestern Louisiana. *Trans. Gulf Coast Assn. Geol. Soc.*, v. 9, pp. 237–259.

Byrne, J. V., G. A. Fowler, and N. J. Maloney, 1966. Uplift of the continental margin and possible continental accretion off Oregon. *Science*, v. 154, pp. 1654–1655.

Caputo, M., G. F. Panza, and D. Postpischl, 1970. Deep structures of the Mediterranean basin. *Jour. Geophys. Res.*, v. 75, no. 26, pp. 4919–4923.

Carlson, P. R., 1967. Marine Geology of Astoria Submarine Canyon. Ph.D. Thesis, Oregon State Univ., Corvallis, mimeo, 257 pp.

Carlson, P. R., and C. H. Nelson, 1969. Sediments and sedimentary structures of Astoria Canyon-Fan system. *Jour. Sed. Petrology*, v. 39, pp. 1269–1282.

Carlson, P. R., T. R. Alpha, and D. S. McCulloch, 1970. The floor of central San Francisco Bay. *Mineral Info. Service* (now *California Geology*), v. 23, no. 5, pp. 97–107.

Carrigy, M. A., and R. W. Fairbridge, 1954. Recent sedimentation, physiography and structure of the continental shelves of Western Australia. *Jour. Roy. Soc. W. Australia*, v. 38, pp. 65–95.

Carruthers, J. N., 1958. A leaning-tube current indicator. *Bull. Inst. Oceanog. Monaco*, no. 1126, 24 pp.

Carsola, A. J., 1954A. Recent marine sediments from Alaskan and Northwest Canadian Arctic. *Amer. Assoc. Petrol. Geol. Bull.*, v. 38, no. 7, pp. 1552–1586.

Carsola, A. J., 1954B. Microrelief on Arctic Sea floor. *Amer. Assoc. Petrol. Geol. Bull.*, v. 38, no. 7, pp. 1587–1601.

Carsola, A. J., R. L. Fisher, C. J. Shipek, and G. A. Shumway, 1961. Bathymetry of the Beaufort Sea. In *Geology of the Arctic*, 1st Intl. Symposium on Arctic Geology, Univ. Toronto Press, Toronto, Canada, pp. 678–689.

Castañares, A. A., and F. B Phleger, eds., 1969. *Coastal Lagoons, A Symposium*. Universidad Nacional Autonoma de Mexico, 686 pp.

Chamberlain, T. K., 1960. Mechanics of Mass Sediment Transport in Scripps Submarine Canyon, California. Ph.D. thesis, Univ. California, Scripps Inst. of Oceanography, mimeo, 200 pp.

Chamberlin, R. T., 1924. The geological interpretation of the coral reefs of Tutuila, American Samoa. *Carnegie Inst. publ. no. 340*, Washington D.C., pp. 147–178.

Chase, T. E., H. W. Menard, and J. Mammerickx, 1971. Topography of the North Pacific map. *Geol. Data Center*, Scripps Inst. Oceanog., and *Inst. Marine Resources*.

Chauveau, J.-C., Guy Deneufroug, and J. A. Sarcia, 1967. Observations sur l'infrastructure de l'atoll de Mururoa (Archipel des Touamotou, Pacifique Sud). *C. R. Acad. Sci. Paris*, v. 265, ser. D, pp. 1113–1116.

Chave, K. E., S. V. Smith, and K. J. Roy (1972). Calcium carbonate production by coral reefs. *Marine Geol.*, v. 12, no. 2, pp. 123–140.

Chmelik, F. B., 1967. Electro-osmotic core cutting. *Marine Geol.*, v. 5, pp. 321–325.

Clark, D. L., 1971. Arctic Ocean ice cover and its Late Cenozoic history. *Geol. Soc. Amer. Bull.,* v. 82, pp. 3313–3324.

Cloud, P. E., Jr., 1958. Nature and origin of atolls. *Proc. 8th Pacific Sci. Congr.,* v. 3-A, Oceanog. Zool., pp. 1009–1024.

Coleman, J. M., 1969. Brahmaputra River: channel processes and sedimentation. *Sediment. Geol.,* v. 3, nos. 2/3, spec. issue, pp. 129–239.

Coleman, J. M., and S. M. Gagliano, 1964. Cyclic sedimentation in the Mississippi River deltaic plain. *Gulf Coast Assn. Geol. Soc. Trans.,* v. 14, pp. 67–80.

Conolly, J. R., 1969. Western Tasman Sea floor. *New Zealand Jour. Geol. and Geophysics,* v. 12, no. 1, pp. 310–343.

Conolly, J. R., and Maurice Ewing, 1965. Ice-rafted detritus as a climatic indicator in Antarctic deep-sea cores. *Science,* v. 150, no. 3705, pp. 1822–1824.

Conolly, J. R., and C. C. von der Borch, 1967. Sedimentation and physiography of the sea floor south of Australia. *Sediment. Geol.,* v. 1, pp. 181–220.

Conolly, J. R., H. D. Needham, and B. C. Heezen, 1967. Late Pleistocene and Holocene sedimentation in the Laurentian Channel. *Jour. Geol.,* v. 75, no. 2, pp. 131–147.

Conolly, J. R., Alan Flavelle, and R. S. Dietz, 1970. Continental margin of the Great Australian Bight. *Marine Geol.,* v. 8, pp. 31–58.

Cooper, L. H. N., and David Vaux, 1949. Cascading over the continental slope of water from the Celtic Sea. *Jour. Marine Biol. Assn. United Kingdom,* v. 28, pp. 719–750.

Correns, C. W., 1939. Pelagic sediments of the North Atlantic Ocean. In *Recent Marine Sediments,* P. D. Trask, ed., Amer. Assoc. Petrol. Geol., Tulsa, Okla., pp. 373–395.

Cotton, C. A., 1954. Deductive morphology and genetic classification of coasts. *Scientific Monthly,* v. 78, no. 3, pp. 163–181.

Cox, Allan, 1969. Geomagnetic reversals. *Science,* v. 163, no. 3864, pp. 237–245.

Cox, Allan, R. R. Doell, and G. B. Dalrymple, 1964. Reversals of the earth's magnetic field. *Science,* v. 144, no. 3626, pp. 1537–1543.

Cox, D. C., and J. F. Mink, 1963. The tsunami of 23 May, 1960 in the Hawaiian Islands. *Bull. Seismol. Soc. Amer.,* v. 53, pp. 1191–1209.

Creager, J. S., 1953. Submarine topography of the continental slope of the Bay of Campeche. *Oceanographic Survey of Gulf of Mexico, Texas A & M Res. Found.,* Project 24, 23 pp.

Creager, J. S., and D. A. McManus, 1967. Geology of the floor of Bering and Chukchi Sea. In *The Bering Land Bridge,* D. H. Hopkins, ed., Stanford Univ. Press, Stanford, Calif., 495 pp.

Crossland, Cyril, 1928. Notes on the ecology of the reef-builders of Tahiti. *Proc. Zool. Soc. London,* Pt. 3, pp. 717–735.

Crowell, J. C., and L. A. Frakes, 1970. Phanerozoic glaciation and the causes of ice ages. *Amer. Jour. Sci.,* v. 268, no. 3, pp. 193–224.

Cullis, C. G., 1904. The mineralogical changes observed in the cores of the Funafuti borings. In *The Atoll of Funafuti,* Royal Society of London, pp. 392–420.

Curray, J. R., 1960. Sediments and history of Holocene transgression, continental shelf, northwest Gulf of Mexico. In *Recent Sediments, Northwest Gulf of Mexico,* F. P. Shepard, F. B Phleger, and Tj. H. van Andel, eds., Amer. Assoc. Petrol. Geol., Tulsa, Okla., pp. 221–266.

Curray, J. R., 1965A. Late Quaternary history, continental shelves of the United States. In *The Quaternary of the United States,* H. E. Wright, Jr. and D. G. Frey, eds., Princeton Univ. Press, Princeton, N.J., pp. 723–735.

Curray, J. R., 1965B. Structure of the continental margin off central California. *Trans. New York Acad. Sci.,* Ser. 2, v. 27, no. 7, pp. 794–801.

Curray, J. R., 1966. Geologic structure on the continental margin, from subbottom profiles, Northern and Central California. *Geol. Northern California Bull.,* 190, pp. 337–342.

Curray, J. R., 1969. Lecture no. 12, in *The New Concept of Continental Margin Sedimentation,* AGI Short Course Notes, Philadelphia, Amer. Geol. Inst., Washington, D.C., JC–XII 1–JC–XII 22.

Curray, J. R., and D. G. Moore, 1964. Pleistocene deltaic progradation of continental terrace, Costa de Nayarit, Mexico. In *Marine Geology of the Gulf of California, A Symposium*, Mem. 3, Tj. H. van Andel and G. G. Shor, Jr., eds., Amer. Assoc. Petrol. Geol., Tulsa, Okla., pp. 193–215.

Curray, J. R., and D. G. Moore, 1971. Growth of the Bengal deep-sea fan and denudation in the Himalayas. *Geol. Soc. Amer. Bull.*, v. 82, pp. 563–572.

Curray, J. R., and R. D. Nason, 1967. San Andreas Fault north of Point Arena, California. *Geol. Soc. Amer. Bull.*, v. 78, pp. 413–418.

Curray, J. R., F. J. Emmel, and P. J. S. Crampton, 1969. Holocene history of a strand plain, lagoonal coast, Nayarit, Mexico. *Lagunas Costeras, un Simposio. Mem. Simp. Intern.* Laguna Costeras. UNAMUNESCO, Mexico, 1967, pp. 63–100.

Curray, J. R., F. P. Shepard, and H. H. Veeh, 1970. Late Quaternary sea-level studies in Micronesia: CARMARSEL Expedition. *Geol. Soc. Amer. Bull.*, v. 81, pp. 1865–1880.

Curry, D., J. B. Hersey, E. Martini, and W. F. Whittard, 1965. The geology of the western approaches of the English Channel. II. Geological interpretation aided by boomer and sparker records. *Philos. Trans. Roy. Soc. London*, ser. B, Biol. Sci., no. 749, v. 248, pp. 315–351.

Daetwyler, C. C., and A. L. Kidwell, 1959. The Gulf of Batabano, a modern carbonate basin. *Fifth World Petroleum Congress*, Sec. 1, Paper 1, pp. 1–21.

Daly, R. A., 1910. Pleistocene glaciation and the coral reef problem. *Amer. Jour. Sci.*, v. 30, pp. 297–303.

Daly, R. A., 1915. The glacial-control theory of coral reefs. *Proc. Amer. Acad. Arts and Sci.*, v. 51, no. 4, pp. 157–251.

Daly, R. A., 1936. Origin of submarine "canyons." *Amer. Jour. Sci.*, ser. 5, v. 31, no. 186, pp. 401–420.

Daly, R. A., 1948. Coral reefs—a review. *Amer. Jour. Sci.*, v. 246, pp. 193–207.

Dana, J. D., 1885. Origin of coral reefs and islands. *Amer. Jour. Sci.*, ser. 3, v. 30, pp. 89–105, 169–191.

Dangeard, Louis, 1962. Observations faites en "Soucoupe Plongeante" au large de Banyuls. *Cahiers Océanog.*, v. 14, no. 1, pp. 19–24.

Dangeard, Louis, and Michel Rioult, 1965. Le domaine de la géologie marine et ses frontières: confrontation de l'océanographe et du géologue. In *Submarine Geology and Geophysics*, W. F. Whittard and R. Bradshaw, eds., Butterworths, London, pp. 93–105.

Dangeard, Louis, Michel Rioult, J.-J. Blanc, and Laure Blanc-Vernet, 1968. Résultats de la plongée en soucoupe n° 421 dans la valée sous-marine de Planier, au large de Marseille. *Bull. Inst. Océanog. Monaco*, v. 67, no. 1384, 21 pp.

D'Anglejan, B. F., 1967. Origin of marine phosphorites off Baja California, Mexico. *Marine Geol.*, v. 5, pp. 15–44.

Dansgaard, W. S. Johnsen, J. Møller, and C. C. Langway, r., 1969. One thousand centuries of climatic record from Camp Century on the Greenland ice sheet. *Science*, v. 166, pp. 377–380.

d'Arrigo, Agatino, 1936. Richerche sul regime dei litorali nel Mediterraneo. *Ric Var. Spiagge ital.*, 14, pp. 1–172.

d'Arrigo, Agatino, 1959. *Premessa Geofisica Alla Ricerca di Sibari*. L'Arte Tipografica, Univ. di Napoli, 192 pp.

Darwin, Charles, 1842. *The Structure and Distribution of Coral Reefs*. Smith, Elder, London, 214 pp. (Reprinted in 1962 by Univ. California Press, Berkeley–Los Angeles, Calif.)

Davis, W. M., 1917. The structure of high-standing atolls. *Proc. Nat. Acad. Sci.*, v. 3, pp. 473–479.

Davis, W. M., 1925. Les côtes et les récifs coralliens de la Nouvelle-Calédonie. *Ann de Georg.*, v. 34, pp. 244–269, 332–359, 423–451, 521–558.

Davis, W. M., 1928. *The Coral Reef Problem*. Amer. Geogr. Soc. Spec. Publ. no. 9, New York, 596 pp.

de Beaumont, Elie, 1845. *Leçons de géologie pratique*. Paris, pp. 223–252.

Debyser, J., 1955. Étude sédimentologique du système lagunaire d'Abidjan (Côte d'Ivoire). *Rev. Int. Francais du Pétrole et Ann. Combustibles Liquides*, v. 10, no. 5, pp. 319–334.

Deffeyes, K. S., 1969. Electron macroprobe for logging deep-sea cores. *Abs. with Programs, 1969 Geol. Soc. Amer.,* pt. 7, 9, p. 264.

Degens, E. T., and D. A. Ross, eds., 1969. *Hot Brines and Recent Heavy Metal Deposits in the Red Sea.* Springer-Verlag, New York, 600 pp.

Degens, E. T., and D. A. Ross, 1970. Oceanographic expedition in the Black Sea. *Die Naturwissensch,* v. 57, no. 7, pp. 349–353.

Delacour, J., and P. Moulin, 1965. Un noveau procede de forage applicable en mer. *Inst. Francais du Pétrole,* ref. 12 212, 16 pp.

Demenitskaya, R. M., and K. L. Hunkins, 1970. Shape and structure of the Arctic Ocean. In *The Sea,* Vol. 4, Pt. II, A. E. Maxwell, ed., Wiley–Interscience, New York, pp. 223–249.

de Roever, W. P., ed., 1969. *Symposium on the Problems of Oceanization in the Western Mediterranean.* Verhandelingen Kon. Ned. Geol. Mijnbouwk Gen., v. 26, 165 pp.

Dewey, J. F., 1969. Evolution of the Appalachian-Caledonian orogen. *Nature,* v. 222, pp. 124–129.

Dewey, J. F., and J. M. Bird, 1970. Mountain belts and the new global tectonics. *Jour Geophys. Res.,* v. 75, pp. 2625–2647.

Dietz, R. S., 1952. Geomorphic evolution of continental terrace, continental shelf and slope. *Amer. Assoc. Petrol. Geol. Bull.,* v. 36, no. 9, pp. 1802–1819.

Dietz, R. S., 1954. Marine geology of northwestern Pacific: description of Japanese bathymetric chart 6910. *Geol. Soc. Amer. Bull.,* v. 65, pp. 1199–1224.

Dietz, R. S., 1961. Continent and ocean basin evolution by spreading of the sea floor. *Nature,* v. 190, no. 4779, pp. 854–857.

Dietz, R. S., 1963. Wave-built, marine profile of equilibrium, and wave-built terraces: a critical appraisal. *Geol. Soc. Amer. Bull.,* v. 74, pp. 971–990.

Dietz, R. S., and J. C. Holden, 1970. Reconstruction of Pangaea: breakup and dispersion of continents, Permian and present. *Jour. Geophys. Res.,* v. 75, no. 26, pp. 4939–4956.

Dietz, R. S., and George Shumway, 1961. Arctic Basin geomorphology. *Geol. Soc. Amer. Bull.,* v. 72, pp. 1319–1330.

Dietz, R. S., and W. P. Sproll, 1970. Fit between Africa and Antarctica: a continental drift reconstruction. *Science,* v. 167, pp. 1612–1614.

Dietz, R. S., K. O. Emery, and F. P. Shepard, 1942. Phosphorite deposits on the sea floor off Southern California. *Geol. Soc. Amer. Bull.,* v. 53, pp. 815–848.

Dietz, R. S., H. J. Knebel, and L. H. Somers, 1968. Cayar Submarine Canyon. *Geol. Soc. Amer. Bull.* v. 79, pp. 1821–1828.

Dietz, R. S., J. C. Holden, and W. P. Sproll, 1970. Geotectonic evolution and subsidence of Bahama platform. *Geol. Soc. Amer. Bull.,* v. 81, pp. 1915–1928.

Dietz, R. S., J. C. Holden, and W. P. Sproll, 1971. Geotectonic evolution and subsidence of Bahama platform: Reply. *Geol. Soc. Amer. Bull.,* v. 82, pp. 811–814.

Dill, R. F., 1964. Contemporary Submarine Erosion in Scripps Submarine Canyon. Ph.D. thesis, Scripps Inst. Oceanog., Univ. California, San Diego, 269 pp.

Dill, R. F., 1967. Military significance of deeply submerged sea cliffs and rocky terraces on the continental slope. *Proc. 4th U.S. Navy Symposium on Military Oceanography,* v. 1, Naval Res. Lab., Washington, D.C., pp. 106–117.

Dill, R. F., 1969A. Submerged barrier reefs on the continental slope north of Darwin, Australia, with Programs for 1969, Geol. Soc. Amer., Pt. 7, pp. 264–266. *Abs.*

Dill, R. F., 1969B. Earthquake effects on fill of Scripps Submarine Canyon. *Geol. Soc. Amer. Bull.,* v. 80, pp. 321–328.

Dill, R. F., and G. A. Shumway, 1954. Geologic use of self-contained diving apparatus. *Amer. Assoc. Petrol. Geol. Bull.,* v. 38, no. 1, pp. 148–157.

Dillon, W. P., and H. B. Zimmerman, 1970. Erosion by biological activity in two New England submarine canyons. *Jour. Sed. Petrology,* v. 40, no. 2, pp. 542–547.

Dingle, R. V., 1970. Preliminary geological map of part of the eastern Agulhas Bank, South African continental margin. *Proc. Geol. Soc. London,* no. 1663, pp. 137–142.

Dixon, W. J., and F. J. Massey, Jr., 1951. *Introduction to Statistical Analysis.* McGraw-Hill, New York, 370 pp.

Dobrin, M. B., 1960. *Introduction to Geophysical Prospecting*, 2nd ed. McGraw-Hill, New York, 446 pp.

Dolan, Robert, 1965. Seasonal variations in beach profiles along the Outer Banks of North Carolina. *Shore and Beach*, Oct.

Dolan, Robert, and J. C. Ferm, 1967. Temporal precision in beach profiling. *Prof. Geographer*, v. 19, no. 1, pp. 12–14.

Dolet, Michel, Pierre Giresse, and Claude Larsonneur, 1965. Sédiments et sédimentation dans la Baie du Mont-Saint-Michel. *Bull. Soc. Linneenne Normandie*, 10 ser., v. 6, pp. 51–65.

Donahue, J. G., R. C. Allen, and B. C. Heezen, 1966. Sediment size distribution profile on the continental shelf off New Jersey. *Sedimentology*, v. 7, pp. 155–159.

Donn, W. L., and Maurice Ewing, 1968. The theory of an ice-free Arctic Ocean. *Meteorol. Monographs*, v. 8, no. 30, pp. 100–105.

Drake, C. L., and H. P. Woodward, 1963. Appalachian curvature, wrench faulting, and offshore structures. *Trans. New York Acad. Sci.*, v. 26, pp. 48–63.

Drake, C. L., Maurice Ewing, and G. H. Sutton, 1959. Continental margins and geosynclines: the East Coast of North America, north of Cape Hatteras. In *Physics and Chemistry of the Earth*, Pergamon, London, v. 3, pp. 110–198.

Drake, C. L., J. Heirtzler, and J. Hirshman, 1963. Magnetic anomalies off eastern North America. *Jour. Geophys. Res.*, v. 68, no. 18, pp. 5259–5275.

Drake, C. L., J. I. Ewing, and Henry Stockard, 1968. The continental margin of the eastern United States. *Canadian Jour. Earth Sci.*, v. 5, pp. 993–1010.

Duane, D. B., 1969. A study of New Jersey and northern New England coastal waters. *Shore and Beach*, Oct.

Duboul-Razavet, C., 1956. *Contribution à l'Étude Géologique et Sedimentologie du Delta du Rhône*. Mem. Soc. Geol. France, v. 76, 234 pp.

Dunham, J. W., 1951. Refraction and diffraction diagrams. *Proc. 1st Conf. Coastal Engr. Council on Wave Res.*, pp. 33–49.

Eaton, J. P., D. H. Richter, and W. U. Ault, 1951. The tsunami of May 23, 1960, on the Island of Hawaii. *Bull. Seis. Soc. Amer.*, v. 51, no. 2, pp. 135–157.

Edgerton, H. E., 1967. The instruments of deep-sea photography. In *Deep-Sea Photography*, J. B. Hersey, ed., Johns Hopkins Press, Baltimore, Md., pp. 47–54.

Edgerton, H. E., Günter Giermann, and Olivier Leenhardt, 1967. Étude structurale de la baie de Monaco en sondage sismique continu. *Bull. Inst. Océanog. Monaco*, v. 67, no. 1377, 6 pp.

Edmondson, C. H., 1928. The ecology of an Hawaiian coral reef. *Bernice P. Bishop Mus. Bull.*, 45, 64 pp.

Ekman S., 1911. Neue apparate zur qualitativen und quantitativen erforschung der bodenfauna der seen. *Intern. Rev. d. ges. Hydrobiol.*, v. 3, pp. 553–561.

Ekman, V. W., 1904. On dead water. In *Norwegian North Polar Expedition 1893–1896, Scientific Results*, v. 5, no. 15.

Elmendorf, C. H., and B. C. Heezen, 1957. Oceanographic information for engineering submarine canyon cable systems. *Bell System Tech. Jour.*, v. 36, no. 5, pp. 1047–1093.

Elsasser, W. M., 1968. Submarine trenches and deformation. *Science*, v. 160, p. 1024.

Embley, R. W., John Ewing, and Maurice Ewing, 1970. The Vidal deep-sea channel and its relationship to the Demerara and Barracuda Abyssal Plains. *Deep-Sea Res.*, v. 17, pp. 539–552.

Emery, K. O., 1944. Beach markings made by sand hoppers. *Jour Sed. Petrology*, v. 14, pp. 26–28.

Emery, K. O., 1945. Entrapment of air in beach sand. *Jour. Sed. Petrology*, v. 15, pp. 39–49.

Emery, K. O., 1952. Continental shelf sediments of Southern California. *Geol. Soc. Amer. Bull.*, v. 63, pp. 1105–1108.

Emery, K. O., 1956A. Deep standing internal waves in California basins. *Limnol. and Oceanog.*, v. 1, no. 1, pp. 35–41.

Emery, K. O., 1956B. Marine geology of Johnston Island and its surrounding shallows, central Pacific Ocean. *Geol. Soc. Amer. Bull.*, v. 67, pp. 1505–1520.

Emery, K. O., 1956C. Sediments and water of Persian Gulf. *Amer. Assoc. Petrol. Geol. Bull.*, v. 40, no. 10, pp. 2354–2383.

Emery, K. O., 1960. *The Sea off Southern California*. Wiley, New York, 366 pp.

Emery, K. O., 1965. Geology of the continental margin off eastern United States. In *Submarine Geology and Geophysics*, W. F. Whittard and R. Bradshaw, eds., Butterworths, London, pp. 1–20.

Emery, K. O., 1966. Atlantic continental shelf and slope of the United States. *U.S. Geol. Survey Prof. Paper* 529-A, pp. 1–23.

Emery, K. O., 1968A. The continental shelf and its mineral resources. *Governor's Conf. on Oceanography*, Oct. 11 and 12, 1967, Rockefeller Univ., New York.

Emery, K. O., 1968B. Relict sediments on continental shelves of the world. *Amer. Assoc. Petrol. Geol. Bull.*, v. 52, no. 3, pp. 445–464.

Emery, K. O., 1968C. Shallow structure of continental shelves and slopes. *Southeastern Geol.*, Duke Univ., v. 9, no. 4, pp. 173–194.

Emery, K. O., 1969A. Distribution pattern of sediments on the continental shelves of western Indonesia. *Tech. Bull. ECAFE*, v. 2, pp. 79–82.

Emery, K. O., 1969B. The continental shelves. In *The Ocean*, Scientific Amer. Book, Freeman, San Francisco, pp. 41–52.

Emery, K. O., 1969C. Continental rises and oil potential. *Oil and Gas Jour.*, v. 67, no. 19, pp. 231–243.

Emery, K. O., and Y. K. Bentor, 1960. The continental shelf of Israel. *Sea Fisheries Res. Sta. Bull.* no. 28, Ministry Dev. *Geol. Survey Bull.* 26, pp. 25–32.

Emery, K. O., and R. S. Dietz, 1941. Gravity coring instrument and mechanics of sediment coring. *Geol. Soc. Amer. Bull.*, v. 52, pp. 1685–1714.

Emery, K. O., and D. Neev, 1960. 1. Mediterranean beaches of Israel. State of Israel, Ministry of Agriculture, Div. Fisheries, *Sea Res. Station Bull.* no. 28, pp. 1–24.

Emery, K. O., and Hiroshi Niino, 1963. Sediments of the Gulf of Thailand and adjacent continental shelf. *Geol. Soc. Amer. Bull.*, v. 74, pp. 541–554.

Emery, K. O., and Hiroshi Niino, 1967. Stratigraphy and petroleum prospects of Korea Strait and the East China Sea. *Geophysical Exploration*, v. 1, no. 1, Geol. Survey Korea, pp. 1–19.

Emery, K. O., and L. C. Noakes, 1968. Economic placer deposits of the continental shelf. *Tech. Bull. ECAFE*, v. 1, pp. 95–111.

Emery, K. O., and J. S. Schlee, 1968. The Atlantic continental shelf and slope, a program for study. *U.S. Geol. Survey Circular* 481, 11 pp.

Emery, K. O., and F. P. Shepard, 1945. Lithology of the sea floor off southern California. *Geol. Soc. Amer. Bull.*, v. 56, pp. 431–478.

Emery, K. O., and R. E. Stevenson, 1957. Estuaries and lagoons—Pt. 1. Physical and chemical characteristics. Pt. 3. Sedimentation in estuaries, tidal flats and marshes. In *Treatise on Marine Ecology and Paleoecology*, J. W. Hedgpeth, ed., Geol. Soc. Amer. Memoir 67, pp. 673–693. 729–749.

Emery, K. O., and R. D. Terry, 1956. A submarine slope of southern California. *Jour. Geol.*, v. 64, pp. 271–280.

Emery, K. O., and Elazar Uchupi, 1965. Structures of Georges Bank. *Marine Geol.*, v. 3, pp. 349–358.

Emery, K. O., and E. F. K. Zarudzki, 1967. Seismic reflection profiles along the drill holes on the continental margin off Florida. *U.S. Geol. Survey Prof. Paper* 581-A, 8 pp.

Emery, K. O., W. S. Butcher, H. R. Gould, and F. P. Shepard, 1952. Submarine geology off San Diego, California. *Jour. Geol.*, v. 60, no. 6, pp. 511–548.

Emery, K. O., J. I. Tracey, Jr., and H. S. Ladd, 1954. Geology of Bikini and nearby atolls. Pt. 1. Geology. *U.S. Geol. Survey Prof. Paper* 260-A, 265 pp.

Emery, K. O., D. S. Gorsline, E. Uchupi, and R. D. Terry, 1957. Sediments of the three bays of Baja California: Sebastian Viscaíno, San Cristobal, and Todos Santos. *Jour. Sed. Petrology*, v. 27, no. 2, pp. 95–115.

Emery, K. O., A. S. Merrill, and J. V. A. Trumbull, 1965. Geology and biology of the sea floor as deduced from simultaneous photographs and samples. *Limnol. and Oceanog.*, v. 10, no. 1, pp. 1–21.

Emery, K. O., B. C. Heezen, and T. D. Allan, 1966. Bathymetry of the eastern Mediterranean Sea. *Deep-Sea Res.,* v. 13, pp. 173–192.

Emery, K. O., Yoshikaza Hayashi, et al., 1969. Geological structure and some water characteristics of the East China Sea and the Yellow Sea. *Tech. Bull. ECAFE,* v. 2 (by Geol. Survey of Japan), pp. 3–43.

Emery, K. O., Elazar Uchupi, et al., 1970. Continental rise off eastern North America. *Amer. Assoc. Petrol. Geol.,* v. 54, no. 1, pp. 44–108.

Emery, K. O., Hiroshi Niino, and Beverly Sullivan, 1971. Post-Pleistocene levels of the East China Sea. In *The Late Cenozoic Glacial Ages,* K. K. Turekian, ed., Yale Univ. Press, New Haven, Conn., pp. 381–390.

Emiliani, Cesare, 1955. Pleistocene temperatures. *Jour. Geol.,* v. 63, no. 6, pp. 538–578.

Emiliani, Cesare, 1966. Isotopic paleotemperatures. *Science,* v. 154, pp. 851–857.

Engel, A. E. J., C. G. Engel, and R. G. Havens, 1965. Chemical characteristics of oceanic basalts and the upper mantle. *Geol. Soc. Amer. Bull.,* v. 76, no. 7, pp. 719–734.

Engel, C. G., and R. L. Fisher, 1969. Lherzolite, anorthosite, gabbro, and basalt dredged from the Mid-Indian Ocean Ridge. *Science,* v. 166, pp. 1136–1141.

Epstein, S., R. P. Sharp, and A. J. Gow, 1971. Climatological implications of stable isotope variations in deep ice cores, Byrd Station, Antarctica. *Antarctic Jour.,* Jan.–Feb., pp. 18–20.

Erickson, B. H., F. P. Naugler, and W. H. Lucas, 1970. Emperor fracture zone: a newly discovered feature in the central North Pacific. *Nature,* v. 225, Jan. 3, pp. 53–54.

Ericson, D. B., and Göesta Wollin, 1968. Pleistocene climates and chronology in deep-sea sediments. *Science,* v. 162, pp. 1227–1234.

Ericson, D. B., and Göesta Wollin, 1970. Pleistocene climates in the Atlantic and Pacific Oceans: a comparison based on deep-sea sediments. *Science,* v. 167, pp. 1483–1485.

Ericson, D. B., Maurice Ewing, Göesta Wollin, and B. C. Heezen, 1961. Atlantic deep-sea sediment cores. *Geol. Soc. Amer. Bull.,* v. 72, no. 2, pp. 193–285.

Evans, Graham, 1965. Intertidal flat sediments and their environments of deposition in the Wash. *Quart. Jour. Geol. Soc. London,* v. 121, pp. 209–245.

Ewing, John, 1963. Elementary theory of seismic refraction and reflection measurements. In *The Sea,* Vol. 3, M. N. Hill, ed., Interscience, New York, pp. 3–19.

Ewing, John, and Maurice Ewing, 1962. Reflection profiling in and around the Puerto Rico Trench. *Jour. Geophys. Res.,* v. 67, no. 12, pp. 4729–4739.

Ewing, John, and Maurice Ewing, 1970. Seismic reflection. In *The Sea,* Vol. 4, Pt. 1, A. E. Maxwell, ed., Wiley–Interscience, New York, pp. 1–52.

Ewing, John, Maurice Ewing, and C. Fray, 1960. Buried erosional terrace on the edge of the continental shelf east of New Jersey. *Abs. Geol. Soc. Amer.,* v. 71, no. 12, pt. 2, p. 1860.

Ewing, John, J. L. Worzel, and Maurice Ewing, 1962. Sediments and ocean structural history of the Gulf of Mexico. *Jour. Geophys. Res.,* v. 67, pp. 2509–2527.

Ewing, John, Manik Talwani, and Maurice Ewing, 1965. Sediment distribution in the Caribbean Sea. *Fourth Caribbean Geol. Conf.,* Trinidad, pp. 317–324.

Ewing, John, Maurice Ewing, and Robert Leyden, 1966A. Seismic-profiler survey of Blake Plateau. *Amer. Assoc. Petrol. Geol. Bull.,* v. 50, no. 9, pp. 1948–1971.

Ewing, John, J. L. Worzel, Maurice Ewing, and Charles Windisch, 1966B. Ages of horizon A and the oldest Atlantic sediments. *Science,* v. 154, no. 3753, pp. 1125–1132.

Ewing, John, R. E. Houtz, and W. J. Ludwig, 1970A. Sediment distribution in the Coral Sea. *Jour. Geophys. Res.,* v. 75, no. 11, pp. 1963–1972.

Ewing, John, Charles Windisch, and Maurice Ewing, 1970B. Correlation of Horizon A with JOIDES bore-hole results. *Jour. Geophys. Res.,* v. 75, no. 29, pp. 5645–5653.

Ewing, John, N. T. Edgar, and J. W. Antoine, 1970C. Structure of the Gulf of Mexico and Caribbean Sea. In *The Sea,* Vol. 4, Pt. II, A. E. Maxwell, ed., Wiley–Interscience, New York, pp. 321–358.

Ewing, Maurice, 1964. Marine geology. In *Ocean Sciences,* Chapt. 2, U.S. Naval Inst., Annapolis, Md., pp. 157–171.

Ewing, Maurice, 1965. The sediments of the Argentine Basin. *Quart. Jour. Roy. Astronomical Soc.*, v. 6, pp. 10–27.

Ewing, Maurice, and W. L. Donn, 1956. A theory of ice ages. *Science*, v. 123, p. 1061.

Ewing, Maurice, and John Ewing, 1964. Distribution of oceanic sediments. In *Studies on Oceanography*. Tokyo Univ. Press, pp. 525–537.

Ewing, Maurice, and E. M. Thorndike, 1964. Suspended matter in deep ocean water. *Science*, v. 147, no. 3663, pp. 1291–1294.

Ewing, Maurice, and Allyn Vine, 1938. Deep-sea measurements without wires or cables. *Trans. Amer. Geophys. Union*, 19th Ann. Meeting, pp. 248–251.

Ewing, Maurice, A. C. Vine, and J. L. Worzel, 1946. Photography of the ocean bottom. *Jour. Optical Soc. Amer.*, v. 36, pp. 307–321.

Ewing, Maurice, D. B. Ericson, and B. C. Heezen, 1958. Sediments and topography of the Gulf of Mexico. In *Habitat of Oil*, Amer. Assoc. Petrol. Geol., Tulsa, Okla., pp. 995–1053.

Ewing, Maurice, B. C. Heezen, and D. B. Ericson, 1959A. Significance of the Worzel deep sea ash. *Proc. Nat. Acad. Sci.*, v. 45, no. 3, pp. 355–361.

Ewing, Maurice, Julius Hirshman, and B. C. Heezen, 1959B. Magnetic anomalies of the Mid-Oceanic Rift. *Intl. Oceanog. Congr. Preprints*, Abs., p. 24.

Ewing, Maurice, W. J. Ludwig, and John Ewing, 1963. Geological investigations in the submerged Argentine coastal plain. *Geol. Soc. Amer. Bull.*, v. 74, pp. 275–292.

Ewing, Maurice, W. J. Ludwig, and John Ewing, 1964. Sediment distribution in the oceans: the Argentine Basin. *Jour. Geophys. Res.*, v. 69, no. 10, pp. 2003–2032.

Ewing, Maurice, T. Saito, and John Ewing, 1966A. Cretaceous and Tertiary sediments from the Walvis Ridge. *Abs. Geol. Soc. Amer.*, Ann. Meeting, San Francisco, p. 64.

Ewing, Maurice, T. Saito, John Ewing, and L. H. Burckle, 1966B. Lower Cretaceous sediments from the northwest Pacific. *Science*, v. 152, pp. 751–755.

Ewing, Maurice, D. E. Hayes, and E. M. Thorndike, 1967. Corehead camera for measurement of currents and core orientation. *Deep-Sea Res.*, v. 14, pp. 253–258.

Ewing, Maurice, Stephen Eittreim, Manek Truchan, and John Ewing, 1969A. Sediment distribution in the Indian Ocean. *Deep-Sea Res.*, v. 16, pp. 231–248.

Ewing, Maurice, R. Houtz, and John Ewing, 1969B. South Pacific sediment distribution. *Jour. Geophys. Res.*, v. 74, no. 10, pp. 2477–2493.

Ewing, Maurice, J. L. Worzel, and C. A. Burk, 1969C. Introduction. In *Initial Reports of the Deep Sea Drilling Project*, Vol. 1, Nat. Sci. Found., Washington, D.C., pp. 3–9.

Fail, J. P., L. Montadert, et al., 1970. Prolongation des zones de fractures de L'Ocean Atlantique dans le Golfe de Guinee. *Earth and Planetary Sci. Letters*, 7, pp. 413–419.

Fairbridge, R. W., 1950A. Landslide patterns on oceanic volcanoes and atolls. *Geogr. Jour.*, v. 115, pp. 82–88.

Fairbridge, R. W., 1950B. Recent and Pleistocene coral reefs of Australia. *Jour. Geol.*, v. 58, no. 4, pp. 330–401.

Fairbridge, R. W., 1961. Eustatic changes in sea level. In *Physics and Chemistry of the Earth*, Vol. 4, Pergamon, New York, pp. 99–185.

Fairbridge, R. W., 1968. Estuary. In *The Encyclopedia of Geomorphology*, R. W. Fairbridge, ed., Reinhold, New York, pp. 325–329.

Fairbridge, R. W., and H. B. Stewart, Jr., 1960. Alexa Bank, a drowned atoll on the Melanesian border plateau. *Deep-Sea Res.*, v. 7, no. 2, pp. 100–116.

Fenner, Peter, Gilbert Kelling, and D. J. Stanley, 1971. Bottom currents in Wilmington Submarine Canyon. *Nature Physical Sci.*, v. 229, no. 2, pp. 52–54.

Fischer, A. G., B. C. Heezen, et al., 1969. Deep Sea Drilling Project: Leg 6. *Geotimes*, v. 14, no. 8, pp. 13–16.

Fischer, A. G., B. C. Heezen, et al., 1970. Geological history of the western North Pacific. *Science*, v. 168, pp. 1210–1214.

Fisher, R. L., 1955. Cuspate spits of St. Lawrence Island, Alaska. *Jour. Geol.*, v. 63, no. 2, pp. 133–142.

Fisher, R. L., 1961. Middle America Trench: topography and structure. *Geol. Soc. Amer. Bull.*, v. 72, no. 5, pp. 703–720.

Fisher, R. L., 1966. The median ridge in the South Central Indian Ocean. The world rift system. *Rept. UMC Symposium,* Ottowa, Sept. 1965, *Geol. Survey Canada Paper* 66–14, pp. 135-147.

Fisher, R. L., and C. G. Engel, 1969. Ultramafic and basaltic rocks dredged from the nearshore flank of the Tonga Trench. *Geol. Soc. Amer. Bull.,* v. 80, pp. 1373-1378.

Fisher, R. L., and H. H. Hess, 1963. Trenches. In *The Sea,* Vol. 3, M. N. Hill, ed., Interscience, New York, pp. 411–436.

Fisher, R. L., and R. R. Revelle, 1966. The trenches of the Pacific. *Scientific Amer.,* v. 194, no. 5, pp. 36–41.

Fisher, R. L., A. J. Carsola, and G. A. Shumway, 1958. Deep-sea bathymetry north of Point Barrow. *Deep-Sea Res.,* v. 5, pp. 1–6.

Fisher, R. L., G. L. Johnson, and B. C. Heezen, 1967. Mascarene Plateau, western Indian Ocean. *Geol. Soc. Amer. Bull.,* v. 78, pp. 1247–1266.

Fisher, R. L., J. G. Sclater, and D. P. McKenzie, 1971. Evolution of the central Indian Ridge, western Indian Ocean. *Geol. Soc. Amer. Bull.,* v. 82, pp. 553–562.

Fisk, H. N., 1952. Geological investigations of the Atchafalaya basin and the problem of Mississippi River diversion. *U.S. Army Corps Engrs., Mississippi River Comm.,* Vicksburg, Miss., 145 pp.

Fisk, H. N., 1959. Padre Island and the Laguna Madre Flats coastal South Texas. In *2nd Coastal Geography Conference,* R. J. Russell, Chm., Coastal Studies Inst., Louisiana State Univ., Baton Rouge, pp. 103–151.

Fisk, H. N., 1961. Bar-finger sands of the Mississippi Delta. In *Geometry of Sand Bodies,* Amer. Assoc. Petrol. Geol., Tulsa, Okla., pp. 29–52.

Fisk, H. N., and Edward McFarlan, Jr., 1955. Late Quaternary deposits of the Mississippi River. *Geol. Soc. Amer. Spec. Paper* 62, pp. 279–302.

Fisk, H. N., Edward McFarlan, Jr., C. R. Kolb, and L. J. Wilbert, Jr., 1954. Sedimentary framework of the modern Mississippi Delta. *Jour. Sed. Petrology,* v. 24, no. 2, pp. 76–99.

Flint, R. F., 1971. *Glacial and Quaternary Geology.* Wiley, New York, 892 pp.

Folk, R. L., 1954. The distinction between grain size and mineral composition in sedimentary-rock nomenclature. *Jour. Geol.,* v. 62, no. 4, pp. 344–359.

Foster, J. H., and N. D. Opdyke, 1970. Upper Miocene to Recent magnetic stratigraphy in deep-sea sediments. *Jour. Geophys. Res.,* v. 75, no. 23, pp. 4465–4473.

Fox, P. J., and B. C. Heezen, 1965. Sands of the Mid-Atlantic Ridge. *Science,* v. 149, no. 3690, pp. 1367–1370.

Fox, P. J., Allen Lowrie, Jr., and B. C. Heezen, 1969. Oceanographer Fracture Zone. *Deep-Sea Res.,* v. 16, pp. 59–66.

Fox, P. J., W. F. Ruddiman, W. B. F. Ryan, and B. C. Heezen, 1970. The geology of the Caribbean crust, I: Beata Ridge. *Tectonophysics,* v. 10, pp. 495–513.

Frakes, L. A., and J. C. Crowell, 1970A. Late Paleozoic glaciation. II. Africa exclusive of Karroo Basin. *Geol. Soc. Amer. Bull.,* v. 81, pp. 2261–2286.

Frakes, L. A., and J. C. Crowell, 1970B. Geologic evidence for the place of Antarctica in Gondwanaland. *Antarctic Jour.,* v. 5, no. 3, pp. 67–69.

Francis-Boeuf, Claude, 1947. Recherches sur le milieu fluvio-marin et les dépôts d'estuaire. *Ann. Inst. Océanog. Monaco,* v. 23, no. 3, pp. 149–344.

Fray, Charles, and Maurice Ewing, 1963. Pleistocene sedimentation and fauna of the Argentine shelf. *Proc. Acad. Nat. Sci.,* Philadelphia, v. 115, no. 6, pp. 113–147.

Frey, Hank, and Paul Tzimoulis, 1968. *Camera Below.* Associated Press, New York, 224 pp.

Frye, J. C., and R. W. Raitt, 1961. Sound velocities at the surface of deep-sea sediments. *Jour. Geophys. Res.,* v. 66, pp. 589–597.

Funkhouser, J. G., I. L. Barnes, and J. J. Naughton, 1968. Determination of ages of Hawaiian volcanoes by K-Ar method. *Pacific Sci.,* v. 22, pp. 369–372.

Gadow, Sibylle, and H.-E. Reineck, 1969. Ablandiger sandtransport dei Sturmfluten. *Senckenbergiana maritima,* (1) 50, pp. 63–78.

Gakkel, Ya. Ya., and V. D. Dibner, 1967. Bottom of the Arctic Ocean. In *International Dictionary of Geophysics,* Pergamon, London, pp. 1–13.

Gardner, J. V., 1970. Submarine geology of the western Coral Sea. *Geol. Soc. Amer. Bull.,* v. 81, pp. 2599–2614.

Garrison, L. E., 1970. Development of continental shelf south of New England. *Amer. Assoc. Petrol. Geol. Bull.*, v. 54, no. 1, pp. 109–124.

Garrison, L. E., and R. L. McMaster, 1966. Sediments and geomorphology of the continental shelf off southern New England. *Marine Geol.*, v. 4, pp. 273–289.

Gealy, Betty Lee, 1955. Topography of the continental slope in northwest Gulf of Mexico. *Geol. Soc. Amer. Bull.*, v. 66, pp. 203–228.

Gennesseaux, Maurice, 1963. Structure et morphologie de la pente continentale de la région niçoise. *Rapports et Procès-verbaux des réunions de C.I.E.S.M.M.*, v. 17 (3), pp. 991–998.

Gennesseaux, Maurice, 1964. L'evolution des fonds sous-marins de la baie des Anges et le delta du Var. *Les Cahiers du C.E.R.B.O.M.*, v. 13, no. 1, pp. 3–17.

Gennesseaux, Maurice, 1966. Prospection photographiques des canyons sous-marins du Var et du Paillon (Alpes Maritimes) au moyen de la Troika. *Rev. Geogr. Physique et Geol. Dynamique*, (2), v. 8, no. 1, Paris. pp. 3–38.

Gibson, R. B., 1962. Crustal structure across the west Florida Escarpment from gravity data. M.S. thesis, A & M College of Texas, 22 pp.

Gibson, T. G., and J. Schlee, 1967. Sediments and fossiliferous rocks from the eastern side of the Tongue of the Ocean, Bahamas. *Deep-Sea Res.*, v. 14, pp. 691–702.

Gibson, W., and H. Nichols, 1953. Configuration of the Aleutian Ridge, Rat Islands— Semisopochnoi I to west of Buldir I. *Geol. Soc. Amer. Bull.*, v. 64, pp. 1173–1181.

Giddings, J. L., 1966. Cross-dating the archeology of northwestern Alaska. *Science*, v 153, no. 3732, pp. 127–135.

Gierloff-Emden, H. G., 1961. Nehrungen und lagunen. *Petermanns Geogr. Mitt.*, Quart. 2, 3, pp. 82–92, 161–176.

Giermann, Günter, 1961. Erläuterungen zur bathymetrischen Karte der Strasse von Gibraltar. *Bull. Inst. Oceanog. Monaco*, no. 116, maps.

Giermann, Günter, 1962. Erläuterungen zur bathymetrischen Karte des westlichen Mittelmeers (zwischen 6°40′W.L. u. 1° öL.). *Bull. Inst. Oceanog. Monaco*, no. 1254 and B, 24 pp.

Giermann, Günter, 1969. The Eastern Mediterranean Ridge. *Rapp. Comm. Intl. Mer. Médit.*, v. 19, no. 4, pp. 605–607.

Gilbert, G. K., 1885. The topographic features of lake shores. *U.S. Geol. Survey 5th Ann. Rept.*, pp. 69–123.

Gilbert G. K., 1914. The transportation of debris by running water. *U.S. Geol. Survey Prof. Paper 83*, 263 pp. Abs. *Washington Acad. Sci.* J4, pp. 154–158.

Ginsburg, R. N., 1956. Environmental relationships of grain size and constituent particles in some south Florida carbonate sediments. *Amer. Assoc. Petrol. Geol. Bull.*, v. 40, no. 10, pp. 2384–2427.

Ginsburg, R. N., 1957. Early diagenesis and lithifaction of shallow-water carbonate sediments in south Florida. In *Regional Aspects of Carbonate Deposition*, R. J. LeBlanc and J. G. Breefing, eds., Soc. Econ. Paleontol. and Mineralog., spec. publ., no. 5, pp. 80–100.

Giresse, P., 1967. Mécanismes de répartition des minéraux argileux des sédiments marins actuels sur le littoral sud du Cotentin. *Marine Geol.*, v. 5, pp. 61–69.

Glangeaud, L., 1968. Les méthodes de la géodynamique et leurs applications aux structures de la Méditerranée occidentale. *Rev. Geographie Physique et Geol. Dynamique*, (2), v. 10, no. 2, pp. 83–135.

Glossary of Geology and Related Sciences, 1960, 2nd ed., Amer. Geol. Inst. under Nat. Acad. Sciences, Nat. Res. Council, Washington, D.C., 325 pp. + 72 pp. supplement.

Goldberg, E. D., and G. Arrhenius, 1958. Chemistry of Pacific pelagic sediments. *Geochim. Cosmochim. Acta.*, v. 13, pp. 153–212.

Golik, Abraham, 1965. Foraminiferal ecology and Holocene history, Gulf of Panama. Ph.D. thesis, Univ. California, San Diego, mimeo., 198 pp.

Goncharov, V. P., and U. P. Neprochnov, 1960. Geomorphology and tectonic problems of the Black Sea. *Twenty-first Intern. Geol. Congr., Reports of Soviet Geologists, Problem 10, Marine Geology.* pp. 94–104.

Goodell, H. G., 1967. The sediments and sedimentary geochemistry of the southeastern Atlantic shelf. *Jour. Geol.*, v. 75, no. 6, pp. 665–692.

Goodell, H. G., and R. K. Garman, 1969. Carbonate geochemistry of Superior deep test well. *Amer. Assoc. Petrol. Geol. Bull.,* v. 53, no. 3, pp. 513–536.

Goodell, H. G., and D. S. Gorsline, 1961. A sedimentologic study of Tampa Bay, Florida. *Twenty-first Sess. Intl. Geol. Congr.,* Norden, Pt. 23, *Proc. Intl. Assoc. Sedimentology,* pp. 75–88.

Goreau, T. F., 1959. The ecology of Jamaican coral reefs. *Ecology,* v. 40, pp. 67–90.

Goreau, T. F., 1961. On the relation of calcification to primary production in reef-building organisms. In *The Biology of Hydra,* H. M. Lenhoff and W. F. Loomis eds., University of Miami Press, pp. 269–285.

Gorsline, D. S., 1963A. Oceanography of Apalachicola Bay, Florida. In *Essays in Marine Geology,* honoring K. O. Emery, Thomas Clements, ed., Univ. Southern California Press, pp. 69–96.

Gorsline, D. S., 1963B. Bottom sediments of the Atlantic shelf and slope off the southern United States. *Jour. Geol.,* v. 71, no. 4, pp. 422–440.

Gorsline, D. S., 1966. Dynamic characteristics of west Florida Gulf Coast beaches. *Marine Geol.,* v. 4, pp. 187–206.

Gorsline, D. S., 1967. Contrasts in coastal bay sediments on the Gulf and Pacific Coasts. In *Estuaries,* G. H. Lauff, ed., Amer. Assoc. Adv. Sci., Washington, D.C., pp. 219–225.

Gorsline, D. S., D. E. Drake, and P. W. Barnes, 1958. Holocene sedimentation in Tanner Basin, California continental borderland. *Geol. Soc. Amer. Bull.,* v 79, pp. 659–674.

Gould, H. R., 1951. Some quantitative aspects of Lake Mead turbidity currents. *Soc. Econ. Paleontol. and Mineralog. Spec. Publ.,* no. 2, pp. 34–52.

Gould, H. R., 1970. The Mississippi Delta complex. In *Deltaic Sedimentation, Modern and Ancient,* J. P. Morgan, ed., Soc. Econ. Paleontol. and Mineralog., spec. publ. no. 15, Tulsa, Okla., pp. 3–30.

Gould, H. R., and E. McFarlan, Jr., 1959. Geologic history of the chenier plain, southwestern Louisiana. *Trans. Gulf Coast Assn. Geol. Soc.,* v. 9. pp. 261–272.

Gould, H. R., and R. H. Stewart, 1955. Continental terrace sediments in northeastern Gulf of Mexico. In *Finding Ancient Shorelines,* Soc. Econ. Paleontol. and Mineralog., New Haven, Conn., pp. 1–19.

Graham, J. R., M. N. A. Peterson, and A. R. McLerran, 1970. Deep sea drilling project, prelude to a decade of ocean exploration. *Marine Technol. Soc. (MTL) Jour.,* v. 4, no. 5, pp. 5–13.

Granelli, N. C. L., 1959. Giant ripples in the Gulf of San Matias, Argentina. Abs. in *Preprints International Oceanographic Congress,* Amer. Assoc. Adv. Sci., Washington, D.C., pp. 616–618.

Grant, A. C., 1966. A continuous seismic profile on the continental shelf off Labrador. *Canadian Jour. Earth Sci.,* v. 3, no. 5, pp. 725–730.

Grant, U. S., 1948. Influence of the water table on beach aggradation and degradation. *Jour. Mar. Res.,* v. 7, no. 3, pp. 655–660.

Graton. L. C., and H. J. Fraser, 1935. Systematic packing of spheres—with particular relation to porosity and permeability. *Jour. Geol.,* v. 43, pp. 785–909.

Greene, H. G., 1970A. Microrelief of an arctic beach. *Jour. Sed. Petrology,* v. 40, no. 1, pp. 419–427.

Greene, H. G., 1970B. Geology of southern Monterey Bay and its relationship to the ground water basin and salt water intrusion. *U.S. Geol. Survey,* open file rept., 50 pp.

Gregory, J. W., 1929. The earthquake south of Newfoundland and submarine valleys. *Nature,* v. 124, p. 945.

Griffin, J. J., Herbert Windon, and E. D. Goldberg, 1968. The distribution of clay minerals in the world ocean. *Deep-Sea Res.,* v. 15, pp. 433–459.

Griggs, G. B., A. G. Carey, Jr., and L. D. Kulm, 1969. Deep-sea sedimentation and sediment-fauna interaction in Cascadia Channel and on Cascadia abyssal plain *Deep-Sea Res.,* v. 16, pp. 157–170.

Grim, M. S., C. L. Drake, and J. R. Heirtzler, 1970. Sub-bottom study of Long Island Sound. *Geol. Soc. Amer. Bull.,* v. 81, pp. 649–666.

Grim, P. J., 1969. Seamap deep-sea channel. *ESSA Tech. Rept.,* ERL 93-Pol 2, Seattle, Washington, pp. 1–27.

Grimsdale, T. F., 1952. *Cycloclypeus* (Foraminifera) in the Funafuti boring, and its geological significance. *The Challenger Soc.*, no. 2, pp. 1–11.

Gripenberg, Stina, 1934. Sediments of the North Baltic and adjoining seas. *Havforskniningstitutes*, Skr. no. 96, Helsingfors, Fennia 60, no. 3.

Gripenberg, Stina, 1939. Sediments of the Baltic Sea. In *Recent Marine Sediments,* Amer. Assoc. Petrol. Geol., Tulsa, Okla., pp. 293–321.

Groba, E., 1959. Geologische Unterwasserkartierung im Litoral der deutschen Ostseeküste. *Acta Hydrophysica,* v. 5, no. 4, pp. 163–200.

Gross, M. G., S. M. Gucluer, J. S. Creager, and W. A. Dawson, 1963. Varved marine sediments in a stagnant fjord. *Science,* v. 141, pp. 918–919.

Gross, M. G., D. A. McManus, and H.-Y. Ling, 1967. Continental shelf sediment, northwestern United States. *Jour. Sed. Petrology,* v. 37, no. 3, pp. 790–795.

Guilcher, Andre, 1951. La formation de la Mer du Nord Pas de Calais et des plaines maritimes cnvironmentes. *Rev. Geogr. de Lyon,* Univ. Lyon, no. 3, pp. 311–329.

Guilcher, Andre, 1954. Morthologie littorale et Sous Marine. *"Orbis" Introduction Études Géographie,* Presses Univ. France, 210 pp.

Guilcher, Andre, 1959. Coastal sand ridges and marshes and their environment near Grand Popo and Ouidah, Dahomey. In *2nd Coastal Geography Conference,* R. J. Russell, Chm., Coastal Studies Inst., Louisiana State Univ., Baton Rouge, pp. 189–212.

Guilcher, Andre, 1965A. Drumlin and spit structures in the Kenmare River, South-west Ireland. *Irish Geography,* v. 5, no. 2, pp. 7–19.

Guilcher, Andre, 1965B. Grand Récif sud Récifs et lagon de Tuo. In *Expédition Française Sur les Récifs Coralliens de la Nouvelle-Caledonie,* Paris, pp. 137–238.

Guilcher, Andre, 1967. Origin of sediments in estuaries. In *Estuaries,* G. A. Lauff, ed., Amer. Assoc. Adv. Sci., Washington, D.C., pp. 149–157.

Guilcher, Andre, and L. Berthois, 1957. Cinq années d'observations sédimentologiques dans quatre estuaires-temoins de l'ouest de la Bretagne. *Rev. Géomorph. Dyn..* 8, pp. 67–86.

Guilcher, Andre, L. Berthois, Y. Le Calvez, R. Battistini, and A. Crosnier, 1965. Les Récifs Coralliens et le Lagon de L'Ile Mayotte. *Office Rech. Sci. et Tech.* Outre-Mer, Orstom, Paris, 210 pp.

Guilcher, Andre, L. Berthois, F. Doumenge, A. Michel, A. Saint-Requier, and R. Arnold, 1969. Les récifs et lagons coralliens de Mopelia et de Bora-Bora (Îles de la Société). *Memoires Orstom,* no. 38, Office Recher. Sci. Tech, Outre-Mer, 103 pp.

Gulliver, F. P., 1896. Cuspate forelands. *Geol. Soc. Amer. Bull.,* v. 7, pp. 399–422.

Gutenberg, Beno, and C. F. Richter, 1954. *Seismicity of the Earth and Associated Phenomena.* Princeton Univ. Press, Princeton, N.J., 2nd ed., 310 pp. (originally published 1949).

Hails, J. R., and J. H. Hoyt, 1968. Barrier development on submerged coasts: problems of sea-level changes from a study of the Atlantic coastal plain of Georgia, U.S.A. and parts of the East Australian coast. *Ann. Geomorphology,* Neue Folge, supp. 7, pp. 24–55.

Hamblin, W. K., 1962. X-ray radiography in the study of structures in homogeneous sediments. *Jour. Sed. Petrology,* v. 32, no. 2, pp. 201–210.

Hamilton, E. L., 1956. Sunken islands of the Mid Pacific Mountains. *Geol. Soc. Amer. Mem.* no. 64, 97 pp.

Hamilton, E. L., 1959. Thickness and consolidation of deep-sea sediment. *Geol. Soc. Amer. Bull.,* v. 70, no. 11, pp. 1399–1424.

Hamilton, E. L., 1960. Ocean basin ages and amounts of original sediments. *Jour. Sed. Petrology,* v. 30, no. 3, pp. 370–379.

Hamilton, E. L., 1967. Marine geology of abyssal plains in the Gulf of Alaska. *Jour. Geophys. Res.,* v. 72, no. 16, pp. 4189–4213.

Hamilton, E. L., 1970A. Prediction of in-situ acoustic and elastic properties of marine sediments. *Geophysics,* v. 36, no. 2, pp. 266–284.

Hamilton, E. L., 1970B. Sound velocity and related properties of marine sediments, North Pacific. *Jour. Geophys. Res.,* v. 75, no. 23, pp. 4423–4446.

Hamilton, E. L., 1971. Elastic properties of marine sediments. *Jour. Geophys. Res.,* v 76, no. 2, pp. 597–604.

Hamilton, E. L., G. A. Shumway, H. W. Menard, and C. J. Shipek, 1956. Acoustic and other physical properties of shallow-water sediments off San Diego. *Jour. Acoustical Soc. Amer.,* v. 28, pp. 1–15.

Hamilton, Warren, and David Krinsley, 1967. Upper Paleozoic glacial deposits of South Africa and southern Australia. *Geol. Soc. Amer. Bull.,* v. 78, pp. 783–800.

Hanna, G. D., 1952. Geology of the continental slope off central California. *Calif. Acad. Sci. Proc.,* v. 27, pp. 325–358.

Hanzawa, Sheshiro, 1940. Micropaleontological studies of drill cores from a deep well in Kita-Daito Zima (N. Borodino Is.). *Jubilee Publication of Prof. H. Yabe's 60th Birthday,* v. 2, pp. 755–802.

Hartwell, A. D., 1970. Hydrography and Holocene sedimentation of the Merrimack River estuary, Massachusetts. *Contrib.* no. 5-CRG, *Dept. Geol., Univ. Massachusetts,* 166 pp.

Hathaway, J. C., P. F. McFarlin, and D. A. Ross, 1970. Mineralogy and origin of sediments from drill holes on the continental margin off Florida. *U.S. Geol. Survey Prof. Paper* 581-E, pp. 1–26.

Hayes, D. E., and Maurice Ewing, 1970. North Brazilian Ridge and adjacent continental margin. *Amer. Assoc. Petrol. Geol. Bull.,* v. 54, no. 11, pp. 2120–2150.

Hayes, D. E., A. C. Pimm, et al., 1971. Deep-Sea drilling project leg 14. *Geotimes,* v. 16, no. 2, pp. 14–17.

Hayes, M. O., 1964. Lognormal distribution of inner continental shelf widths and slopes. *Deep-Sea Res.,* v. 11, pp. 53–78.

Hayes, M. O., 1967. Relationship between coastal climate and bottom sediment type of the inner continental shelf. *Marine Geol.,* v. 5, pp. 111–132.

Hays, J. D., 1965. Radiolaria and the Late Tertiary and Quaternary history of the Antarctic seas. In *Biology of the Antarctic Seas II,* Research Series 5, Amer. Geophys. Union, pp. 125–184.

Heath, G. R., 1969. Carbonate sedimentation in the abyssal equatorial Pacific during the past 50 million years. *Geol. Soc. Amer. Bull.,* v. 80, pp. 689–694.

Heath, G. R., and T. C. Moore, Jr., 1965. Subbottom profile of abyssal sediments in the central equatorial Pacific. *Science,* v. 149, no. 3685, pp. 744–746.

Hedberg, H. D., 1970. Continental margins from viewpoint of the petroleum geologist. *Amer. Assoc. Petrol. Geol. Bull.,* v. 54, no. 1, pp. 3–43.

Hedgpeth, J. W., 1967. The sense of the meeting. In *Estuaries,* G. H. Lauff, ed., Publ. 83, Amer. Assoc. Adv. Sci., Washington, D.C., pp. 707–710.

Heezen, B. C., 1960. The rift in the ocean floor. *Scientific Amer.,* Oct., pp. 98–114.

Heezen, B. C., and C. L. Drake, 1964. Grand Banks slump. *Amer. Assoc. Petrol. Geol. Bull.,* v. 48, no. 2, pp. 221–225.

Heezen, B. C., and Maurice Ewing, 1952. Turbidity currents and submarine slumps, and the Grand Banks earthquake. *Amer. Jour. Sci.,* v. 250, pp. 849–873.

Heezen, B. C., and A. G. Fischer, 1971. Patterns of crustal age. In *Initial Reports of the Deep Sea Drilling Project,* Vol. 6, A. G. Fischer et al., Washington, D.C. (U.S. Govt. Printing Office), pp. 1301–1305.

Heezen, B. C., and Charles Hollister, 1964. Deep-sea current evidence from abyssal sediments. *Marine Geol.,* v. 1, pp. 141–174.

Heezen, B. C., and Charles Hollister, 1971. *The Face of the Deep.* Oxford Univ. Press, New York, London, 659 pp.

Heezen, B. C., and G. L. Johnson, 1965. The South Sandwich Trench. *Deep-Sea Res.,* v. 12, pp. 185–197.

Heezen, B. C., and R. E. Sheridan, 1966. Lower Cretaceous rocks (Neocomian–Albian) dredged from Blake Escarpment. *Science,* v. 154, no. 3757, pp. 1644–1647.

Heezen, B. C., Marie Tharp, and Maurice Ewing, 1959. *The Floors of the Oceans.* I. *The North Atlantic.* Geol. Soc. Amer., spec. pap. 65, 122 pp.

Heezen, B. C., E. T. Bunce, J. B. Hersey, and Marie Tharp, 1964A. Chain and Romanche Fracture Zones. *Deep-Sea Res.,* v. 11, pp. 11–33.

Heezen, B. C., R. J. Menzies, E. D. Schneider, W. M. Ewing, and N. C. L. Granelli, 1964B. Congo Submarine Canyon. *Amer. Assoc. Petrol. Geol. Bull.,* v. 48, no. 7, pp. 1126–1149.

Heezen, B. C., R. D. Gerard, and Marie Tharp, 1964C. The Vema fracture zone in the equatorial Atlantic. *Jour. Geophys. Res.,* v. 69, no. 4, pp. 733–739.

Heezen, B. C., C. D. Hollister, and W. F. Ruddiman, 1966A. Shaping of the continental rise by deep geostrophic contour currents. *Science*, v. 152, no. 3721, pp. 502–508.

Heezen, B. C., Bill Glass, and H. W. Menard, 1966B. The Manihiki Plateau. *Deep-Sea Res.*, v. 13, pp. 445–458

Heezen, B. C., E. D. Schneider, and O. H. Pilkey, 1966C. Sediment transport by the Antarctic Bottom Current on the Bermuda Rise. *Nature*, v. 211, no. 5049, pp. 611–612.

Heirtzler, J. R., 1970. Magnetic anomalies measured at sea. In *The Sea*, Vol. 4, Pt. 1, A. E. Maxwell, ed., Wiley–Interscience, New York, pp. 85–128.

Heirtzler, J. R., and D. E. Hayes, 1967. Magnetic boundaries in the North Atlantic Ocean. *Science*, v. 157, pp. 185–187.

Heirtzler, J. R., G. O. Dickson, E. M. Herron, W. C. Pitman, III, and X. Le Pichon, 1968. Marine magnetic anomalies, geomagnetic field reversals, and motions of the ocean floor and continents. *Jour. Geophys. Res.*, v. 73, no. 6, pp. 2119–2136.

Herman, Yvonne, 1963. Cretaceous, Paleocene, and Pleistocene sediments from the Indian Ocean. *Science*, v. 140, no. 3573, pp. 1316–1317.

Hersey, J. B., 1963. Continuous reflection profiling. In *The Sea*, Vol. 3, M. N. Hill, ed., Interscience, New York, pp. 47–72.

Hersey, J. B., 1965. Sedimentary basins of the Mediterranean Sea. In *Submarine Geology and Geophysics*, W. F. Whittard and R. Bradshaw, eds., Butterworths, London, pp. 75–89.

Hersey, J. B., Ed., 1967. *Deep-Sea Photography*. Johns Hopkins Press, Baltimore, Md., 310 pp.

Hersey, J. B., 1967. The manipulation of deep-sea cameras. In *Deep-Sea Photography*, Johns Hopkins Press, Baltimore, Md., pp. 55–67.

Hess, H. H., 1938. Gravity anomalies and island arc structure. *Proc. Amer. Philos. Soc.*, v. 79, pp. 71–96.

Hess, H. H., 1946. Drowned ancient islands of the Pacific basin. *Amer. Jour. Sci.*, v. 244, pp. 772–791.

Hess, H. H., 1948. Major structural features of the western and north Pacific, an interpretation of H. O. 5985, bathymetric chart, Korea to New Guinea. *Geol. Soc. Amer. Bull.*, v. 59, pp. 417–446.

Hess, H. H., 1962. History of ocean basins. In *Petrologic Studies: a Volume in Honor of A. F. Buddington*. A. E. J. Engel et al., eds., Geol. Soc. Amer., Boulder, Colo., pp. 599–620.

Hickson, R. E., and F. W. Rodolf, 1951. History of Columbia River jetties. *Proc. 1st Conf. Coastal Engineering*, Long Beach, Calif., J. W. Johnson, ed., Council on Wave Res., the Engr. Found., pp. 283–298.

Hilde, T. W. C., J. M. Wageman, and W. T. Hammond, 1969. The structure of Tosa Terrace and Nankai Trough off southeastern Japan. *Deep-Sea Res.*, v. 16, pp. 67–75.

Hill, M. N., 1956. Notes on the bathymetric chart of the N.E. Atlantic. *Deep-Sea Res.*, v. 3, pp. 229–231.

Hill, M N., 1963. Single-ship seismic refraction shooting. In *The Sea*, Vol. 3, M. N. Hill, ed., Interscience, New York, pp. 39–46.

Hill, M. N., and W. B. R. King, 1954. Seismic prospecting in the English Channel and its geological interpretation. *Quart. Jour. Geol. Soc. London*, v. 109, pp. 1–19.

Hinde, G. J., 1904. Report on the materials from the borings at the Funafuti Atoll. *Report of the Coral Reef Committee, Royal Society of London*, W. J. Sollas et al., London, pp. 186–360.

Hiranandani, M. G., and C. V. Gole, 1959. Formation and movement of mudbanks and their effect on southwesterly coast of India. Abs. in *Preprints International Oceanographic Congress*, Amer. Assoc. Adv. Sci., Washington, D.C., pp. 623–624.

Hoare, J. M., 1968. Geologic map of the Hooper Bay quadrangle, Alaska. *U.S. Geol. Survey Misc. Geol. Inv. Map* I-523.

Hoare, J. M., and W. H. Condon, 1966. Geologic map of the Kwiguk and Black quadrangles, western Alaska. *U.S. Geol. Survey Misc. Geol. Inv. Map* I-469.

Hoffmeister, J. E., and H. S. Ladd, 1935. The foundations of atolls; a discussion. *Jour. Geol.,* v. 43, no. 6, pp. 643–665.

Hoffmeister, J. E., and H. G. Multer, 1968. Geology and origin of the Florida Keys. *Geol. Soc. Amer. Bull.,* v. 79, pp. 1487–1502.

Holmes, Arthur, 1945. *Principles of Physical Geology.* Ronald Press, New York, 532 pp.

Holtedahl, Hans, 1955. On the Norwegian continental terrace, primarily outside More-Romsdal: its geomorphology and sediments. With contributions on the Quaternary geology of the adjacent land and on the bottom deposits of the Norwegian Sea. *Univ. Bergen, Årbok, Naturvitenskap. Rekke,* no. 14, 209 pp.

Holtedahl, Hans, 1958. Some remarks on geomorphology of the continental shelves off Norway, Labrador, and southeast Alaska. *Jour. Geol.,* v. 66, no. 4, pp. 461–471.

Holtedahl, Hans, 1959. Geology and paleontology of Norwegian Sea bottom cores. *Jour. Sed. Petrology,* v. 29, no. 1, pp. 16–29.

Holtedahl, Olaf, 1940. The submarine relief off the Norwegian coast. *Norske Videnskaps-Akad.,* Oslo, 43 pp.

Holtedahl, Olaf, 1964. Echo-soundings in the Skagerrak, with remarks on the geomorphology. *Norges Geol. Undersokelse,* no. 223, pp. 139–160.

Holtedahl, Olaf, 1970. On the morphology of the west Greenland shelf with general remarks on the "marginal channel" problem. *Marine Geol.,* v. 8, pp. 155–172.

Hopkins, D. M., 1967. The Cenozoic history of Beringia—a synthesis. In *The Bering Land Bridge,* D. M. Hopkins, ed., Stanford University Press, Stanford, Calif. pp. 451–484.

Hopkins, D. M., D. W. Scholl, et al. (in press). Cretaceous, Tertiary, and Early Pleistocene rocks from the continental margin in the Bering Sea. *Geol. Soc. Amer. Bull.*

Hopkins, T. L., 1964. A survey of marine bottom samplers. In *Progress in Oceanography,* Vol. 2, M. Sears, ed., Macmillan, New York, pp. 215–254.

Horn, D. R., M. N. Delach, B. M. Horn, 1969. Distribution of volcanic ash layers and turbidites in the North Pacific. *Geol. Soc. Amer. Bull.,* v. 80, pp. 1715–1724.

Horn, D. R., M. Ewing, M. N. Delach, and B. M. Horn, 1971. Turbidites of the northeast Pacific. *Sedimentology,* v. 16, pp. 55–69.

Hoskins, Hartley, 1967. Seismic reflection observations on the Atlantic continental shelf, slope, and rise southeast of New England. *Jour. Geol.,* v. 74, no. 5, pp. 598–611.

Houbolt, J. J. H. C., 1957. *Surface Sediments of the Persian Gulf Near the Qatar Peninsula.* Thesis, Univ. Utrecht, Mouton, The Hague, 113 pp.

Houbolt, J. J. H. C., 1968. Recent sediments in the southern bight of the North Sea. *Geol. en Mijbouw.,* v. 47, no. 4, pp. 245–273.

Hough, J. L., 1940. Sediments of Buzzards Bay, Massachusetts. *Jour. Sed. Petrology,* v. 10, no. 1, pp. 19–32.

Hough, J. L., 1956. Sediment distribution in the southern oceans around Antarctica. *Jour. Sed. Petrology,* v. 26, no. 4, pp. 301–306.

Hoyt, J. H., 1967. Barrier island formation. *Geol. Soc. Amer. Bull.,* v. 78, pp. 1125–1136.

Huang, T.-C., and H. G. Goodell, 1967. Sediments of Charlotte Harbor, southwestern Florida. *Jour. Sed. Petrology,* v. 37, no. 2, pp. 449–474.

Hubbert, M. K., 1958. Permeability (fluid). In *Encyclopaedia Britannica,* v. 17, pp. 531–532.

Hubert, J. F., 1964. Textural evidence for deposition of many western North Atlantic deep-sea sands by ocean-bottom currents rather than turbidity currents. *Jour. Geol.,* v. 72, no. 6, pp. 757–785.

Hunkins, Kenneth, 1968. Geomorphic provinces of the Arctic Ocean. In *Arctic Drifting Stations,* J. E. Sater, ed., Arctic Inst. North America, pp. 365–376.

Hunkins, Kenneth, 1969. Arctic geophysics. *Jour. Arctic Inst. North Amer.,* v. 22, no 3, pp. 225–232.

Hunkins, Kenneth, E. M. Thorndike, and Guy Mathieu, 1969. Nepheloid layers and bottom currents in the Arctic Ocean. *Jour. Geophys. Res.,* v. 74, no. 28, pp. 6995–7008.

Hunter, W., and D. W. Parkin, 1960. Cosmic dust in recent deep-sea sediments. *Proc. Roy. Soc. London,* A, v. 255, pp. 382–397.

Hurley, P. M., F. F. M. de Almeida, et al., 1967. Test of continental drift by comparison of radiometric ages. *Science*, v. 157, no. 3788, pp. 495–500.

Hurley, R. J., 1960. The Geomorphology of Abyssal Plains in the Northeast Pacific Ocean. Ph.D. thesis, Univ. California, Scripps Inst. of Oceanog., La Jolla, Calif., 173 pp.

Hurley, R. J., and L. K. Fink, Jr., 1963. Ripple marks show that counter-current exists in Florida Straits. *Science*, v. 139, no. 3555, pp. 603–605.

Hurley, R. J., V. B. Siegler, and L. K. Fink, Jr., 1962. Bathymetry of the Straits of Florida and the Bahama Islands. Pt. 1. Northern Straits of Florida. *Bull. Mar. Sci. Gulf and Caribbean*, v. 12, no. 3, pp. 313–321.

Hvorslev, M. J., and H. C. Stetson, 1946. Free-fall coring tube: a new type of gravity bottom sampler. *Geol. Soc. Amer. Bull.*, v. 57, pp. 935–950.

Hyne, N. J., and H. G. Goodell, 1967. Origin of the sediments and submarine geomorphology of the inner continental shelf off Choctawhatchee Bay, Florida. *Marine Geol.*, v. 5, pp. 299–313.

Ignatius, Heikki, 1958. On the rate of sedimentation in the Baltic Sea. *Comptes Rendus Soc. geol. Finlande*, no. 30, pp. 135–144.

Inderbitzen, A. L., 1963. The Sigma corer, an improved marine sediment sampling device. *ASW & Ocean Systems Organization, Lockheed Rept.* no. 16944, 11 pp.

Inderbitzen, A. L., 1964. Report of an investigation of submarine slope stability. *Lockheed California Co., Lockheed Rept.* no. 18140, 45 pp.

Inman, D. L., 1952. Measures for describing the size distribution of sediments. *Jour. Sed. Petrology*, v. 22, no. 3, pp. 125–145.

Inman, D. L., 1953. Areal and seasonal variations in beach and nearshore sediments at La Jolla, California. *Beach Erosion Board, Corps Engrs. Tech. Memo.* no. 39, 134 pp.

Inman, D. L., 1957. Wave generated ripples in nearshore sands. *Beach Erosion Board, Corps Engrs. Tech. Memo.* no. 79, 43 pp.

Inman, D. L., 1963. Sediments: physical properties and mechanics of sedimentation. In *Submarine Geology*, 2nd ed., F. P. Shepard, Harper & Row, New York, Chap. 5.

Inman, D. L., 1968. *Office of Naval Research Progress Rept.*, p. 24.

Inman, D. L., 1970. Strong currents in submarine canyons. *Abs. Trans. Amer. Geophys. Union*, v. 51, no. 4, p. 319.

Inman, D. L., and R. A. Bagnold, 1963. Littoral processes. In *The Sea*, Vol. 3, N. M. Hill, ed., Interscience, New York, pp. 529–553.

Inman, D. L., and A. J. Bowen, 1963. Flume experiments on sand transport by waves and currents. *Proc. 8th Conf. on Coastal Engr.*, Mexico City, J. W. Johnson, ed., pp. 137–150.

Inman, D. L., and Jean Filloux, 1960. Beach cycles related to tide and local wind wave regime. *Jour. Geol.*, v. 68, no. 2, pp. 226–231.

Inman, D. L., and J. D. Frautschy, 1966. Littoral processes and the development of shorelines. *Coastal Engineering; Santa Barbara Specialty Conf.*, pp. 511–536.

Inman, D. L., and N. Nasu, 1956. Orbital velocity associated with wave action near the breaker zone. *Beach Erosion Board, Corps Engrs. Tech. Memo.* no. 79, 43 pp.

Inman, D. L., and C. E. Norstrom, 1971. On the tectonic and morphologic classification of coasts. *Jour. Geol.*, v. 79, no. 1, pp. 1–21.

Inman, D. L., and G. A. Rusnak, 1956. Changes in sand level on the beach and shelf at La Jolla, California. *Beach Erosion Board, Tech. Memo.* no. 82, 64 pp.

Inman, D. L., W. R. Gayman, and D. C. Cox, 1963. Littoral sedimentary processes on Kauai, a sub-tropical high island. *Pacific Sci.*, v. 17, no. 1, pp. 106–130.

Ippen, A. T., ed., 1966. *Estuary and Coastline Hydrodynamics*. McGraw-Hill, New York, 744 pp.

Isaacs, J. D., and G. B. Schick, 1960. Deep-sea free instrument vehicle. *Deep-Sea Res.*, v. 7, pp. 61–67.

Isaacs, J. D., J. L. Reid, Jr., G. B. Schick, and R. A. Schwartzlose, 1966. Near-bottom currents measured in 4 kilometers depth off the Baja California coast. *Jour. Geophys. Res.*, v. 71, no. 18, pp. 4297–4303.

Isacks, Bryan, Jack Oliver, and L. R. Sykes, 1968. Seismology and the new global tectonics. *Jour. Geophys. Res.*, v. 73, no. 18, pp. 5855–5899.

Iselin, C. O'D., as told to John Lear, 1963. What sank "Thresher," the A-sub? *Saturday Rev.*, Oct. 5, pp. 57–60.

Iversen, H. W., 1953. Waves and breakers in shoaling water. *Proc. 3rd Conf. Coastal Engineering*, J. W. Johnson, ed., Council on Wave Res., pp. 1–12.

Jackson, E. D., E. A. Silver, and G. B. Dalrymple (1972). The Hawaiian-Emperor chain and its relation to Cenozoic circumpacific tectonics. *Geol. Soc. Amer. Bull.*, v. 83, pp. 601–618.

Jarke, J., 1956. Eine neue Bodenkarte der südlichen Nordsee. *Deut. Hydrograph. Z.*, v. 9, 1, pp. 1–8.

Johnson, D. A., 1971. Studies of Deep Sea Erosion Using Deep-Towed Instrumentation. Ph.D. thesis, University of California, Scripps Inst. Oceanog., La Jolla, Calif., 171 pp.

Johnson, D. A., S. D. Einsohn, and G. G. Shor, Jr., 1969. Structure and sediments of the Shatsky Rise. Manuscript, MPL-U-53/69, 13 pp.

Johnson, D. W., 1919. *Shore Processes and Shoreline Development.* Wiley, New York, 584 pp.

Johnson, D. W., 1925. *New England—Acadian Shoreline.* Wiley, New York, 608 pp.

Johnson, D. W., 1939. *The Origin of Submarine Canyons.* Columbia Univ. Press, New York, 126 pp.

Johnson, G. L., and B. C. Heezen, 1967. The Arctic Mid-Ocean Ridge. *Nature*, v. 215, no. 5102, p. 724.

Johnson, Helgi, and B. L. Smith, eds., 1970. *The Megatectonics of Continents and Oceans.* Rutgers Univ. Press, New Brunswick, N.J., 282 pp.

Jones, E. J. W., and John Ewing, 1969. The age of the Bay of Biscay: evidence from seismic profiles and bottom samples. *Science*, v. 166, pp. 102–105.

Jones, O. A., 1966. Geological questions posed by the reef. *Australian Nat. Hist.*, v. 15, no. 8, pp. 245–248.

Jordan, G. F., 1954. Large sink holes in Straits of Florida. *Amer. Assoc. Petrol. Geol. Bull.*, v. 38, no. 8, pp. 1810–1817.

Jordan, G. F., 1958. Pamplona Searidge 1779–1957. *Intl. Hydrographic Rev.*, May, pp. 3–13.

Jordan, G. F., 1961. Erosion and sedimentation of eastern Chesapeake Bay at the Choptank River. *U.S. Coast and Geodetic Survey Tech. Bull.* no. 16, 8 pp.

Jordan, G. F., 1962. Redistribution of sediments in Alaskan bays and inlets. *Geogr. Rev.*, v. 52, no. 4, pp. 548–558.

Jordan, G. F., and H. B. Stewart, Jr., 1959. Continental slope off southwest Florida. *Amer. Assoc. Petrol. Geol. Bull.*, v. 43, no. 5, pp. 974–991.

Jordan, G. F., R. J. Malloy, and J. W. Kofoed, 1964. Bathymetry and geology of Portales Terrace, Florida. *Marine Geol.*, v. 1, pp. 259–287.

Kanaev, V. F., 1959. Kronotsky Bay. *Trudi Inst. Okeanol., Akad. Nauk. USSR*, v. 36, pp. 5–20 (in Russian).

Kanes, W. H., 1970. Facies and development of the Colorado River Delta in Texas. In *Deltaic Sedimentation, Modern and Ancient*, J. P. Morgan, ed., Soc. Econ. Paleontol. and Mineralog., spec. publ. no. 15, Tulsa, Okla., pp. 78–106.

Karig, D. E., 1970A. Ridges and basins of the Tonga–Kermadec Island arc system. *Jour. Geophys. Res.*, v. 75, no. 2, pp. 239–254.

Karig, D. E., 1970B. Kermadec Arc-New Zealand tectonic confluence. *New Zealand Jour. Geol. Geophys.*, v. 13, pp. 21–29.

Karig, D. E., 1971. Structural history of the Mariana Island arc system. *Geol. Soc. Amer. Bull.*, v. 82, pp. 323–344.

Katsumata, Mamory, and L. R. Sykes, 1969. Seismicity and tectonics of the western Pacific: Izu-Mariana-Caroline and Ryukyu-Taiwan regions. *Jour. Geophys. Res.*, v. 74, no. 25, pp. 5923–5948.

Kay, M., 1951. *North American Geosynclines.* Geol. Soc. Amer., Mem. 48, Boulder, Colo., 143 pp.

Kay, M., ed., 1969. *North Atlantic—Geology and Continental Drift.* Amer. Assoc. Petrol. Geol., Tulsa, Okla., 1082 pp.

Kaye, C. A., 1959. Shoreline features and Quaternary shoreline changes Puerto Rico. *U.S. Geol. Survey Prof. Paper* 317-B, pp. 49–140.

Keen, M. J., 1968. *An Introduction to Marine Geology.* Pergamon, London, 218 pp.

Keller, G. H., and A. F. Richards, 1967. Sediments of the Malacca Strait, Southeast Asia. *Jour. Sed. Petrology,* v. 37, no. 1, pp. 102–127.

Kennett, J. P., 1970. Pleistocene paleoclimates and formaminiferal biostratigraphy in subantarctic deep-sea cores. *Deep-Sea Res.,* v. 17, pp. 125–140.

Kennett, J. P., and K. R. Geitzenauer, 1969. Pliocene-Pleistocene boundary in a South Pacific deep-sea core. *Nature,* v. 224, pp. 899–901.

Kenyon, N. H., and A. H. Stride, 1968. The crest length and sinuosity of some marine sand waves. *Jour. Sed. Petrology,* v. 38, March, pp. 255–259.

Kenyon, N. H., and A. H. Stride, 1970. The tide-swept continental shelf sediments between the Shetland Isles and France. *Sedimentology,* v. 14, pp. 159–173.

King, C. A. M., 1953. The relationship between wave incidence, wind direction and beach changes at Marsden Bay, Co. Durham. In *Inst. of British Geographers Transactions and Papers,* Publ. no. 19, pp. 13–23.

King, C. A. M., 1959. *Beaches and Coasts.* Edward Arnold, London, 403 pp.; 2nd ed., 1972, 570 pp.

King, Lester, 1952. The Durham beach problem. *S. African Jour. Sci.,* May, pp. 314–318.

King, Lewis H., 1969. Submarine end moraines and associated deposits on the Scotian Shelf. *Geol. Soc. Amer. Bull.,* v. 80, pp. 83–96.

King, Lewis H., 1970. Surficial geology of the Halifax–Sable Island map area. *Dept. Energy, Mines & Resources, Marine Sci. Branch,* Ottawa, Canada, paper 1, 16 pp.

King, L. H., and Brian MacLean, 1970A. A diapiric structure near Sable Island—Scotian Shelf. *Maritime Sediments,* v. 6, no. 1, pp. 1–4.

King, L. H., and Brian MacLean, 1970B. Observations on Cretaceous outcrop from a submersible—Scotian Shelf. *Canadian Jour. Earth Sci.,* v. 7, no. 1, pp. 188–190.

King, L. H., and Brian MacLean, 1970C. Continuous seismic-reflection study of Orpheus gravity anomaly. *Amer. Assoc. Petrol. Geol. Bull.,* v. 54, no. 11. pp. 2007–2031.

King, L. H., and Brian MacLean, 1970D. Origin of the outer part of the Laurentian Channel. *Canadian Jour. Earth Sci.,* v. 7, no. 6, pp. 1470–1484.

King, L. H., Brian MacLean, G. A. Bartlett, J. A. Jeletzky, and W. S. Hopkins, Jr., 1970. Cretaceous strata on the Scotian Shelf. *Canadian Jour. Sci.,* v. 7, no. 1, pp. 145–155.

Kinsman, B., 1965. *Wind Waves, Their Generation and Propagation on the Ocean Surface.* Prentice-Hall, Englewood Cliffs, N.J., 676 pp.

Klein, G. De V., 1961. *Stratigraphy, Sedimentary Petrology, Structure of Triassic Sedimentary Rocks, Maritime Provinces of Canada.* Yale Univ., New Haven, Conn., 302 pp.

Klenova, M. V., 1940. Sediments of the Barents Sea. *Compt. Rend (Doklady) Acad. Sci. USSR,* v. 26, no. 8, pp. 796–800.

Klenova, M. V., 1948. *Geology of the Sea.* 495 pp. (in Russian).

Klenova, M. V., 1960A. *The Geology of the Barents Sea.* Acad. Sci. USSR, Moscow, 367 pp. (in Russian).

Klenova, M. V., 1960B. Geology of the Barents Sea. English abs. in *Marine Geology,* Intern. Geol. Congr. 21st Sess., Repts. of Soviet Geologists, p. 130.

Knauss, J. A., 1961. The Cromwell Current. *Scientific Amer.,* v. 204, no. 4, pp. 105–116.

Knott, S. T., and H. Hoskins, 1968. Evidence of Pleistocene events in the structure of the continental shelf off the northeastern United States. *Marine Geol.,* v. 6, pp. 5–43.

Kofoed, J. W., and D. S. Gorsline, 1963. Sedimentary environments in Apalachicola Bay and vicinity, Florida. *Jour. Sed. Petrology,* v. 33, no. 1, pp. 205–223.

Kofoed, J. W., and G. F. Jordan, 1964. Isolated fault scarps on the continental slope off southwest Florida. *Southeastern Geol.,* v. 5, no. 2, pp. 69–77.

Kofoed, J. W., and R. J. Malloy, 1965. Bathymetry of the Miami Terrace. *Southeastern Geol.,* v. 6, no. 3, pp. 159–165.

Kögler, F.-C., 1963. Das Kastenlot. *Meyniana,* v. 13, pp. 1–17.

Kolb, C. R., and J. R. Van Lopik, 1958A. Geological investigations of the Mississippi River-Gulf outlet channel. *Misc. Paper 3-259, U.S. Army Engr. Waterway Experimental Sta., Corps of Engrs., Tech. Rept.* No. 3, 120 pp.

Kolb, C. R., and J. R. Van Lopik, 1958B. Geology of the Mississippi River deltaic plain. *U.S. Corps Engrs., Waterways Experimental Sta., Tech. Repts.,* 3–483 and 4–484, 2 vols.

Kolbe, R. W., 1957. Fresh-water diatoms from Atlantic deep-sea sediments. *Science,* v. 126, p. 1053.

Koldewijn, B. W., 1958. *Sediments of the Paria-Trinidad Shelf.* Vol. 3, *Reports of the Orinoco Shelf Expedition.* Thesis, Mouton, The Hague, 109 pp.

Koldijk, W. S., 1968. Bottom Sediments of the Ria de Arosa (Galicia, NW Spain). Ph.D. thesis, Univ. of Leiden, pp. 77–134.

Komar, P. D., 1969. The channelized flow of turbidity currents with application to Monterey deep-sea fan channel. *Jour. Geophys. Res.,* v. 74, no. 18, pp. 4544–4557.

Kornicker, L. S., and D. W. Boyd, 1962. Shallow-water geology and environments of Alacran Reef complex, Campeche Bank, Mexico. *Amer. Assoc. Petrol. Geol. Bull.,* v. 46, no. 5, pp. 640–673

Kotenev, B. N., 1965. Submarine valleys in the zone of the continental slope in the Bering Sea. *Tr. Vses. Nauchn. Issled. Inst. Razved. Okeanol.,* v. 58, pp. 35–44 (in Russian).

Krause, D. C., 1964A. Lithology and sedimentation in the southern continental borderland. In *Papers in Marine Geology,* Shepard Comm. Vol., R. L. Miller, ed., Macmillan, New York, pp. 274–318.

Krause, D. C., 1964B. Guinea fracture zone in the equatorial Atlantic. *Science,* v. 146. no. 3640, pp. 57–59.

Krause, D. C., 1965. East and West Azores fracture-zones in the North Atlantic. In *Submarine Geology and Geophysics,* W. F. Whittard, and R. Bradshaw, eds., Buttersworths, London, pp. 163–173.

Krause, D. C., 1966. Tectonics, marine geology, and bathymetry of the Celebes Sea–Sula Sea region. *Geol. Soc. Amer. Bull.,* v. 77, pp. 813–832.

Krause, D. C., and V. F. Kanaev, 1970. Narrow-beam echo sounding in marine geomorphology. *Intl. Hydrographic Rev.,* v. 47, no. 1, pp. 23–33.

Krause, D. C., and J. G. Schilling, 1969. Dredged basalt from the Reykjanes Ridge, North Atlantic. *Nature,* v. 224, no. 5221, pp. 791–793.

Krause, D. C., H. W. Menard, and S. M. Smith, 1964. Topography and lithology of the Mendocino Ridge. *Jour. Marine Res.,* v. 22, no. 3, pp. 236–250.

Kruit, C., 1955. *Sediments of the Rhone Delta: Grain Size and Microfauna.* Dissert. Rijkuniversiteit te Groningen, Mouton, The Hague, 141 pp.

Krumbein, W. C., 1936. Application of logarithmic moments to size frequency distribution of sediments. *Jour. Sed. Petrology,* v. 6, pp. 35–47.

Krumbein, W. C., 1941. Measurement and geological significance of shape and roundness of sedimentary particles. *Jour. Sed. Petrology,* v. 11, pp. 64–72.

Krumbein, W. C., and Esther Aberdeen, 1937. The sediments of Barataria Bay. *Jour. Sed. Petrology,* v. 7, no. 1, pp. 3–17.

Krumbein, W. C., and L. T. Caldwell, 1939. Areal variation of organic carbon content of Barataria Bay sediments, Louisiana. *Amer. Assoc. Petrol. Geol. Bull.,* v. 23, no. 4, pp. 582–594.

Krumbein, W. C., and G. D. Monk, 1942. Permeability as a function of the size parameters of unconsolidated sand. *Amer. Inst. Mining and Metallurgy Engrs. Tech. Publ.* No. 1492, Petroleum Technol., July, pp. 1–11.

Ku, T.-L., 1968. Protactinium 231 method of dating coral from Barbados Island. *Jour. Geophys. Res.,* v. 73, no. 6, pp. 2271–2276.

Ku, T.-L., and W. S. Broecker, 1966. Atlantic deep-sea stratigraphy: extension of absolute chronology to 320,000 years. *Science,* v. 151, pp. 448–450.

Kudinov, E. I., 1957. Vibro-piston core sampler. *Trudi Inst Okeanol Akad. Nauk USSR,* v. 25, pp. 143–152 (in Russian).

Kuenen, Ph. H., 1933. Geology of coral reefs. In *The Snellius Expedition, 1929–1930, Geological Results.* Vol. 5, pt. 2, Brill, Leiden, 124 pp.

Kuenen, Ph. H., 1942. Bottom samples, collection of the samples and some general aspects. In *The Snellius Expedition, 1929–1930.* Vol. 5, pt. 3, sec. 1, Brill, Leiden.

Kuenen, Ph. H., 1947. Two problems of marine geology: atolls and canyons. *Koninkl. Ned. Akad. Wetenschap., Verh. (Tweede Sec.),* v. 43, no. 3, pp. 1–69.

Kuenen, Ph. H., 1951. Properties of turbidity currents of high density. *Soc. Econ. Paleontol. and Mineralog.,* spec. publ. no. 2, pp. 14–33.

Kuenen, Ph. H., 1953. Origin and classification of submarine canyons. *Geol. Soc. Amer. Bull.,* v. 64, pp. 1295–1314.

Kuenen, Ph. H., 1958. Some experiments on fluviatile rounding. *Koninkl. Ned. Akad. Wetensch. Proc.,* ser. B, v. 61, no. 1, pp. 47–53.

Kuenen, Ph. H., 1959. Turbidity currents a major factor in flysch deposition. *Ecologae Geol. Helv.,* v. 51, no. 3, pp. 1009–1021.

Kuenen, Ph. H., 1964. Bibliography of turbidity currents and turbidites. In *Turbidites. Developments in Sedimentology—3,* A. H. Bouma and A. Brouwer, eds., Elsevier, New York, pp. 222–246.

Kuenen, Ph. H., 1965. Experiments in connection with turbidity currents and clay suspensions. In *Submarine Geology and Geophysics,* W. F. Whittard and R. Bradshaw, eds., Butterworths, London, pp. 47–74.

Kuenen, Ph. H., 1967. Emplacement of flysch-type sand beds. *Sedimentology,* v. 9, pp. 203–243.

Kuenen, Ph. H., and C. I. Migliorini, 1950. Turbidity currents as a cause of graded bedding. *Jour. Geol.,* v. 58, pp. 91–127.

Kullenberg, B., 1947. The piston core sampler. *Sv. Hydr.-Biol. Komm. skr.,* e. ser.: Hydro. Bed. 1, H. 2, Göteborg, 46 pp.

Kullenberg, B., 1955. A new core-sampler. *K. Vet. O. Vitterh. Samh. Handl.,* F. 6, ser. A, v. 6, no. 15, 17 pp.

Kulm, L. D., and J. V. Byrne, 1967. Sediments of Yaquina Bay, Oregon. In *Estuaries,* G. H. Lauff, ed., Amer. Assoc. Adv. Sci., Washington, D.C., pp. 226–238.

Kummel, B., and D. Raup, eds., 1965. *Handbook of Paleontological Techniques.* Freeman, San Francisco, 852 pp.

Laban, André, J.-M. Pérès, and Jacques Piccard, 1963. La photographie sous-marine profonde et son exploitation scientifique. *Bull. Inst. Oceanog. Monaco,* v. 60 no. 1258, pp. 1–32.

LaCoste, L. J. B., N. Clarkson, and G. Hamilton, 1967. LaCoste and Romberg stabilized platform shipboard gravity meter. *Geophysics,* v. 32, pp. 99–109.

Ladd, H. S., 1958. Fossil land shells from western Pacific atolls. *Jour. Paleontol.,* v. 32, no. 1, pp. 183–198.

Ladd, H. S., 1961. Reef building. *Science,* v. 134, no. 3481, pp. 703–715.

Ladd, H. S., and J. E. Hoffmeister, 1936. A criticism of the glacial-control theory. *Jour. Geol.,* v. 44, no. 1, pp. 74–92.

Ladd, H. S., and S. O. Schlanger, 1960. Drilling operations on Eniwetok Atoll, Bikini and nearby atolls, Marshall Islands. *U.S. Geol. Survey Prof. Paper 260-Y,* pp. 863–905.

Ladd, H. S., J. I. Tracey, Jr., J. W. Wells, and K. O. Emery, 1950. Organic growth and sedimentation on an atoll. *Jour. Geol.,* v. 58, no. 4, pp. 410–425.

Ladd, H. S., J. I. Tracey, Jr., and A. G. Gross, 1970. Deep drilling on Midway Atoll. *U.S. Geol. Survey Prof. Paper 680-A,* pp. A1–A22.

LaFond, E. C., 1961. The isotherm follower. *Jour. Marine Res.,* v. 19, no. 1, pp. 33–39.

Land, L. S., F. T. Mackenzie, and S. J. Gould, 1967. Pleistocene history of Bermuda. *Geol. Soc. Amer. Bull.,* v. 78, pp. 993–1006.

Langseth, M. G., Jr., and R. P. Von Herzen, 1970. Heat flow through the floor of the world oceans. In *The Sea,* Vol. 4, Pt. 1, A. E. Maxwell, ed., Wiley–Interscience, New York, pp. 299–352.

Lankford, R. R., and J. R. Curray, 1957. Mid-Tertiary rock outcrop on continental shelf, northwest Gulf of Mexico. *Amer. Assoc. Petrol. Geol. Bull.,* v. 41, no. 9, pp. 2114–2117.

Lankford, R. R., and F. P. Shepard, 1960. Facies interpretation in Mississippi Delta borings. *Jour. Geol.,* v. 68, no. 4, pp. 408–426.

Larson, H. E., 1959. *A History of Self-Contained Diving and Underwater Swimming.* Publ. 469, Nat. Acad. Sci., Nat. Res. Coun., Washington, D.C., 50 pp.

Larson, R. L., 1971. Near-bottom geologic studies of the East Pacific Rise crest. *Geol. Soc. Amer. Bull.,* v. 82, pp. 823–842.

Larson, R. L., and C. G. Chase, 1970. Relative velocities of the Pacific, North America, and Cocos plates in the Middle American region. *Earth Planetary Sci. Letters,* v. 7, pp. 425–428.

Larsonneur, Claude, and Pierre Hommeril, 1967. Sediments et sedimentation dans la partie orientale de la Baie de Seine. *Rev. Soc. Sav. Hte-Normandie, Sci.,* no. 47, pp. 45–75.

Lauff, G. H., Ed., 1967. *Estuaries.* Publ. No. 83, Amer. Assoc. Adv. Sci., Washington, D.C., 757 pp.

Laughton, A. S., 1960. An interplain deep-sea channel system. *Deep-Sea Res.,* v. 7, pp. 75–88.

Laughton, A. S., 1968. New evidence of erosion on the deep ocean floor. *Deep-Sea Res.,* v. 15, pp. 21–29.

Laughton, A. S., 1970. A new bathymetric chart of the Red Sea. *Philos. Trans. Roy. Soc. London,* A, v. 267, pp. 21–22.

Laughton, A. S., and C. Tramontini, 1969. Recent studies of the crustal structure in the Gulf of Aden. *Tectonophysics,* v. 8, pp. 359–375.

Laughton, A. S., R. B. Whitmarsh, and M. T. Jones, 1970A. The evolution of the Gulf of Aden. *Philos. Trans. Roy. Soc. London,* A, v. 267, pp. 227–266.

Laughton, A. S., D. H. Matthews, and R. L. Fisher, 1970B. The structure of the Indian Ocean. In *The Sea,* Vol. 4, Pt. 2, A. E. Maxwell, ed., Wiley–Interscience, New York, pp. 543–586.

Laughton, A. S., W. A. Berggren, et al., 1970C. Deep sea drilling project Leg 12. *Geotimes,* v. 15, no. 9, pp. 10–14.

Lawson, A. C., 1893. The geology of Carmelo Bay. *Univ. California, Dept. Geol. Bull.,* v. 1, pp. 1–59.

Lawson, A. C., 1932. Insular arcs, foredeeps and geosynclinal seas of the Asiatic Coast. *Geol. Soc. Amer. Bull.,* v. 43, pp. 353–381.

LeBlanc, R. J., and W. D. Hodgson, 1959. Origin and development of the Texas shorelines. *Gulf Coast Assn. Geol. Soc. Trans.,* v. 9, pp. 197–220.

LeClaire, Lucien, Jean-Pierre Caulet, and Philippe Bouysse, 1965. Prospection sedimentologique de la marge continentale Nord-Africaine. *Cahiers Oceanog.,* v. 17, no. 7, pp. 467–479.

Leffingwell, E. deK., 1919. The Canning River region, northern Alaska. *U.S. Geol. Survey Prof. Paper* 109, 243 pp.

Lehner, Peter, 1969. Salt tectonics and Pleistocene stratigraphy on continental slope of northern Gulf of Mexico. *Amer. Assoc. Petrol. Geol. Bull.,* v. 53, no. 12, pp. 2431–2479.

Le Pichon, X., 1968. Sea-floor spreading and continental drift. *Jour. Geophys. Res.,* v. 73, pp. 3661–3698.

Le Pichon, X., and J. R. Heirtzler, 1968. Magnetic anomalies in the Indian Ocean and sea-floor spreading. *Jour. Geophys. Res.,* v. 73, pp. 2101–2117.

Levert, C. F., Jr., 1969. Geology of the Flower Garden Banks, northeast Gulf of Mexico. *Trans. Gulf Coast Assn. Geol. Soc.,* 19th Ann. meeting, pp. 89–100.

Lisitzin, A. P., 1966. *Processes of Recent Sedimentation in the Bering Sea.* Acad. Sci. USSR, Dept. Earth Sci. Comm. Sed. Rocks, Inst. Oceanol., Moscow, 574 pp. (in Russian).

Lisitzin, A. P., 1970. Sedimentation and geochemical considerations. In *Scientific Exploration of the South Pacific,* W. S. Wooster, ed., Nat. Acad. of Sci., Washington, D.C., pp. 89–132.

Lisitzin, A. P., 1972. Sedimentation in the world ocean. *Soc. Econ. Paleontol. and Mineralog.* spec. publ. no. 17, 218 pp.

Logan, B. W. et al., 1969. Constituents of carbonate sediments on Yucatan shelf. In *Carbonate Sediments and Reefs, Yucatan Shelf, Mexico,* A. R. McBirney, ed., Amer. Assoc. Petrol. Geol. Mem. no. 11, Tulsa, Okla., pp. 42–68.

Longseth, M. G., and R. P. Von Herzen, 1970. Heat flow through the floor of the world oceans. In *The Sea*, Vol. 4, Pt. 1, A. E. Maxwell, ed., Wiley–Interscience, New York, pp. 299–352.

Loring D. H., 1962. A preliminary study of the soft sediment layers in the Gulf of St. Lawrence and parts of the Scotian and Newfoundland Shelves. *Fisheries Res. Board Canada, Manusc. Rept. Ser.* no. 107, 17 pp.

Lort, J. M., 1971. The tectonics of the eastern Mediterranean: a geophysical review. *Rev. Geophys. and Space Physics*, v. 9, no. 2, pp. 189–216.

Lotze, Franz, 1963. The distribution of evaporites in space and time. In *Problems in Palaeoclimatology*, NATO Paleoclimates Conf. Univ. Newcastle upon Tyne Proc., Interscience, London, New York, pp. 491–507, 531–532.

Louderback, G. D., 1914. Preliminary report upon the bottom deposits in San Francisco Bay. *Univ. California Publs. Zool.*, v. 14, pp. 89–97, 185–196.

Louderback, G. D., 1940. San Francisco Bay sediments. *Sixth Pacific Sci. Congr., 1939 Proc.*, v. 2, pp. 783–793.

Lowenstam, H. A., and S. Epstein, 1957. On the origin of sedimentary aragonite needles of the Great Bahama Bank. *Jour. Geol.*, v. 65, pp. 364–375.

Ludwick, J. C., 1970. Sand waves and tidal channels in the entrance to Chesapeake Bay. *Inst. Oceanog., Old Dominion Univ., Norfolk, Virginia, Tech. Rept.* no. 1, 7 pp.

Ludwick, J. C., and W. R. Walton, 1957. Shelf-edge, calcareous prominences in northeastern Gulf of Mexico. *Amer. Assoc. Petrol. Geol. Bull.*, v. 41, no. 9, pp. 2054–2101.

Ludwig, W. J., 1970. The Manila Trench and west Luzon Trough—III. Seismic-refraction measurements. *Deep-Sea Res.*, v. 17, pp. 553–571.

Ludwig, W. J., J. I. Ewing, and Maurice Ewing, 1968. Structure of Argentine continental margin. *Amer. Assoc. Petrol. Geol. Bull.*, v. 52, no. 12, pp. 2337–2368.

Ludwig, W. J., J. E. Nafe, and C. L. Drake, 1970. Seismic refraction. In *The Sea*, Vol. 4, Pt. 1, A. E. Maxwell, ed., Wiley–Interscience, New York, pp. 53–84.

Ludwig, W. J., S. Murauchi, et al., 1971. Structure of Bowers Ridge, Bering Sea. *Jour. Geophys. Res.*, v. 76, no. 26, pp. 6350–6366.

Luskin, Bernard, B. C. Heezen, M. Ewing, and Mark Landisman, 1954. Precision measurement of ocean depth. *Deep-Sea Res.*, v. 1, pp. 131–140.

Luternauer, J. L., and O. H. Pilkey, 1967. Phosphorite grains: their application to the interpretation of North Carolina shelf sedimentation. *Marine Geol.*, v. 5, pp. 315–320.

Luyendyk, B. P., 1970. Origin and history of abyssal hills in the northeast Pacific Ocean. *Geol. Soc. Amer. Bull.*, v. 81, pp. 2237–2260.

Lyall, A. K., D. J. Stanley, H. N. Giles, and Alvan Fisher, Jr., 1971. Suspended sediment and transport at the shelf-break and on the slope, Wilmington Canyon area, eastern U.S.A. *Marine Technol. Soc. Jour.*, v. 5, no. 1, pp. 15–27.

Lynts, G. W., 1970. Conceptual model of the Bahamian Platform for the last 135 million years. *Nature*, v. 225, no. 5239, pp. 1226–1228.

Mabesoone, J. M., and I. M. Tinoco, 1955–1956. Shelf off Alagoas and Sergipe (northeastern Brazil). *Trans. Inst. Oceanog. Univ. Fed. Pe., Recife*, 7/8, pp. 151–186.

MacCarthy, G. R., 1935. Eolian sands, a comparison. *Amer. Jour. Sci.*, v. 30, no. 176, pp. 81–95.

Macdonald, G. A., 1949. Hawaiian petrographic province. *Geol. Soc. Amer. Bull.*, v. 60, pp. 1541–1596.

MacKenzie, F. T., 1969. Field guide to Bermuda geology. *Bermuda Biol. Sta. for Res.*, spec. publ. no. 4, St. George's, West Bermuda, 14 pp. (mimeo.)

MacNeil, F. S., 1954. Organic reefs and banks and associated detrital sediments. The shape of atolls: an inheritance from subaerial erosion forms. *Amer. Jour. Sci.*, v. 252, pp. 385–401, 402–427.

Mahadevan, C., and M. P. Rao, 1954. Study of ocean floor sediments off the east coast of India. *Andhra Univ. Mem. Oceanog.*, 1, pp. 1–35.

Malaise, Rene, 1969. *A New Deal in Geology, Geography, and Related Sciences.* (Privately printed) Herbyvägen 1, S-181 42 Lidingö, Sweden, 328 pp.

Malfait, B. T., and M. G. Dinkelman, 1972. Circum-Caribbean tectonic and igneous activity and the evolution of the Caribbean plate. *Geol. Soc. Amer. Bull.*, v. 83, no. 2, pp. 251–272.

Malloy, R. J., 1964. Crustal uplift of Montague Island, Alaska. *Science*, v. 146, no. 3647, pp. 1048–1049.

Malloy, R. J., 1968. Depositional anticlines versus tectonic "reverse drag." *Trans. Gulf Coast Assn. Geol. Soc.*, 18th Ann. Meeting, Jackson, Miss., J. O. Snowden, Jr., ed., pp. 114–123.

Mammerickx, Jacqueline, 1970. Morphology of the Aleutian abyssal plain. *Geol. Soc. Amer. Bull.*, v. 81, pp. 3457–3464.

Manheim, F. T., and F. L. Sayles, 1970. Brines and interstitial brackish water in drill cores from the deep Gulf of Mexico. *Science*, v. 170, pp. 57–61.

Margolis, S. V., and J. P. Kennett, 1970. Antarctic glaciation during the Tertiary recorded in sub-Antarctic deep-sea cores. *Science*, v. 170, pp. 1085–1087.

Markl, R. G., G. M. Bryan, and J. I. Ewing, 1970. Structure of the Blake-Bahama outer ridge. *Jour. Geophys. Res.*, v. 75, no. 24, pp. 4539–4555.

Marlow, M. S., D. W. Scholl, et al., 1970. Buldir Depression—a Late Tertiary graben on the Aleutian Ridge, Alaska. *Marine Geol.*, v. 8, pp. 85–108.

Martens, J. H. C., 1931. Beaches of Florida. *Florida State Geol. Survey Ann. Rept.* 21st-22nd., pp. 67–119.

Martin, B. D., and K. O. Emery, 1967. Geology of Monterey Canyon, Calif. *Amer. Assoc. Petrol. Geol. Bull.*, v. 51, no. 11, pp. 2281–2304.

Martin, H. B., and J. E. Kenny, 1971. Recovery of untethered instruments. *Oceanology Intl.*, July, pp. 29–31.

Mascle, J. R., 1968. Contributions a l'étude de la marge continentale et de la plaine abyssale au large de Toulon. Ph.D. thesis, Faculté des Sciences, Paris, 111 pp.

Mathews, W. H., and F. P. Shepard, 1962. Sedimentation of Fraser River Delta, British Columbia. *Amer. Assoc. Petrol. Geol. Bull.*, v. 46, no. 8, pp. 1416–1438.

Matthews, D. J., 1944. *Tables of the Velocity of Sound in Pure Water and Sea Water for Use in Echo-sounding and Sound-Ranging.* 2nd ed., Hydrogr. Dept., Admiralty, H. D. 282, 52 pp.

Maxwell, A. E., ed., 1970. *The Sea*, Vol. 4, Pts. I, II, Wiley–Interscience, New York.

Maxwell, A. E., R. P. Von Herzen, et al., 1970. Deep sea drilling in the South Atlantic. *Science*, v. 168, pp. 1047–1059.

Maxwell, W. G. H., 1968. *Atlas of the Great Barrier Reef.* Elsevier, Amsterdam, 258 pp.

Maxwell, W. G. H., 1970. Deltaic patterns in reefs. *Deep-Sea Res.*, v. 17, pp. 1005–1018.

Maxwell, W. G. H., and W. R. Maiklem, 1964. Lithofacies analysis, southern part of the Great Barrier Reef. *Univ. Queensland Papers*, v. 5, no. 11, pp. 1–21.

Mayor, A. G., 1924. Some posthumous papers relating to work at Tutuila Island and adjacent regions. Papers from *Dept. Marine Biol., Carnegie Inst.*, Washington, D.C., v. 19, pp. 1–90.

McClure, C. D., H. F. Nelson, and W. B. Huckabay, 1958. Marine Sonoprobe system, a new tool for geologic mapping. *Amer. Assoc. Petrol. Geol. Bull.*, v. 42, pp. 701–716.

McCoy, F. W., and R. P. Von Herzen, 1971. Deep-sea corehead camera photography and piston coring. *Deep-Sea Res.*, v. 18, pp. 361–373.

McCulloch, D. S., 1967. Quaternary geology of the Alaskan shore of Chukchi Sea. In *The Bering Land Bridge*, D. M. Hopkins, ed., Stanford Univ. Press, Stanford, Calif., pp. 91–120.

McGeary, D. F. R., 1969. Sediments of the Vema Fracture Zone. Ph.D. Thesis, Univ. California, San Diego, 62 pp. (mimeo.).

McGill, J. T., 1958. Map of coastal landforms of the world. *Geogr. Rev.*, v. 49, no. 3, pp. 402–405.

McHugh, J. L., 1967. Estuarine nekton. In *Estuaries*, G. H. Lauff, ed., Amer. Assoc. Adv. Sci., Washington, D.C., pp. 581–620.

McKee, E. D., 1950. Report on studies of stratification in modern sediments and in laboratory experiments. Project Nonr 164(00), NR 081 123, *Office of Naval Res.*, 61 pp.

McKee, E. D., 1958. Geology of Kapingamarangi Atoll, Caroline Islands. *Geol. Soc. Amer. Bull.,* v. 69, pp. 241–277.

McKee, E. D., 1962. Origin of the Nubian and similar sandstones. *Geol. Rundsch.,* v. 52, pp. 551–587.

McKee, E. D., and T. S. Sterrett, 1961. Laboratory experiments on form and structure of longshore bars and beaches. In *Geometry of Sandstone Bodies,* Amer. Assoc. Petrol. Geol., Tulsa, Okla., pp. 13–28.

McKenzie, D. P., 1970. Plate tectonics of the Mediterranean region. *Nature,* v. 226, no. 5242, pp. 239–243.

McLeod, I. R., 1964. The saline lakes of the Vestfold Hills, Princess Elizabeth Land. In *Antarctic Geology,* Proc. 1st Intl. Symposium on Antarctic Geology, Cape Town, Sept. 1963, North-Holland, Amsterdam, pp. 65–72.

McManus, D. A., R. E. Burns, et al., 1969. Deep Sea Drilling Project Leg 5. *Geotimes,* v. 14, no. 7, pp. 19–20.

McMaster, R. L., 1954. Petrography and genesis of the New Jersey beach sands. *New Jersey State Dept. Conserv. Geol. Ser. Bull.,* 63, 239 pp.

McMaster, R. L., 1960. Sediments of Narragansett Bay system and Rhode Island Sound, Rhode Island. *Jour. Sed. Petrology,* v. 30, no. 2, pp. 249–274.

McMaster, R. L., 1962. Petrography and genesis of recent sediments in Narragansett Bay and Rhode Island Sound, Rhode Island. *Jour. Sed. Petrology,* v. 32, no. 3, pp. 484–501.

McMaster, R. L., and T. P. LaChance, 1968. Seismic reflectivity studies on northwestern African continental shelf: Strait of Gibraltar to Mauritania. *Amer. Assoc. Petrol. Geol. Bull.,* v. 52, no. 12, pp. 2387–2395.

McMaster, R. L., and T. P. LaChance, 1969. Northwestern African continental shelf sediments. *Marine Geol.,* v. 7, pp. 57–67.

McMaster, R. L., T. P. LaChance, and L. E. Garrison, 1968. Seismic-reflection studies in Block Island and Rhode Island Sounds. *Amer. Assoc. Petrol. Geol. Bull.,* v. 52, no. 3, pp. 465–474.

McMaster, R. L., Jelle de Boer, and Asaf Ashraf, 1970A. Magnetic and seismic reflection studies on continental shelf off Portuguese Guinea, Guinea and Sierra Leone, West Africa. *Amer. Assoc. Petrol. Geol. Bull.,* v. 54, no. 1, pp. 158–167.

McMaster, R. L., T. P. LaChance, and Asaf Ashraf, 1970B. Continental shelf geomorphic features off Portuguese Guinea, Guinea, and Sierra Leone (West Africa). *Marine Geol.,* v. 9, pp. 203–213.

McMaster, R. L., J. D. Milliman, and A. Ashraf, 1971. Continental shelf and upper slope sediments off Portuguese Guinea, Guinea, and Sierra Leone, West Africa. *Jour. Sed. Petrology,* v. 41, no. 1, pp. 150–158.

Meisburger, E. P., and D. B. Duane, 1969. Shallow structural characteristics of Florida Atlantic shelf as revealed by seismic reflection profiles. *U.S. Army, Corps Engrs. Coastal Engineering Res. Center,* 1-70, pp. 207–215.

Menard, H. W., 1953. Pleistocene and Recent sediment from the floor of the northwest Pacific. *Geol. Soc. Amer. Bull.,* v. 64, pp. 1279–1294.

Menard, H. W., 1955. Deep-sea channels, topography, and sedimentation. *Amer. Assoc. Petrol. Geol. Bull.,* v. 39, pp. 236–255.

Menard, H. W., 1956. Archipelagic aprons. *Amer. Assoc. Petrol. Geol. Bull.,* v. 40, no. 9, pp. 2195–2210.

Menard, H. W., 1960. Consolidated slabs on the floor of the eastern Pacific. *Deep-Sea Res.,* v. 7, no. 1, pp. 35–41.

Menard, H. W., 1964. *Marine Geology of the Pacific.* McGraw-Hill, New York, 271 pp.

Menard, H. W., 1967A. Sea floor spreading, topography, and the second layer. *Science,* v. 157, no. 3791, pp. 923–924.

Menard, H. W., 1967B. Transitional types of crust under small ocean basins. *Jour. Geophys. Res.,* v. 72, no. 12, pp. 3061–3073.

Menard, H. W., 1969. The deep-ocean floor. In *The Ocean,* Scientific American Book, Freeman, San Francisco, pp. 55–63.

Menard, H. W., and T. M. Atwater, 1968. Changes in direction of sea floor spreading. *Nature,* v. 219, no. 5153, pp. 463–467.

Menard, H. W., and R. S. Dietz, 1952. Mendocino Submarine Escarpment. *Jour. Geol.,* v. 60, no. 3, pp. 266–278.

Menard, H. W., and C. J. Shipek, 1958. Surface concentrations of manganese nodules. *Nature,* v. 182, pp. 1156–1158.

Menard, H. W., and S. M. Smith, 1966. Hypsometry of ocean basin provinces. *Jour. Geophys. Res.,* v. 71, no. 18, pp. 4305–4325.

Menard, H. W., S. M. Smith, and R. M. Pratt, 1965. The Rhône deep-sea fan. In *Submarine Geology and Geophysics,* W. F. Whittard and R. Bradshaw, eds., Butterworths, London, pp. 271–285.

Mero, J. L., 1965. *The Mineral Resources of the Sea.* Elsevier, New York, 312 pp.

Meyerhoff, A. A., 1970. Continental drift, I; implications of paleomagnetic studies, meteorology, physical oceanography, and climatology. *Jour. Geol.,* v. 78, no. 1, pp. 1–51.

Meyerhoff, A. A., and H. A. Meyerhoff, 1972A. "The new global tectonics": major inconsistencies. *Amer. Assoc. Petrol. Geol. Bull.,* v. 56, no. 2, pp. 269–336.

Meyerhoff, A. A., and H. A. Meyerhoff, 1972B. "The new global tectonics": age of linear magnetic anomalies of ocean basins. *Amer. Assoc. Petrol. Geol. Bull.,* v. 56, no. 2, pp. 337–359.

Meyerhoff, A. A., and H. A. Meyerhoff, 1972C. Continental drift, IV: the Caribbean "plate". *Jour. Geol.,* v. 80, no. 1, pp. 34–60.

Meyerhoff, A. A., and Curt Teichert, 1971. Continental drift, III: Late Paleozoic glacial centers, and Devonian-Eocene coal distribution. *Jour. Geol.,* v. 79, no. 3, pp. 285–321.

Middleton, G. V., 1966–1967. Experiments on density and turbidity currents. *Canadian Jour. Earth Sci.,* v. 3, pp. 523–546; v. 3, pp. 627–637; v. 4 (1967), pp. 475–505.

Miller, D. J., 1960. Giant waves in Lituya Bay, Alaska. *U.S. Geol. Survey Prof. Paper 354-C,* pp. 51–86.

Milliman, J. D., F. T. Manheim, R. M. Pratt, and E. F. K. Zarudzki, 1967. Alvin dives on the continental margin off the southeastern United States, July 2-13, 1967. *Woods Hole Oceanog. Inst. Ref. no. 67–80,* 48 pp. (mimeo.).

Mitchell, A. H., and H. G. Reading, 1971. Evolution of island arcs. *Jour. Geol.,* v. 79, no. 3, pp. 253–284.

Moberly, Ralph, Jr., and Theodore Chamberlain, 1964. Hawaiian beach systems. Final rept. Harbors Div., Transport. Dept., Hawaii, Hawaiian Inst. Geophysics, Univ. Hawaii, 177 pp.

Moberly, Ralph, Jr., and F. W. McCoy, Jr., 1966. The sea floor north of the eastern Hawaiian Islands. *Marine Geol.,* v. 4, pp. 21–48.

Moberly, Ralph, Jr., D. C. Cox, Theodore Chamberlain, F. W. McCoy, and J. F. Campbell, 1963. Coastal geology of Hawaii. Final Rept. *Planning and Econ. Development, State Hawaii, Hawaii Inst. Geophysics,* Univ. Hawaii, 216 pp.

Moberly, Ralph, Jr., L. D. Baver, Jr., and Anne Morrison, 1965. Source and variation of Hawaiian littoral sand. *Jour. Sed. Petrology,* v. 35, no. 3, pp. 589–598.

Molengraaff, G. A. F., 1916. The coral reef problem and isostasy. *Proc. Akad. Wetenschap. Amsterdam,* v. 19, pp. 623–624.

Molengraaff, G. A. F., 1921. *De Geologie der Zeeën van Nederlandsch Oost-Indië.* E. J. Brill, Leiden.

Molengraaff, G. A. F., 1930. The recent sediments in the seas of the East Indian Archipelago, with a short discussion on the condition of those seas in former geological periods. *Proc. 4th Pacific Sci. Congr.,* Java, v. 2B, pp. 989–1021.

Molnar, Peter, and L. R. Sykes, 1969. Tectonics of the Caribbean and Middle America regions from focal mechanisms and seismicity. *Geol. Soc. Amer. Bull.,* v. 80, pp. 1639–1684.

Moore, D. G., 1954. Submarine geology of San Pedro shelf. *Jour. Sed. Petrology,* v. 24, pp. 162–181.

Moore, D. G., 1960. Acoustic-reflection studies of the continental shelf and slope off southern California. *Geol. Soc. Amer. Bull.,* v. 71, pp. 1121–1136.

Moore, D. G., 1961A. Submarine slumps. *Jour. Sed. Petrology,* v. 31, no. 3, pp. 343–357.

Moore, D. G., 1961B. The free-corer: sediment sampling without wire and winch. *Jour. Sed. Petrology,* v. 31, no. 4, pp. 627–630.

Moore, D. G., 1969. Reflection profiling studies of the California continental border-land: structure and Quaternary turbidite basins. *Geol. Soc. Amer. Spec. Paper* 107, 142 pp.

Moore, D. G., and J. R. Curray, 1964A. Sedimentary framework of the drowned Pleistocene delta of Rio Grande de Santiago, Nayarit, Mexico. In *Developments in Sedimentology,* Vol. 1, *Deltaic and Shallow Marine Deposits,* L. M. J. U. van Straaten, ed., Elsevier, New York, pp. 275–281.

Moore, D. G., and J. R. Curray, 1964B. Wave-base, marine profile of equilibrium, and wave-built terraces: discussion. *Geol. Soc. Amer. Bull.,* v. 75, pp. 1267–1274.

Moore, D. G., and George Shumway, 1959. Sediment thickness and physical properties: Pigeon Point Shelf, California. *Jour. Geophys. Res.,* v. 64, no. 3, pp. 367–374.

Moore, D. G., J. R. Curray, and R. W. Raitt, 1971. Structure and history of the Bengal deep-sea fan. *Abs. 8th Intl. Sed. Congr., Intl. Assoc. Sedimentologists,* p. 69.

Moore, G. W., 1966. Arctic beach sedimentation. In *Environment of the Cape Thompson Region, Alaska,* N. J. Wilimovsky and J. N. Wolfe, eds., U.S. Atomic Energy Comm., pp. 587–608.

Moore, G. W., 1968. Arctic beaches. In *The Encyclopedia of Geomorphology,* R. W. Fairbridge, ed., Reinhold, New York, pp. 21–22.

Moore, G. W., 1970. Sea-floor spreading at the junction between Gorda Rise and the Mendocino Ridge. *Geol. Soc. Amer. Bull.,* v. 81, pp. 2817–2824.

Moore, G. W., and E. A. Silver, 1968. Geology of the Klamath River Delta, California. *U.S. Geol. Survey Prof. Paper* 600-C, pp. C144–C148.

Moore, J. R. III, 1963. Bottom sediment studies, Buzzards Bay, Massachusetts. *Jour. Sed. Petrology,* v. 33, no. 3, pp. 511–558.

Moore, T. C., Jr., and G. R. Heath, 1966. Manganese nodules, topography and thickness of Quaternary sediments in the central Pacific. *Nature,* v. 212, no. 5066. pp. 983–985.

Moore, T. C., Jr., Tj. H. van Andel, W. H. Blow, and G. R. Heath, 1970. Large submarine slide off northeastern continental margin of Brazil. *Amer. Assoc. Petrol. Geol. Bull.,* v. 54, no. 1, pp. 126–128.

Morelock, Jack, 1969. Shear strength and stability of continental slope deposits, western Gulf of Mexico. *Jour. Geophys. Res.,* v. 74, no. 2, pp. 465–482.

Morgan, J. P., 1963. Louisiana's changing shoreline. *Coastal Studies Inst.,* Louisiana State Univ., Contrib. 63-5, pp. 66–78.

Morgan, J. P., ed., 1970. *Deltaic Sedimentation, Modern and Ancient.* Soc. Econ. Paleontol. and Mineralog., spec. publ. no. 15, Tulsa, Okla., 312 pp.

Morgan, J. P., 1970. Depositional processes and products in the deltaic environment. In *Deltaic Sedimentation, Modern and Ancient,* J. P. Morgan, ed., Soc. Econ. Paleontol. and Mineralog., spec. publ. no. 15, Tulsa, Okla., pp. 31–47.

Morgan, J. P., and W. G. McIntire, 1959. Quaternary geology of the Bengal Basin, East Pakistan and India. *Geol. Soc. Amer. Bull.,* v. 70, pp. 319–342.

Morgan, J. P., L. G. Nichols, and Martin Wright, 1958. Morphological effects of Hurricane Audrey on the Louisiana coast. *Coastal Studies Inst.,* Louisiana State Univ., Tech. Rept. no. 10, 53 pp.

Morgan J. P., J. M. Coleman, and S. M. Gagliano, 1963. Mudlumps at the mouth of South Pass, Mississippi River; sedimentology, paleontology, structure, origin and relation to deltaic processes. *Coastal Studies Inst.,* Louisiana State Univ., ser. 10, 190 pp.

Morgan, W. J., 1968. Rises, trenches, great faults, and crustal blocks. *Jour. Geophys. Res.,* v. 73, pp. 1959–1982.

Morgan, W. J., 1971. Convection plumes in the lower mantle. *Nature,* v. 230, no. 5288, pp. 42–43.

Morgan, W. J., 1972. Deep mantle convection plumes and plate motions. *Amer. Assoc. Petrol. Geol. Bull.,* v. 56/2, pp. 203–213.

Mudie, J. D., W. R. Normark, and E. J. Cray, Jr., 1970. Direct mapping of the sea floor using side-scanning sonar and transponder navigation. *Geol. Soc. Amer. Bull.,* v. 81, pp. 1547–1554.

Müller, German, 1958. Die rezenten sedimente im Golfe von Neapel. I. Die sedimente des Golfes von Pozzuoli. *Geol. Rundschau,* v. 47, no. 1, pp. 117–150.

Müller, German, 1964. Die Korngrössenverteilung in den rezenten sedimenten des Golfes von Neapel. In *Developments in Sedimentology,* Vol. 1. *Deltaic and Shallow Marine Deposits,* L. M. J. U. van Straaten, ed., Elsevier, New York, pp. 282–292.

Müller, German, 1966. Grain size, carbonate content, and carbonate mineralogy of Recent sediments of the Indian Ocean off the eastern coast of Somalia. *Die Naturwissensch.,* v. 53, no. 21, pp. 547–550.

Munk, W. H., 1968. Once again—tidal friction. *Quart. Jour. Roy. Astr. Soc.,* v. 9, pp. 352–375.

Munk, W. H., and M. C. Sargent, 1948. Adjustment of Bikini Atoll to ocean waves. *Trans. Amer. Geophys. Union,* v. 29, no. 6, pp. 855–860.

Murray, G. E., 1961. *Geology of the Atlantic and Gulf Coastal Province of North America,* Harper & Row, New York, 692 pp.

Murray, John, 1880. On the structure of coral reefs and islands. *Proc. Roy. Soc. Edinburgh,* v. 10, pp. 505–518.

Murray, John, and G. V. Lee, 1909. The depth and marine deposits of the Pacific. *Mem. Museum. Comp. Zool.,* v. 38, no. 1, 169 pp.

Murray, John, and A. F. Renard, 1891. *Deep-Sea Deposits Based on the Specimens Collected During the Voyage of H.M.S. 'Challenger' in the Years 1872–1876 'Challenger' Reports.* Longmans, London, 525 pp. (reprint, Johnson, London, 1965).

Nason, R. D., 1968. San Andreas fault at Cape Mendocino. *Proc. Conf. Geol. Problems of San Andreas Fault System,* Stanford Univ. Publ., Geol. Sci., v. 11, pp. 231–241.

National Science Foundation, 1970. Deep sea drilling project, ocean sediment coring program. *Nat. Sci. Found.,* 70-25, Washington, D.C., 14 pp.

Nayudu, Y. R., 1959. Recent Sediments of the Northeast Pacific. Ph.D. Thesis, Univ. Washington, 53 pp.

Nayudu, Y. R., 1964 Volcanic ash deposits in the Gulf of Alaska and problems of correlation of deep-sea ash deposits. *Marine Geol.,* v. 1, pp. 194–212.

Nayudu, Y. R., 1965. Petrology of submarine volcanics and sediments in the vicinity of the Mendocino Fracture Zone. In *Progress in Oceanography—Vol. 3,* Pergamon, Oxford, pp. 207–220.

Nelson, B. W., 1962. Important aspects of estuarine sediment chemistry for benthic ecology. In *Symposium on the Environmental Chemistry of Marine Sediments,* N. Marshall, ed., Occ. Publ. No. 1, Narragansett Marine Lab., Univ. Rhode Island, Kingston, pp. 27–41.

Nelson, B. W., 1970. Hydrography, sediment dispersal, and recent historical development of the Po River Delta, Italy. In *Deltaic Sedimentation, Modern and Ancient,* J. P. Morgan, ed., Soc. Econ. Paleontol. and Mineralog., spec. publ. no. 15, Tulsa, Okla., pp. 152–184.

Nelson, C. H., 1968. Marine Geology of Astoria Deep-Sea Fan. Ph.D. thesis., Oregon State Univ., Corvallis, 287 pp.

Nelson, C. H., L. D. Kulm, P. R. Carlson, and J. R. Duncan, 1968. Mazama ash in the northeastern Pacific. *Science,* v. 161, pp. 47–49.

Nelson, C. H., P. R. Carlson, J. V. Byrne, and T. R. Alpha, 1970. Development of the Astoria Canyon-Fan physiography and comparison with similar systems. *Marine Geol.,* v. 8, nos. 3, 4, pp. 259–291.

Nesteroff, W. D., 1959. Attempt at a synthesis of present-day marine sedimentation along the French Mediterranean coast (eastern part). Abs. in *Preprints International Oceanographic Congress,* Amer. Assoc. Adv. Sci., Washington, D.C., p. 642.

Nesteroff, W. D., 1966. Les dépôts marins actuels de la feuille de Fréjus-Cannes. *Bull. Carte Géol. France,* v. 61, no. 278, pp. 203–211.

Neumann, A. C., 1958. The configuration and sediments of Stetson Bank, northwestern Gulf of Mexico. *Texas A & M College, Dept. Oceanog. and Meteorol.,* Tech. Rept. no. 58-5T, 125 pp.

Neumann, A. C., and M. M. Ball, 1970. Submersible observations in the Straits of Florida; geology and bottom currents. *Geol. Soc. Amer. Bull.,* v. 81, pp. 2861–2874.

Newell, N. D., 1955. Bahamian platforms. *Geol. Soc. Amer. Spec. Paper 62,* pp. 303–315.

Newell, N. D., 1956. Geological reconnaissance of Raroia (Kon Tiki) Atoll, Tuamotu Archipelago. *Amer. Mus. Nat. Hist. Bull.,* v. 109, art. 3, pp. 311–372.

Newell, N. D., 1959. Question of the coral reefs. Pt. I. Biology of the corals. Pt. II. *Nat. Hist.,* v. 68, no. 3, pp. 119–131; no. 4, pp. 226–235.

Newell, N. D., 1971. An outline history of tropical organic reefs. *Amer. Mus. Novitates,* no. 2465, 37 pp.

Newell, N. D., and A. L. Bloom, 1970. The reef flat and "two-meter eustatic terrace" of some Pacific atolls. *Geol. Soc. Amer. Bull.,* v. 81, no. 7, pp. 1881–1894.

Newell, N. D., J. K. Rigby, A. J. Whiteman, and J. S. Bradley, 1951. Shallow-water geology and environments, eastern Andros Island, Bahamas. *Amer. Mus. Nat. Hist. Bull.,* v. 98, art. 1, pp. 1–30.

Newell, N. D., John Imbrie, E. G. Purdy, and D. L. Thurber, 1959. Organism communities and bottom facies, Great Bahama Bank. *Amer. Mus. Nat. Hist. Bull.,* v. 117, art. 4, pp. 177–228.

Newell, N. D., E. G. Purdy, and John Imbrie, 1960. Bahamian oölitic sand. *Jour. Geol.,* v. 68, no. 5, pp. 481–497.

Newman, W. S., and C. A. Munsart, 1968. Holocene geology of the Wachapreague Lagoon, eastern shore peninsula, Virginia. *Marine Geol.,* v. 6, pp. 81–105.

Nichols, Haven, and R. B. Perry, 1966. Bathymetry of the Aleutian Arc, Alaska (six maps), scale 1 :400,000. Dept. Comm., *Environ. Sci. Serv. Adm., Coast Geodetic Survey Monograph,* 3.

Nicholas, Haven, R. B. Perry, and J. W. Kofoed, 1964. Bathymetry of Bowers Bank, Bering Sea. *Surveying and Mapping,* v. 24, no. 3, pp. 443–448.

Nichols, R. L., 1961. Characteristics of beaches formed in polar climates. *Amer. Jour. Sci.,* v. 259, pp. 694–708.

Niino, Hiroshi, and K. O. Emery, 1961. Sediments of shallow portions of the East China Sea and South China Sea. *Geol. Soc. Amer. Bull.,* v. 72, pp. 731–762.

Ninkovich, D., and B. C. Heezen, 1967. Physical and chemical properties of volcanic glass shards from pozzuolana ash, Thera Island, and from upper and lower ash layers in eastern Mediterranean deep sea sediments. *Nature,* v. 213, no. 5076, pp. 582–584.

Noornay, Iraj, and S. F. Gizienski, 1969. Engineering properties of submarine soils, a state-of-the-art-review. *Woodward-Clyde & Assoc.,* San Diego, 67 pp.

Normark, W. R., 1969. Growth Patterns of Deep-Sea Fans. Ph.D. thesis, Univ. California, Scripps Inst. of Oceanog., 165 pp.

Normark, W. R., 1970A. Channel piracy on Monterey Deep-Sea Fan. *Deep-Sea. Res.,* v. 17, pp. 837–846.

Normark, W. R., 1970B. Growth patterns of deep-sea fans. *Amer. Assoc. Petrol. Geol. Bull.,* v. 54, no. 11, pp. 2170-2195.

Normark, W. R., and J. R. Curray, 1968. Geology and structure of the tip of Baja California, Mexico. *Geol. Soc. Amer. Bull.,* v. 79, pp 1589–1600.

Normark, W. R., and D. J. W. Piper, 1969. Deep-sea fan-valleys, past and present. *Geol. Soc. Amer. Bull.,* v. 80, pp. 1859–1866.

Nota, D. J. G., 1958. *Reports of the Orinoco Expedition,* Vol. II. H. Veenman en Zonen, Wageningen, The Netherlands, 98 pp.

Nota, D. J. G., 1968. Geomorphology and sedimentary petrology in the southern Gulf of St. Lawrence. *Geol. en Mijnb.,* v. 47, no. 1, pp. 49–52.

Nota, D. J. G., and D. H. Loring, 1964. Recent depositional conditions in the St. Lawrence River and gulf—a reconnaissance survey. *Marine Geol.,* v. 2, pp. 198–235.

Oba, Tadamichi, 1969. Biostratigraphy and isotopic paleotemperature of some deep-sea cores from the Indian Ocean. *Sci. Repts. Tohoku Univ.,* Sendai, 2nd ser., v. 41, no. 2, pp. 129–195.

Odum, H. T., and E. P. Odum, 1955. Trophic structure and productivity of a windward coral reef community on Eniwetok Atoll. *Ecology Mon.,* v. 25, pp. 291–320.

Off, Theodore, 1963. Rhythmic linear sand bodies caused by tidal currents. *Amer. Assoc. Petrol. Geol. Bull.*, v. 47, no. 2, pp. 324–341.

Olausson, Erik, 1960, 1961. In *Report of the Swedish Deep-Sea Expedition, 1947–48*, 6 (5), 9 (2), 8 (3), 7 (5), 8 (4).

Olausson, Erik, and U. C. Jonasson, 1969. The Arctic Ocean during the Würm and Early Flandrian. *Geol. Föreningens i Stockholm Förhandlingar*, v. 91, pp. 185–200.

Opdyke, N. D., and B. P. Glass, 1969. The paleomagnetism of sediment cores from the Indian Ocean. *Deep-Sea Res.*, v. 16, pp. 249–261.

Opdyke, N. D., B. P. Glass, J. D. Hays, and J. Foster, 1966. Paleomagnetic study of Antarctic deep-sea cores. *Science*, v. 154, no. 3748, pp. 349–357.

Ottmann, Francois, 1962. Sur la classification des côtes. *Bull. Soc. Géol. France*, 7 ser, v. 4, pp. 620–624.

Ottmann, Francois, 1965. *Introduction a la Géologie Marine et Littorale*. Masson, Paris, 259 pp.

Ottmann, Francois, 1968. L'etude des problemes estuariens. *Rev. Geogr. Physique et Geol. Dynamique* (2), v. 10, no. 4, pp. 329–353.

Ottmann, Francois, and C. M. Urien, 1965. Observaciones preliminares sobre la distribución de los sedimentos en la zona externa del Rio de la Plata. *Acad. Brasileira de Ciências*, Rio de Janeiro, pp. 283–288.

Otvos, E. G., Jr., 1970. Development and migration of barrier islands, northern Gulf of Mexico. *Geol. Soc. Amer. Bull.*, v. 81, pp. 241–246.

Paige, H. G., 1955. Phi-millimeter conversion table. *Jour. Sed. Petrology*, v. 25, pp. 285–292.

Pannekoek, A. J., 1966. The ria problem. *Tijdsch. Koninkl. Ned. Aardrijk. Genoots.*, v. 83, no. 3, pp. 289–297.

Pannekoek, A. J., 1969. Uplift and subsidence in and around the western Mediterranean since the Oligocene: a review. *Verhandelingen Kon. Ned. Geol.* v. 26, pp. 53–77.

Parke, M. L., Jr., K. O. Emery, Raymond Szymankeiwicz, and L. M. Reynolds, 1971. Structural framework of continental margin in South China Seas. *Amer. Assoc. Petrol. Geol. Bull.*, v. 55, no. 5, pp. 723–751.

Parker, F. L., and W. H. Berger, 1971. Faunal and solution patterns of planktonic Foraminifera in surface sediments of the South Pacific. *Deep-Sea Res.*, v. 18, no. 1, pp. 73–107.

Parker, F. L., F. B Phleger, and J. F. Peirson, 1953. Ecology of Foraminifera from San Antonio Bay and environs, southwest Texas. *Cushman Found. Foram. Res. Spec. Publ.* 2, Washington, D.C., 63 pp.

Parker, R. H., and J. R. Curray, 1956. Fauna and bathymetry of banks on continental shelf, northwest Gulf of Mexico. *Amer. Assoc. Petrol. Geol. Bull.*, v. 40, no. 10, pp. 2428–2439.

Penck, Albrecht, 1894. *Morphologie der Erdoberfläche*. 2 vols., Stuttgart.

Petelin, V. P., 1960. Bottom sediments in the western part of the Pacific Ocean *Oceanol. Res.*, X sect, I.G.Y. Program, p. 45.

Peterson, M. N. A., 1971. Scientists view deep sea drilling project. *Petroleum Engr. Mag.*, May, p. 14.

Peterson, M. N. A., and E. D. Goldberg, 1962. Feldspar distributions in South Pacific pelagic sediments. *Jour. Geophys. Res.*, v. 67, no. 9, pp. 3477–3492.

Peterson, M. N. A., N. T. Edgar, et al., 1969. Deep sea drilling project Leg 2. *Geotimes*, v. 14, no. 3, pp. 11–13.

Peterson, M. N. A., N. T. Edgar, C. C. von der Borch, and R. W. Rex, 1970. Cruise leg summary and discussion. In *Initial Reports of the Deep Sea Drilling Project*. Vol. 2, Nat. Sci. Found., Washington, D.C., pp. 413–427.

Pettersson, Hans, 1945. Iron and manganese on the ocean floor. *Göteborg Oceanog. Inst. Medd.*, b, 3B, no. 7, pp. 1–37.

Pettersson, Hans, and K. Fredriksson, 1958. Magnetic spherules in deep-sea deposits. *Pacific Sci.*, v. 12, pp. 71–81.

Pfannenstiel, Max, 1960. Erlästerungen zu den bathymetrischen Karten des östlichen Mittelmeeres. *Bull. Inst. Oceanog. Monaco*, No. 1192.

Philippi, E., 1912. Die Grundproben der deutschen Südpolar Expedition. *Deutsche Südpolar Expedition 1901–1903*, v. 2, no. 6, pp. 431–434.

Phillips, J. D., and D. A. Ross, 1970. Continuous seismic reflexion profiles in the Red Sea. *Phil. Trans. Roy. Soc. London*, A, v. 267, pp. 143–152.

Phillips, O. M., 1966. *The Dynamics of the Upper Ocean.* Cambridge Univ. Press, 261 pp.

Phipps, C. V. G., 1963. Topography and sedimentation of the continental shelf and slope between Sydney and Montague Island—N.S.W. *Australasian Oil and Gas Jour.*, Dec., 7 pp.

Phipps, C. V. G., 1966. Gulf of Carpentaria (northern Australia). In *The Encyclopedia of Oceanography*, R. W. Fairbridge, ed., Reinhold, New York, pp. 316–324.

Phipps, C. V. G., 1967. The character and evolution of the Australian continental shelf. *A. P. E. A. Jour.*, pp. 44–49.

Phleger, F. B, 1939. Foraminifera of submarine cores from the continental slope. *Geol. Soc. Amer. Bull.*, v. 50, pp. 1395–1422.

Phleger, F. B, 1960. Sedimentary patterns of microfaunas in northern Gulf of Mexico. In *Recent Sediments, Northwest Gulf of Mexico*, F. P. Shepard, F. B Phleger, and Tj. H. van Andel, eds., Amer. Assoc. Petrol. Geol., Tulsa, Okla., pp. 267–295.

Phleger, F. B, 1965. Sedimentology of Guerrero Negro Lagoon, Baja California, Mexico. In *Submarine Geology and Geophysics*, W. F. Whittard, and R. Bradshaw, eds., Butterworths, London, pp. 205–235.

Phleger, F. B, 1969A. A modern evaporite deposit in Mexico. *Amer. Assoc. Petrol. Geol. Bull.*, v. 53, no. 4, pp. 824–829.

Phleger, F. B, 1969B. Some general features of coastal lagoons. In *Coastal Lagoons*, A. A. Castañares and F. B Phleger, eds., Univ. Nacional Autonoma Mexico, pp. 5–25.

Phleger, F. B, and G. C. Ewing, 1962. Sedimentology and oceanography of coastal lagoons in Baja California, Mexico. *Geol. Soc. Amer. Bull.*, v. 73, no. 2, pp. 145–182.

Phleger, F. B, F. L. Parker, and J. F. Peirson, 1953. North Atlantic core Foraminifera. *Repts. Swedish Deep-Sea Exped.*, v. 7, no. 1, pp. 1–122.

Piccard, Jacques, and R. S. Dietz, 1961. *Seven Miles Down.* Putnam, New York, 249 pp.

Pickett, T. E., and R. L. Ingram, 1969. The modern sediments of Pamlico Sound, North Carolina coast. *Southeastern Geol.*, v. 11, pp. 53–83.

Pierce, J. W., 1968. Sediment budget along a barrier island chain. *Sedimentary Geol.*, v. 3, pp. 5–16.

Pierce, J. W., and D. J. Colquhoun, 1970. Holocene evolution of a portion of the North Carolina coast. *Geol. Soc. Amer. Bull.*, v. 81, no. 12, pp. 3697–3714.

Pigorini, Bruno, 1968. Sources and dispersion of recent sediments of the Adriatic Sea. *Marine Geol.*, v. 6, pp. 187–229.

Pimm, A. C., 1964. Seabed sediments of the South China Sea. *Borneo Region, Malaysia Geol. Survey Ann. Rept.*, pp. 122–146.

Piper, D. J. W., 1970. Transport and deposition of Holocene sediment on La Jolla deep-sea fan, California. *Marine Geol.*, v. 8, nos. 3/4, pp. 211–227.

Piper D. J. W., and N. F. Marshall, 1969. Bioturbation of Holocene sediments on La Jolla deep-sea fan, California. *Jour. Sed. Petrology*, v. 39, no. 2, pp. 601–606.

Pirsson, L. V., 1914. Geology of Bermuda Island; the igneous platform. *Amer. Jour. Sci.*, v. 38, pp. 189–206, 331–334.

Pitman, W. C., 1971. Sea-floor spreading and plate tectonics. *Trans. Amer. Geophys. Union*, v. 52, no. 5, pp. IUGG 130–134.

Pitman, W. C. III, and J. R. Heirtzler, 1966. Magnetic anomalies over the Pacific-Antarctic ridge. *Science*, v. 154, pp. 1164–1171.

Pitman, W. C. III, and M. Talwani, 1972. Sea-floor spreading in the North Atlantic. *Geol. Soc. Amer. Bull.*, v. 83, pp. 619–646.

Plafker, George, 1969. Tectonics of the March 27, 1964 Alaska earthquake. *U.S. Geol. Survey Prof. Paper 543-I*, 174 pp.

Plafker, George, Reuben Kachadoorian, W. B. Eckel, and L. R. Mayo, 1969. Effects of the earthquake of March 27, 1964, on various communities. *U.S. Geol. Survey Prof. Paper 542-G*, 50 pp.

Postma, H., 1957. Size frequency distribution of sands in the Dutch Wadden Sea. In *Archives Neerlandaises Zool.*, v. 12, Brill, Leiden, pp. 319–349.

Postma, H., 1969. Chemistry of coastal lagoons. In *Coastal Lagoons*, A. A. Castañares and F. B Phleger, eds., Univ. Nacional Autonoma Mexico, pp. 421–429.

Potter, P. E., and F. J. Pettijohn, 1963. *Paleocurrents and Basin Analysis*. Academic, New York, 296 pp.

Powers, M. C., 1953. A new roundness scale for sedimentary particles. *Jour. Sed. Petrology*, v. 23, pp. 117–119.

Pratje, Otto, 1948. Die Bodenbedeckung der südlichen und mittleren Ostsee und ihre Bedeutung für die Ausdeutung fossiler Sedimente. *Deutschen Hydrograph. Zeitschr.*, v. 1, nos. 2, 3, pp. 45–61.

Pratje, Otto, 1949. Bodenbedeckung der nordeuropäischen Meer. *Handbuch der Seefischerei Nordsuropas*, v. 1, pt. 3, 23 pp.

Pratt, R. M., and B. C. Heezen, 1964. Topography of the Blake Plateau. *Deep-Sea Res.*, v. 11, pp. 721–728.

Price, W. A., 1951. Barrier island, not "offshore bar." *Science*, v. 113, pp. 487–488.

Price, W. A., 1955. Correlation of shoreline type with offshore bottom conditions. *Texas A & M College, and Geogr. Branch Office Naval Res.*, 8 pp. (mimeo.).

Price, W. A., 1956. Environment and history in identification of shoreline types. *Quaternaria*, 3, pp. 151–166.

Price, W. A., and B. W. Wilson, 1956. Cuspate spits of St. Lawrence Island: a discussion. *Jour. Geol.*, v. 64, no. 1, pp. 94–95.

Psuty, N. P., 1965. Beach-ridge development in Tabasco, Mexico. *Ann. Assoc. Amer. Geographers*, v. 55, no. 1, pp. 112–124.

Psuty, N. P., 1966. The geomorphology of beach ridges in Tabasco, Mexico. *Coastal Studies Inst.*, Louisiana State Univ., Tech. Rept. 30, 51 pp.

Purdy, E. G., 1963. Recent calcium carbonate facies of the Great Bahama Bank—2. Sedimentary facies. *Jour. Geol.*, v. 71, no. 4, pp. 472–497.

Putnam, W. C., D. I. Axelrod, H. P. Bailey, and J. T. McGill, 1960. *Natural Coastal Environments of the World*. Univ. California, Los Angeles, 140 pp.

Pytkowicz, R. M., 1965. Calcium carbonate saturation in the oceans. *Limnol. Oceanog.*, v. 10, no. 2, pp. 220–225.

Raitt, R. W., 1952. The 1950 seismic refraction studies of Bikini and Kwajalein atolls and Sylvana Guyot. *Univ. Calif., Marine Phys. Lab., Scripps Inst. Oceanog.*, SIO Ref. 52-38, pp. 1–25.

Raitt, R. W., 1957. Seismic-refraction studies of Eniwetok Atoll, Bikini and nearby atolls, Marshall Islands. *U.S. Geol. Survey Prof. Paper 260-S*, pp. 685–698.

Raitt, R. W., R. L. Fisher, and R. G. Mason, 1955. Tonga Trench. In *The Crust of the Earth*, A. Poldervaart, ed., Geol. Soc. Amer. Spec. Paper 62, pp. 237–254.

Rajcevic, B. M., 1957. Etudes des conditions de sedimentation dans l'estuaire de la Seine. *Suppl. Ann. Inst. Tech. Batiment et Trav. Publ.*, no. 117.

Rao, M. P., and C. Mahadevan, 1959. Studies in marine geology of Bay of Bengal along the east coast of India. Abs. in *Preprints International Oceanographic Congress*, Amer. Assoc. Adv. Sci., Washington, D.C., pp. 655–656.

Raymond, P. E., and H. C. Stetson, 1932. A calcareous beach on the coast of Maine. *Jour. Sed. Petrology*, v. 2, no. 2, pp. 51–62.

Redfield, A. C., 1958. Preludes to the entrapment of organic matter in the sediments of Lake Maracaibo. In *Habitat of Oil*, L. G. Weeks, ed., Amer. Assoc. Petrol. Geol., Tulsa, Okla., pp. 968–981.

Reimnitz, Erk, 1971. Surf-beat origin for pulsating bottom currents in the Rio Balsas Submarine Canyon, Mexico. *Geol. Soc. Amer. Bull.*, v. 82, pp. 81–90.

Reimnitz, Erk, and Mario Gutierrez-Estrada, 1970. Rapid changes in the head of the Rio Balsas Submarine Canyon system, Mexico. *Marine Geol.*, v. 8, nos. 3/4, pp. 245–258.

Reineck, H.-E., 1958. Kastengreifer und Lotröhre "Schnepfe." *Senckenbergiana leth.*, v. 39, nos. 1/2, pp. 45–48.

Rex, R. W., 1955. Microrelief produced by sea ice grounding in the Chukchi Sea near Barrow, Alaska. *Jour. Arctic Inst. North Amer.*, v. 8, no. 3, pp. 177–186.

Rex, R. W., and E. D. Goldberg, 1958. Quartz contents of pelagic sediments of the Pacific Ocean. *Tellus X*, no. 1, pp. 153–159.

Reynolds, Osborne, 1885. On the dilatancy of media composed of rigid particles in contact. *Philos. Mag.*, ser. 5, v. 20, pp. 469–481.

Richards, H. C., 1940. Results of deep boring operations on the Great Barrier Reef, Australia. *Proc. 6th Pacific Sci. Congr.*, v. 2, p. 857.

Riedel, W. R., 1952. Tertiary Radiolaria in western Pacific sediments. *Medd. Fran Oceanog. Inst. I Göteborg*, Sjatte Foljden, ser. B, v. 6, no. 3, pp. 1–21.

Riedel, W. R., 1959. Siliceous organic remains in pelagic sediments. In *Silica in Sediments*, Soc. Econ. Paleontol. and Mineralog., spec. publ. no. 7, Tulsa, Okla., pp. 94–97.

Riedel, W. R., 1971. The occurrence of the Pre-Quaternary Radiolaria in deep-sea sediments. In *Micropaleontology of Oceans*, B. M. Funnell and W. R. Riedel, eds., Cambridge Univ. Press, pp. 567–594.

Riedel, W. R., and B. M. Funnell, 1964. Tertiary sediment cores and microfossils from the Pacific Ocean floor. *Quart. Jour. Geol. Soc. London*, v. 120, pp. 305–368.

Rigby, J. K., and W. G. McIntire, 1967. The Isla de Lobos and associated reefs, Veracruz, Mexico. *Brigham Young Univ. Geol. Studies*, v. 13, pp. 3–46.

Riley, N. A., 1941. Projection sphericity. *Jour. Sed. Petrology*, v. 11, pp. 94–97.

Rivière, André, and Jean Laurent, 1954. Sur une méthode nouvelle et peu coûteuse de défense contre l'érosion littorale. *Acad. Sci. (Paris) Comptes rendus*, v. 239, pp. 298–300.

Roberson, M. I., 1964. Continuous seismic profiler survey of Oceanographer, Gilbert, and Lydonia Submarine Canyons, Georges Bank. *Jour. Geophys. Res.*, v. 69, no. 22, pp. 4779–4789.

Roberts, D. G., 1970. The Rif-Betic orogen in the Gulf of Cadiz. *Marine Geol.*, v. 9, pp. M31–M37.

Roberts, D. G., and A. H. Stride, 1968. Late Tertiary slumping on the continental slope of southern Portugal. *Nature*, v. 217, no. 5123, pp. 48–50.

Robinson, A. H. W., 1966. Residual currents in relation to shoreline evolution of the East Anglian coast. *Marine Geol.*, v. 4, pp. 57–84.

Robinson, A. H. W., et al., 1953. The storm floods of 1st February, 1953. *Geography*, pp. 132–189.

Rod, Emile, 1956. Strike-slip faults of northern Venezuela. *Amer. Assoc. Petrol. Geol. Bull.*, v. 40, no. 3, pp. 457–476.

Rodgers, John, 1968. The eastern edge of the North American continent during the Cambrian and Early Ordovician. In *Studies of Appalachian Geology: Northern and Maritime*, E-An Zen, W. S. White et al., eds., Interscience, New York, pp. 141–149.

Rodolfo, K. S., 1969. Bathymetry and marine geology of the Adaman Basin, and tectonic implications for Southeast Asia. *Geol. Soc. Amer. Bull.*, v. 80, no. 7, pp. 1203–1230.

Rona, E., and C. Emiliani, 1968. Absolute dating of Caribbean cores P6304-8 and P6304-9. *Science*, v. 163, pp. 66–68.

Rona, P. A., 1969. Linear "lower continental rise hills" off Cape Hatteras. *Jour. Sed. Petrology*, v. 39, no. 3, pp. 1132–1141.

Rona, P. A., 1970A. Comparison of continental margins of eastern North America at Cape Hatteras and northwestern Africa at Cap Blanc. *Amer. Assoc. Petrol. Geol. Bull.*, v. 54, no. 1, pp. 129–157.

Rona, P. A., 1970B. Submarine canyon origin on upper continental slope off Cape Hatteras. *Jour. Geol.*, v. 78, pp. 141–152.

Rona, P. A., 1971A. Bathymetry off central northwest Africa. *Deep-Sea Res.*, v. 18, pp. 321–327.

Rona, P. A., 1971B. Deep-sea salt diapirs. *Geotimes*, v. 16, no. 5, p. 8.

Rona, P. A., E. D. Schneider, and B. C. Heezen, 1967. Bathymetry of the continental rise off Cape Hatteras. *Deep-Sea Res.*, v. 14, pp. 625–633.

Rosalsky, M. B., 1964. Swash and backwash action at the beach scarp. *Rocks and Minerals*, Jan.–Feb., pp. 5–14.

Rosenan, E., 1937. Fisherman's chart—1:100,000 scale in four sheets. *Govt. Palestine, Dept. Agriculture and Fisheries Serv.*

Rosfelder, A. M., 1955. Carte Provisoire au 1/500,000e de la Marge Continentale Algérienne. *La Carte Géologique de l'Algérie (Nfi)*, Bull. no. 5, pp. 57–106.

Rosfelder, A. M., 1966A. Hydrostatic actuation of deep-sea instruments. *Jour. Ocean Technol.*, v. 1, no. 1, pp. 53–64.

Rosfelder, A. M., 1966B. Subsea coring for geological and geotechnical surveys. *Proc. Offshore Exploration Conf.,* 1966, Long Beach, Calif., pp. 709–734.

Rosfelder, A. M., and N. F. Marshall, 1966. Oriented marine cores: a description of new locking compasses and triggering mechanisms. *Jour. Marine Res.,* v. 24, no. 3, pp. 353–364.

Rosfelder, A. M., and N. F. Marshall, 1967. Obtaining large, undisturbed, and oriented samples in deep water. In *Marine Geotechnique,* A. F. Richards, ed., Univ. Illinois Press, Urbana, Ill., pp. 243–263.

Ross, D. A., 1968. Current action in a submarine canyon. *Nature,* v. 218, pp. 1242–1245.

Ross, D. A., 1970A. Source and dispersion of surface sediments in the Gulf of Maine —Georges Bank area. *Jour. Sed. Petrology,* v. 40, no. 3, pp. 906–920.

Ross, D. A., 1970B. Atlantic continental shelf and slope of the United States—heavy minerals of the continental margin from southern Nova Scotia to northern New Jersey. *U.S. Geol. Survey Prof. Paper* 529-G, 40 pp.

Ross, D. A., 1971. The Red and the Black seas. *Amer. Scientist,* v. 59, no. 4, pp. 420–424.

Ross, D. A., and W. R. Riedel, 1967. Comparison of upper parts of some piston cores with simultaneously collected open-barrel cores. *Deep-Sea Res.,* v. 14, pp. 285–294.

Ross, D. A., and G. G. Shor, Jr., 1965. Reflection profiles across the Middle America Trench. *Jour. Geophys. Res.,* v. 70, no. 22, pp. 5551–5572.

Runcorn, S. K., ed., 1962. *Continental Drift.* Academic, New York, 338 pp.

Rusnak, G. A., 1960. Sediments of Laguna Madre, Texas. In *Recent Sediments, Northwest Gulf of Mexico,* F. P. Shepard, F. B Phleger, and Tj. H. van Andel, eds., Amer. Assoc. Petrol. Geol., Tulsa, Okla., pp. 153–196.

Rusnak, G. A., 1967. High-efficiency subbottom profiling. *U.S. Geol. Survey Prof. Paper* 575-C, pp. C81–C91.

Rusnak, G. A., and R. L. Fisher, 1964. Structural history and evolution of Gulf of California. In *Marine Geology of the Gulf of California,* Tj. H. van Andel and G. G. Shor, Jr., eds., Amer. Assoc. Petrol. Geol., Tulsa, Okla., pp. 144–156.

Rusnak, G. A., R. L. Fisher, and F. P. Shepard, 1964. Bathymetry and faults of Gulf of California. In *Marine Geology of the Gulf of California,* Tj. H. van Andel and G. G. Shor, Jr., eds., Amer. Assoc. Petrol. Geol., Tulsa, Okla., pp. 59–75.

Rusnak, G. A., K. W. Stockman, and H. A. Hofmann, 1966. The role of shell material in the natural sand replenishment cycle of the beach and nearshore area between Lake Worth inlet and the Miami ship channel. Final Rept. *Coastal Engr. Res. Center, U.S. Army Corps Engrs.* (DA-49-055-Civ-Eng-63-12), 16 pp.

Russell, R. D., 1939. Effects of transportation on sedimentary particles. In *Recent Marine Sediments,* P. D. Trask, ed., Amer. Assoc. Petrol. Geol., Tulsa, Okla., pp. 32–47.

Russell, R. J., 1936. Physiography of Lower Mississippi River Delta. *Louisiana Geol. Survey, Geol. Bull.* no. 8, 199 pp.

Russell, R. J., 1940. Quaternary history of Louisiana. *Geol. Soc. Amer. Bull.,* v. 51, pp. 1199–1234.

Russell, R. J., 1958. Geological geomorphology. *Geol. Soc. Amer. Bull.,* v. 69. no. 1, pp. 1–22.

Russell, R. J., 1967A. Aspects of coastal morphology. *Geografiska Annal.,* v. 49, ser. A, pp. 299–309.

Russell, R. J., 1967B. River and delta morphology. *Coastal Studies Inst.,* Louisiana State Univ., Rept. 52, 49 pp.

Russell, R. J., and H. V. W. Howe, 1935. Cheniers of southwestern Louisiana. *Geogr. Rev.,* v. 25, pp. 449–461.

Ryan, W. B. F., K. J. Hsu, et al., 1970A. Deep sea drilling project Leg 13. *Geotimes,* v. 15, no. 10, pp. 12–15.

Ryan, W. B. F., D. J. Stanley, et al., 1970B. The tectonics and geology of the Mediterranean Sea. In *The Sea,* Vol. 4, Pt. 2, A. E. Maxwell, ed., Wiley–Interscience, New York, pp. 387–492.

Sachs, P. L., and S. O. Raymond, 1965. Design considerations, development, and trials at sea of a free sampler. *Woods Hole Oceanog. Inst. Ref.* no. 65–3, 16 pp.

Sackett, W. M., 1965. Deposition rates by the protectinium method. In *Symposium on Marine Geochemistry, 1964*. Rhode Island Univ., Narragansett Marine Lab. Occasional Publ. 3, pp. 29–40.

Saito, Tsunemasa, and B. M. Funnell, 1970. Pre-Quaternary sediments and microfossils in the oceans. In *The Sea*, Vol. 4, Pt. 1, A. E. Maxwell, ed., Wiley–Interscience, New York, pp. 183–204.

Sanders, J. E., 1963. North-south trending submarine ridge composed of coarse sand off False Cape, Virginia. *Abs., Amer. Assoc. Petrol. Geol. Bull.*, v. 46, p. 278.

Sasa, Yasuo, and Akira Izaki, 1962. Submarine geology of the Tsugaru Straits. *Proc. Japan Acad.*, v. 38, no. 3, pp. 120–123.

Saski, Tadayoshi, Seichi Watanabe, and Gohachiro Oshiba, 1965. New current meters for great depths. *Deep-Sea Res.*, v. 12, pp. 815–824.

Sayles, R. W., 1931. Bermuda during the ice age. *Proc. Amer. Acad. Arts and Sci.*, v. 66, no. 11, pp. 381–467.

Schatz, C. E., 1971. Observations of sampling and occurrence of manganese nodules. *Offshore Technol. Conf.*, Paper OTC 1364, Dallas, Tex., preprint. pp. I 389–393.

Schnable, J. E., and H. G. Goodell, 1968. Pleistocene-Recent stratigraphy, evolution, and development of the Apalachicola coast, Florida. *Geol. Soc. Amer. Spec. Paper* 12, 72 pp.

Scholl, D. W., 1966. Florida Bay: a modern site of limestone formation. In *The Encyclopedia of Oceanography*, R. W. Fairbridge, ed., Reinhold, New York, pp. 282–288.

Scholl, D. W., and D. M. Hopkins, 1969. Newly discovered Cenozoic basins, Bering Sea shelf, Alaska. *Amer. Assoc. Petrol. Geol. Bull.*, v. 53, no. 10, pp. 2067–2078.

Scholl, D. W., R. von Huene, and J. B. Ridlon, 1968A. Spreading of the ocean floor: undeformed sediments in the Peru-Chile Trench. *Science*, v. 159, pp. 869–871.

Scholl, D. W., E. C. Buffington, and D. M. Hopkins, 1968B. Geologic history of the continental margin of North America in the Bering Sea. *Marine Geol.*, v. 6, pp. 297–330.

Scholl, D. W., E. C. Buffington, D. M. Hopkins, and T. R. Alpha, 1970A. The structure and origin of the large submarine canyons of the Bering Sea. *Marine Geol.*, v. 8, nos. 3/4, pp. 187–210.

Scholl, D. W., M. N. Christensen, R. von Huene, and M. S. Marlow, 1970B. Peru-Chile Trench sediments and sea-floor spreading. *Geol. Soc. Amer. Bull.*, v. 81, pp. 1339–1360.

Schott, Wolfgang, 1935. Die Foraminiferen im aequatorialen Teil des Atlantischen Ozeans. In *Wissenschaftliche Ergebnisse der Deutschen Atlantischen Expedition auf dem Forschungs-und Vermessungsschiff "Meteor," 1925–1927*, Vol. 3, Dritter Teil, Berlin, pp. 43–134.

Schott, Wolfgang, and Ulrich von Stackelberg, 1965. Über rezente sedimentation im Indischen Ozean, ihre Bedeutung für die Entstehung kohlenwasserstoffhaltiger sedimente. *Erdöl und Kohle, Erdgas, Petrochemie*, v. 18, pp. 945–950.

Schwartz, M. L., 1967. Littoral zone tidal-cycle sedimentation. *Jour. Sed. Petrology*, v. 37, pp. 677 683.

Sclater, J. G., J. W. Hawkins, and J. Mammerickx, 1972. Crustal extension between the Tonga and Lau ridges: petrologic and geophysical evidence. *Geol. Soc. Amer. Bull.*, v. 83, pp. 505–518.

Scruton, P. C., 1960. Delta building and the deltaic sequence. In *Recent Sediments, Northwest Gulf of Mexico*, F. P. Shepard, F. B Phleger, Tj. H. van Andel, eds., Amer. Assoc. Petrol. Geol., Tulsa, Okla., pp. 82–102.

Sen Gupta, B. K., and R. M. McMullen, 1969. Foraminiferal distribution and sedimentary facies on the Grand Banks of Newfoundland. *Canadian Jour. Earth Sci.*, v. 6, no. 3, pp. 475–487.

Shaler, N. S., 1889. Geology of the island of Nantucket. *U.S. Geol. Survey Bull.*, 53, 13 pp.

Shepard, F. P., 1930. Fundian faults or Fundian glaciers. *Geol. Soc. Amer. Bull.*, v. 41, pp. 659–674.

Shepard, F. P., 1931. Glacial troughs of the continental shelves. *Jour. Geol.*, v. 39, no. 4, pp. 345–360.

Shepard, F. P., 1932. Sediments of the continental shelves. *Geol. Soc. Amer. Bull.*, v. 43, pp. 1017–1040.

Shepard, F. P., 1937. "Salt" domes related to Mississippi Submarine Trough. *Geol. Soc. Amer. Bull.*, v. 48, pp. 1354–1361.

Shepard, F. P., 1948. *Submarine Geology*, Harper & Row, New York, 348 pp.

Shepard, F. P., 1949. Terrestrial topography of submarine canyons revealed by diving. *Geol. Soc. Amer. Bull.*, v. 60, pp. 1597–1612.

Shepard, F. P., 1951. Mass movements in submarine canyon heads. *Trans. Amer. Geophys. Union*, v. 32, no. 3, pp. 405–418.

Shepard, F. P., 1952. Revised nomenclature for depositional coastal features. *Amer. Assoc. Petrol. Geol. Bull.*, v. 36, no. 10, pp. 1902–1912.

Shepard, F. P., 1953. Sedimentation rates in Texas estuaries and lagoons. *Amer. Assoc. Petrol. Geol. Bull.*, v. 37, no. 8, pp. 1919–1934.

Shepard, F. P., 1955. Delta-front valleys bordering the Mississippi distributaries. *Geol. Soc. Amer. Bull.*, v. 66, no. 12, pp. 1489–1498.

Shepard, F. P., 1957. Northward continuation of the San Andreas Fault. *Seismol. Soc. Amer. Bull.*, v. 47, no. 3, pp. 263–266.

Shepard, F. P., 1960A. Gulf Coast barriers. In *Recent Sediments, Northwest Gulf of Mexico*. F. P. Shepard, F. B Phleger, and Tj. H. van Andel, eds., Amer. Assoc. Petrol. Geol., Tulsa, Okla., pp. 197–220.

Shepard, F. P., 1960B. Mississippi Delta: marginal environments, sediments, and growth. In *Recent Sediments, Northwest Gulf of Mexico*, F. P. Shepard, F. B Phleger, and Tj. H. van Andel, eds., Amer. Assoc. Petrol. Geol., Tulsa, Okla., pp. 56–81.

Shepard, F. P., 1963. *Submarine Geology*, 2nd ed., Harper & Row, New York, 557 pp.

Shepard, F. P., 1964. Sea-floor valleys of the Gulf of California. In *Marine Geology of the Gulf of California—A Symposium*, Mem. No. 3, Tj. H. van Andel and G. G. Shor, Jr., eds., Amer. Assoc. Petrol. Geol., Tulsa, Okla., pp. 157–192.

Shepard, F. P., 1965. Submarine canyons explored by Cousteau's Diving Saucer. In *Submarine Geology and Geophysics*, W. F. Whittard and R. Bradshaw, eds., Butterworths, London, pp. 303–311.

Shepard, F. P., 1966. Meander in valley crossing a deep-ocean fan. *Science*, v. 154, no. 3748, pp. 385–386.

Shepard, F. P., 1967. Submarine canyon origin: based on deep-diving vehicle and surface ship operations. *Rev. Geogr. Physique et Geol. Dynamique*, Paris, (2), v. 9, no. 5, pp. 347–356.

Shepard, F. P., 1970. Lagoonal topography of Caroline and Marshall Islands. *Geol. Soc. Amer. Bull.*, v. 81, pp. 1905–1914.

Shepard, F. P., and C. N. Beard, 1938. Submarine canyons: distribution and longitudinal profiles. *Geogr. Rev.*, v. 28, no. 3, pp. 439–451.

Shepard, F. P., and E. C. Buffington, 1968. La Jolla Submarine Fan-Valley. *Marine Geol.*, v. 6, no. 2, pp. 107–143.

Shepard, F. P., and G. V. Cohee, 1936. Continental shelf sediments off the Mid-Atlantic states. *Geol. Soc. Amer. Bull.*, v. 47, pp. 441–458.

Shepard, F. P., and J. R. Curray, 1967. Carbon-14 determination of sea level changes in stable areas. In *Progress in Oceanography*, Vol. 4, M. Sears, ed., Pergamon, Oxford, pp. 283–291.

Shepard, F. P., and R. F. Dill, 1966. *Submarine Canyons and Other Sea Valleys*. Rand McNally, Chicago, 381 pp.

Shepard, F. P., and K. O. Emery, 1941. *Submarine Topography off the California Coast: Canyons and Tectonic Interpretation.* Geol. Soc. Amer., spec. pap. no. 31, 171 pp.

Shepard, F. P., and U. S. Grant IV, 1947. Wave erosion along the southern California coast. *Geol. Soc. Amer. Bull.*, v. 58, pp. 919–926.

Shepard, F. P., and D. L. Inman, 1950. Nearshore water circulation related to bottom topography and wave refraction. *Trans. Amer. Geophys. Union*, v. 31, no. 2, pp. 196–212.

Shepard, F. P., and R. R. Lankford, 1959. Sedimentary facies from shallow borings in lower Mississippi Delta. *Amer. Assoc. Petrol. Geol. Bull.,* v. 43, no. 9, pp. 2051–2067.

Shepard, F. P., and G. A. Macdonald, 1938. Sediments of Santa Monica Bay, California. *Amer. Assoc. Petrol. Geol. Bull.,* v. 22, no. 2, pp. 201–216.

Shepard, F. P., and N. F. Marshall, 1969. Currents in La Jolla and Scripps Submarine Canyons. *Science,* v. 165, pp. 177–178.

Shepard, F. P., and N. F. Marshall, 1973. Currents along the floors of submarine canyons. *Amer. Assoc. Petrol. Geol. Bull.,* v.57, pp. 244–264.

Shepard, F. P., and D. G. Moore, 1955. Central Texas coast sedimentation: characteristics of sedimentary environment, recent history, and diagenesis. *Amer. Assoc. Petrol. Geol. Bull.,* v. 39, no. 8, pp. 1463–1593.

Shepard, F. P., and D. G. Moore, 1960. Bays of central Texas coast. In *Recent Sediments, Northwest Gulf of Mexico.* F. P. Shepard, F. B Phleger, and Tj. H. van Andel, eds., Amer. Assoc. Petrol. Geol., Tulsa, Okla., pp. 117–152.

Shepard, F. P., and G. A. Rusnak, 1957. Texas bay sediments. *Inst. Marine Sci.,* v. 4, no. 2, pp. 473–481.

Shepard, F. P., and H. R. Wanless, 1971. *Our Changing Coastlines.* McGraw-Hill, New York, 579 pp.

Shepard, F. P., and Ruth Young, 1961. Distinguishing between beach and dune sands. *Jour. Sed. Petrology,* v. 31, no. 2, pp. 196–214.

Shepard, F. P., K. O. Emery, and H. R. Gould, 1949. Distribution of sediments on East Asiatic continental shelf. *Allan Hancock Found. Publ., Occasional Paper* no. 9, 64 pp.

Shepard, F. P., G. A. Macdonald, and D. C. Cox, 1950. The tsunami of April 1, 1946. *Bull. Scripps Inst. Oceanog.,* Univ. Calif. Press, v. 5, no. 6, pp. 391–527.

Shepard, F. P., F. B Phleger, and Tj. H. van Andel, eds., 1960. *Recent Sediments, Northwest Gulf of Mexico.* Amer. Assoc. Petrol. Geol., Tulsa, Okla., 394 pp.

Shepard, F. P., Hiroshi Niino, and T. K. Chamberlain, 1964. Submarine canyons and Trough, east-central Honshu, Japan. *Geol. Soc. Amer. Bull.,* v. 75, pp. 1117–1130.

Shepard, F. P., R. F. Dill, and B. C. Heezen, 1968. Diapiric intrusions in foreset slope sediments off Magdalena Delta, Colombia. *Amer. Assoc. Petrol. Geol. Bull.,* v. 52, no. 11, pp. 2197–2207.

Shepard, F. P., R. F. Dill, and Ulrich von Rad, 1969. Physiography and sedimentary processes of La Jolla Submarine Fan and Fan-Valley, California. *Amer. Assoc. Petrol. Geol. Bull.,* v. 53, no. 2, pp. 390–420.

Sheridan, R. E., and C. L. Drake, 1968. Seaward extension of the Canadian Appalachians. *Canadian Jour. Earth Sci.,* v. 5, no. 3, pt. 1, pp. 337–373.

Sheridan, R. E., C. L. Drake, J. E. Nafe, and J. Hennion, 1966. Seismic-refraction study of continental margin east of Florida. *Amer. Assoc. Petrol. Geol. Bull.,* v. 50, no. 9, pp. 1972–1991.

Sheridan, R. E., R. E. Houtz, C. L. Drake, and Maurice Ewing, 1969A. Structure of continental margin off Sierra Leone, West Africa. *Jour. Geophys. Res.,* v. 74, no. 10, pp. 2512–2530.

Sheridan R F., J. D. Smith, and J. Gardner, 1969B. Rock dredges from Blake Escarpment near Great Abaco Canyon. *Amer. Assoc. Petrol. Geol. Bull.,* v. 53, no. 12, pp. 2551–2558.

Shinn, E. A., 1963. Spur and groove formation on the Florida Reef Tract. *Jour. Sed. Petrology,* v. 33, no. 2, pp. 291–303.

Shor, G. G., Jr., 1959. Reflexion studies in the eastern equatorial Pacific. *Deep-Sea Res.,* v. 5, pp. 283–289.

Shor, G. G., Jr., 1960. Crustal structure of the Hawaiian Ridge near Gardner Pinnacles. *Bull. Seis. Soc. Amer.,* v. 50, no. 4, pp. 563–573.

Shor, G. G., Jr., 1963. Refraction and reflection techniques and procedure. In *The Sea,* Vol. 3, M. N. Hill, ed., Interscience, New York, pp. 20–38.

Shor, G. G., Jr., 1964. Thickness of coral at Midway Atoll. *Nature,* v. 201, no. 4925, p. 1207.

Shor, G. G., Jr., and R. L. Fisher, 1961. Middle America Trench: seismic-refraction studies. *Geol. Soc. Amer. Bull.,* v. 72, no. 5, pp. 721–730.

Shor, G. G., Jr., and R. W. Raitt, 1958. Seismic studies in the southern California continental borderland. *Proc. Congr. Geol. Intl. 20, Geofis. Apl.,* pp. 243–259.

Shor, G. G., Jr., P. Dehlinger, H. R. Kirk, and W. S. French, 1968. Seismic refraction studies off Oregon and northern California. *Jour. Geophys. Res.,* v. 73, no. 6, pp. 2176–2194.

Shouldice, D. H., 1970. Geology of western Canadian continental shelf. *Abs. Amer. Assoc. Petrol. Geol. Bull.,* v. 54/5, pp. 870–871.

Sievers, H. A., 1961. El Maremoto del 22 de Mayo de 1960 en los costas de Chile. Publ. no. 3012, Republica Chile, *Dept. Navegacion e Hidrografia de la Armada,* Valparaiso, Chile, 129 pp.

Sigl, Walter, Ulrich von Rad, Hansjörg Oeltzschner, Karl Braune, and Frank Fabricius, 1969. Diving sled: a tool to increase the efficiency of underwater mapping by scuba divers. *Marine Geol.,* v. 7, pp. 357–363.

Silver, E. A., 1971. Transitional tectonics and Late Cenozoic structure of the continental margin off northernmost California. *Geol. Soc. Amer. Bull.,* v. 82, no. 1, pp. 1–22.

Silverman, M., and R. C. Whaley, 1952. Adaptation of the piston coring device in shallow water. *Jour. Sed. Petrology,* v. 22, no. 1, pp. 11–17.

Singlewald, J. T., Jr., and Turbit Slaughter, 1949. Shore erosion in tidewater Maryland. *Maryland Dept. Geol., Mines, Nat. Resources Bull.,* 6, 128 pp.

Sisoev, N. H., I. E., Milhalsez, et al., 1959. The results of seismo-acoustic investigations of the ocean bottom. Abs. in *Preprints International Oceanographic Congress,* Amer. Assoc. Adv. Sci., Washington, D.C., p. 105.

Sitarz, J. A., 1960. Côtes Africaines. Étude des profils d'équilibre de plages. *Travaux Centre d'Études et Res. Océanog.,* v. 3, no. 4, pp. 43–62.

Sitarz, J. A., 1963. Contribution to the study of beach evolution from the knowledge of equilibrium profiles. *Travaux Centre Rech. d'Études Océanog.,* v. 5, nos. 2–4, pp. 13–14.

Smales, A. A., D. Mapper, and A. J. Wood, 1958. Radioactivation analysis of 'cosmic' and other magnetic spherules. *Geochim. et Cosmochim. Acta.,* v. 13, pp. 123–126.

Smith, A. J., A. H. Stride, and W. F. Whittard, 1965. The geology of the western approaches of the English Channel. IV. A recently discovered Varsican granite west-north-west of the Scilly Isles. In *Submarine Geology and Geophysics,* W. F. Whittard and R. Bradshaw, eds., Butterworths, London, pp. 287–301.

Smith S. V., J. A. Dygas, and K. E. Chave, 1968. Distribution of calcium carbonate in pelagic sediments. *Marine Geol.,* v. 6, pp. 391–400.

Smith, W. S. T., 1902. The submarine valleys of the California coast. *Science,* v. 15, pp. 670–672.

Sonu, C. J., and R. J. Russell, 1966. Topographic changes in the surf zone profile. *Tech. Rept. no. 50, Coastal Studies Inst.,* Louisiana State Univ., pp. 502–524.

Soons, J. M., 1968. Raised submarine canyons: a discussion of some New Zealand examples. *Ann. Assoc. Amer. Geographers,* v. 58, no. 3, pp. 606–613.

Spencer, J. W., 1898. On the continental elevation of the glacial epoch. *Geol. Mag.,* 4 (5), pp. 32–38.

Spencer, Maria, 1967. Bahama deep test. *Amer. Assoc. Petrol. Geol. Bull.,* v. 51, no. 2, pp. 263–268.

Spiess, F. N., and J. D. Mudie, 1970. Small-scale topographic and magnetic features. In *The Sea,* Vol. 4, Pt. 1, A. E. Maxwell, ed., Wiley–Interscience, New York, pp. 205–250.

Spiess, F. N., B. P. Luyendyk, R. L. Larson, W. R. Normark, and J. D. Mudie, 1969. Detailed geophysical studies on the northern Hawaiian Arch using a deeply towed instrument package. *Marine Geol.,* v. 7, pp. 501–527.

Sprigg, R. C., 1963. New structural discoveries off Australia's southern coast. *Australian Oil & Gas Jour.,* v. 9, no. 12, p. 32.

Sproll, W. P., and R. S. Dietz, 1969. Morphological continental drift fit of Australia and Antarctica. *Nature,* v. 222, no. 5191, pp. 345–348.

Squires, D. F., 1963. Modern tools probe deep water. *Natural Hist.,* June–July, pp. 22–29.

Stanley, D. J., 1967. Comparing patterns of sedimentation in some modern and ancient submarine canyons. *Earth and Planetary Sci. Letters,* 3 Amsterdam, pp. 371–380.

Stanley, D. J., 1970. The ten-fathom terrace on Bermuda: its significance as a datum for measuring crustal mobility and eustatic sea-level changes in the Atlantic. *Ann. Geomorphol.,* v. 14, no. 2, pp. 186–201.

Stanley, D. J., and A. E. Cok, 1967. Sediment transport by ice on the Nova Scotian Shelf. In *Ocean Sciences & Engineering of the Atlantic Shelf.* Trans. Nat. Symposium Marine Technol. Soc., pp. 100–125.

Stanley, D. J., and Gilbert Kelling, 1967. Sedimentation patterns in the Wilmington Submarine Canyon area. In *Ocean Sciences & Engineering of the Atlantic Shelf,* Trans. Nat. Symposium Marine Technol. Soc., pp. 127–142.

Stanley, D. J., and Gilbert Kelling, 1970. Interpretation of a levee-like ridge and associated features, Wilmington Submarine Canyon, eastern United States. *Geol. Soc. Amer. Bull.,* v. 81, pp. 3747–3752.

Stanley, D. J., and D. J. P. Swift, 1967. Bermuda's southern aeolianite reef tract. *Science,* v. 157, no. 3789, pp. 677–681.

Stanley, D. J., and D. J. P. Swift, 1968. Bermuda's reef-front platform: bathymetry and significance. *Marine Geol.,* v. 6, pp. 479–500.

Stanley, D. J., Georges Drapeau, and A. E. Cok, 1968. Submerged terraces on the Nova Scotian Shelf. *Ann. Geomorphol.,* Suppl. 7, pp. 85–94.

Starke, G. W., and A. D. Howard, 1968. Polygenetic origin of Monterey Submarine Canyon. *Geol. Soc. Amer. Bull.,* v. 79, pp. 813–826.

Stetson, H. C., 1936. Geology and paleontology of the Georges Bank canyons. I. Geology. *Geol. Soc. Amer. Bull.,* v. 47, pp. 339–366.

Stetson, H. C., 1938. The sediments of the continental shelf off the eastern coast of the United States. *Papers in Physical Oceanog. and Meteorol.,* Massachusetts Inst. Technol. and Woods Hole Oceanog. Inst., v. 5, no. 4, 48 pp.

Stetson, H. C., 1939. Summary of sedimentary conditions on the continental shelf off the east coast of the United States. In *Recent Marine Sediments,* P. D. Trask, ed., Amer. Assoc. Petrol. Geol., Tulsa, Okla., pp. 230–244.

Stetson, H. C., 1949. The sediments and stratigraphy of the East Coast continental margin: Georges Bank to Norfolk Canyon. *Papers in Physical Oceanog. and Meteorol.,* Massachusetts Inst. Technol. and Woods Hole Oceanog. Inst., v. 11, no. 2, 60 pp.

Stetson, T. R., D. F. Squires, and R. M. Pratt, 1962. Coral banks occurring in deep water on the Blake Plateau. *Amer. Mus. Novitates,* no. 2114, 39 pp.

Stetson, T. R., Elazar Uchupi, and J. D. Milliman, 1969. Surface and subsurface morphology of two small areas of the Blake Plateau. *Trans. Gulf Coast Assoc. Geol. Soc.,* v. 19, pp. 131–142.

Stevenson, R. E., 1968. Lagoon. In *The Encyclopedia of Geomorphology,* R. W. Fairbridge, ed., Reinhold, New York, pp. 590–594.

Stevenson, R. E., and K. O. Emery, 1958. Marshlands at Newport Bay, California. *Allan Hancock Found. Publ., Occasional Paper* no. 20, Univ. Southern Calif., Los Angeles, 109 pp.

Stewart, H. B., Jr., 1956. Contorted sediments in modern coastal lagoons explained by laboratory experiments. *Amer. Assoc. Petrol. Geol. Bull.,* v. 40, no. 1, pp. 153–161.

Stewart, H. B., Jr., and G. F. Jordan, 1965. Underwater sand ridges on Georges Shoal. In *Papers in Marine Geology,* Shepard Comm. Vol., R. L. Miller, ed., Macmillan, New York, pp. 102–114.

Stewart, H. B., Jr., G. F. Jordan, and G. G. Salsman, 1959. Underwater sand ridges on Georges Shoal. Abs. in *Preprints International Oceanographic Congress,* Amer. Assoc. Adv. Sci., Washington, D.C., p. 665.

Stewart, H. B., Jr., F. P. Shepard, and R. S. Dietz, 1964. Submarine canyons off eastern Ceylon. *Abs. Program 1964 Ann. Meetings, Geol. Soc. Amer.,* Miami, Fla., p. 197.

Stocks, Theodor, 1933. *Die Echolotprofile Wissenschaftliche Ergebnisse der Deutschen Atlantischen Expedition auf den "Meteor," 1925–1927.* Vol. 2.

Stocks, Theodor, 1956. Der Boden der südlichen Nordsee. 2. Eine neue Tiefenkarte der südlichen Nordsee. *Deutschen Hydrog. Zeitschr.,* v. 9, no. 6, pp. 265–280.

Storr, J. E., 1964. Ecology and oceanography of the coral-reef tract, Abaco Island, Bahamas. *Geol. Soc. Amer. Spec. Papers,* no. 79, 98 pp.

Stride, A. H., 1959A. On the origin of the Dogger Bank, in the North Sea. *Geol. Mag.,* v. 96, no. 1, pp. 33–44.

Stride, A. H., 1959B. A linear pattern on the sea floor and its interpretation. *Jour. Mar. Biol. Ass. U. K.,* v. 38, pp. 313–318.

Stride, A. H., 1960. Recognition of folds and faults on rock surfaces beneath the sea. *Nature,* v. 185, no. 4716, p. 837.

Stride, A. H., 1961A. Geological interpretation of asdic records. *Intl. Hydrogr. Rev.,* v. 38, no. 1, pp. 1–9.

Stride, A. H., 1961B. Mapping the sea floor with sound. *New Scientist,* v. 10, pp. 304–306.

Stride, A. H., 1963A. North-east trending ridges of the Celtic Sea. *Proc. Ussher Soc.,* v. 1, pt. 2, pp. 62–63.

Stride, A. H., 1963B. Current-swept sea floors near the southern half of Great Britain. *Quart. Jour. Geol. Soc. London,* v. 119, pp. 175–199.

Stride, A. H., 1965A. Periodic and occasional sand transport in the North Sea. *Petrole et la Mar,* sec. 1, no. 111, 3 pp.

Stride, A. H., 1965B. Under the North Sea. *Hunting Group Rev.,* no. 3, pp. 20–23.

Stride, A. H., R. H. Belderson, J. R. Curray, and D. G. Moore, 1967. Geophysical evidence on the origin of the Faeroe Bank Channel—I. Continuous reflection profiles. *Deep-Sea Res.,* v. 14, pp. 1–6.

Stride, A. H., J. R. Curray, D. G. Moore, and R. H. Belderson, 1969. Marine geology of the Atlantic continental margin of Europe. *Philos. Trans. Roy. Soc. London,* ser. A, Mathematical and physical sciences, no. 1148, v. 264, pp. 31–75.

Stubbs, A. R., 1963. Identification of patterns on asdic records. *Intl. Hydrogr. Rev.,* v. 40, no. 2, pp. 50–68.

Sverdrup, H. U., M. W. Johnson, and R. H. Fleming, 1942. *The Oceans, Their Physical, Chemistry, and General Biology.* Prentice-Hall, Englewood Cliffs, N. J., 1087 pp.

Swift, D. J. P., 1969. Outer shelf sedimentation: processes and products. In *The New Concepts of Continental Margin Sedimentation,* Lecture no. 5, Amer. Geol. Inst. Short Course Lecture Notes, Philadelphia, Amer. Geol. Inst., Washington, D.C., DS 5, 26 pp.

Swift, D. J. P., and A. K. Lyall, 1968. Origin of the Bay of Fundy, an interpretation from sub-bottom profiles. *Marine Geol.,* v. 6, pp. 331–343.

Sykes, L. R., 1963. Seismicity of the South Pacific Ocean. *Jour. Geophys. Res.,* v. 68, no. 21, pp. 5999–6006.

Sykes, L. R., 1967. Mechanism of earthquakes and nature of faulting on the mid-oceanic ridges. *Jour. Geophys. Res.,* v. 72, no. 8, pp. 2131–2153.

Sykes, L. R., Jack Oliver, and Bryan Isacks, 1970. Earthquakes and tectonics. In *The Sea,* Vol. 4, Pt. 1, A. E. Maxwell, ed., Wiley–Interscience, New York, pp. 353–420.

Tagg, A. R., and H. G. Greene, 1971. Seismic survey locates potential gold deposits in the Bering Sea. *Ocean Industry,* v. 6, no. 8, pp. 40–43.

Tagg, A. R., and Elazar Uchupi, 1967. Subsurface morphology of Long Island Sound, Block Island Sound, Rhode Island Sound, and Buzzards Bay. *U.S. Geol. Survey Prof. Paper 575-C.,* pp. 92–96.

Talwani, Manik, 1970A. Gravity. In *The Sea,* Vol. 4, Pt. 1, A. E. Maxwell, ed., Wiley–Interscience, New York, pp. 251–297.

Talwani, Manik, 1970B. Developments in navigation and measurement of gravity at sea. *Geoexploration,* v. 8, pp. 151–183.

Talwani, Manik, J. L. Worzel, and Maurice Ewing, 1960. Gravity anomalies and structure of the Bahamas. *2nd Caribbean Geol. Conf.,* pp. 156–161.

Talwani, Manik, G. H. Sutton, and J. L. Worzel, 1959. A crustal section across the Puerto Rico Trench. *Jour. Geophys. Res.,* v. 64, no. 10, pp. 1545–1555.

Talwani, Manik, J. Dorman, J. L. Worzel, and G. M. Bryan, 1966. Navigation at sea by satellite. *Jour. Geophys. Res.,* v. 71, no. 24, pp. 5891–5902.

Tanner, W. F., 1960. Florida coastal classification. *Trans. Gulf Coast Assoc. Geol. Soc.,* v. 10, pp. 259–266.

Tanner, W. F., 1961. Offshore shoals in area of energy deficit. *Jour. Sed. Petrology,* v. 31, no. 1, pp. 87–95.

Teichert, Curt, 1958. Cold- and deep-water coral banks. *Amer. Assoc. Petrol. Geol. Bull.,* v. 42, no. 5, pp. 1064–1082.

Teichert, Curt, and R. W. Fairbridge, 1948. Some coral reefs on the Sahul Shelf. *Amer. Geogr. Soc.,* v. 38, no. 2, pp. 222–249.

Terry, R. D., 1966. *The Deep Submersible.* Western Periodicals, North Hollywood, Calif., 456 pp.

Terry, R. D., S. A. Keesling, and Elazar Uchupi, 1956. Submarine geology of Santa Monica Bay, California. *Rept. to Hyperion Engrs. Inc.,* from Geol. Dept., Univ. Southern California, 177 pp. (multilithed).

Terzaghi, Karl, 1956. Varieties of submarine slope failures. *Proc. 8th Texas Conf. Soil Mechanics and Found. Engr.,* 47 pp.

Terzaghi, Karl, and R. B. Peck, 1948. *Soil Mechanics in Engineering Practice.* Wiley, New York, 566 pp.

Thiel, E. C., 1962. The amount of ice on planet Earth. In *Antarctic Research,* American Geophysical Union, Nat. Acad. Sci.—Nat. Res. Council, publ. no. 1036, Geophysical Monograph no. 7, Washington, D.C., pp. 172–175.

Thompson, R. W. (ms.). Recent sediments of Humboldt Bay, Eureka, California. Final Rept. PRF 789-G2, Humboldt State College, Arcata, Calif. 46 pp.

Thompson, W. C., 1955. Sandless coastal terrain of the Atchafalaya Bay area, Louisiana. In *Finding Ancient Shorelines, A Symposium with Discussions,* J. L. Hough and H. W. Menard, eds., Soc. Econ. Paleontol. and Mineralog., spec. publ. no. 3, Tulsa, Okla., pp. 52–77.

Thurber, D. L., W. S. Broecker, R. L. Blanchard, and H. A. Potratz, 1965. Uranium-series ages of Pacific atoll coral. *Science,* v. 149, no. 3679, pp. 55–58.

Tobin, D. G., and L. R. Sykes, 1968. Seismicity and tectonics of the northeast Pacific Ocean. *Jour. Geophys. Res.,* v. 73, pp. 3821–3845.

Tocher, Don, 1956. Earthquakes off the North Pacific Coast of the United States. *Seismol. Soc. Amer. Bull.,* v. 46, pp. 165–173.

Todd, Ruth, and Rita Post, 1954. Bikini and nearby atolls. Pt. 4. Paleontology. Smaller Foraminifera from Bikini drill holes. *U.S. Geol. Survey Prof. Paper 260-N,* pp. 547–568.

Tracey, J. I., Jr., D. P. Abbott, and Ted Arnow, 1961. Natural history of Ifaluk Atoll: physical environment. *Bernice P. Bishop Mus. Bull.,* 222, Honolulu, 75 pp.

Trask, P. D. (assisted by H. E. Hammar and C. C. Wu), 1932. *Origin and Environment of Source Sediments of Petroleum.* American Petroleum Institute Gulf Publ., Houston, Texas, 323 pp.

Trask, P. D., 1952. Source of beach sand at Santa Barbara, California, as indicated by mineral grain studies. *Beach Erosion Board, U.S. Army Corps Engrs. Tech. Memo.* no. 28, 24 pp.

Trask, P. D., 1955. Movement of sand around southern California promontories. *Beach Erosion Board, U.S. Army Corps Engrs. Tech. Memo.* no. 76, 66 pp.

Trask, P. D., and J. W. Rolston, 1951. Engineering geology of San Francisco Bay, California. *Geol. Soc. Amer. Bull.,* v. 62, pp. 1079–1110.

Trefethen, J. M., and R. L. Dow, 1960. Some features of modern beach sediments. *Jour. Sed. Petrology,* v. 30, no. 4, pp. 589–602.

Trumbull, J. V. A., and M. J. McCamis, 1967. Geological exploration in an East Coast submarine canyon from a research submersible. *Science,* v. 158, pp. 370–372.

Trumbull, J. V. A., John Lyman, J. F. Pepper, and E. M. Thomasson, 1958. An introduction to the geology and mineral resources of the continental shelves of the Americas. *U.S. Geol. Survey Bull.* 1067, 92 pp.

Tucker, M. J., and A. R. Stubbs, 1961. Narrow-beam echo-ranger for fishery and geological investigations. *British Jour. Applied Physics,* v. 12, pp. 103–110.

Uchupi, Elazar, 1965. Basins of the Gulf of Maine. *U.S. Geol. Survey Prof. Paper 525-D* pp. D175–D177.

Uchupi, Elazar, 1966A. Shallow structure of the Straits of Florida. *Science,* v. 153, no. 3735, pp. 529–531.

Uchupi, Elazar, 1966B. Structural framework of the Gulf of Maine. *Jour. Geophys. Res.,* v. 71, no. 12, pp. 3013–3027.

Uchupi, Elazar, 1967A. Bathymetry of the Gulf of Mexico. *Trans. Gulf Coast Assn. Geol. Soc.,* 17th Ann. Meeting, pp. 161–172.

Uchupi, Elazar, 1967B. Slumping on the continental margin southeast of Long Island. *Deep-Sea Res.,* v. 14, pp. 635–639.

Uchupi, Elazar, 1968A. Atlantic continental shelf and slope of the United States—physiography. *U.S. Geol. Survey Prof. Paper* 529-C, 30 pp.

Uchupi, Elazar, 1968B. Tortugas Terrace, a slip surface? *U.S. Geol. Survey Prof. Paper* 600-D, pp. 231–234.

Uchupi, Elazar, 1968C. Seismic profiling survey of the east coast submarine canyons. Pt. 1. Wilmington, Baltimore, Washington and Norfolk Canyons. *Deep-Sea Res.,* v. 15, pp. 613–616.

Uchupi, Elazar, 1970. Atlantic continental shelf and slope of the United States—shallow structure. *U.S. Geol. Survey Prof. Paper* 529-I, pp. 1–44.

Uchupi, Elazar, and K. O. Emery, 1963. The continental slope between San Francisco, California and Cedros Island, Mexico. *Deep-Sea Res.,* v. 10, pp. 397–447.

Uchupi, Elazar, and K. O. Emery, 1967. Structure of continental margin off Atlantic Coast of United States. *Amer. Assoc. Petrol. Geol. Bull.,* v. 51, no. 2, pp. 223–234.

Uchupi, Elazar, and K. O. Emery, 1968. Structure of continental margin off Gulf Coast of United States. *Amer. Assoc. Petrol. Geol. Bull.,* v. 52, no. 7, pp. 1162–1193.

Uchupi, Elazar, and R. Gaal, 1963. Sediments of the Palos Verdes shelf. In *Essays in Marine Geology,* honoring K. O. Emery, Univ. Southern California Press, Los Angeles, Calif., pp. 171–189.

Uchupi, Elazar, J. D. Phillips, and K. E. Prada, 1970. Origin and structure of the New England Seamount Chain. *Deep-Sea Res.,* v. 17, pp. 483–494.

Uchupi, Elazar, J. D. Milliman, B. P. Luyendyk, C. O. Bowin, and K. O. Emery, 1971. Structure and origin of the southeastern Bahamas. *Amer. Assoc. Petrol. Geol. Bull.,* v. 55, no. 5, pp. 687–704.

Udden, J. A., 1898. Mechanical composition of wind deposits. *Augustana Library Publ.,* no. 1.

Udintsev, G. B., 1957. Relief of Okhotsk Sea. *Akad. Nauk. USSR,* v. 22, 76 pp. (in Russian).

Udintsev, G. B. 1966. Results of Upper Mantle Project Studies in the Indian Ocean by the survey vessel *Vityaz.* In *The World Rift System,* Report of 1965 UMC Symposium, Ottawa, T. N. Irvine, ed., Geol. Survey Canada Paper 66-14, pp. 148–172.

Ufford, C. W., 1947. Internal waves in the ocean. *Trans. Amer. Geophys Union,* v. 28, pp. 79–86.

Umbgrove, J. H. F., 1945. Different types of island-arcs in the Pacific. *Geogr. Jour.,* v. 106, nos. 5, 6, pp. 198–209.

Umbgrove, J. H. F., 1947. Coral reefs on the East Indies. *Geol. Soc. Amer. Bull.,* v. 58, pp. 729–777.

Umbgrove, J. H. F., 1949. *Structural History of the East Indies.* University Press, Cambridge, England, 62 pp.

Urey, H. C., 1947. The thermodynamic properties of isotopic substances. *Jour. Amer. Chem Soc.,* pp. 562–581.

Vacquier, Victor, 1959. Measurement of horizontal displacement along faults in the ocean floor. *Nature,* v. 183, pp. 452–453.

Vacquier, Victor, and R. P. Von Herzen, 1964. Evidence for connection between heat flow and the Mid-Atlantic Ridge magnetic anomaly. *Jour. Geophys. Res.,* v. 69, no. 6, pp. 1093–1101.

Valentin, Hartmut, 1952. *Die Küsten der Erde.* Justus Perthes Gotha, Berlin, 118 pp.

Valentin, Hartmut, 1955. Die Grenze der Letzten Vereisung im Nordseeraum. *Deutscher Geographentag Hamburg,* Aug., pp. 359–366.

van Andel, Tj. H., 1955. Sediments of the Rhone Delta. II. Sources and deposition of heavy minerals. *Koninkl. Ned. Geol.-Mijnb. Genoot. Verh.,* Geol. Ser., 15, 3e, pp. 516–543.

van Andel, Tj. H., 1964. Recent marine sediments of Gulf of California. In *Marine Geology of the Gulf of California—A Symposium,* Mem. 3, Tj. H. van Andel and G. G. Shor, eds., Amer. Assoc. Petrol. Geol., Tulsa, Okla., pp. 216–310.

van Andel, Tj. H., 1967. The Orinoco Delta. *Jour. Sed. Petrology,* v. 37, no. 2, pp. 297–310.

van Andel, Tj. H., 1969. Recent uplift of the Mid-Atlantic Ridge south of the Vema fracture zone. *Earth Planetary Sci. Letters,* v. 7, pp. 228–230.

van Andel, Tj. H., and C. O. Bowin, 1968. Mid-Atlantic Ridge between 22° and 23° north latitude and the tectonics of mid-ocean rises. *Jour. Geophys. Res.,* v. 73, no. 4, pp. 1279–1298.

van Andel, Tj. H., and S. E. Calvert, 1971. Evolution of sediment wedge, Walvis shelf, southwest Africa. *Jour. Geol.,* v. 79, pp. 585–602.

van Andel, Tj. H., and J. B. Corliss, 1967. The intersection between the Mid-Atlantic Ridge and the Vema Fracture Zone. *Jour. Marine Res.,* v. 25, no. 3, pp. 343–351.

van Andel, Tj. H., and Jacques Laborel, 1964. Recent high relative sea level stand near Recife, Brazil. *Science,* v. 145, no. 3632, pp. 580–581.

van Andel, Tj. H., and D. M. Poole, 1960. Sources of recent sediments in the northern Gulf of Mexico. *Jour. Sed. Petrology,* v. 30, pp. 91–122.

van Andel, Tj. H., and H. Postma, 1954. *Recent Sediments of the Gulf of Paria, Reports of Orinoco Shelf Expedition.* North-Holland, Amsterdam, 245 pp.

van Andel, Tj. H., and P. L Sachs, 1964. Sedimentation in the Gulf of Paria during the Holocene transgression; a subsurface acoustic reflection study. *Jour. Marine Res.,* v. 22, no. 1, pp. 30–50.

van Andel, Tj. H., and G. G. Shor, Jr., eds., 1964. *Marine Geology of the Gulf of California.* Amer. Assoc. Petrol. Geol., Tulsa, Okla., 408 pp.

van Andel, Tj. H., and J. J. Veevers, 1967. Morphology and sediments of the Timor Sea. *Dept. Nat. Development, Australia, Bur. Mineral Resources, Geol. and Geophys. Bull.,* no. 83, 173 pp.

van Andel, Tj. H., G. R. Heath, T. C. Moore, and D. F. R. McGeary, 1967. Late Quaternary history, climate, and oceanography of the Timor Sea, northwestern Australia. *Amer. Jour. Sci.,* v. 265, pp. 737–758.

van Andel, Tj. H., J. D. Phillips, and R. P. Von Herzen, 1969. Rifting origin for the Vema fracture in the North Atlantic. *Earth and Planetary Sci. Letters,* v. 5, pp. 296–300.

van Andel, Tj. H., G. R. Heath, et al., 1971. Deep sea drilling project: Leg 16, *Geotimes,* v. 16, no. 6, pp. 12–14.

van Baren, F. A., and H. Kiel, 1950. Contribution to the sedimentary petrology of the Sunda Shelf. *Jour. Sed. Petrology,* v. 20, no. 4, pp. 185–213.

van den Bussche, H. K. J., and J. J. H. C. Hubolt, 1964. A corer for sampling shallow-marine sands. *Sedimentology,* v. 3, pp. 155–159.

van der Lingen, G. J., 1969. The turbidite problem. *N. Z. Jour. Geol. Geophys.,* v. 12, pp. 7–50.

van der Lingen, G. J., and P. B. Andrews, 1969. Hoof-print structures in beach sand. *Jour. Sed. Petrology,* v. 39, pp. 350–357.

Van Dorn, W. G., 1964. Source mechanism of the tsunami of March 28, 1964 in Alaska. In *Proceedings of the Ninth Conference on Coastal Engineering,* Lisbon, American Society of Civil Engineers, pp. 166–190.

Van Dorn, W. G., 1965. Tsunamis. In *Advances in Hydroscience,* Vol. 2, Academic, New York, pp. 1–48.

van Straaten, L. M. J. U., 1950. Environment of formation and facies of the Wadden Sea sediments. *Tijdschr. van het Koninkl. Ned. Aardrijksk. Genoots.,* v. 47, pt. 3, pp. 94–108.

van Straaten, L. M. J. U., 1953. Rhythmic patterns on Dutch North Sea beaches. *Geol. en Mijnb.,* Nw ser., no. 2, pp. 31–43.

van Straaten, L. M. J. U., 1954A. Composition and structure of recent marine sediments in the Netherlands. *Leidse Geol. Mededeel.,* v. 19, pp. 1–110.

van Straaten, L. M. J. U., 1954B. Sedimentology of recent tidal flat deposits and the Psammites du Condroz (Devonian). *Geol. en Mijnb.,* v. 16, pp. 25–47.

van Straaten, L. M. J. U., 1956. Composition of shell beds formed in tidal flat environment in the Netherlands and in the Bay of Arcachon (France). *Geol. en Mijnb.,* v. 18e, no. 7, pp. 209–226.

van Straaten, L. M. J. U., 1965A. Coastal barrier deposits in South- and North-Holland, in particular in the areas around Scheveningen and IJmuiden. *Mededeel. Geol. Soc.,* Nieuwe ser. no. 17, pp. 41–75.

van Straaten, L. M. J. U., 1965B. Sedimentation in the north-western part of the Adriatic Sea. In *Submarine Geology and Geophysics,* W. F. Whittard and R. Bradshaw, eds., Butterworths, London, pp. 143–162.

van Straaten, L. M. J. U., and Ph. H. Kuenen, 1957. Accumulation of fine grained sediments in the Dutch Wadden Sea. *Geol. en Mijnb.,* NS, v. 19, pp. 329–354.

van Veen, 1936. *Onderzoekingen in de Hoofden.* 'S-Gravenhage, Algemeene Landsdukkerij, 252 pp.

Varney, F. M., and L. E. Redwine, 1937. Hydraulic coring instrument for submarine geologic investigations. Natl. Res. Council Ann. Rept., App. 1, Sedimentation Comm., pp. 107–113.

Vaughan, T. W., 1916. On Recent Madreporaria of Florida, the Bahamas, and the West Indies; and on collections from Murray Island, Australia. *Dept. Marine Biol., Carnegie Inst. Washington Year Book* 14, pp. 220–231.

Vaughan, T. W., and J. W. Wells, 1943. *Revision of the Suborders, Families, and Genera of the Scleractinia.* Geol. Soc. Amer., spec. pap. 44, 363 pp.

Veatch, A. C., and P. A. Smith, 1939. *Atlantic submarine valleys of the United States and Congo Submarine Valley.* Geol. Soc. Amer., spec. pap., no. 7, 101 pp.

Veeh, H. H., 1966. Th^{230}/U^{238} ages of Pleistocene high sea level stand. *Jour. Geophys. Res.,* v. 71, no. 14, pp. 3379–3386.

Veeh, H. H., and J. J. Veevers, 1970. Sea level at −175 m off the Great Barrier Reef 13,600 to 17,000 year ago. *Nature,* v. 226, May, pp. 536–537.

Veltheim, Valto, 1962. On the pre-Quaternary geology of the bottom of the Bothnian Sea. *Bull. Comm. Geol. Finland,* no. 200, 166 pp.

Veltheim, Valto, 1969. On the pre-Quaternary geology of the Bothnian Bay area in the Baltic Sea. *Bull. Comm. Geol. Finland,* no. 239, 56 pp.

Vening Meinesz, F. A., 1929. Results of gravity determination upon the Pacific and the organization of further research. *Proc. 4th Pacific Sci. Congr.,* Java, IIB, p. 665.

Vening Meinesz, F. A., 1955. Plastic buckling of the earth's crust: the origin of geosynclines. In *Crust of the Earth,* A. Poldervaart, ed., Geol. Soc. Amer., spec. pap. 62, pp. 319–330.

Vening Meinesz, F. A., J. H. F. Umbgrove, and Ph. H. Kuenen, 1934. *Gravity Expeditions at Sea 1923–1932,* Vol. 2, J. Waltman, Jr., Delft, 208 pp.

Vernon, J. W., and R. A. Slater, 1964. Submarine tar mounds, Santa Barbara County, California. *Amer. Assoc. Petrol. Geol. Bull.,* v. 47, no. 8, pp. 1624–1627.

Vine, F. J., 1966. Spreading of the ocean floor: new evidence. *Science,* v. 154, pp. 1405–1415.

Vine, F. J., 1968. Magnetic anomalies associated with mid-ocean ridges. In *The History of the Earth's Crust,* R. A. Phinney, ed., Princeton University Press, Princeton, N.J., pp. 73–89.

Vine, F. J., 1970. The geophysical year. *Nature,* v. 227, no. 5262, pp. 1013–1017.

Vine, F. J., and H. H. Hess, 1970. Sea-floor spreading. In *The Sea,* Vol. 4, Pt. 2, A. E. Maxwell, ed., Wiley–Interscience, New York, pp. 587–622.

Vine, F. J., and D. H. Matthews, 1963. Magnetic anomalies over oceanic ridges. *Nature,* v. 199, pp. 947–949.

von der Borch, C. C., 1970. Phosphatic concretions and nodules from the upper continental slope, northern New South Wales. *Jour. Geol. Soc. Australia,* v. 16, pt. 2, pp. 755–759.

von der Borch, C. C., J. R. Conolly, and R. S. Dietz, 1970. Sedimentation and structure of the continental margin in the vicinity of the Otway Basin, southern Australia. *Marine Geol.*, v. 8, pp. 59–83.

von Huene, Roland, and G. G. Shor, Jr., 1969. The structure and tectonic history of the eastern Aleutian Trench. *Geol. Soc. Amer. Bull.*, v. 80, pp. 1889–1902.

Waddell, Hakon, 1932. Volume, shape, and roundness of rock particles. *Jour. Geol.*, v. 40, pp. 443–451.

Waddensymposium, 1950. Tisjchr. Kon. Nederl. Aaedr. k. Gen., Deel 67, Alf. 3, 148 pp.

Wageman, J. M., T. W. C. Hilde, and K. O. Emery, 1970. Structural framework of the West China Sea. *Amer. Assoc. Petrol. Geol. Bull.*, v. 54, no. 9, pp. 1611–1643.

Walker, B. W., 1952. A guide to the grunion. *California Fish and Game*, v. 38, no. 3, pp. 409–420.

Walthier, T. N., A. M. Rossfelder, and C. E. Schatz, 1971. Free-fall bottom sampler. Pat. no. 3,572,129, 4 sheets, *U.S. Patent Office*.

Walton, W. R., 1955. Ecology of living benthonic Foraminifera, Todos Santos Bay, Baja California. *Jour. Paleontol.*, v. 29, no. 6, pp. 952–1018.

Warme, J. E., 1969. Mugu Lagoon, coastal California: origin, sediments, and productivity. In *Coastal Lagoons*, A. A. Castañares and F. B Phleger, eds., Univ. Nacional Autonoma Mexico, pp. 137–154.

Warme, J. E., and N. F. Marshall, 1969. Marine borers in calcareous terrigenous rocks of the Pacific Coast. *Amer. Zoolog.*, v. 9, no. 3, edit. 2, pp. 765–774.

Warme, J. E., T. B. Scanland, and N. F. Marshall, 1971. Submarine canyon erosion: contribution of marine rock burrowers. *Science*, v. 173, no. 4002, pp. 1127–1129.

Washburn, A. L., 1947. *Reconnaissance Geology and Portions of Victoria Island and Adjacent Canada*. Geol. Soc. Amer. Mem. 22, 142 pp.

Wass, R. E., J. R. Conolly, and R. J. MacIntyre, 1960. Bryozoan carbonate sand continuous along Southern Australia. *Marine Geol.*, v. 9, pp. 63–73.

Watkins, N. D., and J. P. Kennett, 1971. Antarctic bottom water: major change in velocity during the Late Cenozoic between Australia and Antarctica. *Science*, v. 173, pp. 813–818.

Weber, K. J., 1971. Sedimentological aspects of oil fields in the Niger Delta. *Geol. en Mijnb.*, v. 50, no. 3, pp. 559–576.

Weeks, L. A., R. N. Harbison, and G. Peter, 1967. Island arc system in Andaman Sea. *Amer. Assoc. Petrol. Geol. Bull.*, v. 51, no. 9, pp. 1803–1815.

Wegener, Alfred, 1912. Die Entstehung der Kontinente. *Geol. Rundsch.*, v. 3, pp. 276–292.

Wegener, Alfred, 1924. *The Origin of Continents and Oceans*. Eng. trans., Dutton, New York, 212 pp.

Weidie, A. E., 1968. Bar and barrier island sands. *Trans. Gulf Coast Assn. Geol. Soc.*, 18th Ann. meeting, v. 18, pp. 405–415.

Wells, A. J., and L. V. Illing, 1964. Present-day precipitation of calcium carbonate in the Persian Gulf. In *Developments in Sedimentology*, Vol. 1, *Deltaic and Shallow Marine Deposits*, L. M. J. U. van Straaten, ed., Elsevier, New York, pp. 429–435.

Wells, J. W., 1954. Recent corals of the Marshall Islands, Bikini, and nearby atolls. II. Oceanography (Biologic). *U.S. Geol. Survey Prof. Paper* 260-I, pp. 385–486.

Wengerd, S. A., 1965. Salt tectonics of the Cuanza Basin, Angola, Portuguese West Africa. *Abs. Amer. Assoc. Petrol. Geol. Bull.*, v. 49, no. 3, pt. I of II, p. 336.

Wentworth, C. K., 1922. A scale of grade and class terms for clastic sediments. *Jour. Geol.*, v. 30, no. 5, pp. 377–392.

White, S. M., 1970. Mineralogy and geochemistry of continental shelf sediments off the Washington-Oregon coast. *Jour. Sed. Petrology*, v. 40, no. 1, pp. 38–54.

Wiegel, R. L., 1953. Waves, tides, currents and beaches: glossary of terms and list of standard symbols. *Council on Wave Res.*, Univ. California, 113 pp.

Wiens, H. J., 1962. *Atoll Environment and Ecology*. Yale University Press, New Haven, Conn., 532 pp.

Wilde, Pat, 1966. Quantitative measurements of deep-sea channels on the Cocos Ridge, east central Pacific. *Deep-Sea Res.*, v. 13, no. 4, pp. 635–640.

Wilhelm, Oscar, and Maurice Ewing, 1972. Geology and history of the Gulf of Mexico. *Geol. Soc. Amer. Bull.,* v. 83, pp. 575–600.

Wilson, J. T., 1962. Cabot Fault, an Appalachian equivalent of the San Andreas and Great Glen Faults and some implications for continental displacement. *Nature,* v. 195, pp. 135–138.

Wilson, J. T., 1965. A new class of faults and their bearing on continental drift. *Nature,* v. 207, pp. 343–347.

Wimberley, C. S., 1955. Marine sediments north of Scripps Submarine Canyon, La Jolla, California. *Jour. Sed. Petrology,* v. 25, no. 1, pp. 24–37.

Windisch, C. C., R. J. Leyden, J. L. Worzel, T. Saito, and John Ewing, 1968. Investigation of horizon beta. *Science,* v. 162, pp. 1473–1479.

Winslow, J. H., 1966. Raised submarine canyons: an exploratory hypothesis. *Ann. Assoc. Amer. Geographers,* v. 56, no. 4, pp. 634–672.

Winslow, J. H., 1968. Stopping at the water's edge: reason or habit? A reply. *Ann. Assoc. Amer. Geographers,* v. 58, no. 3, pp. 614–634.

Winterer, E. L., W. R. Riedel, et al., 1969. Deep sea drilling project: Leg 7. *Geotimes,* v. 14, no. 10, pp. 12–14.

Winterer, E. L., John Ewing, et al., 1971. Deep sea drilling project Leg 17. *Geotimes,* v. 16, no. 9, pp. 12–14.

Wiseman, J. D. H., and C. D. Ovey, 1950. Recent investigations on the deep-sea floor. *Proc. Geol. Assoc.,* v. 61, Pt. 1, pp. 28–84.

Wiseman, J. G. H., and W. R. Riedel, 1960. Tertiary sediments from the floor of the Indian Ocean. *Deep-Sea Res.,* v. 7, pp. 215–217.

Wollin, Göesta, D. B. Ericson, and Maurice Ewing, 1971. Late Pleistocene climates recorded in Atlantic and Pacific deep-sea sediments. In *The Late Cenozoic Glacial Ages,* K. K. Turekian, ed., Yale University, pp. 199–214.

Wong, H. K., E. F. K. Zarudzki, J. D. Phillips, and G. K. F. Giermann, 1971. Some geophysical profiles in the eastern Mediterranean. *Geol. Soc. Amer. Bull.,* v. 82, pp. 91–100.

Woodring, W. P., M. N. Bramlette, and W. S. W. Kew, 1946. Geology and paleontology of Palos Verdes Hills, California. *U.S. Geol. Survey Prof. Paper 207,* 145 pp.

Woodruff, J. L., 1970. A self-deactivating piston for a piston corer. *Ocean Engr.,* v. 1, Pergamon, pp. 597–599.

Worzel, J. L., 1959. Extensive deep-sea sub-bottom reflections identified as white ash. *Proc. Nat. Acad. Sci.,* v. 45, no. 3, pp. 349–355.

Worzel, J. L., and J. C. Harrison, 1963. Gravity at sea. In *The Sea,* Vol. 3, M. N. Hill, ed., Interscience, New York, pp. 134–174.

Worzel, J. L., and G. L. Shurbet, 1955. Gravity interpretations from standard oceanic and crustal sections. In *Crust of the Earth,* A. Poldervaart, ed., Geol. Soc. Amer. spec. pap. 62, pp. 87–100.

Worzel, J. L., R. Leyden, and Maurice Ewing, 1968. Newly discovered diapirs in Gulf of Mexico. *Amer. Assoc. Petrol. Geol. Bull.,* v. 52, pp. 1194–1203.

Wüst, G. 1958. Über Stromgeschwindigkeiten und Strommengen in der Atlantischen Tiefsee. *Geol. Rundsch.,* pp. 187–195.

Yamasaki, Naomasa, 1926. Physiographical studies of the great earthquake of the Kwanto District, 1923. *Jour. Faculty Sci., Univ. Tokyo,* sec. 2, v. 2, pt. 2, pp. 77–119.

Yerkes, R. F., D. S. Gorsline, and G. A. Rusnak, 1967. Origin of Redondo Submarine Canyon, Southern California. *U.S. Geol. Survey Prof. Paper 575-C,* pp. C97–C105.

Yerkes, R. F., H. C. Wagner, and K. A. Yenne, 1969. Petroleum development in the region of the Santa Barbara Channel. *U.S. Geol. Survey Prof. Paper 679-B,* pp. 13–77.

Yonge, C. M., 1940. The biology of reef-building corals. In *Great Barrier Reef Expedition 1928–29, Scientific Reports,* v. 1, no. 13, pp. 353–391.

Yonge, C. M., 1963. The biology of coral reefs. In *Advances in Marine Biology,* F. S. Russell, ed., Academic, New York, pp. 209–260.

Zarudzki, E. F. K., and Elazar Uchupi, 1968. Organic reef alignments on the continental margin south of Cape Hatteras. *Geol. Soc. Amer. Bull.,* v. 79, pp. 1867–1870.

Zeigler, J. M., 1959. Sedimentary environments on the continental shelf of northern South America. Abs. in *Preprints International Oceanographic Congress,* Amer. Assoc. Adv. Sci., Washington, D.C., p. 670.

Zeigler, J. M., 1964. The hydrology and sediments of the Gulf of Venezuela. *Limnol. and Oceanog.,* v. 9, no. 3, pp. 397–411.

Zeigler, J. M., and W. D. Athearn, 1968. The hydrography and sediments of the Gulf of Darien. *4th Caribbean Geol. Conf.,* Trinidad, pp. 335–342.

Zeigler, J. M., and S. D. Tuttle, 1961. Beach changes based on daily measurements of four Cape Cod beaches. *Jour. Geol.,* v. 69, no. 5, pp. 583–599.

Zeigler, J. M., W. D. Athearn, and H. Small, 1957. Profiles across the Peru-Chile Trench. *Deep-Sea Res.,* v. 4, pp. 238–249.

Zeigler, J. M., S. D. Tuttle, G. S. Giese, and H. J. Tasha, 1964. Residence time of sand composing the beaches and bars of outer Cape Cod. *Proc. 9th Conf. Coastal Engineering, Amer. Soc. Civil Engrs.,* pp. 403–416.

Zenkovitch, V. P., 1959. On the genesis of cuspate spits along lagoon shores. *Jour. Geol.,* v. 67, no. 3, pp. 269–277.

Zenkovich, V. P., 1967 (often spelled Zenkovitch). *Processes of Coastal Development.* Eng. trans. by D. G. Fry; J. A. Steers, ed., Wiley–Interscience, New York, 738 pp.

Zhivago, A. V., 1962. Outlines of southern ocean geomorphology. In *Antarctic Research,* Geophysical Monograph No. 7, H. Wexler, M. J. Rubin, and J. E. Caskey, Jr., eds., Amer. Geophys. Union, publ. no. 1036, pp. 74–88.

Zhivago, A. V., 1964. Tectonic and relief maps of the sea floor in the Southern Ocean. In *Antarctic Geology,* R. J. Adie, ed., North-Holland, pp. 715–724.

CONVERSION TABLES

Fathoms to Feet to Meters

Fathoms	Feet	Meters	Fathoms	Feet	Meters
¼	1.5	0.5	6½	39.0	11.9
½	3.0	0.9	6¾	40.5	12.3
¾	4.5	1.4	7	42.0	12.8
1	6.0	1.8	8	48.0	14.6
1¼	7.5	2.3	9	54.0	16.5
1½	9.0	2.7	10	60.0	18.3
1¾	10.5	3.2	11	66.0	20.1
2	12.0	3.7	12	72.0	21.9
2¼	13.5	4.1	13	78.0	23.8
2½	15.0	4.6	14	84.0	25.6
2¾	16.5	5.0	15	90.0	27.4
3	18.0	5.5	16	96.0	29.3
3¼	19.5	5.9	17	102.0	31.1
3½	21.0	6.4	18	108.0	32.9
3¾	22.5	6.9	19	114.0	34.7
4	24.0	7.3	20	120.0	36.6
4¼	25.5	7.8	30	180.0	54.9
4½	27.0	8.2	40	240.0	73.2
4¾	28.5	8.7	50	300.0	91.4
5	30.0	9.1	60	360.0	109.7
5¼	31.5	9.6	70	420.0	128.0
5½	33.0	10.1	80	480.0	146.3
5¾	34.5	10.5	90	540.0	164.6
6	36.0	11.0	100	600.0	182.9
6¼	37.5	11.4			

Statute Miles to Nautical Miles to Kilometers

Statute Miles	Nautical	Kilometers	Statute Miles	Nautical	Kilometers
¼	0.21	0.40	9	7.84	14.50
½	0.43	0.80	10	8.70	16.10
¾	0.65	1.21	20	17.40	32.20
1	0.87	1.61	30	26.16	48.40
2	1.74	3.22	40	34.80	64.50
3	2.61	4.84	50	43.50	80.50
4	3.48	6.45	60	52.20	96.50
5	4.35	8.05	70	61.00	113.00
6	5.22	9.65	80	69.60	129.00
7	6.10	11.30	90	78.40	145.00
8	6.96	12.90	100	87.00	161.00

GAZETTEER

Note: Names are not always given in figures.

Name	Approx. Latitude	Approx. Longitude	Figures
Aden, Gulf of, SW Asia	12°N	47°E	13–13
Agulhas Bank, South Africa	35°30′S	21°00′E	13–13
Aleutian Is., Alaska	52°N	160°W–177°E	14–4
Aleutian Tr., Alaska	50°N	150°W–177°E	13–7, 14–4
Amazon Est., Brazil	1°N	49°30′W	
Amur. R. mouth, Siberia	53°N	141°E	9–29
Andaman Is., SE Asia	12°N	93°E	9–35
Andaman Sea, SE Asia	12°44′N	95°45′E	9–35
Andros Is., Bahamas	24°30′N	78°00′W	11–17
Angola, W. Africa	14°15′S	16°00′E	
Apalachicola Delta, Florida	29°43′N	84°59′W	8–14, 9–13
Arabian Sea	16°N	65°E	13–13
Arafura Depr., Australia	12°S	129°E	9–33
Aricife Alacran, Yucatan	22°30′N	89°40′W	9–13, 12–17
Aru Is., Australia	6°20′S	134°00′E	9–33
Astoria C., Oregon	46°15′N	124°30′W	11–10
Asturias Coast, N. Spain	43°21′N	6°00′W	9–41
Azores Is., N. Atlantic	37°44′N	29°25′W	13–1
Bahama Is.	26°15′N	76°00′W	9–8, 11–17
Balearic Basin, Mediterranean	40°N	7°E	13–18A
Banda Sea, Australia	6°S	128°E	9–33
Banderas Bay, W. Mexico	20°38′N	103°25′W	13–6
Bangka Is., SE Asia	02°S	106°E	
Banquereau, Nova Scotia	44°30′N	58°30′W	9–2, 9–3
Barataria Bay, Louisiana	29°13′N	89°90′W	8–2, 9–13
Barbados Is., West Indies	13°30′N	59°48′W	13–17
Barents Sea	72°14′N	37°28′E	9–44
Barracuda Abyssal Pl., N. Atlantic O.	17°N	56°30′W	13–1
Barrow Sea Valley (Canyon)	72°30′N	154°W	13 16
Bass Strait, Australia	39°40′S	145°40′E	9–33
Belitung Is., Indonesia	3°30′S	107°30′E	13–13
Bell Is., Newfoundland	50°45′N	55°35′W	9–2
Bell Is., Strait of	51°21′N	55°56′W	9–2
Bengal, Bay of	17°30′N	87°00′E	13–15
Bering C., Alaska	54°N	168°W	11–11
Bering Sea, Alaska	60°N	175°W	11–11
Bermuda Is.	32°20′N	65°45′W	13–1
Bermuda Rise	32°30′N	65°W	13–1

Arch., archipelago; C., canyon; Chan., channel; Depr., depression; Est., estuary; Is., island; Penin., peninsula; Plat., plateau; R., river; Tr., trench; Tro., trough.

Name	Approx. Latitude	Approx. Longitude	Figures
Bikini Atoll, Marshall Is.	11°35′N	165°25′E	12–2
Biscay Plain, Atlantic O.	45°N	7°15′W	13–1
Blake Outer Ridge, Atlantic O.	28°–32°N	73°–76°W	9–8, 13–1
Blake Plat.	27°–32°N	77°–79°W	9–8
Block C.	40°N	71°16′W	9–4, 9–6
Block Is.	41°05′N	71°35′W	9–4
Bonaparte Depr., Australia	13°S	128°00′E	9–33
Borodino Is. (Daito)	26°N	132°E	13–5
Bothnia, Gulf of	61°N	19°E	9–42
Bora Bora Is., Society Is.	16°30′S	151°45′W	5–10
Borneo	0°25′N	112°39′E	5–10
Bougainville Is., Solomon Is.	06°S	155°E	5–10
Bowers Bank, Bering Sea	53°N	180°E	9–28
Brahmaputra Delta, Bangladesh	26°45′N	92°45′E	8–6
Brooks Penin., Vancouver Is.	50°N	128°W	13–7
Brown Bank, Nova Scotia	42°40′N	66°05′W	9–2, 9–3
Burwood Bank, Argentina	54°S	60°W	9–21
Buzzards Bay, Massachusetts	41°35′N	70°45′W	9–4
Cabo Rojo, E. Mexico	21°30′N	97°10′W	9–15
Cabot Strait, E. Canada	47°35′N	60°W	9–2
Cadiz, Gulf of, Portugal	36°50′N	7°W	12–1
California, Gulf of	25°N	110°W	13–5
Cambay, Gulf of, W. India	21°05′N	71°58′E	13–13
Campeche Escarp., Mexico	24°N	86°–88°W	9–15
Campeche Shelf, Mexico	22°N	90°W	9–13
Canada Basin, Arctic O.	80°N	145°W	13–16
Canary Is.	28°30′N	15°10′W	13–1
Canton Is.	2°50′S	171°40′W	13–1
Cap Blanc, West Africa	20°39′N	18°08′W	13–1
Cap Breton C., W. France	43°40′N	1°50′W	9–41
Cape Canaveral, Florida	28°30′N	80°23′W	9–4
Cape Espenberg, Alaska	66°34′N	163°40′W	
Cape Hatteras, North Carolina	35°15′N	75°24′W	7–8, 9–4
Cape Kennedy, Florida	28°30′N	80°23′W	9–8
Cape Krusenstern, Alaska	67°10′N	163°50′W	
Cape Mendocino, California	40°25′N	124°22′W	9–26
Cape Romain, South Carolina	33°01′N	79°23′W	7–18, 9–4
Cape St. Elias, Alaska	59°50′N	144°35′W	13–4, 14–4
Cape St. George, Florida	29°30′N	85°20′W	9–13
Cape San Blas, Florida	29°38′N	85°38′W	9–13
Cape San Martin, California	35°30′N	121°30′W	10–2
Cape San Roque, Brazil	5°30′S	35°W	
Cape Verde, West Africa	14°43′N	17°33′W	
Capricorn Chan., Australia	23°S	152°30′E	9–32
Cariaco Tr., Venezuela	10°37′N	65°10′W	13–17
Cariaco, Gulf of, Venezuela	10°33′N	63°47′W	13–17
Carlsberg Ridge, Indian O.	3°S	67°E	13–13
Caroline Ridge (Swell) Pacific O.	8°N	150°E	14–5
Carpentaria, Gulf of, Australia	14°45′S	138°50′E	9–33
Cascadia Chan., Pacific O.	46°N	128°W	12–8
Cayman Trough, Caribbean	19°N	80°W	13–17
Cedros Is., W. Mexico	28°10′N	115°10′W	13–5
Celebes Basin, SE Asia	4°N	122°E	5–10
Celtic Sea, Great Britain	50°–51°N	07°–10°W	9–42
Ceralbo Is., Baja California	24°15′N	109°50′W	
Chagos-Laccadive Ridge, Indian O.	0°–15′N	73°E	13–13
Challenger Deep, Pacific O.	15°N	147°30′E	13–5

Name	Approx. Latitude	Approx. Longitude	Figures
Chandeleur Is., Louisiana	29°53'N	88°35'W	8–2
Charlotte Harbor, Florida	26°47'N	81°58'W	9–14
Cherbourg Penin., W. France	49°39'N	1°43'W	9–41, 13–1
Chukchi Sea, Arctic O.	68°N	168°W	13–16
Clarion Fracture Zone	13°–18°N	117°–160°W	13–5, 13–6
Clipperton Fracture Zone	3°–10°N	105°–153'W	13–5, 13–6
Cocos Ridge, Pacific O.	5°30'N	86°W	13–5
Colorado R. Delta, Texas	29°N	96°W	8–7
Columbia R. Est.	46°20'N	124°W	13–8
Colville R. Delta, Alaska	70°30'N	150°30'W	6–11
Congo C.	6°S	11°50'E	11–13
Congo R. mouth	6°S	12°20'E	11–13
Cook Inlet, Alaska	60°50'N	152°W	13–7
Coral Sea Basin, Pacific O.	14°S	152°E	5–10
Coronado Is., W. Mexico	32°24'N	177°13'W	9–23
Corsair C., New England	41°15'N	66°10'W	9–4
Cortes Bank, California	32°30'N	119°10'W	9–22
Darwin, Australia	12°25'S	131°E	9–33
Darwin Rise, Pacific O.	20°–40°S	130°–150°E	13–5
Delgada C., California	40°N	124°W	9–26
Delgada Fan, California	38°N	128°W	9–26
Demerara Abyssal Pl., Atlantic O.	11°N	48°W	
De Soto C., Florida	29°N	87°30'W	9–13
Dixon Entrance, W. Canada	54°30'N	133°W	13–7, 14–4
Dogger Bank, North Sea	55°N	3°E	9–42, 10–6
Dover Strait (La Manche)	50°50'N	1°15'W	9–42, 10–6
Drake Passage, South America	57°S	65°W	
Dry Tortugas, Florida	24°40'N	82°50'W	9–13
Durban, South Africa	29°48'S	31°E	7–26, 13–13
East Caroline Basin, Pacific O.	5°N	145°E	13–5
East China Sea	30°N	125°E	9–31
Ebro Delta, Spain	40°41'N	0°45'E	13–18A
Elbe R. mouth, Germany	53°47'N	9°20'E	9–43
Eleuthera Is., Bahamas	25°05'N	76°10'W	11–17
Emperor Fracture Zone	40°–50°N	175°W–170°E	13–6
Emperor Seamounts, Pacific O.	30°–50°N	170°E	13–6
Eniwetok, Marshall Is.	12°N	162°E	12–2, 13–5
Eurasia Basin, Arctic	87°N	80°E	13–16
Falkland Is., SE South America	50°45'S	60°W	9–21
Farallon Is., California	37°40'N	123°W	
Fiji Is.	18°50'S	178°E	5–10
Flores Sea	7°09'S	120°30'E	5–10
Florida Keys	24°33'N	81°20'W	9–8, 12–16
Florida, Str. of	24°10'N	81°W	9–8
Fortymile Bank, California	32°39'N	117°58'W	9–22
Funafuti Atoll, Ellice Is.	8°S	179°E	12–6
Fundy, Bay of	44°50'N	66°05'W	9–2, 9–4
Gabes, Gulf of, Tunisia	34°N	10°30'E	13–18A
Galapagos Is.	1°S	91°W	13–5
Galicia Bank, NW Spain	42°40'N	11°45'W	12–1
Ganges Delta, Bangladesh	21°20'N	88°40'E	8–6
Ganges Trough (Swatch of No Ground)	21°15'N	89°28'E	11–22
Georges Bank, New England	41°15'N	67°30'W	9–4, 9–5
Georgia, St. of, Canada	48°56'N	123°06'W	14–4
Gilbert Is.	1°30'S	173°E	13–5
Golfo de Guayaquil, Ecuador	3°03'S	82°12'W	
Golfo San Jorge, Argentina	46°15'S	66°45'W	9–21

Name	Approx. Latitude	Approx. Longitude	Figures
Golfo San Matias, Argentina	41°30'S	63°45'W	9–21
Gorda Escarp., California	40°22'N	125°10'W	9–26
Gorda Rise, Pacific O.	41°N	127°W	13–7
Grand Banks, Newfoundland	45°N	53°W	9–2
Grand Isle, Louisiana	29°12'N	90°W	8–2
Gray Feather Bank, Caroline Is.	08°N	149°E	13–5
Great Abaco C., Bahamas	27°N	77°W	9–8
Great Abaco Is., Bahamas	26°N	77°20'W	11–17
Great Bahama C.	25°40'N	77°W	11–17
Great Barrier Reef, NW Australia	18°S	146°50'E	9–33
Great Kori Bank, W. India	23°N	68°E	13–13
Great Natuna Is., South China Sea	04°N	107°30'E	
Grenadine Is., West Indies	12°37'N	61°35'W	13–17
Guerrero Negro Lagoon, Baja California	28°N	114°W	8–16
Guinea, Gulf of	3°N	2°E	8–5
Guinea Ridge, S. Atlantic	8°S	0°	
Hawaiian Ridge	18°–31°N	153°W–172°E	13–5
Hecate St., Canada	53°34'N	130°53'W	13–7
Heron Is., Great Barrier Reef	23°25'S	151°55'E	9–33
Hogsty Reef, Bahamas	21°41'N	73°48'W	9–4
Hokkaido Is., Japan	43°30'N	143°45'E	9–29
Honshu Is., Japan	36°50'N	135°20'E	13–5
Horizon Guyot, Pacific O.	19°40'N	168°30'W	13–5
Huang Ho R. mouth, China	38°N	118°E	9–31
Hudson C.	39°27'N	72°12'W	9–4, 9–6
Humboldt Bay, California	40°48'N	124°25'W	
Hurd Deep, English Channel	50°N	2°W	9–42
Icy Cape, Alaska	70°20'N	161°40'W	
Indus Trough	23°N	67°E	11–22
Ionian Basin, Mediterranean	36°N	17°E	13–18A
Irish Sea	54°N	5°W	9–42
Irrawaddy Delta, Burma	16°N	95°E	9–35
Jamaica Is., Gulf of Mexico	18°N	77°30'W	13–17
Java Sea	5°S	110°30'E	13–13
Java Tr.	10°30'S	110°E	13–13
Jekyll Is., Georgia	31°03'N	81°25'W	9–8
Johnston Is., N. Pacific O.	16°45'N	169°32'W	13–5
Juan de Fuca Ridge, N. Pacific O.	45°49'N	130°W	5–4, 13–6
Juan de Fuca, Str. of	48°25'N	124°37'W	13–8
Kamchatka, Siberia	55°N	158°E	9–29
Kara Sea (Karskoye Mare), USSR	74°08'N	65°45'E	13–16
Karimata Is., South China Sea	1°08'S	108°10'E	
Kauai, Hawaii	22°09'N	159°15'W	13–5
Kermadec Is., New Zealand	30°30'S	178°30'W	5–10
Klamath R. mouth, California	41°35'N	124°05'W	
Kodiak Is., Alaska	57°24'N	153°32'W	14–4
Krakatau Volcano, SE Asia	6°S	106°E	13–13
Kuril Is., USSR	44°–52°N	148°–156°E	9–29
Kuril Tr., Pacific O.	42°52'N	145°–152°E	9–29
Kuskokwim R. Est., Alaska	60°45'N	162°W	8–7
Kutch, Gulf of, W. India	22°45'N	69°E	13–13
Kwajalein, Marshall Is., Pacific O.	9°15'N	167°30'E	12–4
Laguna Madre, Texas	25°–27°N	97°30'W	9–13
La Jolla C., California	32°53'N	117°17'W	11–2, 11–4
Lake Aheme, Dahomey, W. Africa	06°30'N	2°E	
La Perouse Bank, Canada	48°40'N	125°50'W	13–8
Lau Basin, Pacific O.	20°S	177°W	5–10

Name	Approx. Latitude	Approx. Longitude	Figures
Lau Is., Fiji	19°S	178°30'W	5–10
Lau Ridge, Pacific O.	21°S	178°30'W	5–10
Laurentian Chan., Canada	46°30'N	58°W	9–2
Line Is., Pacific O.	5°N	160°W	13–5
Lituya Bay, Alaska	58°40'N	137°45'W	14–4
Loire R. Est., W. France	47°25'N	02°W	9–41
Lomonosov Ridge, Arctic O.	88°N	140°E	13–16
Los Angeles R. mouth, California	33°46'N	118°14'W	9–24
Lyons, Gulf of, S. France	43°10'N	04°E	13–18A
Mackenzie R. mouth, Yukon Territory	69°N	135°W	13–16
Mafia Is., E. Africa	7°50'S	39°50'E	13–13
Magdalena R. Delta, Colombia	11°N	74°50'W	9–19, 13–17
Magellan Rise, Pacific O.	17°30'N	151°E	13–5
Maine, Gulf of	43°N	68°W	9–4
Malacca Str., SE Asia	2°N	102°E	9–35
Maldive Is., Indian O.	4°30'N	73°E	13–13
Manihiki Plat., Pacific O.	11°S	164°W	13–5
Maracaibo, Gulf of, Venezuela	9°55'N	71°30'W	13–17
Mariana Is.	15°N	146°E	13–5
Marianas Tr., Pacific O.	15°N	147°30'E	13–5
Martaban, Gulf of, S. Asia	16°34'N	96°58'E	9–35
Martha's Vineyard, Massachusetts	41°25'N	70°35'W	9–4
Mascarene Plat., Indian O.	11°S	60°E	13–13
Matagorda Bay, Texas	28°32'N	96°13'W	8–13
Mazatlan, W. Mexico	23°14'N	106°27'W	
Mendocino Fracture Zone	40°N	125°–165°E	13–5
Merrimack R., Massachusetts	42°49'N	70°44'W	9–4
Michaelmas Cay, Great Barrier Reef	16°40'S	146°E	9–33
Mid-America Tr., Pacific O.	17°N	86°–107°W	13–6
Midway Atoll	28°N	177°W	13–5
Mindanao Is., Philippine Is.	7°30'N	125°10'E	
Mississippi Tro.	28°30'N	89°45'W	9–13
Mogami Bank, Caroline Is.	08°30'N	149°E	12–6
Molokai Fracture Zone	18°–26°N	115°–154°W	13–5, 13–6
Monterey Bay, California	36°40'N	122°05'W	11–8
Monterey C., California	36°48'N	122°01'W	11–8
Moorea Is., Society Is.	17°30'S	150°W	5–10
Mugu C., California	34°03'N	119°05'W	
Mugu Lagoon, California	34°06'N	119°05'W	
Murray Fracture Zone	27°–34°N	123°–160°W	13–5, 13–6
Nankai Tro., E. Japan	31°N	134°E	
Nansei Shoto Is. (see Ryukyu Is.)			
Nantucket Is., Massachusetts	41°15'N	70°05'W	9–4
Nares Deep, Atlantic O.	23°30'	63°W	13–1
Narragansett Bay, Rhode Island	41°20'N	71°15'W	
New Caledonia	22°S	165°E	5–10
New England Seamounts	38°N	61°W	13–1
Newfoundland Rise	40°40'N	48°W	9–2
Newport C., California	33°35'N	117°55'W	
Nicobar Is., Indian O.	8°N	94°E	9–35
Ninetyeast Ridge, Indian O.	13°N–10°S	91°–89°E	13–15A
Nome, Alaska	64°30'N	165°20'W	9–27
Northeast Chan., Gulf of Maine	42°N	66°W	9–4
Norton Sound, Alaska	64°N	163°W	8–7
Oman, Gulf of	24°24'N	58°58'E	13–13
Orinoco Delta, South America	9°30'N	61°W	
Owen Fracture Zone, Indian O.	10°N	66°E	13–14

Name	Approx. Latitude	Approx. Longitude	Figures
Padre Is., Texas	27°09′N	97°15′W	8–11, 9–13
Palk St., India	10°N	79°23′E	13–13
Palm Beach, Florida	26°43′N	80°03′W	9–8
Palos Verdes Hills, California	33°48′N	118°24′W	9–22
Pamlico Sound, North Carolina	35°10′N	76°10′W	7–18
Panama, Gulf of	7°45′N	79°20′W	13–17
Paria, Gulf of	10°33′N	62°14′W	9–20
Persian Gulf	27°38′N	50°30′E	13–13
Phuket, Thailand	7°57′N	98°19′E	9–35
Pohai, Gulf of, China	38°N	120°E	9–32
Point Arena, California	38°57′N	123°40′W	9–26
Point Barrow, Alaska	71°20′N	156°W	13–16
Point Conception, California	34°27′N	120°28′W	10–29
Point Dume, California	34°N	118°50′W	
Point Franklin, Alaska	70°53′N	159°W	
Point Hope, Alaska	68°20′N	166°49′W	7–19
Point Reyes Penin., California	38°N	123°W	
Point Sur, California	36°18′N	121°55′W	11–18
Pondicherry, E. India	11°58′N	79°48′E	13–13
Porcupine Bank, Ireland	53°20′N	13°30′W	
Portales Terrace, Florida	24°35′N	82°20′W	9–13
Port Arthur, Texas	29°52′N	93°59′W	9–13
Port Royal, Jamaica	17°50′N	76°45′W	13–17
Pribilof Is., Alaska	57°N	169°20′W	11–11
Puerto Angelo, W. Mexico	15°42′N	96°32′W	13–6
Puerto Rico Tr., Atlantic O.	20°N	66°W	13–1
Puget Sound, Washington	47°49′N	122°26′W	14–4
Punta Gorda, California	40°20′N	124°15′W	9–26
Qatar Penin., Persian Gulf	25°N	52°45′E	13–13
Queen Charlotte Is., Canada	53°40′N	132°50′W	14–4
Queen Charlotte Sound, Canada	51°N	129°W	14–4
Raroia Is., Tuamotu Arch.	16°S	142°30′W	
Rat Is., Aleutians	51°35′N	178°15′E	9–28
Redondo C., California	33°49′N	117°28′W	9–22
Rewa R., Fiji	18°10′S	178°40′E	
Rhone Delta, S. France	43°30′N	04°50′E	13–18A
Rio Balsas C., W. Mexico	18°N	102°W	13–6
Rio de la Plata, Argentina	34°30′S	58°W	9–21
Rio Grande Delta, Texas	25°55′N	97°15′W	8–11, 9–13
Rio Grande Rise, Atlantic O.	31°S	35°W	
Rockall Bank, Atlantic O.	57°30′N	13°50′W	
Romanche Deep, Atlantic O.	0°	17°W	13–3
Romanche Fracture Zone	0°10′S	18°W	13–3
Rongelap Atoll, Marshall Is.	11°30′N	166°45′E	12–6
Ryukyu Is., Pacific O.	26°N	126°E	9–31
Sable Is., Nova Scotia	43°55′N	59°50′W	9–2, 9–3
Sable Is. Bank	43°45′N	60°30′W	9–2, 9–3
Sagami Bay, Japan	35°10′N	139°25′E	13–5
St. Georges Chan., Great Britain	51°45′N	6°30′W	9–42
St. Lawrence Is., Alaska	63°10′N	171°W	9–27
St. Matthew Is., Alaska	60°25′N	172°10′W	9–27
St. Paul Is., Cabot Str., Canada	47°14′N	60°08′W	9–2
Samar Is., Philippine Is.	11°30′N	125°07′W	
San Clemente Is., California	33°02′N	118°36′W	9–22
San Cristobal Bay, Baja California	27°20′N	114°35′W	13–6
San Diego Tro., California	32°40′N	117°35′W	9–22
San Jose C., Baja California	22°45′N	109°35′W	11–6

Name	Approx. Latitude	Approx. Longitude	Figures
San Lucas C., Baja California	22°45'N	109°50'W	11–6
San Miguel Is., California	34°03'N	120°23'W	6–13, 9–22
San Pablo Bay, California	38°04'N	122°25'W	8–19
Santa Catalina Is., California	33°29'N	118°37'W	9–22
Santa Monica Bay, California	33°50'N	118°40'W	9–22
Santa Ynez R. mouth, California	34°35'N	120°35'W	10–2
Sarawak, Borneo	03°N	113°E	
Saya de Malha Bank, Indian O.	10°30'S	61°30'E	13–14
Sea of Azov, USSR	46°N	36°20'E	
Sebastian Vizcaíno Bay, Baja California	28°45'N	115°W	13–6
Seoul, S. Korea	37°35'N	127°03'E	9–31
Severnaya Zemlya Is., Arctic O.	79°N	97°E	13–16
Seychelles Is., Indian O.	5°20'S	55°10'E	13–14
Seymour Narrows, W. Canada	49°50'N	125°W	14–4
Shatsky Rise, Pacific O.	32°30'N	158°E	13–5
Sheba Ridge, Indian O.	13°N	49°E	13–13
Shelter Cove, California	40°03'N	124°04'W	9–26
Shetland Is., Scotland	60°35'N	1°10'W	9–42
Sicily, Str. of, Mediterranean	37°N	12°E	13–18A
Sigsbee Knolls, Gulf of Mexico	23°45'N	92°25'W	9–13
Sigsbee Scarp, Gulf of Mexico	26°N	92°30'W	9–13
Skagerrak Chan., Norway	58°N	9°E	9–42
South China Sea Basin	15°N	115°E	5–10
South Honshu Ridge, E. Japan	24°N	142°E	9–5
Sulu Basin, Indochina	8°N	121°30'E	5–10
Suo Nada, Japan	33°50'N	131°30'E	8–20
Susitna R. mouth, Alaska	61°20'N	150°50'W	13–7
Susquehanna R. mouth, Pennsylvania	39°40'N	76°10'W	8–18
Swain Reefs, Australia	21°40'S	152°15'E	9–33
Swiftsure Bank, W. Canada	48°33'N	125°W	13–8
Tahiti Is., Society Is.	17°S	149°W	5–10, 13–6
Tampa Bay, Florida	27°35'N	82°38'W	9–13
Tanner Bank, California	32°42'N	119°08'W	9–22
Tartary, Gulf of, Siberia	47°N	140°30'E	9–29
Tatar Str., Siberia	51°N	141°45'E	9–29
Terrebonne Is., Louisiana	29°05'N	90°30'W	9–13
The Gully, Nova Scotia	44°15'N	59°15'W	9–2, 9–3
The Wash, England	53°N	0°20'E	9–42
Tijuana R. mouth, California	32°33'N	117°07'W	9–23
Timbalier Is., Louisiana	29°05'N	90°25'W	8–2, 9–13
Tinian Is., Mariana Is.	14°58'N	145°38'E	13–5
Tobago Basin, Caribbean	12°30'N	60°30'W	13–17
Todos Santos Bay, Baja California	31°40'N	116°40'W	
Tokyo C., Japan	35°N	139°30'E	13–5
Tonga Ridge, Pacific O.	15°–27°S	180°–173°W	
Tonga Tr., Pacific O.	16°–26°S	176°–171°W	
Tonkin, Gulf of, North Vietnam	20°30'N	108°10'E	9–32
Torres Str., Australia	10°30'S	141°30'E	9–33
Tortuga Is., Caribbean	10°55'N	76°18'W	13–17
Toulon, S. France	43°09'N	5°54'E	11–15, 13–18A
Trinidad Is., S. Atlantic O.	20°30'S	29°W	
Trinidad Is., West Indies	10°30'N	61°30'W	9–20
Truk Is., Caroline Is.	07°N	152°E	12–11
Tsugaru Str., Japan	41°25'N	140°20'E	9–29
Tsu Shima Is., Korea Str.	34°30'N	129°20'E	9–31
Tufts Plain, Pacific O.	47°N	140°W	13–7
Tunis, Gulf of, N. Africa	37°06'N	10°43'E	13–18A

Name	Approx. Latitude	Approx. Longitude	Figures
Tutuila, Samoa	14°16′S	170°45′W	5–10
Tyrrhenian Sea, Mediterranean	40°N	12°E	13–18A
Vema Fracture Zone, Pacific O.	11°–12°N	38°–49°W	13–1
Vizcaíno Bay (see Sebastian Vizcaíno)			
Vladivostok, Siberia	43°06′N	131°47′E	9–29
Walvis Ridge, Atlantic O.	20°–40°S	10°E–12°W	
Wreck Is., Great Barrier Reef	22°17′S	155°25′E	9–33
Yakutat Bay, Alaska	59°40′N	139°35′W	13–7, 14–4
Yangtze Kiang R. mouth, China	32°N	121°10′E	9–31
Yaquina Bay, Oregon	44°35′N	124°05′W	
Yucatan Penin., Mexico	20°45′N	89°W	9–15
Zambezi R. mouth, E. Africa	18°50′S	36°10′E	13–13
Zanzibar Is., E. Africa	6°S	39°30′E	13–13
Zhemchug Canyon, Bering Sea	58°N	175°W	11–11

Index

Abdel-Gawad, M., 394
Aberdeen, E., 176
Acoustic transponder, 312
Acropora, 350, 364, 368
    related to reef margins, 347
Adams, K. T., 11
Adriatic Basin, 263
Adriatic Sea, 363
    coasts of, 263
    sediment map of, 264
Aerial photographs, 56, 143, 151, 160
Africa
    coastal trend of, 261
    East, rifting of, 90
        continental slope off, 259
        rift valleys off, 89, 300
    South, continental shelf off, 259
    West, 98
        barriers of, 141
        continental terrace off, 259–261
African plate, 403
Afro-trailing-edge coasts, 104, 106, 199, 258, 259
Agassiz, A., 217, 343
Aguijan Island, Marianas, 122
Aguja Canyon, Colombia, 228
Agulhas Bank, 259
Agulhas Plateau, 395
Ahermatypic corals, 214, 346
Airview, Inc., Long Beach, N.J., 58
Akutan Island, Aleutians, 319
Alabama
    continental shelf off, 221–222
    continental slope off, 222
Alacran Reef, Yucatan, 367, 368
Alaska, 149, 172, 273
    coast of, 141
    continental shelf off, 242, 273
    continental slope off, 242
    earthquake of 1964, 54, 57, 108, 109, 242
    Peninsula, 242
    Range, 382
    rivers of, 292
    tsunamis of, 53
    volcanoes of, 411
Albatross, 2
Alboran Basin, 425
Aleutian Abyssal Plain, 382
Aleutian Arc, 244, 245, 319, 389
Aleutian continental shelf, 284
Aleutian Islands, 242, 245, 319
Aleutian Peninsula, 241
Aleutian Trench, 242, 245, 382, 390
    earthquakes in, 54
Alexa Bank, Fiji Islands, 354
Alexandria, Egypt, shelf off, 263
Algal mat, 176, 187
Algeria, 263

Alinat, J., 402
Allen, J. R. L., 141, 167, 167, 259, 260, 261
Alpha, Tau Rho, 234, 246, 316, 318
Alphard Banks, 259
Altemüller, H. J., 28
Aluminaut, 34, 218
Alvin, 33, 33, 34, 325
Amacuro, Venezuela, 169
Amazon Estuary, 195, 229
Amazon River, 61, 169, 192, 195, 229
American Association of Petroleum Geologists, 166, 176, 206, 213
American Geophysical Institute, 197
American Petroleum Institute Project 51, 3, 218
American plate, 93, 241
Amero-trailing-edge coasts, 104, 106, 199, 200
Amur River, USSR, 244
Andaman Islands, 255
    diagram of, 256
Andaman Sea, 255
    diagram of, 256
Anderson, R. S., 24
Andes Mts., 100, 229
    earthquakes of, 94
Andesite line, 388
Andrée, K., 2
Andresen, A. S., 22
Andrews, J. E., 218, 326, 327, 327, 378
Andros Island, 368
Angola, West Africa, 230
Antarctic Rise, 91
Antarctica, 293
    beaches of, 129
    continental shelves off, 284
        deepest in the world, 259
    continental terraces off, 275
    ice cap, melting of, 106
    ocean sediments of, 412
    rifting of, 90, 255
    saline lakes of, 101
Anticlinal ridges, 226, 227
Anticline, sedimentary, 218
Anticosti Island, Gulf of St. Lawrence, 202
Anti-dunes, 86
Antoine, J. W., 220
Apalachicola Bay, Fla., sediments of, 180, 181
Apalachicola Delta, Fla., 180, 220, 221
Appalachee Bay, Fla., 218
Appalachian Mts., system, fit of, 98
Arabia, continental shelf off, 258
Arabian-American Oil Co., 346
Arabian Desert coast, 258
Arabian Sea, 279
Arafura Sea Depression, 252
Aragonite, 182

Arcata Bay, Calif., 185
Archimede, bathyscaph, 33
Archipelagic aprons, 385
Arctic
  beaches, 129
  ice-free hypothesis of, 418
Arctic Coastal Plain, 172
Arctic Mid-Ocean Ridge, 397
Arctic Ocean
  continental shelves of, 292
  continental slopes of, 274
  continental terrace of, 273–275
  physiographic provinces of, *398*
  topography of, 397
Arcuate islands, 242
Arcuate ridges, 388–389
Arcuate trenches, 388–389
Argentina
  continental shelf off, 231
  sedimentary basins off, 231, *232*
Aricife Alcaran, 226
Arnborg, L., 172
Arrhenius, G., 3, 407, 408, 410, 411, 413,
  414, 415
Aru Islands, off New Guinea, 252
Ascension Fan Valley, Calif., 315
Aseismic ridges, 98, 395, 397
Ash fall, worldwide, 411
Asia, southern, continental terrace off, 253–
  258
Askinuk Mts., Alaska, 169
Asphalt domes, 237
Asthenosphere, 89, 98, 429
Astoria Channel, Ore., *384*
Astoria Fan Valley, Ore., 317, 318
Astoria Submarine Canyon, Ore., 190, 315–
  319, *318*, 330
Atchafalaya River, La., sediment deposits of,
  165
Athearn, W. D., 229
Atlantic Basin, 376
Atlantic Ocean
  deep-sea channels of, 376
  fit of sides, 97
  fracture zones of, 372–376
  North. *See* North Atlantic Ocean
  origin of, 90
  South, drilling in, *423*
  topography of, 371–378
  trenches of, 376–377
Atlantis, lost continent of, 372
Atlantis I, 3, *4*, 28
Atlantis II, 3, *5*
Atlas Mt. coast, Africa, 263
Atolls, 348–358
  cross-section of, *350*
  development stages of, 12–14
  drillings into, 357, *357*
  drowned, 354
  emerged, 354
  lagoons of, 175, 354, 360–361
    average depth of, 350, *351*
  origin of, 361
  sea-floor slopes outside of, 354–355

  terraces of, 360–361
  unusual, 352–354
Atterberg limits, 294
Atwater, T. M., *92*, 380, 382
Austral-Gilbert-Marshall Ridge, 98
Austral Ridge, 383
Australia, 252–255
  barrier beaches of, 141
  continental shelves off, 252
    structure of, *253*
  continental slopes off, 255, 279
  continental terrace off, 252–255, 293
  coral off, 293
  marginal basins off, 292
  rifting of, 90, 255
Aves Ridge, 425
Ayala-Castañares, A. *See* Castañares, A. A.
Aymé, J. M., 261
Azores Islands, 374
  tsunami effects of, 54

Backwash, 161
Baffin Bay, 376, *377*
  formation of, 90
Bagnold, R. A., 68, 84
Bahama Banks, 187, 214–218, *219*, 327, 368
  sea-floor spreading, caused by, 97
Bahama block, 299
Bahama Outer Ridge, 413, 422
Bahama Submarine Canyon. *See* Great Ba-
  hama Submarine Canyon
Bahamas, 329
  beaches of, 132
  reefs of, 366, 367
  sand waves off, 271
Baie, L., 277
Baird Inlet, Alaska, 169
Baja California, Mex.
  continental shelves off, 233
  dating of, 293
  fault valleys of, 340
  lagoons of, 182
  rifting of, 380
  submarine canyons off tip of, *312*
  western coast of, 233–234
Balearic Basin, 402, 426
Ball, M. M., 218, 227
Ballard, R. D., 205
Baltic Sea, sediments of, *270*
Baluchistan, South Asia, shelf off, 258
Banda Sea, 252
Banderas Bay, Mexico, 233
Bangka Island, Java Sea, 251
Bangladesh, 168
Banquereau, Nova Scotia, 203, *203*, 284, 325
Banyuls, France, 64
Bar fingers, 165–166
Barataria Bay, La., 176–180, 286
  currents of, 64
Barbados Island, West Indies, 417
Barents Sea, 273, 292
  glacial erosion topography of, *274*
Barnegat Inlet, N.J., *58*
Barr, K. W., 23
Barracuda Abyssal Plain, 376

Barrier islands, *104,* 106, 134, 141, *163,* 165, 176, 182, 187, 286
Barrier reefs, coral, 358–360, 361
  submerged, 214
Barriers, 103, *148,* 165
  Arctic, 141
  beach, 103, 141, 151, *155*
  definition of, 103
  drowned, 210, 286–287
  flats, *145*
  formation of, 143
  islands
    recent, *142,* 143
    reefs, 358–360
    See also Barrier islands
  lagoons of, 185, 360
  origin of, 141, *141*
  prograding, 284
  related to deltas, 145
  spits, 163
  types of, *104*
Barrow Sea Valley, 273
Bars
  crescentic, 127, *128*
  cuspate, 128
  definition of, 103
  development of, *166*
  fingers of, *166*
  longshore, *145*
Basalt, serpentinized, 376
Bascom, W., 124
Basement ridges. *See* Continental shelves
Bass Strait, 252, 254
Bathyal, definition of, 198
Bathyscaph
  Archimede, 33
  Trieste, 306
  See also Deep-diving vehicles
Bathythermograph, 66
Bay of Bengal, 168, 255, 279, 393, 418
  valleys of, 334
Bay of Fundy, 204–205
  tides of, 60
Bay of Pigs. *See* Gulf of Batabano
Beach, 123–161. *See also* Beaches; Sand
  backwash marks, 158, *158,* 160
  bars, 136
  berm, 134, 143, 157
  compartments of, 140, *140*
  cusps on, *151,* 157
  cycles, 134–137, *135*
    grunion related to, 136
  development, mechanics of, 129, 132
  erosion of, 153, *153*
  foreshore, 134
  permeability of, 127
  ridges on, *144, 145,* 149
    elevated gravel, *131*
  rill marks, 161
  ripple marks, 157, 159, *159*
  sands, composition of, 132–133
  scarp, 134
  storm, 143, *144*
  stratification of, 156, *157*
  structures and markings, 156–161

swash marks, *160,* 161
Beach-dune ridges, 143, *145,* 149
  drowned, 287
Beach Erosion Board, 44, *51,* 123, 124, 132
Beaches, 123–161. *See also* Beach
  Arctic, 129, 301
  black sand, 133
  classification of, 126
  cove, 129, *130,* 133
  crescentic, 129
  cuspate, *138*
  engineering structures, related to, 152–• 156
  estuaries, related to, 133
  green sand, 133
  sand, composition of, 132–133
  terminology of, 124–126
  volcanic sand, 132–133
Beagle, H.M.S., 342, 361
Beal, M. A., 81
Beard, C. N., 315
Beaufort Sea, 274
Beckmann, J. P., 402
Bedford Institute drill, 23
Belderson, R. H., 268
Belitung Island, Java Sea, 251
Bell Island, Conception Bay, Newfoundland, 303
Ben Franklin, 34
Bender, M. L., 409
Bengal, Bay of. *See* Bay of Bengal
Bengal Fan, 41, 395, *396, 397*
Benguela Current, 61, 286
Benioff zone, *93,* 389
Bentor, Y. K., 263
Berger, W. H., 408, 410
Berggren, W. A., 422
Beria, Africa, 258
Bering Fan Valley, *320,* 321
Bering Sea, 169, 389
  canyons of, 319–321, *319*
  continental shelf off, 243, 319
  sediments of, 244
  structural contours of, *247*
  continental terrace of, 242–252, *247*
  marginal basins of, 292
Bering Strait, 149, 150, 244, 273, 417
Bering Submarine Canyon, 318, *320*
Bermuda Islands
  drilling into, 355, 357
  land geology of, 352
  resemblance to atolls, 352
Bermuda Lagoon, 354
Bermuda Rise, 371
Bernard, H. A., 141
Berry, R. W., 258
Berthois, L., *192, 193,* 265, 268, *269,* 271
Betic Block, 265
Bezrukov, P. L., 22, 244, *246–247*
Biarritz, France, 268
Biggs, R. B., 188
Bikar Atoll, Marshall Is., *348*
Bikini Atoll
  deep borings in, 343, 355, 357, 358, 363
  substructure of, 355, *356*

Bikini Lagoon, 355, 361
Bimini Island, escarpment east of, 218
Biogenic carbonate detritals, 192
Biogenous sediments, 405, 410–411
Bioherms, 284
Bioturbation, 309, 312, 412
Bird, E. C. F., 111, 124, 141
Birdfoot Delta, Miss., 145, 165, *165*, 167, 176, 222, 224
Biscay Plain, 376
Biscaye, P. E., 409
Bissagos Delta, Africa, 261, *262*
Bissett, L. B., 273
Black, M., 268
Black Sea, 403
Blackett, P. M. S., 89
Blackman, A., 24, 415
Blackwell, D. D., 380
Blake-Bahama Basin, 378
Blake Escarpment, 378
Blake Plateau, 34, 63, 214–218, *215*, 259, 283, 286
    coral reef of, 346
    margin of, *282*
Blanc, J.-J., 263
Blanco Fracture Zone, 382
Block Channel, 210
Block Island, slope off, 212, 325
Block Submarine Canyon, 325
Bloom, A. L., 103, 108, *108*, 346
Board of Geographic Names, 315
Boerma, J. A. K., 21
Böggild, O. B., 2, 274
Boillot, G., 192, 193, 268, *282*
Bonaparte Bay, 252
Bonaparte Depression, 252
Bonatti, E., 408, 410
Bonneville Dam, of Columbia River, 190
Boomer, 40
Bora Bora, Society Is., 358
    lagoon of, 360
Boring organisms, 306, 308
Borneo, 251
Borodino Island (Kito-Daito-Jima), 357
Bougainville Island, 252
Bougainville Trench, 390
Bougis, P., 64
Bouma, A. H., 18, 21, 28, *29*, 172
Bourcart, J., 3, 192, 193, 264, *323*
Boutan, L., 34
Bowen, A. J., 87
Bowers Ridge, *245*, 389
Bowin, C. O., 372, *374*, 376
Box corers (Kastengreifer), 23–25, *24*
Box cores, *24*, 28, *29*, 312, 327
    from La Jolla Submarine Canyon, *311*
Boyd, D. W., 227, *367*, 368
Bradley, W. H., 3, 404, 407, 412, 415
Brahmaputra River, 168
Bramlette, M. N., 3, 358, 404, 407, 412, 415
Brazil, 98, 376, 420
    barriers of, 141
    continental shelf off, 230
    continental slope off, 229–231
Brazos River, Tex., 172

Breakers, 44–49. *See also* Waves
    heights of, 47
    plunging, *46*, 47
    predictions of time, heights, 54
    spilling, *46*, 47
    surf beat, 47
    surging, *46*, 48
    types of, *47*
Breakwaters, 153, *154*
Brenot, R., 268, *269*
Bretz, J. H., 190
Bricker, O. P., 354
Brines, 182, 187
British Columbia
    continental shelf off, 241–242, 283
    continental slope off, 242
    mountains of, 190
    University of, 3
British Isles
    continental shelf off, 268, 271
    continental slope off, 298
British Museum of Natural History, 405
Broecker, W. S., 416, 417, 418
Brooks Peninsula, 241–242
Brooks Range, Alaska, 172
Brown (red) clay, 96, 406, 429
    deposition rate of, 422
Bruun, A. F., 390
Bruun, P. M., 163
Bryan, E. H., Jr., 350
Bryant, W. H., 172
Bryant, W. R., 226, *227*
Bryozoans, 75, *79*, 263
Bucher, W. H., 329
Buddington, E. C., 242, 306, *310*, 331
Bullard, E. C., 3, 8, 42, 89, 97, *97*, 261
Bunce, E. T., 376
Bungo Strait, Sea of Japan, 64, 244
Bureau of Commercial Fisheries, 210
Bureau of Ships, 7
Buried valleys, 210
Burk, C. A., 398
Burke, J. C., 23
Burma, *113*
    continental shelf off, 255
Burns, R. E., 429
Burrowing organisms, as modifying beaches, 161
Burwood Bank, 231
Butler, L. W., 230
Buzzards Bay, Mass., 188
Byrne, J. V., 147, 190, 241

Cable breaks, 66, *322*, 330, 331, *331*
Cabo São Roque, Fla., 106, 226
Cabot Strait, 202, 203
Caera Rise, 424
Calcarenites, 75, 252
Calcium carbonate, production in coral reefs, 347
Calcutta, India, 168
Caldwell, L. T., 176
Caledonian Range, fit of, 98
California
    coastal ranges truncated, *281*

California (*Continued*)
Northern. *See* Northern California
Southern. *See* Southern California
Calvert, S. E., 259, 286
Campbell grab sampler, 36
Campeche Escarpment, 226, 425
Campeche shelf, 368
Canadian Basin, 397
Canadian Hydrographic Service charts, 323
Canary Islands, off Africa, 261
Caño Manamo, of Orinoco River, 192
Canton Island, South Pacific, 352
Canyon walls, upbuilding of, 333–334
Cap Blanc, Africa, 261
Cap Breton, France, 49
Cap Breton Submarine Canyon, 268, 269
Cap Vert. *See* Cape Verde
Cape Cod, Mass., 133, *151*, 209, 292, *294*
Cape Espenberg, Alaska, 149
Cape Hatteras, N.C., 134, 148, 261, 286
continental shelf off, 200, 209–210, 212
continental slope, erosion of, 298
lagoons of, 150
Cape Kennedy (Cape Canaveral), Fla., 149, 212
Cape Krusenstern, Alaska, 149
Cape Lookout, N.C., 134
Cape Lookout Shoals, *288*
Cape Mendocino, Calif., 239
Cape of Good Hope, S. Africa, 259
Cape Range, S. Africa, 98
Cape Romain, S.C., 148
Cape St. Elias, Alaska, 242
Cape St. George, Fla., 221
Cape San Blas, Fla., 149, 220, 221, 286
Cape San Lucas, Baja Calif., Mex., 312
Cape San Martin, Calif., 239
Cape San Roque, Brazil, 229, 230
Cape Verde (Cap Vert), Africa, 150, 261
Cape Verde Basin, Africa, 408
Capricorn Channel, Australia, 252
Caputo, M., 402
Carbon-14 dating, *107*, 149, 261, 416
Cariaco Trench, 229, 425
Caribbean Basin, 229
Caribbean block, 389
Caribbean plate, 277
Caribbean Sea, 229, 398
drillings into, 425
sea-floor relief of, *399*
sea-floor spreading of, 97
Carlsberg Ridge, 91, 393, 395
Carlson, P. R., 192, 315, 318
CARMARSEL Expedition, 358, 361, 364
Carmel Bay, Calif., 315
Carmel Submarine Canyon, Calif., 239, *316*
Carnegie Ridge, 428
Carolina capes, 212
Carolina coastal plain, *148*
Caroline Basin, East, 427
Caroline Islands, 242, 358, 383
atolls of, 350, 354
Caroline Ridge, 427
Carpinteria, Calif., 237
Carrabelle, Fla., 180

Carrigy, M. A., 255
Carruthers, J. N., 39
Carsola, A. J., 273, 275
Carta Litologica Submarina, Portugal, 265
Cascadia Abyssal Plain, *384*
Cascadia Channel, 382, *384*
Castañares, A. A., 163, 174
Catastrophic waves, 53–59, *59*
hurricanes, 56–57, 147
landslide surges, 53, 57–59, *59*
storm surges, 53, 56–57
See also Tsunamis
Cedros Island, Baja Calif., Mex., 234
Celebes Basin, 389
Celtic Sea, 64, 268, 271
Central America, continental slopes off, 233
Central Indian Ridge, 393
Ceralbo Island, Gulf of Calif., 388
Ceylon, slopes off, 257
Chagos Fracture Zone, 393
Chagos-Laccadive Ridge, 393, 395
Chain, R/V, 394
Chain Fracture Zone, *375*
Challenger, H.M.S., 2, 405
Challenger Deep (Trieste), 33, 390
Challenger Expedition, 372, 410, 412
Chamberlain, T. K., 82, 140, 308
Chamberlin, R. T., 358
Chandeleur Islands, off La., 165
Charlotte Harbor, Fla., 180
Chart soundings, 11–12
Charybdis, 64
Chase, R. L., 376
Chase, T. E., 370, *378*
Chauveau, J.-C., 358
Chave, K. E., 347, 352
Cheniers, 127, 145–147, *146*, 169
Cherbourg Peninsula, 268
Cherts
deep-sea, 420
in deep-sea drilling, 429
Chesapeake Bay, 162, 182, 188, *189*, 210, 325
erosion rate of, 188
Chile, 283, 390
tsunamis of, 54
China
continental shelf sediments off, 248, *250*
rivers of, 246
Sea, East. *See* East China Sea
Chmelik, F. B., 28
Choee, G. V., 210
Chukchi Sea, *149*
continental shelf of, 273
sediments of, 273–274
Church, R., 46
Clark, D. L., 418
Clipperton Fracture Zone, 426
Cloud, P. E., Jr., 350, *350*
Coal Oil Point, Calif., 237
Coarse-fraction analyses, *178–179*, 182, *183*
Coastal classifications, Chap. 6
genetic, 109–111
of Johnson, 106, 111
of Ottmann, 109, *110*

Coastal classifications (*Continued*)
  of Shepard, 111–122, *112*
  tectonic, *105*
Coastal Engineering Research Center, 44
Coastal features, nomenclature for, 103
Coastal Research Group, Univ. Mass., 124
Coastal Studies Institute, Louisiana State
    Univ., 124
Coasts
  alluvial fan, *117*
  barrier, 121
  collision, 140
  coral reef, 122
  dendritic, 111
  dip, *121*
  downfaulted, 106
  drowned, 106, 111
    karst topography of, 112
  dune, *118*
  emergent, 106, 108
  fault, 119
  fold, 119
  fossil, 115, *118*
  glacial deposition, 115
  ice, 120
  land erosion, 111
  landslide, 115
  lava-flow, 118
  mangrove, 122
  marine deposition of, 121
  neo-trailing-edge, 104
  neutral, 106
  oyster reef, 122
  primary, 111
  ria, 111
  river deposition, 112
  rockslide, *119*
  secondary, 111, 120
  serpulid, 122
  subaerial deposition of, 112
  submergent, 106, 108
  tephra, 119
  trailing-edge, 104
  trellis erosion of, 111, *113*
  wave-eroded, 120
  wind-prograded, *117*
  upwarping of, 108
  volcanic collapse (explosion), 119
Cocodrie, of Miss. distributary, 165
Cocos Ridge, 383, 428
Coiba Ridge, 428
Cok, A. E., 203, *203*
Coleman, J. M., 176, 334
Collision zone. *See* Subduction zones
Colorado Delta, Tex., 172, *173*
Colorado River, Tex., 66, 233
Columbia Estuary, 190, 241
Columbia River, Ore.–Wash.
  floods of, 190
  mouth of, 190
  transportation of sand, 195
Columbia Submarine Canyon. *See* Astoria
    Submarine Canyon
Colville Delta, Alaska, *116*, 172
Conception Bay, Newfoundland, 303

Condon, W. H., 169
Congo, Africa, continental shelf off, 321
Congo Estuary, 321, *322*
Congo Fan Valley, 321, *322*
Congo River, 321
Congo Submarine Canyon, 259, 321, *322*
Connecticut River, estuary of, 188
Conolly, J. R., 202, 255, 416
Continental borderland, 197, *234*, 235, 244,
    261, 268, 271, 274, 297
  faults and folds on, *296*
  seismic profiles of, *296*
Continental drift, 89, 100
Continental margin, 233, 319, *391*
Continental rises, 198, 279, 323, 332
  distribution of, 298, *298*
  evolution of, 300
  mass movements on, *295*, 296
Continental shelves, 197–275, *282*, 301
  basement ridge on, *239*, 283
  characteristics of, 277–278
  coral growth on, 286–288
  deepest, 259
  economic minerals of, *301*
  glaciation of, 284
  landslide on, *239*
  margin of, *282*
  sediments on, *222*, *225*, 278
  sinkholes on, 212, 286
  structure contours of, *243*, *253*
  terrace on, broad, 258, *285*
  topography of, 235, 277–278, 284–291
    processes influencing, 284–291
  wide, 251
    explanation of, 292
Continental slopes, 204, 251, 252, 255, 257,
    259, 301, 323
  causes of, 298–299
  characteristics of, 279
  coral growth on, 294
  definition of, 197
  erosion of, 297
  irregularities, cause of, 294–298
  mass movements on, 294–296, *295*
  off Mississippi, 222
Continental terraces, Chap. 9
  basic types of, 279–284
  description of, 200
  economic resources of, 300
  off Gulf Coast, 218–226, *220*
  largest, 258
  off northeast United States, 200, *206*
  off northwest Africa, 261
  origin and history of, 276–303
  prograded, *283*
  off southeast Canada, *200*
  off southeast United States, *213*
  types of, 199, 279–284, *280*
Contourites, 411–412
Convection currents, 93
Cooper, L. H. N., 64
Copepods, 347
Coquina banks, 193
Coral
  Acropora, 347, 350

Coral (*Continued*)
  ahermatypic, 214, 346
  banks, 226, 230, 251
  bryozoans, 343
  coralline algae, 343
  fragments of, 132
  growth of, 347
  hermatypic, 343, 346
  hydrocorals, *343*
  largest in world, 252
  mounds, 214
  sands, 132
  Scleractinia, 346
  symbiotic flagellates, 346
Coral reefs, 227, 230, 248, 251, 252, 254–255, 257, 258, 293, Chap. 12
  algal sediments of, 218
  barriers, 180
  calcium production of, 347
  dams of, 283
  ecology of, 346–348
  elevated, 122
  framework of, 343
  geophysical prospecting of, 355
  growth of, 299
  lagoons of, 252. *See also* Lagoons
    average depth of, 350
    floors of, 352
    food supply in, 347
    sinkholes in, 361
    terrace levels in, *360*
  organic, defined, 343
  origin, hypothesis of, 361–363, *362*
  platforms, 368
  raised, 354
  related to sea-level rise, 106, 347, *366*
  salinity limits of, 346
  stages of development, 254, *364*
  surge channels of, 352
  topography related to, 284
  types of, 343, *344*
    atolls, 343
    barriers, 343
    faros, 343, 352, *353*
    fringing, 343
    knolls, 343
    patch, 368
    table, 343
Coral Sea Basin, 389
Cordell Bank, Calif., 238
Cores
  dating of, 416
  long, Quaternary sequence of, 415–418
Coring devices, 19–25
  box corer (Kastengreifer), 23, *24*, *311*
  core retainer, 23
  "flexoforage," 23
  gravity corer, 19, 21
  pinger, 20
  piston corer, 20–21, *20*, 22
  sphincter core retainer, 23
Coring techniques, 21, 26–28
Coriolis force, 61, 317
Coronado, Calif., barrier of, 182
Coronado Islands, Mex., 235

Coronado Strand, Calif., 153
  countercurrent off, 154
Corpus Christi Bay, Tex., 175, 176
Correns, C. W., 404
Corsair Submarine Canyon, 325
Corsican submarine canyons off western Corsica, 321, *323*, 329
Cortes Bank, Calif., 238
Cosmogenous, 405, 410
Cotton, C. A., 102, 111
Council on Wave Research, 123
Cousteau, J.-Y., 3, 31, 32, 33, 36, 306, *307*, 323, *324*
Cove beaches, 129, *130*, 133
Coyote Creek, San Francisco Bay, Calif., 190
Creager, J. S., 226, 244
Crescent City, Calif., tsunami damage at, 54
Cromwell, T., 61
Cromwell Current, 61
Cross-bedding, 172
Crossland, C., 358
Crowell, J. C., 100
Crust
  age, distance from E. Pacific Rise, 428
  continental, 369
  oceanic, 369
Cubmobile, 34
Cullis, C. G., 343
Curray, J. R., 6, *41*, *107*, 141, *144*, 197, 198, 224, *225*, 233, 239, *239*, 240, 277, 279, 283, 289, 291, 300, 312, 334, *335*, 341, 342, 358, 364, 395, *396*, *397*
Current meters
  Isaacs-Schick, 37
  pendulum, 37, *38*
  records from, *332*
  Savonius rotor, 37
Current ripples, 63, *158*, 321, 328
  asymmetrical, 332
Currents, ocean, 43–67, 422
  bottom, 63, 67
    deep, 370–378
    measurements of, 37–39
    related to permanent, 61
  cascading, 64, 333
  contour, *63*, 333
  convection, 93
  countercurrents, 61
  geostrophic, 411
  longshore, *50*, 51, 52
  nearshore, 50–53
  permanent, 61, *62*, 63
  rip, *50*, 51, 52, *53*, *156*
  scour marks of, *409*
  in submarine canyons, 67
  surface, *62*
  tidal, 64, 231
  turbidity. *See* Turbidity currents
  velocities of, *332*
  western boundary of, 377
  wind-drift, 61
Currents, proper names of, *62*
  Benguela, 61, 286
  Cromwell, 61
  Florida, 217

Currents (*Continued*)
Gulf Stream. *See* Gulf Stream
Humboldt, 61
Kuroshio, 61, 63, 248, 251, 291, 321
Labrador, 61, 202
North Equatorial, 165, 286, 291
Oyashio, 61
Pacific Equatorial Undercurrent (Crom-well). *See* Cromwell Current
Curry, D., 268
Cuspate bars, 128, *150*
Cuspate delta, *114*
Cuspate forelands, 121, 148–150, *148, 149,* 151, 212, 226, 284, 286, 287, *288*
Cuspate shorelines, 148
Cuspate spits, 148, 150–152, *150*
Cusps, giants, 148, *151,* 152

D'Abidjan, Africa, 187
D'Anglejan, B. F., 233
D'Arrigo, A., 263, 264
D'Ozouville, L., 268, *282*
da Vinci, Leonardo, 43–44
Daetwyler, C. C., 180
Dahomey, Africa, coast of, 185
Dalhousie University, 3
Daly, R. A., 66, 343, 362, 363
Damietta River, Egypt, 263
Dams of shelf margin, 292
Dana, J. D., 343, 362
Dangeard, L., 2, 268, 323
Darcy's law, 83
Darwin, C., 2, 342, 355
hypothesis of, 355, 361, *362, 363*
Darwin Rise, 385
Dating, 293. *See also* Carbon-14 dating
Davis, W. M., 343, 354, 358, 362, 363
Davis Sea, 275
Daytona Beach, Fla., 132
de Almeida, F. F. M., 98
de Beaumont, E., 141–143, *141*
de Leon, Ponce, 44
de Roever, W. P., 402
De Soto Submarine Canyon, 222
Debyser, J., 187
Decca, 12
Deep-diving vehicles, 31, 33–34, 40, 63, 294, 304, 306, 312, 327
development of, 3, 9
Deep sea
camera, *34*
channels of, 376, *377, 382, 383*
deposits of, 404–430
floor, erosion of, 413
nondepositional areas of, 412–413
sediments. *See also* Deep-sea sediments
distribution of, 413, *414*
stratigraphy, 404–430
Deep Sea Drilling Project, 3, 28, 31, 419–430, *421. See also* JOIDES
operational methods, *30, 31*
Deep-sea sediments, 405–413
authigenic, 413
biogenous, 405, 410–411
cherts, 412, 418, 420

classification of, 407
cosmogenous components of, 405
deposits of, 406–407
distribution of, *414*
erosion areas of, 413
feldspar in, 408
quartz in, 408
sources of, 405
Deep Star, 33, 306, 308, *309,* 312, 325
Deep tow
instruments, 9
methods, 15, *418*
records, 312
Degens, E. T., 394, 403
Delacour, J., 23
Delaware Bay, 210, 325
Delgada Submarine Canyon, Calif., trun-cated head of, *240*
Delmarva Peninsula, Chesapeake Bay, 182
Delta-front troughs, *335–336, 337*
description of, 334–337
origin of, *335,* 336–337
Deltaic foreset beds, 172
Deltaic sediments, 172
Deltas, 164–174
composition of, 172
cuspate, *114*
general characteristics of, 172–174
shifting channels of, 185
Demenitskaya, R. M., 397
Demerara Abyssal Plain, 376
Diamond Shoals, 212
Diapiric intrusions, 226, 284
Diapiric structures, 222
Diapirism, 294, 296
Diapirs, *223,* 226, 230, 283, 398, 402, 424
mud, 224, *228, 229,* 284, 336
Diastrophic movements, as submarine can-yon origin, 329
Diatomites, 75
Dibner, V. D., 397
Dietz, R. S., 8, 21, 33, 88, 89, *91,* 97, 98, 218, 238, 299, 300, 380, 389, 397
Dill, R. F., 31, 32, 63, 237, 251, *265,* 293, 306, 308, *311,* 312, *313,* 321, 328, 331, 332, *339, 341*
Dillon, W. P., 325
Dingle, R. V., 259
Dinkleman, M. G., 402
Discovery II, 3
Discovery III, 6
Diving Saucer (Soucoupe), 33, 306, *307,* 308, *314,* 323
Dixon, W. J., 71, 190
Dixon Entrance, 242
Dobrin, M. B., 40
Dogger Bank, 271
Dolan, R., 134
Dolet, M., 193
Donahue, J. G., 291
Donn, W. L., 416, 418
Dover Strait, 268
Dow, R. L., 159, *159*
Drake, C. L., 201, 202, 212, 239, 299, 300, 330, 374

Drake Passage, currents of, 63, 378
Dredging, methods of, 25–28
   using frame dredge, 25
   using pipe dredge, 25
Drowned barrier islands, 224, 286–287
Drowned beach-dune ridges, 287
Drowned valleys, 106, 163, 188, 193, 329
Dry Tortugas, 350
Duboul-Razavet, C., 264
Dunes, 86
   ridges of, *145*
   sand, 176
   underwater, *208, 219*
Dunham, J. W., *51*
Durban, S. Africa, 153, *155*, 259

E. W. Scripps, R/V, 3, *4*
Earthquakes, 330
   Alaska, in 1964, 54, 108, *109*, 242
   deep-focus, depth of, *92*
   epicenters of, 94, 239
   Grand Banks, 1929, 204, 330, 331, *331*
   Gulf of Kutch, 1819, 257
   Hawaii, in 1946, tsunamis from. See Tsunamis
   Japan, in 1923, 108, 340
   San Francisco, in 1906, 239
East Asia, continental terrace off, 242–252
East Caroline Basin, 427
East China Sea, 246
   structures of, *249*
East Pacific Rise, 233, 408, 426, *428*
   age of crust, *428*
   formation of, 91
   fracture zones of, 241, 379, 380, 381, 382
   heat flow from, 99
Eaton, J. P., 54
Ebro Delta, Spain, 264
Echinoids, 75, *160*, 176
Echo-sounders, narrow-beam, 13, 15
Echo-sounding surveys, profiles of, 9, 15, 271
Echo-soundings, 306
Edgar, N. T., 420, 422
Edgar Tobin Surveys, *145*
Edi Stephan, 2
Edmondson, C. H., 346
Eel River, Calif., sediments from, 185
Egeria, H.M.S., 413
Ekman, S., 23
Ekman, V. W., 66
Elbe River, 185
Electronic position finders, 12, 306
Eleuthera Island, Bahamas, 327
Ellice Islands, Gilbert Is., 355, 383
Elmendorf, C. H., 321
Elsasser, W. M., 390
Embley, R. W., 376
Emerald Basin, Scotian Shelf, 203, *203*
Emerita, 161
Emery, K. O., 2, 5, 8, *18*, 19, 21, 36, 40, 67, *107*, 161, 185, 197, 205, 212, 214, *215*, 220, 224, 226, 233, 235, 236, 237, 238, 246, 248, *249*, *250*, 251, 258, 263, 277, 279, 280, *281*, 283, *283*, 289, *291*, 295,

298, 301, *301*, *302*, 308, 315, 332, 338, 343, *345* *350*, 352, 354, 357, *360*, 361, 402
Emiliani, C., 415, 416
Emperor Fracture Zone, 384
Emperor Seamounts, 383, 384
Emperor Trough, 384
Engel, A. E. J., 372
Engel, C. G., 390, 395
Engineering structures, effects on beaches, 152–156
England, 147
English Channel, 197
   continental shelf off, 268, 281
   sand ridges in, 286, *287*, 288
   tides of, 60, 193, 271
Eniwetok Atoll, Marshall Is., *345*, 355, *356*
   deep borings in, 343, 355, 358, 363
   reef study of, 347
   stages of development, *364*
Ensenada, Baja Calif., Mex., 234
Eolanites, 352
Epstein, S., 132
Erickson, B. H., 384
Ericson, D. B., 325, 404, 406, 412, *414*, 416, *416*, 417
Escarpments, 241, 251, 268
   coral reef, origin of, 220
   greatest vertical extent of, 232
   off West Florida, 220
ESSA (now NOAA), 197
Estuaries, 188–195, 229, 230, 269, 271, 321
   definition of, 175
   general characteristics of, 194–195
   of New England, 188–190
   related to beaches, 133
   typical, *163, 189*
   of western France, 192–193
Eugeosynclines, 298
   buried, *299*, 300
Euphrates River, 258
Eurasian Basin, 397
Eureka, Calif., 185
Europe, western
   continental terraces off, 265
   seismic reflection profiles off, *266*
Eustatic sea-level changes, 103, 106–108. See *also* Sea-level changes
Evans, G., 193
Evaporite belts, 101, 182
Evaporites, historical distribution of, 101
Ewing, G. C., 182
Ewing, J. I., 40, 214, 293, *333*, 376, 385, 389, 397, 398, 418, *419*, 420, 424, 429
Ewing, M., 16, 34, 37, 63, 210, 229, 231, *232*, 330, *337*, 372, 374, 376, 385, 397, 398, 411, 416, 418, 424

Faeroe Islands, 273
Fail, J. P., 261, 374
Fairbridge, R. W., 163, 198, 255, 354
Falkland Islands, 231
Fan valleys, 306–327, *311*, 321, 323, 327
Farallon Islands, Calif., 238, 239, 283
Faros, *353*

Fathogram, *14*
Fault, 220, 242, *296*, *320*, 341
  on continental borderland, *296*
  scarps, 109, 229, 238
    margins of, *282*
  on shelf margins, 280
  transform, 92, *92*, *212*, 227, 382, 389
  troughs, 233, 321
Fecal pellets, 77
Ferromagnesians, 157
  nodules of, 411
Fiji Islands, 346
  atolls of, 350, 354
Filloux, J. H., *154*
Fink, L. K., 218
Fiords, 106, 109, 163, 188, 341
  deepest in world, 231
  stagnant conditions of, 163, 242
Fischer, A. G., 385, 426, 429
Fisher, R. L., 7, 26, 151, 233, 274, 370, 390, 391, *391*, 393, *394*, 395
Fisk, H. N., 106, 165, 166, 167, 175, 176, *177*, 222
Flemish Cap, 202
Flint, R. F., 100
Flores Sea, 252
Florida, 286
  continental shelf off, 283
  continental slope off, 279
  drilling off, *215*
Florida, East, 283
  continental shelf off, 214
Florida, Strait of, 63, 214, 217, 286, 425
Florida, West
  continental shelf sediments on, 218, *222*
  continental slope off, 222
  escarpment off, 220, 222, 226, 283
Florida Bay, 180
Florida Current, 217
Florida Escarpment, 398
Florida Keys, 180, 214, 348, *366*, 368
Florida Panhandle, shoals off, 211
FNRS-I, 33
Fold ridges, 229
Fold valleys, on sea floor, 340–341
Folds
  chevron, 424
  on continental borderland, 269–297, *296*
  slumping due to, 265, *297*
Folk, R. L., *75*
Foraminifera, 75, 76, 132, 238, 263, 325, 352, 415, 422
  benthonic, 76, *79*, 224
  horizons correlated by, *416*
  planktonic, 76, *79*, 224
  sand-sized, 412
Foreset beds, 281
Foresets, 172, 264
Formosa Strait, 248
Fortymile Bank, 303
Fossils
  mammoth bones submerged, 212
  shallow-water, 429
Fox, P. J., 412
Fracture zones, 91–92, 94, *95*, 240, *240*, 241,

261, 370, 374, 376, 379–382, *381*, 384, 424, 426, *428*
Frakes, L. A., 100
Frame dredge, 25–26, *25*
France
  continental shelf off, 268
  submarine canyons off southern, *324*
Francis-Boeuf, C., 192
Franklin, Benjamin, mapping of Gulf Stream, 44
Franklin Shore, 210, 286
Franz Joseph Islands, 273
Fraser, H. J., 81, *81*
Frautschy, J. D., 140, *140*
Fray, C., 231
Fredriksson, K., 410
Freeport, Tex., 172
French Mediterranean, submarine canyons in, 323
French Riviera, 264, 323
Freshwater peat, 289
Frey, H., 36
Frye, J. C., 412
Funafuti Atoll, borings into, 355, 358, 363
Funkhouser, J. G., 98
Funnell, B. M., 411, 412

Gaal, R., 237
Gagliano, S. M., 165
Gagnan, E., 31, 36
Gakkel, Ya. Ya., 397
Galapagos Islands, 428
Galathea, 390
Galicia, West, Spain, rias of, 193
Galicia Bank, 268
Galveston, Tex., 224
  hurricane of 1900, devastation of, *56*
  lagoons near, 176
Ganges-Brahmaputra Delta, 168, *168*, 255, 334, 395
Ganges River, 168, 255
Ganges Trough, 334
Gardner, J. V., 389
Gargano promontory, Italy, 263
Garman, R. K., 368
Garrison, L. E., *209*, 210
Gascony coast, sand-dune belt of, 268
Gealy, B. L., 226, 296
General Oceanographics, 34
Gennesseaux, M., 323
Geological Institute, 62
Geological Society of America, 370
Geophysical methods, 40–42
Georges Bank, New England, 204–205, *208*, *209*, 271, 281, 284, 325
  sediments of, 205
Geosynclines, 252
Geotimes, 420
German South Polar Expedition, 1901–1903, 2
Gibraltar, 54, 329, 376
Gibson, T. G., 218, 325, 333
Gibson, W., 220, *340*
Giddings, J. L., 149
Gierloff-Emden, H. G., 151

Giermann, G., 263, 402
Gilbert, G. K., 86, *141*, 141, 143
Gilbert Islands, 383
Ginsburg, R. N., 180
Gipsiferous deposits, 176
Giresse. P., 192
Gizienski, S. F., 40
Glacial control, hypothesis of, 362
Glacial-marine sediments, 202, 273, 407, 412, 413, 415, 422
Glacial moraine, 271
Glacial outwash, 190, 289
  fans of, 284
Glacial troughs, 201, 244, 273, 284
Glaciation, *99*, 101, 273, 292
Glaciers, 242
Glangeaud, L., 3, 402
Glass, B., 416
Glauconite, 76, *79*, 176, 238, 241, 291
Global Marine Company, 29
Globigerina limestones, 358
Globigerina ooze, 254, 412
Globigerina truncatulinoides, 415
Glomar Challenger, 8, 29, *30*
Godavari Delta, Bay of Bengal, 257
Gold. 132
Goldberg. E. D., 405, 408, 409, 410
Golden Gate, San Francisco, Calif.
  currents at, 64, 191, 238
  sandbar off, 238
  sediments of, 191
Gole, C. V., 257
Goleta, Calif., 237
Golfe du Lion, 64
Golfo de Guayaquil, 231
Golfo San Jorge, 231
Golfo San Matias, 231
Golik, A., 233
Goncharov, V. P., 403
Gondwana, 89
  glaciation of, *100*
Goodell, H. G., 180, 192, 212, 221, *289*, 368
Gorda Escarpment, 239–240, *240*, 380
Gorda Ridge, 380, 382, 429
Goreau, T. F., 343, 346, 352
Gorsline, D. S., 134, 180, *181*, 192, 212, 238
Gould, H. R., 66, 127, *146*, 147, 164, 166, *166*, 219, 222
Grab samplers, 18–19, 36
Graben, 252, 284
Graham, J. R., 29
Grand Banks, 202, 323
  cable breaks at, 66, 330, 331, *331*
  earthquake of 1929 at, 204, 330, 331, *331*
  turbidity currents of, 66, 330–331
Grand Canyon, comparable to submarine canyons, 315
Grand Isle, Barataria Bay, La., 145, 165
Granelli, N. C. L., 231
Grant, A. C., 201
Grant, U. S., 133, 137, *138*
Grapestone facies, 218
Graton, L. C., 81, *81*
Gravimeters, 42
Gravity measurements, 41–42

Gray Feather Bank, Caroline Is., 354
Graywackes, 304
Great Abaco Island, 327
Great Abaco Submarine Canyon, 214
Great Bahama Submarine Canyon, 325–327, *326*, 333
  highest walls in world, 325, *326*
  Northeast Providence Channel of, *326*
  Northwest Providence Channel of, *326*
  photo taken in, *327*
  Tongue of the Ocean of, 325, *326*
Great Barrier Reef
  borings into, 366
  largest, 366
  of Queensland, 366–368
  reef patches on, shapes of, *365*
Great Britain, 271
Great Natuna Island, 251
Great Sole Bank, 269
Great Valley of California, 190, 315
Greene, H. G., 129, 315
Greenland, 90
Gregory, J. W., 202
Griffin, J. J., 408
Grim, M. S., 210
Grim, P. J., 382
Grimsdale, T. F., 355
Groins, *154*, 156, *156*
Gross, M. G., 241, 242
Grunion, related to beach cycles, 136–137
Guerrero Negro Lagoon, Baja Calif., Mex., 182, *184*
  inlet, migration of, 182
Guilcher, A., 3, 185, 192, 193, 271, 342, 358, 359, 360
Guinea, Africa, 261
  geomorphic features off, *262*
Guinea Fracture Zone, 374
Guinea Ridge, 374
Gulf Coast, U.S., 141, 286
  hurricane effects on, 147
Gulf of Aden, 258, 394
Gulf of Alaska, 412
  ash in, 411
Gulf of Batabano (Bay of Pigs), 180–182
Gulf of Bothnia, 108
Gulf of Cadiz, 265
Gulf of California, 3, 233, 283
  origin of, 380
  rifting of, 89, 233
  sediment study of slope gullies, 338
Gulf of Cambay, 257, 258
Gulf of Campeche, 226
Gulf of Cariaco, 229
Gulf of Carpentaria, 252
Gulf of Darien, 229
Gulf of Gabès, 263
Gulf of Guinea, 167
Gulf of Kutch, 257–258
  1819 earthquake in, 258
Gulf of Lyons, 264
Gulf of Maine, 204–205
  ice excavation of, 292
  sediments of, 205
Gulf of Maracaibo, 229

Gulf of Martaban, 255
Gulf of Mexico, 143, 172, 221
  barriers of, *145*
  bathymetry of, *220*
  coastal environments of, 76
  continental terrace of, 218–226
  deltas of, 165
  diapirs in, 226, 283
  drillings in, 398, 424–425
  lagoons of, 174, *174*
  minerals of, 134
  northwest coast of
    continental shelf sediments of, *225*
    seismic reflection profiles of, *226*
  tides of, 60, 64, 286
  topography of, 397–402
Gulf of Oman, shelf of, 258
Gulf of Panama, 233
Gulf of Paria, 169, 192
Gulf of Po Hai, 246
Gulf of St. Lawrence, 106–108, *200*, 202
Gulf of Siam, widest shelves in world, 251
Gulf of Tatary, 244
Gulf of Tehuantepec, 233
Gulf of Tonkin, 249
Gulf of Tunis, 263
Gulf Stream, 34, 44, 61, 150, 202, 214, 217,
  218, 227, 286
Gulliver, F. P., 148
Gully, The, Nova Scotia, 325
Gunter, G., *174*
Gutenberg, B., *391*
Gutierrez-Estrada, M., 332, 339
Guyots, 354, 363, 383–388, *386*

Hails, J. R., 141
Hainan, SE China, 251
Halimeda, 132, 352, 354, 355, 359, 363
Hamblin, W. K., 28
Hamilton, E. L., 39, 40, 82, 382, 388
Hamilton, W., *99*, 100
Hancock Foundation, 234
Hanna, G. D., 238
Hanzawa, S., 352, 357
Hardline Reefs, 366
Harrison, C. G. A., 227
Harrison, J. C., 42
Hartwell, A. D., 190
Hatteras Abyssal Plain, 378, 424
Hawaiian arch, 305
Hawaiian-Emperor Ridge, 98
Hawaiian Islands
  beaches of, 124, 132
  subsidence of, 385
  tides of, 60
  tsunamis of, 53, 54, *55*, *56*
Hawaiian Ridge, 98, 383, 385
Hawkins, J. W., 382
Hayes, D. E., 94, 229, 424
Hayes, M. O., 277, 278
Hays, J. D., 416, 417
Head of Passes, La., 167
Heat flow, measurements of, from ocean
  floors, 42, 99
Heath, G. R., 409, 413, 418, 428

Hecate Strait, 242
Hedberg, H. D., 277, 279, 280, 283, 284
Hedgpeth, J. W., 163
Heezen, B. C., *5*, 7, 8, 36, 37, 63, *63*, 214,
  297, 321, *322*, 330, 332, 370, 372, *373*,
  375, 376, 377, *377*, 378, 385, 386, 393,
  397, 411, 412, 426, 429
Heirtzler, J. R., 42, 94, *95*, 395, 416
Heliopora, 364
Hellenic Trough, 425
Herman, Y., 413
Heron Island, Australia, 366
Hersey, J. B., 36, 40, 370, 402
Hess, H. H., 8, 89, *92*, 385, 388, 390
Hickson, R. E., 190
Hilde, T. W. C., 246
Hill, M. N., 3, 8, 40, 268, 372
Hilo, Hawaiian Is., tsunamis of, 54
Himalaya Mts., 93, 168
Hinde, G. J., 355
Hiranandani, M. G., 257
Hoare, J. M., 169
Hoffmeister, J. E., 342, 343, 346, 363, *366*,
  368 .
Hogsty Reef, Bahamas, atolls of, 350
Hokkaido Island, Japan, 64, 244
Holden, J. C., *91*, 98
Holland (The Netherlands), 147
  barriers of, 141
Hollister, C., 8, 36, 37, 63, 370, 376, 378
Holmes, A., 88
Holocene and Pleistocene boundary, 417
Holtedahl, H., *199*, 210, 273, 412
Holtedahl, O., 273
Hommeril, P., 193
Honduras, 227
Hong Kong, China, 249
Honshu Island, Japan, 64, 244
Hoover Dam, Nev., 66
Hopkins, D. M., 243, *243*, 321
Hopkins, T. L., 18
Horizon, R/V, 354
Horizon Guyot, 429
Horizons
  correlated by foraminifera, *416*
  reflection, in transoceanic profiles, 418–
    419, *419*
Horn, D. R., 382, 411, 412
Hornblende, 134
Hoskins, H., 205, 210, 212
Hot spots, 98
  migration of, *99*
Houbolt, J. J. H. C., 23, 258, 271, *287*
Hough, J. L., 190, 412
Howard, A. D., 315
Howe, H. V. W., 127
Hoyt, J. H., 106, 141, 143
Hsu, K. J., 329, 425
Huang, T.-C., 180
Hubbert, M. K., 83
Hubert, J. F., 412
Hudson Bay, 108, 274
Hudson Channel, 210
Hudson Inlet, 188
Hudson River, 325

Hudson Submarine Canyon, 325
Fan Valley off, 325
profiles of, *333*
Hueneme Submarine Canyon, Calif., 140
Humboldt Bay, Calif., 185
Hunkins, K., *63*, 397, *398*
Hunter, W., 410
Hurd Deep, 268
Hurley, P. M., 98
Hurley, R. J., 218, *326, 327*, 382, *384*
Hurricanes, 56–57
Audrey, 147
Camille, 56
Carla, *57*
catastrophic waves from, 147
locus of, 56
storm surges from, effects of, 147
Hvorslev, M. J., 19, 20
Hwang Ho River, China, 247, 248
Hydrogenous, 405
Hydrophones, 40
Hyne, N. J., 221, *289*

Iberian Peninsula, 265, 268
Iberian Plain, 376
Ice floes, 129
Ice Polygons, *116*
Icefalls into fiords, 59
Icy Cape, Alaska, 149
Ifaluk Atoll, 342
Igneous rocks of sea floor, 132
Illing, L. V., 258
Inderbitzen, A. L., 21, 294
India
continental slope off, 257
continental terrace off, 257
rifting of, 90
Indian Ocean, 255–258, 393–395
atolls of, 352, *353*
cores obtained from, 413
mapping of, 7
geomorphological, *392*
origin of, 90
ridges of, 370
Carlsberg, 91, 393
Central, 393, 394
mid-ocean, 393–395
Southeast Indian Ridge of, 393
Southwest Indian Ridge of, 393
sea-floor spreading in, 94
topography of, 393–395
Indonesia, 388
atolls of, 350, 354
Indus Delta, 257, 258, *336*
Indus River, 334
Indus Trough, 258, 334
Ingram, R. L., 182, *183*
Inman, D. L., 37, *46, 50*, 68, *71, 72, 72*, 73,
87, 103, 104, *105*, 124, 127, 129, 132,
133, *136*, 140, *140*, 141, *142*, 160, 236,
258, 331, 332
Inside Passage (Inland Passage), Alaska, 242
Ionian Basin, 426
Ippen, A. T., 163
Iran, shelf off, 258

Ireland
continental slope off, 271, 273
South, coast of, 268
Irish Channel, 288
Irish Sea, 268–269, 288
Irrawaddy Delta, 255
Isaacs, J. D., 16, 37
Isacks, B., *96*
Iselin, C. O'D., 64
Isla de Lobos, Mex., 368
Isla de Pinos, Cuba, 180
Island arcs, 94, 199, 388
Island barrier reefs. *See* Barrier reefs
Isoletes Columbretes, 265
Isostacy, 89
Isostatic adjustment, 300
Israel
beach sands of, 133
continental shelf off, 263
continental slopes off, 263
Isthmus of Panama, 428
Italian Peninsula
continental shelves off, 264
continental slopes off, 264
Italian Riviera, 264
Iversen, H. W., *47*, 48
Ivory Coast, Africa, 187
Iya Nada, Inland Sea of Japan, 193
Izaki, A., 244

Jackson, E. D., 98, 385
Jacksonville, Fla., 212
Jamaica, coral growth of, 352
Japan, 64
Sea of, 193–194, *248*, 286 ·
sediments off, *194*, 244
Trench off, 246
Japanese islands, 388
Japanese Maritime Safety Board, 244
Jarke, J., 271
Java Sea, 251
sediments of, 251
shelf, 251, 283
Java Trench, 393
Jekyll Island, Ga., 163
Jersey Production Research Company, 180
Jet sampler, 23
Jetties, effects on beaches, 152–154, *153*
Johnson, D. A., 386, *387*, 413
Johnson, D. W., 53, 103, 106, 111, 124, 133,
143, 151, 156, *157*, *204*, 328
Johnson, G. L., 377, 397
Johnson, H., 89
Johnston Island, 352
JOIDES (Joint Oceanographic Institutions for
Deep Earth Sampling), 3, 29–31, *30*,
91, 96, 214, 218, 369, 405, 412, 413,
418-430, *427*
correlation of holes drilled, 422
drilling holes, map of, *421*
drillings in
Atlantic Ocean, 202
Caribbean Sea, 229, 425
Gulf of Mexico, 31, 398, 424–425

JOIDES (Continued)
Mediterranean Sea, 263, 329, 402, 425–426
North Atlantic Ocean, 422
Pacific Ocean, 242, 382, 426–429, 427
evidence from, 298
innovations, reentry funnel, 31
methods of drilling, 8, 29–31, 30
Jonasson, U. C., 417
Jones, B. R., 200
Jones, O. A., 366, 367
Jordan, G. F., 188, 217, 217, 219, 242, 292
Jordan Basin, Gulf of Maine, 204
Juan de Fuca, Strait of, 241
Juan de Fuca Ridge, 380, 382

Kagoshima Bay, Japan, 244–246
Kaimoo, 129, 131
Kamchatka, USSR, 389
Kanaev, V. F., 15
Kaneohe Bay, Oahu, Hawaiian Is., 352
Kanes, W. H., 172
Kara Sea, 273
Karachi, Pakistan, 258
Karig, D. E., 388, 388, 389
Karimata Islands, S. China Sea, 251
Kastengreifer, box corer, 23
Katmai Volcano, Alaska
ash from, 411
eruption of, in 1912, 411, 413
Katsumata, M., 389
Kauai, Hawaiian Is.
beaches of, 133
coastal waves of, 50, 52
tsunami at, 55
Kay, M., 89
classification by, 246
Kaye, C. A., 118
Keen, M. J., 8
Keller, G. H., 255
Kelling, G., 325
Kennett, J. P., 416, 417
Kenny, J. E., 18
Kenya, Africa, 259
Kenyon, N. H., 271
Kermadec Islands, 390
Key West, Fla., 216
Kidwell, A. L., 180
Kiel, H., 251
Kiilerich, A., 390
Kimberley, Australia, 252
King, C. A. M., 102, 124, 133, 135, 137, 147, 153
King, L. H., 202, 203
King, W. B. R., 268
Kinsman, B., 43
Kistna (Krishna) Delta, Bay of Bengal, 257
Klamath River, Calif., 241
Klein, G. De V., 205
Klenova, M. V., 7, 31, 273, 274
Knauss, J. A., 61
Knott,, S. T., 210
Kodiak Island, Alaska, 242
Kofoed, J. W., 180, 181, 216, 217, 217

Kögler, F.-C., 23
Kolb, C. R., 145, 165, 165
Kolbe, R. W., 372
Koldewijn, B. W., 293
Koldijk, W. S., 193
Komar, P. D., 331
Korea, 247
coast of, 248, 250
Gulf of, sand ridges of, 290
Korean Strait, 244, 248
Kori Great Bank, 258
Kornicker, L. S., 227, 367, 368
Kotenev, B. N., 319
Krakatau Volcano, Sunda Strait, 251
eruption of, 411
Krause, D. C., 15, 235, 374, 376, 382, 389
Krinsley, D., 99, 100
Kruit, C., 264
Krumbein, W. C., 69, 70, 74, 78, 83, 84, 176
Ku, T. L., 416, 417, 418
Kudinov, E. I., 22
Kuenen, Ph. H., 2, 66, 81, 159, 185, 331, 333, 334, 347, 354, 357
Kukpuk River, Alaska, 149
Kullenberg, B., 20, 21
Kulm, L. D., 190
Kummel, B., 28
Kure Island, atoll of, 350
Kuril Trench, 391
Kuroshio (Current), 248, 251, 291, 321
Kuskokwim Delta, Alaska, 169, 170
Kuskokwim Estuary, Alaska, 169
Kuskokwim River, Alaska, 321
Kwajalein Atoll, 349, 350
coral reefs of, 355
Kyushu Island, Japan, 58, 246

La Have Basin, Nova Scotia, 203
La Jolla, Calif., 135, 138, 140, 157, 294, 341
Boomer Beach at, 130, 137
Point at, 236
wave-cut bench at, 293
La Jolla Fan Valley, Calif., 306, 309
contour map of, 310
La Jolla Submarine Canyon, Calif., 152, 306–312
box cores from, 309, 311
contour map of, 310
current meter records of, 38, 332, 332
erosion pinnacle in, 309
head of, 307
sediments of, 236, 311
topography of, 294, 308
La Peruse Bank, 241
Laban, A., 36, 36
Laborel, J., 364
Labrador
continental shelf off, 200–201
glacial troughs off, 199
Labrador Current, 202
Labrador Sea, bottom currents of, 422
LaChance, T. P., 261
LaCoste, L. J. B., 42
Ladd, H. S., 342, 343, 346, 357, 358, 361, 363, 364

LaFond, E. C., 66
Lafourche subdelta, 145
Lagoons, 162–195, 252, 347, 350, 352. *See*
  also Atolls
  barriers, terraces of, 360–361, *360*
  cross-section of, *163*
  general characteristics of, 187
  mining of salt in, 182
  sediments of floors of, 182
  sinkholes of, 361
  typical, *163*
  uplifted, 354
Lagoons, coastal, 174–187
  Baja Calif., western coast of, 182
  California, West Coast of, 182–185
  Florida, West Coast of, 180
  Guerrero Negro, Baja Calif., *184*
  Gulf of Guinea, 185
  Gulf of Mexico, 174, *174*
  Texas, central coast of, 176, *178–179*
  United States, central East Coast of, 182
Laguna Madre, Tex., 175–176, *175, 177*
Lake Aheme, Ghana, W. Africa, 187
Lake Geneva (Lac Leman), 66
Lake Mead, Ariz.–Nev., 66
Lake Worth Inlet, Palm Beach, Fla., 152,
  153
Laminar flow, 84
Lamont-Doherty Geological Observatory, 3,
  5, 22, 40, 214, 218, 231, *331*, 413, 418
Land, L. S., 352
Lands End, England, 268
Landslide surges, 57
  at Lituya Bay, Alaska, 57–58, *59*
Landslides, submarine, 239, *239*
Langseth, M. G., Jr., 42
Lankford, R. R., 145, 166, 224, 257
Larson, H. E., 31
Larson, R. L., 380
Larsonneur, C., 193
Lau Basin, 389
Lau Islands, 354
Lau Ridge, 389
Lauff, G. H., 163
Laughton, A. S., 370, 376, *392, 393, 394, 395,*
  422
Laurasia, 89
Laurent, J., 156
Laurentian Channel, 202, 204
Lawson, A. C., 2
Le Calvez, Y., 268, 271
Le Pichon, X., 89, 94, 395
Lebanon, shelf off, 263
LeBlanc, R. J., 141
LeClaire, L., 263
Lee, G. V., 2
Leffingwell, E. deK., 292
"Left hook" fan valleys, 317, *384*
Lehner, P., 224, 226, 283, 296
Levantine Basin, 425
Libya, shelf off, 263
Line Island Ridge, 383
Lisbon, Portugal, tsunami of 1755, 54
Lisitzin, A. P., 7, 244, 405, 413
Lithogenous, 405

Lithosphere, 89, *93*, 98
Lithothamnion ridge, 350, *351*
Lithothamnion rudite, 193
Lituya Bay, Alaska, landslide surge at, 57, *59*
Lituya Glacier, Alaska, *59*
Loen Lake, Norway, 59
Lofoten Islands, Norwegian Sea, 273
Loire Estuary, France, 192, 193
Lomonosov, 7
Lomonosov Ridge, 397
Long Beach, Calif., 182
  continental shelf off, 237
  sediments off, *236, 237*
Long Island, N. Y., 212, 294, 325
Long Island Sound, N. Y., 210
Longshore bars, 127, *145*
  migration of, 136
Longshore currents, 158, *158*
Longshore drift, 143, 157
Loring, D. H., 202
Lort, J. M., 402, 403
Los Angeles River, Calif., 237
Lotze, F., 101
Louderback, G. D., 2, 191
Louisiana, beaches of, 147
  western, continental terrace off, 224–226
Louisiana State University, 218
Lowenstam, H. A., 132
Ludwick, J. C., 134, 188, 222
Ludwig, W. J., 40, 389
Luzon, Philippine Islands, slope valleys off,
  252, 338, *339*
Lyall, A. K., 205, 300, 325
Lynts, G. W., 218
Lysocline, 408

McCamus, M. J., 325
MacCarthy, G. R., 81
Macclesfield Bank, S. China Sea, submerged
  rim of, 354
McClure, C. D., 7
McCoy, F. W., 25, 385
McCulloch, D. S., 150
Macdonald, G. A., 237, 388
McFarlan, E., Jr., 127, *146*, 147, 165
McGeary, D. F. R., 376
McGill, J. T., 104, 111
McHugh, J. L., 163
McIntire, W. G., 168, 257, 368
McKee, E. D., 156, 352
McKenzie, D. P., *102, 103*
MacKenzie, F. T., 352, 354
MacKenzie River, Canada, 273, 292
Mackereth, F. J. H., 19
MacLean, 202, 203
McLeod, I. R., 101
McManus, D. A., 244, 429
McMaster, R. L., 134, 189, *209*, 210, 261, *262*
McMullen, R. M., 202
MacNeil, F. S., 361, 363
Mad River, Calif., sediment from, 185
Mafia Island, E. Africa, 258
Magdalena Delta, Colombia
  sea floor off, *228*
  slope off, 229

Magdalena River, Colombia, 229
Magellan Rise, 385, 427
Magnetic anomaly, 372, 395
 ages of, 96, 429
 measurements, 42
Magnetic bands, 94, *95*
Magnetic belts, matching of, *95*
Magnetic fields, 94
 reversing polarity of, 94
Magnetic poles, 98
Magnetic reversals, 416, *417*
Magnetite, 98
Magnetometer, measurements by, 42
Mahadevan, C., 257
Maiklem, W. R., 254
Malacca Strait, 255
Malaise, R., 372
Malaya, 255
 inner shelf off, 255
Malaysia, broad shelf off, 251
Maldive Islands, atolls of, 352, *353*
Male Atoll, Indian O., *353*
Male Islands, Indian O., *353*
Malfait, B. T., 402
Malloy, R. J., 216, *216*, 217, 242
Mammerickx, J., 370, *378*, 382, 383, *383*
Manar, R. Y. *See* Young, R.
Manganese, 411
Manganese nodules, 76, 301, 386, 409, *409*,
 412, 429
 methods of obtaining, 26
Mangrove coasts, 106, 122
Manihiki Plateau, 385, 386
Map, electronic position finder, 12
Maratoa Island, 357
Marblehead Neck, Mass., *144*
Marginal plateaus, 197–198. *See also* Blake
 Plateau
 off California, 238
 off Queensland, 255
Margolis, S. V., 416
Mariana Islands, *122*, 389
 uplifted lagoons of, 354
Mariana Trench, depth of, 390
Marine Institute, Univ. of Georgia, 124
Marine valleys, types of, 305–306
Markl, R. G., 378
Marlow, M. S., 244
Marsden Bay, England, *135*
Marseille, France, 264
Marshall, N. F., 19, 23, *24*, *32*, 37, *38*, 308,
 332
Marshall Islands, 342, *345*, *348*, *349*, 383
 atolls of, 343, *348*, 350
 drillings in, 342–343, 358
 origin of, 362
 sea-floor slopes off, 354
Marshes, 187
Martens, J. H. C., 132
Martha's Vineyard, Mass., 209
Martin, B. D., 315
Martin, H. B., 18
Mascarene Plateau, 393, 395
Mascle, J. R., 329
Mason, R. G., 42

Massey, F. J., Jr., 71
Matagorda Lagoon, Tex., *173*
Matagorda Peninsula, Tex., effects of Hurri-
 cane Carla, *57*
Mathews, W. H., 296, 337
Matthews, D. H., 94, 372
 tables by, 13
Mauritius Island, 395
Maxwell, A. E., 8, 370, 420, *423*
Maxwell, W. G. H., 252, 254, 342, 347, *365*,
 366
Mayor, A. G., 343, 346, 347
Mazama ash, 330
Mazama Mt., cataclysmic eruption of, 318
Mazatlan, Mex., 233
Meander in sea valley, 315
Mediterranean Ridge, eastern, 402, 425
Mediterranean Sea, 264, 402
 continental terrace of, 263
 currents of, 64
 drillings into, 329, 425–426
 cores from, 426
 eastern, physiographic provinces of, *401*
 lowering of, 329
 tides of, 60
 western, sea-floor relief of, *400*
Mediterraneans, 395–403
Melville, R/V, 3, 6
Menard, H. W., 7, 89, 92, 315, 317, 370, 371,
 378, 379, 380, *381*, *384*, 385, *386*, 388,
 397, 402, 409, 410, 411, 412, *414*
Mendocino Escarpment, 240
Mendocino Fracture Zone, 240, *240*, 382, 426
Mendocino Ridge, 380, 382
Mergui Archipelago, 255
Mero, J. L., 26, 410
Mero Submarine Canyon, Japan, 321
Merrimack River Estuary, Mass., 190
Metastable sediments, 296
Meteor Expedition, 2, 372
Meteorites, 410
Mexico
 anticlinal ridges off eastern, 227
 continental terrace off eastern, 226
Mexico City, 163
Meyerhoff, A. A., 89, 94, 98, 99, 100, 101
Meyerhoff, H. A., 89, 94, 98, 99
Miami, Fla., 180
 continental shelf off, 216
 shells of, 132
Miami Terrace, 216, *216*
Miami University, 3, 197, 218
Michaelmas Cay, Great Barrier Reef, 366
Micronesia, 364
 profiles of, 354
Mid-Atlantic, U.S., continental shelf off, 210
Mid-Atlantic Channel, 376
Mid-Atlantic Ridge, 93, 261, 300, 371–376,
 380
 drillings in, 412, 420, *422*, *423*
 formation of, 91
 heat-flow measurements from, 99
 hot spots of, 98
Mid-Atlantic Rift Valley, 372, *374*

Mid-Ocean Channel l(Mid-Ocean Canyon), 376
Mid-Ocean ridges, 372
Mid-Pacific Mountains, 383
Middle America Trench, 233, 389, 390, 391, 428
Middleton, G. V., 331
Midway Island (Atoll), 98, 350, 355, 358, 364
  drillings at, 255
Midway Lagoon, 355
Migliorini, C. I., 66
Millepora, 368
Miller, D. J., 57, 58, *59*
Milliman, J. D., 214
Mindanao, Philippine Is., 252
Mindanao Trench, 390
Mission Beach, S. Calif., internal waves off, 66
Mississippi, continental shelf off, 221–222
Mississippi Cone, 224
Mississippi Delta, 164–167
  continental slope off, 222, 224, 226, 337
  continental terrace off, 222–224
  environments of, *164*
  lobes of, *165,* 169
    Balize, *165*
  passes of, 145, 165, *165,* 167
  sediments of, 224
  slope gullies off, 337–338, *338*
Mississippi River, 61, 169
  sediment supply related to cheniers, 147
Mississippi Trough, 165, *165,* 224, 334–336, *337*
Mitchell, A. H., *93*
Moberly, R., Jr., 124, 132, 385
Mobile Bay, Ala., 134, 221
Mobile River, Ala., 134
Mogami Bank, Caroline Is., 354
Mohole Project, 420
Mohorovičić discontinuity (Moho), 41, 89, 90
  downbuckling of, 391
Molengraaff, G. A. F., 2
Mollusks, 132, 185
Molnar, 402
Molokai, Hawaiian Is.
  coral reefs off, 347
  land-sea canyons of, 341
Monk, G. D., 74, 83, *84*
Mont Argentario, Italy, 150
Mont-Saint-Michel, France, 103, 193
Montague Island, Alaska, *109*
Monterey Bay, Calif., 315
Monterey Fan Valley, Calif., 331
  meander of, 315, *317*
Monterey Submarine Canyon, 239, 315, *316*
Montgomery, R., 61
Montmorillonite, 193
Moody, D., 287
Moore, D. G., 16, *41,* 172, 176, *178–179,* 197, 233, 235, 237, 238, 239, 277, 279, 283, 294, *296,* 331, 335, 395, 397, 418
Moore, G. W., 129, *131,* 143, *168,* 241, 380
Moore, J. R., III, 190
Moore, T. C., Jr., 229, 409, 418

Moorea, Society Is., 358
  lagoons of, 360
Moran, 12
Morelock, J., 294, 296
Morgan, J. P., *120,* 147, 163, 164, *164,* 166, *168,* 257
Morgan, W. J., 89, 98, *100*
Moriarty, J. R., *80*
Morocco, coast of, 108
Moss Landing, Calif., 315
Moulin, P., 23
Mud blanket, definition of, 198
Mud diapirs, 224, *228,* 229, 284, 336
Mud lump islands, *120*
Mud lumps, 119
Mudbanks, 257
Mudie, J. D., 15, 16, 380
Mugu Lagoon, Calif., 185
Mugu Submarine Canyon, Calif., 141
Müller, G., 259, 264
Multer, H. G., 342, *366,* 368
Munk, W. H., 64, *351,* 352
Munsart, C. A., 182
Murray, G. E., 226
Murray, J., 2, 363, 405, 406, 408, 410, 412
Murray Fracture Zone, 382
Murray Sea Valley, Aleutian Is., *340*
Mururoa Island, drillings in, 355, 358
Musée Océanographique de Monaco, *324*

Nankai Trough, Japan, 246
Nannoplankton, 410
Nansei Shoto Island arc, Japan, 246
Nantucket Island, Mass., 151
  shelf off, 209–210, 212
Nantucket Shoals, 209, 325
Napa Creek, Calif., 190
Naples, Italy, 264
Nares Deep, 376, *377*
Narragansett Bay, Rhode Is., 188
  sediments of, 188–189
Nason, R. D., 240
Nasu, N., 129
Natal, Africa, 259
National Geographic Magazine, 370
National Oceanographic Institute, 3, 6
National Science Foundation, 3, 29
Natural levees
  of fan valley, 309, *317*
  submerged, *115*
Naval Undersea Research and Development Center (formerly Navy Electronics Laboratory), 3, *34,* 39, *117,* 309, 312, *314*
Navigation chart soundings, 11
Nayarit, Mex., *144,* 283
Nayudu, Y. R., 382, 410, 411, 413, *414,* 415
Nazare, Portugal, 49
Necker Ridge, 383
Nekton, 34
Nelson, B. W., 169, *171*
Nelson, C. H., 241, 315, 317, *318,* 330
Neo-trailing-edge coasts, 199, 258
Nepheloid layers, 63
Neprochnov, U. P., 403
Neritic, definition of, 198

Nesteroff, W. D., 264
Neumann, A. C., 218, 224
New Caledonia, coral reefs of, 358, 359
New England, 188
  continental shelf off southern, 210
New England Seamounts, 374, 376
New Jersey
  beach sands of, 134
  continental shelf off, 210, 291
New London, Conn., 210
New York, continental shelf off, 210, 286
New York Bay, 325
New York Harbor, 210
Newell, N. D., 76, 217, 218, 219, 342, 343,
  364, 367, 368
Newfoundland, 303
  continental shelf off, 201–202
  continental slope off, 298
  sea-floor topography off, 200, 201–202
Newfoundland Rise, 376
Newman, W. S., 182
Newport Bay, Calif., 185
Newport Beach, Calif., sea floor off, 90
Newport Harbor, Calif., storm surge at, 156,
  156
Newport Submarine Canyon, Calif., 141
Nicaragua, coral reefs off, 227
Nichols, H., 244, 340, 389
Nichols, R. L., 129
Nichols Shore, 210
Nicobar Islands, Bay of Bengal, 255
Niger Delta, W. Africa, 97, 167, 167, 185,
  261
  continental shelf off, 259, 263, 281, 299
  sediment facies of, 260
Niino, H,, 248, 250, 251
Nile Abyssal Plain, 425
Nile Cone, 426
Nile Delta, continental slope off, 263
Nile River, 133
  continental shelf off, 263, 281
Ninetyeast Ridge, Indian O., 393, 395, 396
Ninkovich, D., 411
NOAA (National Oceanic Atmospheric Ad-
  ministration), 389
Noakes, L. C., 301, 301
Nodules
  ferromanganese, 411
  manganese, 26, 76, 301, 386, 409, 409–
    410, 412
Nome, Alaska
  mining at, 132
  submerged beaches of, 301
Noornay, I., 40
Nordstrom, C. E., 103, 104, 105, 258
Normark, W. R., 312, 315
North Atlantic Ocean
  age of, 422
  physiographic provinces of, 373
North Equatorial Current, 165, 286, 291
North Pacific Ocean
  basement ages of, 427
  bathymetry of, 378
  block movement of, 427

North Sea, 147, 197
  coast of, 193
  sand ridges in, 287
  sand waves of, 272
  sediments of, 185, 270, 271
  shelf glaciation of, 271, 273
  strong currents of, 291
North Slope, Alaska, 172
Northeast Channel, Gulf of Maine, 204, 325
Northeast Providence Channel, Bahamas,
  326
Northern California, 54
  continental shelf off, 282, 297
  waves of 1964 earthquake, 53–54
Northwest Providence Channel, Bahamas,
  326
Norton Sound, Alaska, 169, 170, 244
Norway
  coast of, 64
  continental shelf and slope off, 273
  coral reefs off, 346
Norwegian Basin, 422
Norwegian Sea, cores from, 412
Nota, D. J. G., 202, 229
Nova-Canton Trough, 384
Nova Scotia
  continental shelf off, 203, 203, 204
  continental slope off, 205, 294, 295
  sea floor off, 200, 204
  sediments of, 291
  submarine canyons off, 325
Novaya Zemlya, Russia, 273

Oba, T., 415
Ocean basins, general shape of, 371
Ocean floor
  basement age under, 94–98
  ridges, spreading of, 95
  topography, deep, 369–403
  features defined, 370–371
Oceanic margins, fit of opposite sides, 96–
  98
Oceans Magazine, 189
Odum, E. P., 347, 350
Odum, H. T., 350
Off, T., 288, 290
Office of Naval Research, subsidizing of
  projects, 7
Ojo de Liebre Lagoon, Baja Calif., Mex., 182
Okhotsk Basin, 284
Okhotsk Sea, 244
  contours of, 246
  sediments of, 247
Okinawa, 246
Olausson, E., 3, 417
Olivine crystals, 133
Omega network, positioning, 12
Oölites, 176, 180, 187, 218, 257
Oölitic hematite, 303
Oölitic sands, 220
Oozes
  calcareous, 96, 410
  diatom, 412, 413
  foraminiferal, 422
  Globigerina, 412

Oozes (*Continued*)
  organic, 96
  radiolarian, 411, 413
  siliceous, 96, 429
Opdyke, N. D., 416, 417
Orange-peel buckets, 19
Oregon
  continental shelf off, 241
  submarine canyons off, 315
  waves from 1964 earthquake, 54
Oregon State University, 3, 197, 241, 315
Oriented lakes, *116*
Orinoco Delta, Venezuela-Colombia, 169,
    192, 229, 286, 293
  sediments of, 229, 230
Orinoco River, 192
Ostracods, 75, 76
Ottmann, F., 8, 229, 230
  coastal classification by, 109, *110*
Otvos, E. G., Jr., 134, 141
"Overfalls," 321
Overwash fans, *145*
Ovey, C. D., 270
Ovoid nonskeletal carbonate, 180
Owen, D., *409*
Owen Fracture Zone, 393–394, 395
Oyster reefs, 176, 180, 187

Pacific Basin, 388
Pacific Ocean
  Central, basement age of, *427*
  fracture zones of, 380–382, *381*, *428*
  northeast, sediments of, *415*
  topography of, 379–391
  trenches of, 389–391
    depths of, 390, *391*
Pacific Plate, 242, 385
Pack ice, 129, 172
Padre Island, Tex., 162, 224, 286
Pago Pago, American Samoa, 346
Paige, H. G., 70
Pakistan Navy, 334, *335*, *336*
Palawan, Philippine Is., 251
Paleoclimatic events, dating of, *417*
Paleomagnetic reversal, *417*
Paleomagnetism, 98–99
Palk Strait, India, 257
Palm Beach, Fla., shelf disappears at, 212
Palmer, H. D., 308, *309*
Palos Verdes, Calif.
  continental shelf off, 237
  fault scarp off, 338
Palos Verdes Hills, Calif., 141
Pamlico Sound, N.C., 150, 182
  sediment types of, *183*
Pamplona Searidge, 242
Panama City, Fla., 180
Pangaea
  breakup of, 89, 90, *91*, 395
  glaciation of, 101
  reconstruction of, 100
Pannekoek, A. J., 193, *400*, 402
Paris, University of, 3
Parke, M. L., Jr., 251
Parker, F. L., *79*, 408

Parker, R. H., 224
Parkin, D. W., 410
Pass a Loutre, Mississippi Delta, *120*
Patch reefs, 368
Peck, R. B., 82
Peels, methods of taking, 28
Pelagic sediments, 407
Pellets, mud, 218
Penck, A., 362
Pensacola, Fla., 180
Pensacola Bay, Fla., *150*
Pérès, J.-M., *36*
Perlas Islands, 233
Permafrost, *116*, 169, 172
Permo-carboniferous glaciation, 99–100
Perry, R. B., 244
Perry Company, 34
Persian Gulf, coral reefs of, 346
Peru-Chile Trench, 232, 389
Petaluma Creek, Calif., 190
Petelin, V. P., 22, 390
Petersen grab sampler, 19
Peterson, M. N. A., 408, 420, *422*
Petroleum, 237, 271, 301, 424
  accumulations in reservoir sands, 174
  discovery of, 430
Petroleum industry, 28, 197, 252
  developing drilling techniques, 23
Pettersson, H., 3, 40
Pettijohn, F. J., 156
Pfannenstiel, M., 402
Phi units conversion table, *71*
Philippi, E., 2, 404, 412
Philippine Islands, 388
  continental terraces off, 252
Phillips, J. D., 394
Phillips, O. M., 43
Phillipsite, 413
Phipps, C. V. G., 252, *253*, 255
Phleger, F. B, 3, 163, 174, 182, *184*, 187, 257,
    404
Phleger corer, 19, *19*, 21
Phosphate nodules, 233, 236, 302
Phosphorite, 212, 238, 301
Photography
  aerial, 56, 143, 151, *160*
  bottom, 19, 36, 63, 370
    history of, 34, 36–37
  deep-sea camera, *34*, 36, *36*
  suspended camera, *327*, 327
  underwater, 25, 306
  stereo, *35*, 36
  x-ray, *29*
Phuket, Thailand, 255
Piccard, A., 33
Piccard, J., 33, *36*, 390
Pickett T. E., 182, *183*
Pierce, J. W., 134
Pigeon Point, Calif., 238, 239
Piggot, C. S., 412
Piggot transatlantic cores, *407*, 415
Pigorini, B., 263, *264*, 264
Pimm, A. C., 251, 424
Pipe dredge, 25
Piper, D. J. W., 309, 312

Pirsson, L. V., 357
Piston corer, *20*, 21, 415
Piston cores, 21, *22*, 229
Pitman, W. C., III, 94
Placer deposits, 301
Plafker, G., 54, *109*, 242
Planet, 2
Plate tectonics, 7, 88–101, 103, 104, 233, 393,
    426, 429. *See also* Sea-floor spreading
  evidence for, 94–101
  hypothesis of, 89, 92
Plateau, marginal, 238, 255, 259
Plates, major, 92
Platform reefs, 368
Pleistocene–Pliocene boundary, 417
Plum Island, Mass., 190
Po Delta, Italy, 169, 264
  history of, 169, *171*
Point Arena, Calif., 238
Point Barrow, Alaska, 149, 150, 273, 274
  continental shelf off, 273
Point Conception, Calif., 134, 139, 140, 235,
    237
Point Dume, Calif., *136*, 141
Point Franklin, Alaska, 149
Point Hope, Alaska, 149, *149*
Point Reyes Peninsula, Calif., 341
Point Sal, Calif., *139*
Point Sur, Calif., 239
Pondicherry, India, 257
Poole, D. M., 134
Porcupine Bank, 271
Porites, 347, 350
Port Arthur, Tex., 224
Port Royal, Jamaica, tsunami of, 54
Portales Terrace, 216
Portugal
  continental slope off, 298
  continental terrace off, 265
  slump folds off, *297*
Poseidonia beds, 264, *265*
Positioning at sea, 10
  sampling, *27*
Post, R., 257
Postglacial, 364
  rebound, 108, *108*, *131*
  sea-level rise, 106, *108*, 133, 195, 222
Postma, H., 185, 192
Potomac River, Va.–W. Va., 188
Potter, R. E., 156
Powers, M. C., 78, *80*
Powers models, 80, *80*
Pratje, O., *270*, 271
Pratt, R. M., 378
Precision depth recorders (PDR), 13
Pribilof Islands, Alaska, 243, 284
Pribilof Submarine Canyon, Bering Sea, *320*
Price, W. A., 103, 111, 151
Provincetown, Mass., 133
Psuty, N. P., 143
Puerto Rico Trench, 376, 391
  cores from, 412
Puget Sound, Wash., 241
Punta Banda, Baja Calif., Mex., 234
Punta Gorda, Calif., 239, 380

Purdy, E. G., 218
Putnam, W. C., 111, *121*, *122*
Pyrite, 187
Pytkowicz, R. M., 410

Qatar Peninsula, Arabia, 258
Quartz, *79*, 132, 192, 408
Queen Charlotte Bank, 230
Queen Charlotte Islands, 242
Queen Charlotte Sound, 242
Queensland shelf, coral reefs of, 252, 286,
    347, 366–368
Quicksand, 193

Radar, 12
Radiography, 28. *See also* X-ray photographs
Radiolarians, 75
Raitt, R. W., *90*, 355, *356*, 391, 412
Rajcevic, B. M., 192
Rann, Gulf of Kutch, 257
Rao, M. P., 257
Raroia Island, 343
Rat Islands, Aleutians, 389
Raup, D., 28
Raydist, 12
Raymond, P. E., 132
Raymond, S. O., 16
Reading, H. G., *93*
Red clay, 406, 413, 422
Red Sea
  atolls of, 350
  continental shelf of, 258
  rift valley of, 258
  rifting of, 89
  sediments of, 394
Redfield, A. C., 229
Redondo, Calif.
  breakwater at, 153
  jetty at, 152
Redondo Submarine Canyon, 141
Redwine, L. E., 22
Reef patches, *365*
Reefs. *See* Coral Reefs
Reflectors, 418, 420
  M-, 426
Reimnitz, E., 331, 332, 339
Reineck, H.-E., 23
Relict sands, 237
Relict sediments, 198, 241, 289
  zones, 288–291, *291*
Renard, A. F., 405, 406, 408, 412
Resources for future off U.S., *302*
Revelle, R., *391*
Rewa River, Fiji Is., 346
  coral reefs off, 346
Rex, R. W., 273, 408
Reynard, A. F., 2
Reynolds, O., 82, 84
Reynolds' number, definition of, 84
Reynolds Submarine Services, 34
Rhine River, 185
Rhizome mat, 182
Rhode Island University, 3
Rhone Delta, France, 264
  turbidity currents of, 66

Ria de Arosa, Spain, sediments of, 193
Rias, 111, 188, 193
Richards, A. F., 255
Richards, H. C., 254
Richter, C. F., *391*
Riedel, W. R., 21, 385, 404, 411, 412, 413, *414*
Rift valleys, 229, 233, 384, 394
 of Africa, 89
Rigby, J. K., 368
Riley, N. A., 78
Rio Balsas Delta, 338–339
Rio Balsas Submarine Canyon, Mex., 331
Rio de Janeiro, 230
Rio de la Plata Estuary, Argentina, 230
Rio Grande Delta, continental shelf off, 224
Rio Grande deltaic plain, 175
Rio Grande Ridge, 98
Rio Grande Rise, 374, 420
Rio Grande River, 226
Rio São Francisco del Norte, Brazil, 230
Rip current channels, 158
Rip currents, 50, 51, 52, *53*, 78, *156*
Ripple marks, 63, 157–159, *158*, 218, *314*,
 323, 325, 327, 328, 411, 412
 giant, 158, *159*, 218, 231, 234, 378
 wavelengths of, 87
Rivière, A., 156
Roberson, M. I., 205
Roberts, D. G., 265, *297*
Robinson, A. H. W., 147, 271
Rockall Bank, 271
Rockport, Tex., bays of, 176, 180
Rodolf, F. W., 190
Rodolfo, K. S., 255, *256*
Rodriquez Ridge, 393
Rolston, J. W., 192
Romanche Deep, 374
Romanche Fracture Zone, 261, 374, *375*
Rona, P. A., 63, 261, 298, 333, 377, 424
Rongelap Lagoon, Marshall Is., 360
Rosalsky, M. B., 156, 161
Rosetta River, Egypt, 263
Rosfelder, A. M. (also Rossfelder), 16, 19,
 22, 23, *24*, 263
Ross, D. A., 21, 204, 205, 233, 325, 394, 403
Ross, Sir James, 13
Rowley shelf, western Australia, 255
Roy, K. J., 347
Roy, R. F., 380
Ruivo, M., 64
Runcorn, S. K., 89
Rusnak, G. A., 132, 140, 175, *175*, 176, 224,
 233
Russell, R. D., 81
Russell, R. J., 106, *115*, 127, 134, 145, 165,
 229
Ryan, W. B. F., 329, 402, 425
Ryukyu Island, 246, 284, 388

Sable Island, Nova Scotia, 203, 325
Sable Island Bank, Nova Scotia, 203, *203*,
 284, 325
Sachs, P. L., 16, 192
Sackett, W. M., 416
Sacramento River, Calif., 190

Sagami Bay, Japan, 321, 340
 earthquake of 1923, effects of, 108, 340
Sagami Bay Fault Trough, 321
Saguenay River, Quebec, 202
Sahara Desert, 408
Sahul shelf, 252
St. Bernard subdelta, *115*, 145, 165
St. Elias Range, Alaska, 242
St. Georges Channel, 268
St. Lawrence Island, Alaska, 243
St. Matthew Island, Alaska, 243
St. Paul Island, Gulf of St. Lawrence, 202
Saito, T., 412
Sale Cypremont, distributary of Mississippi,
 165
Saline lakes, 101
Salt
 cores, 424
 discovery of, 430
 domes, 224, 226, 258, 261, 283, 301, 336,
 398, 402, 424, 426
 intrusions, 223, 226
 marshes, 185
 mining of, 182
Saltation, 85
Samar, Philippines, 252
Samoa, 358, 383
Samplers
 box core, 23–25, *24*
 dredges, 25, 26–28
 grab, 18–19
 Campbell, *18*, 19, 36
 Petersen, 19
 jet, 25
 orange peel bucket, 19
 Phleger corer, 19, *19*, 21
 piston corer, 20–21, *20*
Samples
 bottom, 16
 impregnation of, 25
 methods of handling, 16, 28
 positioning vessels for sampling, 27
 x-ray radiography, 28, *29*
San Andreas Fault, 92, 240, *240*, 380, 382
San Antonio Bay, Tex., 162
San Blas, W. Mex., 233
San Clemente Island, Calif., 238
 slope off, 338
San Cristobal Bay, Baja Calif., Mex., 233
San Diego, Calif., 293
 continental shelf off, 235
 Point Loma at, 153
 sediments off, 235, *235*, 236
San Diego Harbor, Calif., 182, 235
San Diego Trough, 14, *15*, 308
 contour map of, *310*
 profile across, *296*
 sediments of, 238, 309
San Francisco, Calif.
 continental shelf off, 238
 continental slope off, 239
 earthquake of 1906, 239
San Francisco Bay, Calif., 64, 190–192, *191*
 sediments of, *191*
San Joaquin River, Calif., 190

San Jose Submarine Canyon, Baja Calif., Mex., *313*, 315
San Lucas Submarine Canyon, Baja Calif., Mex., 312–315, *313*
  granite blocks on floor of, *314*
  ripple-marked sands in, *314*
  underwater sandfall in, *314*
San Miguel Island, Calif., *117*
San Onofre, Calif., *135*
San Pablo Bay, Calif., 190, 191
San Pedro, Calif., 182
Sand. *See also* Beaches
  algal, 220
  backwash marks on, 158, *158*, 160
  bars, 136, 165
  black, 133
  blanket, 198
  coral, 132
  domes, 161
  drips, 161
  erosion of, 153, *153*
  falls, 312, *314*
  flows, 140, 308, 312
  green, 133
  oölitic, 220
  patterned, 161
  permanent loss of, 137–141
  quartz in, 132
  ribbons, 271
  ridges. *See* Sand ridges
  rill marks on, 161
  sea-floor highs from, 412
  sources of, 133
  swash marks on, *160*, 161
  volcanic, 132, 133
  waves, 86, 188, 218, 221, 271, *272*
Sand barriers, 180
Sand crabs *(Emerita)*, 161
Sand-dune belts, 268
Sand hoppers, 161
Sand Island, 358
Sand islands, 57
Sand ridges, 209, 210, *211*, 218, 219, 224, 269, 271, *287*, 288, *289*, 290
  on continental shelves, 286–288
Sand ripples, 86, *314*
Sand swales, 210
Sandbar channels, 190
Sanders, J. E., 287
Sandfall, underwater, *314*
Sandwich Islands, 377
Sangamon shoreline, 150
Santa Ana River, Calif., 141, 185
Santa Barbara, Calif.
  breakwater at, 133, 134, 139, 152, *154*
  coast of, 133, 141
  continental shelf off, 235, 237, 238
  oil spill at, 424
Santa Catalina Island, Calif., 238
Santa Lucia Mountain coast, Calif., slope off, 239
Santa Maria River, Calif., 133, *139*
Santa Maria Submarine Canyon, Baja Calif., Mex., 29
Santa Marta, Colombia, sea floor off, 228

Santa Monica, Calif., breakwater at, 7, 23, 24, 152, *154*
Santa Monica Bay, Calif.
  oil and gas seepages in, 237
  sediments of, *236*, 237
Santa Monica Mts., 141
Santa Ynez Mts., Calif., 141
Santa Ynez River, Calif., 133
Santander, Spain, 268
Santorini Volcano, 411
Sarawak, Malaysia, 251
Sargent, M. C., *351*, 352
Sasa, Y., 244
Saski, T., 38
Satellite navigation, 10, 12
Satellite observations, 42
Savonius rotor current meters, 37, 39
Saya da Malha Bank, 395
Sayles, R. W., 352
Sayner, D. B., *362*
Scandinavia, 273
Schatz, C. E., 16, 18
Scheelite, 132
Schick, G. B., 16, 37
Schlanger, S. O., 343
Schlee, J. S., *18*, 19, 218, 325, 333
Schnabel, J. E., 221
Scholl, D. W., 180, 231, 232, 243, *243*, 244, 319, *319*, *320*, 321, 390
Schorre, 193
Schott, W., 257, 259, 404
Sclater, J. G., 389
Scotia Sea, bottom currents of, 298, 378
Scotia Shelf, 203–204, *203*
Scotland, 273
Scripps Institution of Oceanography, 3, *5*, 37, 76, 218, 309, 312, 354, *397,·* 412
  beach at, 137, 157, 234, 418
  Hydraulic Laboratory at, 124
  pier at, 137
  waves at, 47
Scripps Submarine Canyon, 37, 141, 306–312, *307*
  heads of, 308
  sand chutes of, 308
  sediments of, 237, 308
  topography of, *307*, 308, 323
  vertical walls of, 306
Scruton, P. C., 106, 165
Scuba diving, 31–33, *32*, 36, 40, 237, 304, 306, 312
  development of, 3
Sea-floor highs, sands from, 412
Sea-floor observations by scientists, 31–32
Sea-floor spreading, 88–101, 218, 233, 241, 242, 255, 380–382, 389, 426, 429. *See also* Plate tectonics
  evidence for, 94–101, 261, 299
  hypothesis of, 7, 89, 374, 390, 419
  measuring of, 42
  of ocean ridges, *95*
  resulting from, 300
Sea-floor trenches, 233, 246, 389–391, *391*
Sea-floor valleys, types of, *305*

Sea-level changes, 103, 198, 221, 261, 284, 364, *366. See also* Eustatic sea-level changes
  coral reefs related to, 347
  history of, 106–108, *107*
  low stages of, 182, 224, 252, 261, 271, 286, 288, 292–294, 296
  lowering effects of, 198, 284–286
  maximum low due to glaciation, 292–294
  rising, barriers drowned by, 287–288
Sea of Azov, USSR, 187
Sea of Japan, Inland, 64, 193–194, *194*, 244, *248*, 286
Sea of Okhotsk, USSR, 244, *246*, *247. See also* Okhotsk Sea
Seacoasts
  classifications of, 102–122
    genetic, 109–111, *110*
  nomenclature for, 103
  origin of, 102–122
  See also Coasts
Sealab II, III, 32
Seamounts, 212, 383
Sebastian Vizcaíno Bay, Baja Calif., Mex., 182, 233, 234
Sedge peat, 182
Sediment, 68–87
  basins filled with, 251
  chemical properties of, 68
  coarse-fraction analysis of, 76–78, *77, 79*
  deposition of, 52
  dilatation of, 81–82
  distribution of, 70–74, 413
  glacial marine, 202, 273, 407, 412, 413, 415, 422
  grading of, 3
    scales of, 69
  kurtosis, 70, 73, 74
  logarithmic scale of, 69
  packing of, 81–82, *81*
  palimest, 198
  particles, constituents of, 75–78
  permeability of, 83–84, *84*
  phi deviation, 72
  phi median diameter, 73
  phi skewness of, 73–74
  phi unit, conversion table, 70
  plume, 87
  porosity of, 81–82
  proportion of, 68–69
  relict, 198, 212, 288–291, *291*
  roundness, 80, 81, 143
    classes of, *80*
    definition of, 78
  shape of, 78–81
  size of, 69–74
    descriptive measures of, 70–74, *72*
    distribution of, 74
  skewness, 69, 73, 74
  sphericity of, *78*
  spontaneous liquefaction of, 82
  standard deviation of, 72
  terrigenous, 413
    defined by mineral composition, 75
  transportation, modes of, 84–87, *85*

  unmixing of, 198
  volcanic, 411
Sediment particles, constituents of, 75–78
  arkose, 75
  authigenic sands, 76
  biogenous sands, 75
  bryozoans, 75, 76, *79*
  calcarenites, 75
  diatomites, 75, 76
  echinoids, 75, 76, 132
  foraminiferal, 75, 76
  glauconite, 76, *79*
  graywacke, 75
  ilmenite, 132
  magnetite, 132
  manganese nodules, 76
  mica, 76, 78, 132
  orthoquartzite, 75
  ostracods, 75
  phosphate nodules, 76
  pyrite, 76
  radiolarites, 75
  scheelite, 132
  terrigenous sands, 75
  wolframite, 132
Sediment plume, 87
Sedimentary basins, 251
Seibold, E., 3
Seiches, 151
Seine Estuary, France, 192, 193
Seismic reflection profiles, 40, *41*, 214, *228*, 242, 243, *246*, 255, 261, *266*, 268, 273, 284, 292, 296, *296*, 332, 418, 420
  buried valleys found by, 210
  slumping shown by, *295*
  transoceanic, 418–419, *419*
Seismic refraction profiles, 40–41, 300, 355
Sen Gupta, B. K., 202
Seoul, S. Korea, 247
Serpents Mouth, Gulf of Paria, 192
Severnaya Zemlya Islands, USSR, 273
Seychelles Islands, 393, 395
Seymour Narrows, British Columbia, 64
Shaler, N. S., 150
Shatsky Rise, 385, 427, 429
  profiles across, *387*
Sheba Ridge, 394
Sheet flow, 86
Shelter Cove, Calif., 239
Sheridan, R. E., 201, 202, 214, 217, 261
Shetland Islands, *273*
Shimbara Bay, Japan, catastrophic waves at, 58
Shinn, E. A., 352, 368
Shipek, C. J., *34*, 409
Shor, G. G., Jr., 3, 40, 41, *90*, 233, 319, 355, 385, 390, 391
Shore, definition of, 103
Shore processes, 123–161. *See also* Beaches
Shoreline, definition of, 103
Shouldice, D. H., 241
Shumway, G. A., 31, 238, 239, 397
Shurbet, G. L., *90*, 391
Siberia, shelf off, 244, 273–274
Siboga, 2

Siboga Expedition, 2
Sicily, Strait of, 263, 402
Side scanning, 9, 16, *17*, 271, *272*
Sierra de Pillahuinco Range, Argentina, 98
Sierra Leone, Africa, 261, 374
Sierra Nevada Mts., 190
Sievers, H. A., 54
Sigl, W., 32
Sigsbee Abyssal Plain, 397–398
Sigsbee Basin, 425
Sigsbee Deep, 424
Sigsbee Knolls, 398, 424
Sigsbee Scarp, 226
Silver, E. A., 241, *282*, 382
Silverman, M., 472
Singewald, J. T., Jr., 188
Sinkholes, 361
  on continental shelf, 212, 286
  drowned, 216
Sisoev, N. N., 391
Sitarz, J., 134
Sivash Lagoon, USSR, 187
Skagerrak Channel, Norway, 271
Skagerrak Trough, Norway, 273
Slater, R. A., 237
Slaughter, T., 188
Slide blocks, 296
Slikke, 193
Slope gullies, *305*, 306
  Mississippi Delta, off, *338*
  origin of, 337–338
Slope valleys, off Luzon, 252, 338, *339*
Slump folds, *297*
Slumping, *295*, 308, *320*
Smales, A. A., 410
Smith, A. J., 268
Smith, B. L., 89
Smith, P. A., 2, 210, 321
Smith, S. M., 371, 379, 397
Smith, S. V., 347
Smith, W. S. T., 2
Snellius Expedition, 2
Society Islands, 383
Soil Mechanics methods, 39–40
Soldado Bank, Gulf of Paria, 192
Somali Republic, Africa, 259
Somayajulu, B. L. K., 415
Sonar (Asdic), 271
Sonu, C. J., 134
Soons, J. M., 341
Soucoupe (Diving Saucer), Cousteau's, 33, 323
Soundings, correction of, 13–14
South Africa, diamonds on shelf off, 301
South America
  continental terraces off eastern, 229–231
  continental terraces off northern, 227–229
  continental terraces off western, 231–233
South American plate, 227
South Atlantic Ocean, drilling legs of, *423*
South China Sea, 251
South China Sea Basin, 251, 389
South Honshu Ridge, 246
South Natuna Island, 251

South Pole, plates movement away from, 101
South Sandwich arc, 389
South Sandwich Trench, 377
Southeast Channel, New England, 205
Southeast Indian Ridge, 393
Southern California, continental borderland off, *234*, *236*, *237*, *296*, *297*
Southwest Indian Ridge, 393
Southwest Pass, La., bars development at mouth of, 165, *166*
Soviet Union, 7, *7*, 185, 197, 242, 292
Spain
  continental shelf off, 264
  continental terrace off, 268
Sparker (arcer), 40
Spencer, J. W., 2
Spencer, M., 218, 333
Sphalerite crystals, 424
Spiess, F. N., 15
Spitsbergen Islands, Norway, 273
Sproll, W. P., 97, 98
Squires, D. F., 214, 346
Stanley, D. J., 197, 203, *203*, 325, 352, 354
Starke, G. W., 315
Sterrett, T. S., 156
Stetson, H. C., 3, 20, 132, 210, 212, 325
Stetson, T. R., 214
Stevenson, R. E., 164, 185, 187
Stewart, H. B., Jr., 157, *208*, 219, 220, 257, 354
Stewart, R. H., *222*
Stocks, T., 2, 271, 372
Storm surges, 53
  effects of, 56–57
Storr, J. E., 368
Strait of Belle Isle, 202
Strait of Florida, 63, 214, 217, 286, 425
  currents of, 150
Strait of Georgia, Canada–U.S., 241
Strait of Gibraltar, 261, 263
Strait of Magellan, 231
Strait of Messina, tides of, 64
Strait of Sicily, 263, 402
Stride, A. H., 3, 16, *17*, 265, *266–267*, 268, 269, 271, *272*, 273, 277, 297, *297*, 341
Strike-slip faults, 92, 98, 227, 233, 382
Stroup, R., 61
Stubbs, A. R., 16
Subaerial erosion, contributing to cause of canyon origin, 329–330
Subdeltas, 165
Subduction
  of lithosphere, *93*, *93*
  special types of, 263
  of trenches, 94, 199
Subduction coasts, 104, 242
Subduction zones, *93*, 241, 299, 382, 389. *See also* Benioff zone
Submarine canyons, Chap. 11, 304–341
  cable breaks in, 322
  current-meter records from, *38*, *332*
  currents in, 67
  deepest, 325, *326*
  description of, 306–327

Submarine canyons (*Continued*)
  heads of, 257
  largest, 319
  longest, 244
  origin of, 327–334
    artesian springs, 328
    diastrophism, 329
    drowned river valleys, 329
    mass movements, 332
    river cut, 329–330
    subaerial erosion, 329–330
    turbidity currents, 328
  related to beaches, 140, *140*
  turbidity currents in, *317*, 318, 321, *322*,
    330–331, 332
  uplifted, 341
  walls of
    upbuilding of, 333–334
    vertical, 306
Submarine landslides, 54, 57–59
Submarine valleys, types of. *See* Marine
  valleys
Subsidence, 385
Suffolk, England, coastal cliffs of, 147
Sulu Basin, 389
Sumatra, 251, 255
  coral reefs of, 346
Suo Nada, Inland Sea of Japan, 193
Surf beat, 332
Surge channels, *351*
Surveyor Fracture Zone, 426
Susquehanna River, Pa., 188
Sverdrup, H. U., 413
Swain Reefs, 252, 254
Swash, 127
Swatch of No Ground, *335*, 336
Swedish Deep Sea Expedition, 3, 412
Swift, D. J. P., 198, 205, 210, 221, 277, 287,
  291, 352
Swiftsure Bank, 241
Sydney, Australia, 252
Sykes, L. R., 8, 89, 94, 374, 382, 389, 402
Sylvania Seamount, 355

Tabasco, Mex., 143
Tagg, A. R., 210
Tahiti, 358, 360
Taiwan, 246
  continental shelf off, 249
  continental terrace off, 248
  sediments off, *250*
Talwani, M., 12, 42, 94, 217, 391
Tampa Bay, Fla., 192
Tangue, 193
Tanner, W. F., 111, 221
Tanner Bank, Calif., 238
Tanzania, Africa, 259
Tar mounds, 237
Tasmania, 252, 255
Tatar Strait, USSR, 244
Teche, distributary of Mississippi River, 165
Tectonic dams, 280, 283
Teichert, C., 89, 100, 346
Terrace, wave-cut, *109*
Terry, R. D., 33, *33*, 237, 338

Terzaghi, K., 82, 172, 296, 338
Tethys Sea, 100
Texas A. & M. University, 3, 197, 218
Texas coast
  beaches of, 134
  continental slope off, 296
  drilling on, 301
  lagoons of, *178*
  sediments of, *179*
Texas–Louisiana shelf, 224–226, 296
Thailand, broad shelf off, 251
Theil, E. C., 106
Thermal flux, measuring of, 42
Thermistors, 99
Thompson, R. W., 185
Thompson, W. C., 292
Thorndike, E. M., 63
Thresher, U.S. submarine, 64
Thurber, D. L., 363
Tia Juana River, Mex.–Calif., 154
  sediments off, 154, 235
Tiber Delta, *114*
Tidal currents, 64, 231
Tidal deltas, *58*, 143
Tides, 59–60
  cause of, 59
  definition of, 59
  erosion by, *65*
  neap, 59, 127, 136
  spring, 59, 136
  types of curves, *60*
    diurnal, *60*
    mixed, *60*
    semidiurnal, *60*
Tigris River, 258
Timbalier Island, 145
Timor Sea, 252, *254*
Tinian Island, uplifted lagoon of, 354
Tobago Basin, *399*
Tobin, D. G., 382
Tocher, D., 239
Todd, R., 357
Todos Santos Bay, Baja Calif., Mex., 233, 234
Tokyo Bay, 321
Tokyo Submarine Canyon, 321
Tokyo University, 3
Tombolo, 150, *159*, 261, 312
Tonga–Kermadec arc, *388*
Tonga Ridge, 389
Tonga Trench, 390, 391
Tongue of the Ocean, Bahamas, *325*, *326*
Torres Strait, 252
Tortugas Terrace, 217, *217*
Toulon Submarine Canyon, France, 329
Tracey, J. I., Jr., 242, *350*
Tramontini, C., 394
Transducer, towed electric, 40
Transform faults, 92, *92*, 212, 227, 382, 389
Trask, P. D., 2, 3, 71, 133, 134, 139, 141, 192
Trefethen, J. M., 159, *159*
Trenches, 233, 246, 389–391, *391*
Trieste, 33, 306, 390
Trincomalee Bay, Ceylon, 257
Trindad Island, off Venezuela, 192
Trinidad Islands, off Brazil, 374

Tristan da Cunha Island, 98
Troika, 32, 36, *36*, 323
Troughs, 214, 231, 246, 273, 284, 341
  fault, 232, 241, 321
Truk Islands, 359
  lagoon of, 358, *359*
Trumbull, J. V. A., 303, 325
Tsugaru Strait, Japan, 244
  currents of, 64, *65*
Tsunamis, 53–56, *55*, 147, 389. *See also*
  Earthquakes
  as canyon origin, 329
  characteristics of, 54
  Hawaii, 1946, 54, *55*, *160*
  destruction by, 56
  height of, at Kauai, *56*
Tsushima Island, 244
Tuamotu Islands, 383
  atolls of, 350, 358
Tuamotu-Line Island Ridge, 98
Tucker, M. J., 16
Tufts Plain, 382
Tungsten, 132
Turbidites, 299, 304, 318, 382, 407, 411, 412,
  424, 425, 429
Turbidity currents, 140, 198, 218, 308, 309,
  312, *317*, 318, 321, *322*, 328, 330–331,
  332, 383, 411–412, 429
  cable breaks caused by, *322*, 330, 390
  causes of, 66
  definition of, 66
  first observed at, 66
Turbulent flow, 84
Tuttle, S. D., 134
Tutuila Island, Samoa, 358
Tyrrhenian Basin, 426
Tyrrhenian Sea, 426
Tzimoulis, P., 36

Uchupi, E., 8, 97, 197, 200, *202*, 204, 205,
  210, *211*, 212, 214, 217, 220, *220*, 223,
  224, *226*, 235, 237, 277, 282, 295, 323,
  325, 374
Udden, J. A., 69
Udden's table, *70*
Udintsev, G. B., 390, 395
Ufford, C. W., 66
Umbgrove, J. H. F., 251, 346, 388
Undertow, 52
United States
  future resources off East Coast of, *302*
  submarine canyons off northeastern, 323–
  325
U.S. Army Corps of Engineers, 44, 123, 139
  152, 172, 190
U.S. Coast and Geodetic Survey, 2, 14, *48*,
  57, *142*, 210, 241, 315
U.S. Geological Survey, 3, 19, *109*, *116*, 169,
  197, 210, 218, *234*, 242, *245*, 323, 357
U.S. Hydrographic Office (now Oceano-
  graphic Office), 3
U.S. Navy, 3, 7, 33, *160*, 234, 242
  subsidizing of projects, 32
Ural Mts., formation of, 94
Urey, H. C., 415

Urien, C. M., 230
Uruguay, continental slope off, 230

Vacquier, V., 372
Valencia Trough, 426
Valentin, H., 102, 106, 271
van Andel, Tj. H., 3, 134, 169, 192, 229, *230*,
  233, 252, *254*, 259, 264, 286, 364, 372,
  374, *374*, 428
van Baren, F. A., 251
Van den Bussche, H. K. J., 23
van der Lingen, G. J., 157, 330
Van Dorn, W. G., 53, 54, 55
Van Lopik, J. R., 145, 165, *165*
van Straaten, L. M. J. U., 141, 156, 185, 263
van Veen, J., 271
Vancouver Island, Canada, 241
Varney, F. M., 22
Varsican Range, fit of, 98
Varved layers, 242
Vaughan, T. W., 343, 346, 347
Vaux, D., 64
Veatch, A. C., 2, 210, 321
Veeh, H. H., 293, 417
Veevers, J. J., 252, *254*, 293
Vema, 2, *5*, 28, 231
Vema Fracture Zone, 374, 376, 424
Venezuela
  Cariaco Trench off, 229
  continental shelf off, 227
  Gulf of, 229
  sediments off, 229
Venice, Italy, 169
Vening Meinesz, F. A., 41, 388, 391
Vernon, J. W., 237
Victoria, Australia, 255
Vidal Channel, Brazil, 376
Vietnam, 249, 251
Vigia Submarine Canyon, off Baja Calif.,
  Mex., *313*
Vine, F. J., 8, 16, 89, 92, 94, 372, 395
Vitiaz, 22, *35*
Vladivostok, USSR, 244
Volcanic activity related to continental
  slopes, 294
Volcanic basements, 355
Volcanic coasts, 109, 118, 244
Volcanic glass, 411
Volcanic islands, 362
Volcanic ridges, 383–388
Volcanic sand, 133
Volcanic sediments, 411
Volcanoes, 261
  andesite, 389
  glass of, 411
  sediment source of, 246
Volta Delta, Africa, 261
von der Borch, C. C., 255
Von Herzen, R. P., 25, 42, 372, 420, *423*
von Huene, R., 390
von Rad, U., *311*
von Stackelberg, U., 257, 259

Wachapreague Lagoon, Va., 182
Waddell, H., 78

Wadden Sea, The Netherlands, 185, *186*
Wageman, J. M., 246
Waikiki, Oahu, Hawaiian Is., 346
Walker, B. W., 137
Walsh, Lt. D., 390
Walthier, T. N., 16
Walton, W. R., 222, 234
Walvis Ridge, 98, 374
Wanless, H. R., 57, 124, 133, 147, 148, 153, 165, 169, *170*, 172, *173*
Warme, J. E., 185, 308
Wash, The, England, 193
Washburn, A. L., *131*
Washington
  continental shelf off, 241
  submarine canyons off, 315
Washington, University of, 3, 197
Washover sand, 176
Wass, R. E., 255
Wave-built terraces, 277, 280, 281
Wave convergence, 49–50, *49*
Wave-cut bench, 292, *293*
Wave-cut terrace, 106, *109*, 280
Wave diffraction, 49–50, *51*
Wave divergence, 49, *49*
Wave reflection, 49–50, *52*
Wave refraction, 49–50
  diagram of, *49*
Waves, 43–67, *48*
  Airy, 44, *45*
  backwash of, 52
  breakers, 44
    predictions of heights and time of, 54
  catastrophic. *See* Catastrophic waves
  internal, 66, 332
  sea, 44
  sinusoidal, *45*
  solitary, 44, *59*
  Stokes, 44
  storm, 53, 56–57
  swell, 44
  wind-generated, 44, 53
Weber, K. J., 167
Weeks, L. A., 255
Wegener, A., 329
Wegener hypothesis, 7, 88, 89, 93, 100
Weidie, A. E., 141
Wells, A. J., 258, 261
Wells, J. W., 343, 346, 347
Wentworth, C. K., 69
Wentworth scale, 69, 70, 73, 74
West Corsica, 321–323, *323*, 329
West Florida, continental shelf off, 218, 222
West Florida Escarpment, 220, 222, 226, 283
West Indian arc, 389
West Indies, 54, 376
Westinghouse Electric Company, 33, 306
Whaley, R. C., 472
White, S. M., 241
Wiegel, R. L., 103, 124
Wiens, H. J., 342
Wilde, P., 383

Wilhelm, O., 398
Wilkinson Basin, Gulf of Maine, 204
Williamson turn, 27
Wilmington Submarine Canyon, East Coast U.S., 325
Wilson, B. W., 151
Wilson, J. T., 89, 98
Wimberley, C. S., 236
Windisch, C. C., 418
Winslow, J. H., 341
Winterer, E. L., 384, 385, 426, *427*, *428*, 429
Wire soundings, methods of, 12
Wisconsin glacial stage, 278, 293
Wiseman, J. G. H., 370, 413
Wolframite, 132
Wollin, G., 415, 416, 417
Wong, H. K., *401*, 402
Woodring, W. P., 237
Woodruff, J. L., 21
Woods Hole Oceanographic Institution, 3, 4, *5*, 19, 33, 189, 197, 204, 210, 214, 218, 323, 403
Woodward, H. P., 374
Worzel, J. L., 42, *90*, 226, 391, 398, 411
Wreck Island, Great Barrier Reef, 366
Wüst, G., 63

X-ray photographs, 28, *29*

Yakutat Bay, Alaska, 242, 279
Yalu River, Korea, 247
Yamasaki, N., 340
Yangtze Kiang River, China, 247
Yap Trench, 390
Yaquina Bay, Ore., 190
Yaquina River, Ore., minerals from, 190
Yellow Sea, 246, 248, 292
  structure of, *249*
Yonge, C. M., 343
Young, R., *79*, 80, *80*, 81, 143
Yucatan Peninsula, continental shelf off, 226, 227, 286
Yugoslavia, 264
Yukon Delta, Alaska, 169, *170*
Yukon River, *170*, 321
Yukon Territory, Canada, 172

Zacatecas Fracture Zone, 226
Zambezi River, Africa, 258
Zanzibar Island, E. Africa, 258
Zarduzki, E. F. K., 214, *215*
Zeigler, J. M., 133, 134, 229, 231
Zenkevitch, N. L., *35*
Zenkovich (or Zenkovitch), V. P., 7, 124, 133, 141, 150, 151
Zeolites (phillipsite), 408–409, 424
Zeugogeosynclines, 246, 251
Zhemchug Submarine Canyon, Bering Sea, 319, 320
Zhivago, A. V., 7, 275
Zimmerman, H. B., 325

Printer and Binder: The Murray Printing Company

78 79 80   10 9 8 7 6